The Theory and Practic
Revenue Management

Recent titles in the
INTERNATIONAL SERIES IN
OPERATIONS RESEARCH & MANAGEMENT SCIENCE
Frederick S. Hillier, Series Editor, *Stanford University*

Ramík, J. & Vlach, M. / *GENERALIZED CONCAVITY IN FUZZY OPTIMIZATION AND DECISION ANALYSIS*
Song, J. & Yao, D. / *SUPPLY CHAIN STRUCTURES: Coordination, Information and Optimization*
Kozan, E. & Ohuchi, A. / *OPERATIONS RESEARCH/ MANAGEMENT SCIENCE AT WORK*
Bouyssou et al. / *AIDING DECISIONS WITH MULTIPLE CRITERIA: Essays in Honor of Bernard Roy*
Cox, Louis Anthony, Jr. / *RISK ANALYSIS: Foundations, Models and Methods*
Dror, M., L'Ecuyer, P. & Szidarovszky, F. / *MODELING UNCERTAINTY: An Examination of Stochastic Theory, Methods, and Applications*
Dokuchaev, N. / *DYNAMIC PORTFOLIO STRATEGIES: Quantitative Methods and Empirical Rules for Incomplete Information*
Sarker, R., Mohammadian, M. & Yao, X. / *EVOLUTIONARY OPTIMIZATION*
Demeulemeester, R. & Herroelen, W. / *PROJECT SCHEDULING: A Research Handbook*
Gazis, D.C. / *TRAFFIC THEORY*
Zhu, J. / *QUANTITATIVE MODELS FOR PERFORMANCE EVALUATION AND BENCHMARKING*
Ehrgott, M. & Gandibleux, X. /*MULTIPLE CRITERIA OPTIMIZATION: State of the Art Annotated Bibliographical Surveys*
Bienstock, D. / *Potential Function Methods for Approx. Solving Linear Programming Problems*
Matsatsinis, N.F. & Siskos, Y. / *INTELLIGENT SUPPORT SYSTEMS FOR MARKETING DECISIONS*
Alpern, S. & Gal, S. / *THE THEORY OF SEARCH GAMES AND RENDEZVOUS*
Hall, R.W./*HANDBOOK OF TRANSPORTATION SCIENCE - 2nd Ed.*
Glover, F. & Kochenberger, G.A. / *HANDBOOK OF METAHEURISTICS*
Graves, S.B. & Ringuest, J.L. / *MODELS AND METHODS FOR PROJECT SELECTION: Concepts from Management Science, Finance and Information Technology*
Hassin, R. & Haviv, M./ *TO QUEUE OR NOT TO QUEUE: Equilibrium Behavior in Queueing Systems*
Gershwin, S.B. et al/ *ANALYSIS & MODELING OF MANUFACTURING SYSTEMS*
Maros, I./ *COMPUTATIONAL TECHNIQUES OF THE SIMPLEX METHOD*
Harrison, T., Lee, H. & Neale, J./ *THE PRACTICE OF SUPPLY CHAIN MANAGEMENT: Where Theory And Application Converge*
Shanthikumar, J.G., Yao, D. & Zijm, W.H./ *STOCHASTIC MODELING AND OPTIMIZATION OF MANUFACTURING SYSTEMS AND SUPPLY CHAINS*
Nabrzyski, J., Schopf, J.M., Węglarz, J./ *GRID RESOURCE MANAGEMENT: State of the Art and Future Trends*
Thissen, W.A.H. & Herder, P.M./ *CRITICAL INFRASTRUCTURES: State of the Art in Research and Application*
Carlsson, C., Fedrizzi, M., & Fullér, R./ *FUZZY LOGIC IN MANAGEMENT*
Soyer, R., Mazzuchi, T.A., & Singpurwalla, N.D./ *MATHEMATICAL RELIABILITY: An Expository Perspective*

* A list of the early publications in the series is at the end of the book *

THE THEORY AND PRACTICE OF REVENUE MANAGEMENT

KALYAN T. TALLURI
Department of Economics and Business
Universitat Pompeu Fabra
Barcelona

GARRETT J. VAN RYZIN
Graduate School of Business
Columbia University
New York

 Springer

Library of Congress Cataloging-in-Publication Data

A C.I.P. Catalogue record for this book is available from the Library of Congress

Tallury & Van Ryzin/ *THE THEORY AND PRACTICE OF REVENUE MANAGEMENT*

ISBN 1-4020-7701-7 (hardcover)   0-387-24376-3 (paperback)   1-4020-7701-7 (eBook)

First paperback printing © 2005 Springer Science+Business Media, Inc.

© 2004 Springer Science+Business Media, Inc.
All rights reserved. This work may not be translated or copied in whole or in part without the written permission of the publisher (Springer Science+Business Media, Inc., 233 Spring Street, New York, NY 10013, USA), except for brief excerpts in connection with reviews or scholarly analysis. Use in connection with any form of information storage and retrieval, electronic adaptation, computer software, or by similar or dissimilar methodology now know or hereafter developed is forbidden.
The use in this publication of trade names, trademarks, service marks and similar terms, even if the are not identified as such, is not to be taken as an expression of opinion as to whether or not they are subject to proprietary rights.

Printed in the United States of America.

9 8 7 6 5 4 3 2 1   SPIN 11374756

springeronline.com

To Cristina and Uma
for the love and joy,
K.T.

To Mary Beth,
Stephanie, Claire and
Andrea with love and
thanks, and to the
memory of my father
John R. van Ryzin,
G.V.

Contents

Dedication		v
List of Figures		xvii
List of Tables		xxi
Preface		xxv
Acknowledgments		xxxi

1. **INTRODUCTION**   1
   - 1.1 What Is "RM"?   1
     - 1.1.1 Demand-Management Decisions   2
     - 1.1.2 What's New About RM?   4
   - 1.2 The Origins of RM   6
     - 1.2.1 Airline History   6
     - 1.2.2 Consequences of the Airline History   10
   - 1.3 A Conceptual Framework for RM   11
     - 1.3.1 The Multidimensional Nature of Demand   11
     - 1.3.2 Linkages Among Demand-Management Decisions   12
     - 1.3.3 Business Conditions Conducive to RM   13
     - 1.3.4 Industry Adopters Beyond the Airlines   16
   - 1.4 An Overview of a RM System   17
   - 1.5 The State of the RM Profession   18
   - 1.6 Chapter Organization and Reading Guide   20
     - 1.6.1 Chapter Organization   20
     - 1.6.2 Reading Guide   22
   - 1.7 Notes and Sources   23

Part I Quantity-based RM

2. SINGLE-RESOURCE CAPACITY CONTROL   27
   2.1 Introduction   27
      2.1.1 Types of Controls   28
      2.1.2 Displacement Cost   32
   2.2 Static Models   33
      2.2.1 Littlewood's Two-Class Model   35
      2.2.2 n-Class Models   36
      2.2.3 Computational Approaches   41
      2.2.4 Heuristics   44
   2.3 Adaptive Methods   50
      2.3.1 Adaptive Algorithm   50
      2.3.2 A Numerical Comparison with EMSR and Censored Forecasting   52
   2.4 Group Arrivals   56
   2.5 Dynamic Models   57
      2.5.1 Formulation and Structural Properties   58
      2.5.2 Optimal Policy   59
   2.6 Customer-Choice Behavior   62
      2.6.1 Buy-Up Factors   62
      2.6.2 Discrete-Choice Models   64
   2.7 Notes and Sources   75

3. NETWORK CAPACITY CONTROL   81
   3.1 Introduction   81
      3.1.1 The Promise and Challenge of Network Control   82
      3.1.2 Types of Controls   83
   3.2 The Theory of Optimal Network Controls   87
      3.2.1 The Structure of Optimal Controls   88
      3.2.2 Bid Price Controls   89
      3.2.3 Nonoptimality of Bid-Price Controls   90
      3.2.4 Evidence in Support of Bid Prices   91
      3.2.5 Bid Prices and Opportunity Cost   91
   3.3 Approximations Based on Network Models   92
      3.3.1 The Deterministic Linear Programming Model   93
      3.3.2 The Probabilistic Nonlinear Programming Model   95
      3.3.3 The Randomized Linear Programming Model   98

3.4	Approximations Based on Decomposition	100
	3.4.1 OD Factors Method	101
	3.4.2 Prorated EMSR	102
	3.4.3 Displacement-Adjusted Virtual Nesting (DAVN)	103
	3.4.4 Dynamic Programming Decomposition	107
	3.4.5 Iterative Decomposition Methods	108
3.5	Stochastic Gradient Methods	111
	3.5.1 Continuous Model with Gradient Estimates	112
	3.5.2 Discrete Model with First-Difference Estimates	116
3.6	Asymptotic Analysis of Network Problems	118
	3.6.1 Asymptotic Optimality of Partitioned Controls	118
	3.6.2 Asymptotic Optimality of Bid-Price Controls	120
	3.6.3 Comments on Asymptotic Optimality	120
3.7	Decentralized Network Control: Airline Alliances	121
3.8	Notes and Sources	122

4. OVERBOOKING 129

4.1	Business Context and Overview	130
	4.1.1 A History of Legal Issues in Airline Overbooking	131
	4.1.2 Managing Denied-Service Occurrences	135
	4.1.3 Lessons Beyond the Airline Industry	137
4.2	Static Overbooking Models	138
	4.2.1 The Binomial Model	139
	4.2.2 Static-Model Approximations	147
	4.2.3 Customer Class Mix	149
	4.2.4 Group Cancellations	150
4.3	Dynamic Overbooking Models	152
	4.3.1 Exact Approaches	152
	4.3.2 Heuristic Approaches Based on Net Bookings	154
4.4	Combined Capacity-Control and Overbooking Models	155
	4.4.1 Exact Methods for No-Shows	156
	4.4.2 Class-Dependent No-Show Refunds	158
	4.4.3 Exact Methods for Cancellations	159
	4.4.4 Class-Dependent Cancellation Refunds	160
4.5	Substitutable Capacity	161
	4.5.1 Model and Formulation	162
	4.5.2 Joint Optimal Overbooking Levels	164
4.6	Network Overbooking	166

4.7	Notes and Sources	168

Part II Price-based RM

5.	DYNAMIC PRICING			175
	5.1	Introduction and Overview		175
		5.1.1	Price versus Quantity-Based RM	176
		5.1.2	Industry Overview	177
		5.1.3	Examples of Dynamic Pricing	179
		5.1.4	Modeling Dynamic Price-Sensitive Demand	182
	5.2	Single-Product Dynamic Pricing Without Replenishment		187
		5.2.1	Deterministic Models	188
		5.2.2	Stochastic Models	200
	5.3	Single-Product Dynamic Pricing with Replenishment		209
		5.3.1	Deterministic Models	209
		5.3.2	Stochastic Models	212
	5.4	Multiproduct, Multiresource Pricing		215
		5.4.1	Deterministic Models Without Replenishment	216
		5.4.2	Deterministic Models with Replenishment	218
		5.4.3	Stochastic Models	219
		5.4.4	Action-Space Reductions	220
	5.5	Finite-Population Models and Price Skimming		223
		5.5.1	Myopic Customers	223
		5.5.2	Strategic Customers	226
	5.6	Promotions Optimization		229
		5.6.1	An Overview of Promotions	229
		5.6.2	Retailer Promotions	232
		5.6.3	Trade-Promotion Models	234
	5.7	Notes and Sources		235
6.	AUCTIONS			241
	6.1	Introduction and Industry Overview		241
		6.1.1	An Overview of Auctions in Practice	242
		6.1.2	Types of Auctions	245
	6.2	Independent Private-Value Theory		247
		6.2.1	Independent Private-Value Model and Assumptions	247
		6.2.2	An Informal Analysis of Sealed-Bid, First- and Second-Price Auctions	248

	6.2.3	Formal Game-Theoretic Analysis	254
	6.2.4	Revenue Equivalence	257
	6.2.5	Optimal Auction Design	259
	6.2.6	Relationship to List Pricing	262
	6.2.7	Departures from the Independent Private-Value Model	266
6.3	Optimal Dynamic Single-Resource Capacity Auctions		272
	6.3.1	Formulation	272
	6.3.2	Optimal Dynamic Allocations and Mechanisms	274
	6.3.3	Comparisons with Traditional RM	278
6.4	Optimal Dynamic Auctions with Replenishment		280
	6.4.1	Dynamic Programming Formulation	281
	6.4.2	Optimal Auction and Replenishment Policy	282
	6.4.3	Average-Profit Criterion	284
	6.4.4	Comparison with a List-Price Mechanism	285
6.5	Network Auctions		288
	6.5.1	Problem Definition and Mechanism	289
	6.5.2	Equilibrium Analysis	290
	6.5.3	Relationship to Traditional Auctions	291
	6.5.4	Relationship to Traditional Network RM	291
	6.5.5	Revenue Maximization and Reserve Prices	293
6.6	Notes and Sources		294

Part III   Common Elements

7.	CUSTOMER-BEHAVIOR AND MARKET-RESPONSE MODELS		301
7.1	The Independent-Demand Model		301
7.2	Models of Individual Customer Choice		303
	7.2.1	Reservation-Price Models	303
	7.2.2	Random-Utility Models	304
	7.2.3	Customer Heterogeneity and Segmentation	308
7.3	Models of Aggregate Demand		310
	7.3.1	Demand Functions and Their Properties	311
	7.3.2	Multiproduct-Demand Functions	320
	7.3.3	Common Demand Functions	321
	7.3.4	Stochastic-Demand Functions	327
	7.3.5	Rationing Rules	330
7.4	Notes and Sources		330

xii THE THEORY AND PRACTICE OF REVENUE MANAGEMENT

8. **THE ECONOMICS OF RM** — 333
   - 8.1 Introduction — 333
   - 8.2 Perfect Competition — 336
     - 8.2.1 Perfectly Competitive Markets — 336
     - 8.2.2 Firm-Level Decisions Under Perfect Competition — 338
     - 8.2.3 Precommitment and Demand Uncertainty — 338
     - 8.2.4 Peak-Load Pricing Under Perfect Competition — 341
     - 8.2.5 Identifiable Peak Periods — 341
     - 8.2.6 Uncertainty over the Timing of Peak Loads — 343
     - 8.2.7 Advance Purchases in Competitive Markets — 345
   - 8.3 Monopoly Pricing — 349
     - 8.3.1 Single-Price Monopoly — 350
     - 8.3.2 Monopoly with Capacity Constraints — 351
     - 8.3.3 Multiple-Price Monopoly and Price Discrimination — 352
     - 8.3.4 Strategic Customer Behavior — 363
     - 8.3.5 Optimal Mechanism Design for a Monopolist — 369
     - 8.3.6 Advance Purchases and Peak-Load Pricing Under Monopoly — 372
   - 8.4 Price and Capacity Competition in an Oligopoly — 375
     - 8.4.1 Static Models — 376
     - 8.4.2 Dynamic Models — 388
     - 8.4.3 Product Differentiation — 395
   - 8.5 Notes and Sources — 402

9. **ESTIMATION AND FORECASTING** — 407
   - 9.1 Introduction — 407
     - 9.1.1 The Forecasting Module of RM Systems — 408
     - 9.1.2 What Forecasts Are Required? — 410
     - 9.1.3 Data Sources — 412
     - 9.1.4 Design Decisions — 415
   - 9.2 Estimation Methods — 419
     - 9.2.1 Estimators and Their Properties — 420
     - 9.2.2 MSE Estimators — 422
     - 9.2.3 Maximum-Likelihood (ML) Estimators — 425
     - 9.2.4 Method of Moments and Quantile Estimators — 427
     - 9.2.5 Endogeneity, Heterogeneity, and Competition — 428
   - 9.3 Forecasting Methods — 433
     - 9.3.1 Ad-Hoc Forecasting Methods — 434
     - 9.3.2 Time-Series Forecasting Methods — 439

	9.3.3	Stationary Time-Series Models	442
	9.3.4	Nonstationary Time-Series Models	447
	9.3.5	Box-Jenkins Identification Process	449
	9.3.6	Bayesian Forecasting Methods	450
	9.3.7	State-Space Models and Kalman Filtering	458
	9.3.8	Machine-Learning (Neural-Network) Methods	464
	9.3.9	Pick-up Forecasting Methods	470
	9.3.10	Other Methods	472
	9.3.11	Combining Forecast Methods	472
9.4	Data Incompleteness and Unconstraining		473
	9.4.1	Expectation-Maximization (EM) Method	474
	9.4.2	Gibbs Sampling	481
	9.4.3	Kaplan-Meir Product-Limit Estimator	483
	9.4.4	Plotting Procedures	484
	9.4.5	Projection-Detruncation Method	485
9.5	Error Tracking and System Control		486
	9.5.1	Estimation Errors	487
	9.5.2	Forecasting Errors and System Control	496
9.6	Industry Models of RM Estimation and Forecasting		499
	9.6.1	Airline No-Show and Cancellations Forecasting	499
	9.6.2	Groups Demand and Utilization Forecasting	502
	9.6.3	Sell-Up and Recapture Forecasting	504
	9.6.4	Retail Sales Forecasting	505
	9.6.5	Media Forecasting	508
	9.6.6	Gas-Load Forecasting	510
9.7	Notes and Sources		511
10. INDUSTRY PROFILES			**515**
10.1	Airlines		515
	10.1.1	History	515
	10.1.2	Customers, Products, and Pricing	516
	10.1.3	RM Practice	521
10.2	Hotels		524
	10.2.1	Customers, Products, and Pricing	524
	10.2.2	RM Practice	526
10.3	Rental Car		531
	10.3.1	Customers, Products, and Pricing	531
	10.3.2	RM Practice	532

10.4	Retailing	533
	10.4.1 Customers, Products, and Pricing	534
	10.4.2 RM Practice	541
10.5	Media and Broadcasting	542
	10.5.1 Customers, Products, and Pricing	543
	10.5.2 RM Practice	545
10.6	Natural-Gas Storage and Transmission	546
	10.6.1 Customers, Products, and Pricing	547
	10.6.2 RM Practice	550
10.7	Electricity Generation and Transmission	551
	10.7.1 Industry Structure	552
	10.7.2 Customers, Products, and Pricing	554
	10.7.3 RM Practice	554
10.8	Tour Operators	555
	10.8.1 Customers, Products, and Pricing	556
	10.8.2 Capacity Management and Base-Price Setting	556
	10.8.3 RM Practice	558
10.9	Casinos	559
	10.9.1 Customers, Products, and Pricing	559
	10.9.2 RM Practice	559
10.10	Cruise Ships and Ferry Lines	560
	10.10.1 Customers, Products, and Prices	560
	10.10.2 RM Practice	561
10.11	Passenger Railways	561
	10.11.1 Customers, Products, and Pricing	561
	10.11.2 RM Practice	563
10.12	Air Cargo	563
	10.12.1 Customers, Products, and Pricing	563
	10.12.2 RM Practice	563
10.13	Freight	564
	10.13.1 Customers, Products, and Pricing	565
	10.13.2 RM Practice	566
10.14	Theaters and Sporting Events	567
	10.14.1 Customers, Products, and Pricing	567
	10.14.2 Ticket Scalping and Distribution	567
	10.14.3 RM Practice	571
10.15	Manufacturing	574

	10.15.1 Customers, Products, and Pricing	574
	10.15.2 RM Practice	575
10.16	Notes and Sources	576

11. IMPLEMENTATION 579

11.1	Segmentation and Product Design	579
	11.1.1 Segmentation	580
	11.1.2 Product Design	585
11.2	System Architecture, Hardware, Software, and Interfaces	594
	11.2.1 Hardware Requirements	594
	11.2.2 User-Interface Design	594
	11.2.3 GDS, CRS, and PMS Interfaces	598
	11.2.4 Retail Management Systems	605
11.3	Revenue-Opportunity Assessment and Revenue-Benefits Measurement	608
	11.3.1 Revenue-Opportunity Assessment	608
	11.3.2 Revenue-Benefits Measurement	610
11.4	RM Simulation	611
	11.4.1 Generating Aggregate Number of Customers	613
	11.4.2 Generating the Customer-Arrival Pattern	613
11.5	Customer Perceptions and Reactions	614
	11.5.1 RM Perception Problems	614
	11.5.2 Managing Perceptions	618
	11.5.3 Overbooking Perceptions	619
11.6	Cultural, Organizational, and Training Issues	620
	11.6.1 Changes in Responsibility by Function	620
	11.6.2 Project and Organizational Structure	623
	11.6.3 Training	627
11.7	Notes and Sources	628

Appendices		631
A	Notation	631
B	Probability	635
C	Convexity and Optimization	643
D	Dynamic Programming	651
E	The Theory of Choice	657
F	Game Theory	667

References 671
Index 709

List of Figures

1.1	A firm's demand landscape	11
1.2	RM process flow	19
2.1	Booking limits b_j, protection levels y_j, and bid prices $\pi(x)$	29
2.2	Optimal protection level y_j^* in the static model	39
2.3	Monte Carlo estimates of optimal protection levels for 50 simulated data points	44
2.4	Adaptive-method example 1	54
2.5	Adaptive-method example 2	55
2.6	Optimal protection level $y_j^*(t)$ in the dynamic model	61
2.7	Scatter plot of $Q(S)$ and $R(S)$	69
3.1	Network examples	82
3.2	Comparison of DLP, RLP, and dynamic programming decomposition	100
3.3	A network example	115
3.A.1	Stochastic gradient calculation for nested booking limits: gradient equal to zero	127
3.A.2	Stochastic gradient calculation for nested booking limits: gradient nonzero	127
4.1	Overbooking notification statement	134
4.2	Overbooking limits over time	140
4.3	Overbooking for the multiclass and binomial models	165
4.4	Network overbooking	168
5.1	Sample price path at a discount air carrier	181

5.2	The maximum concave envelope produced by discrete prices	196
5.3	Sales volume example	198
5.4	Effect of markdowns	200
5.5	An example of the optimal price path in the stochastic case	204
5.6	A six-node, two-hub airline network	218
5.7	Optimal price-skimming solution for myopic customers	225
6.1	Perturbing the bid v_i in a second-price auction	249
6.2	Illustration of the direct-revelation mechanism	257
6.3	Optimal allocations in the dynamic auction model	275
6.4	Optimal allocations in the dynamic-auction model with replenishment	284
7.1	Individual demand and the aggregate demand	311
7.2	Revenue and marginal-revenue curves	319
7.3	Some common aggregate price-response functions	323
8.1	Revenue from selling a product at multiple prices	355
8.2	The equilibrium bipartite graph	382
8.3	Response functions for a duopoly RM game	383
8.4	The two cases for the function $g(\cdot)$	400
8.5	Best response cycle example	400
8.6	No-purchase probabilities causing a best-response cycle	401
9.1	Forecasting module in a RM system	409
9.2	Wedge-shaped bookings data	416
9.3	Time series components	435
9.4	Exponential smoothing with different smoothing parameters	437
9.5	Sample ACF and PACF functions	443
9.6	ACF and PACF examples	446
9.7	The hierarchial Bayes model	454
9.8	Kalman filter smoothing	463
9.9	Neural network example	465
9.10	Neural network activation functions	467
9.11	Incremental booking data	471
9.12	Over-fitting example	495
9.13	Booking curve with cancellations and no-shows	500
9.14	Cancellation probabilities as a function of booking time	501

List of Figures

9.15	Induction tree on cancellations data	503
9.16	Neural network for gas-load forecasting	510
10.1	Pricing an air travel itinerary	519
10.2	Revenue sources and revenue drivers for a hotel	526
10.3	A rental car RM system implementation	533
10.4	Store type breakdown for the top 200 retailers	535
10.5	Growth of department store markdowns	537
10.6	Gas pipeline network	551
10.7	Electricity industry structures	553
10.8	Capacity planning at a tour operator	557
10.9	Purchase plan and price setting at a tour operator	557
10.10	RM process for tour-operators	558
10.11	Washington Opera Kennedy Center layout	573
11.1	Advance-purchase and max-stay restrictions for an airline trip represented by a grid	590
11.2	Products and customers utility reduction modeling	591
11.3	Nightly batch processing	595
11.4	Forecasting under the independent-class model	596
11.5	RM process flow	597
11.6	Quantity-based RM user interface	598
11.7	Quantity-based RM user interface	599
11.8	GDS reservation processing	600
11.9	Seamless availability	604
11.10	Generating arrivals over time	613
11.11	RM organization charts	625

List of Tables

2.1	Static single-resource model and protection levels	48
2.2	Revenue performance for Example 2.3	49
2.3	Static single-resource model data for Example 2.4	49
2.4	Revenue performance for Example 2.4	50
2.5	Fares, demand statistics, and protection levels for adaptive-method numerical examples	53
2.6	Starting values of protection levels for adaptive-method numerical examples	53
2.7	Fare-product revenues and restrictions for Example 2.5	65
2.8	Segments and their characteristics for Example 2.5	65
2.9	Choice probabilities $P_j(S)$, probability of purchase $Q(S)$, and expected revenue $R(S)$ for Example 2.5	66
2.10	Illustration of nested policy for Example 2.5	72
2.11	The different segment choices in Example 2.5 if all classes are open and resulting demand for a population size of 20	74
2.12	Inputs to the EMSR-b model	75
2.13	Protections for the EMSR-b model without and with buy-up factors	75
2.14	Simulation results comparison between choice dynamic program and EMSR-b with buy-up	75
3.1	Example of a bid price table for a single resource based on remaining time and remaining capacity	90
3.2	Problem data for the bid price counterexample	90
3.3	Data for the iterative proration-method example	111

3.4	Example of convergence of the iterative proration method ($t = 1$)	111
3.5	Data for the Williamson [566] network example	115
3.6	Initial protection levels produced by DAVN	116
3.7	Improved protection levels produced by the stochastic gradient algorithm	116
4.1	U.S. major airline denied-boarding rates, 1990-2000	133
4.2	Binomial and normal approximation overbooking probabilities	143
4.3	Comparison of normal and Gram-Charlier (G-C) approximations	149
4.4	Empirical distribution of group sizes	150
5.1	Allocations of capacity between periods 1 and 2 and the marginal values and total revenue	190
5.2	Example of the marginal-allocation algorithm	192
5.3	Example of discrete prices and revenues	195
5.4	Solution of a linear program for the discrete-price example	195
5.5	Results of different markdown policies on 60 markdown styles	200
5.6	Example performance of the deterministic price heuristic	206
5.7	Demand-function parameters, itineraries, and optimal solution for Example 5.6	219
5.8	Empirical generalizations on promotions	231
6.1	Dynamic auction revenues for different concentrations of customers	280
6.2	DLPCC suboptimality gaps relative to a dynamic auction for different demand to capacity ratios	280
6.3	Dynamic auction and replenishment profits for different numbers of customers	287
6.4	Dynamic auction and replenishment profits for different holding costs	288
6.5	Network auction simulation results: average revenues as a function of reserve price	294
7.1	Attribute weights x_m^j for attributes $m = 1, 2$ in alternative $j = 1, 2, 3$	311
7.2	Estimated elasticities (absolute values) for common products	314

List of Tables

7.3	Common demand functions	322
8.1	Prices and capacities for Example 8.2 without an advance-purchase market	347
8.2	Prices and capacities for Example 8.3 with an advance-purchase market	349
8.3	Revenue and variance calculations for Example 8.9 with a single-price policy	361
8.4	Revenue and variance calculations for Example 8.9 with a multiple-price policy	361
9.1	Assumptions of ordinary least-squares (OLS) estimation	424
9.2	Means and covariances of some stationary time-series processes	445
9.3	Results of the AR(2) forecasting example	448
9.4	EM algorithm iterations on constrained data	478
10.1	An example of airline fare codes, classes, and their restrictions	522
10.2	Features of a hotel property management system (PMS)	528
10.3	World's top 10 retailers, store types, and their revenues for the year 2002	534
10.4	U.S. Apparel sales by channel	537
10.5	U.S. apparel sales by category	538
10.6	Inventory definitions in television, radio, and print media	544
10.7	A sample advertising purchase plan	544
10.8	An example of a pipeline delivery contract	547
10.9	Sample pipeline tariffs	548
10.10	Sample natural gas transportation and storage products	549
10.11	Amtrak accommodation and fare types	562
10.12	Sample freight product differentiation	565
10.13	An example of ticket categories for a broadway show	568
10.13	(continued) An example of ticket categories for a broadway show	569
10.14	Washington Opera Kennedy Center pricing (2003–2004 season)	573
10.15	The Mets four-tier pricing plan (year 2002)	574
11.1	Customer segments and subsegments by industry	581
11.2	Classification of segment bases	582

11.3	Some common segment bases used in RM	583
11.3	(continued) Some common segment bases used in RM	584
11.4	Attributes and their levels for a hotel application	586
11.5	Major global distribution systems (GDSs) as of 1998	601
11.6	An availability request message as software code and the same request as a message	602
11.7	Typical data tables provided by a hotel PMS	602
11.8	Table BIDPRICE	603
11.9	RMS pricing and inventory functions	607
11.10	Functionality of EDI for the travel and tourism industries	609
11.11	Commonly tracked RM system performance measures	612
11.12	Task list for a RM implementation	624

Preface

Revenue management (RM) has gained attention recently as one of the most successful application areas of operations research (OR). The practice has grown from its origins as a relatively obscure practice among a handful of major airlines in the post-deregulation era in the U.S. (circa 1978) to its status today as a mainstream business practice with a growing list of industry users from Walt Disney Resorts to National Car Rental and a supporting industry of software and consulting firms. Major airlines, hotel chains, and car rental companies have large staffs of developers and analysts working on RM, and major consulting and software firms also employ large numbers of RM professionals.

There are now several major industry RM conferences each year: The Airline Group of the International Federation of Operational Research Societies (AGIFORS) sponsors an annual reservation and yield management conference that attracts has attracted up to 200 professionals, and The International Air Travel Association (IATA) hosts an annual RM conference that has drawn up to 800 attendees in recent years. The Professional Pricing Society also hosts professional conferences that address science-based pricing methods and technologies and general pricing strategy.

Over this same period, academic and industry research on the methodology of RM has also grown rapidly. The number of published papers on RM has increased dramatically in the last ten years. INFORMS, the leading professional society of OR, has started a Pricing and RM Section, which has now hosted several annual conferences on RM, each drawing in excess of 100 researchers and professionals. And several universities now offer specialized RM courses, at both the M.B.A and Ph.D. levels.

Despite this explosion of both professional and scholarly activity, no book has comprehensively covered the field of RM. For any area in such a mature state of development and with such widespread industry usage, such a reference is desirable. However, for RM the need is particularly acute for several reasons:

- RM is very much a professional practice and as such there is a considerable amount of "institutional" knowledge surrounding it that is relatively inaccessible to those outside the profession.

- Many of the early and even some more recent seminal ideas do not appear in published journals. Even those that have been published sometimes appear in relatively obscure sources such as AGIFORS proceedings, industry newsletters, and standard industry practice.

- The terminology, concepts, and notation have not been standardized to date, so it is often confusing for an outsider to reconcile the various contributions of the extant literature.

- There is often a considerable gap between practitioners and academics in the field. Academics are often not aware of the real world complexities faced by practitioners of RM, and industry practitioners are often not aware of the more recent advances in the academic literature.

Our aim in writing this book is to meet this need. The book seeks - as its title indicates—to cover both the *theory* and the *practice* of RM. Fundamentally, RM is an applied discipline; its value and significance ultimately derive from the business results it achieves. At the same time it has strong elements of an applied science, and the technical elements of the subject deserve rigorous treatment. Both these practical and theoretical elements of the field reinforce each other, and to a large extent this is what makes the topic exciting. It is this constructive interplay of theory and practice that we have strived to capture in this book.

Audience

We have two primary audiences in mind for this book—(1) analytically trained (or at least "analytically tolerant") practitioners in industry and (2) academic researchers and teachers. We view our core reader as someone who has the equivalent of a master's degree or higher in a technical subject such as engineering, operations research, statistics, or economics. However, significant portions of the text are accessible to general or business readers, particularly the introduction, Chapter 10 on industry profiles, and Chapter 11 on implementation issues. In addition, the introductions to the technical chapters provide high-level overviews of each chapter, which are designed to provide a qualitative understanding of the main topics covered and their business context, and give the reader a sense of the essence—if not the details—of the material.

For experienced practitioners this book serves as a single-source reference for the major theory and application issues involved in RM. The key technical results in the field are organized and presented precisely and in consistent notation, so that practitioners can easily refer to relevant models, formulas, and algorithms as needed. For new employees in the RM industry our book also serves as a useful primer on the subject, allowing them to "get up to speed" on the details of the field quickly through a consistent presentation of the material. For the technically oriented user it serves as an unbiased, noncommercial source for understanding the competing methodologies available for RM and their relative strengths and weaknesses.

We view the academic audience for the book as consisting of the many researchers now working on various RM-related topics, as well as those who work in related areas (such as supply-chain management), who may want a single-source, accessible overview of the main theory and practice components of the field. Academics who teach management science or operations management courses may also find the book useful, either directly as a supplementary text or simply for the instructor's personal use as a reference on the subject. Our experiences with colleagues outside the field has suggested that most are curious about RM but perhaps not confident enough about the theory and practice to introduce the subject in their classes. This book should help "demystify" the subject for them.

Finally, a growing number of courses have specifically focused on RM. Though not designed particularly as a textbook, the book should serve as a useful reading and reference in such courses. While we have not put in homework exercises, we did include many small, technically uncluttered examples throughout the book that illustrate the core concepts being discussed.

We forewarn the reader that the material in some places in the book has an airline bias. This is as it should be in our opinion; airline RM practice remains an important topic in its own right. In addition, a large number—indeed the vast majority—of RM practitioners and researchers working in the field today are involved directly in airline RM practices. So airline RM is deserving of rigorous and careful coverage, which is one of our goals in writing this book.

At the same time, not every industry is like the airline industry and "airlinelike" conditions are not, in our view, that necessary to apply RM ideas. Therefore, we have attempted to present RM in as generic terms as possible and included several topics and chapters that generalize beyond the airline industry. We have tried to be somewhat forward looking in this regard, while at the same time not venturing too far into the realm of pure speculation.

Content and Style

As for the choices of material, we have aimed for an applied technical (engineering) level in our treatment of the subject. For example, we have chosen to present all problems in discrete time. This eliminates several technical complications, while still allowing us to address a wide range of problems in a simple, yet rigorous way. Moreover, continuous-time models and methods are not frequently used in practice, so the focus on discrete-time methods is well justified from a practical standpoint.

Similarly, we have not included a large number of proofs. This is both consistent with the applied orientation of the field and reflects our view that RM models and theory do not share enough in common to justify a highly formalistic, deductive approach to the subject. In a few cases we provide proofs of the theoretical results, but even these are relegated to appendices. When proofs are omitted, we provide references to the original sources and if possible give either informal arguments or intuition about the results.

In addition, the bodies of each chapter do not contain a large number of literature references. This is because we want the reader to "see the material for what it is" and not be sidetracked by a lot of discussion of the literature. Where ideas are strongly associated with specific papers and people, we, of course, point this out. Detailed references to the literature and a discussion of sources are collected in a Notes and Sources section provided at the end of each chapter. To further assist the reader, appendices containing basic results on probability theory, continuous optimization, dynamic programming, and game theory are provided to make the technical material in the book as self-contained as possible.

We tried to be comprehensive in our coverage of RM, covering both quantity- and price-based RM as well as the supporting topics of forecasting and economics. While we might have risked over-extending ourselves in this regard, we believe such a comprehensive approach is necessary to fully understand the subject. Indeed, a key contribution of the book is to unify the various forms of RM and to link them closely to each other and to the supporting fields of statistics and economics. The topics and coverage do, however, reflect our own personal choices about what is and is not important to understand RM. While we have tried to be as comprehensive, fair, and balanced as possible in arriving at these choices, undoubtedly our choices have resulted in some biases. However,

PREFACE

the benefit to the reader is that the text has a point of view and is not merely an uncritical inventory of all research results to date in the field.

Finally, we have also tried to come up with a notation that is generic and consistent across all the chapters. Much of this notation will not coincide with the notation found in the original papers in the field, which is by and large quite inconsistent anyway. A summary of notation is provided in Appendix A for reference. The consistency of notation and presentation, we believe, makes reading the book much easier than looking at the corresponding collection of original-source articles, and it also highlights the connections among topics.

Acknowledgments

This book has been a long time in the making, and writing it would not have been possible without strong support from our institutions, colleagues and family.

Jointly we would like to acknowledge the following colleagues for their time in reading, either sections, or significant parts of the book, and providing us with valuable feedback: Antonio Cabrales (UPF), James Dana (Northwestern), Srinivas Bollapragada (NBC), and the graduate OM class of UPF (2002); Gustavo Vulcano (NYU), Itir Karaesmen (U. of Maryland), Sanne de Boer (MIT), Michael Harrison (Stanford) (and doctoral students in Mike's 2002 Ph.D. seminar on RM), Costis Maglaras (Columbia), Serguei Netessine (Wharton), Qian Liu (Columbia), Yannis Paschalidis (Boston U.) and Andy Philpott (U. of Auckland) and the graduate students of the seminar taught at the Auckland University in 2002. The book has benefited greatly from their comments Certainly, all remaining errors and obfuscations are our responsibility.

Interactions with many industry colleagues over the years, especially those with Surain Adyanthaya, Andy Boyd, Sebastian Ceria, Ren Curry, Mark Diamond, Kevin Geraghty, Craig Hopperstad, Bob Philips, Anand Rao, John Salch, Barry Smith, and Ben Vinod, have also greatly benefited the book. Dr. Rama Ramakrishnan of Profitlogic was kind enough to provide screenshots of markdown pricing software for use in the implementation chapter.

Kalyan Talluri would like to thank the Department of Economics and Business of the Universitat Pompeu Fabra for their support and healthy research environment, and the Deming Center of Columbia Business School for funding many trips to New York to work on the book. He also would like to acknowledge that his knowledge of RM, and his research, benefited from his long collaboration with the Pricing and RM department at Iberia airlines, specifically working for many years with

Fernando Castejon and Juan Magaz. On the personal side, the stress and labors of writing a long book like this, he would like to acknowledge, were vastly mitigated by the love and joy of companionship of Cristina Ferrer and Uma Talluri Ferrer, both of whom no doubt greet this book with a big sigh of relief.

Garrett van Ryzin would like to thank Columbia Business School for supporting this project over many years, and in particular the Deming Center and its director, Nelson Fraiman, who provided travel funds and research support which helped make writing this book possible. Significant portions of this book were written during a sabbatical visit to the University of Auckland in 2001-2002, and the support of the MSIS Department and especially its then head-of-department, Justo Diaz, is greatfully acknowledged. A course taught at Auckland also helped improve early drafts of the book, as did input and discussions with Andy Philpott of the Engineering Science Department at Auckland. Much of the content of this book is the result of research collaborations with a number of colleagues, including Guillermo Gallego, Aliza Heching, Itir Karaesmen, Costis Maglaras, Siddharth Mahajan, Jeff McGill and Gustavo Vulcano. It has been a privilege to work with such a talented group of colleagues, and this book has benefited greatly from their collective contributions. Finally, writing this book would not have been possible without the patience, love and support of Mary Beth, Stephanie, Claire and Andrea—who generously (if not joyfully) tolerated Dad's many long hours of isolation in his office. Like Cristina and Uma, they too very much deserve to celebrate the completion of this book.

Chapter 1

INTRODUCTION

This chapter provides an introduction to the topic of revenue management (RM). We begin with an explanation of RM and its history and origins. We then provide a conceptual framework for understanding the objectives of RM, the types of business conditions under which it is applied, and the ways RM systems work. Finally we conclude by giving an outline of the remaining chapters of the book.

1.1 What Is "RM"?

Every seller of a product or service faces a number of fundamental decisions. A child selling lemonade outside her house has to decide on which day to have her sale, how much to ask for each cup, and when to drop the price (if at all) as the day rolls on. A homeowner selling a house must decide when to list it, what the asking price should be, which offer to accept, and when to lower the listing price—and by how much—if no offers come in. A stamp dealer selling on an Internet auction site has to select the duration of the auction, what reserve price to set (if any), and so on.

And anyone who has ever faced such decisions knows the uncertainty involved. You want to sell at a time when market conditions are most favorable, but who knows what the future might hold? You want the price to be right—not so high that you put off potential buyers and not so low that you lose out on potential profits. You would like to know how much buyers value your product, but more often than not you must just guess at this number.

Indeed, it is hard to find anyone who is entirely satisfied with their pricing and selling decisions. Even if you succeed in making a sale, you often wonder whether you should have waited for a better offer or whether you accepted a price that was too low.

Businesses face even more complex selling decisions. For example, how can a firm segment buyers by providing different conditions and terms of trade that profitably exploit their different buying behavior or willingness to pay? How can a firm design products to prevent cannibalization across segments and channels? Once it segments customers, what prices should it charge each segment? If the firm sells in different channels, should it use the same price in each channel? How should prices be adjusted over time based on seasonal factors and the observed demand to date for each product? If a product is in short supply, to which segments and channels should it allocate the products? How should a firm manage the pricing and allocation decisions for products that are complements (seats on two connecting airline flights) or substitutes (different car categories for rentals)?

RM is concerned with such *demand-management* decisions[1] and the methodology and systems required to make them. It involves managing the firm's "interface with the market" as it were—with the objective of *increasing revenues*. RM can be thought of as the complement of *supply-chain management* (SCM), which addresses the *supply decisions* and processes of a firm, with the objective (typically) of *lowering the cost* of production and delivery.

Other roughly synonymous names have been given to the practice over recent years—*yield management* (the traditional airline term), *pricing and revenue management, pricing and revenue optimization, revenue process optimization, demand management, demand-chain management* (favored by those who want to create a practice parallel to supply-chain management)—each with its own nuances of meaning and positioning. However, we use the more standard term revenue management to refer to the wide range of techniques, decisions, methods, processes, and technologies involved in demand management.

1.1.1 Demand-Management Decisions

RM addresses three basic categories of demand-management decisions:

---

[1] These can be referred to as either *sales decisions* (we are making decisions on where and when to sell and to whom and at what price) or *demand-management decisions* (we are estimating demand and its characteristics and using price and capacity control to "manage" demand). We use the latter consistently and use the shorter *demand management* whenever appropriate.

Introduction 3

- Structural decisions: Which selling format to use (such as posted prices, negotiations or auctions); which segmentation or differentiation mechanisms to use (if any); which terms of trade to offer (including volume discounts and cancellation or refund options); how to bundle products; and so on.

- Price decisions: How to set posted prices, individual-offer prices, and reserve prices (in auctions); how to price across product categories; how to price over time; how to markdown (discount) over the product lifetime; and so on.

- Quantity decisions: Whether to accept or reject an offer to buy; how to allocate output or capacity to different segments, products or channels; when to withhold a product from the market and sale at later points in time; and so on.

Which of these decisions is most important in any given business depends on the context. The timescale of the decisions varies as well. Structural decisions about which mechanism to use for selling and how to segment and bundle products are normally strategic decisions taken relatively infrequently. Firms may also have to commit to certain price or quantity decisions, for example, by advertising prices in advance or deploying capacity in advance, which can limit their ability to adjust price or quantities on a tactical level. The ability to adjust quantities may also be a function of the technology of production—the flexibility of the supply process and the costs of reallocating capacity and inventory. For example, the use of capacity controls as a tactic in airlines stems largely from the fact that the different "products" an airline sells (different ticket types sold at different times and under different terms) are all supplied using the same, homogeneous seat capacity. This gives airlines tremendous quantity flexibility, so quantity control is a natural tactic in this industry. Retailers, in contrast, often commit to quantities (initial stocking decisions) but have more flexibility to adjust prices over time. The ability to price tactically, however, depends on how costly price changes are, which can vary depending on the channel of distribution such as online versus catalog.

Whether a firm uses quantity or price-based RM controls varies even across firms within a given industry. For instance, while most airlines commit to fixed prices and tactically allocate capacity, low-cost carriers tend to use price as the primary tactical variable.

Firms can also find innovative ways to increase their ability to make price or quantity recourse decisions. For example, retailers may hold back some stock in a centralized warehouse and then make a mid season replenishment decision rather than precommit all their stock to stores up front. Some major airlines have experimented with movable partitions that allow them to reallocate seats from coach to business cabins on a short-term basis. And other major airlines have recently experimented with a practice called *demand-driven dispatch* (D^3), in which aircraft of different sizes are dynamically assigned to each flight departure in response to fluctuations in demand, and are not precommitted to flights [50]. Car rental companies also may reallocate their fleet from one city to another. In terms of pricing, using online channels or advertising products without price ("call for our low price") provides firms with more price flexibility. All these innovations increase the opportunity for quantity and price-based RM.

Broadly speaking, RM addresses all three categories of demand-management decisions—structural, pricing, and quantity decisions. We qualify RM as being either *quantity-based RM* or *price-based RM* if it uses (inventory- or) capacity-allocation decisions or prices as the primary tactical tool respectively for managing demand. Both the theory and practice of RM differ depending on which control variable is used, and hence we use this dichotomy as necessary.

1.1.2 What's New About RM?

In one sense, RM is a very old idea. Every seller in human history has faced RM-type decisions. What price to ask? Which offers to accept? When to offer a lower price? And when to simply "pack up one's tent" as it were and try selling at a later point in time or in a different market. In terms of business practice, the problems of RM are as old as business itself.

In terms of theory, at a broad level the problems of RM are not new either. Indeed, the forces of supply and demand and the resulting process of price formation—the "invisible hand" of Adam Smith—lie at the heart of our current understanding of market economics. They are embodied in the concept of the "rational" (profit-maximizing) firm, and define the mechanisms by which market equilibria are reached. Modern economic theory addresses many advanced and subtle demand-management decisions, such as nonlinear pricing, bundling, segmentation, and optimizing in the presence of asymmetric information between buyers and sellers.

What *is* new about RM is not the demand-management decisions themselves but rather *how* these decisions are made. The true innovation of RM lies in the *method* of decision making—a technologically sophisticated, detailed, and intensely operational approach to making demand-management decisions.

Introduction

This new approach is driven by two complementary forces. First, scientific advances in economics, statistics, and operations research now make it possible to model demand and economic conditions, quantify the uncertainties faced by decision makers, estimate and forecast market response, and compute optimal solutions to complex decision problems. Second, advances in information technology provide the capability to automate transactions, capture and store vast amounts of data, quickly execute complex algorithms, and then implement and manage highly detailed demand-management decisions. This combination of science and technology applied to age-old demand management is the hallmark of modern RM.

And both the science and technology used in RM are quite new. Much of the science used in RM today (demand models, forecasting methods, optimization algorithms) is less than fifty years old, most of the information technology (large databases, personal computers, Internet) is less than twenty years old, and most of the software technology (Java, object-oriented programming) is less than five years old. Prior to these scientific developments, it would have been unthinkable to accurately model real world phenomena and demand-management decisions. Without the information technology, it would be impossible to *operationalize* this science. These two capabilities *combined* make possible an entirely new approach to decision making—one that has profound consequences for demand management.

The first consequence is that science and technology now make it possible to manage demand on a *scale and complexity* that would be unthinkable through manual means (or would require a veritable army of analysts to achieve). A modern large airline, for example, can have thousands of flights a day and provide service between hundreds of thousands of origin-destination pairs, each of which is sold at dozens of prices—and this entire problem is replicated for hundreds of days into the future! A similar complexity is found at most large retail chains, which can have tens of thousand of SKUs[2] sold in hundreds of stores and over the Web with prices monitored and updated on a daily basis. The sheer scale and complexity of the decision-making task in these cases is beyond the ability of human decision makers. And if not automated, the task has to be so highly aggregated and simplified that significant opportunities for incremental gains—on particular products, at particular locations, at specific points in time—are simply lost.

---

[2] A *SKU* (stock-keeping unit) is the lowest level at which we identify inventory—such as men's Arrow blue Oxford shirts, long sleeves, size medium.

The second consequence of science and technology is that they make it possible to improve the *quality* of demand-management decisions. The management tasks that are involved—quantifying the risks and rewards in making demand-management decisions under uncertainty; working through the often subtle economics of pricing; accurately interpreting market conditions and trends and reacting to this information with timely, accurate, and consistent real-time decisions; optimizing a complex objective function subject to many constraints and business rules—are tasks most humans, even with many years of experience, are simply not good at. Models and systems are better at separating market signals from market noise, evaluating complex tradeoffs, and optimizing and producing consistent decisions. The application of science and technology to demand decisions often produces an improvement in the quality of the decisions, resulting in a significant increase in revenues.

Of course, even with the best science and technology, there will always be decisions that are better left to human decision makers. Models can detect only what's in the data. They cannot reason through the consequences of a demand shock, new technologies, a sudden shift in consumer preferences, or the surprise price war of a competitor. These higher-level analyses are best left to experienced, human analysts. Most RM systems recognize this fact and parse the decision-making task, with models and systems handling routine demand-management decisions on an automated basis and human analysts overseeing these decisions and intervening (based on flags or alerts from the system) when extraordinary conditions arise. Such man-machine interaction offers a firm the best of both human and automated decision making.

The process of managing demand decisions with science and technology—implemented with disciplined processes and systems, and overseen by human analysts (a sort of "industrialization" of the entire demand-management process)—defines modern RM.

1.2 The Origins of RM

Where did RM come from? In short, the airline industry. There are few business practices whose origins are so intimately connected to a single industry. Here we briefly review the history of airline RM and then discuss the implications of this history for the field.

1.2.1 Airline History

The starting point for RM was the Airline Deregulation Act of 1978. With this act, the U.S. Civil Aviation Board (CAB) loosened control of airline prices, which had been strictly regulated based on standardized

Introduction

price and profitability targets. Passage of the act led to rapid change and a rash of innovation in the industry. Established carriers were now free to change prices, schedules, and service without CAB approval. Large airlines accelerated their development of computerized reservation systems (CRSs) and global distribution systems (GDSs), and the GDS business became profitable in its own right. The majors developed hub-and-spoke networks, which allowed them to offer service in many more markets than was possible with point-to-point service but also made pricing and operations more complex.

At the same time, new low-cost and charter airlines entered the market. Many of these upstarts—because of their lower labor costs, simpler (point-to-point) operations, and no-frills service—were able to profitably price much lower than the major airlines. These new entrants tapped into an entirely new and vast market for discretionary travel—families on a holiday, couples getting away for the weekend, college students visiting home—many of whom might otherwise have driven their cars or not traveled at all. It turned out (quite surprisingly to some at the time) that air travel was quite price elastic; with prices sufficiently low, people switched from driving to flying, and demand from this segment surged.

The potential of this market was embodied in the rapid rise of PeopleExpress, which started in 1981 with cost-efficient operations and fares 50 to 70% lower than the major carriers. By 1984, its revenues were approaching $1 billion, and for the year 1984 it posted a profit of $60 million, its highest profit ever (Cross [137]).

While these developments resulted in a significant migration of price-sensitive discretionary travelers to the new, low-cost carriers, the major airlines had strengths that these new entrants lacked. They offered more frequent schedules, service to more city pairs and an established brand name and reputation. For many business travelers, schedule convenience and service was (and still is) more important than price, so the threat posed by low-cost airlines was less acute in the business-traveler segment of the market. Nevertheless, the cumulative losses in revenue from the shift in traffic were badly damaging the profits of major airlines.

A strategy to recapture the leisure passenger was needed. Yet, for the majors, a head-to-head, across-the-board price war with the upstarts was deemed almost suicidal; with their much lower costs, airlines like PeopleExpress could earn a profit at the new low prices, while most majors would lose money at a staggering rate.

Robert Crandall, American Airline's vice president of marketing at the time, is widely credited with the breakthrough in solving this problem. He recognized that his airline was already producing seats at a marginal cost near zero because most of the costs of a flight (capital costs, wages, fuel) are fixed. As a result, American could in fact afford to "compete on cost" with the upstarts using its surplus seats.

However, two problems had to be solved to execute this strategy. First, American had to have some way of identifying the "surplus" seats on each flight. The scheme would not be profitable if a sale of a low-price seats displaced high-paying business customers.[3] Second, they had to ensure that American's business customers did not switch and buy the new low-price products it offered to discretionary, leisure customers.

American solved these problems using a combination of *purchase restrictions* and *capacity-controlled fares*. First, they designed discounts that had significant restrictions for purchase: they had to be purchased 30 days in advance of departure, were nonrefundable, and required a seven day minimum stay. These restrictions were designed to prevent most business travelers from utilizing the new low fares. At the same time, American limited the number of discount seats sold on each flight: they *capacity-controlled* the fares. This combination provided the means to compete on price with the upstart airlines without damaging their core business-traveler revenues.

The new pricing scheme was launched in 1978 as American Super-Saver Fares. The fares were quite successful at stemming the tide of defections of discretionary travelers to the low-cost airlines.

Despite this initial success, American experienced some significant problems implementing its new strategy. Initially, American's capacity controls were based on setting aside a fixed portion of seats on each flight for the new low-fare products. But as American gained experience with its Super-Saver fares, it realized that not all flights were the same. Flights on different days and at different times had very different patterns of demand. Some had many excess seats and could profitably support a higher allocation of discount seats; others had sufficient demand for regular-priced seats and warranted very little if any allocation to the new, discounted products.

American realized that a more intelligent approach was needed to realize the full potential of capacity-controlled discounts. It therefore embarked on the development of what became known as the *Dynamic Inventory Allocation and Maintenance Optimizer* system (DINAMO). These efforts on DINAMO represent, in many ways, the first large-scale RM system development in the industry. (Though on a more modest scale, the capacity-control problem dates back to the mid-1970s, and other airlines and the Boeing Aircraft Company were experimenting with

---

[3] As we show in the chapters that follow, a notion of this sort of *displacement cost* is central to the theory of RM.

similar ideas at the time.) The DINAMO system was large and complex and took several years to develop and refine.

DINAMO was implemented in full in January 1985 along with a new fare program entitled Ultimate Super-Saver Fares, which matched or undercut the lowest discount fares available in every market American served.

DINAMO made all this possible. American could now be much more aggressive on price. It could announce low fares that spanned a large swath of individual flights, confident in its capability to accurately capacity-control the discounts on each individual departure. If a rival airline advertised a special fare in one of American's markets, American could immediately match the offer across the board, knowing that the DINAMO system would carefully control the availability of this fare on the thousands of departures affected by the price change. Moreover, the competition could not observe American's capacity controls unlike prices themselves, which, thanks to GDSs, instantly became public information. This feature of pricing aggressively and competitively at an aggregate, market level, while controlling capacity at a tactical, individual-departure level still characterizes the practice of RM in the airline industry today.

The effect of this new capability was dramatic. PeopleExpress was especially hard hit as American repeatedly matched or beat their prices in every market it served. PeopleExpress's annual profit fell from an all-time high in 1984 (the year prior to implementation of DINAMO) to a loss of over $160 million by 1986 (one year after DINAMO was implemented). It soon went bankrupt as a result of mounting losses, and in September 1986 the company was sold to Continental Airlines.

Donald Burr, CEO of PeopleExpress, summarized the reasons behind the company's failure [137]:

> We were a vibrant, profitable company from 1981 to 1985, and then we tipped right over into losing $50 million a month. We were still the same company. What changed was American's ability to do widespread Yield Management in every one of our markets. We had been profitable from the day we started until American came at us with Ultimate Super Savers. That was the end of our run because they were able to under-price us at will and surreptitiously.
>
> Obviously PeopleExpress failed ... We did a lot of things right. But we didn't get our hands around Yield Management and automation issues. ... [If I were to do it again,] the number one priority on my list every day would be to see that my people got the best information technology tools. In my view, that's what drives airline revenues today more than any other factor—more than service, more than planes, more than routes."

This story was played out in similar fashion throughout the airline industry in the decades following deregulation. And airlines that did not have similar RM capabilities scrambled to get them.

As a result of this history, the practice of RM in the airline industry today is both pervasive and mature, and RM is viewed as critical to running a modern airline profitably. For example, American Airlines' estimates that its RM practices generated $1.4 billion in additional incremental revenue over a three-year period starting around 1988 [477]. Many other carriers also attribute similar improvements in their revenue due to RM.

1.2.2 Consequences of the Airline History

The intimate connection of RM to the airline industry is both a blessing and a curse for the field of RM. The blessing is that RM can point to a major industry in which the practice of RM is pervasive, highly developed, and enormously effective. Indeed, a large, modern airline today would just not be able to operate profitably *without* RM. By most estimates, the revenue gains from the use of RM systems are roughly comparable to many airlines' total profitability in a good year (about 4 to 5% of revenues).[4] And the scale and complexity of RM at major airlines is truly mind-boggling. Therefore, the airline success story validates both the economic importance of RM and the feasibility of executing it reliably in a complex business environment. This is the good-news story for the field from the airline experience.

The bad news—the curse if you will—of the strong association of RM with airlines is that it has created a certain myopia inside the field. Many practitioners and researchers view RM solely in airline-specific terms, and this has at times tended to create biases that have hampered both research and implementation efforts in other industries.

A second problem with the airline-specific association of RM is that airline pricing has something of a bad reputation among consumers. While on the one hand customers love the very low fares made possible by RM practices, the fact that fares are complex, are available one minute and gone the next, and can be drastically different for two people sitting side by side on the same flight, has led to a certain hostility toward the way airlines price. As a result, managers outside the industry are at times, quite naturally, somewhat reluctant to try RM practices for fear of engendering a similar hostile reaction among their customers.

---
[4]Many skeptics point to Southwest Airlines as a counterexample, but Southwest does use RM systems. However, because its pricing structure is simpler than most other airlines the use of RM is less obvious to consumers and casual observers.

Introduction

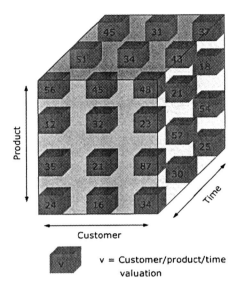

Figure 1.1. Conceptual view of a firm's demand landscape

Yet the reality is that, in most cases, applying RM does *not* involve radically changing the structure of pricing and sales practices; rather, it is a matter of making more intelligent decisions.

1.3 A Conceptual Framework for RM

So if airlinelike conditions aren't strictly necessary for RM, then exactly where *does* it apply? A short answer is: in any business where tactical demand management is important and the technology and management culture exists to implement it. But this in turn begs the question: when do *these* conditions arise? To answer this question, it helps to begin with a conceptual framework for thinking about the demand management process.

1.3.1 The Multidimensional Nature of Demand

A firm's demand has multiple dimensions, including (1) the different products it sells, (2) the types of customers it serves, their preferences for products, and their purchase behaviors, and (3) time. Other dimensions (such as, locations or channels) also affect the nature of a firm's demand, but these three dimensions—products, customers, and time—suffice to illustrate the idea.

Figure 1.1 shows these three demand dimensions. A single cell in the figure indicates a particular customer's valuation for a particular product at a particular point in time. RM addresses the structural, price, timing and quantity decisions a firm makes in trying to exploit the potential of this multidimensional demand landscape.

For example, some RM problems look at exploiting heterogeneity in valuations among customers for a single product at a single point in time: they fix the product and time dimension and try to optimize over the customer dimension. This problem is characteristic of the classical auction-design problems discussed in Chapter 6 and classical price-discrimination problems discussed in Chapter 8. Other RM problems look at dynamically pricing a single product to heterogeneous customers over time: they fix the product dimension and optimize over the customer and time dimensions. Such problems are addressed in Chapter 5. Others, such as the network problems in Chapter 3, address managing demand decisions for multiple products over multiple time periods, and the customer-behavior dimension is not explicitly considered. Of course, all three dimensions are important factors in practice. However, methodologically one often has to decompose and simplify the problem to develop implementable solutions.

1.3.2 Linkages Among Demand-Management Decisions

If the decisions affecting the demand landscape in Figure 1.1 were independent, then the decision-making problem would be considerably simpler. However, typically one or more of the following three factors link the demand across these dimensions.

First, multiple products may share production capacity or have joint production costs. In such cases, the demand-management decision for different products or for a given product in different periods of time are interrelated. For example, because of joint capacity constraints, accepting demand from a customer for a particular product at a specific point in time may mean giving up opportunities to accept demand at later points in time, or because lowering the price of one product increases its demand, this may reduce the capacity available for producing other products.

Second, even if production constraints do not link demand decisions, customer behavior often does. Customers may choose among substitute products at any given point in time, or customers may strategize over their timing in purchasing a given product. As a result, the price or quantity decisions that a firm makes about one product may affect demand for related products—or may affect the future demand for the same product.

Finally, demand decisions for different products, customers, and time-periods may also be linked in terms of the information the firm gains. The most common link is over time; observed demand to date may reveal information about future demand. Thus, a decision about price today may affect the information we gain about demand sensitivity, which will affect future pricing decisions. Also, a firm selling the same product in geographically separated markets or in different channels may gain information in one location or channel as a result of observing demand that impacts its decisions in other locations and channels. Or the observed purchase decisions of a given customer may reveal information about that customer's future purchase decisions. Such linkages complicate demand-management decisions, and managing the often subtle tradeoffs they create is a key motivation for RM.

1.3.3 Business Conditions Conducive to RM

Given this conceptualization of the demand-management problem, one can begin to gain insights into conditions in which RM is likely to be beneficial. Here, we discuss a few such conditions.

1.3.3.1 Customer Heterogeneity

If all customers value a product identically and exhibit similar purchase behavior, then the customer dimension of Figure 1.1 is essentially lost. As a result, there is less potential to exploit variations in willingness to pay, variations in preference for different products, and variations of purchase behavior over time. Therefore, the more heterogeneity in customers, the more potential there is to exploit this heterogeneity strategically and tactically to improve revenues.

Customers in the airline and hotel industries certainly exhibit this characteristic. They have widely varying patterns of usage and behavior in terms of when they purchase and how flexible their plans are, and they place very different valuations on the need to travel.

1.3.3.2 Demand Variability and Uncertainty

The more demand varies over time (due to seasonalities, shocks and so on) and the more uncertainty one has about future demand (the more variance there is along the time dimension in Figure 1.1) the more difficult the demand-management decisions become. Hence, the potential to make bad decisions rises, and it becomes important to have sophisticated tools to evaluate the resulting complex tradeoffs.

Consider the demand for air travel. It exhibits significant variations (by season, time of day, day of week, holidays) and even correcting for this predictable seasonal variation is highly uncertain for a given flight.

1.3.3.3 Production Inflexibility

As mentioned, joint production constraints and costs complicate the demand-management problem. If a firm can "absorb" variations in *demand* easily and costlessly through variations in *supply*, then the complexity of managing demand diminishes; you just supply enough to meet demand. However, the more inflexible the production—the more delays involved in producing units, the more fixed costs or economies of scale involved in production, the more the switch-over costs, the more capacity constraints—the more difficult or costly it becomes to match demand variations with supply variations. As a result, inflexibility leads to more interaction in the demand management at different points in time, between different segments of customers, across different products of a product line, and across different channels of distribution (the different cells in Figure 1.1). The complexity increases and the consequences of poor decisions become more acute. Hence, RM becomes more beneficial.

Again, the airline industry is one in which production is very inflexible. Essentially, when committing to fly a flight from A to B, an airline both *fixes* the level of its output (the number of seats) and, for all practical purposes, the total cost of that output—independent of how many customer actually fly on the flight. Its unit cost per seat sold, therefore, varies tremendously with the volume of sales, and once the capacity constraint is reached, no more production is possible. Worse yet, like all services, output cannot be inventoried, so production of air transport output in one period cannot be used to satisfy demand in later periods (an unsold seat on Monday cannot be used to supply the need of an excess passenger on Tuesday). All these factors combine to create extreme inflexibility in the technology of air transport service, and this is one of the key driving factors in the importance of RM in this industry.

1.3.3.4 Price as a Signal of Quality

The extent to which price is a signal of quality is also a factor. For example, people buy a $10,000 Patek Philippe watch partly for its aesthetics and functionality but also, to a large extent, because they want the exclusivity of a $10,000 watch. The price is a key feature of the watch, as it is with most luxury goods. They are status symbols, and to lower or manipulate the price risks damaging this status.

A more subtle case is observed in situations where it is hard to assess quality through other, objective means. For example, the hourly rate of a prominent attorney or consultant, the tuition at an Ivy League university, and the price of a bottle of wine on a dinner menu—all play important roles as signals of quality to consumers. Again, tampering with prices for tactical reasons in such settings jeopardizes the signaling value of prices. Therefore, RM is more suited to products where price is *not* a status symbol and *not* a significant signal of value—where price and quality are decoupled in the consumer's mind.

Airlines are arguably a good example. While different airlines position themselves differently with respect to price and quality (e.g., no-frills discount carriers and full-service, mainstream carriers), consumers generally do not associate the price of an airline ticket with the quality of the particular flight. We do not expect a "nicer" flight when paying $300 more because we booked our ticket at the last minute or because we booked our flight on a holiday weekend as opposed to a normal weekday. Moreover—despite what some airline marketers might like to believe—most consumer do not have strong quality preferences among airlines, at least not sufficient to outweigh even relatively small differences in schedule and price. This is one of the main motivations behind the introduction of loyalty schemes in the industry, which are really an attempt to "synthesize" a high level of brand loyalty among a group of consumers who innately have very little of it.

1.3.3.5 Data and Information Systems Infrastructure

To operationalize RM requires data to accurately characterize and model demand. It also requires systems to collect and store the data and to implement and monitor the resulting real-time decisions. In most industries it is usually feasible—in theory, at least—to collect and store demand data and automate demand decisions. However, attempting to apply RM in industries that do not have databases or transactions systems in place can be a time-consuming, expensive, and risky proposition. RM, therefore, tends to be more suited to industries where and transaction-processing systems are already employed as part of incumbent business processes.

Again, the airline industry is a perfect case in point. It is an industry whose pricing and distribution processes were largely automated with the introduction of GDSs in the 1960's and 1970's. In fact, it is one of the earliest industries to move almost entirely to electronic selling and distribution—decades before the advent of e-commerce. This long history of using information systems to automate business processes meant that it was quite natural to implement RM in the airline industry when the time came.

1.3.3.6 Management Culture

RM is a technically complex and demanding practice. There is a risk, therefore, that a firm's management may simply not have sufficient familiarity with—or confidence in—science and technology to make implementing a RM system a realistic prospect. The culture of the firm may not be receptive to innovation or may value more intuitive approaches to problem solving. This is often due to the culture of the industry and its managers: their educational backgrounds, their professional experiences and responsibilities en route to leadership positions, and the skills required to succeed in the industry.

Again, the airline industry serves as a good example. Modern airlines cannot run without information systems: systems for ticketing and reservations, scheduling crews and aircraft, handling baggage, planning meals and operational control (rerouting aircraft because of delays and breakdowns, and so on). Also, airline managers are accustomed to applying scientific methods in managing these various operations. In fact, long before RM was practiced in the industry, most large airlines had staffs of operations researchers working on complex problems of scheduling and fleet assignment. When RM came along, the management and culture in the industry were therefore well conditioned to accept it.

1.3.4 Industry Adopters Beyond the Airlines

What do these conditions imply for adopters of RM technology? Chapter 10 reviews specific industry adopters in detail, so here we only briefly mention some of them.

The production-inflexibility characteristics of airlines are shared by many other service industries, such as hotels, cruise ship lines, car rental companies, theaters and sporting venues, and radio/TV broadcasters, to name a few. Indeed, RM is strongly associated with service industries.

Retailers have recently begun to adopt RM, especially in the fashion apparel, consumer electronics, and toy sectors. Retail demand is highly volatile and uncertain, consumers' valuations change rapidly over time, and with short selling seasons and long production and distribution lead times, supply is quite inflexible. On the technology front, the introduction of bar codes and point-of-sale (POS) technology has resulted in a high degree of automation of sales transactions for most major retailers.

The energy sector has been a recent adopter of RM methods as well, principally in the area of managing the sale of pipeline capacity for gas transportation. Again, energy demands are volatile and uncertain, and the technology for generating and transmitting electricity and gas can be inflexible. Also, thanks to deregulation in the industry, there has

Introduction

been a lot of experimentation and innovation in the pricing practices of energy, gas, and transmission markets.

Manufacturing is potentially a vast market for RM methods, though to date relatively few instance of the practice have been documented. To a large extent this is due to the fact that supply is more flexible, and, for durable goods, customers have more flexibility in their purchase timing. This somewhat diminishes the impact of RM and creates unique challenges for the methodology as well. Still, there is immense interest in RM in manufacturing. Enterprise resource planning (ERP), supply-chain management (SCM), and customer-relationship management (CRM) systems are commonplace in the industry, and most manufacturers have huge amounts of data and heavily automated business processes, which could form the foundations for RM. Indeed, in the auto industry Ford Motor Corporation recently completed a high-profile implementation of RM technology [135].

What about future adopters of RM? Given the criteria outlined above, one can argue that many industries are potential candidates. Almost all businesses must deal with demand variability, uncertainty, and customer heterogeneity. Most are subject to some sort of supply or production inflexibility. Finally, thanks largely to the wave of enterprise software and e-commerce innovation of late, many firms have now automated their business processes. All of these factors bode well for the future of RM.

Nevertheless, as with any technological and business-practice innovation, the case for RM ultimately boils down to a cost-benefit analysis for each individual firm. For some, the potential benefit will simply never justify the costs of implementing RM systems and business processes. However, we believe that for the majority of firms, RM will eventually be justified once the technology and methodology in their industry matures. Indeed, the history of RM in industries such as airlines, hotels, and retail suggests that once the technology gains a foothold in an industry, it spreads quite rapidly. As a result, we would not be surprised to see RM systems (or systems performing RM functions under a different label) become as ubiquitous as ERP, SCM, and CRM systems are today.

1.4 An Overview of a RM System

Here, we give a brief description of the generic operations of a RM system. This introduces the key components and gives an overview of the information flows, controls, and design of a RM system. The details of the science and systems involved in each component are covered in later chapters.

RM generally follows four steps:

1. Data collection: Collect and store relevant historical data (prices, demand, causal factors).

2. Estimation and forecasting: Estimate the parameters of the demand model; forecast demand based on these parameters; forecast other relevant quantities like no-show and cancellation rates, based on transaction data.

3. Optimization: Find the optimal set of controls (allocations, prices, markdowns, discounts, overbooking limits) to apply until the next re-optimization.

4. Control: Control the sale of inventory using the optimized control. This is done either through the firm's own transaction-processing systems or through shared distribution systems (such as GDSs).

The RM process typically involves cycling through these steps at repeated intervals. The frequency with which each step is performed is a function of many factors such as the volume of data, the speed that business conditions change, the type of forecasting and optimization methods used, and the relative importance of the resulting decisions. For example, most RM systems in airline and hotel applications stagger the dates—data collection points (DCPs)—when they collect data, reforecast, and reoptimize, with the cycle occurring more frequently (at least daily) as the service time nears. This is because in these industries, a substantial portion of the reservations occurs during the last few days before the time of service.

Figure 1.2 shows the process flow in a RM system. Data is fed to the forecaster; the forecasts become input to the control optimizer; and finally the controls are uploaded to the transaction-processing system, which controls actual sales.

1.5   The State of the RM Profession

On the practice side, the profession can be divided into *users* (the firms and individuals who use RM methods to manage their business) and *vendors* (the firms and individuals who develop and supply technology and consulting services to users). Of course, this division is not always sharp. Many users of RM, especially in the airline industry, have research and development organizations that provide significant components of their firm's RM technology. Still, most users—even those with their own RM staff—rely on vendors in part or whole for their technology. Often, the role of a user's R&D staff is to serve as in-house technology advisers and consultants, helping senior management evaluate new technologies and manage the relationships with the firm's technology vendors.

Introduction

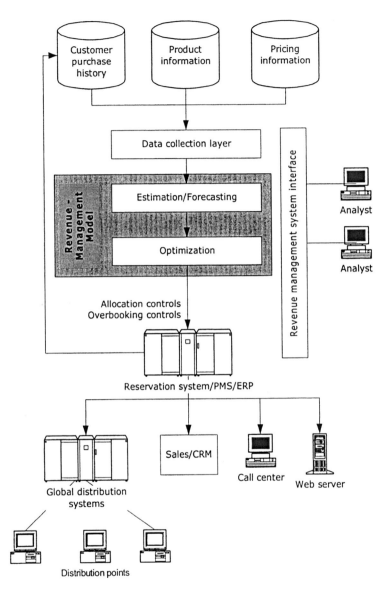

Figure 1.2. RM process flow.

In most user organizations, the vast majority of RM staff are involved in day-to-day RM operational activities: training and supervision of field staff, managing pricing and capacity controls in individual markets, overseeing automated decisions and intervening where necessary or maintaining computer systems. The typical senior management titles in such organizations are *VP of RM* or *VP of pricing and inventory control*. The organization typically has a corporate staff that is responsible for overall RM strategy, policy and systems and line management and staff responsible for RM processes in specific business units and markets.

Except for a few large airlines that develop their own systems, RM software is developed by a handful of RM vendors, many of whom specialize in a particular industry. Most of these firms have both a scientific staff to develop models and algorithms (operations researchers, marketing scientists, statisticians, economists), an IT and software staff to develop the associated software and systems, and a consulting staff to provide training and implementation services. The resulting products are usually customized for each user's particular business conditions. Vendors also provide training and consulting in the use of the systems. To this list of RM-specific vendors, one ought to add the major enterprise software and technology companies and general IT consulting and software companies that, though not specifically identified as RM vendors, nevertheless provide some RM products and services.

1.6 Chapter Organization and Reading Guide

We next describe the organization of the book and then provide our suggestions for how to approach the material.

1.6.1 Chapter Organization

The book is divided into three main parts. Part I addresses *quantity-based RM*, in which the primary demand-management decisions concern product rationing and availability control—how much to sell to whom, whether to accept or reject requests for products, and so on. These are the core set of problems behind traditional airline RM and closely related industries like hotels and rental car industries. Part I is comprised of these chapters:

- Chapter 2, Single-Resource Capacity Control: This chapter looks at capacity controls for a single resource (seats on a single flight, hotel rooms on a single night) that is sold to differentiated demand classes—the so-called single-leg problem in airline RM parlance. We provide a comprehensive treatment of the classic exact and heuristic approaches to this problem, as well as a number of more recent advances.

- Chapter 3, Network Capacity Control: This chapter looks at the same capacity-control decisions, but in a setting in which products require multiple resources—called the *network problem*. The main motivation is controlling availability of discount classes at an origin-destination (O&D) level in an airline network. However, hotels face a similar network problem when they control capacity by length of stay. Because the network capacity-control problem is significantly more complex than the single-resource problem, most of the methods in this chapter are based on approximations.

- Chapter 4, Overbooking: This chapter looks at the practice of overbooking—accepting more reservations than physical capacity as a hedge against cancellations. The topic is somewhat specialized to reservation-based industries like airlines, hotels, and car rentals. While in a sense overbooking is a demand-management decision, it is somewhat different from the pricing and allocation decisions of the other chapters. However, overbooking is intimately connected to RM in the airline and hotel industries and is almost always implemented in conjunction with RM capacity controls. It is also extremely important economically in these industries and forms a significant and visible part of RM.

Part II of the book examines *price-based RM*, in which the primary demand decisions are prices—how to price to various customer groups or how to vary prices over time. Both posted price and auction mechanisms are considered. These price-based RM problems are more typical of retail and manufacturing RM. Part II has two chapters:

- Chapter 5, Dynamic Pricing: In this chapter we look at a problem in which the principle demand decision is how to adjust prices over time, subject to demand variability and uncertainty and various constraints or costs on re-supply. Many of the retail RM systems are based on the types of models discussed in this chapter.

- Chapter 6, Auctions: Auctions are an important and long-standing pricing mechanism in many industries and, with the rise of e-commerce, have gained popularity as a alternatives to posted pricing. The basic types of auctions are discussed along with the theory of optimal auction design. We discuss the implications of this theory for dynamic pricing in general and look at classical auctions, dynamic auctions and network auctions.

Finally, the five chapters in Part III of the book examine components of RM that are common to both quantity and price-based RM:

- Chapter 7, Customer Behavior and Market-Response Models: This chapter summarizes the core demand-modeling theory and methodology underlying RM. We discuss the basic theory of consumer behavior and develop several of the demand models used in both quantity and price-based RM. Both individual customer choice and aggregate market-demand models are covered.

- Chapter 8, The Economics of RM: Here we discuss the economic theory of RM. We briefly survey classical monopoly and oligopoly pricing theory as well as the theory of price discrimination, peak-load pricing, and pricing under demand uncertainty, all of which are particularly relevant to understanding the strategies and tactics used in RM practice.

- Chapter 9, Estimation and Forecasting: This chapter addresses the broad range of issues involved in estimating models from data and building forecasts of future demand. We survey the main estimation and forecasting methods commonly used in practice. The coverage is not intended to be as in-depth as specialized books on these topics but rather to review the basic assumptions and theory of each method and its role in RM practice.

- Chapter 10, Industry Profiles: This chapter provides detailed descriptions of several industries practicing RM, including information on consumers, products, sales practice and technology—all of which impact the real world practice of RM. For experienced industry insiders, much of this material may be well-known. However, for new employees in an industry, for academics, and for industry practitioners looking at a different industry, the chapter provides useful information on the institutional context in which RM is practiced.

- Chapter 11, Implementation: This chapter discusses issues involved in implementing a RM system, including product design, organizational and technology-management issues, all factors critical in making a RM system effective in application.

1.6.2   Reading Guide

Some readers will not want to read the book in strict sequential order. It is certainly possible to read Parts I and II independently of one another. Readers who are interested primarily in traditional quantity-based RM should begin with Part I, while those interested primarily in price-based RM problems could begin with Part II and then look at Part I afterwards. However, within Parts I and II chapters are interrelated, with later chapters building on ideas developed in earlier chapters.

Each chapter provides a comprehensive introduction as well, so readers may wish to begin by looking through each of the chapter introductions to get a sense of the scope of each one and then read individual chapters in detail according to their level of interest.

Parts I and II can also be read largely independently of Part III, though the material in Part III provides useful background. While some readers may choose to use Part III only as a reference, in our view each chapter in Part III is also of significant independent interest. Readers interested in the theory underlying RM will find Chapter 7 on demand modeling and Chapter 8 on economics of particular interest. Those interested primarily in the applied elements of RM will find Chapter 9 on forecasting methods and Chapters 10 and 11 on industry profiles and implementation (respectively) most useful.

The chapters in Part III are not strongly interrelated and may be read in any order. However, the material in Part III is best understood in the context of the topics covered in Parts I and II; hence, we recommend at least skimming the introductions of chapters in Parts I and II before reading Part III in detail.

1.7   Notes and Sources

The 1997 book by Robert Cross, *RM: Hard Core Tactics for Market Domination* [137] was influential in popularizing the story of airline RM and introducing the concept of RM to the general business community. Bob Cross was then chairman and CEO of Aeronomics, a RM consultancy and software firm. It is a nontechnical and lively book for a general audience , and is informative reading, providing nice descriptions of the early history of RM in the airline industry, many practical anecdotes, and insights into the philosophy and challenges of implementing RM. Several other books on RM have been published recently. One is an edited volume by Ingold, McMahon-Beattie, and Yeoman [263] that focuses primarily on the hotel industry. Another, Daudel and Vialle [146], focuses on air transportation. Both, however, deal more with practical and conceptual issues and do not cover the scientific methods of RM in much depth. The book by Nagel and Holden [400] provides a comprehensive overview of many managerial issues involved in pricing and is useful reading. However, it does not address tactical RM decision making in depth.

Several survey articles provide general coverage of RM. The *Handbook of Airline Economics* edited by Jenkins [268] provides several good practice-oriented articles on RM in the airline industry. Kimes [301] provides a conceptual introduction to RM with a hotel RM focus. Smith et al. [477] provide a nice description of the practice of RM at American Airlines and the DINAMO system.

As for guides to the research literature, Weatherford and Bodily [556] propose a taxonomy for classifying the sets of assumptions used in many traditional RM models, although the taxonomy itself is little used. McGill and van Ryzin [374] provide a comprehensive overview and annotated bibliography of the published academic literature in the field through 1998. Elmaghraby and Keskinocak [177] provide a survey on research in the area of dynamic pricing.

PART I

QUANTITY-BASED RM

Chapter 2

SINGLE-RESOURCE CAPACITY CONTROL

2.1 Introduction

In this chapter, we examine the problem of quantity-based revenue management for a single resource; specifically, optimally allocating capacity of a resource to different classes of demand. Two prototypical examples are controlling the sale of different fare classes on a single flight leg of an airline and the sale of hotel rooms for a given date at different rate classes. This is to be contrasted with the multiple-resource—or network—problems of Chapter 3, in which customers require a bundle of different resources (such as two connecting flights or a sequence of nights at the same hotel). In reality, many quantity-based RM problems are network RM problems, but in practice, they are still frequently solved as a collection of single-resource problems (treating the resources independently). For this reason, it is important to study single-resource RM models. Moreover, single-resource models are useful as building blocks in heuristics for the network case.

We assume that the firm sells its capacity in n distinct classes[1] that require the same resource. In the airline and hotel context, these classes represent different discount levels with differentiated sale conditions and restrictions. In the early parts of this chapter, we assume that these products appeal to distinct and mutually exclusive segments of the market: the conditions of sale segment the market perfectly into n segments—one for each class. Customers in each segment are eligible for or can afford only the class corresponding to their segment. Later in

---

[1] In the case of airlines, these are called *fare classes*. Terms like *rate products*, *rate classes*, *revenue classes*, *booking classes* and *fare products* are also used. We shall use the generic term *class* in this chapter.

the chapter, we look at models that do not assume that customers are perfectly segmented, but instead that they choose among the n classes.

The units of capacity are assumed homogeneous, and customers demand a single unit of capacity for the resource. The central problem of the chapter is how to optimally allocate the capacity of the resource to the various classes. This allocation must be done dynamically as demand materializes and with considerable uncertainty about the quantity or composition of future demand. The remainder of the chapter focuses on various models and methods for making these capacity-allocation decisions.

2.1.1 Types of Controls

In the travel industry, reservation systems provide different mechanisms for controlling availability. These mechanisms are usually deeply embedded in the software logic of the reservation system and, as a result, can be quite expensive and difficult to change. Therefore, the control mechanisms chosen for a given implementation are often dictated by the reservation system. The details of reservations systems and the constraints they impose are discussed in greater detail in Chapters 10 and 11. Here, we focus on the control mechanisms themselves.

2.1.1.1 Booking Limits

Booking limits are controls that limit the amount of capacity that can be sold to any particular class at a given point in time. For example, a booking limit of 18 on class 2 indicates that at most 18 units of capacity can be sold to customers in class 2. Beyond this limit, the class would be "closed" to additional class 2 customers. This limit of 18 may be less than the physical capacity. For example, we might want to protect capacity for future demand from class 1 customers.

Booking limits are either *partitioned* or *nested*: A *partitioned booking limit* divides the available capacity into separate blocks (or *buckets*)—one for each class—that can be sold only to the designated class. For example, with 30 units to sell, a partitioned booking limit may set a booking limit of 12 units for class 1, 10 units for class 2, and 8 units for class 3. If the 12 units of class 1 capacity are used up, class 1 would be closed regardless of how much capacity is available in the remaining buckets. This could be undesirable if class 1 has higher revenues than do classes 2 and 3 and the units allocated to class 1 are sold out.

With a *nested booking limit*, the capacity available to different classes overlaps in a hierarchical manner—with higher-ranked classes having access to all the capacity reserved for lower-ranked classes (and perhaps more). Let the nested booking limit for class j be denoted b_j. Then

Single-Resource Capacity Control

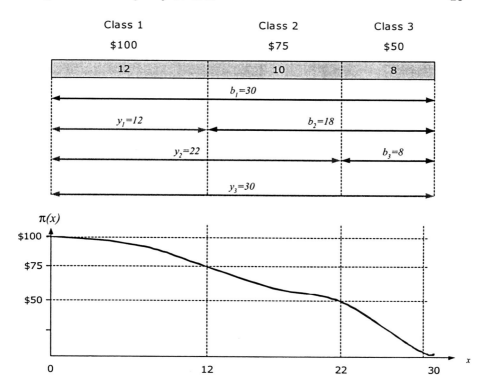

Figure 2.1. The relationship between booking limits b_j, protection levels y_j, and bid prices $\pi(x)$.

b_j is the maximum number of units of capacity we are willing to sell to classes j and lower. So in Figure 2.1, the nested booking limit on class 1 and lower (all classes) would be $b_1 = 30$ (the entire capacity), the nested booking limit on classes 2 and 3 combined would be $b_2 = 18$, and the nested booking limit on class 3 alone would be $b_3 = 8$. We would accept at most 30 bookings for classes 1, 2, and 3, at most 18 for classes 2 and 3 combined, and at most 8 for class 3 customers. Effectively, this logic simply allows any capacity "left over" after selling to low classes to become available for sale to higher classes.

Nesting booking limits in this way avoids the problem of capacity being simultaneously unavailable for a high class yet available for lower classes. Most reservations systems that use booking-limit controls quite sensibly use nested rather than partitioned booking limits for this reason.

2.1.1.2 Protection Levels

A *protection level* specifies an amount of capacity to reserve (protect) for a particular class or set of classes. Again, protection levels can be *nested* or *partitioned*. A partitioned protection level is trivially equivalent to a partitioned booking limit; a booking limit of 18 on class 2 sales is equivalent to protecting 18 units of capacity for class 2.

In the nested case, protection levels are again defined for sets of classes—ordered in a hierarchical manner according to class order. Suppose class 1 is the highest class, class 2 the second highest, and so on. Then the protection level j, denoted y_j, is defined as the amount of capacity to save for classes $j, j-1, \ldots, 1$ combined—that is, for classes j and higher (in terms of class order). Continuing our example, we might set a protection level of 12 for class 1 (meaning 12 units of capacity would be protected for sale only to class 1), a protection level of 22 for classes 1 and 2 combined, and a protection level of 30 for classes 1, 2, and 3 combined. (Though frequently no protection level is specified for this last case since it is clear that all the capacity is available to at least one of the classes.)

Figure 2.1 shows the relationship between protection levels and booking limits. The booking limit for class j, b_j is simply the capacity minus the protection level for classes $j-1$ and higher. That is,

$$b_j = C - y_{j-1}, \quad j = 2, \ldots, n,$$

where C is the capacity. For convenience, we define $b_1 = C$ (the highest class has a booking limit equal to the capacity) and $y_n = C$ (all classes combined have a protection level equal to capacity).

2.1.1.3 Standard Versus Theft Nesting

The standard process for using booking limits or nested protection levels proceeds as follows. Starting with C units of capacity, we begin receiving bookings. A bookings for class j is accepted provided (1) there is capacity remaining and (2) the total number of requests accepted for class j to date is less than the booking limit b_j (equivalently, the current capacity remaining is more than the protection level y_{j-1} for classes higher than j). This is called *standard nesting*, and it is the most natural and common way to implement nested-capacity controls.

Another alternative, which is less prevalent though still encountered occasionally in practice, is called *theft nesting*. In theft nesting, a booking in class j not only reduces the allocation for class j but also "steals" from the allocation of all lower classes. So when we accept a request for class j, not only is the class j allocation reduced by one but so are the allocations for classes $j+1, j+2, \ldots, n$. This is equivalent to keeping y_j

units of capacity protected for *future* demand from class j and higher. In other words, even though we just accepted a request for class j, under theft nesting we continue to reserve y_j units for class j and higher, and to do so requires reducing the allocation for classes $j+1, j+2, \ldots, n$. Under standard nesting, in contrast, when we accept a request from class j we effectively reduce by one the capacity we protect for future demand from class j and higher.

The rationale for standard nesting is that the capacity protected for, say, class 1 is based on a forecast of future demand for class 1. Once we observe some demand for class 1, we then *reduce* our estimate of future demand—and hence the capacity we protect for class 1. Standard nesting does this by reducing the capacity protected for future class 1 demand on a one-for-one basis after each arriving request is accepted (and similarly for other classes as well). To illustrate, suppose in our example demand for class 1 is deterministic and equal to the protection level $y_1 = 12$. Then if we receive 5 requests for class 1, we know for certain that future demand for class 1 will be only 7 and hence that it makes sense to reduce the capacity we protect for future demand from 12 to 7, which is precisely what standard nesting does. Theft nesting, in contrast, intuitively corresponds to an assumption of "memorylessness" in demand. In other words, it assumes the demand to date for class 1 does not affect our estimate of *future* demand for class 1. Therefore, we continue to protect y_1 units of capacity for class 1 (and hence must reduce the allocation for classes $2, 3, \ldots, n$).

The two forms of nesting are in fact equivalent if demand arrives strictly in low-to-high class order; that is, the demand for class n arrives first, followed by the demand for class $n-1$, and so on.[2] This is what the standard (static) single-resource models assume, so for these static models, the distinction is not important. However, in practice demand rarely arrives in low-to-high order, and the choice of standard versus theft nesting matters. With mixed order of arrivals, theft nesting protects more capacity for higher classes (equivalently, allocates less capacity to lower classes). Again, however, standard nesting is the norm in RM practice.

2.1.1.4 Bid Prices

What distinguishes bid-price controls from both booking limits and protection levels is that they are revenue-based rather than class-based controls. Specifically, a bid-price control sets a threshold price (which

---

[2] It is easy to convince oneself of this fact by tracing out the accept/deny decisions under both forms of nesting, and doing so is an instructive exercise.

may depend on variables such as the remaining capacity or time), such that a request is accepted if its revenue exceeds the threshold price and rejected if its revenue is less than the threshold price. Bid-price controls are, in principle, simpler than booking-limit or protection-level controls because they require only storing a single threshold value at any point in time—rather than a set of capacity numbers, one for each class. But to be effective, bid prices must be updated after each sale—and possibly also with time as well—and this typically requires storing a table of bid price values indexed by the current available capacity, current time, or both.

Figure 2.1 shows how bid prices can be used to implement the same nested-allocation policy as booking limits and protection levels. The bid price $\pi(x)$ is plotted as a function of the remaining capacity x. When there are 12 or fewer units remaining, the bid price is over \$75 but less than \$100, so only class 1 demand is accepted. With 13 to 22 units remaining, the bid price is over \$50 but less than \$75 so only classes 1 and 2 are accepted. With more than 22 units of capacity available, the bid price drops below \$50 so all three classes are accepted.

Bid-price control is criticized by some as being "unsafe"—the argument being that having a threshold price as the only control means that the RM system will sell an unlimited amount of capacity to any class whose revenues exceed the bid price threshold. But this is true only if the bid price is not updated. As shown in Figure 2.1, if the bid price is a function of the current remaining capacity, then it performs exactly like a booking limit or protection level, closing off capacity to successively higher classes as capacity is consumed. Without this ability to make bid prices a function of capacity, however, a simple static threshold is indeed a somewhat dangerous form of control.

One potential advantage of bid-price controls is their ability to discriminate based on revenue rather than class. Often (see Section 10.1.3.1) a number of products with different prices are booked in a single class. RM systems then use an average price as the price associated with a class. However, if actual revenue information is available for each request, then a bid-price control can selectively accept only the higher revenue requests in a class, whereas a control based on class designation alone can only accept or reject all requests of a class. Of course, if the exact revenue is not observable at the time of reservation, then this advantage is lost.

2.1.2 Displacement Cost

While the mathematics of optimal capacity controls can become complex, the overriding logic is simple. First, capacity should be allocated to a request if and only if its revenue is greater than the value of the capacity required to satisfy it. Second, the value of capacity should be measured by its (expected) *displacement cost*—or *opportunity cost*—which is the expected loss in future revenue from using the capacity now rather than reserving it for future use.

Theoretically, the displacement-cost idea is captured by using a *value function*, $V(x)$, that measures the optimal expected revenue as a function of the remaining capacity x. The displacement cost is then the difference between the value function at x and the value function at $x-1$, or $V(x)-V(x-1)$. Much of the theoretical analysis of the capacity controls boils down to analyzing this value function. But conceptually, the logic is simply to compare revenues to displacement costs to make the accept or deny decision.

2.2 Static Models

In this section, we examine one of the first models for quantity-based RM, the so-called *static*[3] single-resource models.

The static model makes several assumptions that are worth examining in some detail. The first is that demand for the different classes arrives in nonoverlapping intervals in the order of increasing prices of the classes.[4] In reality, demand for the different classes may overlap in time. However, the nonoverlapping-intervals assumption is a reasonable approximation (for example, advance-purchase discount demand typically arrives before full-fare coach demand in the airline case). Moreover, the optimal controls that emerge from the model can be applied—at least heuristically—even where demand comes in arbitrary order (using either bid prices or the nesting policies, for example). As for the strict low-before-high assumption, this represents something of a worst-case scenario; for instance, if high-revenue demand arrives before low-revenue demand, the problem is trivial because we simply accept demand first come, first serve.

The second main assumption is that the demands for different classes are independent random variables. Largely, this assumption is made for analytical convenience because to deal with dependence in the demand structure would require introducing complex state variables on the history of observed demand. We can make some justification of the

---

[3]The term *static* here is somewhat of a misnomer because demand does arrive sequentially over time, albeit in stages ordered from low-revenue to high-revenue demand. However, this term is now standard and helps distinguish this class of models from *dynamic* models that allow arbitrary arrival orders.
[4]Robinson [445] generalizes the static model to the case where demand from each class arrives in nonoverlapping intervals but the order is not necessarily from low to high revenue.

assumption by appealing to the forecast inputs to the model. That is, to the extent that there are systematic factors affecting *all* demand classes (such as seasonalities), these are often reflected in the forecast and become part of the *explained* variation in demand in the forecasting model (for example, as the differences in the forecasted means and variance on different days). The randomness in the single-resource model is then only the residual, unexplained variation in demand. So, for example, the fact that demand for all classes may increase on peak flights does not in itself cause problems provided the increase is predicted by the forecasting method. Still, one has to worry about possible residual dependence in the *unexplained variation* in demand, and this is a potential weakness of the independence assumption.

A third assumption is that demand for a given class does not depend on the capacity controls; in particular, it does not depend on the availability of other classes. Its only justification is if the multiple restrictions associated with each class are so well designed that customers in a high revenue class will not buy down to a lower class and if the prices are so well separated that customers in a lower class will not buy up to a higher class if the lower class is closed. However, neither is really true in practice. There is considerable porousness (imperfect segmentation) in the design of the restrictions, and the price differences between the classes are rarely that dispersed. The assumption that demand does not depend on the capacity controls is therefore a weakness, though in Section 2.6 we look at models that handle imperfect segmentation.

Fourth, the static model suppresses many details about the demand and control process within each of the periods. This creates a potential source of confusion when relating these models to actual RM systems. In particular, the static model assumes an aggregate quantity of demand arrives in a single stage and the decision is simply how much of this demand to accept. Yet in a real reservation system, we typically observe demand sequentially over time, or it may come in batch downloads. The control decision has to be made knowing only the demand observed to date and is usually implemented in the form of prespecified controls uploaded to the reservation system. These details are essentially ignored in the static model. However, fortunately (and perhaps surprisingly), the form of the optimal control is not sensitive to this assumption and can be applied quite independently of how the demand is realized within a period (all at once, sequentially, or in batches). The simplicity and robustness of the optimal control is in fact a central result of the theory for this class of models.

A fifth assumption of the model is that either there are no groups, or if there are group bookings, they can be partially accepted. Group bookings cause significant methodological problems, and these are discussed in Section 2.4.

Finally, the static models assume risk-neutrality. This is a reasonable assumption in practice, since a firm implementing RM typically makes such decisions for a large number of products sold repeatedly (for example, daily flights, daily hotel room stays, and so on). Maximizing the average revenue, therefore, is what matters in the end. While we do not cover this case here, some researchers have recently analyzed the single-resource problem with risk-averse decision makers (Feng and Xiao [187]).

We start with the simple two-class model in order to build some basic intuition and then examine the more general n-class case.

2.2.1 Littlewood's Two-Class Model

The earliest single-resource model for quantity-based RM is due to Littlewood [347]. The model assumes two product classes, with associated prices $p_1 > p_2$. The capacity is C, and we assume there are no cancellations or overbooking. Demand for class j is denoted D_j, and its distribution is denoted by $F_j(\cdot)$. Demand for class 2 arrives first. The problem is to decide how much class 2 demand to accept before seeing the realization of class 1 demand.

The two-class problem is similar to the classic *newsboy problem* in inventory theory, and the optimal decision can be derived informally using a simple marginal analysis: Suppose that we have x units of capacity remaining and we receive a request from class 2. If we accept the request, we collect revenues of p_2. If we do not accept it, we will sell unit x (the marginal unit) at p_1 if and only if demand for class 1 is x or higher. That is, if and only if $D_1 \geq x$. Thus, the expected gain from reserving the xth unit for class 1 (the *expected marginal value*) is $p_1 P(D_1 \geq x)$. Therefore, it makes sense to accept a class 2 request as long as its price exceeds this marginal value, or equivalently, if and only if

$$p_2 \geq p_1 P(D_1 \geq x). \tag{2.1}$$

Note the right-hand side of (2.1) is decreasing in x. Therefore, there will be an optimal protection level, denoted y_1^*, such that we accept class 2 if the remaining capacity exceeds y_1^* and reject it if the remaining capacity is y_1^* or less. Formally, y_1^* satisfies

$$p_2 < p_1 P(D_1 \geq y_1^*) \quad \text{and} \quad p_2 \geq p_1 P(D_1 \geq y_1^* + 1).$$

If a continuous distribution $F_1(x)$ is used to model demand (as is often the case), then the optimal protection level y_1^* is given by the simpler expressions

$$p_2 = p_1 P(D_1 > y_1^*), \quad \text{equivalently}, \quad y_1^* = F_1^{-1}(1 - \frac{p_2}{p_1}), \qquad (2.2)$$

which is known as *Littlewood's rule*. Setting a protection level of y_1^* for class 1 according to Littlewood's rule is an optimal policy. Equivalently, setting a booking limit of $b_2^* = c - y_1^*$ on class 2 demand is optimal. Alternatively, we can use a bid-price control with the bid price set at $\pi(x) = p_1 P(D_1 > x)$.

We omit a rigorous proof of Littlewood's rule since it is a special case of a more general result proved below. However, to gain some insight into it, consider the following example:

Example 2.1 Suppose D_1 is normally distributed with mean μ and standard deviation σ. Then by Littlewood's rule, $F_1(y_1^*) = 1 - p_2/p_1$, which implies the optimal protection level can be expressed as

$$y_1^* = \mu + z\sigma,$$

where $z = \Phi^{-1}(1 - p_2/p_1)$ and $\Phi(\cdot)^{-1}$ denotes the inverse of the standard normal c.d.f. Thus, we reserve enough capacity to meet the mean demand for class 1, μ, plus or minus a factor that depends both on the revenue ratio and the demand variation σ. If $p_2/p_1 > 0.5$, the optimal protection level is less than the mean demand; and if $p_2/p_1 < 0.5$, it is greater than the mean demand. In general, the lower the ratio p_2/p_1, the more capacity we reserve for class 1. This makes intuitive sense because we should be willing to take very low prices only when the chances of selling at a high price are lower.

2.2.2 n-Class Models

We next consider the general case of $n > 2$ classes. Again, we assume that demand for the n classes arrives in n stages, one for each class, with classes arriving in increasing order of their revenue values. Let the classes be indexed so that $p_1 > p_2 > \cdots > p_n$. Hence, class n (the lowest price) demand arrives in the first stage (stage n), followed by class $n-1$ demand in stage $n-1$, and so on, with the highest price class (class 1) arriving in the last stage (stage 1). Since there is a one-to-one correspondence between stages and classes, we index both by j. Demand and capacity are most often assumed to be discrete, but occasionally we model them as continuous variables when it helps simplify the analysis and optimality conditions.

2.2.2.1   Dynamic Programming Formulation

This problem can be formulated as a dynamic program in the stages (equivalently, classes), with the remaining capacity x being the state variable. At the start of each stage j, the demand $D_j, D_{j-1}, \ldots, D_1$ has not been realized. Within stage j, the model assumes that the following sequence of events occurs:

1. The realization of the demand D_j occurs, and we observe its value.

2. We decide on a quantity u of this demand to accept. The amount accepted must be less than the capacity remaining, so $u \leq x$. The optimal control u^* is therefore a function of the stage j, the capacity x, and the demand D_j, $u^* = u^*(j, x, D_j)$, though we often suppress this explicit dependence on j, x and D_j in what follows.

3. The revenue $p_j u$ is collected, and we proceed to the start of stage $j - 1$ with a remaining capacity of $x - u$.

This sequence of events is assumed for analytical convenience; we derive the optimal control u^* "as if" the decision on the amount to accept is made *after* knowing the value of demand D_j. In reality, of course, demand arrives sequentially over time, and the control decision has to be made *before* observing all the demand D_j. However, it turns out that optimal decisions do not use the prior knowledge of D_j as we show below. Hence, the assumption that D_j is known is not restrictive.

Let $V_j(x)$ denote the value function at the start of stage j. Once the value D_j is observed, the value of u is chosen to maximize the current stage j revenue plus the revenue to go, or

$$p_j u + V_{j-1}(x - u),$$

subject to the constraint $0 \leq u \leq \min\{D_j, x\}$. The value function entering stage j, $V_j(x)$, is then the expected value of this optimization with respect to the demand D_j. Hence, the Bellman equation is[5]

$$V_j(x) = E\left[\max_{0 \leq u \leq \min\{D_j, x\}} \{p_j u + V_{j-1}(x - u)\}\right], \quad (2.3)$$

with boundary conditions

$$V_0(x) = 0, \quad x = 0, 1, \ldots, C.$$

The values u^* that maximize the right-hand side of (2.3) for each j and x form an optimal control policy for this model.

---

[5] Readers familiar with dynamic programming may notice that this Bellman equation is of the form $E[\max\{\cdot\}]$ and not $\max E[\cdot]$ as in many standard texts. The relationship between these two forms is explained in detail in Appendix D. But essentially, the $\max E[\cdot]$ form can be recovered by considering the demand D_j to be a state variable along with x. While the two forms can be shown to be equivalent, the $E[\max\{\cdot\}]$ is simpler to work with in many RM problems. In our case, this leads to the modeling assumption that we optimize "as if" we observed D_j.

2.2.2.2 Optimal Policy: Discrete Demand and Capacity

We first consider the case where demand and capacity are discrete. To analyze the form of the optimal control in this case, define

$$\Delta V_j(x) \equiv V_j(x) - V_j(x-1).$$

$\Delta V_j(x)$ is the *expected marginal value of capacity* at stage j—the expected incremental value of the xth unit of capacity. A key result concerns how these marginal values change with capacity x and the stage j (See Appendix 2.A for proof.):

PROPOSITION 2.1 *The marginal values $\Delta V_j(x)$ of the value function $V_j(x)$ defined by (2.3) satisfy $\forall x, j$:*
(i) $\Delta V_j(x+1) \leq \Delta V_j(x)$
(ii) $\Delta V_{j+1}(x) \geq \Delta V_j(x)$.

That is, at a given stage j the marginal value is decreasing in the remaining capacity, and at a given capacity level x the marginal value increases in the number of stages remaining. These two properties are intuitive and greatly simplify the control. To see this, consider the optimization problem at stage $j+1$. From (2.3) and the definition of $\Delta V_j(x)$, we can write

$$V_{j+1}(x) = V_j(x) + E\left[\max_{0 \leq u \leq \min\{D_{j+1}, x\}} \{\sum_{z=1}^{u}(p_{j+1} - \Delta V_j(x+1-z))\}\right],$$

where we take the summation above to be empty if $u = 0$. Since $\Delta V_j(x)$ is decreasing in x by Proposition 2.1(i), it follows that the terms in the sum $p_{j+1} - \Delta V_j(x+1-z)$ are decreasing in z. Thus, it is optimal to increase u (keep adding terms) until the terms $p_{j+1} - \Delta V_j(x+1-z)$ become negative or the upper bound $\min\{D_{j+1}, x\}$ is reached, whichever comes first.

The resulting optimal control can be expressed in terms of optimal protection levels y_j^* for $j, j-1, \ldots, 1$ (class j and higher in the revenue order) by

$$y_j^* \equiv \max\{x : p_{j+1} < \Delta V_j(x)\}, \quad j = 1, \ldots, n-1. \quad (2.4)$$

(Recall the optimal protection level $y_n^* \equiv C$ by convention.) The optimal control at stage $j+1$ is then

$$u^*(j+1, x, D_{j+1}) = \min\{(x - y_j^*)^+, D_{j+1}\}, \quad (2.5)$$

where the notation $z^+ = \max\{0, x\}$ denotes the positive part of z. The quantity $(x - y_j^*)^+$ is the remaining capacity in excess of the protection

Single-Resource Capacity Control

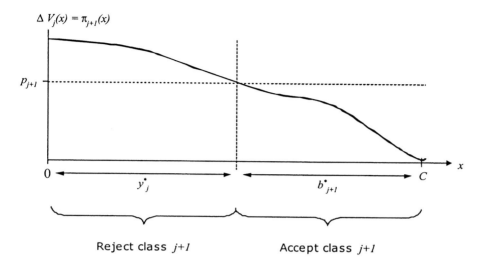

Figure 2.2. The optimal protection level y_j^* in the static model.

level, which is the maximum capacity we are willing to sell to class $j+1$. The situation is shown in Figure 2.2.

In practice, we can simply post the protection level y_j^* in a reservation system and accept requests first come, first serve until the capacity threshold y_j^* is reached or the stage ends, whichever comes first. Thus, the optimal protection-level control at stage $j+1$ requires no information about the demand D_{j+1}, yet it produces the same optimal decision "as if" we knew D_{j+1} exactly at the start of stage $j+1$. The reason for this is that knowledge of D_{j+1} does not affect the future value of capacity, $V_j(x)$. Deciding to accept or reject each request simply involves comparing current revenues to the marginal cost of capacity, and this comparison does not depend on how many stage-$(j+1)$ requests there are in total.

Proposition 2.1(ii) implies the nested protection structure

$$y_1^* \leq y_2^* \leq \cdots \leq y_n^*.$$

This fact is easily seen from Figure 2.2. If p_{j+1} increases with j and the curve $\Delta V_j(x)$ decreases with j, then the optimal protection level y_j^* will shift to the left (decrease). Together, this ordering produces the nested protection-level structure.

One can also use booking limits in place of protection levels to achieve the same control. Optimal nested booking limits are defined by

$$b_j^* \equiv C - y_{j-1}^*, \quad j = 2, \ldots, n, \tag{2.6}$$

with $b_1^* \equiv C$. The optimal control in stage $j+1$ is then to accept

$$u^*(j+1, x, D_{j+1}) = \min\{(b_{j+1} - (C-x))^+, D_{j+1}\}.$$

Note that $C - x$ is the total capacity sold prior to stage $j + 1$ and b_{j+1} is the booking limit for class $j+1$, so $(b_{j+1} - (C-x))^+$ is the remaining capacity available for class $j+1$. The optimal booking limit is also shown in Figure 2.2.

Finally, the optimal control can also be implemented through a table of bid prices. Indeed, if we define the stage $j + 1$ bid price by

$$\pi_{j+1}(x) \equiv \Delta V_j(x), \qquad (2.7)$$

then the optimal control is

$$u^*(j+1, x, D_{j+1}) = \begin{cases} 0 & \text{if } p_{j+1} < \pi_{j+1}(x) \\ \max\{z : p_{j+1} \geq \pi_{j+1}(x-z)\} & \text{otherwise.} \end{cases}$$

In words, we accept the zth request in stage $j + 1$ if the price p_{j+1} exceeds the bid price value $\pi_{j+1}(x - z)$ of the zth unit of capacity that is allocated. In practice, we can store a table of bid prices and process requests by sequentially comparing the price of each product to the table values corresponding to the remaining capacity.

We summarize these results in the following theorem:

THEOREM 2.1 *For the static model defined by (2.3), the optimal control can be achieved using either*
(i) Nested protection levels defined by (2.4),
(ii) Nested booking limits defined by (2.6), or
(iii) Bid price tables defined by (2.7).

How to compute these various policies is discussed in Section 2.2.3.

2.2.2.3 Optimality Conditions for Continuous Demand

Next, consider the case where capacity is continuous and demand at each stage has a continuous distribution. In this case, the dynamic program is still given by (2.3); however D_j, x, and u are now continuous quantities. The analysis of the dynamic program is slightly more complex than it is in the discrete-demand case, but many of the details are quite similar. Hence, we only briefly describe the key differences.

The main change is that the marginal value $\Delta V_j(x)$ is now replaced by the derivative of $V_j(x)$ with respect to x, $\frac{\partial}{\partial x} V_j(x)$. This derivative is still interpreted as the marginal expected value of capacity. And an argument nearly identical to that in the proof of Proposition 2.1 shows that the marginal value $\frac{\partial}{\partial x} V_j(x)$ is decreasing in x (equivalently, $V_j(x)$ is concave in x).

Therefore, the optimal control in stage $j+1$ is to keep increasing u (keep accepting demand) as long as

$$p_{j+1} \geq \frac{\partial}{\partial x} V_j(x-u)$$

and to stop accepting once this condition is violated or the demand D_{j+1} is exhausted, whichever comes first. Again, this decision rule can be implemented with optimal protection levels, defined by

$$y_j^* \equiv \max\{x : p_{j+1} < \frac{\partial}{\partial x} V_j(x)\}, \quad j = 1, \ldots, n-1.$$

One of the chief virtues of the continuous model is that it leads to simplified expressions for the optimal vector of protection levels $\mathbf{y}^* = (y_1^*, \ldots, y_n^*)$. We state the basic result without proof (see Brumelle and McGill [91] for a proof).

First, for an arbitrary vector of protection levels $\mathbf{y}$ and vector of demands $\mathbf{D} = (D_1, \ldots, D_n)$, define the following $n-1$ *fill events*

$$B_j(\mathbf{y}, \mathbf{D}) \equiv \{D_1 > y_1, D_1 + D_2 > y_2, \quad (2.8)$$
$$\ldots, D_1 + \cdots + D_j > y_j\}, \quad j = 1, \ldots, n-1.$$

$B_j(\mathbf{y}, \mathbf{D})$ is the event that demand to come in stages $1, 2, \ldots, j$ exceeds the corresponding protection levels. A necessary and sufficient condition for $\mathbf{y}^*$ to be an optimal vector of protection levels is that it satisfy the $n-1$ equations

$$P(B_j(\mathbf{y}^*, \mathbf{D})) = \frac{p_{j+1}}{p_1}, \quad j = 1, 2, \ldots, n-1. \quad (2.9)$$

That is, the jth fill event should occur with probability equal to the ratio of class $(j+1)$ revenue to class 1 revenue. As it should, this reduces to Littlewood's rule (2.2) in the $n = 2$ case, since $P(B_1(\mathbf{y}^*, \mathbf{D})) = P(D_1 > y_1^*) = p_2/p_1$.

Note that

$$B_j(\mathbf{y}, \mathbf{D}) = B_{j-1}(\mathbf{y}, \mathbf{D}) \cap \{D_1 + \cdots + D_j > y_j\},$$

so the event $B_j(\mathbf{y}, \mathbf{D})$ can occur only if $B_{j-1}(\mathbf{y}, \mathbf{D})$ occurs. Also, if $y_j = y_{j-1}$ then $B_j(\mathbf{y}, \mathbf{D}) = B_{j-1}(\mathbf{y}, \mathbf{D})$. Thus, if $p_j < p_{j-1}$, we must have $y_j^* > y_{j-1}^*$ to satisfy (2.9). Thus, the optimal protection levels are strictly increasing in j if the revenues are strictly decreasing in j.

2.2.3 Computational Approaches

At first glance it may appear that the optimal nested allocations are difficult to compute. However, computing these values is in fact quite easy and efficient algorithmically. There are two basic approaches: dynamic programming and Monte Carlo integration.

42   THE THEORY AND PRACTICE OF REVENUE MANAGEMENT

2.2.3.1   Dynamic Programming

The first approach is based on using the dynamic programming recursion (2.3) directly and requires that demand and capacity are discrete—or in the continuous case that these quantities can be suitably discretized. The inner optimization in (2.3) is simplified by using the optimal protection levels y_{j-1}^* from the previous stage. Thus, substituting (2.5) into (2.3) we obtain the recursion

$$V_j(x) = E\left[p_j \min\{D_j, (x - y_{j-1}^*)^+\} + V_{j-1}(x - \min\{D_j, (x - y_{j-1}^*)^+\})\right], \quad (2.10)$$

where y_j^* is determined using (2.4), and we define $y_0^* = 0$. This procedure is repeated starting from $j = 1$ and working backward to $j = n$.

For discrete-demand distributions, computing the expectation in (2.10) for each state x requires evaluating at most $O(C)$ terms since $\min\{D_j, (x - y_{j-1}^*)^+\} \leq C$. Since there are C states (capacity levels), the complexity at each stage is $O(C^2)$. The critical values y_j^* can then be identified from (2.4) in $\log(C)$ time by binary search as $\Delta V_j(x)$ is nonincreasing. Indeed, since we know $y_j^* \geq y_{j-1}^*$, the binary search can be further constrained to values in the interval $[y_{j-1}^*, C]$. Therefore, computing y_j^* does not add to the complexity at stage j. Since these steps must be repeated for each of the $n-1$ stages (stage n need not be computed as mentioned above), the total complexity of the recursion is $O(nC^2)$.

2.2.3.2   Monte Carlo Integration

The second approach to computing optimal protection levels is based on using (2.9) together with Monte Carlo integration. It is most naturally suited to the case of continuous demand and capacity, though the discrete case can be computed (at least heuristically) with this method as well.

The idea is to simulate a large number K of demand vectors, $\mathbf{d}^k = (d_1^k, \ldots, d_n^k)$, $k = 1, \ldots, K$, from the forecast distributions for the n classes. We then progressively sort through these values to find thresholds $\mathbf{y}$ that approximately satisfy (2.9).

In what follows, it is convenient to note that

$$P(B_j(\mathbf{y}, \mathbf{D})) = P(D_1 + \cdots + D_j > y_j | B_{j-1}(\mathbf{y}, \mathbf{D})) P(B_{j-1}(\mathbf{y}, \mathbf{D}))$$

Thus, (2.9) implies that the optimal $\mathbf{y}^*$ must satisfy

$$\begin{aligned} P(D_1 + \cdots + D_j > y_j^* | B_{j-1}(\mathbf{y}^*, \mathbf{D})) &= \frac{1}{P(B_{j-1}(\mathbf{y}^*, \mathbf{D}))} \frac{p_{j+1}}{p_1} \\ &= \frac{p_{j+1}}{p_j}, \quad j = 1, \ldots, n-1, \end{aligned}$$

Single-Resource Capacity Control

since $P(B_{j-1}(\mathbf{y}^*, \mathbf{D})) = p_j/p_1$. The following algorithm computes the optimal $\mathbf{y}^*$ approximately using the empirical conditional probabilities estimated from the sample of simulated demand data:

STEP 0: Generate and store K random demand vectors $\mathbf{d}^k = (d_1^k, \ldots, d_n^k)$.
For $k = 1, \ldots, K$ and $j = 1, \ldots, n-1$, compute the partial sums
$$S_j^k = d_1^k + d_2^k + \cdots + d_j^k$$
and form the vector $\mathbf{S}^k = (S_1^k, \ldots, S_n^k)$.
Initialize a list $\mathcal{K} = \{1, \ldots, K\}$ and counter $j = 1$.

STEP 1: Sort the vectors $\mathbf{S}^k, k \in \mathcal{K}$ by their j^{th} component values, S_j^k. Let $[l]$ denote the lth element of $\mathcal{K}$ in this sorted list so that
$$S_j^{[1]} \leq S_j^{[2]} \leq \cdots \leq S_j^{[|\mathcal{K}|]}.$$

STEP 2: Set $l^* = \lfloor \frac{p_{j+1}}{p_j} |\mathcal{K}| \rfloor$. Set $y_j = \frac{1}{2}(S_j^{[l^*]} + S_j^{[l^*+1]})$.

STEP 3: Set $\mathcal{K} \leftarrow \{k \in \mathcal{K} : S_j^k > y_j\}$, and $j \leftarrow j + 1$.
IF $j = n - 1$ STOP
ELSE GOTO STEP 1.

The complexity of this method is $O(nK \log K)$, which is nearly linear in the number of simulated demand vectors K. Thus, it is relatively efficient even with large samples. It is also quite simple to program and can be used with any general distribution. The following example illustrates the method:

Example 2.2 Consider a three-class example, where the prices are $p_1 = 100$, $p_2 = 70$, and $p_3 = 42$. The demand for each class is normally distributed. Class 1 has a mean of 20 and standard deviation of 9; class 2 has a mean of 45 and standard deviation of 12. (The statistics for class 3 do not affect the calculation.)

Figure 2.3 shows a plot of the partial sums $S_1 = D_1$ and $S_2 = D_1 + D_2$ for 50 simulated data points of this problem. Since the first ratio $p_2/p_1 = 70/100 = 0.7$, the Monte Carlo algorithm starts by finding a value y_1 such that 70% of these points (35 points) have S_1 values above y_1. The result is shown in Figure 2.3 by the vertical line at $y_1 = 16.3$.

In the next iteration of the algorithm, the 35 points to the right of this vertical line are sorted again by their S_2 value and y_2 is chosen so that a fraction $p_3/p_2 = 42/70 = 0.6$ of the points (21 points) lie above y_2. This occurs at $y_2 = 68.3$.

The algorithm then terminates with the estimates $y_1^* \approx 16.3$ and $y_2^* \approx 68.3$.

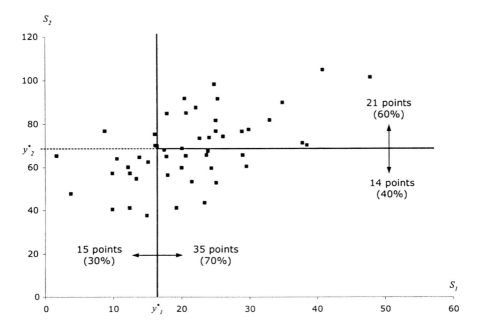

Figure 2.3. A plot of the partial sums S_1, S_2 and resulting Monte Carlo estimates of optimal protection levels for fifty simulated data points.

2.2.4 Heuristics

As we have seen, computing optimal controls for the static single-resource model is not particularly difficult. Despite this fact, exact optimization models are not widely used in practice. Indeed, most single-resource airline RM systems use one of several heuristics to compute booking limits and protection levels.

There are two main reasons for this state of affairs. The first is simply a case of practice being one step ahead of the underlying theory. As mentioned, in the airline industry the practice of using capacity controls to manage multiple classes quickly gained popularity following deregulation in the mid 1970s. But this predates the theory of optimal controls by more than a decade. The only known optimal controls in the 1970s were Littlewood's results for the two-class problem. As a result, heuristics were developed for the general n-class problem. During the decade following deregulation, RM software embedded these heuristics, and people grew accustomed to thinking in terms of them. The inertia generated from this early use of the heuristics is one reason for their continued popularity today.

Heuristics are also widely used because they are simpler to code, quicker to run, and generate revenues that in many cases are close to optimal. Indeed, many practitioners in the airline industry simply believe that even the modest effort of computing optimal controls is not worth the benefit they provide in improved revenue performance. Proponents of heuristics argue that the potential improvement from getting better revenue data and improving demand forecasts swamps the gains from using optimal controls—reflecting the philosophy that it is better to be "approximately right" than it is to be "precisely wrong."

While these points are well taken, such criticisms are somewhat misdirected. For starters, using optimal controls does not mean one has to give up on improvements in other areas, such as forecasting. These activities are not mutually exclusive, though a understaffed development group might very well consider refining optimization modules a low-priority task. Yet given the very modest cost of coding and computing optimal controls, the strong objections to the use of optimal controls are often not entirely rational.

Regardless of one's view on the use of heuristics, it is important to understand them. They remain widely used in practice and can also help develop useful intuition.

We next look at the two most popular heuristics: EMSR-a and EMSR-b, both of which are attributed to Belobaba [38–40]. Both heuristics are based on the n-class, static, single-resource model defined above in Section 2.2. They differ only in how they approximate the problem. Static model assumptions apply: classes are indexed so that $p_1 > p_2 > \cdots > p_n$, $F_j(x)$ denotes the c.d.f. of class j demand, and low-revenue demand arrives before high-revenue demand in stages that are indexed by j as well. Moreover, for ease of exposition we assume that capacity and demand are continuous and that the distribution functions $F_j(x)$, $j = 1, \ldots, n$, are continuous as well, though these assumptions are easily relaxed.

2.2.4.1 EMSR-a

EMSR-a (*expected marginal seat revenue–version a*) is the most widely publicized heuristic for the single-resource problem. Despite this fact, it is less popular in practice than its close cousin, EMSR-b, which surprisingly is not well documented in the literature. Generally, EMSR-b provides better revenue performance, and it is certainly more intuitive, though EMSR-a is important to know just the same.

EMSR-a is based on the idea of adding the protection levels produced by applying Littlewood's rule to successive pairs of classes. Consider stage $j+1$ in which demand of class $j+1$ arrives with price p_{j+1}. We are interested in computing how much capacity to reserve for the remaining classes, $j, j-1, \ldots, 1$; that is, the protection level, y_j, for classes j and higher. To do so, let's consider a single class k among the remaining classes $j, j-1, \ldots, 1$ and compare k and $j+1$ *in isolation*. Considering only these two classes, we would use Littlewood's rule (2.2) and reserve capacity y_k^{j+1} for class k, where

$$P(D_k > y_k^{j+1}) = \frac{p_{j+1}}{p_k}. \tag{2.11}$$

Repeating for each future class $k = j, j-1, \ldots, 1$, we could likewise compute how much capacity to reserve for each such class k in isolation. The idea of EMSR-a, then, is simply to add up these individual protection levels to approximate the total protection level y_j for classes j and higher. That is, set the protection level y_j as

$$y_j = \sum_{k=1}^{j} y_k^{j+1}, \tag{2.12}$$

where y_k^{j+1} is given by (2.11). One then repeats this same calculation for each stage j.

EMSR-a is certainly simple and has an intuitive appeal. For a short while it was even believed to be optimal, but this notion was quickly dispelled once the published work on optimal controls appeared.

Indeed, it is not hard to see intuitively that EMSR-a can be excessively conservative and produce protection levels that are larger than optimal in certain cases. This is because adding the individual protection levels y_k^{j+1} ignores the statistical averaging effect (*pooling effect*) produced by aggregating demand across classes. For example, for the sake of illustration, suppose that at stage $j+1$ all future demand has the same revenue, i.e., $p_j = p_{j-1} = \cdots = p_1 = p$. Then EMSR-a will set protection levels so that

$$P(D_k > y_k^{j+1}) = \frac{p_{j+1}}{p}, \quad k = 1, \ldots, j.$$

However, it is clear that all these future classes should be aggregated since they have identical revenues, in which case we can apply Littlewood's rule (2.2) to obtain the optimal protection level, y_j^*, using

$$P(\sum_{k=1}^{j} D_k > y_j^*) = \frac{p_{j+1}}{p}.$$

Since for any random variables and any numbers y_k^{j+1}

$$P(\sum_{k=1}^{j} D_k > \sum_{k=1}^{j} y_k^{j+1}) \leq \sum_{k=1}^{j} P(D_k > y_k^{j+1}),$$

the optimal protection level y_j^* is less than the EMSR-a protection level, $\sum_{k=1}^{j} y_k^{j+1}$.[6] This behavior suggests that EMSR-a may perform badly when there are large numbers of classes whose revenues are close together.

2.2.4.2 EMSR-b

EMSR-b (*expected marginal seat revenue–version b*) is an alternative single-resource heuristic that avoids the lack-of-pooling defect in EMSR-a mentioned above. EMSR-b is again based on an approximation that reduces the problem at each stage to two classes, but in contrast to EMSR-a, the approximation is based on aggregating *demand* rather than aggregating *protection levels*. Specifically, the demand from future classes is aggregated and treated as one class with a revenue equal to the weighted-average revenue.

Consider stage $j + 1$ in which we want to determine protection level y_j. Define the aggregated future demand for classes $j, j - 1, \ldots, 1$ by

$$S_j = \sum_{k=1}^{j} D_k,$$

and let the weighted-average revenue from classes $1, \ldots, j$, denoted $\bar{p}_j$, be defined by

$$\bar{p}_j = \frac{\sum_{k=1}^{j} p_k E[D_k]}{\sum_{k=1}^{j} E[D_k]}. \qquad (2.13)$$

Then the EMSR-b protection level for class j and higher, y_j, is chosen by Littlewood's rule (2.2) so that

$$P(S_j > y_j) = \frac{p_{j+1}}{\bar{p}_j}. \qquad (2.14)$$

It is common when using EMSR-b to assume demand for each class j is independent and normally distributed with mean μ_j and variance σ_j^2, in

---

[6] To see this, consider a sample realization of demand, and note that if $\sum_{k=1}^{j} D_k > \sum_{k=1}^{j} y_k^{j+1}$, then $D_k > y_k^{j+1}$ for at least one k but the converse need not be true. So the probability of the event $\sum_{k=1}^{j} D_k > \sum_{k=1}^{j} y_k^{j+1}$ cannot exceed the sum (over $k = 1, \ldots, j$) of the probabilities of the events $D_k > y_k^{j+1}$.

which case
$$y_j = \mu + z_\alpha \sigma,$$
where $\mu = \sum_{k=1}^{j} \mu_k$ is the mean and $\sigma^2 = \sum_{k=1}^{j} \sigma_k^2$ is the variance of the aggregated demand to come at stage $j+1$ and $z_\alpha = \Phi^{-1}(1 - p_{j+1}/\bar{p}_j)$ (recall $\Phi^{-1}(x)$ is the inverse of the standard normal c.d.f.). Again, one repeats this calculation for each j. (See Section 11.A for approximation formulas for the normal and inverse normal distributions.)

EMSR-b clearly captures the *pooling*—or *statistical averaging*—effect that is lacking in EMSR-a. This is an advantage of EMSR-b over EMSR-a. However, using the weighted-average revenue is a somewhat crude approximation that can distort the resulting protection levels. In practice EMSR-b is more popular and seems to generally perform better than EMSR-a, though studies comparing the two have at times shown mixed results. Belobaba [41] reports studies in which EMSR-b is consistently within 0.5 percent of the optimal revenue, whereas EMSR-a can deviate by nearly 1.5 percent from the optimal revenue in certain cases, though with mixed order of arrival and frequent reoptimization, he reports that both methods perform well. However, another recent study by Polt [425] using Lufthansa airline data showed more mixed performance, with neither method dominating the other.

2.2.4.3 Numerical Examples

A few simple numerical examples give some sense of the protection levels and revenues produced by these two approximations. The example we consider is based on a slightly modified instance of the data reported by Wollmer [576]:

Example 2.3 There are four classes, and demand is assumed to be normally distributed. Table 2.1 shows the demand data and Table 2.2 the protection levels produced by EMSR-a, EMSR-b, and the optimal policy.

Table 2.1. Static single-resource model and protection levels.

		Demand Statistics		Protection Levels (y_j)		
j	p_j	μ_j	σ_j	OPT	EMSR-a	EMSR-b
1	1050	17.3	5.8	16.7	16.7	16.7
2	567	45.1	15.0	42.5	38.7	50.9
3	534	39.6	13.2	72.3	55.6	83.1
4	520	34.0	11.3			

Table 2.2. Revenue performance for Example 2.3.

C	DF	OPT Rev.	EMSR-a Rev.	EMSR-a % Sub. Opt.	EMSR-b Rev.	EMSR-b % Sub. Opt.
80	1.70	49,666	49,515	0.30%	49,463	0.41%
90	1.51	54,846	54,721	0.23%	54,560	0.52%
100	1.36	60,063	59,985	0.13%	59,786	0.46%
110	1.24	65,112	65,078	0.05%	64,881	0.35%
120	1.13	69,916	69,904	0.02%	69,764	0.22%
130	1.05	73,975	73,973	0.00%	73,898	0.10%
140	0.97	77,177	77,174	0.00%	77,143	0.04%
150	0.91	79,544	79,544	0.00%	79,534	0.01%

Note from Table 2.1 that (in this example) there is a considerable discrepancy between the protection levels computed under the heuristics, both compared with each other and with the optimal protection levels.

The revenue performance of the methods from a simulation study are shown in Table 2.2. Capacity is varied from 80 to 150 to create demand factors (ratio of total mean demand to capacity) in the range 1.7 to 0.9. The percentage suboptimality is also reported (one minus the ratio of heuristic revenues to optimal revenues). Note for this example that EMSR-a is slightly better than EMSR-b, though both perform quite well; the suboptimality gap of EMSR-b reaches a high of 0.52%, while the maximum suboptimality of EMSR-a is only 0.30%.

Example 2.4 The demand statistics are the same as in Example 2.3, but the revenue values are more evenly spaced. The data and resulting protection levels are shown in Tables 2.3 and 2.4.

Table 2.3. Static single-resource model data for Example 2.4.

j	p_j	Demand Statistics μ_j	Demand Statistics σ_j	Protection Levels (y_j) OPT	Protection Levels (y_j) EMSR-a	Protection Levels (y_j) EMSR-b
1	1050	17.3	5.8	9.7	9.8	9.8
2	950	45.1	15.0	54.0	50.4	53.2
3	699	39.6	13.2	98.2	91.6	96.8
4	520	34.0	11.3			

The revenue performance of the heuristics in Example 2.4 is shown in Table 2.4. In this case, there is less discrepancy among the protection levels computed under the heuristics and the optimal policy. The performance of both heuristics is also very good, especially under EMSR-b, which is for all practical purposes optimal. Performance such as this helps explain why these heuristics are popular in practice.

Table 2.4. Revenue performance for Example 2.4.

C	DF	OPT Rev.	EMSR-a Rev.	% Sub. Opt.	EMSR-b Rev.	% Sub. Opt.
80	1.70	67,512	67,462	0.07%	67,516	-0.01%
90	1.51	74,003	73,950	0.07%	74,000	0.00%
100	1.36	79,429	79,164	0.33%	79,426	0.00%
110	1.24	84,884	84,554	0.39%	84,862	0.03%
120	1.13	89,879	89,668	0.23%	89,875	0.00%
130	1.05	95,054	94,899	0.16%	95,045	0.01%
140	0.97	99,072	99,004	0.07%	99,068	0.00%
150	0.91	102,346	102,339	0.01%	102,346	0.00%

2.3 Adaptive Methods

We next examine an adaptive algorithm for determining optimal protection levels for the static, single-resource model of Section 2.2.2. The optimal protection levels in this case are determined by the conditions (2.8). Typically, application of the optimality conditions (2.8) requires three steps. First, historical demand data are studied to determine suitable models for the demand distributions. Second, forecasting techniques are applied to estimate the parameters of these distributions. Third, the forecasts are passed to an optimization routine that solves for protection levels $\mathbf{y}^*$. The resulting controls are then used to make individual accept or deny decisions as reservations come in. In practice, bookings from similar resources are fed back into the forecasting system, and the process is repeated cyclically over time.

In this section, we look at a method for *directly* updating booking policy parameters for the next resource usage based on observations of the performance of the parameters on previous instances, without recourse to the complex cycles of forecasting and optimization. We show how to construct a simple adjustment scheme of this sort that is based on stochastic approximation methods (the Robbins-Monro algorithm [443]) and that provably converges to an optimal policy with repeated application. The convergence, however, is guaranteed only for the case of stationary, independent demand and may be quite slow, requiring a large number of adjustments to reach a near-optimal set of protection levels.

2.3.1 Adaptive Algorithm

Our starting point in developing an adaptive algorithm is condition (2.9), which states that for $\mathbf{y}^*$ to be an optimal set of protection levels, it must satisfy

$$P(B_j(\mathbf{y}^*, \mathbf{D})) = \frac{p_{j+1}}{p_1}, \quad j = 1, 2, \ldots, n-1.$$

For example, if the revenue ratio $\frac{p_{j+1}}{p_1}$ is 0.6, then this condition says that the fill event $B_j(\mathbf{y}, \mathbf{D})$ defined by (2.8) should occur on 60% of the service instances, on average, if $\mathbf{y} = \mathbf{y}^*$. Note that it is easy to determine the frequency of the fill events $B_j(\mathbf{y}, \mathbf{D})$ from historical records, since it is necessary only to observe if demand reached the protection levels, not the degree to which it exceeded them.[7]

To develop the algorithm, for $j = 1, \ldots, n-1$ define

$$H_j(\mathbf{y}, \mathbf{D}) = \frac{p_{j+1}}{p_1} - \mathbf{1}(B_j(\mathbf{y}, \mathbf{D})),$$

where $\mathbf{1}(E)$ denotes the indicator function of event E (a function that is 1 if event E occurs and is zero otherwise). The quantity $H_j(\mathbf{y}, \mathbf{D})$ will be negative if the jth fill event occurs and positive otherwise. If protection levels are being adjusted, an occurrence of the jth fill event (all of classes 1 through j reached their protection levels) suggests that the protection level y_j should be adjusted upward. Thus $-H_j(\mathbf{y}, \mathbf{D})$ can be viewed as an *adjustment direction* for protection level y_j. The corresponding *adjustment vector* is $\mathbf{H}(\mathbf{y}, \mathbf{D}) = (H_1(\mathbf{y}, \mathbf{D}), \ldots, H_{n-1}(\mathbf{y}, \mathbf{D}))$. Define

$$h_j(\mathbf{y}) = \frac{p_{j+1}}{p_1} - P(B_j(\mathbf{y}, \mathbf{D})) = E[H_j(\mathbf{y}, \mathbf{D})], \quad j = 1, \ldots, n-1,$$

and let $\mathbf{h}(\mathbf{y}) = (h_1(\mathbf{y}), \ldots, h_{n-1}(\mathbf{y}))$. Thus, $-\mathbf{h}(\mathbf{y})$ can be properly viewed as the *expected adjustment vector* for protection levels given current levels $\mathbf{y}$. The optimality condition (2.9) stipulates that we should seek a $\mathbf{y}^*$ such that the expected adjustment for all protection levels is zero; or, $\mathbf{h}(\mathbf{y}^*) = 0$.

The Robbins-Monro [443] algorithm (generalized for vector quantities) constructs a sequence of parameter estimates, $\{\mathbf{y}^{(1)}, \mathbf{y}^{(2)}, \ldots\}$, from a sequence of independent instances, $\{\mathbf{D}^{(1)}, \mathbf{D}^{(2)}, \ldots\}$, using the recursion

$$\mathbf{y}^{(k+1)} = \mathbf{y}^{(k)} - \gamma_k \mathbf{H}(\mathbf{y}^{(k)}, \mathbf{D}^{(k)}), \tag{2.15}$$

---

[7]There are two important exceptions to this statement. First, if y_j happens to exceed the maximum number of seats available for sale (usually the physical capacity plus an overbooking pad), then the event $D_1 + \cdots + D_j > y_j$ is not observable (unless the rejected sales are recorded). Second, if protection levels are revised during the booking period prior to the usage of the resource, it can easily happen that a new protection level exceeds the *remaining* capacity (a problem similar to the first one) that earlier, high protection levels constrained demand during part of the booking period in one or more discount classes. In this case, total demand is not observed (a variant of censorship of the demand data).

where γ_k is a sequence of nonnegative step sizes satisfying

$$\sum_k \gamma_k = +\infty \quad \text{and} \quad \sum_k \gamma_k^2 < +\infty. \tag{2.16}$$

The simplest example of a suitable step size sequence is defined by $\gamma_k = \frac{1}{k}$; however, this simple averaging sequence takes small steps early in the procedure, which can delay convergence. In the development to follow, we use a sequence of the form $a/(k+b)$, where a and b are constants chosen to give larger early steps.

The directions $\mathbf{H}(\mathbf{y}^{(k)}, \mathbf{D}^{(k)})$ can be determined after the completion of each instance k (each departure in the airline case). If the jth fill event occurs, $H_j = p_{j+1} - 1 < 0$, and the protection level $y_j^{(k)}$ is increased by $\gamma_k(1 - p_{j+1})$; if not, then $H_j = p_{j+1} > 0$, and y_j^k is reduced by $\gamma_k p_{j+1}$. Thus protection levels are stepped up when high demand is observed and stepped down when low demand is observed, with the step size becoming smaller as the algorithm progresses.

Some relatively mild regularity conditions ensure that the procedure (2.15) does converge (a.s.) to a value $\mathbf{y}^*$ satisfying $h(\mathbf{y}^*) = 0$. (See van Ryzin and McGill [526] for exact conditions as well as bounds on the rate of convergence.)

2.3.2 A Numerical Comparison with EMSR and Censored Forecasting

We next look at a brief numerical example of the performance of the adaptive algorithm compared with a procedure that combines censored forecasting with EMSR-b protection levels. These comparisons are based on simulated data in an idealized stationary setting, but do provide some insight into the performance of each method.

The combined forecasting/EMSR-b (F/EMSR) scheme constructs a demand forecast from censored data based on the Kaplan-Meier estimator of the survivor function $S(x) = P(D > x)$. (See Section 9.4.3.) Protection levels are set using EMSR-b.

The test problem has four classes (three protection levels). Demand is modeled using a normal distribution. The distribution data, along with optimal and EMSR-b protection levels, are shown in Table 2.5. The protection level y^* is the optimal level when demand is normally distributed, while y-EMSR is the protection level computed by the EMSR-b heuristic. To illustrate convergence, starting protection levels are set far from optimal, corresponding to cases of very high and very low starting values (Table 2.6). There are two demand scenarios. The high-demand scenario has a starting inventory of 124 seats, corresponding to a 125% demand factor (ratio of expected total demand to capacity) and approx-

Table 2.5. Fares, demand statistics, and protection levels for adaptive-method numerical examples.

Class	Fare ($)	Mean	Std. Dev.	y-EMSR	y^*
1	1,050	17.3	5.8	16.7	16.7
2	567	45.1	15.0	51.5	44.6
3	527	73.6	17.4	131.4	134.0
4	350	19.8	6.6	n.a.	n.a.

Table 2.6. Starting values of protection levels for adaptive-method numerical examples.

	y_1	y_2	y_3
Low	0	15	65
High	35	110	210

imately a 95% load factor (ratio of average number of seats sold to capacity) under optimal protection levels. The low-demand scenario starts with 164 seats, resulting in a demand factor of 95% and a load factor of approximately 90% under optimal protection levels. For the stochastic approximation procedure, the step size sequence is $\gamma_k = 200/(10+k)$.

Figure 2.4 shows three graphs of the data for the case of low demand and high starting values. The top graph of Figure 2.4 shows the average cumulative revenue as a percentage of the optimal revenue for the two methods as a function of the number of iterations (flights). The error bars show the 95% confidence intervals about these averages. The middle graph shows the average protection levels over time for the stochastic approximation (SA) procedure. The horizontal dotted lines are the optimal protection levels. The lowest line corresponds to y_1^*, the middle line to y_2^*, and the top line to y_3^*. The solid lines are the corresponding average protection levels produced by the stochastic approximation (SA) method. The error bars on the solid lines give the 25-percentile and 75-percentile values for each protection level at each iteration, which provides some sense of the variability in protection levels across sample paths. The bottom graph shows the identical plot of protection levels for the F/EMSR method.

Figure 2.4 shows that the F/EMSR procedure converges more quickly than the SA procedure. In this case, the faster convergence of the F/EMSR has a significant impact on the cumulative revenue performance: F/EMSR generates about 2 to 3% higher revenue on average in the early iterations. With low demand and low starting values, the two methods perform comparably.

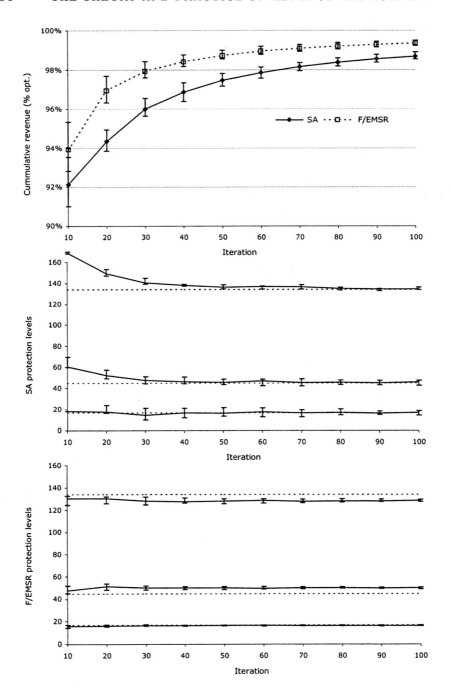

Figure 2.4. Adaptive-method example: low demand, high start, normal distribution.

Single-Resource Capacity Control

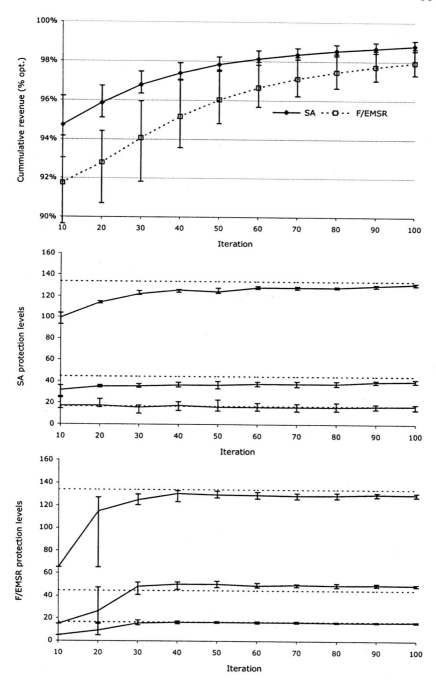

Figure 2.5. Adaptive-method example: high demand, low start, normal distribution.

The results are quite different in the high demand factor case. Figure 2.5 shows the simulation results for low starting protection values and high load factor. Note, as indicated by the error bars in the bottom graphs, that the F/EMSR procedure is very volatile and somewhat slow to converge in the early iterations. The revenue effect of this behavior is quite significant, with F/EMSR generating cumulative revenues roughly 8% lower than optimal and 2 to 3% lower than SA in the early iterations. However, the performance and protection levels of F/EMSR improve after about 30 iterations. In contrast, the SA procedure is considerably more stable, and it converges faster in the early iterations, which accounts for its superior revenue performance. F/EMSR performs badly in this case because the forecasting procedure suffers from the frequent censoring caused by a combination of low protection levels and high demand. For high starting protection levels, the F/EMSR performs better, since there is less initial censoring of data.

This behavior suggests that adaptive methods may be useful as a means of automatically adjusting protection levels in cases where very little demand information is available (such as with new products) and forecasting is difficult due to a high degree of censoring. In such cases, adaptive methods provide a robust way to adjust protection levels and may also help speed up the forecasting method itself by nudging protection levels in the right direction, thereby reducing the amount of censoring.

2.4 Group Arrivals

Group arrivals can pose additional complications. A group request is a single request for multiple units of capacity (such as a family of four traveling together). We briefly describe this case but omit any detailed formulations because the basic ideas follow readily from what we have seen thus far (and the more complicated ideas are beyond the scope of this text).

If groups can be *partially accepted*—that is, given a request for $m \geq 1$ units, we can sell any quantity u in the range $0 \leq u \leq m$ (and more important, the customer is willing to buy any amount in this range, something that is not unusual among tour operators)—then there is little impact on the single-resource models discussed above. Indeed, the static model (2.3) can be thought of as a group model where in each period one "large group" of size D_j arrives because we can sell u units, where $0 \leq u \leq \min\{D_j, x\}$ and x is the total available capacity. Thus, with groups that can be partially accepted, we need only to keep track of the aggregate demand for each class and the formulations are essentially the same as those of the static case in Section 2.2.2.

The real complication in group arrivals occurs when groups must be accepted on an *all-or-none* basis—that is, given a request for $m > 1$ units we can sell only all m units or none at all. This seemingly modest change has a profound impact on the structure of optimal allocation policies. First, we must specify the distribution of group sizes to model how much demand we have from groups of various sizes. But this in itself does not pose too much of a theoretical difficulty. The bigger problem is that the value function may not be concave (the marginal value of capacity may in fact increase), so using booking limits, protection levels, or bid prices may not be optimal.

An example illustrates what can go wrong. Consider a static model with only two stages. Suppose that in the last stage (stage $j = 1$), only groups of size two arrive. In the first stage (stage $j = 2$), groups of varying sizes can arrive. Now suppose that we have $x = 2$ units of capacity remaining. Note the marginal value of the last unit of capacity in stage 1 is zero; that is, $\Delta V_1(1) = V_1(1) - V_1(0) = 0$, because we only get demand for groups of size 2 in the last stage, and therefore having only one unit of capacity is of no value. On the other hand, provided we have some positive probability of demand for a group of size 2 in stage 1, then the second unit of capacity will have a positive marginal value; that is, $\Delta V_1(2) = V_1(2) - V_1(1) > 0$. Hence, the marginal value of capacity $\Delta V_j(x)$ is no longer decreasing in x. As a result, in first stage it can be optimal to reject a request for a single unit of some class when there are two units of capacity remaining but optimal to accept the same request when there is only one unit of capacity remaining. So the notion that there is a booking limit above which we will not sell to a class is no longer valid.

Essentially, the requirement to completely accept or reject groups creates a combinatorial (*bin-packing*) phenomenon in allocating capacity. The resulting nonmonotone value function means that optimal policies are considerably more complex than in the case where groups can be partially accepted. Intuitively, one might expect that if a sufficiently large fraction of demand is from small groups (size one or two) and the capacities are reasonably large, then these combinatorial effects could be ignored and the nongroup models may be a good approximation. This is the implicit assumption in most single-resource RM systems used in practice.

2.5 Dynamic Models

Dynamic models relax the assumption that the demand for classes arrives in a strict low-to-high revenue order. Instead, they allow for an arbitrary order of arrival, with the possibility of interspersed arrivals of several classes. While at first this seems like a strict generalization of the static case, the dynamic models require the assumption of Markovian (such as Poisson) arrivals to make them tractable. This puts restrictions on modeling different levels of variability in demand. Indeed, this limitation on the distribution of demand is the main drawback of dynamic models in practice. In addition, dynamic models require an estimate of the pattern of arrivals over time (called the *booking curve*), which may be difficult to estimate in certain applications. Thus, the choice of dynamic versus static models essentially comes down to a choice of which set of approximations is more acceptable and what data is available in any given application.

Other assumptions of the static model are retained. Demand is assumed independent between classes and over time and also independent of the capacity controls. The firm is again assumed to be risk-neutral. The justifications (or criticisms) for these assumptions are the same as in the static-model case.

2.5.1 Formulation and Structural Properties

In the simplest dynamic model, we have n classes as before with associated prices $p_1 \geq p_2 \geq \cdots \geq p_n$. There are T total periods and t indexes the periods, with the time index running forward ($t = 1$ is the first period, and $t = T$ is the last period; this is in contrast to the static dynamic program, where the stages run from n to 1 in the dynamic programming recursion). Since there is no longer a one-to-one correspondence between periods and classes, we use separate indices—t for periods and j for classes.

In each period we assume, by a sufficiently fine discretization of time, that at most one arrival occurs.[8] The probability of an arrival of class j in period t is denoted $\lambda_j(t)$. The assumption of at most one arrival per period implies that we must have

$$\sum_{j=1}^{n} \lambda_j(t) \leq 1.$$

In general, the periods need not be of the same duration. For example, early in the booking process when demand is low we might use a period of several days whereas during periods of peak booking activity we might use a period of less than an hour. Note also the arrival probabilities may vary with t, so the mix of classes that arrive may vary over time. In

---

[8]The assumption of one arrival per period can be generalized as shown by Lautenbacher and Stidham [330], but both theoretically and computationally it is a convenient assumption.

Single-Resource Capacity Control

particular, we do not require lower classes to arrive earlier than higher classes.

2.5.1.1 Dynamic Program

As before, let x denote the remaining capacity and $V_t(x)$ denote the value function in period t. Let $R(t)$ be a random variable, with $R(t) = p_j$ if a demand for class j arrives in period t and $R(t) = 0$ otherwise. Note that $P(R(t) = p_j) = \lambda_j(t)$. Let $u = 1$ if we accept the arrival (if there has been one) and $u = 0$ otherwise. (We suppress the period subscript t of the control as it should be clear from the context.) We want to maximize the sum of current revenue and the revenue to go, or

$$R(t)u + V_{t+1}(x - u).$$

The Bellman equation is therefore

$$\begin{aligned} V_t(x) &= E\left[\max_{u \in \{0,1\}} \{R(t)u + V_{t+1}(x - u)\}\right] \\ &= V_{t+1}(x) + E\left[\max_{u \in \{0,1\}} \{(R(t) - \Delta V_{t+1}(x))u\}\right], \end{aligned} \quad (2.17)$$

where $\Delta V_{t+1}(x) = V_{t+1}(x) - V_{t+1}(x-1)$ is the expected marginal value of capacity in period $t+1$. The boundary conditions are [9]

$$V_{T+1}(x) = 0, \quad x = 0, 1, \ldots, C,$$

and

$$V_t(0) = 0, \quad t = 1, \ldots, T.$$

2.5.2 Optimal Policy

An immediate consequence of (2.17) is that if a class j request arrives, so that $R(t) = p_j$, then it is optimal to accept the request if and only if

$$p_j \geq \Delta V_{t+1}(x).$$

Thus, the optimal control can be implemented using a bid-price control where the bid price is equal to the marginal value,

$$\pi_t(x) = \Delta V_t(x). \quad (2.18)$$

---

[9] The second boundary condition can be eliminated if we use the control constraint $u \in \{0, \min\{1, x\}\}$ instead of $u \in \{0, 1\}$. However, it is simpler conceptually and notationally to use the $x = 0$ boundary conditions instead.

Revenues that exceed this threshold are accepted; those that do not are rejected.

As in the static case, an important property of the value function is that it has decreasing marginal value $\Delta V_t(x) = V_t(x) - V_t(x-1)$. (See Appendix 2.A for proof.)

PROPOSITION 2.2 *The increments $\Delta V_t(x)$ of the value function $V_t(x)$ defined by (2.17) satisfy $\forall x, t$:*
(i) $\Delta V_t(x+1) \leq \Delta V_t(x)$,
(ii) $\Delta V_{t+1}(x) \leq \Delta V_t(x)$.

This theorem is natural and intuitive since one would expect the value of additional capacity at any point in time to have a decreasing marginal benefit and the marginal value at any given remaining capacity x to decrease with time (because as time elapses, there are fewer opportunities to sell the capacity).

As a consequence, the optimization on the right-hand side of (2.17) can also be implemented as a nested-allocation policy, albeit one that has time-varying protection levels (or booking limits). Specifically, we can define time-dependent optimal protection levels

$$y_j^*(t) = \max\{x : p_{j+1} < \Delta V_{t+1}(x)\}, \quad j = 1, 2, \ldots, n-1 \qquad (2.19)$$

that have the usual interpretation that $y_j^*(t)$ is the capacity we protect for classes $j, j-1, \ldots, 1$. Then the protection levels are nested, $y_1^*(t) \leq y_2^*(t) \leq \cdots \leq y_{n-1}^*(t)$, and it is optimal to accept class j if and only if the remaining capacity exceeds $y_{j-1}^*(t)$. The situation is illustrated in Figure 2.6.

Time-dependent nested booking limits can also be defined as before by

$$b_j^*(t) \equiv C - y_{j-1}^*(t), \quad j = 2, \ldots, n, \qquad (2.20)$$

That the booking limits and protection levels depend on time in this case essentially stems from the fact that the demand to come varies with time in the dynamic model. The change in demand to come as time evolves effects the opportunity cost and therefore the resulting booking limit and protection levels.

As a practical matter, because the value function is not likely to change much over short periods of time, fixing the protection levels or booking limits computed by a dynamic model and updating them periodically (as is done in most RM systems in practice) is usually close to optimal. Still, the time-varying nature of the protection levels remains a key distinction between static and dynamic models.

We summarize these results in the following theorem:

Single-Resource Capacity Control

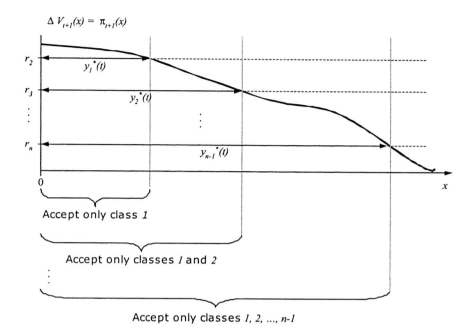

Figure 2.6. Optimal protection level $y_j^*(t)$ in the dynamic model.

THEOREM 2.2 *For the dynamic model defined by (2.17), the optimal control can be achieved using either:*
(i) Time-dependent nested protection levels defined by (2.19),
(ii) Time-dependent nested booking limits defined by (2.20), or
(iii) Bid-price tables defined by (2.18).

2.5.2.1 Computation

Computationally, the dynamic model is solved by substituting the optimal policy into (2.17). This yields the recursion

$$V_t(x) = V_{t+1}(x) + E\left[(R(t) - \Delta V_{t+1}(x))^+\right]$$
$$= \sum_{j=1}^n \lambda_j(t)(p_j - \Delta V_{t+1}(x))^+, \quad t = 1, \ldots, T.$$

Starting with the boundary condition $V_{T+1}(x) = 0$, $\forall x$, we proceed with the recursion backward in time t. Each stage t requires $O(nC)$ operations, so the overall complexity is $O(nCT)$. Usually, the value of T is $O(C)$ because in most practical problems the total expected demand is the same magnitude as the capacity and the periods are typically chosen so that there is $O(1)$ arrival per period. So the complexity in terms of

2.6 Customer-Choice Behavior

A key assumption in the models that we have described thus far is that demand for each of the classes is completely independent of the capacity controls being applied by the seller. That is, it is assumed that the likelihood of receiving a request for any given class does not depend on which other classes are available at the time of the request. Needless to say, this is a somewhat unrealistic assumption. For example, in the airline case the likelihood of selling a full-fare ticket may very well depend on whether a discount fare is available at the same time and the likelihood that a customer buys at all may depend on the lowest available fare. When customers buy a higher fare when a discount is closed it is called *buy-up* (from the firm's point of view, this is also called *sell-up*); when they choose another flight when a discount is closed it is called *diversion*.

Clearly, such customer behavior could have important RM consequences and ought to be considered when making control decisions. We next look at some heuristic and exact methods for incorporating customer-choice behavior in single-resource problems.

2.6.1 Buy-Up Factors

One approach to modeling customer-choice behavior that works with the two-class model is to include buy-up probabilities—also called *buy-up factors*—in the formulation.

The approach works as follows. Consider the simple two-class static model, and recall that Littlewood's rule (2.2) (slightly restated) is to accept demand from class 2 if and only if

$$p_2 \geq p_1 P(D_1 > x), \qquad (2.21)$$

where x is the remaining capacity—that is, if the revenue from accepting class 2 exceeds the marginal value of the unit of capacity required to satisfy the request. Now suppose that there is a probability q that a customer for class 2 will buy class 1 if class 2 is closed. The net benefit of accepting the request is still the same, but now rather than losing the request when we reject it, there is some chance the customer will buy up to class 1. If so, we earn a net benefit of $p_1 - p_1 P(D_1 > x)$ (the class 1 revenue minus the expected marginal cost). Thus, it is optimal to accept class 2 now if $p_2 - p_1 P(D_1 \geq x) \geq q p_1(1 - P(D_1 > x))$ or equivalently if

$$p_2 \geq (1-q) p_1 P(D_1 > x) + q p_1. \qquad (2.22)$$

Note that the right-hand side of the modified rule (2.22) is strictly larger than the right-hand side in Littlewood's rule (2.21), which means that the modified rule (2.22) is more likely to reject class 2 demand. This is intuitive because with the possibility of customers upgrading to class 1, we should be more eager to close class 2.

The difficulty with this approach is that it does not extend to more than two classes—at least not in an exact way—because the probability that a customer buys class i given that class j is closed depends not only on i and j but also on which other classes are also available. In other words, with more than two classes the customer faces a *multinomial* choice rather than a *binary* choice.

However, one can at least heuristically extend the buy-up factor idea to EMSR-a or EMSR-b because these heuristics approximate the multi-class problem using the two-class model.

For example, EMSR-b can be extended to allow for a buy-up factor by modifying the equation for determining the protection level y_j, (2.14), as follows:

$$p_{j+1} = (1 - q_{j+1})\bar{p}_j P(S_j > y_j) + q_{j+1}\hat{p}_{j+1}, \qquad (2.23)$$

where q_{j+1} is the probability that a customer of class $j+1$ buys up to one of the classes $j, j-1, \ldots, 1$; $\bar{p}_j$ is the weighted-average revenue from these classes as defined by (2.13); and $\hat{p}_{j+1} > p_{j+1}$ is an estimate of the average revenue received given that a class $j+1$ customer buys up to one of the classes $j, j-1, \ldots, 1$ (for example, $\hat{p}_{j+1} = p_j$ if customers are assumed to buy up to the next-highest price class). Again, the net result of this change is to increase the protection level y_j and close down class $j+1$ earlier than one would do under the traditional EMSR-b rule.[10]

While this modification to EMSR-b provides a simple heuristic way to incorporate choice behavior, it is a somewhat ad hoc adjustment to an already heuristic approach to the problem. Beyond the limitations of the model and its assumptions, there are some serious difficulties involved in estimating the buy-up factors. Indeed, in current applications of the model, they are often simply made-up, reasonable-sounding numbers. Moreover, the assumptions of the model can clash with unconstraining and recapture procedures that are subsequently applied, resulting in double counting of demand. Despite these limitations, buy-up factors have proved useful as a rough-cut approach for incorporating choice behavior in practice.

---

[10]That it increases the protection level about the usual EMSR-b value can be seen by noting that $p_{j+1} = \bar{R}_j P(S_j > y_j)$ in the usual EMSR-b case and $\hat{p}_{j+1} > p_{j+1}$; thus, y_j has to increase to satisfy the equality (2.23).

2.6.2 Discrete-Choice Models

We next look at a single-resource problem in which customer-choice behavior is explicitly modeled using a general discrete-choice model. In contrast to the heuristic approach of buy-up factors, this model provides a more theoretically sound approach to incorporating choice behavior. It also provides insights into how choice behavior affects the optimal availability controls. The theory is first developed for the general choice model case and then applied to some special demand models.

2.6.2.1 Model Definition

As in the traditional dynamic model of Section 2.5, time is discrete and indexed by t, with the indices running forward in time ($t = T$ is the period of resource usage). In each period there is at most one arrival. The probability of arrival is denoted by λ, which we assume, for ease of exposition, is the same for all time-periods t. There are n classes, and we let $\mathcal{N} = \{1, \ldots, n\}$ denote the entire set of classes. We let choice index 0 denote the no-purchase choice; that is, the event that the customer does not purchase any of the classes offered. Each class $j \in \mathcal{N}$ has an associated price p_j, and without loss of generality we index classes so that $p_1 \geq p_2 \geq \cdots \geq p_n \geq 0$. We let $p_0 = 0$ denote the revenue of the no-purchase choice.

Customer purchase behavior is modeled as follows. In each period t, the seller chooses a subset $S_t \subseteq \mathcal{N}$ of classes to offer. When the set of classes S_t is offered in period t, the probability that a customer chooses class $j \in S_t$ is denoted $P_j(S_t)$. $P_0(S_t)$ denotes the no-purchase probability.

The probability that a sale of class j is made in period t is therefore $\lambda P_j(S_t)$, and the probability that no sale is made is $\lambda P_0(S_t) + (1 - \lambda)$. Note that this last expression reflects the fact that having no sales in a period could be due either to no arrival at all or an arrival that does not purchase. This leads to an incomplete-data problem when estimating the model, as discussed in Section 9.4.1.2.

The only condition we impose on the choice probabilities $P_j(S)$ is that they define a proper probability function. That is, for every set $S \subseteq \mathcal{N}$, the probabilities satisfy

$$P_j(S) \geq 0, \quad \forall j \in S$$
$$\sum_{j \in S} P_j(S) + P_0(S) = 1.$$

Single-Resource Capacity Control

This includes most choice models of practical interest (see Ben-Akiva and Lerman [48]) and even some rather pathological cases.[11] The following running example will be used to illustrate the model and analysis:

Example 2.5 An airline offers three fare products—Y, M, and K. These products differ in terms of revenues and conditions as shown in Table 2.7. The airline has

Table 2.7. Fare-product revenues and restrictions for Example 2.5.

Fare Product (Class)	SA Stay	21-Day Adv.	Revenue
Y	No	No	$800
M	No	Yes	$500
K	Yes	Yes	$450

five segments of customers—two business segments and three leisure segments. The segments differ in terms of the restrictions that they qualify for and the fares they are willing to pay. The data describing each segment are given in Table 2.8. The second

Table 2.8. Segments and their characteristics for Example 2.5.

		Qualifies for Restrictions?		Willing to Buy?	
Segment	Prob.	SA Stay	21-Day Adv.	Y Class	M Class
Bus. 1	0.1	No	No	Yes	Yes
Bus. 2	0.2	No	Yes	Yes	Yes
Leis. 1	0.2	No	Yes	No	Yes
Leis. 2	0.2	Yes	Yes	No	Yes
Leis. 3	0.3	Yes	Yes	No	No

column of Table 2.8 gives the probability that an arriving customer is from each given segment.

Given this data for Example 2.5, the first four columns of Table 2.9 give the choice probabilities that would result.[12]

---

[11] For example, some psychologists have shown that customers can be overwhelmed by more choices, and they may become more reluctant to purchase as more options are offered (see Iyengar and Lepper [265]). Such cases would be covered by a suitable choice of $P_j(S)$ that results in the total probability of purchase, $\sum_{j \in S} P_j(S)$, being decreasing in S.

[12] To see how the probabilities in Table 2.9 are derived, consider the set $S = \{Y, K\}$. If $S = \{Y, K\}$ is offered, segments Business 1 and Business 2 buy the Y fare because they cannot qualify for both the SA stay and 21-day advance-purchase restrictions on K, so $P_Y = 0.1 + 0.2 = 0.3$. Similarly, Leisure 1 cannot qualify for the SA stay restriction of K and is not willing to purchase Y, so these customers do not purchase at all. Segments Leisure 2 and 3, however, qualify for both restrictions on K and purchase K. Hence, $P_K = 0.2 + 0.3 = 0.5$. Class M is not offered, so $P_M = 0$. The other rows of Table 2.9 are filled out similarly.

66   THE THEORY AND PRACTICE OF REVENUE MANAGEMENT

Table 2.9. Choice probabilities $P_j(S)$, probability of purchase $Q(S)$, and expected revenue $R(S)$ for Example 2.5.

S	$P_Y(S)$	$P_M(S)$	$P_K(S)$	$Q(S)$	$R(S)$	Efficient?[a]
$\{Y\}$	0.3	0	0	0.3	240	Yes
$\{M\}$	0	0.4	0	0.4	200	No
$\{K\}$	0	0	0.5	0.5	225	No
$\{Y,M\}$	0.1	0.6	0	0.7	380	No
$\{Y,K\}$	0.3	0	0.5	0.8	465	Yes
$\{M,K\}$	0	0.4	0.5	0.9	425	No
$\{Y,M,K\}$	0.1	0.4	0.5	1	505	Yes

[a] Efficient sets are defined in Section 2.6.2.3.

This particular method of generating choice probabilities is only for illustration. Other choice models could be used and in general any proper set of probabilities could be used to populate Table 2.9.

2.6.2.2 Formulation

As before, let C denote the total capacity, T the number of time-periods, t the current period, and x the number of remaining inventory units. Define the value function $V_t(x)$ as the maximum expected revenue obtainable from periods $t, t+1, \ldots, T$ given that there are x inventory units remaining at time t. Then the Bellman equation for $V_t(x)$ is

$$V_t(x) =$$
$$\max_{S \subseteq \mathcal{N}} \left\{ \sum_{j \in S} \lambda P_j(S)(p_j + V_{t+1}(x-1)) + (\lambda P_0(S) + 1 - \lambda)V_{t+1}(x) \right\}$$
$$= \max_{S \subseteq \mathcal{N}} \left\{ \sum_{j \in S} \lambda P_j(S)(p_j - \Delta V_{t+1}(x)) \right\} + V_{t+1}(x), \quad (2.24)$$

where $\Delta V_{t+1}(x) = V_{t+1}(x) - V_{t+1}(x-1)$ denotes the marginal cost of capacity in the next period, and we have used the fact that for all S,

$$\sum_{j \in S} P_j(S) + P_0(S) = 1.$$

The boundary conditions are

$$V_{T+1}(x) = 0, \quad x = 0, 1, \ldots, C \quad (2.25a)$$
$$V_t(0) = 0, \quad t = 1, \ldots, T. \quad (2.25b)$$

Note one key difference in this formulation compared to our analysis of the traditional independent-class models of Section 2.2.2 and Section 2.5—we assume the seller *precommits* to the open set of classes S

Single-Resource Capacity Control

in each period, while in the traditional models, we assume the seller observes the class of the request and then makes an accept or deny decision based on the class. The reason for the difference is that in the traditional models the class of an arriving request is completely independent of the controls, so it doesn't matter whether we precommit to the set of open classes or not. However, in the choice-based model, the class that an arriving customer chooses depends (through the choice model $P_j(S)$) on which classes S we report as being open. Hence, the formulation (2.24) reflects this fact (we are taking $\max E[\cdot]$ in 2.24 instead of $E[\max(\cdot)]$); we must choose S prior to seeing the realization of the choice decision.

2.6.2.3 Structure of the Optimal Policy

The problem (2.24) at first seems to have very little structure, but a sequence of simplifications provides a good characterization of the optimal policy. The first simplification is to write (2.24) in more compact form as

$$V_t(x) = \max_{S \subseteq \mathcal{N}} \{\lambda(R(S) - Q(S)\Delta V_{t+1}(x))\} + V_{t+1}(x), \quad (2.26)$$

where

$$Q(S) = \sum_{j \in S} P_j(S) = 1 - P_0(S)$$

denotes the total probability of purchase, and

$$R(S) = \sum_{j \in S} P_j(S) p_j$$

denotes the total expected revenue from offering set S. Table 2.9 gives the values $Q(S)$ and $R(S)$ for our Example 2.5. For theoretical purposes, we also consider allowing the seller to randomize over the sets S that are offered at the beginning of each time-period, but this relaxation is not strictly needed since there is always at least one set S that achieves the maximum in (2.26).

The second simplification is to note that not all $2^n - 1$ subsets need to be considered when maximizing the right-hand side of (2.26). Indeed, the search can be reduced to only those sets that are efficient as defined below:

DEFINITION 2.1 *A set T is said to be **inefficient** if there exist probabilities $\alpha(S), \forall S \subseteq \mathcal{N}$ with $\sum_{S \subseteq \mathcal{N}} \alpha(S) = 1$ such that*

$$Q(T) \geq \sum_{S \subseteq \mathcal{N}} \alpha(S) Q(S) \quad \text{and} \quad R(T) < \sum_{S \subseteq \mathcal{N}} \alpha(S) R(S),$$

*A set is said to be **efficient** if no such probabilities $\alpha(S)$ exist.*

In words, a set T is inefficient if we can use a randomization of other sets S to produce an expected revenue that is strictly greater than $R(T)$ with no increase in the probability of purchase $Q(T)$.

The significance of inefficient sets is that they can be eliminated from consideration:

PROPOSITION 2.3 *An inefficient set is never an optimal solution to (2.24).*

The proof is omitted, but the fact that such sets should be eliminated from consideration is quite intuitive from (2.26); an inefficient set T provides strictly less revenue $R(T)$ than do other sets and incurs at least as high a probability of consuming capacity $Q(T)$ (and hence incurs at least as high an opportunity cost $Q(S)\Delta V_{t+1}(x)$ in (2.26)).

For Example 2.5, Table 2.9 shows which sets are efficient—namely, the sets $\{Y\}$, $\{Y, K\}$, and $\{Y, K, M\}$. That these sets are efficient follows from inspection of Figure 2.7, which shows a scatter plot of the values $Q(S)$ and $R(S)$ for all subsets S. Note from this figure and Definition 2.1 that an efficient set is a point that is on the "efficient frontier" of the set of points $\{Q(S), R(S)\}, S \subseteq \mathcal{N}$. Here, "efficiency" is with respect to the tradeoff between expected revenue $R(S)$ and probability of sale $Q(S)$.

The third simplification is to note that the efficient sets can be easily ordered. Indeed, let m denote the number of efficient sets. These sets can be indexed $S_1, \ldots, S_m$ such that both the revenues and probabilities of purchase are monotone increasing in the index. That is, if the collection of m efficient sets is indexed such that $Q(S_1) \leq Q(S_2) \leq \cdots \leq Q(S_m)$, then $R(S_1) \leq R(S_2) \leq \cdots \leq R(S_m)$ as well. The proof of this fact is again omitted, but it is easy to see intuitively from Figure 2.7. Note from Table 2.9 that there are $m = 3$ efficient sets $\{Y\}$, $\{Y, K\}$, and $\{Y, K, M\}$. These can be ordered $S_1 = \{Y\}$, $S_2 = \{Y, K\}$, and $S_3 = \{Y, K, M\}$ with associated probabilities of purchase $Q_1 = 0.3$, $Q_2 = 0.8$, and $Q_3 = 1$ and prices $p_1 = \$240$, $p_2 = \$465$, and $p_3 = \$505$ as claimed.

Henceforth, we assume the efficient sets are denoted $S_1, \ldots, S_m$ and are indexed in increasing revenue and probability order. Also, to simplify notation we let $R_k = R(S_k)$ and $Q_k = Q(S_k)$ and note R_k and Q_k are both increasing in k. So the Bellman equation can be further simplified to

$$V_t(x) = \max_{k=1,\ldots,m} \{\lambda(R_k - Q_k \Delta V_{t+1}(x))\} + V_{t+1}(x). \qquad (2.27)$$

The final simplification is to show that when expressed in terms of the sequence $S_1, \ldots, S_m$ of efficient sets, the optimal policy has a simple form as stated in the following theorem:

Single-Resource Capacity Control

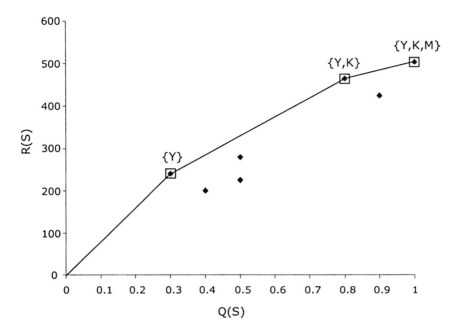

Figure 2.7. Scatter plot of $Q(S)$ and $R(S)$ for Example 2.5 (efficient points are enclosed in squares and labeled).

THEOREM 2.3 *An optimal policy for (2.24) is to select a set k^* from among the m efficient, ordered sets $\{S_k : k = 1, \ldots, m\}$ that maximizes (2.27). Moreover, for a fixed t, the largest optimal index k^* is increasing in the remaining capacity x, and for any fixed x, k^* is increasing in time t.*

The proof of this theorem is involved but derives from the fact that the marginal value $\Delta V_{t+1}(x)$ is decreasing in x (see Appendix 2.A for a proof) and the fact that the optimal index k^* is decreasing in this marginal value.

This characterization is significant for several reasons. First, it shows that the optimal sets can be reduced to only those that are efficient, which in many cases significantly reduces the number of sets we need to consider. Moreover, it shows that this limited number of sets can be sequenced in a natural way and that the more capacity we have (or the less time remaining), the higher the set we should use in this sequence.

For example, applying Theorem 2.3 to Example 2.5, we see that the efficient sets $S_1 = \{Y\}$, $S_2 = \{Y, K\}$, and $S_3 = \{Y, K, M\}$ would be used as follows. With very large amounts of capacity remaining, S_3 is optimal: all three fare classes are opened. As capacity is consumed, at some point we switch to only offering S_2: class M is closed, and only Y and K are offered. As capacity is reduced further, at some point we close class K and offer only class Y (set S_1 is used).

Note what's odd here; it can be optimal to offer the highest fare Y and the lowest fare K, but not the middle fare M. This is because opening M causes some buy-down from Y to M, whereas K is sufficiently restricted to prevent buy-down. Only when capacity is plentiful is M opened.

2.6.2.4 Identifying Efficient Sets

Finding the efficient sets is, in general, computationally complex. The naive approach is to enumerate all $2^n - 1$ subsets of $\mathcal{N}$ and for each set T solve a linear program (in variables $\alpha(S), S \subseteq \mathcal{N}$) to test for efficiency using the conditions in Definition 2.1.

However, a more efficient alternative is to use the following *largest marginal revenue* procedure. First, let $S_0 = \emptyset$. Then successive sets can be found by the following recursion. Let S_i be the ith efficient set. Then the $(i+1)^{\text{st}}$ efficient set, S_{i+1}, is found by checking among the sets S with $Q(S) \geq Q(S_i)$ and $R(S) \geq R(S_i)$ for the one that maximizes the marginal revenue ratio

$$\frac{R(S) - R(S_i)}{Q(S) - Q(S_i)}.$$

(Note that this is simply the increase in expected revenue per unit increase in expected demand.) The procedure starts with $i = 0$ and stops as soon as no sets S with $Q(S) \geq Q(S_i)$ and $R(S) \geq R(S_i)$ exist; it returns the complete sequence $S_1, \ldots, S_m$. Since there are $O(2^n)$ subsets to check at each step, the recursion has complexity $O(m2^n)$, where m is the number of efficient sets (which in the worst case could be $O(2^n)$ itself).

For small numbers of classes, this largest marginal revenue procedure is practical, especially since it can be performed off line. But it is still exponential in the number of classes n. For large numbers of classes, heuristic or analytic approaches can be used to reduce the complexity of identifying efficient sets. For example, one could enumerate a limited collection of subsets S rather than all $2^n - 1$ subsets and apply the largest marginal revenue procedure to determine which subsets in the collection are efficient relative to other sets in the collection. In some special cases, as shown below, one can identify which subsets are efficient analytically, thus eliminating the need to enumerate all possible subsets.

Single-Resource Capacity Control

2.6.2.5 Optimality of Nested-Allocation Policies

The optimization results above also have important implications for the optimality of nested-allocation policies. Indeed, Definition 2.1 and Theorem 2.3 can be used to provide a complete characterization of cases in which nested-allocation polices are optimal. They also can be used to provide conditions under which the optimal nesting is by revenue order.

We begin with a precise definition of a nested-allocation policy in the context of the choice model:

DEFINITION 2.2 *A control policy is called a **nested policy** if there is an increasing family of subsets* $S_1 \subseteq S_2 \subseteq \cdots \subseteq S_m$ *and an index* $k_t(x)$ *that is increasing in* x, *such that set* $S_{k_t(x)}$ *is chosen at time* t *when the remaining capacity is* x.

Though this is a somewhat abstract definition of a nested policy, it is in fact a natural generalization of nested allocations from the traditional single-resource models of Section 2.2.2 and 2.5 and implies an ordering of the classes based on when they first appear in the increasing sequence of sets S_k. That is, class i is considered "higher" than class j in the nesting order if class i appears earlier in the sequence. Returning to Example 2.5, we see that the efficient sets are indeed nested according to this definition because $S_1 = \{Y\}$, $S_2 = \{Y, K\}$, and $S_3 = \{Y, K, M\}$ are increasing. Class Y would be considered the highest in the nested order, followed by class K and then class M.

If the optimal policy is nested in this sense, then we can define optimal protection levels $y_k^*(t), k = 1, \ldots, m$, such that classes lower in the nesting order than those in S_k are closed if the remaining capacity is less than $y_k^*(t)$, just as in the traditional single-resource case. The optimal protection levels for $k = 1, 2, \ldots, m - 1$ are defined by

$$y_k^*(t) = \max\{x : R_k - Q_k \Delta V_{t+1}(x) > R_{k+1} - Q_{k+1} \Delta V_{t+1}(x)\}.$$

Nested booking limits can also be defined in the usual way, $b_k(t) = C - y_{k-1}(t)$.

We again return to Example 2.5 to illustrate this concept. Table 2.10 shows the objective function value $R_k - Q_k \Delta V_{t+1}(x)$ for each of the three efficient sets $k = 1, 2, 3$, for a particular marginal value function $\Delta V_{t+1}(x)$, which we assume is given in this example. Capacities are in the range $x = 1, 2, \ldots, 20$. The last column of Table 2.10 gives the index, $k_t^*(x)$, of the efficient set that is optimal for each capacity x.

Note that for capacities $1, 2$, and 3, the set $S_1 = \{Y\}$ is the optimal set, so class Y is the only open fare. Once we reach 4 units of remaining capacity, set $S_2 = \{Y, K\}$ becomes optimal and we open class K in addition to class Y. When the remaining capacity reaches 13, set $S_3 =$

$\{Y, K, M\}$ becomes optimal, and we open M in addition to Y and K. As a result, the optimal protection level for set S_1, is $y_1^* = 3$, and the protection level for set S_2 is $y_2^* = 12$. S_3 has a protection level equal to capacity.

Table 2.10. Illustration of nested policy for Example 2.5.

x	$\Delta V_{t+1}(x)$	\multicolumn{3}{c}{$R_k - Q_k \Delta V_{t+1}(x)$}	$k_t^*(x)$
		$k=1$	$k=2$	$k=3$	
1	780.00	6.00	-159.00	-275.00	1
2	624.00	52.80	-34.20	-119.00	1
3	520.00	84.00	49.00	-15.00	1
4	445.71	106.29	108.43	59.29	2
5	390.00	123.00	153.00	115.00	2
6	346.67	136.00	187.67	158.33	2
7	312.00	146.40	215.40	193.00	2
8	283.64	154.91	238.09	221.36	2
9	260.00	162.00	257.00	245.00	2
10	240.00	168.00	273.00	265.00	2
11	222.86	173.14	286.71	282.14	2
12	208.00	177.60	298.69	297.00	2
13	195.00	181.50	309.00	310.00	3
14	183.53	184.94	318.18	321.47	3
15	173.33	188.00	326.33	331.67	3
16	164.21	190.74	333.63	340.79	3
17	156.00	193.20	340.20	349.00	3
18	148.57	195.43	346.14	356.43	3
19	141.82	197.45	351.55	363.18	3
20	135.65	199.30	356.48	369.35	3

2.6.2.6 Nesting by Revenue Order

Revenues provide a natural nesting ordering, and, as described earlier in this chapter, this is traditionally how most quantity-based RM systems have been conceived and implemented. From a practical standpoint, therefore, it is important to understand when a particular choice model leads to nesting by revenue order. Yet Example 2.5 makes clear that nesting by revenue order need not be the optimal policy in general.

Talluri and van Ryzin [500] provide conditions that guarantee a given choice model will always have this property. The results show, for example, why the optimal control for the traditional independent-demand model is nested by revenue order. Talluri and van Ryzin [500] also show that if the choice probabilities follow the multinomial-logit (MNL) choice model (See Section 7.2.2.3.), then the optimal policy is always nested by revenue order. A similar result holds if customers' choice behavior is such that they always buy the lowest-price open class. However, in general nesting by revenue order need not be optimal.

2.6.2.7 Comparisons of Optimality Conditions

The optimality conditions in the nested-by-revenue-order case also provide some intuition into how choice-based controls differ from traditional controls. Let $C_k = \{1, \ldots, k\}$ denote the set of the k highest classes (in revenue order). Then one can show it is optimal to open class $k+1$ if and only if

$$P_{k+1}(C_{k+1})\left(r_{k+1} - \Delta V_{t-1}(x)\right) \geq \sum_{j=1}^{k} \Delta P_j(C_k)(r_j - \Delta V_{t-1}(x)), \quad (2.28)$$

where

$$\Delta P_j(C_k) = P_j(C_k) - P_j(C_{k+1})$$

is the change (usually an increase for most choice models) in purchase probability for class j as the result of *not* offering class $k+1$. This expression is intuitive: The left-hand side is the probability of selling class $k+1$ times the "net gain" from selling it; that is, the revenue we get from class $k+1$ minus the opportunity cost, $\Delta V_{t-1}(x)$, of using a unit of capacity. The right-hand side is the net gain (loss) among the other classes caused by eliminating (adding) class $k+1$ (the sum over all the other class j in C_k of the change in purchase probability times the net gain from selling j). Therefore, the condition (2.28) simply says that if the expected gain on class $k+1$ exceeds the incremental loss on the other classes caused by adding $k+1$, then it pays to open $k+1$; otherwise, $k+1$ should be closed.

The expression (2.28) should be compared with the optimality condition for independent-demand model; namely, it is optimal to open class $k+1$ if and only if

$$r_{k+1} - \Delta V_{t-1}(x) \geq 0.$$

(Indeed, note that (2.28) reduces to the above expression for the independent-demand model since for this model $\Delta P_j(C_k) = 0$ for all j, k.) Note that the right-hand side above is zero while the right-hand side of (2.28) is nonzero (typically positive). This happens because in the independent-demand model, if we close class $k+1$, we lose all demand for that class. Therefore, it is optimal to accept class $k+1$ whenever r_{k+1} exceeds the opportunity cost $\Delta V_{t-1}(x)$. However, in the choice-based model, if we close class $k+1$, customers choose from among the other classes that are offered (e.g., from C_k). Hence, the threshold on the right-hand side of (2.28) is nonzero. This difference reflects the fact

74   THE THEORY AND PRACTICE OF REVENUE MANAGEMENT

that customers may buy up to a higher class, so there is some nonzero benefit to rejecting a request for $k+1$.

2.6.2.8   A Numerical Comparison to EMSR-b with Buy-Up

We next look at the results of a small simulation study comparing choice-based methods to a traditional EMSR-b method with buy-up as described in Section 2.6.1. While the EMSR-b model is developed under the static-model assumptions, it is frequently used as a heuristic in the dynamic case by simply aggregating the total demand to come for each class. Also, it is one of the few models available that incorporates some type of choice behavior. For these reasons, it serves as a useful benchmark for comparison.

This simulations are based on our running three-class example, Example 2.5. We compare the optimal control given by the choice dynamic program with the traditional EMSR-b (buy-up) recommendations. (Recall that for Example 2.5 the optimal dynamic programming policy uses the fare classes in the order Y, K, M, whereas the EMSR-b uses them in the fare order Y, M, K.)

The capacity $C = 20$, and there are three population sizes: $15, 20,$ and 25. The fares, restrictions, and customer segments are as given in Table 2.8. For a population size of 20 this results in an unconstrained mean and variance as shown in Table 2.11. These statistics are used to create inputs for EMSR-b. The buy-up factors are computed as shown

Table 2.11. The different segment choices in Example 2.5 if all classes are open and resulting demand for a population size of 20 (N.E. stands for not eligible).

Segments	%	Sat-Night	21-Day	Can Buy Y?	Can Buy M?	Eligible for K?	Choice	Mean	Var.
B1	0.1	No	No	Yes	N.E	N.E.	Y	2	1.8
B2	0.2	No	Yes	Yes	Yes	N.E.	M	4	3.2
L1	0.2	No	Yes	No	Yes	N.E.	M	4	3.2
L2	0.2	Yes	Yes	No	Yes	Yes	K	4	3.2
L3	0.3	Yes	Yes	No	No	Yes	K	6	4.2

in Table 2.12, which lists the unconstrained choices and demands when all classes are open. (N.E. signifies that the segment is not eligible.) This estimate roughly mimics the traditional practice of unconstraining and forecasting demand in each class. Table 2.12 then sums the unconstrained demands for each fare class. The buy-up factors are estimated as follows. The buy-up factor for K is given by the percentage K customers who would buy up to M if we go from offering $\{Y, M, K\}$ to $\{Y, M\}$. Similarly the buy-up factor for M is the fraction who buy up

Table 2.12. Inputs to the EMSR-b model.

Class	Mean	S.D.	Buy-Up Factors
Y	2	1.34	
M	8	2.52	0.33
K	10	2.72	0.40

to Y when going from offering $\{Y, M\}$ to Y. Table 2.12 shows these demands and buy-up factors for a population size of 20, and Table 2.13 shows the computed EMSR-b protection levels for this population size. The results for the three load factors are summarized in Table 2.14.

Table 2.13. Protections for the EMSR-b model without and with buy-up factors.

Class	EMSR-b	EMSR-b with Buy-Up
Y	1.57	2.20
Y + M	7.55	8.71

Note that the choice dynamic program shows significant improvements

Table 2.14. Simulation results comparison between choice dynamic program and EMSR-b with buy-up.

Population Size	EMSR-b Revenue	Choice DP Revenue	% gain from Choice DP	99% Conf. Int. Error on % Gain
15	7,466	7,466		
20	9,825	10,129	3.10	± 0.0050
25	10,142	11,301	11.48	± 0.0079

on this example, achieving an 11.5% improvement in revenue in the high-demand case. Again, part of this improvement can be attributed to the fact that the choice dynamic program uses a different sequence of classes (only the efficient sets $\{Y\}, \{Y, K\}, \{Y, M, K\}$).

2.7 Notes and Sources

The notion of theft versus standard nesting is not well documented and is part of the folklore of RM practice. Our understanding, however, greatly benefited from discussions with our colleagues Peter Belobaba, Sanne de Boer, and Craig Hopperstad.

The earliest paper on the static models of Section 2.2 is Littlewood [347]. Another early applied paper is Bhatia and Parekh [64]. But there are close connections to earlier work on the stock-rationing problem in the inventory literature by Kaplan [288] and Topkis [515];

see also Gerchak [209, 210] and Ha [246]. Indeed, Topkis's [515] results can be used to show the optimality of nested-allocation policies.

Optimal policies for the $n > 2$ case were obtained in close succession (using slightly different methods and assumptions) in papers by Brumelle and McGill [91], Curry [139], Robinson [445], and Wollmer [576]. See also McGill's thesis [375]. Robinson [445] also analyzed the case where the order of arrival is not the same as the revenue order. Brumelle et al. [90] analyzed a two-class static model with dependent demand.

The Monte Carlo method for computing optimal overbooking limits presented in Section 2.2.3.2 is due to Robinson [445].

The dynamic model of Section 2.5 was first analyzed by Lee and Hersh [336]. However, the proofs of Propositions 2.1 and 2.2 in Appendix 2.A are due to Lautenbacher and Stidham [330], who provided a unified analysis of both the static and dynamic single-resource models. Walczak and Brumelle [543] relate this problem to a dynamic pricing problem using a Markov model of demand that allows for partial information on the revenue values or customer types. See Liang [342] for an analysis of a continuous-time version of the dynamic model.

The EMSR-a and EMSR-b heuristics are both due to Belobaba. The most detailed coverage of EMSR-a is contained in Belobaba's 1987 thesis [39], but see also the published articles from it [38] and [40]. EMSR-b was introduced in [41]; see also Belobaba and Weatherford [37].

The problem of group or batch request in Section 2.4 was addressed by Lee and Hersh [336] for the partially accepted case. For the more complex case where groups must be completely accepted, see Walczak and Brumelle [544], Kleywegt and Papastavrou [307], and Van Slyke and Young [530].

The adaptive algorithm in Section 2.3 is due to van Ryzin and McGill [526]. The buy-up heuristics in Section 2.6.1 are due to Belobaba [39, 38, 40]. See also Belobaba and Weatherford [37], Weatherford [555], and the simulation study of Bohutinsky [81]. See Titze [514] for a discussion of passenger behavior in the simple two-class model. The material in Section 2.6.2 on choice-based models is from Talluri and van Ryzin [500]; see also Algers and Besser [7] and Andersson [18] for an application of discrete-choice models at SAS. For a good reference on discrete-choice modeling, see Ben-Akiva and Lerman [48]. De Boer [155] is another recent work that addresses customer choice in a single-resource problem.

APPENDIX 2.A: Monotonicity Proofs

The proofs of monotonicity are based on a lemma of Stidham [487] originally developed to analyze queueing-control problems. The lemma was adapted to provide

APPENDIX 2.A: Monotonicity Proofs

a convexity proof for a general single-resource problem, which includes both the static and dynamic models, in work by Lautenbacher and Stidham [330]. The proof here follows theirs.

We begin with a definition:

DEFINITION 2-2.A.3 *A function defined on the set of nonnegative integers, $g : \mathcal{Z}_+ \to \Re$ is **concave** if it has nonincreasing differences. That is, $g(x+1) - g(x)$ is nonincreasing in $x \geq 0$.*

LEMMA 2-2.A.1 *Suppose $g : \mathcal{Z}_+ \to \Re$ is concave. Let $f : \mathcal{Z}_+ \to \Re$ be defined by*

$$f(x) = \max_{a=0,1,\ldots,m} \{ap + g(x-a)\}$$

for any given $p \geq 0$ and nonnegative integer $m \leq x$. Then $f(x)$ is concave in $x \geq 0$ as well.

Proof
First, note that by changing variables to $y = x - a$ we can write $f(x) = \hat{f}(x) + rx$, where

$$\hat{f}(x) = \max_{x-m \leq y \leq x} \{-yp + g(y)\}.$$

We first analyze $\hat{f}(x)$. Let $y^* = \arg\max_{y \geq 0}\{-yp + g(y)\}$. Since $g(y)$ is concave, $-yp + g(y)$ is also concave, and moreover is nondecreasing for values of $y \leq y^*$ and nonincreasing for values of $y > y^*$. Therefore, for a given m and p,

$$\hat{f}(x) = \begin{cases} -xp + g(x) & x \leq y^* \\ y^*p + g(y^*) & y^* \leq x \leq y^* + m \\ -(x-m)p + g(x-m) & x \geq y^* + m. \end{cases}$$

Therefore, in the range $x < y^*$ and using the fact that $g(x)$ is concave

$$\begin{aligned} \hat{f}(x+1) - \hat{f}(x) &= -p + g(x+1) - g(x) \\ &\leq -p + g(x) - g(x-1) \\ &= \hat{f}(x) - \hat{f}(x-1). \end{aligned}$$

For $y^* \leq x < y^* + m$, it follows that $\hat{f}(x+1) - \hat{f}(x) = 0$, so $\hat{f}(x)$ is trivially concave in this range.

Finally, for $x \geq y^* + m$, from the concavity of $g(x)$

$$\begin{aligned} \hat{f}(x+1) - \hat{f}(x) &= -p + g(x+1-m) - g(x-m) \\ &\leq -p + g(x-m) - g(x-1-m) \\ &= \hat{f}(x) - \hat{f}(x-1). \end{aligned}$$

Thus, $\hat{f}(x)$ is concave in $x \geq 0$ and since $f(x) = \hat{f}(x) + rx$, $f(x)$ is concave in $x \geq 0$ as well. QED

Proof of Proposition 2.1
We first prove part (i) of Proposition 2.1—namely, that the marginal value $\Delta V_j(x)$ is nonincreasing in x (that $V_j(x)$ is concave in x). The proof is by induction on the stages. Note that in the terminal stage (stage 0), $V_{n+1}(x) = 0$ for all x, which is

trivially concave. Now assume that $V_{j-1}(x)$ is concave in x and consider $V_j(x)$. Note that

$$V_j(x) = E\left[\max_{0 \leq u \leq \min\{D_j, x\}} \{p_j u + V_{j-1}(x - u)\}\right].$$

The inner maximization is precisely of the form given by Lemma 2-2.A.1 with $m = \min\{D_j, x\}$. Hence, it follows that for any realization of D_j, the function

$$H(D_j, x) = \max_{0 \leq u \leq \min\{D_j, x\}} \{p_j u + V_{j-1}(x - u)\}$$

is concave in x. Since $V_j(x) = E_{D_j}[H(D_j, x)]$, it is a weighted average (nonnegative weights) of concave functions, and hence it follows that $V_j(x)$ is concave as well.

Part (ii) of the Proposition 2.1 says that the marginal value at a given capacity x at stage j is less than at stage $j+1$. This is shown as follows:

$$\Delta V_j(x) =$$
$$E\left[p_j \min\{D_j, (x - y_{j-1}^*)^+\} + V_{j-1}(x - \min\{D_j, (x - y_{j-1}^*)^+\})\right]$$
$$-E\left[p_j \min\{D_j, (x - 1 - y_{j-2}^*)^+\} + V_{j-1}(x - 1 - \min\{D_j, (x - y_{j-2}^*)^+\})\right]$$
$$\geq E\left[p_j \min\{D_j, (x - y_{j-2}^*)^+\} + V_{j-1}(x - \min\{D_j, (x - y_{j-2}^*)^+\})\right]$$
$$-E\left[p_j \min\{D_j, (x - 1 - y_{j-2}^*)^+\} + V_{j-1}(x - 1 - \min\{D_j, (x - y_{j-2}^*)^+\})\right]$$
$$\geq E\left[\Delta V_{j-1}(x - \min\{D_j, (x - y_{j-1}^*)^+\})\right]$$
$$\geq \Delta V_{j-1}(x),$$

where the first inequality follows because y_{j-1}^* is the optimal protection level at stage $j-1$, the second inequality follows from the nonnegativity of $\min\{D_j, (x - y_{j-2}^*)^+\} - \min\{D_j, (x - 1 - y_{j-2}^*)^+\}$, and the last inequality follows from the fact that $\Delta V_{j-1}(x)$ is decreasing in x. QED

Proof of Proposition 2.2 Similarly, we can use Lemma 2-2.A.1 to show that the increments $\Delta V_t(x)$ of the value function $V_t(x)$ defined by (2.17) are nonincreasing as well: $V_t(x)$ is concave in x. The proof is by induction on t. First, note $V_{T+1}(x) = 0$ for all x, so $\Delta V_{T+1}(x)$ is trivially decreasing in x. Next, assume $V_{t+1}(x)$ is concave and consider period t. The Bellman equation (2.17) is

$$V_t(x) = E\left[\max_{u \in \{0,1\}} \{R(t)u + V_{t+1}(x - u)\}\right].$$

The inner maximization is again in the form of Lemma 2-2.A.1 with $m = 1$. Hence, the function

$$G(R(t), x) = \max_{u \in \{0,1\}} \{R(t)u + V_{t+1}(x - u)\}$$

is concave in x for any realization of $R(t)$. Since $V_t(x) = E_{R(t)}[G(R(t), x)]$, it follows that $V_t(x)$ is concave in x as well.

To show monotonicity in t, note that

$$\Delta V_t(x) = E\left[(R(t) - \Delta V_{t+1}(x))^+ + V_{t+1}(x)\right]$$
$$-E\left[(R(t) - \Delta V_{t+1}(x - 1))^+ + V_{t+1}(x - 1)\right]$$
$$= \Delta V_{t+1}(x) + E\left[(R(t) - \Delta V_{t+1}(x))^+ - (R(t) - \Delta V_{t+1}(x - 1))^+\right]$$
$$\geq \Delta V_{t+1}(x),$$

APPENDIX 2.A: Monotonicity Proofs

where the last inequality follows from the fact that $\Delta V_{t+1}(x) \leq \Delta V_{t+1}(x-1)$ and hence for any realization $R(t)$,

$$(R(t) - \Delta V_{t+1}(x))^+ \geq (R(t) - \Delta V_{t+1}(x-1))^+.$$

QED

Proof of Monotonicity of Marginal Values from the Choice-Based Model

We next show for completeness that the marginal values in the choice-based models are also decreasing in remaining capacity. Namely,

PROPOSITION 2-2.A.4 *For the choice-based single-resource problem defined by (2.24), the value function satisfies*

$$\Delta V_t(x) \leq \Delta V_t(x-1), \quad \forall t, x$$

and

$$\Delta V_t(x) \geq \Delta V_{t-1}(x), \quad \forall t, x$$

Proof

The proof is by induction on t. First, the statement is trivially true for $t = T+1$ by the boundary conditions (2.25a). Assume it is true for period $t+1$ and consider period t. Let $S_t^*(x)$ denote the optimal solution to (2.24) and note

$$\begin{aligned}
\Delta V_t(x) - \Delta V_t(x-1) &= (\Delta V_{t+1}(x) - \Delta V_{t+1}(x-1)) \\
&+ \sum_{j \in S_t^*(x)} \lambda P_j(S_t^*(x))(p_j - \Delta V_{t+1}(x)) \\
&- \sum_{j \in S_t^*(x-1)} \lambda P_j(S_t^*(x-1))(p_j - \Delta V_{t+1}(x-1)) \\
&- \sum_{j \in S_t^*(x-1)} \lambda P_j(S_t^*(x-1))(p_j - \Delta V_{t+1}(x-1)) \\
&+ \sum_{j \in S_t^*(x-2)} \lambda P_j(S_t^*(x-2))(p_j - \Delta V_{t+1}(x-2))
\end{aligned}$$

(2.A.1)

From the optimality of the set defined by $S_t^*(\cdot)$, the following inequalities hold:

$$\sum_{j \in S_t^*(x-1)} \lambda P_j(S_t^*(x-1))(p_j - \Delta V_{t+1}(x-1)) \geq$$

$$\sum_{j \in S_t^*(x)} \lambda P_j(S_t^*(x))(p_j - \Delta V_{t+1}(x-1))$$

and

$$\sum_{j \in S_t^*(x-1)} \lambda P_j(S_t^*(x-1))(p_j - \Delta V_{t+1}(x-1)) \geq$$

$$\sum_{j \in S_t^*(x-2)} \lambda P_j(S_t^*(x-2))(p_j - \Delta V_{t+1}(x-1))$$

Substituting into (2.A.1) we obtain

$$\Delta V_t(x) - \Delta V_t(x-1) \leq \Delta V_{t+1}(x) - \Delta V_{t+1}(x-1)$$
$$+ \sum_{j \in S_t^*(x)} \lambda P_j(S_t^*(x))(p_j - \Delta V_{t+1}(x))$$
$$- \sum_{j \in S_t^*(x)} \lambda P_j(S_t^*(x))(p_j - \Delta V_{t+1}(x-1))$$
$$- \sum_{j \in S_t^*(x-2)} \lambda P_j(S_t^*(x-2))(p_j - \Delta V_{t+1}(x-1))$$
$$+ \sum_{j \in S_t^*(x-2)} \lambda P_j(S_t^*(x-2))(p_j - \Delta V_{t+1}(x-2))$$

Rearranging and canceling terms yields

$$\Delta V_t(x) - \Delta V_t(x-1) \leq (1 - \sum_{j \in S_t^*(x)} \lambda P_j(S_t^*(x)))(\Delta V_{t+1}(x) - \Delta V_{t+1}(x-1))$$
$$+ \sum_{j \in S_t^*(x-2)} \lambda P_j(S_t^*(x-2))(\Delta V_{t+1}(x-1) - \Delta V_{t+1}(x-2))$$

By induction, $\Delta V_{t+1}(x) - \Delta V_{t+1}(x-1) \leq 0$ and $\Delta V_{t+1}(x-1) - \Delta V_{t+1}(x-2) \leq 0$. Therefore, $\Delta V_t(x) - \Delta V_t(x-1) \leq 0$.

To show the marginal values are monotone increasing in the remaining time, note that

$$\Delta V_t(x) = V_t(x) - V_t(x-1)$$
$$= \max_k \{\lambda(R_k - Q_k \Delta V_{t-1}(x))\} - \max_k \{\lambda(R_k - Q_k \Delta V_{t-1}(x-1))\} + \Delta V_{t-1}(x).$$

From the monotonicity in x we have that $\Delta V_{t-1}(x) \leq \Delta V_{t-1}(x-1)$, and therefore for any value k,

$$\lambda(R_k - Q_k \Delta V_{t-1}(x)) - \lambda(R_k - Q_k \Delta V_{t-1}(x-1)) \geq 0.$$

Hence

$$\max_k \{\lambda(R_k - Q_k \Delta V_{t-1}(x))\} - \max_k \{\lambda(R_k - Q_k \Delta V_{t-1}(x-1))\} \geq 0$$

as well, and it follows that $\Delta V_t(x) \geq \Delta V_{t-1}(x)$. **QED**

Chapter 3

NETWORK CAPACITY CONTROL

3.1 Introduction

In this chapter, we examine quantity-based revenue management of multiple resources—commonly referred to as *network* RM. This class of problems arises in airline, railway, cruise-line, and hotel RM—and in general, whenever customers buy bundles of resources in combination under various terms and conditions. In the airline case, the problem is managing the capacities of a set of connecting flights across a network, where flights can have a mix of connecting and local traffic A product in this case is an *origin-destination itinerary fare class* combination—or ODIF as it is called in airline terminology. In the hotel RM case, the problem is managing room capacity on consecutive days when customers stay multiple nights, where a mix of customers with different lengths of stay share the capacity on any given day. (See Figure 3.1.)

When products are sold as bundles, the lack of availability of any one resource in the bundle limits sales. This creates interdependence among the resources, and hence, to maximize total revenues, it becomes necessary to jointly manage (coordinate) the capacity controls on all resources.[1] In the airline industry, this problem is also called *the passenger-mix problem* or *O&D (origin-destination) control*, and in the hotel industry, *length-of-stay control*. We use the now-standard term *network RM*—though the term is something of a misnomer because the theory and methodology do not require an explicit network structure as such.

---

[1] As in all quantity-based RM, we assume here that prices are fixed for all the products and that we manage only the *allocation* of the resources to the different products.

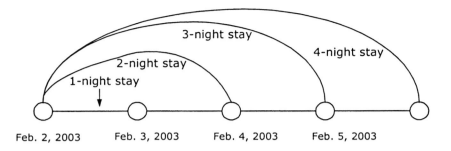

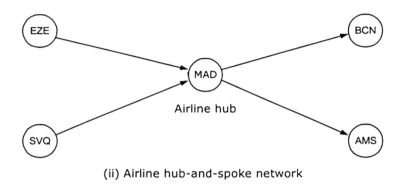

Figure 3.1. Examples of network RM: (i) hotel length-of-stay network and (ii) airline hub-and-spoke network.

3.1.1 The Promise and Challenge of Network Control

There can be significant revenue benefits in taking a network approach to RM. Indeed, simulation studies of airline hub-and-spoke networks by various researchers have demonstrated notable revenue benefits from using network methods over single-resource methods [36, 43, 565, 566]. While precise improvement numbers are dependent on the type of network, the mix and variance of demand, the load factor, etc., improvements on the order of 1.5% are not uncommon at moderate load factors, and gains can be as high as 3% or more at higher load factors.[2] In terms of practice, the potential improvements have been sufficient to justify

---

[2]Such gains have been observed in simulations on airline data, for networks with up to 100 flights, and under an independent-class assumption similar to that of Section 2.2.2.

significant investments in network RM systems within the airline and hotel industries and elsewhere.

While the potential benefit may be high, network RM poses significant implementation and methodological challenges. On the implementation side, network RM vastly increases the complexity and volume of data that one must collect, store, and manage. The resulting data and information systems requirements make the transition from single-resource to network RM often a difficult and expensive one.

Network RM also creates organizational challenges. Since revenue decisions and their effects now span an entire network, it is no longer easy to assign revenue responsibility for a resource to a single analyst; revenue losses at one point in the network may be offset by gains elsewhere in the network. As a result, creating organizations and incentive structures that support the objectives of network RM is a challenging task. Moreover, the transition to a network-oriented RM organization can be difficult for employees who are used to a simpler view of the world.

Network RM creates methodological and operational challenges as well. On the forecasting side, it requires a massive increase in the scale of the forecasting system, which now must forecast demand for each individual network product at each point in the booking process. In addition, the sparsity of the data at this finer level of detail creates numerical and estimation problems, and one has to wrestle with the problem of combining forecasts at various levels of aggregation.

Optimization is more complex as well. In the case of a single-resource problem, there are many exact optimization methods. However, in the network case, exact optimization is for all practical purposes impossible. Therefore, optimization methods use approximations of various types. Achieving a good balance between the quality of the approximation and the efficiency of the resulting algorithms becomes the primary challenge.

3.1.2 Types of Controls

As with single-resource problems, in network allocation problems there are a variety of ways one can control the availability of capacity. The objective is to find an effective and easy-to-implement mechanism to allocate capacity to requests for products that require a combination of resources. The technological constraints imposed by the distribution systems, the revenue performance achievable by the method, and the overall robustness of the control scheme are important criteria for choosing the control mechanism.

We next look at the major categories of network controls. Most are network versions of the controls used in single-resource RM (Section 2.1.1). But others—virtual nesting, in particular—are specific to network RM.

3.1.2.1 Partitioned Booking Limits

Partitioned booking-limit control for network RM is an extension of the partitioned control for single-resource RM (Section 2.1.1.1). In the network case, partitioned booking limits allocate a fixed amount of capacity on each resource for every product that is offered. These allocated amounts of capacity are non-overlapping—or partitioned; demand for a product has exclusive access to its allocated capacity, and no other product may use this capacity.

Partitioned booking limits for the network case have all the defects mentioned in Section 2.1.1.1 and then some. Allocating fixed amounts of capacity to each product results in dividing the capacity of each resource into a very large number of small allocations. This fragmenting of capacity can result in tremendous inefficiencies when demand is stochastic, and for this reasons partitioned booking limits are seldom used in practice.[3]

Partitioned booking limits do have an important role to play both theoretically and computationally, however. Theoretically, they are used to provide bounds on the optimal network revenue. Computationally, they are used in many approximate models (such as the deterministic linear programming method discussed below).

3.1.2.2 Virtual Nesting Controls

Nested booking limits, of the type we saw in Section 2.1.1 for the single-resource case, are difficult to translate directly into a network setting. One difficulty is that the ordering of classes is no longer straightforward.[4] A second difficulty is that the different capacities of the resources involved makes it difficult to specify protection levels or booking limits for products that are consistent across the resources in the network. However, the ability of nested controls to dynamically share the capacity of a resource—and thereby recover the pooling economies lost in partitioned controls—is an attractive feature.

---

[3]Though Ciancimino et al. [118] report study a RM model for the Italian railroad in which partitioned allocations worked well (at least relative to first come, first serve), though in their railroad problem the number of products using each resource was small (3 to 6).

[4]Of course, we've seen in Chapter 2 that for even the single-resource case, the ordering is not always simply the order of the prices. For example, when refunds are class dependent or one models consumer choice behavior, nesting may not be optimal and the ordering of classes may be different from the simple fare order.

Virtual nesting control—a hybrid of network and single-resource controls—provides one solution. American Airlines is credited with developing this control scheme (circa 1983) as a way of incorporating some degree of network control within the single-resource nested-allocation structure of American's single-resource-based (at that time) reservation system.

Virtual nesting uses single-resource nested-allocation controls for each resource in the network. However, the classes used in these nested allocations are *virtual classes*, which group together sets of products that use a given resource. Products are assigned to a virtual class through a process known as *indexing*, which essentially provides a table that maps every product to a virtual class on each resource. This indexing, which attempts to cluster products based on their "network value," may be updated periodically as network demand patterns change.

Nested booking limits (or protection levels) for each resource are then computed using these virtual classes. To decide on a request for a network product, the system checks for availability of its corresponding virtual classes on each resource. If *all* the virtual classes are available, the request is accepted; if *any one* of the virtual classes is closed, the request is rejected.

Virtual nesting has a few notable disadvantages. For one, the indexing process can create potential difficulties in data collection and forecasting. Specifically, if data is collected at the virtual class level, then reindexing can alter the parameters of demand in a virtual class by changing which products are mapped into each virtual class. As a result, the virtual class demand statistics may shift dramatically even when the underlying product-level demand is unchanged. In this way, indexing can introduce its own "noise" into the data and forecasts. Virtual classes can also cause confusion for analysts, who may not be able to easily interpret virtual class demand. Finally, both the indexing process and the control logic of mapping products to virtual classes create complexity.

At the same time, virtual nesting has several advantages. It preserves the booking-class control logic of most existing GDSs, so no major infrastructure changes are required, yet it allows a RM system to incorporate some network information. It therefore provides an acceptable compromise between single-resource and full network control and has proven to be quite effective and popular in practice, especially in the airline industry.

Virtual nesting was a genuine breakthrough in RM when it was introduced by American Airlines and still has a strong contingency of industry and academic supporters. However, its main benefit—its ability to preserve the single-resource, nested-allocation structure of control—is somewhat particular to the travel and hospitality industry, where such controls are part of the industry's legacy distribution systems. For industries without this legacy infrastructure, bid-price controls provide simpler, more intuitive, and powerful means of network control. Even within the airline industry, bid-price control is gaining popularity as GDS and RM system vendors upgrade their technologies.

3.1.2.3 Bid-Price Controls

While nested allocations are difficult to extend directly to network RM control, network bid-price control is a simple extension of the single-resource version described in Chapter 2. In a network setting, a bid-price control sets a threshold price—or *bid price*—for each resource in the network. This bid price is normally interpreted as an estimate of the marginal cost to the network of consuming the next incremental unit of the resource's capacity. When a request for a product comes in, the revenue of the request is compared with the *sum* of the bid prices of the resources required by the product. If the revenue exceeds the sum of the bid prices, the request is accepted; if not, it is rejected.

Bid-price controls have many advantages. First, even in a network setting the structure of the control remains simple: we have to specify only a single value for each resource (not each product), so the number of parameters involved is minimal.[5] Second, evaluating a request for a product requires only a simple comparison of revenue to the sum of bid prices for the requested resources, so the transaction-processing task is quick. Third, bid prices are intuitive and have a natural economic interpretation as the marginal cost to the network of each resource. Finally, if implemented correctly, bid prices have very good revenue performance and can even be shown to be theoretically near-optimal under certain conditions.

Despite these advantages, bid-price controls have generated considerable controversy among both RM practitioners and academics alike. The debate about bid prices being "unsafe" was discussed in Section 2.1.1.4 in the single-resource context. The issues are essentially the same in the network case. Also, in the airline industry, as simple as the bid-price control logic may be, it is a vast departure from the control logic used in existing GDSs, and their acceptance has not been immediate. Bid-price control requires the GDS to support seamless availability (Sec-

---

[5]Though to work properly, bid prices must be updated with changes in capacity and time. This often necessitates storing a table of bid price values and the values are retrieved based on the current remaining capacity and the current remaining time. (See Table 3.1.) Still, only one such table is required for each resource, and the look-up is quick.

tion 11.2.3.2) and other real-time capabilities to connect to the host CRS and instantly respond to requests—technologies that have only recently become widespread. Hotel RM systems, of more recent vintage and without the legacy-system burdens of airlines, have been in a better position to adapt bid-price control, and it has become the dominant form of control for network RM in the hotel industry.

We next look at the theory behind network controls. Network RM is complex and difficult to analyze, and few exact theoretical results are available. However, asymptotic analysis giving some insight into the optimal form of control is sometimes possible, and we review it in Section 3.6.

3.2 The Theory of Optimal Network Controls

We begin with a basic model of the problem. The network has m resources, and the firm sells n products. (Each network product is a combination of a bundle of the m resources sold with certain purchase terms and restrictions at a given price.) Let $a_{ij} = 1$ if resource i is used by product j and $a_{ij} = 0$ otherwise. Define the *incidence matrix* $\mathbf{A} = [a_{ij}]$. Thus, the jth column of $\mathbf{A}$, denoted $\mathbf{A}_j$, is the *incidence vector* for product j; the ith row, denoted $\mathbf{A}^i$, has an entry of one in column j corresponding to a product j that uses resource i.[6] We let A_j denote the set of legs used by product j and A^i to denote the set of products that use resource i, so that the notation $i \in A_j$ indicates that resource i is used by product j, and $j \in A^i$ indicates that product j uses resource i.

The state of the network is described by a vector $\mathbf{x} = (x_1, \ldots, x_m)$ of resource capacities. If product j is sold, the state of the network changes to $\mathbf{x} - \mathbf{A}_j$. To simplify our analysis at this stage, we will ignore cancellations and no-shows. We will also omit the possibility of booking requests for multiple units of capacity (say, corresponding to a group or batch request) to simplify the presentation.

Time is discrete, there are T periods, and the index t represents the current time (with the time indices running forward, so $t = T$ is the time of service). Within each time-period t, as we did for the dynamic single-resource case, we assume that at most one request for a product can arrive; that is, the discretization of time is sufficiently fine so that the probability of more than one request is negligible.

---

[6]Note that there will be multiple identical columns if there are multiple ways of selling a given bundle of resources. For example, in the airline case there may be many fare classes for the same itinerary. Each would have an identical column in the matrix $\mathbf{A}$, but they could have different revenue values and different demand patterns.

Demand in period t is modeled as the realization of a *single* random vector $\mathbf{P}(t) = (P_1(t), \ldots, P_n(t))$. If $P_j(t) = p_j > 0$, this indicates a request for product j occurred and that its associated price is p_j; if $P_j(t) = 0$, this indicates there is no request for product j. A realization $\mathbf{P}(t) = 0$ (all components equal to zero) indicates that no request from *any* product occurred at time t.

For example, if we have $n = 3$ products, then a value $\mathbf{P}(t) = (0, 0, 0)$ indicates no requests arrived, a value $\mathbf{P}(t) = (120, 0, 0)$ indicates a request for product 1 with a price of \$120, a value of $\mathbf{P}(t) = (140, 0, 0)$ indicates a request for product 1 with a price of \$140, and a value $\mathbf{P}(t) = (0, 70, 0)$ indicates a request for product 2 with a price of \$70, and so on.

The sequence $\{\mathbf{P}(t); t \geq 1\}$ is assumed to be independent across time t with known joint distribution in each period t. Let the prices associated with the products be $\mathbf{p} = (p_1, \ldots, p_n)$. This is the network version of the independence by class assumption that we make in Chapter 2. (See also Section 7.1.) We make this assumption for all models in this chapter.

Given the current time t, the current remaining capacity $\mathbf{x}$, and the current request $\mathbf{P}(t)$, the quantity-based RM decision is as follows: Do we or do we not accept the current request?

Let the n-vector $\mathbf{u}(t)$ denote this decision, where $u_j(t) = 1$ if we accept a request for product j in period t, and $u_j(t) = 0$ otherwise. The decision to accept, $u_j(t)$, is a function of the remaining capacity vector $\mathbf{x}$ and the price p_j of product j—that is, $u_j(t) = u_j(t, \mathbf{x}, p_j)$—and hence $\mathbf{u}(t) = \mathbf{u}(t, \mathbf{x}, \mathbf{p})$. Since we can accept at most one request in any period and resources cannot be oversold, if the current seat inventory is $\mathbf{x}$, then $\mathbf{u}(t)$ is restricted to the set $\mathcal{U}(\mathbf{x}) = \{\mathbf{u} \in \{0, 1\}^n : \mathbf{A}\mathbf{u} \leq \mathbf{x}\}$.

3.2.1 The Structure of Optimal Controls

To formulate a dynamic program to determine optimal decisions $\mathbf{u}^*(t, \mathbf{x}, \mathbf{p})$, let $V_t(\mathbf{x})$ denote the maximum expected revenue to go, given remaining capacity $\mathbf{x}$ in period t. Then $V_t(\mathbf{x})$ must satisfy the Bellman equation

$$V_t(\mathbf{x}) = E\left[\max_{\mathbf{u} \in \mathcal{U}(\mathbf{x})} \left\{\mathbf{P}(t)^\top \mathbf{u}(t, \mathbf{x}, \mathbf{p}) + V_{t+1}(\mathbf{x} - \mathbf{A}\mathbf{u})\right\}\right] \quad (3.1)$$

with the boundary condition

$$V_{T+1}(\mathbf{x}) = 0, \quad \forall \mathbf{x}. \quad (3.2)$$

Network Capacity Control

It is easy to show that $V_t(\mathbf{x})$ is finite for all finite $\mathbf{x}$. Moreover, an optimal control $\mathbf{u}^*(\cdot)$ satisfies

$$u_j^*(t, \mathbf{x}, p_j) = \begin{cases} 1 & \text{if } p_j \geq V_{t+1}(\mathbf{x}) - V_{t+1}(\mathbf{x} - \mathbf{A}_j) \text{ and } \mathbf{A}_j \leq \mathbf{x} \\ 0 & \text{otherwise.} \end{cases}$$

(3.3)

The control (3.3) says that an optimal policy for accepting requests is of the form: accept a request for product j (at price p_j) if and only if we have sufficient remaining capacity and

$$p_j \geq V_{t+1}(\mathbf{x}) - V_{t+1}(\mathbf{x} - \mathbf{A}_j),$$

where $P_j(t) = p_j$ is the price of product j. This reflects the rather intuitive notion that we accept a booking request for a product j only if its price exceeds the *opportunity cost* of the reduction in resource capacities required to satisfy the request.

3.2.2   Bid Price Controls

The displacement cost intuition of (3.3) leads naturally to bid-price controls for network RM, and indeed the analogy to the role of dual prices in deterministic optimization motivated the early development of bid-price control schemes by Simpson [476] and Williamson [565, 566]. If we suppose, in a heuristic sense, that the value function $V_{t+1}(\mathbf{x})$ has a gradient $\nabla V_{t+1}(\mathbf{x}))$, then the condition for accepting product j can be approximated by

$$\begin{aligned} p_j &\geq V_{t+1}(\mathbf{x}) - V_{t+1}(\mathbf{x} - \mathbf{A}_j) \\ &\approx \nabla \mathbf{V}_{t+1}^\top(\mathbf{x}) \mathbf{A}_j \\ &= \sum_{i \in A_j} \pi_i(t, \mathbf{x}), \end{aligned}$$

(3.4)

where $\pi_i(t, \mathbf{x}) = \frac{\partial}{\partial x_i} V_{t+1}(\mathbf{x})$. This line of reasoning motivates the following definition:

DEFINITION 3.1 *A control* $\mathbf{u}(t, \mathbf{x}, \mathbf{p})$ *is a **bid-price control** if there exist real-valued functions* $\boldsymbol{\pi}(t, \mathbf{x}) = (\pi_1(t, \mathbf{x}), \ldots, \pi_m(t, \mathbf{x}))$, $t = 1, 2, \ldots, T$, *(called **bid prices**) such that*

$$u_j(t, \mathbf{x}, p_j) = \begin{cases} 1 & \text{if } p_j \geq \sum_{i \in A_j} \pi_i(t, \mathbf{x}), \ \mathbf{A}_j \leq \mathbf{x} \\ 0 & \text{otherwise.} \end{cases}$$

(3.5)

In other words, a bid-price control specifies a set of bid prices for each resource, at each point in time, and for each capacity level, such that we accept a request for a particular product if and only if there is available capacity and the price exceeds the sum of the bid prices for all the resources used by the product. Bid-price control can be implemented by specifying a bid price table for each resource. Table 3.1 gives an example of such a table.

Table 3.1. Example of a bid price table for a single resource, based on remaining time and remaining capacity.

Capacity	t	$t+1$	$t+2$	...
C	147.3	146.5	145.8	...
C-1	149.5	148.7	147.9	...
C-2	151.8	150.8	150.1	...
⋮	⋮	⋮	⋮	⋱

3.2.3 Nonoptimality of Bid-Price Controls

Despite the generality of Definition 3.1, bid prices are not always an optimal form of control. The following example shows that there are cases where no set of bid prices is able to achieve the optimal control decision (3.3):

Example 3.1 Consider a network with two resources. There is one unit of capacity on each resource and two time-periods. The product data are shown in Table 3.2. In

Table 3.2. Problem data for the bid price counterexample.

Period (t)	Product	($\mathbf{A}_j$)	Fare	Probability
1	1	(1 0)	$250	0.3
	2	(0 1)	$250	0.3
	3	(1 1)	$500	0.4
2	3	(1 1)	$500	0.8
	No arrival			0.2

period 1, demands for all three products can arrive: products 1 and 2 require resources 1 and 2, respectively, and each has a price of $250 and a probability of arrival 0.3. Product 3 requires both resources and has a price of $500 and probability of arrival 0.4.

In the last period (period 2), there is demand only for product 3 with the same $500 price and a probability of arrival of 0.8. Arrivals in each period are mutually exclusive (only one product per period arrives).

It is not hard to see by inspection that an optimal policy is to reject product 1 and 2 in period 1 and accept product 3 in both periods. This is true because accepting products 1 or 2 in period 1 yields $250 in revenue but prevents us from accepting product 3 in period 2, which results in an opportunity cost of $400 (0.8 × $500). So it is optimal to *reject* both products 1 and 2 in period 1. However, we clearly want to accept product 3 in period 1, since we cannot gain more than $500 in period 2.

This implies that the bid prices, π_1 and π_2, in period 1 must satisfy $\pi_1 > 250$, $\pi_2 > 250$, and $\pi_1 + \pi_2 \leq 500$, which is, of course, impossible. Therefore, no bid price policy can produce an optimal decision in period 1.

Indeed, it is not hard to show that the best a bid price policy can do in this example is to reject *all* demand in period 1 and accept only product 3 (if it arrives) in period 2, yielding a $400 expected revenue. The optimal policy, in contrast, generates an expected revenue of $440—fully 10% more expected revenue than by the best possible bid price policy.[7]

This counterexample illustrates the two main reasons that bid-price controls may not be optimal. One reason that bid prices in this example fail is that selling a seat is a "large" change in the capacity of a resource. Large relative changes in capacity on several resources simultaneously cannot, in general, be expected to produce the same revenue effect as the sum of the individual changes. So gradient-based reasoning falls short in explaining bid price optimality.

The second reason that bid prices may fail to capture the opportunity cost is that future revenues may depend in a highly nonlinear way on the displaced capacity. In the counterexample, note that the *minimum* capacity on the two resources determines future expected revenues. Hence, the opportunity cost of using a single resource exactly equals the opportunity cost of using both resources simultaneously. This phenomenon is analogous to degeneracy in mathematical programming, and it can occur in the optimal value function or in various approximations to the optimal value function, as shown below in Section 3.3.

3.2.4 Evidence in Support of Bid Prices

While most of these results on bid price optimality are negative, Section 3.6 provides some theoretical results showing that bid-price controls are asymptotically optimal under certain "large-number" scalings of the network RM problem. Thus, there are conditions under which bid prices are provably good. In addition, numerical simulation studies have largely confirmed that bid prices (properly implemented) are an effective control mechanism for network RM.

3.2.5 Bid Prices and Opportunity Cost

One interesting observation about bid prices is that there need not be a one-to-one correspondence between optimal bid prices and the oppor-

---

[7]The $440 optimal revenue is obtained by accepting product 3 only in both periods 1 and 2, in which case the expected revenue is $0.4 \times \$500 + (1 - 0.4) \times 0.8 \times \$500 = \$440$.

tunity cost of capacity. That is, one can generate examples of bid prices that give optimal accept or deny decisions but that at the same time are very poor approximations to the marginal value of resource capacity.

Consider, for example, a single-resource problem in which high-revenue products arrive strictly *before* low-revenue products. In this case, it is clear that it is optimal to accept arrivals in a first-come, first-serve (FCFS) order. Therefore, a constant bid price of zero is one optimal solution. On the other hand, the opportunity cost $\Delta V_t(x) = V_t(x) - V_t(x-1)$ at each point in time t will in general not be zero.

In other words, while it is *sufficient* to compare the revenue with the opportunity cost at each time t to make optimal accept or deny decisions, it is not *necessary* to do so; other threshold values may produce the same accept or deny decision and the same optimal revenues, as in our example. One might argue that this difference is not worth worrying about as long as we can identify one bid-price value that is optimal. However, in practice a good estimate of opportunity cost is desirable, since opportunity-cost numbers from a RM system often support other decision making, such as capacity planning or pricing. Thus, bid prices that accurately approximate the opportunity cost of capacity are more valuable than ones that do not, even if both perform equally well in a RM system.

3.3 Approximations Based on Network Models

For any network of realistic size, computing the value function $V_t(x)$ in (3.1) exactly is essentially hopeless because of the large dimensionality of the state space (e.g., a network with $m = 20$ resources and capacities of $C = 100$ on each resource has 100^{20} states). Instead, one must rely on approximations of various types. Most approximation methods proposed to date follow one of two basic (not necessarily mutually exclusive) strategies. The first, which we look at in this section, is to use a simplified network model—for example, posing the problem as a static mathematical programming problem. The second strategy, which we look at in the next section, is to decompose the network problem into a collection of single-resource problems. Whichever method is used, it is useful to view all such methods as producing different approximations of the optimal value function.

The most useful output of an approximation method are estimates of displacement costs—or bid prices. These are used either directly in bid-price control mechanisms or indirectly in other mechanisms like virtual nesting. Given an approximation method M that yields an estimate of the value function $V_t^M(\mathbf{x})$, we can approximate the displacement cost of accepting product j by

$$V_t^M(\mathbf{x}) - V_t^M(\mathbf{x} - \mathbf{A}_j) \approx (\nabla V_t^M(\mathbf{x}))^\top \mathbf{A}_j,$$

where $\nabla V_t^M(\mathbf{x})$ is the gradient of the value function approximation $V_t^M(\mathbf{x})$, assuming the gradient exists. The bid prices are then defined by

$$\pi_i^M(t, \mathbf{x}) = \frac{\partial}{\partial x_i} V_t^M(\mathbf{x}).$$

If the gradient does not exist, then $\nabla V_t^M(\mathbf{x})$ is typically replaced (at least implicitly) by a subgradient of $V_t^M(\mathbf{x})$. If the approximation is discrete, then first differences are used in place of partial derivatives.

Using a sum of bid prices to approximate displacement cost, rather than the difference $V_t^M(\mathbf{x}) - V_t^M(\mathbf{x} - \mathbf{A}_j)$ directly, is done mainly to simplify the control. As mentioned above, one of the appeals of bid-price control is that it requires only m resource-level bid prices (one for each resource) rather than n product-level controls—and in large networks, m can be several orders of magnitude smaller than n. This makes additive bid prices appealing on a practical level. Still, as shown by Example 3.1, such additive approximations can be inaccurate in some cases, and non-additive bid prices may be needed. (Berstsimas and Popescu [61] examine this issue of additive versus non-additive bid-price controls.)

Clearly, one objective for an approximation method is to produce a good estimate of the value function—and more important, a good estimate of the displacement costs or bid prices. On the other hand, speed of computation matters as well. The approximation $V_t^M(\mathbf{x})$ may be a static approximation that must be re-solved quite frequently in practice to account for changes in remaining capacity $\mathbf{x}$ and remaining time t. A static method that is accurate but computationally complex will therefore be of little use in practice. An approximation that produces dynamic estimates of displacement cost—for example, bid prices that vary with capacity and time—can be solved less frequently, but still changes in forecasts and fare input data inevitably occur and necessitate recomputing the model. Thus, one should always keep these two criteria—accuracy and speed—in mind when judging network approximation methods.

3.3.1 The Deterministic Linear Programming Model

Let the aggregate demand to come at time t for each product j be denoted D_j (demand over the periods $t, t+1, \ldots, T$) with mean μ_j. Let $\mathbf{D} = (D_1, \ldots, D_n)$ and $\boldsymbol{\mu} = E[\mathbf{D}]$ denote the vector of demands and mean demands, respectively. The *deterministic linear programming* (DLP) method uses the approximation

$$V_t^{LP}(\mathbf{x}) = \max \ \mathbf{p}^\top \mathbf{y} \qquad (3.6)$$
$$\text{s.t.} \ \mathbf{Ay} \leq \mathbf{x}$$
$$0 \leq \mathbf{y} \leq \boldsymbol{\mu}.$$

The decision variables $\mathbf{y} = (y_1, \ldots, y_n)$ represent the partitioned allocation of capacity for each of the n products. The approximation effectively treats demand as if it were deterministic and equal to its mean $\boldsymbol{\mu}$ and makes an optimal partitioned allocation accordingly. Using Jensen's inequality, one can show that $V_t^{LP}(\mathbf{x})$ is in fact an upper bound on the optimal value function [111, 112, 131].

Occasionally, the optimal primal solution to (3.6) is used to construct a partitioned control directly. More often, the primal allocations are discarded, and one uses only the optimal dual variables, denoted $\boldsymbol{\pi}^{LP}$, associated with the constraints $\mathbf{Ay} \leq \mathbf{x}$ as bid prices. If the optimal solution is not degenerate, that is, if the active constraints are linearly independent at the optimal solution, then $\nabla V_t^{LP}(\mathbf{x})$ exists and is given by the unique vector of optimal dual prices, $\boldsymbol{\pi}^{LP}$; if the optimal solution is degenerate, then there are multiple optimal dual price vectors, each of which is only a subgradient of the function $V_t^{LP}(\mathbf{x})$.

The main advantage of the DLP model is that it is computationally very efficient to solve. Due to its simplicity and speed, it is popular in practice. The weakness of the DLP approximation is that it considers only the mean demand, ignoring all uncertainty in the forecasts, and models only a partitioned control mechanism. One consequence of this is that the dual values will be zero on any resource that has a mean demand less than capacity, a behavior that can cause problems in practice.

Despite this deficiency, simulation studies (with the independent-class assumption) by several researchers have shown that with frequent reoptimization, the performance of DLP bid prices is rather respectable, producing higher revenue than both the probabilistic nonlinear programming model of Section 3.3.2 and a variety of leg-based EMSR heuristics [36, 43, 565, 566], though other studies have reported more mixed results [42]. In general, the performance of the DLP method, like many network methods, depends heavily on the type of network, the variance in the demand forecasts, the order in which fare products arrive, and the frequency of reoptimization.[8]

---

[8]Cooper [131] provides a counterexample in which more frequent reoptimization of a DLP model (using bid-price controls) results in worse revenue performance, though the preponderance of numerical evidence suggests that frequent reoptimization is beneficial.

One variation in the use of DLP proposed by Bertsimas and Popescu [61] is not to use the dual prices produced by the model but to use discrete differences as an estimate of the marginal value of capacity. Specifically, for each product j, we compute the difference in the value functions

$$V_t^{LP}(\mathbf{x}) - V_t^{LP}(\mathbf{x} - \mathbf{A}_j)$$

and then use this value as a bid price for product j. The motivation here is both to eliminate the potential problem of degeneracy and also to better reflect the discrete change in capacity that occurs when accepting a request. In numerical experiments, this modification appears to offer a modest improvement over using additive bid prices based on the dual variables π^{LP}. A disadvantage of this approach, however, is that to determine controls for all n products, one must solve a separate linear program for each product j, which increases the computational effort substantially relative to DLP-based bid price controls.[9]

3.3.2 The Probabilistic Nonlinear Programming Model

The *probabilistic nonlinear programming* (PNLP) method uses the approximation

$$V_t^{PNLP}(\mathbf{x}) = \max \sum_{j=1}^n p_j E[\min\{D_j, y_j\}] \qquad (3.7)$$
$$\text{s.t.} \quad \mathbf{Ay} \leq \mathbf{x}$$
$$\mathbf{y} \geq 0,$$

where D_j and p_j are defined as in the DLP case. As in the DLP, the decision variable y_j represents a partitioned allocation of capacity to product j, and the term $E[\min\{D_j, y_j\}]$ is the expected sales of product j under this partitioned allocation. Since partitioned allocations are certainly a feasible policy for the network problem, $V_t^{PNLP}(\mathbf{x})$ is in fact a lower bound on the optimal value function [111].

While this model results in a nonlinear program, it is a relatively easy one; the objective function is concave and separable in the variables y_j, and the constraints are linear. A variety of specialized algorithms for separable concave problems make the PNLP model feasible for large networks.

If demand is discrete, one can also convert the model to a linear program, albeit one with many more variables. This is achieved by assigning

---
[9]Though "warm starting" the LP solver using the current solution for $\mathbf{x}$ reduces the computation time significantly.

a variable to each product *and* to each unit of capacity. Specifically, define variable z_{jd} to be the dth unit of capacity allocated to product j. Then the PNLP model can be written

$$V_t^{PNLP}(\mathbf{x}) = \max \sum_{j=1}^n p_j \sum_{d=1}^{M_j} z_{jd} P(D_j \geq d) \quad (3.8)$$

$$\text{s.t.} \quad y_j = \sum_{d=1}^{M_j} z_{jd}$$

$$\mathbf{Ay} \leq \mathbf{x}$$

$$0 \leq z_{jd} \leq 1, \quad j = 1, \ldots, n, \quad d = 1, \ldots, M_j,$$

where M_j is an upper bound on the capacity allocated to product j (for example, the minimum capacity remaining among the resources used by product j). The equivalence of the formulations (3.7) and (3.8) follows from the fact that (for discrete distributions) $E[\min\{D_j, y_j\}] = \sum_{d=1}^{y_j} P(D_j \geq d)$ and that $P(D_j \geq d)$ is decreasing in d.

Formulation (3.8) is sometimes referred to as the *probabilistic linear programming model*, but it is in fact equivalent to the PNLP for discrete demand and is a close approximation to the PNLP for continuous demand distributions. Because of the large number of variables introduced in converting the PNLP model to a linear program—and the simple structure of the nonlinear version of the PNLP model—it is often more efficient to solve the PNLP model directly as a nonlinear program. Another alternative is to attempt to reduce the size of the PNLP by approximating demand distributions with a coarser discretization, which de Boer et al. [154] show does not affect the performance significantly.

Whichever approach is used to solve the PNLP, one can obtain bid price values from the resulting dual variables. If the active constraints are linearly independent at the optimal solution, then $\nabla V_t^{PNLP}(\mathbf{x})$ exists and is given by the unique vector of optimal dual prices associated with these constraints; if the active constraints are dependent, then multiple optimal dual vectors are subgradients of the function $V_t^{PNLP}(\mathbf{x})$.

3.3.2.1 Qualitative Behavior of PNLP

On the surface, the PNLP approximation appears somewhat better than the DLP approximation, in that the term $E[\min\{D_j, y_j\}]$ in the objective function captures the randomness in demand. For example, unlike in the DLP approximation, the PNLP dual price of a resource can be positive even when mean demand for the resource is strictly less than capacity, reflecting the fact that there is some probability that demand will exceed capacity. However, the assumption of partitioned allocations of capacity to each product can lead to poor behavior, and simulation studies have generally confirmed that PNLP bid prices result in lower revenues than DLP bid prices [154, 565, 566].

To understand the weakness of the PNLP approximation, consider the following example:

Example 3.2 A network has $m = 1$ leg and n products on the leg, each of which is identical. That is, each product j has demand $D_j \sim N(\mu, \sigma)$, and each product has the same revenue p. The PNLP formulation is then

$$V_t^{PNLP}(x) = \max \sum_{j=1}^{n} pE[\min\{D_j, y_j\}]$$

$$\text{s.t.} \sum_{j=1}^{n} y_j \leq x$$

$$y \geq 0.$$

Now, by symmetry the optimal solution is $y_j = x/n$, $j = 1, \ldots, n$, and hence the Kuhn-Tucker conditions imply the optimal dual price π satisfies

$$\pi = pP(nD_j > x) = p(1 - \Phi(\frac{x - n\mu}{n\sigma})), \tag{3.9}$$

where $\Phi(\cdot)$ is the c.d.f. of the standard normal distribution. π then forms an estimate of the marginal value of the xth seat.

However, since all products are identical, the marginal value in this problem should be unchanged if we aggregate all n products into one product with mean $n\mu$ and variance $n\sigma^2$. Aggregating and applying the PNLP, we find the optimal dual multiplier in this case satisfies

$$\pi = pP(\sum_{j=1}^{n} D_j > x) = p(1 - \Phi(\frac{x - n\mu}{\sqrt{n}\sigma})). \tag{3.10}$$

If $x \neq n\mu$ and n is large, (3.9) and (3.10) give very different estimates of the marginal revenue. Of course, from first principles, the opportunity cost should be independent of how we aggregate (or disaggregate) the identical products.

Example 3.2 shows how the partitioned allocation assumption of the PNLP model can potentially lead to poor estimates of opportunity cost. While on the surface this example seems contrived, it is not uncommon to observe similar behavior in large airline networks. For example, in hub-and-spoke networks if many uncongested in-bound legs have connecting passengers traveling on a single congested out-bound leg and passengers pay comparable revenues, then the situation is similar to the example above; one would like to treat all passengers as a single class ("pool" the products), but the PNLP allocates space to each one separately, resulting in a distorted estimate of the marginal value. The DLP method does not suffer this lack-of-pooling defect as shown below:

Example 3.3 Consider the deterministic linear programming version of Example 3.2. If we define the problem for the original n products, we get the linear program

$$\max\{\sum_{j=1}^{n} py_j : \sum_{j=1}^{n} y_j \leq x, 0 \leq y \leq E[D]\}.$$

If we aggregate all the demand in to one product, we get the linear program

$$\max\{ry_1 : y_1 \leq x, 0 \leq y_1 \leq \sum_{j=1}^{n} E[D_j]\}.$$

It is not hard to see that these both have the same optimal dual solution and hence produce the same estimate of the marginal revenue.

Essentially, the DLP makes one particularly bad assumption: demand is deterministic. But once this assumption is made, there is no further loss in optimality by formulating the model using partitioned allocations. In the PNLP case, the demand assumption is more realistic, but under this demand assumption, using partitioned allocations becomes highly suboptimal.

3.3.3 The Randomized Linear Programming Model

Randomized linear programming (RLP) is another approach for incorporating stochastic information into the DLP method based on replacing the expected value of $\mathbf{D}$ in the constraint (3.6) by the random vector $\mathbf{D}$ itself. The expected value of the resulting optimal solution then forms an approximation to the value function. That is, define

$$\begin{aligned} H_t(\mathbf{x}, \mathbf{D}) = \quad &\max \quad \mathbf{p}^\top \mathbf{y} \\ &\text{s.t.} \quad \mathbf{Ay} \leq \mathbf{x} \\ & \quad\quad 0 \leq \mathbf{y} \leq \mathbf{D}. \end{aligned} \quad (3.11)$$

The optimal value $H_t(\mathbf{x}, \mathbf{D})$ is a random variable. Let $\boldsymbol{\pi}(\mathbf{x}, \mathbf{D})$ denote an optimal vector of dual prices for the set of constraints $\mathbf{Ay} \leq \mathbf{x}$, and note that $\boldsymbol{\pi}(\mathbf{x}, \mathbf{D})$ is a random vector.

Next, consider approximating the value function by the expected value of $H_t(\mathbf{x}, \mathbf{D})$,

$$V_t^{RLP}(\mathbf{x}) = E[H_t(\mathbf{x}, \mathbf{D})]. \quad (3.12)$$

Note the right-hand side corresponds to a "perfect-information" approximation because it reflects a case in which *future* allocations (and revenues) are based on perfect knowledge of the realized demand $\mathbf{D}$; however, at time t, $\mathbf{D}$ is not yet realized so the right-hand side is the expected value of this perfect-information solution. Assuming that the gradient exists, we then use $\nabla_\mathbf{x} E[H_t(\mathbf{x}, \mathbf{D})]$ as a vector of bid prices.

Network Capacity Control

3.3.3.1 Estimating the Gradient via Simulation

The RLP approximation (3.12) requires a method to efficiently compute $\nabla_\mathbf{x} E[H_t(\mathbf{x}, \mathbf{D})]$. One simple approach is to consider interchanging differentiation and expectation. There are formal conditions under which such an interchange is justified,[10] but assuming it is, we have

$$\nabla_\mathbf{x} E[H_t(\mathbf{x}, \mathbf{D})] = E[\nabla_\mathbf{x} H_t(\mathbf{x}, \mathbf{D})]. \quad (3.13)$$

This interchange, in turn, suggests a procedure for estimating $\nabla_\mathbf{x} E[H_t(\mathbf{x}, \mathbf{D})]$. Simply simulate k independent samples of the demand vector, $\mathbf{D}^{(1)}, \ldots, \mathbf{D}^{(k)}$, and solve (3.11) for each sample. Then estimate the gradient using

$$\boldsymbol{\pi}^{RLP} = \frac{1}{k} \sum_{i=1}^{k} \boldsymbol{\pi}(\mathbf{x}, \mathbf{D}^{(i)}). \quad (3.14)$$

In other words, simply *average* the dual prices from k perfect-information allocation solutions on randomly generated demands. This idea is called the *randomized linear programming* (RLP) method.

The RLP method has several appealing features. First, it is a simple modification to the DLP method, so it can be easily incorporated into RM systems based on DLP. Second, it has the flexibility to model very general demand distributions, since we need only the ability to simulate demand to apply the method (for example, one could allow for different coefficients of variation or correlations among the components of $\mathbf{D}$). In addition, the quality and complexity of the model can be controlled by varying the number of random samples k. Finally, unlike the DLP, it incorporates distributional information on demand.

3.3.3.2 Comparisons to DLP

To understand qualitatively how RLP compares with DLP, consider the following simple example:

Example 3.4 Let $m = 1$ (a single resource) and $\mathbf{A} = [1]$ (one product), so we have scalar values p, x, D, and $\pi(x, D)$. It is easy to see that $\pi(x, D) = r$ if $D > x$ and $\pi(x, D) = 0$ if $D < x$, so if D has a continuous distribution, then

$$\frac{d}{dx} E[H_t(x, D)] = E[\pi(x, D)] = pP(D > x),$$

which is precisely Littlewood's rule (2.2) for the expected marginal value of capacity. So the RLP method produces the exact expected marginal value in this case. In

---

[10] The conditions essentially ensure that the linear program has a unique dual solution with probability one, so that the gradient with respect to $\mathbf{x}$ is well defined. However, they are somewhat difficult to verify in general. See Talluri and van Ryzin [499] for details.

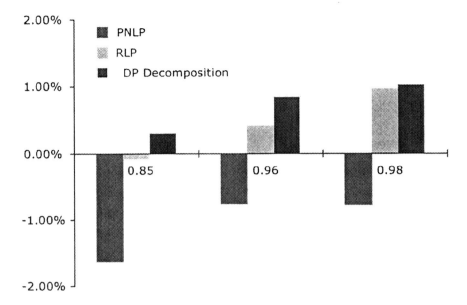

Figure 3.2. Comparison of DLP, RLP, and dynamic programming decomposition: hotel network, 42 days, 497 length-of-stays, 5 classes, 2 optimizations/day, 12 days, 450 rooms.

contrast, the DLP method produces a marginal value of zero if $E[D] < x$ and a marginal value of p if $E[D] > x$—neither of which is correct.

RLP also retains the desirable pooling property of the DLP method. Indeed, consider splitting up the demand D in this example into n classes all with revenue p such that $D_1 + \cdots + D_n = D$. Then it is not hard to see that for each realization of demand, the dual price is again $\pi(x, D) = r$ if $D_1 + \cdots + D_n = D > x$ and $\pi(x, D) = 0$ if $D_1 + \cdots + D_n = D < x$. Thus, the expected dual price is unchanged.

Example 3.4 suggests why the RLP method tends to produce a somewhat better approximation of the marginal value of capacity than the DLP method.

Figure 3.2 show some simulation comparisons of DLP and RLP on a hotel network. In these simulations, RLP provides a small but significant benefit in revenue over DLP at high occupancy. Nevertheless, methods like dynamic programming decomposition that we cover in Section 3.4.4 tend to dominate both DLP and RLP.

3.4 Approximations Based on Decomposition

Another strategy for generating network controls is to decompose the problem into m single-resource problems, each of which may incorporate some network information but are solved essentially independently. Formally, one can think of such a decomposition method as follows. An approximation method decomposes the network problem into m single-resource problems and applies a single-resource method M_i on each resource i, with value functions $V_t^{M_i}(x_i)$, that depends on the time to go t and the remaining capacity x_i of resource i. These may be constructed by incorporating some static, network information into the estimates. Then the total value function is approximated by

$$V_t(\mathbf{x}) = \sum_{i=1}^{m} V_t^{M_i}(x_i).$$

Typically, such approximations are discrete and yield bid prices

$$\pi_i(t, \mathbf{x}) = \Delta V_t^{M_i}(x_i), \quad i = 1, \ldots, m,$$

where $\Delta V_t^{M_i}(x_i) = V_t^{M_i}(x_i) - V_t^{M_i}(x_i - 1)$ is the marginal expected value produced by the single-resource method applied to resource i.

Decomposition approximations have several advantages relative to network approximations. First, because they are based on single-resource models, the displacement costs and bid prices are typically dynamic, varying as a function of capacity and time, and can be represented as a table of outputs (in the case of dynamic programming models) or simple formulas (in the case of EMSR approximations). Thus, it is easy to quickly determine the effect of changes in both the remaining time t and remaining capacity $\mathbf{x}$ on the bid prices. This should be contrasted with network models, which must be re-solved to determine the effects of such changes. Second, the simpler single-resource models used in decomposition approaches allow for more realistic assumptions, such as discrete demand and capacity, sequential decision making over time and stochastic dynamic demand.

The primary disadvantage of decomposition methods is that in the process of separating the problem by resources one can lose important network information. Nevertheless, as we show below, hybrids of the two approaches can be used to try to achieve the benefits of both network and decomposition methods.

3.4.1 OD Factors Method

Origin-destination factors method, or *OD factors* method (also called *heuristic bid price*), is a simple type of decomposition approximation. In this method, one solves independent single-resource problems for each resource i without making any prior adjustment to the input data for these problems. Typically, the total revenue of every product is used on each resource to solve a single-resource problem for that resource. The method may be either heuristic (such as EMSR-b) or exact (such as the dynamic program of Section 2.5). Let the value functions from the problem on resource i be denoted $V_t^{ODF_i}(x_i)$.

When evaluating a request for product j on resource i, the price p_j is compared with a bid price of the form

$$\Delta V_t^{ODF_i}(x_i) + \gamma \sum_{l \in A_j, l \neq i} \Delta V_t^{ODF_l}(x_l),$$

where $\Delta V_t^{ODF_i}(x_i)$ is the marginal value from the resource-level approximation method i and γ is a global "OD factor" that tries to capture the displacement costs on resources other than i. Typically, $\gamma < 1$ and is tuned by simulation. If p_j exceeds the OD factor bid price, then product j is accepted on resource i. A similar calculation is performed on all other resources used by product j. Product j is accepted if it is accepted on all these resources. In other words, p_j should be greater than or equal to the maximum of the bid prices on the resources it consumes.

Note the effect of this strategy. If product j uses only resource i, then the decision is the same as that under the single-resource problem i. However, if product j uses resources other than i, the bid price is higher than it would be under problem i alone. Intuitively, this is as it should be since accepting product j causes displacement at other resources $l \in A_j$ in the network. The reason for the OD factor γ is that each single-resource problem l tends to overestimate its own displacement cost as each problem uses the total revenue of each product, yet displacing a product that uses more than one resource frees up capacity elsewhere in the network, and the loss in gross revenue is partially offset by the marginal value of capacity that is freed up elsewhere in the network.

The motivation for this OD factor approach is simply to take single-resource methods that one might already have in place and, in a simple way, convert their outputs into estimates of network displacement costs. Thus, it has the advantage of not requiring any new modeling or estimation. With proper tuning of the OD factor γ, it can produce revenue benefits over simple, single-resource controls. Indeed, this later claim is almost immediate since $\gamma = 0$ recovers the single-resource strategy and another choice, of course, can potentially improve revenues further. Nevertheless, the choice of the γ parameter is quite ad hoc, and the alternative strategies that we present next almost always perform better.

3.4.2 Prorated EMSR

A slightly more sophisticated method for decomposition is to use prorated revenues in a collection of single-resource problems. The idea was first investigated by Williamson [566] with EMSR heuristics and is called the *prorated expected marginal seat revenue* (PEMSR) scheme.

The PEMSR schemes involve allocating a portion of the revenue of each product to the resources used by the product. One then solves m single-resource problems using an EMSR heuristic, though other single-resource methods can very well be substituted. The resulting marginal values from each resource are then used as bid prices in a bid-price control scheme, or the allocations are used directly in resource-level nested allocation controls.

Specifically, let $\boldsymbol{\alpha} = (\alpha_1, \ldots, \alpha_m)$ be a nonnegative real vector. For each product j define new revenues, one for each resource i used by the product, by

$$\bar{p}_{ij} = \frac{\alpha_i}{\sum_{l \in A_j} \alpha_l} p_j, \quad i \in A_j.$$

Next, treat each resource i independently as if it received demand D_j but with reduced revenue $\bar{p}_{ij}$, and solve the corresponding EMSR problem. The approximation to the value function is then

$$V_t^{PEMSR}(x) = \sum_{i=1}^{m} V_t^{EMSR_i}(x_i, \boldsymbol{\alpha}),$$

where $V_t^{EMSR_i}(x_i, \boldsymbol{\alpha})$ denotes the expected revenue of problem i under the allocation $\boldsymbol{\alpha}$.

Williamson [566] investigated several methods for determining the allocation weights $\boldsymbol{\alpha}$ in airline problems, including prorating based on mileage, number of resources, and the relative revenue value of local demand on each resource. Her conclusion is that none of these fixed allocations is very robust in general. Indeed, it is not hard to see that if one resource used by a product is highly congested and all others have abundant capacity, then the revenue of the product should be entirely allocated to the congested resource. Depending on the forecast of demand, however, the congested resource could be any of the resources used by the product; hence, no fixed allocation scheme can be expected to work well in all cases. Indeed, the iterative prorating schemes of Section 3.4.5 below are partially motivated by this insight.

3.4.3 Displacement-Adjusted Virtual Nesting (DAVN)

Displacement-adjusted virtual nesting (DAVN) starts with a set of static bid prices—or marginal value estimates—which we denote by $\boldsymbol{\pi} = (\pi_1, \ldots, \pi_m)$. These estimates may be obtained, for example, from one of the various network heuristic models presented in Section 3.3. Given the bid prices $\boldsymbol{\pi}$, one then solves a single-resource problem on each resource i as follows.

First, for all products j that use i, a *displacement-adjusted revenue* $\bar{p}_{ij}$ is computed using

$$\bar{p}_{ij} = p_j - \sum_{l \in A_j, l \neq i} \pi_l. \qquad (3.15)$$

That is, the revenue of product j on resource i is reduced by the static bid price values of the other resources used by product j. This adjustment is intended to approximate the net benefit of accepting product j on resource i.[11]

The next step is clustering, or *indexing*. In this step, the displacement-adjusted revenue values on each resource are clustered into a specified number $\bar{c}+1$ of *virtual classes*, or *buckets*. (The number of virtual classes, $\bar{c}+1$, is a design parameter, but is typically on the order of 10. It may also vary across resources.)

The indexing from product j to virtual class c on each resource can be performed using a variety of algorithms. For example, one can simply look at the range of displacement-adjusted revenue values on a given resource and split this range into $\bar{c}+1$ equal intervals (equal-revenue-width partitioning). All products with displacement-adjusted revenues in the first interval are assigned to virtual class 1, all those with displacement-adjusted revenues in the second interval are assigned to virtual class 2, and so on. Because one would like the displacement-adjusted revenues in each virtual class to be roughly the same, more sophisticated clustering methods attempt to assign products to virtual classes so as to minimize some measure of the variation of displacement-adjusted revenue values within each virtual class. The particular indexing method and clustering criteria are also design decisions and vary from implementation to implementation.

The following is an example of an advanced indexing scheme used by American Airlines (see Vinod [537, 538]):

Example 3.5 (DYNAMIC PROGRAMMING BASED INDEXING) Consider a given resource i. For ease of exposition, we assume that products are renumbered so that the displacement-adjusted revenue for product j, $\bar{p}_{ij}$, are ordered as follows:[12]

$$\bar{p}_{i1} \geq \bar{p}_{i2} \ldots \geq \bar{p}_{in}.$$

Let μ_j denote the mean demand for product j (again under the revised numbering of products).

---

[11] Observe that the displacement-adjusted revenue could be negative. In this case, product j is never accepted on resource i, and typically we either eliminate product j from the problem on resource i or (equivalently) set the displacement-adjusted revenue value to zero.
[12] This can be achieved by simply sorting the displacement-adjusted revenue values while keeping tracking of the original product numbers for each sorted value.

We want to partition these products into $\bar{c}+1$ virtual classes. To do so, we first assume that the partitions are contiguous so that if $\bar{p}_{il} \geq \bar{p}_{ik}$ and both products l and k belong to the same partition, then products $l+1, l+2, \ldots, k-1$ also belong to the same partition. For a given contiguous partition (l, k), $l \leq k$, define the squared deviation of revenue within partition (l, k) by

$$c_{lk} = \sum_{j=l}^{k} \mu_j (\bar{p}_{ij} - m_{kl})^2,$$

where

$$m_{lk} = \frac{\sum_{j=l}^{k} \mu_j \bar{p}_{ij}}{\sum_{j=l}^{k} \mu_j}$$

is the weighted-average displacement adjusted revenue for partition (l, k). We would then like to find $\bar{c}+1$ contiguous partitions that minimize the total sum of the squared deviations (minimizes the total *within-group variation*).

This can be solved by dynamic programming as follows. Let $V_c(k)$ denote the minimum squared deviation possible when partitioning the products $1, 2, \ldots, k$ into c contiguous groups. Then

$$V_c(k) = \min_{1 \leq l \leq k} \{c_{lk} + V_{c-1}(l-1)\} \quad (3.16)$$

with boundary conditions

$$\begin{aligned} V_c(0) &= 0, \quad c = 0, 1, \ldots, \bar{c}+1 \\ V_c(k) &= 0, \quad c = k, k+1, \ldots, \bar{c}+1 \\ V_1(k) &= c_{1k}, \quad k = 1, \ldots, \bar{c}+1. \end{aligned}$$

Starting with $c = 1$ groups, we solve (3.16) for successive values of c until we reach $c = \bar{c}+1$. Then $V_{\bar{c}+1}(n)$ gives the minimum total squared deviation, and by backtracking in the recursion (3.16) we can find the corresponding optimal partition. The complexity of this DP is $O(n^2)$, which is much more efficient than complete enumeration.

Vinod [537] reports that this DP-based indexing method significantly improved the performance of American Airline's virtual nesting implementation (by about 0.5%) relative to their prior indexing scheme, which was based on a simple equal-revenue-width partitions. This suggests that the performance of DAVN is sensitive to the choice of indexing method. Indexing, however, is not a real-time process because changes to the virtual class mappings can be quite disruptive for both host reservation systems and human RM analysts alike; therefore it is usually performed only periodically.

Once the virtual classes are formed and the indexing is determined, we compute a representative revenue value for each class—which is usually the (mean demand) weighted average displacement-adjusted revenue. Then, we compute the distribution of aggregate demand in a virtual class by adding the means and variances of demand to come for all products that are indexed to that virtual class. Finally, we solve a single-resource problem (using any standard single-resource method) based on this data to determine $\bar{c}$ protection levels on each resource. We call this problem i. This procedure yields a set of booking limits (or protection levels) for the virtual classes on each resource i and a value-function estimate $V_t^{DAVN_i}(x_i)$.

The control strategy then proceeds as follows. A request for product j is converted into a request for the corresponding virtual class on each resource i required by product j using the indexing scheme. If the virtual class on each resource is available, the request is accepted. If the virtual class on one or more resources is closed, the request is rejected. Thus, once the indexing from products to virtual classes is performed, the control logic is an independent, nested allocation class-level control on each resource in the network. This is the primary appeal of—and motivation for—the method in the airline industry because it produces the sort of booking-class-level controls that are widely used by GDSs.

An important point to note here is that even if the original product demand arrives in low-to-high *revenue* order, it is unlikely that the demand for virtual classes arrives in low-to-high *displacement-adjusted revenue* order. Thus, the revenue-order assumptions of the standard static single-resource model are likely to be violated under DAVN.

Finally, to gain some intuition into DAVN, it is helpful to examine the control decisions that it produces on each resource. Assume, for illustration, that an exact static model is solved for each resource. Let R_{ic} denote the representative revenue value for virtual class c on resource i (the weighted average or the median, depending on the implementation). By the optimality condition (2.7) for the single-resource model, a request on resource i in virtual class c is accepted if and only if

$$R_{ic} \geq \Delta V_t^{DAVN_i}(x_i).$$

Now if the indexing is done well, any product j that is mapped into virtual class c on resource i should have a displacement-adjusted revenue close to R_{ic}; hence, $\bar{p}_{ij} \approx R_{ic}$. Substituting this approximation above and using the definition of $\bar{p}_{ij}$ (3.15), we then obtain (approximately)

$$p_j \geq \Delta V_t^{DAVN_i}(x_i) + \sum_{l \in A_j, l \neq i} \pi_l.$$

Observe the similarity of this condition to (3.4). In accepting a product j in virtual class c on resource i, we are (approximately) comparing its revenue p_j to the sum of displacement costs on all resources it uses, where $\Delta V_t^{DAVN_i}(x_i)$ is a dynamic (time and state-dependent) approximation of the displacement cost on resource i itself and the values π_l are static approximations of the displacement cost on the other resources $l \neq i$, $l \in A_j$. This heuristic connection to (3.4) illustrates the basic intuition behind the DAVN approach.

3.4.4 Dynamic Programming Decomposition

Dynamic programming decomposition is similar in spirit to DAVN in that it uses displacement-adjusted revenues to decompose the network into a series of resource-level problems. Indeed, the only real difference is that while DAVN takes displacement-adjusted revenues and aggregates them into a small number of virtual classes, in dynamic programming decomposition, the revenue and demand remains disaggregated. As with other decomposition methods, there are several possible variations to the basic approach. However, for purposes of illustration we focus on one special case here—specifically, the dynamic single-resource model in which demand for product j arrives in period t with probability $\lambda_j(t)$. (Using the dynamic model, one also does not have to worry about making a low-before-high revenue-order assumptions, which as in DAVN is likely to be violated as a result of the displacement-adjusted revenue step.)

We start the decomposition as in DAVN with a static vector of bid price $\bar{\pi}$. Again, this may be computed in a variety of ways—say, using one of the network math programming models of Section 3.3. Then for each resource i, we solve a single-resource dynamic program based on displacement-adjusted revenues. That is, for each product j that uses resource i, compute the displacement-adjusted revenue

$$\bar{p}_{ij} = p_j - \sum_{l \in A_j, l \neq i} \bar{\pi}_l.$$

Then formulate a dynamic single-resource problem for resource i (problem i) with arrival rates $\lambda_j(t)$ and revenues $\bar{p}_{ij}$. Let the resulting value function be denoted $V_t^{DPD_i}(x_i)$. The total value function approximation is then

$$V_t^{DPD}(\mathbf{x}) = \sum_{i=1}^{m} V_t^{DPD_i}(x_i),$$

and the bid prices are given by

$$\pi_i^{DPD}(t, \mathbf{x}) = \Delta V_t^{DPD_i}(x_i), \quad i = 1, \ldots, m,$$

where $\Delta V_t^{DPD_i}(x_i)$ is the marginal value from problem i.

Because dynamic programming decomposition is so similar to DAVN, the choice of which one to use is most often dictated by the control strategy one wants to use in the end. If the objective is to construct

108   THE THEORY AND PRACTICE OF REVENUE MANAGEMENT

virtual nesting controls, then aggregating and indexing as in DAVN will accurately match the control strategy. If one is using bid-price controls, then the aggregating and indexing of DAVN is not necessary, and a dynamic programming decomposition will tend to yield more accurate bid price approximations, and a bid price table (see Table 3.1) for better control.

3.4.5   Iterative Decomposition Methods

Iterative decomposition methods are also closely related to DAVN. The methods were originally based on the EMSR heuristics for the single-resource problem and so were called *iterative prorated EMSR* methods, but there are several variations of the general idea, and it can be used with any single-resource model. We look at two basic versions that differ primarily in the way they convert product revenues into resource-level revenues.

3.4.5.1   Iterative DAVN

Iterative DAVN is essentially a method for computing the static bid prices $\bar{\pi}$ used by DAVN. The motivation for the approach is heuristic but intuitively appealing. Suppose that $\bar{\pi}$ indeed represents the marginal displacement cost vector. Then once DAVN is solved, for consistency we should have that

$$\bar{\pi}_i = \pi_i^{DAVN}(t, \mathbf{x}), \quad i = 1, \ldots, m,$$

where $\pi_i^{DAVN}(t, \mathbf{x}) = \Delta V_t^{DAVN_i}(x_i)$, the marginal value generated by problem i at the current time t. That is, the marginal costs produced by the DAVN decomposition should match our static estimate $\bar{\pi}$.

A natural question arises then. What happens if $\pi_i^{DAVN}(t, \mathbf{x}) \neq \bar{\pi}_i$? The idea of iterative DAVN is that if these values do not match, we simply feed back the estimates $\pi_i^{DAVN}(t, \mathbf{x})$ as new static bid prices into the procedure and recompute the DAVN models.

Formally, let $\bar{\pi}^{(k)}$ denote the static bid price at iteration k in the algorithm. The algorithm proceeds as follows:

STEP 0 (Initialize):
   $k = 1$. Initialize $\bar{\pi}^{(k)}$ to an arbitrary starting value (say, zero).

STEP 1 (Compute new displacement-adjusted revenues):
   FOR $j = 1, \ldots, n$, DO:
   Compute the displacement-adjusted revenues for each product j on

resource $i \in A_j$ using

$$\bar{p}_{ij} = p_j - \sum_{l \in A_j, l \neq i} \bar{\pi}_l^{(k)}, \quad i \in A_j.$$

FOR $i = 1, \ldots, m$, DO:
Cluster and index the products into virtual classes on resource i and solve a single-resource problem to generate the value-function estimates $V_t^{DAVN_i}(x_i)$ and the marginal value estimates

$$\Delta V_t^{DAVN_i}(x_i) = V_t^{DAVN_i}(x_i) - V_t^{DAVN_i}(x_i - 1).$$

STEP 2 (Check for Convergence):
IF $|\Delta V_t^{DAVN_i}(x_i) - \bar{\pi}_i^{(k)}| < \epsilon$ for $i = 1, \ldots, m$, STOP.
ELSE
$$\bar{\pi}_i^{(k+1)} \leftarrow \Delta V_t^{DAVN_i}(x_i), \quad i = 1, \ldots, m$$
and $k \leftarrow k + 1$. GOTO STEP 1.

Abstractly, this algorithm produces a mapping Ψ from the space of bid price vectors $\bar{\pi}$ onto itself. That is, $\bar{\pi}^{(k+1)} = \Psi(\bar{\pi}^{(k)})$. The algorithm terminates if it finds a fixed point $\bar{\pi}^*$ of this map, $\bar{\pi}^* = \Psi(\bar{\pi}^*)$. However, whether Ψ is a contraction mapping or not is not known, so convergence of this method is not always guaranteed.

3.4.5.2 Iterative Prorated EMSR

Iterative prorated EMSR—denoted IEMSR—is nearly equivalent to iterated DAVN except that it uses prorated rather than displacement-adjusted revenues. It was originally proposed using the EMSR-b approximation, but other single-resource methods could be used in place of EMSR-b.

Iterative prorated EMSR simply replaces Step 1 of the iterative DAVN algorithm with the following:

STEP 1 (Modified):
FOR $j = 1, \ldots, n$, DO:
Compute the prorated revenues on each resource i used by product j as

$$\bar{p}_{ij} = p_j \frac{\bar{\pi}_i^{(k)}}{\sum_{l \in A_j} \bar{\pi}_l^{(k)}}, \quad i \in A_j.$$

FOR $i = 1, \ldots, m$ DO:
Cluster and index the products into virtual classes and use the EMSR-b method for the decomposed problem on resource i to generate the value-function estimates $V_t^{IEMSR_i}(x_i)$ and the marginal value estimates $\Delta V_t^{IEMSR_i}(x_i)$.[13]

That is, the revenue used for product j in problem i is a prorated revenue based on the static bid price vector $\bar{\pi}^{(k)}$. New static bid prices are then formed using the marginal values from an EMSR algorithm, and the procedure is repeated. This continues until the marginal values produced by the EMSR algorithm equal the static bid prices $\bar{\pi}^{(k)}$.

Again, this iteration can be viewed as a mapping of the form $\bar{\pi}^{(k+1)} = \Psi(\bar{\pi}^{(k)})$. However, unlike the mapping produced by iterative DAVN, Bratu [87] has shown that the mapping produced by iterative prorated EMSR is a contraction mapping under certain conditions; namely, when the demand is left disaggregated (there is no indexing to virtual classes) and the EMSR-b heuristic is used. Thus, this version of the algorithm always converges to a set of bid prices $\bar{\pi}^*$ that satisfy $\bar{\pi}^* = \Psi(\bar{\pi}^*)$.

3.4.5.3 Comments on Convergence of Iterative Methods

While convergence guarantees are somewhat reassuring, one has to be careful not to read too much into this fact. Indeed, despite the intuitive appeal and simplicity of iterative methods, there are counterexamples that show that the resulting convergent bid prices can be quite bad. Consider the following example:

Example 3.6 Consider a three-resource line network, with nodes A, B, C, and D. Each of the three resources AB, BC, and CD has one unit of capacity. Suppose there are $T = 2$ periods and we have data for itinerary arrivals as shown in Table 3.3. If we start with static bid price $\bar{\pi}_{AB} = \bar{\pi}_{BC} = \bar{\pi}_{CD}$, prorate the revenues in period 1 by these weights and then compute the exact expected marginal value of each leg, we get a new static set of bid prices $\bar{\pi}_{AB} = 250$, $\bar{\pi}_{BC} = 500$, and $\bar{\pi}_{CD} = 250$. We can then prorate by these new bid prices and repeat the procedure.

The results of repeated applications of this procedure are shown in Table 3.4. Note that the bid prices converge to $\bar{\pi}_{AB} = 0$, $\bar{\pi}_{BC} = 1000$ and $\bar{\pi}_{CD} = 0$. However, by inspection of the data in Table 3.3, it is clear that we want to reject both of the itineraries arriving in period 1, so we need $\bar{\pi}_{AB} > 100$ and $\bar{\pi}_{CD} > 100$. Such a policy

---

[13] How to compute the EMSR-b marginal value is somewhat nonobvious, but it can be obtained easily using the functions $\text{EMSR}_j(x) = \bar{p}_j P(S_j > x)$ where, as defined in Section 2.2.4, $\bar{p}_j$ is the weighted-average revenue for classes j and higher and S_j is the aggregate demand for class j and higher. Taking the maximum provides a (heuristic) marginal-revenue curve for the EMSR-b model.

Table 3.3. Data for the iterative proration-method example.

Time (t)	Product ($\mathbf{A}_j$)	Revenue (p_j)	Probability
1	AB	$100	0.5
	CD	$100	0.5
2	ABC	$1000	0.5
	BCD	$1000	0.5

Table 3.4. Example of convergence of the iterative proration method ($t = 1$).

Iteration k	$\bar{\pi}_{AB}$	$\bar{\pi}_{BC}$	$\bar{\pi}_{CD}$
0	250.0	500.0	250.0
1	166.7	666.7	166.7
2	100	800	100
	.	.	.
	.	.	.
	.	.	.
∞	0	1000	0

yields an expected revenue of $1,000. Because the iterative proration scheme produces zero bid prices for legs AB and CD, it accepts both products AB and CD in period 1, generating an expected revenue of only $100—only 10% of the optimal revenue.

This example illustrates that iterative methods, even though they converge, may not converge to values that yield good approximations of the network opportunity cost. Despite such examples, the methods have worked well in simulation studies.

3.5 Stochastic Gradient Methods

Another class of methods for network capacity control uses simulation to optimize over a parametric family of control policies. Here, we look at optimizing the nested protection levels of a virtual nesting scheme. The idea is to first fix an indexing scheme and a nesting order on each resource (for example, using DAVN) and then to set nested protection levels (or booking limits) for each resource based on this nesting order.

Using a network-level simulation, one can generate samples of demand and compute stochastic gradients—"noisy" estimates of the partial derivatives of the network revenue with respect to the protection-level parameters of the control policy. This gradient information can then be used in a steepest decent algorithm to search for a network-optimal (rather than resource-level optimal) set of protection-level parameters.

One version of the approach is based on a continuous-demand, continuous-capacity model that allows for gradient calculations. We look at this method in detail, since it is the simplest and most direct method. Discrete versions of the approach, which rely on first difference rather than gradient sensitivity estimates, are also discussed briefly afterwards.

3.5.1 Continuous Model with Gradient Estimates

Assume continuous demands and capacities in the network model. Let a realization of a sample path of network demand, denoted ω, be defined by a sequence of customer requests $\omega = \{(j_t, q_t) : t = 1, \ldots, N\}$, where N is the total number of customers in the sequence (a random variable), t indexes the individual customers (in order of arrival), j_t is the product requested by customer t, and q_t is the quantity of product j_t requested. In general, q_t could be any nonnegative quantity (such as a request for one unit or a batch request for five units). We can generate these sample paths by essentially any demand model we like, with no significant restrictions on the distribution (the only one being that N is finite w.p.1). For example, there could be arbitrary order of arrivals, demands could be correlated or coefficients of variation could be arbitrary. This level of generality in the demand model is one of the primary advantages of simulation-based optimization methods.

We assume there are $\bar{c} + 1$ virtual classes on each resource. Let y_{ic} denote the protection level for virtual class c and higher on resource i. (We assume for notational convenience that $y_{i0} = 0$ for all i.) Let $\mathbf{y} = (y_{11}, \ldots, y_{1\bar{c}}, \ldots, y_{m1}, \ldots, y_{m\bar{c}})$ denote the vector of all $m\bar{c}$ protection levels. The protection levels are nested on each resource, so we require

$$0 \leq y_{i1} \leq y_{i2} \cdots \leq y_{i\bar{c}} \leq x_i, \quad i = 1, \ldots, m,$$

where x_i is the capacity of resource i. Let the set of all $\mathbf{y}$ satisfying these constraints be denoted Θ. A request for product j is mapped to virtual class $c_i(j)$ on each resource i used by product j according to the fixed indexing scheme.

Request t for an amount $q_t = q$ of product $j_t = j$ is processed as follows. First, let $\mathbf{x}(t)$ denote the vector of remaining capacities at the time request t arrives. The available capacity for product j on each resource $i \in A_j$ is then $(x_i(t) - y_{c_i(j)-1})^+$. That is, the remaining capacity $x_i(t)$ on i minus the protection level for virtual classes higher than the virtual class of j or zero if this difference is negative. The amount of the request accepted, denoted $u_j(\mathbf{x}(t), \mathbf{y}, q)$, is given by the minimum available capacity among all the resources required by product j or q if there are at least q units of capacity on all of these resources. Formally,

$$u_j(\mathbf{x}(t), \mathbf{y}, q) = \min\{q, (x_i(t) - y_{i,c_i(j)-1})^+ : i \in A_j\}. \quad (3.17)$$

Note here that we are allowing the system to partially accept a request if the available capacity is positive but less than the quantity q that is requested. This assumption is necessary to make the revenue on the sample path, denoted $R(\mathbf{y},\omega)$, a smooth function of the protection levels $\mathbf{y}$. Indeed, one can show that $R(\mathbf{y},\omega)$ is continuous and piecewise linear in $\mathbf{y}$ and always has right and left derivatives with respect to each y_{ic}.

Our objective is to maximize the expected revenue, $E[R(\mathbf{y},\omega)]$, over the set of feasible protection levels Θ:

$$\max_{\mathbf{y}\in\Theta} E[R(\mathbf{y},\omega)]. \qquad (3.18)$$

One method for solving this problem is to use an iterative gradient projection method of the form

$$\mathbf{y}^{(k+1)} = \text{Proj}_\Theta \left(\mathbf{y}^{(k)} + \gamma_k \nabla_\mathbf{y} E[R(\mathbf{y}^{(k)},\omega)] \right),$$

where $\text{Proj}_\Theta(\mathbf{x})$ denotes a projection of the point $\mathbf{x}$ onto the set Θ, $\nabla_\mathbf{y} E[R(\mathbf{y},\omega)]$ is the gradient of $E[R(\mathbf{y},\omega)]$, and γ_k are appropriately selected step sizes. The difficulty is computing $\nabla_\mathbf{y} E[R(\mathbf{y},\omega)]$. However, if we can interchange expectation and differentiation, so that

$$\nabla_\mathbf{y} E[R(\mathbf{y},\omega)] = E[\nabla_\mathbf{y} R(\mathbf{y},\omega)],$$

the gradient $\nabla_\mathbf{y} E[R(\mathbf{y},\omega)]$ can be approximated by the *stochastic gradient* $\nabla_\mathbf{y} R(\mathbf{y},\omega)$, the gradient of the revenue with respect to $\mathbf{y}$ on the single sample path ω, which is a random vector. One can use either an average of a sample of such stochastic gradients as an estimate, or just a single sample. Using the latter leads to the following stochastic gradient method:

$$\mathbf{y}^{(k+1)} = \text{Proj}_\Theta \left(\mathbf{y}^{(k)} + \gamma_k \nabla_\mathbf{y} R(\mathbf{y}^{(k)},\omega) \right). \qquad (3.19)$$

If the step sizes are nonnegative and satisfy $\sum_k \gamma_k = +\infty$ and $\sum_k \gamma_k^2 < +\infty$ and the stochastic demand sequence satisfies a bounded variance condition,[14] then the algorithm can be shown to converge (w.p.1) to at least a stationary point of the optimization problem (3.18).[15]

3.5.1.1 Sample Path Gradient Calculation

Of course, we must be able to compute the sample path gradient $\nabla_\mathbf{y} R(\mathbf{y},\omega)$ efficiently. For the continuous model presented here this can

---
[14] For example, $\gamma_k = 1/k$ will do, but other more effective, adaptive step-size rules are available.
[15] These claims are deliberately informal. The reader is refereed to Bertsekas and Tsitsiklis [55] and Ermoliev [178] for precise conditions and convergence results on stochastic, iterative algorithms of this type. These references also discuss step-size selection and stopping criteria.

be accomplished through a recursive calculation that—although somewhat messy algebraically—is not much more complex than simulating the acceptance policy itself.

To proceed, define $R_t(\mathbf{x}(t), \mathbf{y}, \omega)$ to be the revenue to go over arrivals $t, t+1, \ldots, N$ starting with a vector $\mathbf{x}(t)$ of remaining capacities and protection levels $\mathbf{y}$. We then have the following set of recursive *forward equations* for determining the revenues

$$R_t(\mathbf{x}(t), \mathbf{y}, \omega) = p_{j_t} u_{j_t}(\mathbf{x}(t), \mathbf{y}, q_t) + R_{t+1}(\mathbf{x}(t+1), \mathbf{y}, \omega) \quad (3.20)$$
$$\mathbf{x}(t+1) = \mathbf{x}(t) - \mathbf{A}_{j_t} u_{j_t}(\mathbf{x}(t), \mathbf{y}, \omega), \quad (3.21)$$

for $t = 1, \ldots, N$, with boundary condition $\mathbf{x}(1) = \mathbf{x}$ and $R_{N+1}(\mathbf{x}, \mathbf{y}, \omega) = 0$. The total sample path revenue is given by $R(\mathbf{y}, \omega) = R_1(\mathbf{x}, \mathbf{y}, \omega)$. By taking derivatives through this recursion, one can come up with a gradient $\nabla_{\mathbf{y}} R(\mathbf{y}, \omega)$. The details of this computation are summarized in Appendix 3.A, but it results in a simple and efficient recursion for determining the gradients based on a sample simulation of the network demand.

The main advantage of this simulation-based optimization approach is that the resulting gradients accurately estimate the true network revenue effects of perturbing $\mathbf{y}$. Moreover, as mentioned, the procedure is at least locally convergent, so given good starting values it can potentially find the network-optimal set of virtual nesting protection levels. At a minimum, it is at least guaranteed to improve on any given initial set of protection levels $\mathbf{y}$. The disadvantage of the approach, however, is that it can become computationally intensive.

3.5.1.2 Numerical Example

A numerical example from [528] illustrates the behavior of the stochastic gradient algorithm. Here we consider an airline network with five cities (nodes) and eight legs (resources) with 10 round-trip itineraries and four fare classes per itinerary, producing 80 products as shown in Figure 3.3. The data in Table 3.5 is from Williamson [566]. Demands are assumed normally distributed with a standard deviation equal to the square root of the mean. Each leg has the same capacity, which is varied to generate different load factors. A version of DAVN as described in Section 3.4.3 is used to define the indexing and find the initial set of protection levels. The results shown below are for 5,000 instances (sample paths) run with the stochastic gradient method to improve on the initial protection levels. Tables 3.6 and 3.7 show an example of the protection levels produced by DAVN and the improved protection levels produced by the stochastic gradient algorithm. Note that protection levels are increased on some legs and decreased on others. Some virtual classes

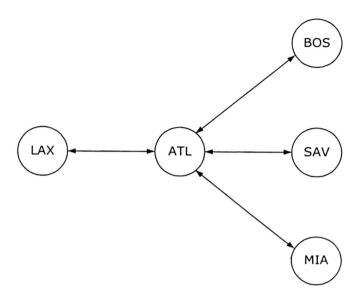

Figure 3.3. A network example (Williamson [566]).

Table 3.5. Data for the Williamson [566] network example.

	Itinerary Fare Class Values							
	Revenues				Mean Demand			
Itinerary	Y	M	B	Q	Y	M	B	Q
ATLBOS/BOSATL	310	290	95	69	12	9	11	17
ATLLAX/LAXATL	455	391	142	122	8	4	11	27
ATLMIA/MIAATL	280	209	94	59	20	9	7	14
ATLSAV/SAVATL	159	140	64	49	25	7	5	13
BOSLAX/LAXBOS	575	380	159	139	9	7	11	24
BOSMIA/MIABOS	403	314	124	89	11	5	14	20
BOSSAV/SAVBOS	319	250	109	69	5	7	11	27
LAXMIA/MIALAX	477	239	139	119	17	11	7	14
LAXSAV/SAVLAX	502	450	154	134	5	7	7	32
MIASAV/SAVMIA	226	168	84	59	13	4	7	25

are, in fact, "merged" (virtual classes 1 and 2 are merged on the first two legs; virtual classes 1, 2, and 3 are merged on leg 4). Virtual class 4 "disappears" on legs 1, 2, 5, and 6, meaning that there is no specific capacity reserved for it. The same thing happens to virtual class 3 on legs 6 and 7. In these examples, the improved protection levels produce dramatically increased revenues (on the order of 5%), though this was largely because the DAVN policy used theft nesting without any reop-

Table 3.6. Initial protection levels produced by DAVN.

Resource	Protection Levels y_{ic}						
	1	2	3	4	5	6	7
1	7	34	41	47	57	70	82
2	7	34	41	47	57	70	82
3	7	18	29	56	68	95	-
4	7	18	29	56	68	95	-
5	3	10	16	36	60	75	89
6	3	10	16	36	60	75	89
7	26	51	60	75	92	101	132
8	26	51	60	75	92	101	132

Table 3.7. Improved protection levels produced by the stochastic gradient algorithm.

Resource	Protection Levels y_{ic}						
	1	2	3	4	5	6	7
1	0	34	39	39	43	70	76
2	0	34	39	39	40	71	76
3	2	3	16	44	70	101	-
4	0	0	10	47	70	99	-
5	3	9	10	10	56	75	81
6	3	10	11	11	58	75	82
7	20	39	39	75	92	101	148
8	18	42	42	73	92	101	147

timizations. More frequent reoptimization improves both methods and narrows the performance gap, but the stochastic gradient method still improves the performance of DAVN significantly.

3.5.2 Discrete Model with First-Difference Estimates

Using simulation-based optimization of nested allocation controls first appeared in a method proposed by Bertsimas and de Boer [60]. We only outline their method because it is somewhat involved and is similar to the continuous model discussed above. There are two key differences in their approach, however. First, it is based on a discrete demand, discrete capacity, and discrete parameter model, where each request t is assumed to be for a single unit of capacity ($q_t = 1$ for all t) and the requests are either fully accepted or rejected. The protection levels **y** are also integral. The resulting model is more realistic, but one must use first-difference (rather than gradient) estimates, which cannot be computed as efficiently as in the continuous model. In addition, because it is a discrete problem, even local convergence is not guaranteed. Second, because of the complexity of propagating first-difference estimates, Bertsimas and de Boer use an approximation of the revenue to go to simplify the computation, so the method provides only approximate estimates of the first differences. This approximation, however, also accounts for the value of reoptimization (it is a true value-function estimate), which adds another refinement to the simulation-based approach. Here, however, we focus only on the simulation-based optimization portion of their overall approach.

As before, let $R(\mathbf{y}, \omega)$ represent the revenue from a sample path and let $R_t(\mathbf{x}(t), \mathbf{y}, \omega)$ be the revenue to go over arrivals $t, t+1, \ldots, N$ starting with a vector $\mathbf{x}(t)$ of remaining capacities and protection levels $\mathbf{y}$. Again, we have a set of recursive forward equations for determining the revenues

$$R_t(\mathbf{x}(t), \mathbf{y}, \omega) = p_{j_t} u_{j_t}(\mathbf{x}(t), \mathbf{y}) + R_{t+1}(\mathbf{x}(t+1), \mathbf{y}, \omega)$$
$$\mathbf{x}(t+1) = \mathbf{x}(t) - \mathbf{A}_{J_t} u_{j_t}(\mathbf{x}(t), \mathbf{y}, \omega),$$

where $u_{j_t}(\mathbf{x}(t), \mathbf{y})$ is either 0 or 1. Since the model is discrete, one must look at first-difference estimates of sensitivity—that is,

$$\Delta_{y_{ic}} R(\mathbf{x}(t), \mathbf{y}, \omega) = R(\mathbf{x}(t), \mathbf{y} + \mathbf{e}_{ic}, \omega) - R(\mathbf{x}(t), \mathbf{y}, \omega),$$

where $\mathbf{e}_{ic}$ is the unit vector corresponding to parameter y_{ic} in the vector $\mathbf{y}$.

However, in contrast to the differential changes in the continuous model, propagating discrete changes using the above recursion cannot be done in parallel for all parameters $\mathbf{y}$. Rather, one must trace the effects of the change for each parameter y_{ic} in a separate calculation, though only protection levels that are binding need to be calculated.

Since this is too computationally complex to be practical for large networks, one can approximate the effects of a discrete change by simulating for a small number of arrivals s and using an approximation $\tilde{R}_t(\mathbf{x}(t), \mathbf{y}, \omega)$ to estimate the revenue effects from time s onward. Bertsimas and de Boer [60] propose using a piecewise linear approximation to the revenue function, based on recursively applying the stochastic first-difference estimates over a successive number of periods.

That is, divide the problem into periods. In the last period, the first-difference estimates are computed exactly for a sample of simulation runs. The resulting averaged first differences are then used to form a piecewise linear approximation of the revenue function in the last period. In the second-to-last period, the simulation estimates of the first differences are run again, but the approximation of the revenue function is used to estimate the revenue effects over the last period. The resulting first-difference estimates for the second-to-last period are then used to form a piecewise approximation of the revenue function in that period, and so on.

Again, the complete method is somewhat involved, and the reader is referred to Bertsimas and de Boer [60] for details. However, in their tests, the method was computationally manageable and effective at improving the revenue performance of the initial protection levels. The improvements increased with load factor and ranged from 0.0 to 0.75%. These results again suggest the potential to improve protection-level policies using simulation-based optimization methods.

3.6 Asymptotic Analysis of Network Problems

Some asymptotic results have been proven for network problems, which we briefly review here. The analysis essentially looks at the performance of different policies as the capacity and demand are scaled up by the same factor. It gives a rigorous—albeit somewhat crude—characterization of optimal network policies. The first set of results relates to partitioned controls, and the second to bid-price controls.

3.6.1 Asymptotic Optimality of Partitioned Controls

We first consider simple partitioned allocation controls generated by the deterministic linear programming approximation of Section 3.3.1. Our analysis here is based on Cooper [131]. Consider the linear program (3.6), which we restate here:

$$\bar{v} = \max \; \mathbf{p}^\top \mathbf{y} \quad (3.22)$$
$$\text{s.t.} \; \mathbf{A}\mathbf{y} \leq \mathbf{x}$$
$$0 \leq \mathbf{y} \leq \boldsymbol{\mu}.$$

$\bar{v}$ provides an upper bound on the expected revenue from any dynamic allocation policy. This follows because any policy will have a cumulative vector of product sales $\mathbf{y}$ that satisfies $\mathbf{A}\mathbf{y} \leq \mathbf{x}$ and $\mathbf{y} \leq \mathbf{D}$ pathwise. Thus, $E[\mathbf{N}]$ is a feasible solution to the linear program (3.22). This, together with Jensen's inequality (see Appendix B) shows that $\bar{v}$ is an upper bound on v^*.

Moreover, an optimal solution $\mathbf{y}^*$ defines partitioned allocations y_j^* for each product j. These partitions may not be integral, but in the limiting case we consider here this is not significant, and hence we ignore integrality. The (random) revenue produced by this allocation, denoted

$V^{PA}(\mathbf{D})$, is given by

$$V^{PA}(\mathbf{D}) = \sum_{j=1}^{n} p_j \min\{y_j^*, D_j\}.$$

Let θ be a positive integer and consider a sequence of such problems indexed by $\theta = 1, 2, \ldots$ with capacities $\theta \mathbf{x}$ and demand vectors $\mathbf{D}_\theta$ such that $E[\mathbf{D}_\theta] = \theta \boldsymbol{\mu}$ and

$$\frac{1}{\theta} \mathbf{D}_\theta \xrightarrow{D} \boldsymbol{\mu},$$

where $\xrightarrow{D}$ denotes convergence in distribution. For example, $\mathbf{D}_\theta$ could be constructed by adding θ i.i.d. vectors $\mathbf{D}$, in which case the above follows by the law of large numbers. In this sequence of problems, demand and capacity are scaled up proportionately by a factor θ.

We can define a sequence of linear programs denoted $PA(\theta)$ analogous to (3.22) by

$$\bar{v}(\theta) = \max \ \mathbf{p}^\top \mathbf{y}$$
$$\text{s.t.} \ \mathbf{Ay} \leq \theta \mathbf{x}$$
$$0 \leq \mathbf{y} \leq \theta \boldsymbol{\mu},$$

and $\bar{v}(\theta)$ is again an upper bound on the optimal expected revenue, $v^*(\theta)$, in problem $PA(\theta)$. Also, note that $\bar{v}(\theta) = \theta \mathbf{p}^\top \mathbf{y}^*$ and that $\theta \mathbf{y}^*$ is the optimal solution to the scaled problem. Finally, the random revenue produced by this solution is

$$V^{PA}(\theta) = \sum_{j=1}^{n} p_j \min\{\theta y_j^*, D_j(\theta)\}.$$

Now a simple scaling of these expressions by $1/\theta$ shows that

$$\frac{1}{\theta} V^{PA}(\theta) = \frac{1}{\theta} \sum_{j=1}^{n} p_j \min\{\theta y_j^*, D_j(\theta)\}$$

$$= \sum_{j=1}^{n} p_j \min\{y_j^*, \frac{1}{\theta} D_j(\theta)\}$$

$$\xrightarrow{D} \sum_{j=1}^{n} p_j \min\{y_j^*, \mu_j\} = \bar{v},$$

where the last convergence follows from the continuity of the $\min\{\cdot\}$ function.

Thus, the scaled revenue of the partitioned allocation policy converges in distribution to a constant; namely, the upper bound, $\bar{v}$. Hence, the revenue of the partitioned allocation policy converges almost surely to the optimal revenue.

3.6.2 Asymptotic Optimality of Bid-Price Controls

One can also show that bid-price controls have good asymptotic properties in the same scaling of the problem. The details of the analysis are somewhat involved, but here we provide a brief description of the main results.

Consider a sequence of scaled deterministic linear programs as in (3.22). It is not hard to show that the dual prices for each program in this sequence of problems are constant because the capacity and demand are multiplied by the same factor θ. Let $\boldsymbol{\pi}$ denote this vector of dual prices. Using this vector of dual prices, one can define a fixed bid-price heuristic in which we accept a request for product j with revenue p_j if and only if there is sufficient capacity and $p_j > \boldsymbol{\pi}^\top \mathbf{A}_j$.

One can then show via upper and lower bounds that the expected revenue produced by this fixed bid-price heuristic converges (in ratio) to the optimal expected revenue as the scale parameter θ increases.[16] In particular, the expected revenue of the fixed-price heuristic is within $O(\sqrt{\theta})$ of the upper bound $\bar{v}(\theta)$ on the optimal expected revenue given by (3.22). Since the $\bar{v}(\theta)$ is $O(\theta)$, this shows the relative suboptimality of the fixed bid-price heuristic tends to zero as θ increases.

For one, this result shows that, asymptotically, optimal bid prices are constant. That is, a set of constant bid prices will be near optimal in this scaling of the problem when demand and capacity are large. This suggests that even though bid prices may be volatile, one should not expect to see much of a drift in bid-price values over time, at least for large problem instances. In addition, because the asymptotically optimal bid prices are based on the DLP model, the results provide some theoretical support for using DLP to compute bid prices.

3.6.3 Comments on Asymptotic Optimality

While these asymptotic results provide some theoretical insights, their practical value is debatable. What these asymptotic results provide is, essentially, a characterization of the first-order properties of a given pol-

---

[16] In this case, it is convergence of the expected values and not convergence in probability as in the partitioned allocation case.

icy. That is, the analysis *averages out* the randomness in the problem and leaves only the mean demand effects. (Note that the asymptotic bounds above depend only on the mean demand.) So we can conclude that bid prices based on partitioned allocations of the deterministic linear program (3.22) will produce the optimal first-order revenues in this sense.

However, this asymptotic first-order optimality is a rather crude measure of performance. One would certainly be suspicious of a policy that does *not* have this property; but the fact that a policy does achieve first-order optimality does not mean it is necessarily effective in practical settings. For example, the partitioned allocation policy typically performs poorly in practice, despite the fact that it is first-order optimal.

3.7 Decentralized Network Control: Airline Alliances

Thus far, we have looked at network control from the standpoint of a single firm that manages the capacity of all network resources. However, there are many instances where no single firm controls all of a network's resources. For example, in hotel chains or car rental operations, individual locations may be owned and operated by a franchisee who operates as a profit center, yet there may be demands for multilocation stays or one-way rentals that impact more than one of these locations. In the airline industry, alliances between major carriers create integrated networks that jointly market products—code shared flights or connections between the carriers' networks, for example—but each firm has a separate profit motive. The management of such *decentralized networks* is a relatively new problem area in RM. Here we briefly review this topic, focusing on the case of airline alliances.

Alliances provide a variety of benefits in a network industry like air transport. For one, airlines can cross-list each other's flights through *code-sharing* arrangements, in which the same physical flight is listed under different flight numbers, one for each partner airline in the alliance. When two networks are combined, connecting flights between one carrier's network and another are created, which again can be cross-listed and marketed by each carrier. Alliances also create economies of scale in providing ground services such as checking, baggage handling, and passenger lounges. Finally, the consolidation of loyalty programs benefits consumers. Indeed, in many ways airlines alliances achieve many of the benefits of an outright merger without the attendant labor and regulatory difficulties.

Yet airline alliances have created new challenges for RM. On an operational level, one must find some mechanism for controlling the capacity on shared flights. The two broad categories of mechanisms used by airlines are the following:

- **Blocked-seat allotment** In a blocked-seat allotment, the carriers agree to partition the capacity of shared flights, with each carrier having individual control over its allotment of seats. The allotments may be updated periodically in various ways, but fundamentally the seat inventory is managed as if it were two "virtual" flights, each one controlled independently by its respective alliance partner.

- **Free sale** Under free sale, capacity is not partitioned. Rather, the carrier that is operating a flight provides some form of dynamic availability information to its alliance partners. For example, it may provide information about which booking classes are available and how much capacity is available in each class. The nonoperating alliance partners can then book seats subject to the capacity controls set by the operating carrier. Alternatively, it may provide bid-price information, in essence providing a spot price for the capacity at a given point in time. The alliance partner may then sell the seat under some revenue and cost-sharing terms that depend on the bid price.

While blocked-seat allotment is simpler operationally, it is not hard to see, just as with partitioned controls in a centralized network, that it can be quite suboptimal to divide capacity this way: one firm may have excess capacity in its allotment and no demand, while another may be sold out and be turning away requests. At the same time, partitioned allocations can usually be accommodated within the current RM practices of alliance partners. Free sale, while offering more flexibility and the potential to provide enhanced revenues, is considerably more complex to manage.

But beyond the simple mechanics of sharing capacity, firms in an alliance must find coordinated policies for managing their shared resources. Moreover, they must do this in a way that provides proper incentives for each party involved. For example, Boyd [86] argues that using bid prices as a transfer price for charging alliance partners for seat capacity can coordinate alliance revenues. His analysis is based on linear programming theory.

3.8 Notes and Sources

The earliest reports and articles on network problems in the airline industry are D'Sylva [163], Glover et al. [215], and Wang [550]. Dror et al. [162] also analyze a network flow model that allows for deterministic cancellations on the arcs. See also network models by Phillips et al. [418], Wong [578], Wong et al. [577], and Wysong [584].

Bid-price controls were first introduced by Smith and Penn [478] and Simpson [476]. Williamson's thesis [565, 566] provides variations of bid-price controls and provides detailed simulation comparisons with other control methods. See also Phillips [419] for an analysis of using and computing marginal values.

The basic theoretical properties and asymptotic analysis of bid-price controls in Section 3.2 and Section 3.6 are from Talluri and van Ryzin [498] as is the bid-price counterexamples. The nonoptimality of bid prices was first raised by Curry [140], who used an analogy to the Taylor series expansion of the value function to argue that second-order "interaction" terms may be significant in determining optimal revenue thresholds. The asymptotic analysis of partitioned controls is due to Cooper [131]. For path-wise theoretical bounds, see Cooper [130].

As for the approximation methods in Section 3.3, the deterministic linear program (DLP) was among the first models analyzed in the early work of D'Sylva [163], Glover et al. [215], Dror et al. [162], and Wong [578, 577]. Wollmer [575] proposed the first linear programming version of the PNLP model. A specialized nonlinear programming algorithm for the PNLP model is provided by Ciancimino et al. [118]. Chen et al. [111, 112] investigate a bid-price approximation that combines both the DLP and PNLP value function approximations using multivariate adaptive regression splines (MARS). The randomized linear programming method was first discussed in Smith and Penn [478] but was investigated in detail in Talluri and van Ryzin [499]. A cutting plane method for joint pricing and allocation in network RM is provided in Garcia-Diaz and Kuyumcu [201].

Descriptions of virtual nesting can be found in Belobaba's thesis [39], Smith and Penn [478], Williamson [565, 566], and Vinod [537, 538]. Vinod [537] provided an algorithm for the indexing step of virtual nesting. Curry [139] analyzes a similar scheme in which classes are nested by origin and destination.

The iterative methods for computing bid prices in Section 3.4.5 can be found in several sources, including Williamson [565, 566] and Phillips [419]. Bratu's master's thesis [87] provides an analysis of the iterative EMSR method and related network approximations.

The discrete stochastic gradient method in Section 3.5.2 for updating virtual nesting booking parameters was first proposed by Bertsimas and de Boer [60]; the continuous gradient model in Section 3.5.1 is due to van Ryzin and Vulcano [528].

The discussion of airline alliances in Section 3.7 is based on Boyd [86]. See also Brueckner and Spiller [89].

The network RM model with passenger routing control and customer-choice behavior was investigated by Talluri [501]. Gallego, Iyengar and Phillips [197] propose a more elaborate model with an airline offering flexible products (at a lower price) and customers self-selecting according to a discrete-choice model.

Finally, we note that in this section we largely ignore the stochastic programming (Birge and Louveauz [68]) approach, as there have been, to our knowledge, no RM implementations that use such techniques directly. However, the basic ideas behind the DLP and RLP approximations are in fact standard ones in stochastic programming. de Boer et al. [154] also apply stochastic programming ideas (i.e., scenario aggregation) to network RM.

APPENDIX 3.A: Computation of Sample Path Gradients

To compute the sample path by gradients, we start by differentiating (3.20) and (3.21) with respect to $\mathbf{y}$ and $\mathbf{x}$, respectively, to obtain the set of *backward equations* for the derivatives with respect to y_{ic}:

$$\frac{\partial}{\partial y_{ic}} R_t(\mathbf{x}(t), \mathbf{y}, \omega) = \left(p_{j_t} - \sum_{k \in \mathbf{A}_{j_t}} \frac{\partial}{\partial x_k} R_{t+1}(\mathbf{x}(t+1), \mathbf{y}, \omega) \right)$$

$$\times \frac{\partial^+}{\partial y_{ic}} u_{j_t}(\mathbf{x}(t), \mathbf{y}, q_t)$$

$$+ \frac{\partial}{\partial y_{ic}} R_{t+1}(\mathbf{x}(t+1), \mathbf{y}, \omega), \quad \forall i, c, t, \quad (3.\text{A}.1)$$

Similarly, a set of backward equations for the derivative with respect to x_i is:

$$\frac{\partial}{\partial x_i} R_t(\mathbf{x}(t), \mathbf{y}, \omega) = \left(p_{j_t} - \sum_{k \in \mathbf{A}_{j_t}} \frac{\partial}{\partial x_k} R_{t+1}(\mathbf{x}(t+1), \mathbf{y}, \omega) \right)$$

$$\times \frac{\partial^-}{\partial x_i} u_{j_t}(\mathbf{x}(t), \mathbf{y}, q_t)$$

$$+ \frac{\partial}{\partial x_i} R_{t+1}(\mathbf{x}(t+1), \mathbf{y}, \omega), \quad \forall i, t; \quad (3.\text{A}.2)$$

with boundary conditions

$$\frac{\partial}{\partial y_{ic}} R_{N+1}(\mathbf{x}(N+1), \mathbf{y}, \omega) = 0, \quad \forall i, c;$$

$$\frac{\partial}{\partial x_i} R_{N+1}(\mathbf{x}(N+1), \mathbf{y}, \omega) = 0, \quad \forall i.$$

(The reason for the use of the right and left derivatives above is somewhat subtle and is explained in detail in [528].)

APPENDIX 3.A: Computation of Sample Path Gradients

Finally, from (3.17) we have that for all i and c,

$$\frac{\partial^+}{\partial y_{ic}} u_j(\mathbf{x}, \mathbf{y}, q) = \begin{cases} -1 & \text{if} \quad \begin{array}{l} (i) \ i \in A_j \\ (ii) \ x_i - y_{i,c_i(j)-1} \leq x_k - y_{k,c_k(j)-1}, \ \forall k \in A_j \\ (iii) \ 0 < x_i - y_{i,c_i(j)-1} \leq q \\ (iv) \ c < c_i(j) \\ (v) \ y_{ic} = y_{i,c_i(j)-1} \end{array} \\ 0 & \text{otherwise.} \end{cases}$$

(3.A.3)

In words, the quantity of demand accepted from a request for product j in state $\mathbf{x}$ is reduced (one for one) by a slight increase in the protection level y_{ic} if and only if all of the following hold: (1) resource i is used by j, (2) the capacity available on resource i is a binding constraint, (3) the amount accepted is positive but constrained by the protection levels, (4) class c is higher in the nesting order than the virtual class of product j, and (5) the protection level for class c is binding. In all other cases, a small change in y_{ic} does not affect the amount of product j we accept.

These conditions are further illustrated in Figures 3.A.1 and 3.A.2, which show a request at time t for q units of product j that uses resources 1, 2, and 3. The height of the bars represents the capacity remaining at time t, and the quantities $y_{i,c_i(j)-1}$ represent the protection levels for product j on each resource i. The unshaded areas therefore represent the capacity available for product j on each of the three resources. Note in Figures 3.A.1 that there is sufficient capacity available on all three resources to fully satisfy the request for product j so $u_j = q$. Thus, a small increase in $y_{i,c_i(j)-1}$ will not affect the quantity of product j accepted in this time period, and therefore $\frac{\partial^+}{\partial y_{i,c_i(j)-1}} u_j(\mathbf{x}, \mathbf{y}, q) = 0$ for all i. However, as shown in Figure 3.A.2, the requested quantity exceeds the available capacity on resources 1 and 3, and the quantity accepted is constrained by the protection-level constraint on resource 3, so $u_j = x_3 - y_{3,c_3(j)-1}$. In this case, a small increase in $y_{3,c_3(j)-1}$ will reduce the amount of product j that we accept in this period, so $\frac{\partial^+}{\partial y_{3,c_3(j)-1}} u_j(\mathbf{x}, \mathbf{y}, q) = -1$. An increase in any of the other protection levels on resources 1 and 2, however, will not affect the quantity accepted because these protection levels are not binding. This example illustrates the conditions leading to (3.A.3).

A similar reasoning provides the left derivatives with respect to x_i:

$$\frac{\partial^-}{\partial x_i} u_j(\mathbf{x}, \mathbf{y}, q) = \begin{cases} 1 & \text{if} \quad \begin{array}{l} (i) \ i \in A_j \\ (ii) \ x_i - y_{i,c_i(j)-1} < x_k - y_{k,c_k(j)-1}, \\ \quad \ \forall k \in A_j, k \neq i \\ (iii) \ 0 < x_i - y_{i,c_i(j)-1} \leq q \end{array} \\ 0 & \text{otherwise.} \end{cases}$$

(3.A.4)

The recursions (3.20)–(3.21) and (3.A.1)–(3.A.2) together with (3.A.3) and (3.A.4) provide the basis for an efficient method for computing the sample path gradient. First, generate a sample sequence of demand ω, and use (3.20)-(3.21) to compute the sequence of acceptance decisions $u_{j_t}(\mathbf{x}(t), \mathbf{y}, q_t)$ and remaining capacities $\mathbf{x}(t)$ by simulating the system forward in time. With these data in hand, starting from N and working backward in time down to $t = 1$, use (3.A.1)-(3.A.2) to compute the derivatives backward in time. The resulting set of derivative values for $t = 1$ give the gradient estimate of $R(\mathbf{y}, \omega)$ with respect to both $\mathbf{y}$ and $\mathbf{x}$. The complexity of the backward pass for computing the gradient is the same as simulating the allocation decisions in the forward pass.

APPENDIX 3.A: Computation of Sample Path Gradients 127

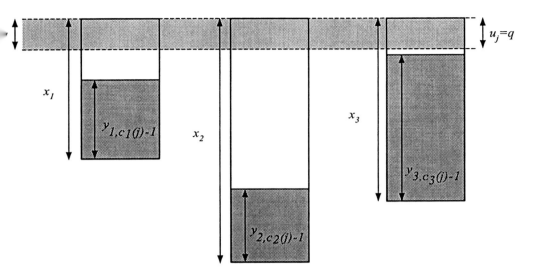

Figure 3.A.1. Illustration of stochastic gradient calculation for nested booking limits: gradient equal to zero.

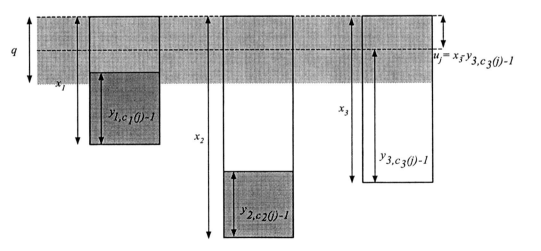

Figure 3.A.2. Illustration of stochastic gradient calculation for nested booking limits: gradient nonzero.

Chapter 4

OVERBOOKING

Overbooking is somewhat distinct from the core pricing and capacity-control problems of revenue management. RM is mainly concerned with how best to price or allocate capacity—how to achieve the best *mix* of demand, in essence. In contrast, overbooking is concerned with increasing capacity utilization in a reservation-based system when there are significant cancellations.[1] Its focus is increasing the total *volume* of sales in the presence of cancellations rather than optimizing customer mix. The problems of optimizing demand mix and volume are quite related, however, and both are considered integral parts of RM.

Indeed, from a historical standpoint, overbooking is the oldest—and, in financial terms, among the most successful—of RM practices. In the airline industry it is estimated that approximately 50% of reservations result in cancellations or no-shows[2] and about 15% of all seats would go unsold without some form of overbooking. [477] This is to be compared to fare-class allocation, which by most estimates leads to incremental revenues to the order of 5%. Despite its economic importance, many researchers consider overbooking a somewhat mature area, and it has

---

[1] We note, however, that reservations are not used in all quantity-based RM industries. In certain advertising markets, for example, one advertiser is allowed to preempt another if it'is willing to pay more for the same ad slot. This effectively produces an auction in which the current highest bidder has claim to the capacity.
[2] A *cancellation* is defined as a reservation that is terminated by a customer strictly *prior* to the time of service. A no-show, in contrast, occurs when a customer does not cancel his reservation but rather just fails to show up at the time of service. The distinction is important because the firm has some opportunity to compensate for a cancellation by accepting more reservations after the fact, while no such opportunities exist when a customer no-shows.

received less attention in the recent research literature than fare-class allocation or pricing.

As a business practice, the biggest challenges in overbooking are managing the negative effects of denying service on customer relations and dealing with the resulting legal and regulatory issues. On a planning level, overbooking involves controlling the level of reservations to balance the potential risks of denied service against the rewards of increased sales. Theoretically, this involves controlling parameters of a probability distribution, which introduces somewhat unique methodology that is not encountered in other areas of RM.

4.1 Business Context and Overview

A reservation is essentially a forward contract between a customer and the firm. Reservations give customers the right to use a service in the future at a fixed price and often also the option to opt out (perhaps with a penalty) before the time of service.

Customers value reservations whenever the costs of unavailability at the desired time of consumption are higher than the costs of unavailability prior to the time of consumption. For instance, because customers travel to attend business meetings, visit family members, or take vacations, they must coordinate their travel with hotel arrangements, business appointments, scheduling of vacation days, and so on. Since it is generally more costly to change or renege on these contingent arrangements at the time of service than it is to change or renege on them in advance, customers value reservations for travel services.

Yet committing to purchasing in advance has its own risks. Uncertain future events (such as clients rescheduling meetings, illness, or more attractive vacation opportunities) may make it impossible or undesirable to use the service. Therefore, customers also value the option to cancel reservations. Indeed, a reservation with a cancellation option gives customers the best of both worlds—the benefit of locking-in availability in advance and the flexibility to renege should their plans or preferences change.

While advance reservations with a cancellation option are highly valued by customers, they require a firm to take a two-sided risk—to honor the reservation when customers show up (or provide suitable compensation if it cannot honor the reservation), and in cases where customers cancel or do not show up, to bear the opportunity cost of wasted capacity. Firms try to manage this risk through a combination of cancellation penalties and overbooking.

Cancellation and no-show penalties effectively allow customers and the firm to share the risks of cancellations. In practice, penalties for cancellations and no-shows range from zero to full price and are most often implemented as a sale condition of the product. In fact, the potential for abuse without such penalties is substantial; customers may make multiple reservations to preserve various options and then cancel all of them except the one they want, not an uncommon practice in certain wholesale international air travel markets. Some minimal penalty is necessary to curb such abuses. (Though firms in a surprising number of industries—the restaurant industry for one—do not penalize for cancellations.) On the other hand, if the penalties are too large, the cancellation option has little value or effectively becomes a nonoption for customers.

To further reduce the costs of cancellations, a firm may also adopt a strategy of accepting more reservations than it has capacity to serve, taking the chance that the number of surviving reservations will be within capacity. This is the essence of planned overbooking.

A firm that chooses a strategy of planned overbooking is immediately faced with several important problems. One is confronting the legal and regulatory implications of failing to honor the reservation contract. Even if the firm is on safe legal ground, it must have operational policies and procedures in place to deal with the situation in which service must be denied. Once these basic structural and policy elements are determined, it must develop methodology to control the level of overbooking on an operational basis. We look at each of these issues in turn.

4.1.1 A History of Legal Issues in Airline Overbooking

Legally, overbooking involves the risk of failing to deliver on a contract to provide service. While there are somewhat different legal requirements in each industry in this regard, it is instructive to look at the evolution of airline overbooking regulations in the United States as an example of the legal issues involved.

Prior to 1961, intentional overbooking was practiced somewhat clandestinely by U.S. airlines and was not acknowledged publicly. Despite this fact, Rothstein [449] reports that as director of Operations Research at American Airlines he, "found much publicly available evidence that all the major airlines were deliberately overbooking." In 1961, the Civil Aeronautics Board (CAB) reported a no-show rate of 1 out of every 10 passengers booked among the 12 leading carriers at that time. The CAB acknowledged that this situation created real economic problems for the airlines. As a result, the CAB implemented a no-show penalty of 50% of the ticket price. At the same time, they explicitly required airlines to pay a penalty of 50% of the ticket price to passengers who were denied boarding. However, the CAB still did not officially sanction overbooking practices. The no-show penalty was abandoned in 1963 largely because airline management felt that the penalties created ill will among passengers and might be discouraging air travel in general.

The CAB conducted another study of overbooking during 1965-66. They found that the denied-boarding rate at that time was approximately 7.69 per 10,000 passengers boarded [119]. Their conclusion was issued in a 1967 docket [119]:

> There is a substantial reservation turnover before flight time from cancellations and no-shows. The airlines are engaging in deliberate or controlled overbooking to compensate for it. Through carefully controlled overbooking, the airlines can reduce the number of empty seats and at the same time serve the public interest by accommodating more passengers.
>
> The present reservation systems of the carriers greatly benefit the traveling public. The Board is not prepared, therefore, to require changes in these systems.

Thus, as of 1965, overbooking was an officially sanctioned practice, provided it was "carefully controlled," a criterion that was never precisely defined by the CAB.

In parallel, the CAB also increased the denied-boarding penalty to 100% of the coupon. Airlines controlled the percentage of denied boardings, and the CAB carefully monitored the denied-boarding performance of each airline. The involuntary denied-boarding rate is still carefully monitored in the United States by the Department of Transportation (DOT) and currently hovers around 0.5 to 1.5 involuntary and 15 to 20 voluntary denied boardings per 10,000 passengers (see Table 4.1).

Despite this progress in formalizing the practice of planned overbooking, the traveling public was still largely unaware of its existence. This was to change in 1972 when Ralph Nader, the well-known U.S. consumer advocate, was denied boarding on an Allegheny Airlines flight. Rather than accept the standard compensation, he sued Allegheny, won, and was awarded $25,000 in punitive damages. The judge's ruling was based on the fact that Allegheny did not advise passengers of its practice of deliberate overbooking. The case was appealed all the way to the U.S. Supreme Court, but the ruling was upheld. As Rothstein [448] noted at the time:

> ... public policy may very well force the airlines into the position that a reservation *involves a definite, legal claim on a seat*. And if this happens, most of the operations research carried out on this problem will have to be discarded and eventually redone.

Table 4.1. U.S. major airline denied-boarding rates, 1990-2000.[a,b,c]

	1990	1991	1992	1993	1994	1995	1996	1997	1998	1999	2000	2001	2002
Boarded ($\times 10^6$)	421	428	445	449	457	460	481	503	514	523	543	498	467
Total DB ($\times 10^3$)	628	646	764	683	824	843	957	1,072	1,126	1,070	1,119	941	837
Voluntary	561	599	718	632	771	794	899	1,018	1,081	1,024	1,062	898	803
Involuntary	67	47	46	51	53	49	58	54	45	46	55	43	34
VDB rate (per 10^4)	14.92	15.09	17.17	15.21	18.03	18.33	19.90	21.31	21.03	19.58	20.60	18.89	17.92
IDB rate (per 10^4)	1.59	1.10	1.03	1.14	1.16	1.07	1.21	1.07	0.87	0.88	1.05	0.86	0.72

[a] Data are for nonstop scheduled service flights between points within the United States (including territories) by the 10 largest U.S. air carriers—that is, those with at least 1% of total domestic scheduled-service passenger revenues (Alaska, America West, American, Continental, Delta, Northwest, Southwest, TWA, United, and US Airways). Before 1994, carriers included both major and national airlines—that is, airlines with over $100 million in revenue.

[b] Statistics are the number of passengers who hold confirmed reservations and are denied boarding ("bumped") from a flight because it is oversold. These figures include only passengers whose oversold flight departs without them; they do not include passengers affected by cancelled, delayed, or diverted flights.

[c] Source: U.S. Department of Transportation, Office of the Secretary, Air Travel Consumer Report (Washington, DC: Annual March issues).

> NOTICE: OVERBOOKING OF FLIGHTS
> Airline flights may be overbooked, and there is a slight chance that a seat will not be available on a flight for which a person has a confirmed reservation. If the flight is overbooked, no one will be denied a seat until airline personnel first ask for volunteers willing to give up their reservation in exchange for a payment of the airline's choosing. If there are not enough volunteers, the airline will deny boarding to other persons in accordance with its particular boarding priority. With few exceptions, persons denied boarding involuntarily are entitled to compensation. The complete rules for the payment of compensation and each airline's boarding priorities are available at all airport ticket counters and boarding locations. Some airlines do not apply these consumer protections to travel from some foreign countries, although other consumer protections may be available. Check with your airline or your travel agent.

Figure 4.1. Overbooking notification statement required by Department of Transportation on domestic U.S. airline tickets [523].

Rothstein proposed that airlines charge for reservations as a possible solution:

> In other words, the *reservation* itself, as opposed to the physical seat on the plane, is to be considered of value ... the reservation itself is a commodity to be purchased for an amount of money and possibly to be relinquished for a different amount of money.

In the wake of Mr. Nader's suit—and after much debate—the CAB revised its rules concerning overbooking as follows:

- Denied-boarding compensation was doubled again to 200% of the coupon.

- Airlines were required to seek volunteers first before denying boarding to any passenger involuntarily.

- The traveling public was to be notified of the deliberate overbooking practices of the airlines.

- A statement warning passengers that their flight may be overbooked and informing them of their rights was to be printed on every ticket.

As a result of this ruling, the DOT requires an overbooking notification statement on all U.S. airline tickets (see Figure 4.1).

These basic regulations are still in existence today in the U.S. Since deregulation in 1974, airlines have increasingly relied on vouchers and payments to attract volunteers to give up their seats on oversold flights. As a result, involuntary denied boardings are much less frequent today than they were in the days when overbooking was a clandestine practice.

4.1.2 Managing Denied-Service Occurrences

Managing the compensation and selection of customers in the event of oversales can have a significant impact on denied-service costs and customer perceptions of overbooking. We next briefly look at the main issues involved in managing oversales.

4.1.2.1 Compensation for Denied Service

While legally mandated compensation often specifies payment of monetary damage, this is often viewed as inadequate in the eyes of customers. A car rental customer who is planning to take a tour of the California coast would most likely find the prospect of getting a full refund plus 50% of the contracted rental rate as poor compensation for a ruined vacation. It is often more effective to offer customers a substitute service (such as an upgrade) plus ancillary services that may make the short-run disruption in their schedule more palatable.

To illustrate, the same car-rental customer may be much more satisfied with an offer to provide a ride to a competing rental company, a free upgrade to a luxury car, plus a voucher for future rentals. Compensation that is targeted to substitute for the denied service and perhaps enhance it somewhat is frequently less expensive for a firm and more effective in the eyes of the customer than pure monetary compensation.

4.1.2.2 Selection Criteria

Selecting which customers are to be denied service also can have a significant impact on both the firm's direct costs and customer goodwill. From a legal standpoint, such selection must not be discriminatory. For example, for airlines, current DOT regulations state that [523]

> Every carrier shall establish priority rules and criteria for determining which passengers holding confirmed reserved space shall be denied boarding on an oversold flight in the event that an insufficient number of volunteers come forward.
>
> Such rules and criteria shall not make, give, or cause any advantage to any particular person or subject any particular person to any unjust or unreasonable prejudice or disadvantage in any respect whatsoever.

The default option for allocating service to customers is usually to do it on a first-come, first-serve (FCFS) basis. While a FCFS allocation is perceived as fair and encourages customers to arrive on time, there are many business situations where this allocation is quite undesirable.

A good example is hotel overbooking. Using a FCFS allocation for a hotel means that the customers who are denied service are those who arrive very late in the evening. This creates two difficulties. First, it is usually much more disruptive to relocate a customer who arrives late at night. These customers are often tired and irritable and simply want to go to bed as soon as possible. A customer who arrives in the late afternoon, in contrast, may be willing to sightsee around the town for several hours or go out for a meal while alternative accommodations are secured and their baggage is transported. Second, late-arriving hotel customers are typically business travelers who pay the highest rates, travel often, and therefore represent the most profitable segment for most hotels. In terms of the lifetime value, these customers are the most costly to lose.

Hotels therefore do not always allocate rooms on a FCFS basis. Rather, they monitor arrival rates and occupancy throughout the day to anticipate a potential oversold condition. If at some point managers expect an oversale, they may find alternative arrangements for early-arriving customers to avoid denying service to customers due to check in very late.

In other service situations, it is sometimes possible to select among a pool of customers when allocating service. For example, in airline boarding, customers usually gather to the gate before departure. This gives gate agents a chance to see which passengers have arrived for the flight and to selectively target specific passengers for denied-boarding offers. Indeed, we are aware of one Australian airline that trains its gate agents to solicit young, student travelers ("backpackers") as volunteers. The airline found that customers in this segment are eager to receive a nice hotel room and a good meal in exchange for taking a flight the following day.

4.1.2.3 Oversale Auctions

An alternative method of managing oversales is to conduct an auction to attract volunteers to give up their reservations in exchange for monetary or other compensation. While this practice is now widespread in airlines and familiar to most travelers, the idea was not well received initially.

In 1968, economist Julian Simon proposed what he called "an almost practical solution to airline overbooking, " in which airlines would conduct a sealed-bid "reverse auction" to find passengers willing to accept monetary compensation for being bumped. Simon predicted (rightly so, as initial responses to his letters to airline executives later indicated) that the airlines would object to the scheme

> because such an auction does not seem decorous; it smacks of the pushcart rather than the one-price store; it is "embarrassing" and "crass," i.e., frankly commercial, like 'being in trade' in Victorian England ([472]).

Simon cites this tongue-in-cheek reaction from airline executive Blaine Cooke:

> I greatly fear that your Overbooking Auction Plan suffers from a flawed premise and a fatal defect. The flawed premise is that you assume that airline management and regulation is a rational exercise. It is not. It is more accurately described as an exercise in applied insanity. The effect is your plan offers a market-sensitive and sensible solution to a real problem but a solution not conceived by an airline. Accordingly, the idea must be disallowed since it is well established in airline marketing that only ideas which originate within the airline industry are permissible.

Simon wrote many letters to executives, regulators, policy makers, and customer groups arguing for his "oversale-auction" idea. Despite these efforts, he failed to get even one airline to experiment with it on even a single flight. Even prominent fellow economists questioned the practicality of the idea. Simon [474] quotes a letter from Milton Friedman:

> If the plan is as good as you and I think it is, I am utterly baffled by the unwillingness of one or more of the airlines to experiment with it. I conclude that we must be overlooking something. I realize that you have tested this quite exhaustively, and I have no reason to question your results; yet I find it even harder to believe that opportunities for large increments of profit are being rejected for wholly irrational reasons.

The scheme continued to flounder until 1977 when Alfred Kahn, an economist, was appointed to head the CAB. Simon wrote to Kahn about his proposal and Kahn liked and largely adopted it under the heading of a "volunteer" denied-boarding plan, as mentioned above. At the same time, Kahn increased penalties for involuntary denied boardings.

Simon [474] quotes an American Airlines internal newsletter from April 27, 1979:

> The happiest result of the volunteer plan is that airlines now have a fair and efficient way to avoid denying seats to people who for business or personal reasons have a pressing need to make their flights as planned. VP, Passenger Services, Robert H. Phillips points out that the voluntary program has twin virtues: "It enables us to reduce costs while maintaining customer goodwill and thereby protecting future revenue".

Given the success of this volunteer denied-boarding plan, it appears that airline management has indeed inched closer to the notion of a "rational exercise."

4.1.3 Lessons Beyond the Airline Industry

There are several broader lessons to be learned from this history of airline overbooking. One is that it takes time for customers to get used to and accept overbooking practices, and providers in turn have to learn how to develop strategies and operational practices that make overbooking as painless as possible for customers. In the airline industry, this process took decades to develop. A second lesson is that some seemingly fanciful techniques—in particular the oversale auction—can in fact prove to be surprisingly popular and effective in practice, which serves as a caution for those who are quick to criticize such innovations. Finally, while there is no denying that overbooking is a well-developed and refined practice in the airline and hotel industries, it nevertheless remains a primary source of dissatisfaction for customers. Overbooking is frequently cited in customer complaints, both to individual firms and to government regulators. So even at its best, overbooking is a somewhat awkward compromise between economic efficiency and service quality.

4.2 Static Overbooking Models

We next look at the methodology for making overbooking decisions. The simplest and most widely used methodology is based on *static overbooking models*. In static models, the dynamics of customer cancellations and new reservation requests over time are ignored. Rather, the models simply determine the maximum number of reservations to hold at the current time given estimates of cancellation rates from the current time until the day of service. This maximum number of reservations, or *overbooking limit*, is then recomputed periodically prior to service to reflect changing state and cancellation probabilities over time. While more sophisticated dynamic overbooking models have been developed and are discussed in Section 4.3, the simplicity, flexibility, and robustness of the simpler static models have made them more popular in practice.

Two types of events impact the overbooking decision—*cancellations* and *no-shows*—with the difference simply related to the timing of the events. (Again, a cancellation is a reservation that is withdrawn by a customer strictly prior to the time of service; a no-show is someone who does not cancel and does not show up at the time of service.) While both result in a situation where a reservation does not "survive" to the time of service, with a cancellation, the firm has an opportunity to possibly replace the cancelled reservation; in contrast, there is little recourse available to compensate for a no-show. Under a static model, the distinction between the two is unnecessary, since a static model assumes a static overbooking limit is set without recourse to adjust it. Thus, all that matters is the probability that a reservation survives to the time of service (the *show demand*, as it is called). In dynamic overbooking models, however, the distinction between no-shows and cancellations is important.

In airline and hotel practice, static models are used to compute overbooking limits—also called *virtual capacities* or *overbooking authorization levels* in the airline industry—which are, in turn, used as inputs to capacity-allocation models. These static overbooking models are typically re-solved periodically to account for changes in the cancellation and no-show probabilities over time, resulting in overbooking limits that vary (typically decline) over time. The current overbooking limit gives the maximum number of reservations one will accept at any time.

The situation is illustrated in Figure 4.2. The top, wide line is the overbooking limit over time. Solving a static model gives one point on this curve. Overbooking limits are set high initially because the probability of a reservation cancelling prior to the time of service or no-showing is usually higher the longer the time till service. As the time of service (T) approaches, the overbooking limits fall. At the same time, reservations are being accumulated in the system over time. The dark line in Figure 4.2 shows that with overbooking in place, the reservations in the system can exceed the capacity C, and we don't stop accepting reservations until the overbooking limit is reached. At that point reservations are rejected. The resulting show demand (demand that shows up finally) at time T is ideally close to the capacity C. The lower line shows the same trajectory of reservations without overbooking. In this case, the reservations in the system are truncated at the capacity C early on in the booking process. As a result, once reservations start to cancel and no-show, the show demand is significantly less than capacity.

4.2.1 The Binomial Model

The simplest static model is based on a *binomial model* of cancellations in which no-shows are lumped together with cancellations (that is, a no-show is treated simply as a cancellation that occurs at the day of service). The following assumptions are made:

- Customers cancel independently of one another.

- Each customer has the same probability of cancelling.

- The cancellation probability is Markovian; it depends only on the time remaining to service and is independent of the age of the reservation.

Let t denote the time remaining until service, C denote the physical capacity, y denote the number of reservations on hand, and q the probability that a reservation currently on hand shows up at the time of service ($1 - q$ is the probability that customers cancel prior to the time

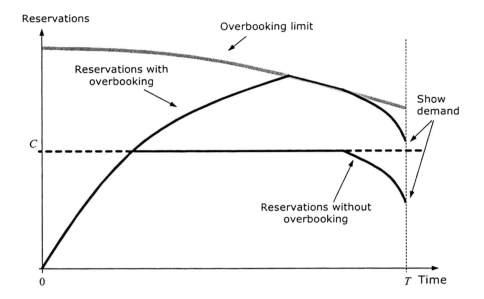

Figure 4.2. Illustration of overbooking limits and reservations over time.

of service). Note that q is really a function of the time remaining, since in general the more time remaining the more likely it is that customers cancel before the time of service. However, to keep the notation simple we suppress the dependence of q on t. Also, in practice estimates of q may be based on the ratio of show demand to reservations on hand (the net bookings) rather than on individual customer cancellation rates; this approach is discussed further in Section 4.3.2.

Under the assumptions stated above, the number of customers who show up at the time of service given there are y reservations on hand, denoted $Z(y)$ (the *show demand*), is binomially distributed with p.m.f.:

$$\begin{aligned} P_y(z) &= P(Z(y) = z) \\ &= \binom{y}{z} q^z (1-q)^{y-z}, \quad z = 0, 1, \ldots, y, \end{aligned} \quad (4.1)$$

and with c.d.f.:

$$\begin{aligned} F_y(z) &= P(Z(y) \leq z) \\ &= \sum_{k=0}^{z} \binom{y}{k} q^k (1-q)^{y-k}, \end{aligned} \quad (4.2)$$

with mean $E[Z(y)] = qy$ and variance $Var(Z(y)) = yq(1-q)$. It is convenient to work with the complement of the distribution F_y, denoted by $\bar{F}_y$, which is defined by

$$\bar{F}_y(z) = P(Z(y) > z) = 1 - F_y(z).$$

Several airline industry studies have validated this binomial model of cancellations. For example, in one of the earliest investigations of overbooking, Thompson [508] considers data from 59 flights from Auckland to Sydney operated by Tasman Empire Airways. He eliminated groups of six or more since they exhibited much lower cancellation rates and although rare (11 total booking on the 59 flights), can significantly distort the cancellation rate on the flights involved. Parties of six or fewer constituted 99.6% of all bookings; 81% of the remaining were singles; 15% were paired and 4% were parties of three to six. (See Table 4.4.) While the results showed that group-cancellation behavior does invalidate the binomial model for certain cabins on certain flights, overall he concluded that the binomial model adequately fits the data. (Group-cancellation effects are discussed further in Section 4.2.4.)

4.2.1.1   Overbooking Based on Service-Level Criteria

One measure of service is the probability of oversale at the time of service, which we call the *Type 1 service level*. We assume the firm uses an *overbooking-limit policy* to control the number of reservations that are accepted. The overbooking limit is denoted x. In other words, we assume the firm continues to accept reservations as long as the number of reservations on hand y is less than x and stops accepting reservations once $y = x$.[3]

The Type 1 service level is denoted $s_1(x)$ and is given by

$$s_1(x) = \bar{F}_x(C) = P(Z(x) > C).$$

That is, if we assume that the number of reservations on hand y reaches our overbooking limit x, then $s_1(x)$ will be the probability that we have to deny service to one or more customers. Hence, setting an overbooking limit of x guarantees that the probability of oversale will not exceed $s_1(x)$.

An arguably more natural measure of service is the long-run fraction of customers who are denied service, which we call the *Type 2 service*

---

[3]Whether such a threshold policy is in any sense optimal for the dynamic decision-making problem is addressed in Section 4.3. Here we simply assume such an overbooking policy is used.

level denoted by $s_2(x)$. This fraction is given by[4]

$$s_2(x) = \frac{E[(Z(x) - C)^+]}{E[Z(x)]}$$
$$= \frac{\sum_{k=C+1}^{x}(k - C)P_x(k)}{x(1 - q)}.$$

Setting an overbooking limit of x ensures that, at most, a fraction $s_2(x)$ of customer will be denied service. Through some algebraic simplification, one can show that

$$s_2(x) = \bar{F}_{x-1}(C - 1) - \frac{C}{qx}\bar{F}_x(C), \tag{4.3}$$

which is a more convenient formula for computations.

Table 4.2 shows the Type 1 and Type 2 service levels for an example with $C = 150$, $q = 0.85$, and varying overbooking limit x. In practice, we first specify a service level and then numerically search for the largest booking level x^* satisfying the specified service level. The resulting x^* is the overbooking limit. The quantity $x^* - C$ (the excess over capacity) is referred to as the *overbooking pad*.

Example 4.1 Suppose we want no more than 0.1% of customers to be denied service and our capacity is $C = 150$ and $q = 0.85$. From Table 4.2, we should accept at most 168 reservations ($x^* = 168$), since this is the largest value for which $s_2(x)$ less than .001, (though 169 has a service level only slightly over the standard and might be a candidate as well). Reservations would then be accepted as long as the number of bookings on hand is less than 168. The overbooking pad would be $168 - 150 = 18$.

Note that if we do not receive at least x^* requests for reservations, the service levels will in fact be higher than $s_1(x^*)$ and $s_2(x^*)$. In other words, these measures predict the service level for instances in which demand exceeds x^*, but the service level will be higher if demand is strictly less than x^*. So effectively, we are considering a worst-case service level (demand exceeding the overbooking limit) rather than an average-case service level.

---

[4] As a technical aside, note that one may be tempted to define the Type 2 service level as

$$E\left[\frac{(Z(x) - C)^+}{Z(x)}\right],$$

the average fraction denied service, rather than by (4.3). This is wrong, however, because it does not account for the varying number of customers served. For example, if $C = 100$, then it would count a day in which $Z(x) = 1$ and a day in which $Z(x) = 100$ equally as two days with denied service fractions of zero, when in reality the second day represents 100 times as many customers. The renewal-reward theorem leading to (4.3) provides the correct measure of the long-run fraction of customers who are denied service.

Table 4.2. Binomial and normal approximation overbooking probabilities with $C = 150$, $q = 0.85$.

	Binomial Model			Normal Approximation	
x	$\bar{F}_{x-1}(C-1)$	$s_1(x)$	$s_2(x)$	$s_1(x)$	$s_2(x)$
160	0.00021	0.00019	0.00000	0.00097	0.00001
161	0.00053	0.00048	0.00001	0.00185	0.00002
162	0.00122	0.00111	0.00001	0.00340	0.00003
163	0.00262	0.00239	0.00003	0.00601	0.00006
164	0.00523	0.00480	0.00006	0.01022	0.00011
165	0.00979	0.00904	0.00012	0.01676	0.00020
166	0.01726	0.01603	0.00022	0.02652	0.00033
167	0.02883	0.02691	0.00039	0.04053	0.00053
168	0.04577	0.04294	0.00066	0.05989	0.00084
169	0.06935	0.06539	0.00107	0.08566	0.00127
170	0.10062	0.09533	0.00166	0.11873	0.00188
171	0.14025	0.13351	0.00247	0.15966	0.00270
172	0.18838	0.18015	0.00355	0.20855	0.00377
173	0.24449	0.23484	0.00494	0.26496	0.00514
174	0.30743	0.29654	0.00668	0.32785	0.00685
175	0.37549	0.36365	0.00879	0.39565	0.00891
176	0.44654	0.43411	0.01127	0.46635	0.01134
177	0.51828	0.50565	0.01414	0.53773	0.01415
178	0.58842	0.57600	0.01736	0.60753	0.01732
179	0.65493	0.64309	0.02093	0.67366	0.02084
180	0.71616	0.70520	0.02479	0.73442	0.02467
181	0.77096	0.76110	0.02891	0.78856	0.02877
182	0.81870	0.81006	0.03325	0.83539	0.03310
183	0.85919	0.85182	0.03776	0.87472	0.03761
184	0.89270	0.88657	0.04241	0.90681	0.04227
185	0.91975	0.91477	0.04715	0.93225	0.04703

The average service level is, however, easy to calculate for a given distribution of demand. To illustrate, consider the Type 2 service level. Let the random variable D denote the demand (unrestricted by capacity). Then by the renewal-reward theorem, the average Type 2 service for an overbooking level x, denoted $\bar{s}_2(x)$, is given by

$$\bar{s}_2(x) = \frac{E[\min\{D,x\} s_2(\min\{D,x\})]}{E[\min\{D,x\}]}$$
$$= \frac{\sum_{y=0}^{x-1} y P(D=y) s_2(y) + x P(D \geq x) s_2(x)}{\sum_{y=0}^{x-1} y P(D=y) + x P(D \geq x)}$$

One then searches for the largest value of x that provides an average service level $\bar{s}_2(x)$ that is within a given limit.

The problem with using average service levels is that customers who make reservations only on congested days will experience service levels closer to $s_2(x)$ than to $\bar{s}_2(x)$. This is a form of the *inspection paradox* of probability theory, in which customers who book only on busy days experience worse-than-average service. Thus, a service standard based on $s_2(x)$ is often justified since it guarantees that all customers, regardless of their patterns of usage, will experience at least the target service standard.

4.2.1.2 Overbooking Based on Economic Criteria

An alternative to setting overbooking limits based on service levels is to use an economic criterion. This approach requires an estimate of the revenue loss from not accepting additional reservations and an estimate of the cost of denied service. We first develop the details of the economic model and then discuss some of the issues involved in estimating the revenue loss and cost inputs.

Model and Basic Results Suppose z customers show up on the day of service (the show demand), and let $c(z)$ denote the denied-service cost. We shall assume $c(z)$ is an increasing convex function of z. For example, a common assumption in practice is that each denied-service costs the firm a constant marginal amount h, in which case

$$c(z) = h(z - C)^+. \tag{4.4}$$

An arguably more realistic assumption is to assume strictly increasing marginal costs, reflecting the need to offer higher levels of compensation (or incur higher goodwill costs) as each additional customer is denied service.

Let p denote the marginal revenue generated by accepting an additional reservation. One could also allow this marginal revenue to vary, but it is a common simplification in practice to consider it fixed. (We discuss this issue further below.) Then the total expected profit from having y reservations on hand is given by

$$V(y) = py - E[c(Z(y))], \tag{4.5}$$

where, as before, the random variable $Z(y)$ denotes the number of customers who show up on the day of service out of y reservations. One can show for the binomial model that if $c(\cdot)$ is convex, then $V(\cdot)$ is concave,[5] in which case, since $V(y)$ is concave, it follows that it is optimal to accept the yth reservation as long as the marginal profit $\Delta V(y) = V(y) - V(y-1)$

---

[5]This follows from stochastic convexity arguments; see Appendix B for a discussion.

Overbooking 145

is positive and to continue accepting reservations until this marginal profit turns negative. Thus, the optimal booking limit x^* is the largest value of x satisfying

$$\Delta V(x) = E[c(Z(x))] - E[c(Z(x-1))] \leq p.$$

For the binomial model with constant marginal costs and parameter q, this condition reduces to

$$hqP(Z(x-1) \geq C) \leq p. \qquad (4.6)$$

This expression can be argued intuitively by noting that when we accept the xth reservation, we incur a marginal denied-boarding penalty of h if and only if (1) the current reservations on hand consume all the capacity ($Z(x-1) \geq C$), and (2) the xth customer shows up. The left-hand side of (4.6) is simply the marginal penalty multiplied by the probability of this event or equivalently the expected marginal cost. Then x^* is the largest value of x for which the expected marginal cost is less than the marginal revenue.

We can express (4.6) as

$$\bar{F}_{x-1}(C-1) \leq \frac{p}{qh}. \qquad (4.7)$$

Note that this is equivalent to setting a fixed Type 1 service level for a capacity of $C - 1$. For large capacities C, $\bar{F}_x(C-1) \approx \bar{F}_x(C)$, so using economic criteria with constant marginal costs corresponds approximately to specifying a particular Type 1 service level. This fact provides one justification for using Type 1 service levels.

To illustrate (4.7), consider the following example:

Example 4.2 Suppose $C = 150$, $q = 0.85$, the overbooking cost is $h = \$500$, and the marginal revenue is $p = \$100$. Then $p/qh = 0.235$. From Table 4.2 and (4.7), we see that the optimal overbooking limit is then $x^* = 172$ since this is the largest value of x for which $\bar{F}_{x-1}(C-1) \leq 0.235$. The overbooking pad is then $172 - 150 = 22$.

Cost and Revenue Parameters While overbooking based on economic criteria is conceptually appealing, it requires good estimates of the marginal revenues and costs. The marginal revenue is usually the easier of the two to determine. If there is only one class, the marginal revenue is simply the common price but with multiple classes determining the marginal revenue is more complex. A heuristic approach is to use the weighted-average revenue. However, as shown in Chapter 2, the marginal revenue produced by an additional unit of capacity is not, in general, equal to the weighted-average revenue. Moreover, the marginal revenue is typically decreasing in the available capacity, so the linear marginal revenue assumption is violated. Both these factors complicate the estimation of marginal revenue in practice.

Estimating the denied-boarding cost involves other complications. Some elements of this cost are clear: in particular, any refund of the purchase price or monetary compensation or both is easy to quantify in most cases. But if auctions are used to determine compensation, then this compensation must be estimated. Vouchers for free service in the future require a more careful accounting of the actual cost of providing the service, as this is often less than the face value of the voucher.

Most difficult of all to quantify is the goodwill loss of upsetting a customer. In principle, this can be taken to be equal to the discounted potential revenue stream of future purchases from the customer (the so-called lifetime value of the customer). This is rather difficult quantify, but it is usually worth an attempt to make this calculation to at least get the correct order of magnitude of goodwill losses.

One useful idea to get around these estimation problems is to compute imputed costs based on subjective service-level criteria rather than specifying a denied-service cost a priori. To obtain an imputed cost h from a given overbooking limit x^* set according to Type 2 service levels, one simply rearranges (4.7) to obtain

$$h = \frac{p}{q\bar{F}_{x^*-1}(C-1)}.$$

The following example illustrates the use of this formula:

Example 4.3 We saw above that if the service standard is 0.1% ($s_2(y) \leq 0.001$), we should accept at most $x^* = 168$ reservations. Since $\bar{F}_{x^*-1}(C-1) = 0.02883$, if the marginal revenue is $p = \$100$, this implies an imputed cost of denied service of

$$h = \frac{100}{0.85 \times 0.02883} = \$4,080.$$

The figure looks rather large relative to the $100 revenue, so one might question if a 0.1% Type 2 service level is economically justified.

Often, these imputed cost numbers provide useful feedback, since they translate service levels, which are tangible and somewhat easier to specify, into economic penalties, which most people find harder to quantify. The economic costs, in turn, serve as a useful "sanity check" on the reasonableness of a given service level by giving the magnitude of the implied costs.

4.2.2 Static-Model Approximations

While the binomial model is quite simple, it is often desirable to have simpler, closed-form expressions for the overbooking limits. We next look briefly at such approximations.

4.2.2.1 Deterministic Approximation

The *deterministic approximation* simply sets the overbooking limit so that the average show demand is exactly equal to the capacity; namely,

$$x^* = \frac{C}{q}.$$

As simplistic as this approximation is, we have seen several RM implementations that use it. The approximation is not completely unjustified, however, as illustrated by the following example:

Example 4.4 Consider our continuing example where $C = 150$ and $q = 0.85$. The deterministic approximation yields an overbooking limit of $x^* = C/q = 176.5$. From Table 4.2, one can see that both service measures $s_1(y)$ and $s_2(y)$ begin to change rapidly in the range of $y = 170$ to $y = 180$, which is approximately centered around the deterministic level 176. So lacking detailed service standards or cost information, a value around the deterministic level is not an unreasonable heuristic to use.

4.2.2.2 Normal Approximation

In practical implementations, it is common to use a continuous approximation to the binomial model to simplify computations. One popular choice is the *normal approximation*, in which the distribution $F_x(\cdot)$ is replaced by the normal distribution with mean μ_x and variance σ_x^2 chosen to match the binomial, viz.,

$$\mu_x = xq$$
$$\sigma_x^2 = xq(1-q).$$

The Type 1 service level is then approximated by

$$s_1(x) \approx 1 - \Phi(z_x),$$

where

$$z_x = \frac{C - \mu_x}{\sigma_x}$$

and $\Phi(\cdot)$ is the c.d.f. of the standard normal distribution.

The Type 2 service level is then approximated by[6]

$$s_2(x) \approx \frac{\sigma_x}{\mu_x}\left[\phi(z_x) - z_x(1 - \Phi(z_x))\right]. \tag{4.8}$$

Table 4.2 shows the estimates produced by the normal approximation for our example with $C = 150$ and $q = 0.85$. As can be seen, they are reasonably close to the values of the binomial model.

The economic overbooking limit (4.7) for the constant marginal-cost function (4.4) is approximated by choosing x^* to satisfy

$$\Phi_{x^*}(C) = 1 - \frac{p}{qh}. \tag{4.9}$$

4.2.2.3 Gram-Charlier Series Approximation

The *Gram-Charlier series* improves on the normal approximation of the binomial distribution by allowing for skewness of the distribution. The standardized density function for this distribution is

$$f(z) = \frac{1}{\sqrt{2\pi}} e^{\frac{-z^2}{2}} \left(1 + \frac{1}{6}\beta(z^3 - 3z)\right),$$

where

$$\beta^2 = \frac{E^2[(Z - E[Z])^3]}{E^3[(Z - E[Z])^2]}$$

is the squared coefficient of skewness. If $\beta = 0$, this reduces to the standard normal distribution. For the binomial model, the coefficient of skewness is given by

$$\beta = \frac{1 - q}{\sqrt{q(1-q)x}}.$$

Letting $z_x = \frac{C - \mu_x}{\sigma_x}$ denote the standardized booking level as before, the fraction of overbooked passengers is approximated by

$$s_2(x) \approx \frac{\sigma_x}{\mu_x}\left[1 + \frac{1}{6}z_x\beta\phi(z_x) - z_x(1 - \Phi(z_x))\right], \tag{4.10}$$

where, as above, $\phi(z)$ and $\Phi(z)$ are the standard normal density and distribution, respectively.

---

[6]This follows from the fact that if Z is a normal random variable with mean μ and variance σ^2, then

$$E[(Z - C)^+] = \sigma\left(\phi(z) - z(1 - \Phi(z))\right),$$

where $z = (C - \mu)/\sigma$.

Table 4.3 shows some numerical comparisons of the normal and Gram-Charlier approximations of the binomial model. In general, the normal approximation tends to overestimate the fraction of denied boardings. The Gram-Charlier approximation also overestimates, but less so.

Table 4.3. Comparison of normal and Gram-Charlier (G-C) approximations.[a]

Capacity	Booking Limit x	Cancellation q	Type-2 Service Level		
			Exact	Normal	G-C Series
36	37	0.05	0.0047	0.0064	0.0051
		0.10	0.0006	0.0016	0.0010
	41	0.20	0.0030	0.0044	0.0034
		0.25	0.0006	0.0010	0.0007
	44	0.25	0.0058	0.0073	0.0052
		0.30	0.0013	0.0018	0.0014
		0.35	0.0002	0.0004	0.0003
12	13	0.25	0.0024	0.0059	0.0037
		0.30	0.0011	0.0026	0.0018
	14	0.35	0.0025	0.0046	0.0034
		0.40	0.0011	0.0018	0.0016
	15	0.40	0.0036	0.0049	0.0044

[a] *Note:* Data as reported by Taylor [504].

4.2.3 Customer Class Mix

One important practical issue that arises in overbooking is that different classes may have quite different cancellation rates. For example, in the airline case, full-coach customers often have no cancellation penalty, while discount-class customers typically incur a significant fee for cancelling a reservation. As a result, the two classes exhibit very different rates of cancellation. Thus, the cancellation rate observed in a collection of reservations may be highly dependent on the mix of classes.

Exact methods to handle this class-mix problem involve keeping track of the inventory of each class as a separate state variable and then making overbooking decisions based on this complete vector of state variables. Such an approach is described in detail for a multiclass model in Section 4.5 below.

The difficulty with such exact approaches, however, is that they result in significantly more complicated overbooking models and methodology. As a result, most often in practice, one of several heuristic approaches is used to account for customer class mix.

The most common practice is to use a cancellation probability that is empirically estimated for each resource separately. In this way, one can capture at least the historical mix of customer segments booked on each resource. Another approach is to estimate the cancellation probability for each class and then use estimates of the class mix on each resource to construct a weighted average cancellation probability for each resource. Compared with straight estimation of the resource-level cancellation rate, this method has the advantage of reducing the variance in the estimates and allows for a more rapid adjustment of cancellation rates as the class mix changes over time.

4.2.4 Group Cancellations

The presence of groups also has an important effect on cancellation models in practice. If a group decides to cancel, then all reservations are cancelled simultaneously. The resulting positive correlation in cancellations increases the variance of the show demand. When dealing with large numbers of reservations, it is often possible to ignore the effect of groups, but with small numbers of reservations, group effects can result in significant deviations from the binomial model.

To gain some sense of the presence of groups in RM bookings, Table 4.4 provides an empirical distribution of group sizes over approximately half a million airline reservations. About half of the reservations are individual reservations, while the other half are from groups of two or more.

Table 4.4. Empirical distribution of group sizes.[a]

Number in Party	Count of Passengers	Percentage	Cumulative Percentage
1	198,056	45.1	45.1
2	114,418	26.0	71.1
3	28,641	6.5	77.6
4	25,688	5.8	83.5
5	17,930	4.1	87.6
6	7,134	1.6	89.2
7	2,135	0.5	89.7
8	2,896	0.7	90.3
9	1,125	0.3	90.6
10	4,960	1.1	91.7
>10	36,375	8.3	100.0
Total	439,358	100.0	

[a] *Note:* Data reported by Rothstein and Stone [446].

One simple technique used in practice to adjust for group size is to simply inflate the variance of the show demand by a factor that accounts for group size. For example, if we are using the normal approximation to the binomial model as described in Section 4.2.2.2, then the estimate of the mean show demand, μ_x, is unchanged but the variance estimate, σ_x^2, is modified as follows:

$$\mu_x = xq$$
$$\sigma_x^2 = kxq(1-q),$$

where k is a factor to account for group cancellations—for example, the average group size.[7]

A more refined technique for adjusting for groups is based on moment-generating functions. Recall that the moment-generating function of a random variable Z is $\psi_Z(t) = E[e^{tZ}]$. We can then find the moments of Z using the fact that

$$E[Z^n] = \left.\frac{d^n}{dt^n}\psi_Z(t)\right|_{t=0}.$$

Let x denote the overbooking limit and x_k denote the number of groups of size k. We will assume that

$$x_k = \frac{\alpha_k x}{k},$$

where α_k is the historical fraction of reservations that are from groups of size k. As an approximation, we allow x_k to be nonintegral. Let q_k denote the probability that a group of size k survives to the time of service (called the *utilization ratio* in the airline industry). Then the moment-generating function for $Z(x)$, the number of survivals from x total reservations, is

$$\psi_{Z(x)}(t) = \prod_{k=1}^{n}(1 - q_k + q_k e^t)^{x_k},$$

from which one can find the first three central moments of show demand,

$$E[Z(x)] = \sum_{k=1}^{n} x_k k q_k$$
$$E[(Z(x) - E[Z(x)])^2] = \sum_{k=1}^{n} x_k k^2 q_k (1 - q_k)$$
$$E[(Z(x) - E[Z(x)])^3] = \sum_{k=1}^{n} x_k k^3 q_k (1 - q_k)(1 - 2q_k).$$

---

[7]Setting k equal to the average group size is obtained by assuming that all reservations are in groups of exactly size k, in which case with x reservations on hand, there are x/k groups of size k, so the variance of the show demand is $k^2(x/k)q(1-q) = kxq(1-q)$.

These three moments can then be used in the Gram-Charlier series approximation (4.10); alternatively, the first two moments can be used in the normal approximation (4.8).

4.3 Dynamic Overbooking Models

The static models do not explicitly account for the dynamics of arrivals, cancellations, and decision making over time. Here we look at models of overbooking that account for such intertemporal effects. We first look at an exact *dynamic overbooking model* and then discuss heuristic approaches.

4.3.1 Exact Approaches

The model presented here is a simplification of one due to Chatwin [109]. The state variables are time, $t = 1, \ldots, T$, and the number of reservations on hand, y. Let the value function be denoted $V_t(y)$. The terminal costs are

$$V_{T+1}(y) = \begin{cases} 0 & y \leq C \\ -c(y - C) & y > C, \end{cases} \quad (4.11)$$

where C is the fixed capacity and $c(\cdot)$ is a convex cost function penalizing denied service. The revenue received from accepting a new reservation in period t is denoted $p(t) \geq 0$. If a reservation is cancelled in period t, the firm pays a refund of $r(t) \geq 0$. (Note that in this model the refund depends only on the time the reservation is cancelled and not on the time-period in which the reservation was made; this (somewhat unrealistic) assumption is necessary to simplify the state space.)

Let $Z_t(y)$ denote the number of surviving reservations at the end of period t, so that given y reservations are on hand at the end of period t, $Z_t(y)$ is the random number surviving to the start of period $t + 1$. We assume that $Z_t(y)$ has a binomial distribution with survival probability q_t. Let D_t denote the random number of new reservation requests in period t (the demand in period t). D_t is assumed independent across time and independent of $Z_t(y)$.

The order of events in a period is as follows: (1) there are y reservations on hand, and D_t new reservation requests arrive; (2) booking decisions are made for the new reservation requests, raising the booking level to x where $y \leq x \leq y + D_t$; then (3) cancellations are observed at the end of the period. The dynamic programming recursion is then

$$\begin{aligned} v_{t+1}(x) &= E[V_{t+1}(Z_t(x)) - (x - Z_t(x))r(t)] \\ V_t(y) &= E[\max_{y \leq x \leq y + D_t} \{v_{t+1}(x) + (x - y)p(t)\}]. \end{aligned}$$

We then have the following result:

PROPOSITION 4.1 *If the denied-service cost $c(\cdot)$ is convex, then an overbooking-limit policy is optimal. That is, in each period t there exists a critical value $x^*(t)$ such that it is optimal to continue accepting new reservations until the total number of reservations on hand reaches $x^*(t)$.*

This result provides some theoretical support for the use of booking limit policies. The following proposition in turn provides sufficient conditions for optimal booking limits to be monotone in time:

PROPOSITION 4.2 *Suppose the denied-service cost $c(\cdot)$ is convex and the survival probabilities q_t, the revenues $p(t)$, and the refunds $r(t)$ satisfy*

$$q_t(p(t) - p(t+1)) + (1 - q_t)(p(t) - r(t)) \geq 0.$$

Then the optimal overbooking limit $x^*(t)$ (or greatest optimal booking limits if more than one optimal limit exists) decline with time. That is, $x^*(1) \geq x^*(2) \geq \cdots \geq x^*(T)$.

This declining-booking-limit situation corresponds to the overbooking curve shown in Figure 4.2. Note that this condition is satisfied whenever the revenues are decreasing over time t ($p(t) \geq p(t+1)$) and refunds paid in period t do not exceed the price in period t ($p(t) \geq r(t)$).[8]

Another important monotonicity result concerns how overbooking limits are affected by the magnitude of future demand. In particular, let θ be a parameter of the distribution of arrivals so that $D_t = D_t(\theta)$. Then we have the following

PROPOSITION 4.3 *Suppose the denied-service cost $c(\cdot)$ is convex and $D_t(\theta)$ is stochastically increasing in θ. Then the optimal booking limits $x^*(t)$ (or greatest optimal booking limits if more than one optimal limit exists) are nonincreasing in θ.*

---

[8]To gain some intuition for this condition, note it can be written as

$$p(t) \geq q_t p(t+1) + (1 - q_t) r(t).$$

Roughly, this can be interpreted as follows. Suppose we are willing to accept a booking in period $t+1$ in state y. Then $p(t+1)$ must exceed the opportunity cost of an additional reservation in state y in period $t+1$. Now consider state y in period t. If we accept an additional booking, we collect revenue $p(t)$. If this reservation cancels at the end of period t, we pay a refund $r(t)$; this occurs with probability $1 - q_t$, and the state (after the cancellation) returns to y. If the reservation survives to period $t+1$, it creates an opportunity cost analogous to accepting a request in period $t+1$, which by the argument above is at most $p(t+1)$ (since we accept a request in state y in period $t+1$); this occurs with probability q_t. So if the revenue $p(t)$ exceeds the "average cost" of these outcomes, then we should be willing accept an arrival in period t in state y as well. Hence, the booking limit in period t is at least as large as that in period $t+1$.

This result says that as demand to come increases (stochastically), it is better to be less aggressive in overbooking at any given point in time. The intuition is that if we have more opportunities to book seats in the future, we do not need to take as great an overbooking risk in the current period. The result also highlights the fact that the optimal overbooking limits in general do depend on future demand, which is something that the static overbooking models ignore. In particular, note that the calculation of costs in the static overbooking model effectively assumes there are no opportunities to replace cancelled reservations with new reservations.[9] Since the degenerate case of no future demand is always stochastically smaller than any nontrivial distribution of future demand, Proposition 4.3 implies that static overbooking models will produce overbooking limits that are higher than optimal.

4.3.2 Heuristic Approaches Based on Net Bookings

While dynamic overbooking models provide some nice insights, they are not used very often in practice. This is due partly to their added complexity and partly because to their similarity to the more general combined capacity control and overbooking models that we look at in Section 4.4 below.

In RM practice, the dynamics of cancellations and new reservations and arrivals are more commonly accounted for by using relative changes in bookings on hand (so-called *net bookings*) rather than customer-level cancellation probabilities as a basis for estimating cancellation rates in a static overbooking model. The idea of net bookings can be best illustrated by going back to Figure 4.2, which shows a sample of the level of booking on hand over time. The change in bookings on hand from one time-period to the next depends on both the number of cancellations and the number of new reservations that are accepted. Looking at changes in the bookings on hand gives us a measure of the net bookings. Quite often in practice, net-bookings data is in fact the only data available for use in estimating overbooking parameters.

Since net bookings reflect both cancellations and new reservations, they can be used to provide an alternative estimate of the cancellation

---

[9]More precisely, the assumption in the static model is that the show demand when the booking limit x is reached is $Z(x)$, a binomial random variable representing the number of surviving reservations out of x total reservations. Hence, show demand consists *only* of those reservations that survive from the current time until the time of service. If new reservations are accepted to replace cancelled reservations, then the show demand will be larger than $Z(x)$, which is what the dynamic model accounts for.

rate, which one can interpret as an approximation to an exact dynamic model. More precisely, one can estimate the survival "rate" q_t as the average ratio of show demand to the number of bookings on hand in time t (or to the number of peak bookings on hand if t is before the peak; see Figure 4.2.). This net-bookings approach to estimating cancellation rates is again quite prevalent in RM practice and seems to lead to better approximations of real world service levels and costs.

4.4 Combined Capacity-Control and Overbooking Models

Thus far, we have analyzed the overbooking problem in isolation without considering the interaction of overbooking decisions with capacity controls. We next look at both exact and approximate methods to model cancellations and no-shows together with the class allocations of quantity-based RM.

Incorporating no-shows or cancellations in either the static or dynamic single-resource model is not too difficult theoretically, provided one makes the following (not entirely satisfying, but analytically useful) set of assumptions:

ASSUMPTION 4.1
(i) The cancellation and no-show probabilities are the same for all customers.
(ii) Cancellations and no-shows are mutually independent across customers.
(iii) Cancellations and no-shows in any period are independent of the time the reservations on hand were accepted.
(iv) The refunds and denied-service costs are the same for all customers.

The assumptions imply that the number of no-shows and the costs incurred are only a function of the total number of reservations on hand. As a result, we need only to retain a single state variable, and the resulting dynamic programs are only slightly more complex than those presented in Chapter 2.

The most restrictive of these assumptions in practice are (i) and (iv): cancellation options and penalties are often linked directly to a class, so cancellation and no-show rates and costs can vary significantly from one class to the next. Ideally, these differences should be accounted for when making allocation decisions. However, this significantly complicates the problem, as we show below. As already mentioned, Assumption 4.1(ii) is often unrealistic because reservations from people in groups typically cancel at the same time. Assumption 4.1(iii) is less of a problem in practice and has some empirical support [508].

In most implementations, the overbooking problem is separated from the capacity-allocation problem. Often, an approximate static overbooking model can be solved that is able to relax (at least heuristically) some or all parts of Assumption 4.1. However, given Assumption 4.1, the overbooking and capacity-allocation problems can be combined exactly, as we show next.

4.4.1 Exact Methods for No-Shows Under Assumption 4.1

We first consider only no-shows and assume that there are no cancellations prior to the time of service. Let q_0 denote the probability that a customer with a reservation shows up for service ($1 - q_0$ is the no-show probability). Assumption 4.1(i) says this probability is assumed to be the same for all customers, and Assumption 4.1(iii) that it is independent of when the reservation was made.

Let $Z_i = 1$ if customer i shows up for service and $Z_i = 0$ otherwise. Given there are y reservations on hand just prior to the time of service, the number of customers who show up at time zero (the show demand), denoted $Z(y)$, is then

$$Z(y) = \sum_{i=1}^{y} Z_i,$$

and by Assumption 4.1(iii) $Z_0(y)$ is a binomial(q_0, y) random variable, with

$$P(Z_0(y) = z) = \binom{y}{z} q_0^z (1 - q_0)^{y-z}, z = 0, 1, \ldots, y.$$

By Assumption 4.1(iv), the total cost of denied service is only a function of the show demand z. Let $c(z)$ denote the overbooking cost given z. We will require that $c(z)$ be increasing and convex with $c(0) = 0$. Convexity in cost is quite natural since the marginal cost of denying service to customers tends to increase with the number denied. For example, we could have a simple linear cost h per denied customer in which case $c(z) = h(z - C)^+$ where, as before, C is the capacity.

Given this no-show model, $V_0(y)$, the expected cost of service given that there are y reservations on hand at the time of service, is given by

$$V_0(y) = E[-c(Z(y))], y \geq 0. \tag{4.12}$$

Stochastic convexity arguments (see Appendix B) show that $V_0(y)$ is concave in y if $c(\cdot)$ is convex. The above expression then replaces the boundary conditions of the dynamic program for the static and dynamic models.

4.4.1.1 Static Model

Consider the static model of Section 2.2, where the classes are ordered by prices $p_1 > p_2 > \cdots > p_n$, and we assumed classes arrive in the order of lowest to highest revenue. Classes and stages are indexed by j. The state variable is now defined to be the number of reservations on hand y rather than the remaining capacity x as in Section 2.2.

The Bellman equation (2.3) for the static model is then modified to account for no-shows as follows

$$V_j(y) = E\left[\max_{0 \leq u \leq D_j} \{p_j u + V_{j-1}(y+u)\}\right], \tag{4.13}$$

with boundary conditions (4.12) and $V_j(0) = 0$ for all j, where $V_j(y)$ is now interpreted as the expected *net benefit* (expected revenue minus the expected terminal cost) of operating the system from stage j onward given that there are y reservations on hand.[10]

Given the concavity of $V_0(y)$, the same argument as in Proposition 2.1 from Chapter 2 shows that the value function $V_j(y)$ in (4.13) is concave in y for all j and y. Since there is no hard capacity constraint in this case, it is more meaningful to express the optimal policy in terms of booking limits. The optimal nested booking limits are given by

$$b_j^* = \min\{y \geq 0 : p_j < \Delta V_{j-1}(y)\}, \quad j = 1, \ldots, n-1,$$

where $\Delta V_{j-1}(y) \doteq V_{j-1}(y) - V_{j-1}(y+1)$ now has the interpretation as the marginal cost of holding another reservation in stage $j-1$ and is increasing in y. It is then optimal to accept class j if and only if the number of reservations on hand y is strictly less than b_j^*.

4.4.1.2 Dynamic Model

Similarly, the optimality equations (2.17) for the dynamic model of Section 2.5 are modified to account for no-shows as follows:

$$V_t(y) = E\left[\max_{u \in \{0,1\}} \{R(t)u + V_{t+1}(y+u)\}\right], \tag{4.14}$$

---

[10]Note that in this case $V_j(y)$ is a decreasing function of y, since the more reservations we have on hand now, the fewer the future opportunities to collect revenue or the higher the expected future terminal costs.

where, recall, $R(t)$ is the random revenue in period t, equal to p_j with probability $\lambda_j(t)$. The boundary conditions are

$$V_{T+1}(y) = E[-c(Z(y))], \quad y \geq 0 \tag{4.15}$$

and $V_t(0) = 0$ for all t. It is optimal to accept an arrival of class j if and only if

$$p_j \geq \Delta V_{t+1}(y),$$

where again $\Delta V_{t+1}(y) = V_{t+1}(y) - V_{t+1}(y+1)$ is interpreted as the marginal cost of accepting another reservation.

Note that under this model, one can always justify accepting a sufficiently high revenue p_j, provided the marginal cost $\Delta V_{t+1}(y)$ is finite. This makes perfect economic sense since we should in principle be willing to accept an almost certain denied-service cost if some customer is willing to pay enough to compensate us for this cost. For example, if the overbooking cost is linear of the form $c(z) = h(z - C)^+$, then the marginal cost is never more than h, so any request with revenue greater than h will always be accepted.

This property of not having an explicit limit on the number of reservations (rather, just an economic limit) has been called *infinite overbooking* by some in the airline industry, since it is in sharp contrast to the usual practice of setting a hard overbooking limit. Also, it highlights the potential suboptimality of using fixed overbooking limits.

4.4.2 Class-Dependent No-Show Refunds

If one relaxes one or more parts of Assumption 4.1, then the problem becomes considerably more difficult. The difficulty stems from the fact that if no-show rates or costs depend on customer class or the time of purchase or both, then one must retain a state variable for each class or each time-period or both. The resulting increase in dimensionality of the dynamic program makes it essentially intractable. However, it turns out that class-dependent refunds can be readily incorporated through an appropriate change of variable.

Suppose customers of class j who no-show in period zero are given a refund h_{j0} that is strictly less than the revenue we receive from them, $h_{j0} < p_j$. However, all other assumptions in Assumption 4.1 hold. A naive formulation of this class-dependent refund feature would require keeping track of each class separately so that refunds can be properly awarded at the time of service. However, whether a given customer no-shows is completely independent of all other decisions and events in the system. Thus, one can in fact charge for the expected refund at the time the reservation is accepted rather than at the time of service, with no

Overbooking

resulting difference in total expected revenues and costs. (This is merely a bookkeeping change.)

More precisely, if we accept a reservation from a customer of class j, it will yield a reduced revenue of

$$\hat{p}_j = p_j - (1 - q_0)h_{j0}$$

independent of everything else in the system. Therefore, we simply use $\hat{p}_j$ in place of p_j in either (4.13) or (4.14) to modify the problem formulation. Note, however, that depending on the refund, the ordering of $\hat{p}_j$ may be different from the ordering of p_j. For example, customers in the high revenue class who receive a full refund if they no-show may yield a lower net revenue $\hat{p}_j$ than customers of a lower class who get no refund if they no-show. Since the nested protection levels are now based on the net revenue $\hat{p}_j$ rather than the gross revenue p_j, the optimal policy may reject the high gross-revenue customer in favor of the high net-revenue one.

4.4.3 Exact Methods for Cancellations Under Assumption 4.1

Cancellations complicate the dynamic program a little more than no-shows, but they are still quite manageable under Assumption 4.1. Again, we look at the static and dynamic models in turn.

4.4.3.1 Static Model

Let q_j denote the probability that a reservation in the system at the start of stage j survives to stage $j-1$ (recall that in the single-resource static model of Section 2.2 stages go from N to 0). So $1 - q_j$ is the probability that a reservation cancels in stage j. By Assumption 4.1 (i), (ii), and (iii), these probabilities are the same and independent for all customers as well as the age of their reservations. Let $Z_j(y)$ denote the number of reservations that survive stage j given that there are y reservations on hand in stage j (so $y - Z_j(y)$ are the number of cancellations in stage j).

The Bellman equation (2.3) for the static model is then modified to account for cancellations as follows

$$V_j(y) = E\left[\max_{0 \leq u \leq D_j} \{p_j u + H_{j-1}(y+u)\}\right], \qquad (4.16)$$

with boundary conditions (4.12), where

$$H_{j-1}(y) = E[V_{j-1}(Z_j(y))] = \sum_{z=0}^{y} \binom{y}{z} q_j^z (1-q_j)^{y-z} V_{j-1}(z)$$

is the expected value function after cancellations in stage j. Again, stochastic convexity arguments show that if $V_{j-1}(z)$ is concave in z, then $H_{j-1}(y)$ is concave in y, and hence a modification of the argument in Proposition 2.1 shows that the value function $V_j(y)$ defined by (4.16) is concave in y.

Nested booking limits are optimal with the optimal booking limits given by

$$b_j^* = \min\{y \geq 0 : p_j < H_{j-1}(y) - H_{j-1}(y+1)\}, \quad j = 1, \ldots, n-1,$$

where we accept class j if and only if the number of reservations on hand y is strictly less than b_j^*.

4.4.3.2 Dynamic Model

Let q_t denote the probability that a reservation in the system at the start of period t survives to period $t+1$, so by Assumption 4.1 (i), (ii), and (iii) the number of surviving reservations $Z_t(y)$ is again binomial. The optimality equations for the dynamic model with cancellations become

$$V_t(y) = E\left[\max_{u \in \{0,1\}} \{R(t)u + H_{t+1}(y+u)\}\right], \qquad (4.17)$$

where

$$H_{t+1}(y) = E[V_{t+1}(Z_t(y))] = \sum_{z=0}^{y} \binom{y}{z} q_t^z (1-q_t)^{1-z} V_{t+1}(z)$$

is the expected value function after cancellations in period t. The boundary conditions are given by (4.15).

As a result, it is optimal to accept an arrival of class j if and only if

$$p_j \geq H_{t+1}(y) - H_{t+1}(y+1).$$

4.4.4 Class-Dependent Cancellation Refunds

Again, relaxing the fact that cancellation rates or costs depend on the class or the time of purchase (or both) requires expanding the state space and is not practical if one has more than two classes. However, as with no-shows, a change of accounting can be used to allow for class-dependent refunds. We illustrate the idea for the dynamic model only, but a similar idea applies for the static model.

Suppose a customer of class j who cancels in period t is given a refund h_{jt} which is strictly less than the revenue we receive, $h_{jt} < p_j$. All other assumptions in Assumption 4.1 hold. As in the no-show case, one can charge for the expected refund at the time the reservation is accepted rather than at the time of service, with no resulting difference in total expected revenues and costs.

This is accomplished as follows. Let $G_t(j)$ denote the expected refund given to a class j reservation from period t through to the time of service. We can solve for $G_t(j)$ recursively using

$$G_t(j) = (1 - q_t)h_{jt} + q_t G_{t-1}(j), \quad t = 1, 2, \ldots, T$$

with boundary condition

$$G_0(j) = (1 - q_0)h_{j0}.$$

We then form the reduced revenue

$$\hat{p}_{jt} = p_j - G_t(j)$$

and simply use $\hat{p}_{jt}$ in place of p_j in (4.17) to modify the problem formulation. Note, as in the case of no-show refunds, that the ordering of $\hat{p}_{jt}$ may be different from the ordering of p_j.

Again, the key practical insight here is that the reduced-revenue $\hat{p}_{jt}$ should be used in evaluating the economic benefit of accepting a class j customer—not the gross-revenue p_j. This is because even if a class gives a higher current revenue, much of that revenue may be forfeited on average, so the net benefit of accepting it can be quite different from the gross revenue.

4.5 Substitutable Capacity

We next look at an overbooking problem with multiple classes and multiple resources (types of capacity). Here, we assume classes correspond to different products a customer can purchase, while resources are physically different, albeit related, types of capacity. The multiple capacity types may be used to satisfy the demand of a given class—or multiple classes may use a single capacity type. A prominent example is overbooking jointly in multiple cabins of an aircraft (coach and business class), where the first-class cabin serves as a substitute capacity if the coach cabin is oversold. Another example is overbooking on back-to-back scheduled flights between a pair of cities, where customers booked on an early flight can be served (perhaps at a cost) on a later flight. Hotels with multiple room types and car rental fleets with multiple car types are further examples. The case of a single resource with multiple classes can be applied to a traditional single-resource problem to control overbooking when cancellation rates differ across classes (for example, to determine separate overbooking limits for each class based on the joint vector of reservations on hand for each class). All these problems share the feature that capacity of a different resource (such as a later flight, an alternative room type, or a vehicle type) can serve as a substitute in the case of oversales.

In the presence of such substitution effects, the overbooking decisions for the resources are related. For example, we might tolerate a higher level of overbooking in the coach cabin of an aircraft if we know that the number of bookings in the first-class cabin is low, and conversely we would be more conservative about overbooking the coach cabin if the first-class cabin was fully booked. Therefore, the key question in such situations is how to jointly determine optimal overbooking levels.

4.5.1 Model and Formulation

One approach to joint overbooking across resources is to approximate this problem as a two-period optimization problem. In the first period (the *reservation period*), we assume reservations are accepted given only probabilistic knowledge of cancellations. In the second period (the *service period*), cancellations are realized, and surviving customers are assigned to the various resources to maximize the net benefit of assignments (for example, minimize downgrading penalties). This gives us essentially a multiclass version of the traditional static overbooking model.

Let n denoted the number of classes and m denote the number of resources. In the reservation period, assume that for each class j we currently have y_j reservations on hand. (This is the current "state.") The decision variables are the maximum number of reservations we are willing to hold after the reservation period is over, denoted by x_j, $j = 1, \ldots, n$. These decision variables have to satisfy $x_j \geq y_j$ for all $j = 1, \ldots, n$, since the maximum number of reservations after the reservation period must be at least as large as the number at the start of the reservation period. (There are no cancellations *during* the reservation period.)

In the service period, cancellations and no-shows are realized, and all remaining customers are either assigned to one of m resources, indexed by i, or they are denied service. This assignment of customers to resources is modeled as a deterministic network-flow problem. The following notation is used:

h_{ji}—The net benefit of assigning a customer of class j to resource i during service period (objective function coefficients).

C_i— The capacity of resource i.

z_j—The number of customers of class j that show up at the service period (number of survivals).

z_{ji}—The number of customers among the z_j who showed up assigned to resource i during the service period (decision variables).

One can add a virtual resource, type $i = 0$, to account for denied service. This resource has finite but very high capacity, and assigning a customer to it means that the customer is denied service. The assignment variables corresponding to the virtual resource are z_{j0}, and the objective function coefficients h_{j0} take into account the loss-of-goodwill cost incurred by denying service to customers of reservation class j, as well as any other direct compensation costs.

Let $\mathbf{z}$ denote the n-vector of show demand and $\mathbf{C}$ denote the $(m+1)$-vector of resource capacities (including the denied-service, virtual-resource capacity, C_0). The maximum value obtained during the service period is denoted by $V(\mathbf{z}, \mathbf{C})$. The allocation problem can be represented as

$$(TP) \quad V(\mathbf{z}, \mathbf{C}) = \max \sum_{j=1}^{n} \sum_{i=0}^{m} h_{ji} z_{ji}$$

$$\text{s.t.} \quad \sum_{i=0}^{m} z_{ji} = z_j \quad j = 1, \ldots, n \quad (4.18)$$

$$\sum_{j=1}^{n} z_{ji} \leq C_i \quad i = 0, 1, \ldots, m \quad (4.19)$$

$$z_{ji} \geq 0; \quad j = 1, \ldots, n; \quad i = 0, 1, \ldots, m.$$

(TP) is a transportation problem in which the supplies are customers requesting service and demands are the available capacities. Let the dual variables associated with constraints (4.18) and (4.19) in (TP) be $\boldsymbol{\mu} = (\mu_1, \ldots, \mu_n)$ and $\boldsymbol{\lambda} = (\lambda_0, \ldots, \lambda_m)$, respectively.

To formulate the reservation-period problem, let Z_j be the show demand for customers from class j. This show demand is, of course, a function of the number of accepted reservations, so $Z_j = Z_j(x_j)$. We let q_j denote the probability that a class j reservation shows up in the service period. Several models can be used for this show demand, most naturally the binomial model discussed in Section 4.2.1. But it is useful theoretically and computationally to approximate the binomial with a Poisson distribution, in which case, booking limits can be treated as continuous variables.

Let $\mathbf{y} = (y_1, \ldots, y_n)$, $\mathbf{x} = (x_1, \ldots, x_n)$, and $\mathbf{Z}(\mathbf{x}) = (Z_1(x_1), \ldots, Z_n(x_n))$. Let the price and refund (on cancellation) vectors for the classes be denoted by $\mathbf{p}$ and $\mathbf{s}$, where we assume $\mathbf{p} \geq \mathbf{s}$. Finally, let $G(\mathbf{x})$ be the expected value of future revenues and costs (*net revenue*) as a function of the final overbooking level, $\mathbf{x}$.

The reservation period problem is, then,

$$\max_{\mathbf{x} \geq \mathbf{y}} G(\mathbf{x}), \qquad (4.20)$$

where

$$G(\mathbf{x}) = \mathbf{p}^\top (\mathbf{x} - \mathbf{y}) - E[\mathbf{s}^\top (\mathbf{x} - \mathbf{Z}(\mathbf{x}))] - E[V(Z(\mathbf{x}), \mathbf{C})] \qquad (4.21)$$

and the expectation above is with respect to the random vector of survivals $Z(\mathbf{x})$.

4.5.2 Joint Optimal Overbooking Levels

The following proposition shows how the overbooking levels for the classes are related if show demand is modeled as a Poisson random variable:

PROPOSITION 4.4 *If for each $j = 1, \ldots, n$, $Z_j(x_j)$ is a Poisson distributed random variable with mean $q_j x_j$, then the function $G(\mathbf{x})$ defined by (4.21) is component-wise concave in each x_j, $j = 1, \ldots, n$, and submodular in $\mathbf{x}$. That is, letting $\mathbf{e}_j$ denote the jth unit vector, for all j, the first differences*

$$G(\mathbf{x} + \mathbf{e}_j) - G(\mathbf{x}),$$

are decreasing in x_j, $j = 1, \ldots, n$.

The component-wise concavity of the expected net revenue function implies that there are critical booking levels for each class j beyond which the expected value does not increase, provided booking levels of other classes are kept constant. The submodularity property implies that the optimal booking limit for class j is nonincreasing in the booking limit for any other class $i \neq j$. These are natural and intuitive properties. They simply reflect the fact that low reservation levels in one class mean that capacity will be less constrained in the service period, and this in turn reduces the potential costs of overbooking in other classes because more (or at least less costly) substitution options will be available.

In Appendix 4.A we give a stochastic gradient method for computing the optimal joint overbooking limits. It solves (4.20) in the case of the Poisson cancellation model using a simulation-based, stochastic gradient algorithm. The following example illustrates how this method compares with the independent binomial model:

Example 4.5 There are four consecutive flights between the same city pair. For simplicity, assume that all four flights serve one class each and each flight has the same capacity of 100. Flights are ordered in time (the earliest flight is flight 1). Overbooking leads to substitution forward in time, so customers denied boarding on

Overbooking

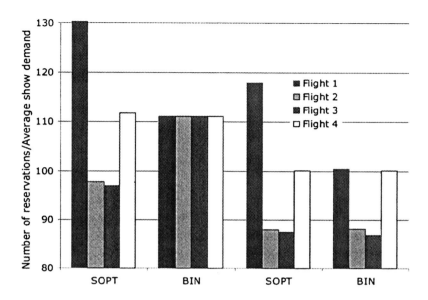

Figure 4.3. Overbooking limits and show demand for the multiclass and binomial models.

an oversold flight can take later flights with some loss of goodwill. Denying service completely to a customer results in a higher cost compared with the cost of goodwill due to delays. Delaying a customer by one flight costs $300; delaying by two flights costs $400; delaying by three flights is $500. The cost of denying service to a customer on any flight is $1,000. The unit revenue for reservations is $500, which is fully refundable on cancellation. There are 10 reservation periods in the planning horizon, and the survival probabilities are $0.81, 0.82, \ldots, 0.90$ from the first period to the last. Flights 1 and 4 receive 30% of reservation demand, while flights 2 and 3 receive only 20% each.

Figure 4.3 shows the overbooking limits and final average show demand for the multiclass, stochastic gradient method (SOPT) and binomial model (BIN) for this example. Note that the SOPT procedure is much more aggressive in overbooking flight 1 than flight 4 even though they have the same demand. This is natural, since overselling flight 1 is less costly because passengers can be put onto later flights; oversold passengers on flight 4 must be denied service. Indeed, SOPT in a sense deliberately "plans" oversales on flight 1, since delayed customers on these flights generate more revenue than penalties. This results in nearly 5% of passengers being delayed for one or more flights, while with the BIN procedure only 0.5% of passengers are delayed. Nevertheless, the multiclass model produces a 1.4% increase in revenues (net of penalty costs) over the independent binomial model, as the increased revenues more than compensate for the increase in delay penalties.

While the parameters of this example are not the most realistic, the example illustrates how coordinated overbooking policies for related resources may differ from those computed using independent models.

4.6 Network Overbooking

We next consider how to set overbooking levels on a network. The capacities of network resources are key inputs to capacity-control problems. Using overbooking, these capacities may be inflated—defining virtual capacities for each resource that exceed the physical capacity. This increase in capacity, in turn, affects the accept or reject decisions of the capacity-control method. On the other hand, capacity-control decisions clearly influence the opportunity cost of capacity, which is a key input to economic overbooking models. Hence, the total revenue for a network (net of penalties) is affected both by overbooking and seat inventory-control practices. Despite the strong interdependence of these decisions, the two problems are typically separated in practice.

In this section, we look at one model for coordinating network-capacity controls and overbooking decisions. The method combines the deterministic linear programming model of Section 3.3.1 with a single-period overbooking model, though it can be adapted to other network approximations as well (such as PNLP and RLP).

As in Chapter 3, consider a network with n products and m resources. We divide the time horizon into two periods: a reservation period, and a service period. The reservation period spans $(0, T]$ and is the period the reservations can be made for any of the n products. The reservation period is followed by the service period, during which the customers with reservations show up or become no-shows. During the service period, the firm may deny service to customers who show up in case of insufficient capacity, in which case it pays a penalty.

The demand or reservation requests arrive according to a stochastic process during $(0, T]$. As before, let $\mathbf{p} = (p_1, \ldots, p_n)$ denote the vector of prices and $\mathbf{C} = (C_1, \ldots, C_m)$, the vector of resource capacities. There is a denied-service cost on each resource given by the vector $\mathbf{h} = (h_1, \ldots, h_m)$. The denied-service cost may differ from one resource to another, but it does not vary with time or product type. The matrix $\mathbf{A} = [a_{ij}]$ is the usual network incidence matrix with $a_{ij} = 1$ if resource i is used by product j, and $a_{ij} = 0$, otherwise. Recall $\mathbf{A}^i$ denotes the ith row and $\mathbf{A}_j$ the jth column of matrix $\mathbf{A}$. For simplicity, we ignore refund for cancellations or no-shows, but this can be included easily in this model.

One way to formulate this overbooking problem is as a two-stage, static model that combines the DLP model and the cost-based overbooking models. The same formulation applies to a variety of network bid-price methods, though we focus on the DLP method for simplicity.

The decision variables are **x**, the vector of overbooking levels (virtual capacities), and **y**, the vector of primal allocations. (Note here we are changing our running definition of **y** to conform with Chapter 3; that is, **y** is now a vector of capacity allocations not a vector of reservations on hand.) The formulation is as follows:

$$\max_{\mathbf{x},\mathbf{y}} \quad \mathbf{p}^\top \mathbf{y} - E[\mathbf{h}^\top (\mathbf{Z}(\mathbf{x}) - \mathbf{C})^+] \quad (4.22)$$
$$\text{s.t.} \quad \mathbf{A}\mathbf{y} \leq \mathbf{x}$$
$$0 \leq \mathbf{y} \leq E[\mathbf{D}]$$
$$\mathbf{x} \geq \mathbf{C}.$$

The problem parameters are $E[\mathbf{D}]$, the vector of expected demand to come for the n classes. The objective function is the total revenue-to-come, net of denied-service costs.

Note that the show demand for resource i in this formulation is approximated by the random variable $Z_i(x_i)$. The actual show demand, however, will be less, since the show demand for i is $Z_i(x_i)$ *only* if the overbooking limit x_i is reached. (Recall the discussion after Section 4.2.1.1.) Otherwise, the number of reservation on resource i will be less than x_i, and so the show demand will be less than $Z_i(x_i)$. However, this approximation greatly simplifies the model and is a good approximation in the important case where demand is high.

We let $H(\mathbf{x}) = E[\mathbf{h}^\top (\mathbf{Z}(\mathbf{x}) - \mathbf{C})^+]$ denote the overbooking-cost function and $F(\mathbf{y}) = \mathbf{p}^\top \mathbf{y}$ denote the revenue function in (4.22). The overbooking-cost function H is a nondecreasing and convex function of the overbooking limit **x** if the random variable associated with the number of survivors for leg i, $Z_i(x_i)$ is assumed to follow the binomial or Poisson model with survival probability q_i. Thus, the objective function of problem (4.22) is jointly concave in **y** and **x** under these two models of cancellation. One can use a general-purpose nonlinear programming method to solve (4.22), but Appendix 4.B provides an algorithm specialized to this problem's structure. The following numerical example from [292] shows the performance of this method:

Example 4.6 The example here is based on the same network of Williamson [566] as shown in Figure 3.3 of Chapter 3. The itinerary revenue values and base-case mean demand values are show in Table 3.5 of Chapter 3 as well. The cancellation rate is assumed to be 15%, and the denied service penalty is assumed to be $1,000 on all legs. Different load factors, proportions of local versus through traffic, and arrival order were simulated to create 4 variations of the problem from the base case. A version of the network overbooking model using binomial, rather than Poisson, assumptions of cancellations (denoted BIN) was computed to find joint overbooking levels and corresponding DLP solutions. The resulting overbooking limits and dual prices were then tested by simulation.

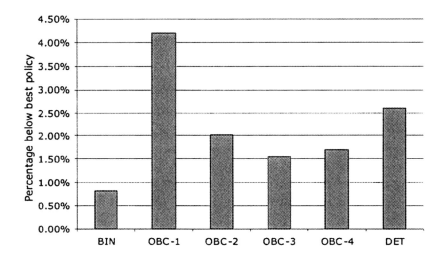

Figure 4.4. Network overbooking example: numerical comparison of policies.

This BIN policy is compared with several versions of ad-hoc overbooking rules. Four of these are cost-based overbooking models, denoted OBC-1 to OBC-4. They differ only in terms of the revenue values used to compute the overbooking limits. Finally, a deterministic overbooking limit (DET) based on the approximation described in Section 4.2.2.1 was also tested. Thus overall five methods are compared with BIN. Once overbooking levels were determined, a DLP model was solved, and the resulting bid prices were used to allocate capacity.

Since no exact methods are known for this problem, the deviation from the best of the six methods was used as the performance metric. That is, the maximum expected revenue (net of penalties) from all the policies is computed and for each individual policy, and the percentage deviation from this maximum is recorded.

Figure 4.4 from [292] shows the average percentage deviation of the six methods. Note that BIN is not always the best method (its average percentage deviation is slightly positive), but it is better than all the other methods.

Similar behavior is observed in other examples in [292], where BIN is not uniformly better than the ad-hoc overbooking mechanisms but is never very far from the best policy and moreover is significantly more robust than any of the ad-hoc methods. These and other tests of the method show the importance of network overbooking; the deviations between the best and worst policy can be quite large—several percentage points of difference in net revenues.

4.7 Notes and Sources

Much of the material in Section 4.1.1 comes from the carefully documented work of Rothstein [447–449] on the development of overbooking in the airline industry. For papers on overbooking from a policy perspective, see Falkson [180] and Ruppenthal [450]. Overview articles on overbooking include Bodily and Pfeifer [80], Dunleavy [165] and Rothstein [447, 449].

There was also much lively debate surrounding the oversale-auction idea, captured in a series of articles by Simon [472–475]. Vickery [534] proposed using his second-price auction mechanism for the problem as well.

The static overbooking problem of Section 4.2 first appeared in a pair of papers by Beckmann [31, 32]. Other early treatments of the static problem are Taylor [504], Thompson [508], and Rothstein and Stone [446]. See also Bierman and Thomas [66] and Shlifer and Vardi [465]. Martinez and Sanchez [362] test the memoryless property of the binomial model empirically.

The Gram-Charlier approximation in Section 4.2.2.3 is due to Taylor [504] as is the moment-generating-function method presented in Section 4.2.4.

The material in Section 4.3 on dynamic overbooking is from Chatwin's thesis ([107]) and subsequent published articles [108, 109].

The material on combined allocation and overbooking problem of Section 4.4 is from Subramanian et al. [494], who also developed the cost transformation technique of Sections 4.4.2 and 4.4.4.

The multiclass overbooking model with substitution (and associated optimization algorithm) in Section 4.5 are from Karaesmen and van Ryzin [290]. The network overbooking model and algorithm presented in Section 4.6 is from Karaesmen and van Ryzin [291]; see also Karaesmen's thesis [292]. Ladany [320] analyzes a two-class version of this problem for hotels using dynamic programming.

Some papers on practical considerations in hotel overbooking include Lambert et al. [326] and Lefever [337]. The latter discusses handling oversales situations in hotels. For models of hotel overbooking, see Ladany [320, 321] and Liberman and Yechiali [343, 344]. Bitran and Gilbert [71] analyze the problem of sequentially determining when to deny service to arriving customers based on the relative costs of denying service early and late in the evening.

APPENDIX 4.A: Computations for the Substitutable Capacity Model

The optimization problem (4.20) with Poisson cancellations and continuous booking limits can be solved numerically using a simulation-based optimization (stochastic gradient) method. To do so we need an estimator of the gradient of the objective function $G(\mathbf{x})$. Let the vector $\mathbf{H}^{(k)}$ denote a gradient estimator at the kth iteration of the algorithm. (How this estimator is constructed is discussed below.) The algorithm requires a sequence of step sizes, $\{\gamma_k\}$ satisfying $\sum_{k=1}^{\infty} \gamma_k = +\infty$ and $\sum_{k=1}^{\infty} \gamma_k^2 < +\infty$; for example $\gamma_k = 1/k$. Then, the algorithm proceeds as follows:

STEP 0: Initialize: $k = 1$ and $\mathbf{x}^{(k)} := \mathbf{y}$.

STEP 1: Get the next stochastic gradient:

- Randomly generate a new vector $\mathbf{Z}(\mathbf{x}^{(k)})$.
- Compute the gradient estimate $\mathbf{H}^{(k)}$ (discussed below).

STEP 2: Compute
$$\mathbf{x}^{(k+1)} = \Pi(\mathbf{x}^{(k)} + \gamma_k \mathbf{H}^{(k)}) \qquad (4.A.1)$$
where $\Pi(.)$ projects $\mathbf{x}^{(k)} + \gamma_k \mathbf{H}^{(k)}$ onto $\{\mathbf{x} : \mathbf{x} \geq \mathbf{y}\}$

STEP 3: Set $k := k + 1$ and GOTO STEP 1.

An estimator for $\mathbf{H}^{(k)}$ can be constructed using the random function
$$\Delta V_j(\mathbf{Z}(\mathbf{x})) = q_j(V_0(\mathbf{Z}(\mathbf{x}) + \mathbf{e}_j) - V_0(\mathbf{Z}(\mathbf{x}))). \qquad (4.A.2)$$
Letting
$$\Delta V(\mathbf{Z}(\mathbf{x})) = (\Delta V_1(\mathbf{Z}(\mathbf{x})), \ldots, \Delta V_n(\mathbf{Z}(\mathbf{x}))),$$
one can show that
$$\nabla_\mathbf{x} E[V(\mathbf{Z}(\mathbf{x}))] = E[\Delta V(\mathbf{Z}(\mathbf{x}))],$$
so $\Delta V(\mathbf{Z}(\mathbf{x}))$ is an unbiased estimator of the gradient of $E[V(\mathbf{Z}(\mathbf{x}))]$. The estimator $\Delta V(\mathbf{Z}(\mathbf{x}))$ can be obtained easily by simulating $\mathbf{Z}(\mathbf{x})$ and solving a network-flow problem to obtain $V(\mathbf{Z}(\mathbf{x}))$. Then each estimate $V(\mathbf{Z}(\mathbf{x}) + e_j)$, $j = 1, \ldots, n$ can be determined by perturbing $\mathbf{Z}(\mathbf{x})$ and re-solving the network problem.

Let $\mathbf{z}^{(k)}$ be a realization of show demand when the number of reservations on hand is $\mathbf{x}^{(k)}$ at the kth iteration of the stochastic gradient algorithm. Then the gradient of the objective function at that time is given by the vector $\mathbf{H}^{(k)} = (H_1^{(k)}, \ldots, H_n^{(k)})$, where
$$\begin{aligned} H_j^{(k)} &= p_j - s_j(1 - q_j) + \Delta V_j(\mathbf{Z}(\mathbf{x}^{(k)})) \\ &= p_j - s_j(1 - q_j) + q_j(V(\mathbf{z}^{(k)} + \mathbf{e}_j) - V(\mathbf{z}^{(k)})) \end{aligned} \qquad (4.A.3)$$
for $j = 1, \ldots, n$.

APPENDIX 4.B: Alternating-Direction Method for Network Overbooking

To determine the optimal solution $(\mathbf{y}^*, \mathbf{x}^*)$ for the model (4.22), one can use an alternating-direction method for the function. This method efficiently exploits the structure of the problem.

APPENDIX 4.B: Alternating-Direction Method for Network Overbooking

Define the set $\Omega = \{\mathbf{y} : 0 \leq \mathbf{y} \leq E[\mathbf{D}]\}$ and $\Upsilon = \{\mathbf{x} : \mathbf{x} \geq \mathbf{C}\}$. The augmented Lagrangian function is

$$L(\mathbf{y}, \mathbf{x}, \boldsymbol{\pi}) = F(\mathbf{y}) - H(\mathbf{x}) - \boldsymbol{\pi}^T(\mathbf{A}\mathbf{y} - \mathbf{x}) - \frac{\gamma}{2}\|\mathbf{A}\mathbf{y} - \mathbf{x}\|^2,$$

where γ is a positive (scalar) parameter. An alternating-direction method can be used to find the maximizers of the augmented Lagrangian. The method proceeds at iteration $k+1$ as follows:

$$\mathbf{y}^{(k+1)} = \arg\max_{\mathbf{y} \in \Omega}\{F(\mathbf{y}) - (\boldsymbol{\pi}^{(k)})^T\mathbf{A}\mathbf{y} - \frac{\gamma}{2}\|\mathbf{A}\mathbf{y} - \mathbf{x}^{(k)}\|^2\} \quad (4.B.1)$$

$$\mathbf{x}^{(k+1)} = \arg\max_{\mathbf{x} \in \Upsilon}\{-H(\mathbf{x}) + (\boldsymbol{\pi}^{(k)})^T\mathbf{x}$$
$$- \frac{\gamma}{2}\|\mathbf{A}\mathbf{y}^{(k+1)} - \mathbf{x}\|^2\} \quad (4.B.2)$$

$$\boldsymbol{\pi}^{(k+1)} = \boldsymbol{\pi}^{(k)} + \gamma(\mathbf{A}\mathbf{y}^{(k+1)} - \mathbf{x}^{(k+1)}). \quad (4.B.3)$$

The parameter $\gamma > 0$, initial vectors $\mathbf{x}^{(0)} \geq \mathbf{C}$ and $\boldsymbol{\pi}^{(0)} \geq 0$ are arbitrary. Let $\gamma = 1$. One can show that a sequence $\{\mathbf{y}^{(k)}, \mathbf{x}^{(k)}, \boldsymbol{\pi}^{(k)}\}$ generated by the algorithm (4.B.1), (4.B.2), and (4.B.3) is bounded and every limit point of $\{\mathbf{y}^{(k)}, \mathbf{x}^{(k)}\}$ is an optimal solution to the original problem (4.22). Furthermore, $\{\boldsymbol{\pi}^{(k)}\}$ converges to the optimal dual variable associated with the virtual capacity constraints. A proof of this fact and more details on the method are provided in Bertsekas and Tsitsiklis [56].

To apply this algorithm, we have to solve two different nonlinear programming problems.

Finding $\mathbf{y}^{(k+1)}$, requires solving the following problem:

$$\max F(\mathbf{y}) - (\boldsymbol{\pi}^{(k)})^T\mathbf{A}\mathbf{y} - \frac{1}{2}\|\mathbf{A}\mathbf{y} - \mathbf{x}^{(k)}\|^2 \quad (4.B.4)$$

for $\mathbf{y} \in \Omega$. For the DLP model, this is equivalent to the following quadratic program:

$$(QP) \quad \max \quad (\mathbf{p} - \mathbf{A}^T\boldsymbol{\pi}^{(k)} + \mathbf{A}^T\mathbf{x}^{(k)})^T\mathbf{y} - \frac{1}{2}\mathbf{y}^T\mathbf{A}^T\mathbf{A}\mathbf{y} \quad (4.B.5)$$

$$\text{s.t.} \quad 0 \leq \mathbf{y} \leq E[\mathbf{D}]. \quad (4.B.6)$$

Problem (QP) can be solved by any standard nonlinear programming method specialized to quadratic programming.

Finding $\mathbf{x}^{(k+1)}$ requires solving

$$(SP) \quad \min_{\mathbf{x} \geq \mathbf{C}} H(\mathbf{x}) - (\boldsymbol{\pi}^{(k)})^T\mathbf{x} + \frac{1}{2}\|\mathbf{A}\mathbf{y}^{(k+1)} - \mathbf{x}\|^2 \quad (4.B.7)$$

$$= \mathbf{h}^T E[(\mathbf{Z}(\mathbf{x}) - \mathbf{C})^+] - (\boldsymbol{\pi}^{(k)} + \mathbf{A}\mathbf{y}^{(k+1)})^T\mathbf{x}$$
$$+ \frac{1}{2}\mathbf{x}^T\mathbf{x} + \frac{1}{2}(\mathbf{y}^{(k+1)})^T\mathbf{A}^T\mathbf{A}\mathbf{y}^{(k+1)}.$$

The function in (4.B.7) is separable, convex, and differentiable under the Poisson model of cancellation, and can therefore be solved with a simple line-search method.

We summarize the steps of the algorithm:

STEP 0: Initialize: $\gamma = 1$, $\mathbf{x}^{(0)} = 0$, $\pi^{(0)} = 0$, $k = 1$.
STEP 1: Solve problem (QP) and get $\mathbf{y}^{(k)}$.
STEP 2: Solve problem (SP) and get $\mathbf{x}^{(k)}$.
STEP 3: Compute $\pi^{(k)}$ using (4.B.3).
STEP 4: Set $k \leftarrow k+1$ and GOTO STEP 1 if $\mathbf{y}^{(k)}, \mathbf{x}^{(k)}, \pi^{(k)}$ do not meet a stopping criterion.

There are several options for the stopping criteria: (1) check that $\mathbf{y}^{(k)}, \mathbf{x}^{(k)}, \pi^{(k)}$ satisfy the KKT conditions, (2) check that $\mathbf{y}^{(k)}, \mathbf{x}^{(k)}, \pi^{(k)}$ are not significantly different from the values of $\mathbf{y}^{(k-1)}, \mathbf{x}^{(k-1)}, \pi^{(k-1)}$, or (3) reach a preset number of iterations; this can be done if one has prior experience with the algorithm and the problems. Karaesmen and van Ryzin [291] show that the algorithm is quite fast and stable on many examples.

PART II

PRICE-BASED RM

Chapter 5

DYNAMIC PRICING

5.1 Introduction and Overview

In this chapter, we look at settings in which prices rather than quantity controls are the primary variables used to manage demand. While the distinction between quantity and price controls is not always sharp (for instance, closing the availability of a discount class can be considered equivalent to raising the product's price to that of the next highest class), the techniques we look at here are distinguished by their explicit use of price as the control variable and their explicit modeling of demand as a price-dependent process.

In terms of business practice, varying prices is often the most natural mechanism for revenue management. In most retail and industrial trades, firms use various forms of dynamic pricing—including personalized pricing, markdowns, display and trade promotions, coupons, discounts, clearance sales, and auctions and price negotiations (request for proposals and request for quotes—RFP/RFQ processes)—to respond to market fluctuations and uncertainty in demand. Exactly how to make such price adjustments in a way that maximizes revenues (or profits, in the case where variable costs are involved) is the subject of this chapter.

Dynamic pricing is as old as commerce itself. Firms and individuals have always resorted to price adjustments (such as haggling at the bazaar) in an effort to sell their goods at a price that is as high as possible yet acceptable to customers. However, the last decade has witnessed an increased application of scientific methods and software systems for dynamic pricing, both in the estimation of demand functions and the optimization of pricing decisions.

5.1.1 Price versus Quantity-Based RM

Some industries use price-based RM (retailing), whereas others use quantity-based RM (airlines). Even in the same industry, firms may use a mixture of price- and quantity-based RM. For instance, many of the RM practices of the new low-cost airlines more closely resemble dynamic pricing than the quantity-based RM of the traditional carriers. What explains these differences?

It is hard to give a definitive answer, and indeed Chapter 8 is devoted to different theoretical explanations of RM practice. But in essence, it boils down to a question of the extent to which a firm is able to vary quantity or price in response to changes in market conditions. This ability, in turn, is determined by the commitments a firm makes (to price or quantity), its level of flexibility in supplying products or services, and the costs of making quantity or price changes.

Consider airlines, for example. While arguably less true today than in the past, airlines normally commit to prices for their various fare products in advance of taking bookings. This is due to advertising constraints (such as the desire to publish fares in print media and fare tariff books), distribution constraints, and a desire to simplify the task of managing prices. For these marketing and administrative reasons, most airlines advertise and price fare products on an aggregate origin-destination market level, for a number of flights over a given interval of time, and do not price on a departure-by-departure basis. This limits their ability to use price to manage the demand on any given departure, demand that varies considerably by flight and is quite uncertain at the time of the price posting. At the same time, the supply of the various classes is almost perfectly flexible between the products (subject to the capacity constraint of the flight), since all fare products sold in the same cabin of service share a homogeneous seat capacity. It is this combination of price commitments together with flexibility on the supply side that make quantity-based RM an attractive tactic in the airline industry. Hotels, cruise ships, and rental cars—other common quantity-based RM industries—share many of these same attributes.

In other cases, however, firms have more price flexibility than quantity flexibility. In apparel retailing, for example, firms commit to order quantities well in advance of a sales season—and may even commit to certain stocking levels in each store. Often, it is impossible (or very costly) to reorder stock or reallocate inventory from one store to another. At the same time, it is easier (though not costless) for most retailers to change prices, as this may require only changing signage and making data entries into a point-of-sale system. Online retailers in particular enjoy tremendous price flexibility because changing prices is almost costless.

Business-to-business sales are often conducted through a RFP/RFQ process, which allows firms to determine prices on a transaction-by-transaction basis. In all these situations, price-based RM is therefore a more natural practice. Of course, the context could dictate a different choice even in these industries. For example, if a retailer commits to advertised prices in different regional markets yet retains a centralized stock of products, it might then choose to manage demand by tactically allocating its supply to these different regions—a quantity-based RM approach.

However, given the choice between price- and quantity-based RM, one can argue that price-based RM is the preferred option. The argument is as follows (see Gallego and van Ryzin [199]). Quantity-based RM operates by rationing the quantity sold to different products or to different segments of customers. But rationing, by its very nature, involves reducing sales by *limiting* supply. If one has price flexibility, however, rather than reducing sales by *limiting supply*, we can reduce sales by *increasing price*. This achieves the same quantity-reducing function as rationing, but does it more profitably because by increasing price we both reduce sales *and* increase revenue at the same time. In short, price-based "rationing" is simply a more profitable way to limit sales than quantity-based rationing.

In practice, of course, firms rarely have the luxury of choosing price and quantity flexibility. Therefore, practical business constraints dictate which tactical response—price- or quantity-based RM (or a mixture of both)—is most appropriate in any given business context.

5.1.2 Industry Overview

To give a sense of the scope of activity in the area of dynamic pricing, we next review pricing innovations in a few industries.

5.1.2.1 Retailing

Retailers, especially in apparel and other seasonal-goods sectors, have been at the forefront in deploying science-based software for pricing, driven primarily by the importance of pricing decisions to retailers' profits. For example, Kmart alone wrote off $400 million due to markdowns in one quarter of 2001, resulting in a 40% decline in its net income [194].

Several software firms specializing in RM in retailing have recently emerged. Most of this software is currently oriented toward optimizing markdown decisions. Demand models fit to historical point-of-sale data together with data on available inventory serve as inputs to optimization models that recommend the timing and magnitude of markdown decisions.

Major retailers—including Gymboree, J. C. Penney, L. L. Bean, Liz Claiborne, Safeway, ShopKo, and Walgreen's—are experimenting with this new generation of software [194, 214, 270, 379]. Many have reported significant improvements in revenue from using pricing models and software. For example, ShopKo reported a 24% improvement in gross margins as a result of using its model-based pricing software [270] and other retailers report gains in gross margins of 5% to 15% [194]. Academic studies based on retail data have also documented significant improvements in revenues using model-based markdown recommendations [70, 247].

5.1.2.2 Manufacturing

Scientific approaches to pricing are gaining acceptance in the manufacturing sector as well. For example, Ford Motor Co. reported a high-profile implementation of pricing-software technology to support pricing and discounts for its products [135]. The project, started in 1995, focused on identifying features that customers were most willing to pay for and changing salesforce incentives to focus on profit margins rather than unit-sale volumes. Ford then applied pricing models developed by an outside consulting firm to optimize prices and dealer and customer incentives across its various product lines. In 1998, Ford reported that the first five U.S. sales regions using this new pricing approach collectively beat their profit targets by $1 billion, while the 13 that used their old methods fell short of their targets by about $250 million [135].

5.1.2.3 E-business

E-commerce has also had a strong influence on the practice of pricing [529]. Companies such as eBay and Priceline have demonstrated the viability of using innovative pricing mechanisms that leverage the capabilities of the Internet. E-tailers can discount and markdown on the fly based on customer loyalty and click-stream behavior. Since a large e-tailer like Amazon.com has to make a large number of such pricing decisions based on real-time information, automating decision making is a natural priority. The success of these e-commerce companies—inconsistent and volatile as it may appear at times—is at least partly responsible for the increased interest among traditional retailers in using more innovative approaches to pricing.

On the industrial side, e-commerce pricing has been influenced by the growth of business-to-business (B2B) exchanges and other innovations in using the Internet to gain trading efficiencies. While this sector too has had its ebbs and flows, it has produced an astounding variety of new pricing and trading mechanisms, some of which are use regularly for the sale products such as raw materials, generic commodity items and excess inventory. For example, Freemarkets has had significant success in providing software and service for industrial-procurement auctions, and as of this writing claims to have facilitated over $30 billion in trade since its founding in 1999. Covisint—an exchange jointly funded by Daimler-Chrysler, Ford Motor Company, and General Motors—while slow to develop, looks nevertheless to become a permanent feature of the auto-industry procurement market. Most infrastructure software for B2B exchanges—sold by firms such as Ariba, i2, IBM, and Commerce One—also has various forms of dynamic pricing capabilities built in.

For all these reasons, e-commerce has given price-based RM a significant boost in recent years.

5.1.3 Examples of Dynamic Pricing

We next examine three specific example of dynamic pricing and the qualitative factors driving price changes in each case.

5.1.3.1 Style-Goods Markdown Pricing

Retailers of style and seasonal goods use markdown pricing to clear excess inventory before the end of the season. This type of price-based RM is most prevalent in apparel, sporting goods, high-tech, and perishable-foods retailing. The main incentive for price reductions in such cases is that goods perish or have low salvage values once the sales season is over; hence, firms have an incentive to sell inventory while they can, even at a low price, rather than salvage it.

But apart from inventory considerations, there are other proposed explanations for markdown pricing. One explanation, proposed by Lazear [332] (see Examples 8.11 and 8.12) and investigated empirically in Pashigan [415] and Pashigan and Bowen [414], is that retailers are uncertain about which products will be popular with customers. Therefore, firms set high prices for all items initially. Products that are popular are the ones for which customers have high reservation prices, so these sell out at the high initial price. The firm then identifies the remaining items as low-reservation-price products and marks them down. In this explanation, markdown pricing serves as a form of demand learning.

A second explanation for markdowns is that customers who purchase early have higher willingness to pay, either because they can use the product for a full season (a bathing suit at the start of summer) or because there is some cache to being the first to own it (a new dress style or electronic gadget). Markdown pricing then serves as a segmentation mechanism to separate price-insensitive customers from those price-sensitive customers willing to defer consumption to get a lower price.

Warner and Birsky [554] give yet another explanation, with empirical evidence, for markdown pricing. On holidays and during peak-shopping periods (such as before Christmas), customers can search for the lowest prices more efficiently because they are actively engaged in search, making many shopping trips over a concentrated period of time. Even those customers who normally do not spend much time searching for the best price change their behavior during these peak shopping periods and become more vigilant. The result is that demand during peak periods is more price-sensitive and retailers respond by running "sales" during these periods.

5.1.3.2 Discount Airline Pricing

Not all dynamic pricing involves price reductions, however. As we mentioned earlier, discount airlines use primarily price-based RM, but with prices often going up over time. These airlines (some examples are easyJet and Ryanair in Europe and jetBlue in the U.S.) typically offer only one type of ticket on each flight, a non-refundable, one-way fare without advance-purchase restrictions. However, they offer these tickets at different prices for different flights, and moreover, during the booking period for each flight, vary prices dynamically based on capacity and demand for that specific departure. To quote from one practitioner of this type of dynamic pricing (Easyjet website, 2003):

> The way we structure our fares is based on supply and demand and prices usually increase as seats are sold on every flight. So, generally speaking, the earlier you book, the cheaper the fare will be. Sometimes, however, due to market forces our fares may be reduced further. Our booking system continually reviews bookings for all future flights and tries to predict how popular each flight is likely to be.

Figure 5.1 shows the evolution of prices for a particular European discount airline flight as a function of the number of weeks prior to departure. Note that prices are highest in the last few weeks prior to departure.

There are some fundamental differences between air travel and style- and seasonal-goods products that explain this increasing price pattern. For one, the value of air travel to customers does not necessarily go down as the deadline approaches. Conversely, the value of a ticket earlier on is lower for customers as customers multiply the value by the probability that they will indeed use the ticket (especially for a non-refundable ticket). Somewhat related to these points, additionally, although customers purchase tickets at different points of time, all customers consume

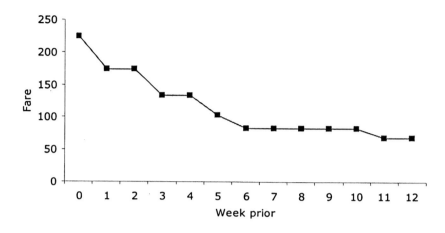

Figure 5.1. Prices as a function of weeks prior to departure at a European low-cost discount air carrier.

the product (fly the flight) at the same time. So two factors come into play. Customers who purchased early may get upset to see prices drop while they are still holding a reservation; indeed, many airlines give a price guarantee to refund the difference if there is a price drop (to encourage passengers to book early), making it costly for the firms to lower prices. And in the travel business, high-valuation high-uncertainty customers tend to purchase closer to the time of service. Hence, demand is less price-sensitive close to the time of service.

5.1.3.3 Consumer-Packaged Goods Promotions

In contrast to markdown and discount airline pricing, promotions are short-run, temporary price reductions. Promotions are the most common form of price-based RM in the consumer packaged-goods (CPG) industry (soap, diapers, coffee, yogurt, and so on).

The fact that customers purchase CPG products repeatedly has important implications for pricing and promotions. Specifically, customers are aware of past prices and past promotions, so running promotions too frequently may condition customers to view the brand as a frequently discounted product, cutting into brand equity in the long run. Because customers are aware of past prices, promotions impact their subjective "reference price"—or sense of the "fair" price—for products. And customers may *stockpile* products, so short-run increases in demand due to promotions may come at the expense of reduced future demand.

The institutional structure of promotions is also more complicated. There are three parties involved—manufacturers, retailers, and end customers. Promotions are run either by a manufacturer as discounts to retailers (trade promotions), which may or may not be passed on to the customers by the retailers (retailer pass-thru), or by retailers (retail promotions or consumer promotions). In some forms of promotion (e.g., mail-in coupons) manufacturers gives a discount directly to the end customer.

The motivations of the manufacturer and the retailer are different as well. While a manufacturer is interested in increasing sales or profits for its brand, retailers are interested in overall sales or profits for a category constituting multiple brands from multiple manufacturers. For a retailer, discounting a particular brand may increase sales for that brand but dilute overall category profits as customers switch from high-margin brands to the discounted brand. So in designing optimal promotions structures, one has to consider complex incentive compatibility constraints.

5.1.4 Modeling Dynamic Price-Sensitive Demand

Any dynamic-pricing model requires a model of how demand—either individual or aggregate—responds to changes in price. The basic theory of consumer choice and the resulting market-response models are covered in Chapter 7. We draw on these models in this chapter.

However, in dynamic-pricing problems some additional factors must be considered. The first concerns how individual customers behave over time—what factors influence their purchase decisions and how sophisticated their decision-making process is, and so on. The second concerns the state of market conditions—specifically the level of competition and the size of the customer population. We next look at each of these assumptions qualitatively.

5.1.4.1 Myopic- Versus Strategic-Customer Models

One important demand-modeling assumption concerns the level of sophistication of customers. Most of the models we consider in this chapter assume *myopic customers*—those who buy as soon as the offered price is less than their willingness to pay. Myopic customers do not adopt complex buying strategies, such as refusing to buy in the hope of lower prices in the future. They simply buy the first time the price drops below their willingness to pay. Models that incorporate *strategic customers*, in contrast, allow for the fact that customers will optimize their own purchase behavior in response to the pricing strategies of the firms.

Of course, the strategic-customer model is more realistic. However, such a demand model makes the pricing problem essentially a strategic game between the customers and the firm, and this significantly complicates the estimation and analysis of optimal pricing strategies—often making the problem intractable. In contrast, the myopic-customer model is much more tractable and hence is more widely used. The issue in practice is really a matter of how "bad" the myopic assumption is in any given context. In many situations, customers are sufficiently spontaneous in making decisions that one can ignore their strategic behavior. Moreover, customers often do not have the sufficient time or information to behave very strategically. However, the more expensive and durable the purchase, the more important it becomes to model strategic customer behavior (for example, automobile buyers waiting to purchase at the end of a model year).

One common defense of the myopic assumption is the following. The forecasting models that use observations of past customer behavior in a sense reflects the effects of our customers' strategic behavior. For example, if the customers who are most price-sensitive tend to adopt a strategy of postponing their purchases until end-of-season clearance sales, then the estimated price sensitivity in these later periods will tend to appear much higher than in earlier periods. Therefore, even though we do not model the strategic behavior directly, our forecasting models indirectly capture the correct price response.

This view is plausible if the pricing strategies obtained from a model are roughly similar to past policies, so that they can be viewed as "perturbations" or "fine tuning" of a historical pricing strategy—a strategy that customers have already factored into their behavior. On the other hand, if optimized pricing recommendations are radically different in structure from past pricing strategies, then it is reasonable to expect that customers will adjust their buying strategies in response. If this happens, the predictions of myopic models that are fit to historical data may be very bad indeed.

Yet even when the myopic approach works (in the sense of correctly predicting price responses), it runs the risk of reinforcing "bad equilibrium" pricing strategies. For example, a myopic model fit to past data may reconfirm the "optimality" of lowering prices significantly at the end of a sales season or running periodic holiday sales because it estimates, based on historical data, that demand is especially price-sensitive in these periods. But this price sensitivity may be due to the fact that customers have learned not to buy at other times, because they know prices will be cut at the end of the season or during holidays. If the firm was to adopt a constant price strategy—and customers were convinced that the firm was sticking to this strategy—then the observed price sensitivity might shift. The resulting equilibrium might be more profitable, but it is one that the firm would not discover using a myopic-customer model.

Despite these limitations and potential pitfalls of the myopic model, it is practical, is widely used, and provides useful insight into dynamic pricing. We therefore focus on the myopic case for the most part in this chapter. However, we consider strategic customers in Section 5.5.2 below and in considerably more depth in Chapter 6, where we look at auctions, the analysis of which is entirely based on strategic-customer models.

5.1.4.2 Infinite- Versus Finite-Population Models

Another important assumption in demand modeling is whether the population of potential customers is finite or infinite. Of course, in reality, every population of customers is finite; the question is really a matter of whether the number and type of customers that have already bought changes one's estimate of the number or type of future customers.

In an infinite-population model, we assume that we are sampling *with replacement* when observing customers. As a result, the distribution of the number of customers and the distribution of their willingness to pay is not affected by the past history of observed demand. This is often termed the *nondurable-goods assumption* in economics because we can view this as a case where customers immediately consume their purchase and then reenter the population of potential customers (say, for a can of Coke). This assumption is convenient analytically because one does not need to retain the history of demand (or a suitable sufficient statistic) as a state variable in a pricing-optimization problem.

The finite-population model assumes a random process *without replacement*. That is, there are a finite (possibly random) number of customers with heterogeneous willingness to pay values. If one of the customers in the population purchases, the customer is removed from the population of potential customers, and therefore future purchases only occur from the remaining customers. This is termed the *durable-goods assumption* in economics because we can consider it as a case where the good being purchased is consumed over a long period of time (for example, an automobile) and hence once a customer purchases, he effectively removes himself from the population of potential customers.

For example, suppose we assume a price $p(t)$ is offered in period t and all customers who value the item at more than $p(t)$ purchase in period t (myopic behavior). Then, under a finite-population model, we know that after period t, the remaining customers all have valuations less than $p(t)$. In particular, the future distribution of willingness to pay is conditioned on the values being less than $p(t)$. As a result, in formulating a dynamic-pricing problem, we have to keep track of past pricing decisions and their effect on the residual population of customers.

Which of these models is most appropriate depends on the context. While often the infinite-population model is used simply because it is easier to deal with analytically, the key factors in choosing one model over the other are the number of potential customers relative to the number that actually buy and the type of good (durable versus nondurable). Specifically, the infinite-population model is a reasonable approximation when there is a large population of potential customers and the firm's demand represents a relatively small fraction of this population because in such cases the impact of the firm's past sales on the number of customers and the distribution of their valuations is negligible. It is also reasonable for consumable goods. However, if the firm's demand represents a large fraction of the potential pool of customers or if the product is a durable good, then past sales will have a more significant impact on the statistics of future demand, and the finite-population assumption is more appropriate.

Qualitatively, the two models lead to quite different pricing policies. Most notably, finite-population models typically lead to *price skimming* as an optimal strategy, in which prices are lowered over time in such a way that high-valuation customers pay higher prices earlier while low-valuation customers pay lower prices in later periods. Effectively, this creates a form of second-degree price discrimination, segmenting customers with different values for the good and charging differential prices over time. In infinite-population models, there is no such price-skimming incentive. Provided the distribution of customer valuations does not shift over time, the same price that yields a high revenue in one period will yield a high revenue in later periods, and thus a firm has no incentive to deviate from this revenue-maximizing price.

5.1.4.3 Monopoly, Oligopoly, and Perfect-Competition Models

Another key assumption in dynamic-pricing models concerns the level of competition the firm faces. Many pricing models used in RM practice are *monopoly models*, in which the demand a firm faces is assumed to depend only on its own price and not on the price of its competitors. Thus, the model does not explicitly consider the competitive reaction to a price change. Again, one makes this assumption primarily for tractability and is not always realistic.

As with the myopic-customer model, the monopoly model can be partly justified on empirical grounds—namely, that an observed historical price response has embedded in it the effects of competitors' responses to the firm's pricing strategy. So for instance, if a firm decides to lower its price, the firm's competitors might respond by lowering their prices. With market prices lower, the firm and its competitors see an increase in demand. The observed increase in demand is then measured empirically and treated as the "monopoly" demand response to the firm's price change in a dynamic-pricing model—even though competitive effects are at work.

Again, while such a view is pragmatic and reflects the conventional wisdom behind the pricing models used in practice, there are some dangers inherent in it, paralleling those of the myopic-customer model. The price-sensitivity estimates may prove wrong if the optimized strategy deviates significantly from past strategies because then the resulting competitive response may be quite different from the historical response. Also, the practice runs the risk of reinforcing "bad" equilibrium responses. Despite these risks, monopoly models have still proved to be valuable for decision support.

It is worth noting that oligopoly models, in which the equilibrium-price response of competitors is explicitly modeled and computed, also have their pitfalls. Most notably, the assumption that firms behave rationally (or quasi-rationally, if heuristics are used in place of optimal strategies) may result in a poor predictor of their actual price response. These potential modeling errors together with the increased complexity of analyzing oligopoly models—and the difficulty in collecting competitor data to estimate the models accurately—has made them less popular in practice. Shugan [468] provides a good summary of this point of view; he notes that "the strong approximating assumption of no competitive response is sometimes better than the approximating assumption of preexisting optimal behavior." However, properly designed and validated, oligopoly models can provide valuable insights on issues of pricing strategy.

Finally, one can also consider perfectly competitive models—in which many competing firms supply an identical commodity. As described in Section 8.2, the output of each firm is assumed to be small relative the market size, and this, combined with the fact that each firm is offering identical commodities, means that a firm cannot influence market prices.[1] Therefore, each firm is essentially a *price taker*—able to sell as much as it wants at the prevailing market price but unable to sell anything at higher prices. Despite the importance of perfect-competition models in economic theory, the assumption that firms have no pricing power means that the results are not that useful for price-based RM. Nevertheless, they do play a role in quantity-based RM. For example, one can interpret the capacity-control models of Chapters 2 and 3 as stemming from competitive, price-taking models; firms take the price for their various products as given (set by competitive market forces), and control only the quantity they supply (the availability or allocation) at these competitive prices. As our focus in the chapter is on price-based RM, we do not consider this model of competition further in this chapter.

5.2 Single-Product Dynamic Pricing Without Replenishment

The first problem we look at is dynamic pricing of a single product over a finite sales horizon given a fixed inventory at the start of the sales horizon. We assume that the firm is a monopolist, customers are myopic, and there is no replenishment of inventory.

The models are representative of the type used in style and seasonal goods retail RM. For such retailers, production and ordering cycles are typically much larger than the sales season, and the main challenge is to determine the price path of a particular style at a particular store location, given a fixed set of inventory at the beginning of the season.

At one level, such models are simplistic: they consider only a single product in isolation and assume customers are myopic, and therefore demand is a function solely of time and the current price (although other factors such as inventory depletion are sometimes included). They therefore ignore competition, the impact of substitution, and the possible strategic behavior of customers over time. Despite these simplifications, the models provide good rough-cut approximations and are useful in practice. In addition, by decomposing the problem and treating products independently, it is possible to solve such models efficiently even when there are hundreds of thousands of product-location combinations. Finally, even with the simplifying assumptions, the analysis can

---

[1]This is in contrast to the Cournot model of quantity competition discussed in Section 8.4, in which there are only a small number of firms whose quantity decisions do affect the market price. Roughly speaking, Cournot competition approaches perfect competition as the number of firms in the industry tends to infinity.

still become complex if we allow stochastic demand and put constraints on prices.

Since we consider only a single product, there is a single (scalar) price decision at each time t, denoted $p(t)$, which induces a unique (scalar) demand rate $d(t,p)$. The set of allowable prices is denoted Ω_p, and Ω_d denotes the set of achievable demand rates. We assume that these functions satisfy the regularity conditions in Assumptions 7.1, 7.2, and 7.3 unless otherwise specified. These include several regularity properties, which we summarize here:

- The demand functions are continuously differentiable and strictly decreasing, $d'(t,p) < 0$, on Ω_p. Hence, they have an inverse, denoted $d(t,p)$.

- The demand functions are bounded above and below and tend to zero for sufficiently high prices—namely,
$$\inf_{p \in \Omega_p} d(t,p) = 0.$$

- The revenue functions $r(t,p) = pd(t,p)$ (equivalently $r(t,d) = dp(t,d)$) are finite for all $p \in \Omega_p$ and have a finite maximizer interior to Ω_p.

- The marginal revenue as a function of demand, d, defined by
$$J(t,d) \equiv \frac{\partial}{\partial d} r(t,d) = p(t,d) + dp'(t,d),$$
is strictly decreasing in d. (Assumption 7.2)

Readers who are not familiar with demand functions are encouraged to review Section 7.3 for more discussion of these and other related properties of demand functions. As discussed in Section 7.3, the demand function can also be expressed as $d(t,p) = N_t(1 - F(t,p))$, where N_t is the market-size parameter and $F(t,p)$ is the fraction of the market with willingness to pay less than p. We let $x(t)$ denote the inventory at time $t = 1, \ldots, T$, where T is the number of periods in the sale horizon. The initial inventory is $x(0) = C$.

5.2.1 Deterministic Models

The simplest deterministic pricing model is formulated in discrete time as follows. Given an initial inventory $x(0) = C$, select a sequence of prices $p(t)$ (inducing demand rates of $d(t,p(t))$) that maximize total revenues. Formulating the problem in terms of the demand rates $d(t)$, the optimal rates $d^*(t)$ must solve

$$\max \sum_{t=1}^{T} r(t, d(t)) \tag{5.1}$$

$$\text{s.t.} \sum_{t=1}^{T} d(t) \leq C$$

$$d(t) \geq 0.$$

Let π^* be the Lagrange multiplier on the inventory constraint, and recall that $J(t,d) = \frac{\partial}{\partial d} r(t,d)$ denotes the marginal revenue. Then the first-order necessary conditions for the optimal rates $d^*(t)$ and multiplier π^* are

$$J(t, d^*(t)) = \pi^*, \tag{5.2}$$

subject to the complementary slackness condition

$$\pi^*(C - \sum_{t=1}^{T} d^*(t)) = 0 \tag{5.3}$$

and the multiplier nonnegativity constraint $\pi^* \geq 0$. Under Assumption 7.2, $J(t,d)$ is decreasing in d and so $r(t,d)$ is concave; hence, these conditions are also sufficient.

The optimality conditions are quite intuitive. The Lagrange multiplier π^* has the interpretation as the marginal opportunity cost of capacity. The condition $J(t, d^*(t)) = \pi^*$ says that the marginal revenue should equal the marginal opportunity cost of capacity in each period. This makes sense because if marginal revenues and costs are not balanced, we can increase revenues by reallocating sales (by adjusting prices) from a period of low marginal revenue to a period of higher marginal revenue. Finally, the complementary slackness condition says that the opportunity cost cannot be positive if there is an excess of stock. If the opportunity cost is zero ($\pi^* = 0$), then if we maximize revenue without a constraint in every period (pricing to the point where marginal revenue is zero), we will still not exhaust the supply. This means it can be optimal—even in the absence of any costs for capacity—not to sell all the available supply.

Note that this problem is essentially equivalent to the problem of optimal third-degree price discrimination (see Section 8.3.3.2) if we consider customers in each period t to be different segments who are offered discriminatory prices $p(t)$. Another way of viewing the above argument is that the firm, faced with a capacity constraint, decides how much to sell in each period, and its optimal allocation of capacity occurs when the marginal revenue in all the periods are the same. The following example illustrates the idea:

Example 5.1 Consider a two-period selling horizon, where during the first period demand is given by $d_1 = -p_1 + 100$ and in period 2 demand is given by $d_2 = -2p_2 + 120$. (Customers in the second period are more price-sensitive than those in the first period.) Purchase behavior is assumed to be myopic. Considered separately, the revenue-maximizing price for the first period (maximizing $r_1 = p_1(-p_1 + 100)$) is given by $p_1^* = 50$ and $d_1^* = 50$, and in the second period by $p_2^* = 30, d_2^* = 60$ (maximizing $r_2 = p_2(-2p_2 + 120)$).

Intertemporal effects come into play if the firm has only a limited number of items to sell (less than 50+60). Suppose the firm's capacity is 40. How should it divide the sale between the two periods?

Note that here, $J(1, d_1) = -2d_1 + 100$ and $J(2, d_2) = -d_2 + 60$. Consider the table of marginal values, Table 5.1, at various allocations and the corresponding revenues. The total revenue is maximized at the point where the marginal values for the two periods are approximately the same (when $d_1 = 27, d_2 = 13$), conforming to our intuition; if they were not equal, the firm would reallocate capacity to the higher marginal-value period.

Table 5.1. Allocations of capacity between periods 1 and 2 and the marginal values and total revenue.

d_1	d_2	$J(1, d_1)$	$J(2, d_2)$	r
22	18	56	42	2634
23	17	54	43	2646.5
24	16	52	44	2656
25	15	50	45	2662.5
26	14	48	46	2666
27	**13**	**46**	**47**	**2666.5**
28	12	44	48	2664
29	11	42	49	2658.5
30	10	40	50	2650
31	9	38	51	2638.5
32	8	36	52	2624
33	7	34	53	2606.5

To see qualitatively how prices will change over time, we can write the optimality condition (5.2) as

$$\frac{p^*(t) - \pi^*}{p^*(t)} = \frac{1}{|\epsilon(t, p^*)|},$$

where $\epsilon(t, p)$ is the elasticity of demand in period t, defined by

$$\epsilon(t, p) \equiv \frac{p}{d(t, p)} \frac{\partial d(t, p)}{\partial p}.$$

See Section 7.3.1.3 for a further discussion of price elasticity. Thus, more elastic demand in period t implies a lower optimal price $p^*(t)$.

For example, if customers that buy toward the end of the sales horizon are more price-sensitive than those that buy early, then optimal prices will decline over time. If customers early on are price-sensitive, and those buying later are less price-sensitive, then optimal prices will increase over time. This observation offers one explanation for why in some industries (such as apparel retailing) prices tend to decline over time, while in others (such as airlines) prices increase over time. Chapter 8 provides additional explanations for intertemporal price patterns.

5.2.1.1 Computational Approaches

Problem (5.1) is a rather simple nonlinear program to solve. Each value π implies a value $d^*(t)$ by (5.2). If the value π is too low, these demand rates will be too high, and the constraint $\sum_{t=1}^{T} d^*(t) \leq C$ will be violated. If π is too high, total demand will not exhaust supply, and (5.3) will be violated. Of course, if $\pi = 0$ results in a total demand that is less than C, then this is the optimal dual value. Using these rules, it is straightforward to derive a search procedure to find the optimal π^*.

Another computational approach is to apply a greedy allocation algorithm, based on the observation that the marginal revenues in all periods are equal at optimality. Specifically, discretize the capacity C into M units of size δ each, so that $C = M\delta$. The greedy algorithm then proceeds by allocating demand in discrete amounts δ so as to equalize the marginal revenue:

STEP 0 (Initialize): Initialize solution $d(t) = 0$, $t = 1, \ldots, T$. Initialize counter $k = 0$.

STEP 1 (Evaluate marginal revenues): IF $\max_t\{J(t, d(t))\} > 0$, THEN DO:
Increment the demand of this highest marginal revenue period t^*:

$$d(t^*) \leftarrow d(t^*) + \delta.$$

ELSE, IF $\max\{J(t, d(t))\} \leq 0$ STOP (Current solution optimal).

STEP 2 (Check capacity constraint and repeat): IF $k = M$, STOP;
ELSE $k \leftarrow k + 1$ and GOTO STEP 1.

This algorithm takes $O(M \log T)$ time and is quite simple to program. Provided the marginal revenue is decreasing in each period, this greedy procedure produces an optimal (discretized) solution. (See Federgruen and Groenvelt [182].) The following example illustrates the algorithm:

Example 5.2 Consider a two-period problem with inverse-demand functions $p(1, d_1) = 10 - d_1$ and $p(2, d_2) = 10 - 2d_2$. The corresponding marginal revenue functions are

$$J(1, d_1) = 10 - 2d_1 \quad \text{and} \quad J(2, d_2) = 10 - 4d_2.$$

There are $C = 6$ units of capacity and we let the increment $\delta = 1$. The algorithm then proceeds as shown in Table 5.2. At the start, both periods have the same marginal revenue of 10. We break ties arbitrarily by assigning demand to period 1, so we assign the first unit to period 1. After this assignment, the marginal revenue in period 1 drops to 8 while the marginal revenue in period 2 is still 10, so we assign the next unit to period 2. The process continues as shown in Table 5.2, assigning units to the period with highest marginal revenue until all six units are used up. The algorithm terminates with $d_1^* = 4$ and $d_2^* = 2$; all six units are allocated and the marginal revenues are equalized $J(1, d_1^*) = J(2, d_2^*) = 2$.

Table 5.2. Example of the marginal-allocation algorithm.

k	d_1	d_2	$J(1, d_1)$	$J(2, d_2)$
0	0	0	10	10
1	1	0	8	10
2	1	1	8	6
3	2	1	6	6
4	3	1	4	6
5	3	2	4	2
6	4	2	2	2

5.2.1.2 Solution in the Time-Homogenous Case

A few additional observations can be made from this model when demand is time-homogenous, i.e., $d(t, p) = d(p)$ for all t. In this case, the optimal price p^*, given by $J(p^*) = \pi^*$, is the same in each period. This shows that prices fluctuate from period to period in the deterministic model (5.1) only as a result of changes in the demand function over time.

The optimal static price will either be the price that causes the supply to run out exactly at the end of the horizon (if $\pi^* > 0$) or the price at which the unconstrained revenue is maximized (if $\pi^* = 0$—that is, the revenue-maximizing price). Specifically, let p^0 be defined to be the value at which marginal revenue is zero, $J(p^0) = 0$, called the *revenue-maximizing price*. Let $\bar{p}$ denote the value at which $d(\bar{p}) = C/T$, which we call the *stock-clearing price*. Then

$$p^* = \max\{p^0, \bar{p}\},$$

so the optimal solution reduces to using the maximum of the revenue-maximizing price and the stock-clearing price. Simply, one cannot do better than pricing at p^0 at all times. If $Td(p^0) \leq C$, this price is feasible because demand is less than supply. If not, $Td(p^0) > C$, and demand at p^0 exceeds supply. We then have to raise the price, and $\bar{p} > p^0$ is the highest price at which we can still manage to sell all C units.

5.2.1.3 Discrete Prices

Often, in practice, we would like to choose prices from a discrete set. For example, prices close to convenient whole dollar amounts (such as $24.99 or $149.99), or fixed percentage markdowns (such as 25% off or 50% off) are often used because they are familiar to customers and easy to understand. In such cases, it may be desirable as a matter of policy to constrain prices to a finite set of k discrete price points, so that $p(t) \in \Omega_p$, where $\Omega_p = \{p_1, \ldots, p_k\}$. Equivalently, the sales rate $d(t)$ is constrained to a discrete set $d(t) \in \Omega_d(t)$ (time-varying in this case if the demand function is time-varying), where $\Omega_d(t) = \{d_1(t), \ldots, d_k(t)\}$, and $d_i(t) = d(t, p_i)$ denotes the sales rate at time t when using the price p_i.

The discreteness of the prices imposes technical complications when attempting to solve the dynamic pricing problem (5.1) because the problem is no longer continuous or convex. However, one can overcome this difficulty by relaxing the problem to allow the use of convex combinations of the discrete prices (or demand rates). In most periods, the optimal solution will be to use only one of the discrete prices; in the remaining periods, the solution has the interpretation of allocating a fraction of time to each of several prices.

To see this, define a vector of new variables $\alpha_i(t)$ for each t, $\boldsymbol{\alpha}(t) = (\alpha_1(t), \ldots, \alpha_k(t))$, which represent convex weights: they are nonnegative and sum to one. Next, in each period replace the variable $d(t)$ with the convex combination

$$d(t) = \sum_{i=1}^{k} \alpha_i(t) d_i(t),$$

and replace the constraint $d(t) \in \Omega_d(t)$ with the constraint

$$\boldsymbol{\alpha}(t) \in W \equiv \{\alpha \in \Re^k : \sum_{i=1}^{k} \alpha_i = 1, \alpha \geq 0\}.$$

The optimization problem is then

$$\max_{\boldsymbol{\alpha}(t) \in W} \sum_{t=1}^{T} \sum_{i=1}^{k} r_i(t) \alpha_i(t) \tag{5.4}$$

$$\text{s.t.} \quad \sum_{t=1}^{T} \sum_{i=1}^{k} \alpha_i(t) d_i(t) \leq C,$$

where $r_i(t) = p_i d_i(t)$ is the revenue rate at price p_i. This is a linear program in the variables $\boldsymbol{\alpha}(t)$, so it is easy to solve numerically.

To relate the solution to the unconstrained price case, introduce a dual variable π^* on the capacity constraint as before. The optimal solution $\boldsymbol{\alpha}^*(t)$ in each period is then characterized by solving

$$\max_{\boldsymbol{\alpha}(t) \in W} \left\{ \sum_{i=1}^{k} \alpha_i(t)(r_i(t) - \pi^* d_i(t)) \right\}, \tag{5.5}$$

where $\pi^* \geq 0$ and $\boldsymbol{\alpha}^*(t)$ are convex weights satisfying the complementary slackness condition

$$\pi^* \left(\sum_{t=1}^{T} \sum_{i=1}^{k} \alpha_i^*(t) d_i(t) - C \right) = 0. \tag{5.6}$$

Since the objective function of (5.5) is linear in $\alpha(t)$, if there is a unique index i^* for which $r_{i^*}(t) - \pi^* d_{i^*}(t)$ is greatest, then the optimal solution is simply $\alpha_{i^*}(t) = 1$, which corresponds to using the discrete price p_{i^*}. If there is more than one such value i^*, then there will be multiple solutions to (5.5), and determining which is optimal can be resolved by appealing to the complementary slackness condition (5.6). Of course, such a choice could result in a fractional solution in which $\alpha_i(t) > 0$ for two or more values i. However, this can be interpreted as saying that we should use the price i for a fraction $\alpha_i(t)$ of period t. Hence, the solution of (5.4) can be converted in practice into a discrete-price recommendation. The following example illustrates the calculation.

Example 5.3 Consider a two week selling season in which there is a linear-demand function $d(1,p) = 100 - p$ in week 1 and a demand function $d(2,p) = 100 - 1.4p$ in week 2. The firm is constrained to offer prices in the set $\{40, 50, 70\}$. The demand and revenues are then given in Table 5.3. Solving the linear program (5.4) for different value of the initial inventory C, we obtain the results in Table 5.4. For example, with an initial inventory of 50, the solution has $\alpha_{70}(1)$ and $\alpha_{50}(2) = 0.64$ and $\alpha_{70}(2) = 0.36$. This corresponds to pricing at \$70 for all of week 1 and 36% of week 2 then lowering the price to \$50 for the remainder of week 2. Similarly, when the initial inventory is 70, the solution calls for pricing at \$70 for half of week 1 and then lowering the price to \$50 for the remainder of the selling season. At very high levels of inventory (110 and 120), it is optimal to charge a price of \$50 in week 1 and a price of \$40 in week 2.

Dynamic Pricing

Table 5.3. Example of discrete prices and revenues.

p	$d(1,p)$	$r(1,p)$	$d(2,p)$	$r(2,p)$
20	60	2,400	44	1,760
30	50	2,500	30	1,500
50	30	2,100	2	140

Table 5.4. Solution of a linear program for the discrete-price example.

Inv. (C)	$\alpha_{40}(1)$	$\alpha_{50}(1)$	$\alpha_{70}(1)$	$\alpha_{40}(2)$	$\alpha_{50}(2)$	$\alpha_{70}(2)$	Total Sold
50	0.00	0.00	1.00	0.00	0.64	0.36	50
60	0.00	0.00	1.00	0.00	1.00	0.00	60
70	0.00	0.50	0.50	0.00	1.00	0.00	70
80	0.00	1.00	0.00	0.00	1.00	0.00	80
90	0.00	1.00	0.00	0.71	0.29	0.00	90
100	0.00	1.00	0.00	1.00	0.00	0.00	94
110	0.00	1.00	0.00	1.00	0.00	0.00	94
120	0.00	1.00	0.00	1.00	0.00	0.00	94

5.2.1.4 Maximum Concave Envelope

In the discrete-price problem, certain discrete prices may never be optimal to use and can in fact be eliminated from the problem. Indeed, suppose that for a give price p_j there exist convex weights $\alpha_i(t)$ such that

$$\sum_{i=1}^{k} \alpha_i(t) r_i(t) > r_j(t) \qquad (5.7)$$

$$\sum_{i=1}^{k} \alpha_i(t) d_i(t) \leq d_j(t).$$

Then the price p_j is never optimal at time t. Intuitively, this follows since a convex combination of other prices produces strictly higher revenue yet consumes no more capacity than using p_j. This is in fact the same notion of efficiency described in Section 2.6.2 for the discrete-choice model of demand in the single-resource capacity-control problem. All such inefficient prices j can be eliminated from consideration at time t. The remaining efficient prices define the *maximum concave envelope* of the pairs of values $\{(d_i(t), r_i(t)) : i = 1, \ldots, k\}$ as shown in Figure 5.2.

5.2.1.5 Inventory-Depletion Effect

Another practical factor affecting dynamic pricing in many retailing contexts is the adverse effects of low inventory levels. This is sometimes

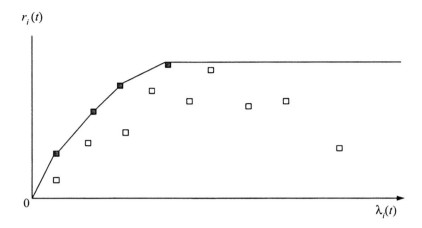

Figure 5.2. The maximum concave envelope produced by discrete prices (scatter plot of pairs $(r_i(t), d_i(t))$: efficient points are shaded).

referred to in retailing as a *broken-assortment effect*. For example, if the inventory-pricing model is applied at an aggregate item level, where an item contains several SKUs—such as color-size combinations in apparel retailing—then when inventories run low, certain SKUs may be out of stock even though there is a positive inventory for the item as a whole (for example, if a color or size runs out). The resulting reduction in alternatives naturally reduces the sales rate at any given price. Indeed, empirical studies have confirmed a positive correlation between inventory levels and sales rates [65].

These inventory-depletion effects can be modeled by making the demand rate a function of inventory as well as of price and time, so that the demand rate becomes $d(t, p(t), x(t))$. We can use a variety of functional forms to represent this inventory-depletion effect. For example, one proposed model is the following multiplicative form [480]:

$$\hat{d}(t, x(t)) = d(t)g(x(t)), \quad (5.8)$$

where $g(\cdot)$ is a depletion-effect term. We will call $d(t)$ the *unadjusted sales rate* (the rate of sales if inventory were unlimited) and $\hat{d}(t, x(t))$ the *adjusted sales rate* (the rate adjusted for inventory-depletion effects). One choice for g is

$$g(x) = 1 - \gamma \max\{0, 1 - x/x_0\},$$

where x_0 is the minimum *full-fixture inventory* and $0 \leq \gamma \leq 1$ is a sensitivity parameter. Both x_0 and γ can be estimated from historical data. Note that $g(x)$ is concave in x.

Another possible form is

$$g(x) = e^{-\gamma \max\{0, 1-x/x_0\}},$$

where γ and x_0 have the same interpretation (see Smith and Achabal [480]).

For this model with inventory depletion one must keep track of the inventory at each time t in the optimization problem. For example, assuming the multiplicative inventory-depletion model of (5.8) and formulating the problem in terms of the unadjusted sales rate $d(t)$, the inventory evolves according to the state equation

$$x(t+1) = x(t) - d(t)g(x(t)),$$

and the revenue-maximization problem can be formulated as

$$\max_{d(t) \geq 0} \sum_{t=1}^{T} r(t, d(t)) g(x(t)) \tag{5.9}$$

$$\text{s.t.} \quad x(t+1) = x(t) - d(t)g(x(t)), \quad t = 1, \ldots, T$$
$$x(T) \geq 0,$$
$$x(0) = C,$$

where $r(t, d(t)) = p(t, d(t))d(t)$ is the unadjusted revenue-rate function.

While somewhat more complex than the case without inventory-depletions effects, this is still a relatively simple nonlinear program to solve because the objective function is separable and the constraints are linear. (The objective function, however, is not necessarily jointly concave even if $r(t, d(t))$ and $g(x)$ are both concave.)

One qualitative impact of this inventory-depletion phenomenon is that optimal prices may decline over time even though the unadjusted revenue-rate function is time-invariant. (Recall that in the problem without inventory-depletion effects, a time-invariant revenue-rate function implied a time-invariant optimal price.) For example, Smith and Achabal [480] show, for the continuous-time version of this model, that if the unadjusted revenue-rate function is constant and the inventory-depletion effect is multiplicative, then optimal prices decline over time in such a way that the adjusted sales rate $g(x(t))d(t)$ is constant; that is, as inventory depletion reduces demand, the optimal prices fall to exactly compensate for the drop in sales due to inventory depletion.

5.2.1.6 A Retail Markdown Application

Here we look at the study of Heching et al. [247] that applied deterministic pricing models of the sort discussed above to analyze markdown

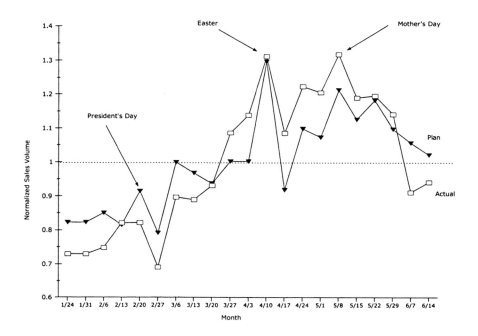

Figure 5.3. Sales volume for spring 1993: actual and planned.

pricing at an apparel retailer. The study provides an example of how such models can be applied and gives an indication of their potential impact.

The firm studied by Heching et al. [247] was a women's specialty apparel retailer with approximately 50 stores in the United States. The firm sold primarily its own private-label products and generally stocked items once at the beginning of the season. It then used markdowns to clear slow-selling merchandize.

The data set included the majority of the firm's sales over the spring 1993 season, spanning 184 styles in 25 groups (a collection of related styles). Weekly sales were obtained for each style sold during this period. Of all the styles in the data set, the firm took markdowns on 60 (the *markdown styles*). The remaining 124 styles had no price changes. While representing only one-third of all styles, markdown styles accounted for 42% of gross sales revenue.

There were strong seasonalities in sales due to major holidays and traditional shopping seasons as shown in Figure 5.3. Total weekly sales ranged from roughly 70% of average in slow weeks to 130% of average in the strongest weeks. The data also indicated that demand was indeed price-sensitive. After adjusting for seasonalities, the conditional probability of a sales increase, given a markdown, was 85%, while the unconditional probability of an increase was only 38%.

Sales of nearly all styles also tended to decline over time. Figure 5.4(i) plots weekly sales for one style that maintained the same price over its entire 12-week selling season. The weekly sales figures have been adjusted to eliminate any seasonality that can be attributed to traditional shopping seasons. Figure 5.4(ii) plots weekly sales for one of the markdown styles. A 28.6% markdown was implemented in week 6. The graph indicates a decline in sales over the weeks prior to the price change, as well as a decline in sales after the markdown price is implemented. Explanations for this declining-sales phenomenon include saturation of the customer base, loss of customers to competitors, a decline in the perceived value of an item as the selling season progresses and depletion of inventories of individual stock-keeping units.

The following demand model was used to model these features,

$$d(t,p) = w_t(a + bp)e^{-\alpha(t-t_0)},$$

where w_t is a seasonality factor, α is an age factor, and a and b are demand-function parameters. The seasonality factor was estimated from aggregate chainwide data. The age factor was estimated at the group level, while the demand function coefficients a and b were estimated using regression at the individual-style level. While there were significant errors in the prediction of individual weekly sales using this simple model, the average error in total revenues at the style level was only 1.2%; the error in the total revenue of all 60 markdown styles was only 0.53%.

The model was then used to estimate the effects of changes in the firm's markdown policy on the 60 markdown styles. The firm's markdown policy was compared with the markdowns recommended by a RM model that combined a simple online forecasting method with a deterministic dynamic-pricing model. Each week the demand function was reestimated, and an optimal price was computed based on this demand estimate. The new price was implemented if it was at least 20% lower than the initial price (a minimum markdown of 20%). The results are shown in Table 5.5. Note that model-based policy marks down only 33 of the 60 styles and that its average markdown week is much earlier than the firm's, though the average markdown is approximately the same. The estimated increase in revenue is 4.8%. This gain is due to (1) a better selection of which styles to markdown and (2) taking earlier markdowns on the styles that were marked down.

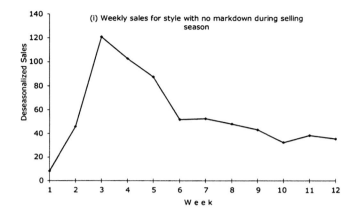

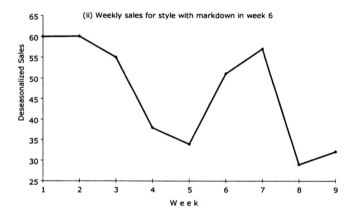

Figure 5.4. Effect of markdowns on two sample styles (sales adjusted for seasonality).

Table 5.5. Results of different markdown policies on 60 markdown styles.

	Model-Based Policy	*Firm's Policy*
Number of markdowns	33	60
Average markdown	25.3%	25.8%
Average markdown week	4.3	8.6
Revenue increase	4.8%	—

5.2.2 Stochastic Models

Next, we look at the case where the price-sensitive demand is stochastic. We separate the case of continuous-demand models from the case of Bernoulli (discrete Poisson) demand, though qualitatively the two cases

Dynamic Pricing

are similar. We assume that the stochastic regularity Assumption 7.6 (namely, that the demand has bounded variance) holds throughout.

5.2.2.1 Continuous Demand

Here we assume demand in each period is a continuous random variable $D(t, p, \xi_t)$ of the form discussed in Section 7.3.4 with expectation $d(t,p) = E[D(t, p, \xi_t)]$. Capacity is continuous as well. Also, we assume initially that prices in every period have no constraint other than being nonnegative.

As in the deterministic case, we assume that the demand function $d(t,p)$ has an inverse $p(t,d)$. As a result, there is a one-to-one correspondence between prices p and mean demand d in each period, so we can express the random demand as a function of d. That is, $D(t, d, \xi_t)$ is the demand in period t where $E[D(t, d, \xi_t)] = d$. In this way, we can view the mean demand d as our decision variable. We require the following convexity assumption on the random demand:

ASSUMPTION 5.1 *For all t, the random demand $D(t, d, \xi_t)$ is convex and increasing in d on the set $\{d : d \geq 0\}$ for every value ξ_t. That is,*

$$D(t, \alpha d_1 + (1-\alpha)d_2, \xi_t) \leq \alpha D(t, d_1, \xi_t) + (1-\alpha)D(t, d_2, \xi_t),$$

for all $d_1 \geq 0, d_2 \geq 0$ and for all $0 \leq \alpha \leq 1$.

This simply says the demand function is convex in d for each realization of the random-noise term ξ_t. (Such a random function is called strongly stochastically convex; see Appendix B.) Note that both the additive- and the multiplicative-demand models satisfy this convexity assumption, as do combinations of the two models.

We also define the following truncated expected revenue function:

$$r^+(t, d, x) = p(t, d) E\left[\min\{D(t, d, \xi_t), x\}\right]. \tag{5.10}$$

This is interpreted as follows. Given a remaining capacity x and a price $p(t,d)$ in period t, then $r^+(t,d,x)$ is the expected revenue received, since what we sell is the minimum of the demand $D(t, d, \xi_t)$ and the capacity available x. We make the following additional assumption:

ASSUMPTION 5.2 *For all t and for every value ξ_t, both the inverse-demand function $p(t, d)$ and the random revenue $p(t, d)D(t, d, \xi_t)$ are concave in d on the set $\Omega_d(t)$.*

While somewhat restrictive, one can show that this assumption holds for both the additive- and the multiplicative-demand models provided the inverse-demand function $p(t, d)$ and revenue function $r(t, d)$ are concave, which is true, for example, for the linear- and log-linear-demand functions.

The optimization problem can then be formulated as follows:

$$\begin{aligned} V_t(x) &= \max_{d \geq 0} E\left[p(t,d) E\left[\min\{D(t,d,\xi_t), x\}\right] + V_{t+1}(x - D(t,d,\xi_t))\right] \\ &= \max_{d \geq 0} \left\{r^+(t,d,x) + G_{t+1}(x,d)\right\}, \end{aligned} \qquad (5.11)$$

with boundary conditions are $V_{T+1}(x) = 0$ for all x and $V_t(0) = 0$ for all t, where we define

$$G_{t+1}(x, d) \equiv E[V_{t+1}(x - D(t, d, \xi_t))].$$

This function is like the value function, in that it gives the expected revenue to go in stage $t + 1$ as a function of certain state variables—in this case, the current inventory x and the demand rate decision d. The difference is that it replaces the future inventory state x_{t+1} in the value function $V_{t+1}(x_{t+1})$ by the two variables that determine x_{t+1}—namely, x and d.

The following proposition characterizes the properties of the functions $V_t(x)$ and $G_t(x, d)$:

PROPOSITION 5.1 *If Assumptions 5.1 and 5.2 hold, then for all t,*
(i) $G_t(x, d)$ is jointly concave in x and d,
(ii) $V_t(x)$ is concave in x, and
(iii) $\frac{\partial}{\partial d} G_t(x, d)$ is increasing in x and decreasing in d.

This proposition is proved in Appendix 5.A and has important consequences for the optimal pricing policy. First, under Assumption 5.2 $r^+(t, d, x)$ is concave in d (it is the minimum of two concave functions), and from Proposition 5.1(i) we know that G_{t+1} is concave in d as well. Therefore, a necessary and sufficient condition for an optimal d^* is obtained by differentiating the term inside the maximization in the dynamic program (5.11) and setting the result to zero, which yields

$$\frac{\partial}{\partial d} r^+(t, d, x) = -\frac{\partial}{\partial d} G_{t+1}(x, d^*).$$

By Proposition 5.1(iii), the right-hand side above is decreasing in x, and since $\frac{\partial}{\partial d} r^+(t, d, x)$ is decreasing in d, this means that higher inventory levels x imply a higher optimal sales rate d^*—and consequently a lower optimal price p^*—in any period t. That higher inventories lead to lower optimal prices is certainly intuitive.

5.2.2.2 Bernoulli Demand

If the random demand is Bernoulli (discrete Poisson), then a different analysis is required. Here we assume there is only one customer per period and the customer in period t has a willingness to pay v_t; that is, a random variable with distribution $F(t,v) = P(v_t \leq v)$. Therefore, if the firm offers a price of p in period t, it will sell exactly one unit if $v_t > p$ (with probability $1 - F(t,p)$). Letting $d(t,p) = 1 - F(t,p)$ denote the (average) demand rate, we can define an inverse-demand function, $p(t,d) = F_t^{-1}(1 - d(t))$ and revenue-rate function, $r(t,d) = dp(t,d)$, as before. The inventory and demand in this case are both assumed to be discrete.

Letting $V_t(x)$ denote the optimal expected revenue to go, the problem can be formulated in terms of demand rates $d(t)$ using the Bellman equation:

$$\begin{aligned} V_t(x) &= \max_{d \geq 0} \{d(p(t,d) + V_{t+1}(x-1)) + (1-d)V_{t+1}(x)\} \\ &= \max_{d \geq 0} \{r(t,d) - d\Delta V_{t+1}(x)\} + V_{t+1}(x) \end{aligned} \quad (5.12)$$

with boundary conditions $V_{T+1}(x) = 0$ for all x and $V_t(0) = 0$ for all t, where $\Delta V_t(x) = V_t(x) - V_t(x-1)$ is the expected marginal value of capacity. Under the monotonicity Assumption 7.2 and assuming an interior solution, necessary and sufficient conditions for the optimal rate d^* are

$$J(t, d^*) = \Delta V_{t+1}(x), \quad (5.13)$$

which again, as in the deterministic case of equation (5.2), has the interpretation that we set the marginal revenue equal to the marginal opportunity cost in every period t. One can show

PROPOSITION 5.2 *If Assumption 7.2 holds, then the expected marginal value of capacity, $\Delta V_t(x)$, of the dynamic program (5.12) is decreasing in t and x—that is, $\forall x, t$*
(i) $\Delta V_{t+1}(x) \leq \Delta V_t(x)$ and (ii) $\Delta V_t(x+1) \leq \Delta V_t(x)$.

Again, this monotonicity has intuitive implications for the optimal price. Consider, for simplicity, the case where the marginal revenue is not time dependent, so $J(t,d) = \frac{\partial r(t,d)}{\partial d} = J(d)$. Note that (5.13) and Assumption 7.2 (that $J(d)$ is decreasing) together imply that higher marginal values correspond to lower optimal-demand rates—and hence higher optimal prices. Thus, Proposition 5.2(i) above says that with more time remaining, the marginal value of capacity increases and there-

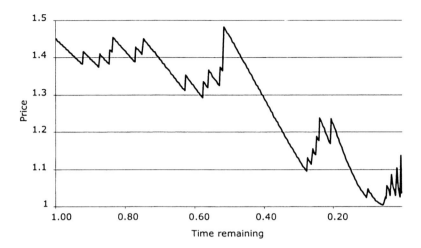

Figure 5.5. An example of the optimal price path in the stochastic case (25 units; exponential demand: $a = -1.1$, $b = 1$).

fore the optimal price increases as well.[2] Conversely, if time elapses without any sales taking place, the optimal price will fall. Proposition 5.2(ii) says the opposite is true of capacity; the more capacity remaining at any given point in time, the lower the optimal price. A numerical example illustrates this behavior:

Example 5.4 Consider a problem with $T = 333$ time-periods, an initial inventory of $C = 25$ units, and a time-homogeneous, exponential-demand function $d(t, p) = ae^{-bp}$ in each period t with parameters $a = 1.1$ and $b = 1$. A sample of the optimal-price path is shown in Figure 5.5. The time axis is normalized to one and represents the fraction of total time remaining. The points at which the price jumps correspond to sales; each sale results in a step increase in price. As time elapses without any sales taking place, prices decline. This is exactly the behavior implied by Proposition 5.2.

5.2.2.3 Comparing the Deterministic and Stochastic Models

One fact that is useful theoretically and computationally is that the deterministic model (5.1) provides an upper bound on the expected revenue from the stochastic model (5.11). This can be shown in a variety of ways. For example, by relaxing the capacity constraint in the stochastic

---

[2] Note this behavior does not necessarily hold if the marginal revenue varies with time, since in such cases whether the condition $J(t, d^*) = \Delta V_{t+1}(x)$ results in d^* rising or falling over time for a fixed x depends on how both $J(t, d)$ and $\Delta V_{t+1}(x)$ vary with time.

Dynamic Pricing

problem with a multiplier $\pi \geq 0$, we can form the relaxed problem

$$\max_{d(t)} E\left[\sum_{t=1}^{T} D(t, d(t), \xi_t) p(t, d(t)) + \pi(C - \sum_{t=1}^{T} D_t(d(t)))\right], \quad (5.14)$$

where the random variable $D_t(d(t)) = 1$ if there is an arrival in period t using the control $d(t)$, and $D_t(d(t)) = 0$ otherwise. Note an optimal policy for the original stochastic problem (5.11) satisfies $x(0) \geq \sum_{t=1}^{T} D_t(d(t))$ (a.s.); therefore, since $\pi \geq 0$ if we evaluate the objective function of (5.14) for such an optimal policy, it will give an upper bound on the optimal expected revenue for the original problem (5.11). Hence, maximizing (5.14) over all $d(t)$ certainly provides an upper bound as well. However, this function is separable in time t, so we can choose the control in each period to maximize $E[D_t(d(t))(p(t, d(t)) - \pi)]$ at each time t. Since $E[D_t(d(t))] = d(t)$, (5.14) is equivalent to maximizing

$$\sum_{t=1}^{T}[r(t, d(t)) - \pi d(t)] + \pi C.$$

This is solved as in the deterministic case by setting the marginal revenue $J(t, d) = \pi$ in each period t. Since this upper bound is valid for any $\pi \geq 0$, we can take $\pi = \pi^*$, the optimal dual price in the deterministic problem. This results in $d(t) = d^*(t)$, the optimal solution of the deterministic problem. Moreover, by the complementary slackness condition (5.3), the optimal dual price satisfies $\pi^*(C - \sum_{t=1}^{T} d^*(t)) = 0$, so the bound (5.14) becomes

$$\sum_{t=1}^{T} d^*(t) p(t, d^*(t)),$$

which is exactly the optimal deterministic revenue. Hence, the optimal deterministic revenue is an upper bound on the optimal expected stochastic revenue.

It's also possible to show that the solution produced by the deterministic dynamic-pricing problem is a reasonably good heuristic for the stochastic-pricing problem. Numerically, it performs well, and theoretically it can be shown to be asymptotically optimal for problems with large demand volumes (such as a large number of time-periods) and large initial inventories (see Gallego and van Ryzin [198]). Such properties provide support for using deterministic models as an approximation. The following example illustrates the deterministic approximation:

Example 5.5 Consider a variation of Example 5.4, where we have $T = 333$ periods and a time-homogenous exponential demand function $d(t, p) = ae^{-bp}$ with parameters

$a = 1$ and $b = 2.718$. Since the demand function is not time varying, the optimal deterministic price is constant. We denote this p^{DET}. The unconstrained revenue-maximizing price is $p^0 = 1$. Starting inventories C range from 1 to 20. Table 5.6 shows the prices and resulting revenues for this problem. As Table 5.6 shows, the

Table 5.6. Example performance of the deterministic-price heuristic.

C	p^{DET}	$V_T(C)$	$\frac{V_T^{DET}(C)}{V_T(C)}$
1	3.30	2.40	0.871
2	2.61	4.11	0.926
3	2.20	5.43	0.945
4	1.92	6.47	0.954
5	1.69	7.30	0.956
6	1.51	7.96	0.956
7	1.35	8.49	0.952
8	1.22	8.89	0.946
9	1.11	9.22	0.937
10	1.00	9.46	0.925
11	1.00	9.64	0.951
12	1.00	9.77	0.970
13	1.00	9.85	0.982
14	1.00	9.91	0.990
15	1.00	9.95	0.995
16	1.00	9.97	0.997
17	1.00	9.99	0.999
18	1.00	9.99	0.999
19	1.00	10.00	1.000
20	1.00	10.00	1.000

relative performance of the deterministic heuristic is poorest at $C = 1$ (13% below the optimal revenue) and $C = 10$ (7.5% below the optimal revenue) but otherwise performs reasonably well, especially in the very unconstrained case of initial inventory approaching 20. Note that $Td(p^0) = 10$, so $C = 10$ is the boundary between the constrained and unconstrained regions of the deterministic problem (the constrained region is where the multiplier $\pi^* > 0$ and the stock-clearing price $\bar{p}$ is used; the unconstrained region is where $\pi^* = 0$ and the revenue-maximizing price p^0 is used).

Intuitively, the deterministic prices perform well because they capture the correct "first-order" effect. That is, they maximize revenue subject to the constraint that the mean demand is within the capacity constraint. The stochastic policy does this as well but also adjusts prices dynamically to respond to fluctuations about the mean demand. In addition, the stochastic policy has a tendency to price higher earlier in the sales process (Figure 5.5), which reflects the *option value* of keeping initial prices high in the event realized demand is stronger than average. These two "second-order" adjustments result in the improvement in revenue exhibited in Table 5.6.

Hence, there are really two separate benefits to dynamic pricing. The first is simply to exploit the time-varying price sensitivity of customers; if the demand function $d(t,p)$ varies with t, then even the optimal deterministic price will vary with t due to the optimality condition (5.2). But in addition if demand is stochastic, dynamic pricing helps compensate for random fluctuations in demand and the option value of holding rather than selling units. This is seen in Figure 5.5, where the optimal stochastic prices vary despite the fact that the optimal deterministic prices for this example are constant. In general, both factors will be present in practical problems, but it is useful to distinguish the different forces at work in each case.

5.2.2.4 Prices Constrained to a Discrete Set

Just as in the deterministic case, it may be desirable in practice to constrain the prices to a finite set, $p(t) \in \Omega_p$, where $\Omega_p = \{p_1, \ldots, p_k\}$. Equivalently, the sales rate d are constrained to a discrete set, $d(t) \in \Omega_d(t)$, where as before, $\Omega_d(t) = \{d_1(t), \ldots, d_k(t)\}$, $d_i(t) = d(t, p_i)$ denotes the sales rate at time t when using the price p_i, and $r_i(t) = p_i d_i(t)$ denotes the corresponding revenue rate. For simplicity, we consider only the Bernoulli demand case here (see Section 7.3.4.3).

Computationally, using discrete prices is not a difficult change and in fact reduces the complexity of the dynamic program (5.12) because the search at each stage is now reduced to a finite set of prices. As in the deterministic case, the finite set of prices can be further reduced to only those prices defining the maximum concave envelope (the efficient prices) by using the efficiency criteria (5.7). The reasoning is identical to the deterministic case; inefficient prices produce less expected revenue and have a higher probability of consuming capacity than done by efficient prices (or mixtures of efficient prices) and therefore are never an optimal choice.

Theoretically, the analysis of the discrete problem can again be relaxed to put it in a form similar to the unconstrained price case by allowing the firm to randomize over the discrete set of prices. As in the deterministic case, define new variables $\alpha_i(t)$ that represent convex weights, and in each period replace the variable $d(t)$ in (5.12) with the convex combination

$$d(t) = \sum_{i=1}^{k} \alpha_i(t) d_i(t)$$

and replace the constraint $d(t) \in \Omega_d(t)$ with the constraint

$$\boldsymbol{\alpha}(t) \in W \equiv \{\boldsymbol{\alpha} : \sum_{i=1}^{k} \alpha_i = 1, \boldsymbol{\alpha} \geq 0\}.$$

The dynamic program then becomes

$$V_t(x) = \max_{\boldsymbol{\alpha}(t) \in W} \left\{ \sum_{i=1}^{k} [r_i(t) - d_i(t)\Delta V_{t+1}(x)] \alpha_i(t) \right\} + V_{t+1}(x)$$

with the usual boundary conditions $V_{T+1}(x) = 0$, $\forall x$ and $V_t(0) = 0$ for all t. In the stochastic case, a fractional solution $\boldsymbol{\alpha}(t)$ can be directly interpreted as a randomization of the prices in $\Omega_p = \{p_1, \ldots, p_k\}$. Also, one can eliminate inefficient prices using the maximum concave envelope as in the deterministic case.

We can put this in a form similar to (5.12) by noting there is a correspondence between the optimal choice of $\boldsymbol{\alpha}(t)$ and the optimal value of $d(t) = \sum_{i=1}^{k} \alpha_i(t)d_i(t)$, since for any fixed sales rate d, the optimal $\boldsymbol{\alpha}(t)$ that achieves this sales rate must maximize the expected revenue—that is, it solves the linear program

$$\hat{r}(t, d) = \max \sum_{i=1}^{k} \alpha_i r_i(t)$$

$$\text{s.t.} \quad \sum_{i=1}^{k} \alpha_i d_i(t) = d$$

$$\sum_{i=1}^{k} \alpha_i d_i(t) = 1$$

$$\alpha \geq 0.$$

The resulting $\hat{r}(t, d)$ in fact will define the maximum concave envelope of the fixed set of prices. Hence, the optimization problem can be formulated as

$$V_t(x) = \max_d \{\hat{r}(t, d) - d\Delta V_{t+1}(x)\} + V_{t+1}(x), \tag{5.15}$$

which has exactly the same form as (5.12) except that the maximum concave envelope function $\hat{r}(t, d)$, though continuous and concave, is no longer differentiable. (Like all objective functions of a maximization linear program, $\hat{r}(t, d)$ is a concave and piecewise linear function of the right-hand side d.) Thus, the optimality condition (5.13) must be replaced by

$$\Delta V_{t+1}(x) \in \partial r(t, d^*(t)), \tag{5.16}$$

where $\partial r(t, d^*(t))$ denotes the set of subgradients (the subdifferential) of $r(t, d^*(t))$ at the value $d^*(t)$. (See Appendix C for a definition and discussion of subgradients and nondifferentiable optimization.)

Practically speaking, the above condition implies that the optimal d will most often be at a corner point of the function $\hat{r}(t, d)$ (or there will be multiple optimal solutions along an interval containing two adjacent corner points) and we can always find one of the fixed prices that is optimal without randomizing. However, by formulating the problem this way, we preserve the concavity of $\hat{r}(t, d)$. Therefore, the structure of the optimal solution to (5.15) is the same as that of (5.12), and Proposition 5.1 continues to hold for this case.

5.3 Single-Product Dynamic Pricing with Replenishment

We next consider situations in which inventory can be replenished at a cost in each period, as in many production and supply-chain-management contexts. In such cases, both pricing and inventory decisions need to be made; pricing decisions are used to control demand, while replenishment decisions are used to control supply. The central problem is to optimally coordinate these demand and supply decisions.

As in the finite-supply case, we first look at deterministic models of this problem and then examine stochastic models.

5.3.1 Deterministic Models

We assume a single good with an end-of-period inventory, denoted $x(t)$, that can be replenished over time. There is a per-unit holding cost h_t for inventory in period t and a unit cost for replenishment c_t. We let $y(t)$ denote the amount ordered in period t. As in the finite-supply case, we can formulate the problem in terms of the sales rate $d(t)$, in which case we let $r(t, d(t))$ and $J(t, d)$ denote, respectively, the revenue rate and marginal revenue as before. Again, we assume that these functions satisfy the regularity conditions in Assumptions 7.1, 7.2, and 7.3 unless otherwise specified.

5.3.1.1 Unconstrained Capacity

We first consider the case where is no capacity constraint on the amount ordered in each period. The problem can be formulated as finding a set of rates $d^*(t)$ and reorder quantities $y^*(t)$ that solve

$$\max \sum_{t=1}^{T} r(t, d(t)) - h_t x(t) - c_t y(t) \qquad (5.17)$$

s.t. $\quad x(t) = x(t-1) - d(t) + y(t), \quad t = 1, \ldots, T$
$\quad\quad d(t), x(t), y(t) \geq 0, \quad t = 1, \ldots, T,$

where we assume the initial inventory $x(0) = 0$ for simplicity.

The problem as stated above is not difficult to solve. Indeed, for $s \leq t$, define the coefficients

$$\gamma_{st} = c_s + \sum_{k=s}^{t-1} h_k,$$

and note that α_{st} is the cost of satisfying demand in period t with supply from period s. Let

$$\gamma^*(t) = \min_{s \leq t} \{\gamma_{st}\},$$

denote the lowest cost for supplying period t, and let $s^*(t)$ denote an index that achieves the minimum on the right-hand side above.

The optimal sales rate in any period t, $d^*(t)$, is then determined by equating the marginal revenue to this lowest marginal cost,

$$J(t, d^*(t)) = \gamma^*(t), \quad t = 1, \ldots, T.$$

And the optimal quantity to order in period s is simply determined by adding up the sales rates from later periods t whose lowest-cost supply is from period s,

$$y^*(s) = \sum_{t: s^*(t) = s} d^*(t), \quad s = 1, \ldots, T.$$

An interesting observation for this problem is that even if the demand functions are time-invariant ($r(t, d) = r(d)$ for all t), the optimal price can still vary over time due to changes in the cost of supply. In other words, because the optimality conditions equate marginal revenue to marginal cost, $J(d^*(t)) = \gamma^*(t)$, changes in the costs $\gamma^*(t)$ over time will lead to time-varying prices, even though the marginal revenue function is time-invariant.

5.3.1.2 Capacity Constraints on Ordering

The problem becomes somewhat more complex when there are capacity constraints on the order quantities of the form

$$y(t) \leq b_t, \quad t = 1, \ldots, T.$$

Such constraints, for example, could be due to limited production, transportation, or handling capacity. While (5.17) can be solved as a nonlinear program with these added capacity constraints, there is a simpler approach. If one discretizes the sales quantities, we can solve the problem using a greedy algorithm under the assumption that the marginal revenue in each period is decreasing. (See Chann, Simchi-Levi, and Swann [105] for a proof.)

The greedy algorithm proceeds as follows. For a fixed vector of rates $\mathbf{d} = (d(1), \ldots, d(T))$, define

$$f(\mathbf{d}) = \sum_{t=1}^{T} r(t, d(t)) - g(\mathbf{d}), \qquad (5.18)$$

where $g(\mathbf{d})$ is the minimum cost for meeting the sales rates $\mathbf{d}$, defined by fixing $\mathbf{d}$ and solving the following optimization problem in the variables $x(t), y(t), t = 1, \ldots, T$:

$$g(\mathbf{d}) = \min \sum_{t=1}^{T} h_t x(t) + c_t y(t) \qquad (5.19)$$

$$\text{s.t.} \quad x(t) = x(t-1) - d(t) + y(t), \quad t = 1, \ldots, T$$

$$y(t) \leq b_t, \quad t = 1, \ldots, T$$

$$x(t), y(t) \geq 0, \quad t = 1, \ldots, T.$$

Thus, $f(\mathbf{d})$ is the optimal profit given the demand rates $\mathbf{d}$. Computing $f(\cdot)$ is efficient because the minimization problem to determine $g(\cdot)$, (5.19), is simply a minimum-cost network-flow problem.

For notational convenience, let $\mathbf{e}_t$ denote the tth unit vector (the vector with a 1 in the tth component and a zero in all other components), and let δ denote the discretization increment (all components of the vector d are assumed to be integral multiples of δ).

The greedy algorithm is as follows.

STEP 0 (Initialize): Initialize solution

$$\mathbf{d} = (d(1), \ldots, d(T)) = (0, \ldots, 0).$$

Calculate $f(\mathbf{d})$ using (5.18).

STEP 1 (Compute marginal values): FOR $t = 1, \ldots, T$, DO:
Compute $f(\mathbf{d} + \delta \mathbf{e}_t)$ from (5.18).

STEP 2 (Find largest marginal increase): Chose the index t^* for which the marginal gain $f(\mathbf{d} + \delta \mathbf{e}_t) - f(\mathbf{d})$ is largest.
IF $f(\mathbf{d} + \delta \mathbf{e}_{t^*}) - f(\mathbf{d}) \leq 0$, STOP (optimal solution found);
ELSE, update $\mathbf{d}$:

$$\mathbf{d} \leftarrow \mathbf{d} + \delta \mathbf{e}_t$$

and GOTO STEP 1.

In words, at each stage the algorithm simply adds an increment δ of demand to the period t that yields the largest net gain $f(\mathbf{d}+\delta \mathbf{e}_t) - f(\mathbf{d})$ and stops when no period produces a positive net gain. Biller et al. [67] report a test of this model and algorithm on data from the automobile industry.

5.3.2 Stochastic Models

A stochastic version of the dynamic-pricing problem with replenishment can also be formulated as follows: As in Section 5.2.2, let $x(t)$ denote the inventory at the end of period t and T be the number of periods in the horizon. (We consider an infinite-horizon, stationary version of the problem in Section 5.3.2.2.) Because demand is random, it is possible that demand in a period can exceed the available inventory. In such cases, we assume that the firm can back-order demand, and this is represented by a negative inventory $x(t)$.

As before, we represent demand in each period as a random variable $D(t, p, \xi_t)$, of the form discussed in Section 7.3.4, with expectation $d(t, p) = E[D(t, p, \xi_t)]$, with a unique inverse $p(t, d)$. We assume that the quantities and demand are continuous. Also, we assume that prices in every period are unconstrained (with $p \geq 0$ the only requirement). Finally, we assume that the demand $D(t, d, \xi_t)$ satisfies the regularity condition in Assumption 7.6 and the convexity condition in Assumption 5.1. The random revenue in each period is $R(t, d, \xi_t) = p(t, d) D(t, d, \xi_t)$.

The inventory after ordering is denoted $y(t)$, and hence the quantity ordered is $y(t) - x(t)$. For notational convenience, we use $y(t)$ as the quantity-decision variable. We assume that we cannot dispose of items, so $y(t) \geq x(t)$.

There is a per-unit ordering cost c_t in period t and a convex cost $h_t(x)$ on the ending inventory x in period t. This cost typically will penalize both positive inventories (due to capital costs, storage costs, and so on), and negative inventories (due to lost goodwill or penalties for late delivery). For example, a function of the form

$$h(x) = ax^+ + bx^-$$

is commonly used, where $x^+ = \max\{x, 0\}$, $x^- = \max\{-x, 0\}$, a is the cost of holding a unit, and b is the penalty cost for back-ordering a unit.

5.3.2.1 Finite-Horizon Problem

In the multi-period case, the optimization problem can then be formulated as follows:

$$V_t(x) = \max_{y \geq x, d \geq 0} E\left[R(t, d, \xi_t) - c_t(y - x) - h_t(y - D(t, d, \xi_t))\right]$$

Dynamic Pricing

$$\begin{aligned}&\quad\quad\quad\quad\quad +V_{t+1}(y - D(t,d,\xi_t))]\\ &= \max_{y \geq x, d \geq 0} \{r(t,d) - c_t(y-x) + G_{t+1}(y,d)\},\end{aligned} \quad (5.20)$$

where we define

$$G_{t+1}(y,d) \equiv E\left[V_{t+1}(y - D(t,d,\xi_t)) - h_t(y - D(t,d,\xi_t))\right].$$

Using arguments that are essentially the same as those in Proposition 5.1, one can show the following:

PROPOSITION 5.3 *(i)* $G_t(y,d)$ *is jointly concave in* y *and* d.
(ii) $V_t(x)$ *is concave in* x.
(iii) $\frac{\partial}{\partial d} G_t(y,d)$ *is increasing in* y.
(iv) $\frac{\partial}{\partial y} G_t(y,d)$ *is increasing in* d.

Proposition (5.3) (iii) and (iv) imply that G_t is a *supermodular function*. (See Appendix C for a definition of the supermodularity property.) These properties allow us to characterize the optimal pricing and ordering policy.

Specifically, let $y^0(t)$ and $d^0(t)$ denote the values that maximize (5.20) without the constraint $y \geq x$; that is, they solve

$$\max_{d \geq 0, y} \{r(t,d) - c_t(y-x) + G_{t+1}(y,d)\}.$$

Further, for simplicity assume an interior optimal solution for d and y so that, by joint concavity of G_t, the necessary and sufficient conditions for $y^0(t)$ and $d^0(t)$ are then

$$\begin{aligned}J(t,d^0(t)) &= -\frac{\partial}{\partial d}G_{t+1}(y^0(t), d^0(t))\\ c_t &= \frac{\partial}{\partial y}G_{t+1}(y^0(t), d^0(t)).\end{aligned}$$

(If there are two or more sets of values satisfying these conditions, take the pair $(y^0(t), -d^0(t))$ that is lexicographically the largest.)

It follows, then, that if $x \leq y^0(t)$, the optimal policy in period t is to order up to $y^0(t)$ and set the demand rate at $d^0(t)$ (that is, $y^* = y^0(t)$ and $d^* = d^0(t)$), since the unconstrained optimal solution $(d^0(t), y^0(t))$ is feasible. However, if $x > y^0(t)$, then one can show that it is optimal to order nothing (for example, set $y^* = x$) and choose a demand rate d^* that is higher than $d^0(t)$. Equivalently, set the price lower than $p(t, d^0(t))$. Moreover, the higher the inventory x, the higher the optimal rate d^* (equivalently, the lower the optimal price $p(t, d^*)$).[3] The resulting policy

---

[3] To see this, we can argue informally as follows. Suppose that the optimal y^* and d^* satisfy $y^* > x > y^0(t)$ and $d^* < d^0(t)$. Then since the constraint $y \geq x$, is not binding, these values

is called a *base-stock, posted-price policy*. If inventory is less than the *base-stock* level $y^0(t)$, then order up to this level, and price at the *posted price* $p(t, d^0(t))$. If inventory exceeds the optimal base-stock level $y^0(t)$, then order nothing, and discount the price below the posted price, with the discount being larger the more the inventory exceeds the optimal base-stock level.

5.3.2.2 Infinite-Horizon, Stationary Problem

One can also extend this same analysis to an infinite-horizon setting. We assume that all the parameters of the problem are time-invariant and profits are discounted by a factor $0 < \beta < 1$ in each period.[4]

The value function in this case is also time-homogenous. The formulation is

$$V(x) = \max_{y \geq x, d \geq 0} \left\{ r(d) - c(y - x) + \beta G(y, d) \right\},$$

where

$$G(y, d) \equiv E\left[V(y - D(d, \xi)) - h(y - D(d, \xi))\right].$$

In this infinite-horizon case, one can show that a time-invariant, base-stock, posted-price policy is optimal. That is, there exist values y^0 and d^0 such that if $x \leq y^0$, it is optimal to order up to y^0 and price at $p^0 = p(d^0)$. If $x > y^0$, we order nothing, and the optimal demand rate d^* is greater than d^0 and increasing in x. Note that in this infinite-horizon case, once we reach a point where $x < y^0$, then in all remaining periods we simply price at the posted price p^0 and order up to y^0. In other words, we use dynamic pricing only to clear inventory that is higher than the optimal base stock y^0. However, since such high inventory levels are only transient, in the long run, the policy ends up using a constant price.

---

must satisfy the first-order condition,

$$J(t, d^*) = -\frac{\partial}{\partial d} G_{t+1}(y^*, d^*)$$

$$c_t = \frac{\partial}{\partial y} G_{t+1}(y^*, d^*).$$

But this contradicts the fact the $(y^0(t), -d^0(t))$ are the lexicographically largest pair of values satisfying the first-order conditions. Therefore, we must have $y^* = x$ and $d^* \geq d^0(t)$. Since $y^* = x$, the fact that d^* is increasing in x now follows from the fact that $J(t, d^*) = -\frac{\partial}{\partial d} G_{t+1}(x, d^*)$, that $J(t, d)$ is decreasing in d, and that $-\frac{\partial}{\partial d} G_{t+1}(x, d)$ is decreasing in y. See Federgruen and Heching [183] for a complete proof of these properties.

[4]Similar results hold for the case of the average profit criteria by considering the discounted problem with $\beta \to 1$. See Federgruen and Heching [183].

5.3.2.3 Fixed Costs

Another variation of the problem is to include a fixed cost for ordering. The finite-horizon version of this problem was studied by Chen and Simchi-Levi [113]. In this case, the cost function becomes

$$c_t(x) = \begin{cases} K_t + c_t x & \text{if } x > 0 \\ 0 & x = 0 \end{cases}$$

This results in a significantly more complex value function. However, one can show, in certain cases, that properties of the optimal policy are similar to those of classical inventory theory. For example, when the demand function has additive uncertainty, then the optimal ordering policy is of the (s_t, S_t) form, wherein we order only in period t if the inventory $x(t)$ drops below s_t, and in this case we order enough to restore the inventory to the target level S_t (order an amount $S_t - x$). However, this property does not hold for other stochastic-demand functions.[5]

Moreover, the optimal state-dependent price is quite a bit more complex. For example, Chen and Simchi-Levi [113] show that, as a function of the current inventory, the optimal price may not be decreasing in the inventory level between ordering epochs. This is because while there is an incentive to decrease price to reduce inventory, there is also an incentive to increase price to delay reordering and postpone incurring the fixed-ordering cost.

5.4 Multiproduct, Multiresource Pricing

Multiproduct, multiresource—or network—versions of dynamic pricing problems arise in many applications. Two fundamental factors typically link the pricing decisions for multiple products. First, demand for products may be correlated. For example, when products are substitutes or complements, the price charged for one product effects the demand for other related products. Then, a firm jointly managing the pricing of a family of such products must consider these cross-elasticity effects when determining its optimal pricing policy. Second, products may be linked by joint capacity constraints. For example, two products may require the same resource, which is available in limited supply. Even if there are no cross-elasticity effects between the two products, the pricing decision for one product will need to account for the joint effect on demand for the other product that uses the limited resource.

As in the case of capacity controls, most problems in real life are multiproduct problems, either because of cross-elasticity effects or because

---
[5] A somewhat more complex variant of this (s, S) policy does hold more generally, however; see Chen and Simchi-Levi [113].

of joint capacity constraints, or both. For example, a grocery store that is pricing brands in a food category—say, salty snacks—needs to consider the cross-elasticity effects of its pricing decision for all products in the category. An increase in the price of a packet of potato chips will not just cause a drop in demand for potato chips but will likely also increase the demand for corn chips. At the same time, these products may occupy the same limited shelf space, so stocking more of one product may require stocking less (or none) of other products.

We can model such situations using multiproduct-demand functions and joint capacity constraints on resources. However, like the network problems of capacity control, such formulations quickly become difficult to analyze and solve, which is the reason that many commercial applications of dynamic-pricing models make the simplifying assumption of unrelated products and independent demands and solve a collection of single-product models as an approximation.

Yet in cases where cross-elasticity or resource-constraint effects are strong—for example, when products are only slightly differentiated, customers are very price-sensitive, or joint capacity constraints are tight—then ignoring multiproduct effects can be severely suboptimal. In such cases, we must solve a pricing problem incorporating these effects—or at least approximating them in some fashion. In this section, we look as such multiproduct, multiresource models and methods.

5.4.1 Deterministic Models Without Replenishment

Under a deterministic-demand assumption, it is relatively straightforward to formulate a multiproduct, multiresource version of dynamic pricing similar to those described in Section 5.2. There are n products, indexed by j, and m resources, indexed by i. There is a horizon of T periods, with each period indexed by t. As in Section 7.3.2, let $\mathbf{d} = (d_1, \ldots, d_n)$ denote the demand rate for the n products and $\mathbf{p}(t, \mathbf{d})$ denote the inverse-demand function in period t. We further assume that the revenue-rate function $r(t, \mathbf{d})$ satisfies the regularity conditions of Assumption 7.4.

Product j uses a quantity a_{ij} of resource i. The matrix $\mathbf{A} = [a_{ij}]$ therefore describes the *bill of materials* for all n products. We assume there are limited capacities $\mathbf{C} = (C_1, \ldots, C_m)$ of the m resources.

The dynamic-pricing problem can then be formulated as finding a sequence of demand vectors $d^*(t)$ that maximizes the firm's total revenue

Dynamic Pricing 217

subject to the capacity constraints C:

$$\max \quad \sum_{t=1}^{T} r(t, \mathbf{d}(t)) \tag{5.21}$$

$$\text{s.t.} \quad \sum_{t=1}^{T} \mathbf{A}\mathbf{d}(t) \leq \mathbf{C}$$

$$\mathbf{d}(t) \geq 0, \quad t = 1, \ldots, T.$$

By Assumption 7.4, $r(t, \mathbf{d})$ is concave in $\mathbf{d}$, and therefore, the following Kuhn-Tucker conditions are necessary and sufficient for characterizing an optimal solution $\mathbf{d}^*(t)$ to (5.21):

$$J(t, \mathbf{d}^*(t)) = \mathbf{A}^\top \boldsymbol{\pi}^* \tag{5.22a}$$

$$\boldsymbol{\pi}^{*\top}(\mathbf{C} - \sum_{t=1}^{T} \mathbf{A}\mathbf{d}(t)) = 0 \tag{5.22b}$$

$$\boldsymbol{\pi}^* \geq 0, \tag{5.22c}$$

where $J(t, \mathbf{d}) = \nabla_d r(t, \mathbf{d})$ is the marginal-value vector and $\boldsymbol{\pi}^*$ is the optimal dual price on the joint-capacity constraints, having the usual interpretation as the vector of marginal opportunity costs (marginal values) for the m resources. Condition (5.22a) says that at the optimal sales rate, the marginal revenue for each product j should equal the marginal opportunity cost of the resources used by product j, or $\boldsymbol{\pi}^{*\top}\mathbf{A}_j$. Condition (5.22b) says that the marginal opportunity cost of resource i can be positive only if the corresponding capacity constraint for resource i is binding. Finally, (5.22c) requires that the marginal opportunity costs be nonnegative.

The nonlinear program (5.21) is relatively easy to solve numerically, since the objective function is concave and the constraints are linear. (See Bertsekas [58, 59] for specific techniques.)

Example 5.6 Consider the six-node airline network shown in Figure 5.6. Nodes 2 and 3 are "hub" nodes. (Leg seat capacities are as indicated in the figure.) For a given path j on the network, the revenue function is time homogeneous and log-linear

$$d_j(p_j) = a_j e^{-\epsilon_j(p_j/\bar{p}_j - 1)},$$

where $\bar{p}_j$ is interpreted as a reference price for itinerary j, a_j is the demand rate at the reference price, and ϵ_j is the magnitude of the elasticity of demand at the reference price. Demand-function parameters for all O-D pairs are shown in Table 5.7 along with the path (itinerary) used by each O-D pair.

Because the demand functions are time-homogeneous, optimal prices are constant over time. The optimal O-D prices and demand are shown in the last two columns in Table 5.7. The solution gives a total revenue of $661,200$ across all O-D pairs.

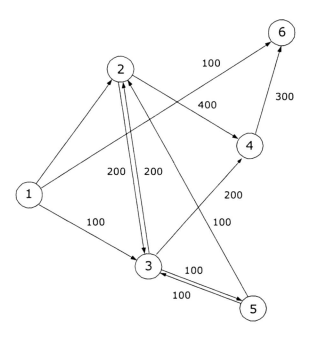

Figure 5.6. A six-node, two-hub airline network.

5.4.2 Deterministic Models with Replenishment

We can formulate deterministic multiproduct models with replenishment, analogous to those in Section 5.3 as follows:

$$\max \quad \sum_{t=1}^{T} r(t, \mathbf{d}(t)) - \mathbf{h}_t^\top \mathbf{x}(t) - \mathbf{c}_t^\top \mathbf{y}(t) \quad (5.23)$$
$$\text{s.t.} \quad \mathbf{x}(t) = \mathbf{x}(t-1) - \mathbf{A}\mathbf{d}(t) + \mathbf{y}(t), \quad t = 1, \ldots, T$$
$$\mathbf{y}(t) \leq \mathbf{b}_t, \quad t = 1, \ldots, T$$
$$\mathbf{d}(t) \geq 0, \quad t = 1, \ldots, T$$
$$\mathbf{x}(t), \mathbf{y}(t) \geq 0, \quad t = 1, \ldots, T,$$

where $\mathbf{x}(t)$ is an m-vector of inventory levels at the end of period t, $\mathbf{y}(t)$ is an m-vector of order quantities in period t, $\mathbf{h}_t$ is a vector of holding costs, $\mathbf{c}_t$ is a vector of ordering costs, and $\mathbf{b}_t$ is a vector of capacity constraints on the order quantities.

The introduction of an inventory-state variable makes this a more difficult problem to solve. However, in certain specialized cases the greedy allocation algorithm of the type described in Section 5.3.1.2 can be used to solve it exactly. (See Swann [497] for details.) This greedy algorithm can also be used as a heuristic in more general cases.

Table 5.7. Demand-function parameters, itineraries, and optimal solution for Example 5.6.

Market		Demand Function				Optimal Solution	
O	D	a_j	ϵ_j	$\bar{p}_j$	Path	d_j^*	p_j^*
1	2	300	1.0	220	1–2	135	$396.62
1	3	300	1.2	220	1–3	67	$495.86
1	4	300	2.0	400	1–2–4	165	$520.11
1	5	300	1.0	250	1–3–5	33	$752.04
1	6	300	0.8	200	1–6	100	$525.58
2	3	300	1.0	230	2–3	168	$364.28
2	4	300	0.9	200	2–4	143	$365.74
2	5	300	2.0	200	2–3–5	32	$423.79
2	6	300	1.0	200	2–4–6	92	$436.80
3	2	300	1.0	200	3–2	200	$281.76
3	4	300	2.0	230	3–4	131	$325.30
3	5	300	2.0	120	3–5	35	$249.51
3	6	300	2.0	150	3–4–6	14	$378.60
4	6	300	1.0	150	4–6	162	$243.30
5	2	300	1.0	200	5–2	100	$420.39
5	3	300	2.0	150	5–3	47	$289.90
5	4	300	1.0	160	5–3–4	21	$585.20
5	6	300	1.0	230	5–3–4–6	32	$748.50

5.4.3 Stochastic Models

Stochastic multiproduct pricing problems, like stochastic multiproduct capacity-allocation problems, are quite difficult to solve exactly. While in principle they can be formulated as dynamic programs, the size of the state space is often prohibitively large. Therefore, approximations offer the only practical hope to solve such problems.

One natural approach for a stochastic multiproduct problem is to approximate it by its deterministic equivalent problem, which as we've seen in Section 5.2.2.3 are reasonably easy to solve. As in the case of the single-product problem discussed in Section 5.2.2.3, one can indeed show that deterministic solutions are asymptotically optimal (in the same fluid scaling of the problem) in certain cases. That is, suppose the revenue in period t, $R(t, \mathbf{d}, \boldsymbol{\xi}_t)$, is random and we consider a deterministic problem that replaces this random demand by its mean, $r(t, \mathbf{d}) = E[R(t, \mathbf{d}, \boldsymbol{\xi}_t)]$. Then the optimal deterministic price trajectory from the resulting deterministic problem, when applied as an open-loop control for the stochastic problem, produces an expected revenue that is provably close to the optimal stochastic expected revenue.

For example, Gallego and van Ryzin [199] show that for a continuous time version of the multiproduct pricing problem of Section 5.4.1

220   *THE THEORY AND PRACTICE OF REVENUE MANAGEMENT*

with Poisson uncertainty, the solution to the equivalent deterministic problem is asymptotically optimal for the stochastic problem as the capacities and time horizon are scaled up proportionally. The arguments and formal definition of the scaling are similar to the asymptotic analysis of network capacity control problems presented in Section 3.6.2 and 3.6 and are omitted. However, the result does provide some intuition into the connection between these two problems.

5.4.4   Action-Space Reductions

One simplification that is useful for multiproduct dynamic pricing problems is to express the problem in terms of resource-consumption rates rather than the demand rates **d**. This yields an equivalent formulation with often a greatly reduced dimensionality that can be much easier to solve. The approach is due to Maglaras and Meissner [354].

To illustrate the main idea, consider the case of the deterministic model (5.21) where there is only $m = 1$ resource but $n > 1$ products. For example, this could be a situation similar to the traditional single-resource problem of Chapter 2 but one in which we control the demand for each product j, d_j, by adjusting its price p_j. The deterministic problem (5.21) in this case is then

$$\max \sum_{t=1}^{T} r(t, \mathbf{d}(t)) \tag{5.24}$$

$$\text{s.t.} \quad \sum_{t=1}^{T} \sum_{j=1}^{n} d_j(t) \leq C$$

$$\mathbf{d}(t) \geq 0, \quad t = 1, \ldots, T.$$

To reduce the dimensionality of this problem, we express the problem in terms of the aggregate-demand rate rather than the individual demand rates **d**. To this end, define the aggregate-demand rate

$$\hat{d} = \sum_{j=1}^{n} d_j,$$

and for a given $\hat{d}$ define the maximized revenue-rate function by

$$\hat{r}(t, \hat{d}) = \max \quad r(t, \mathbf{d}) \tag{5.25}$$

$$\text{s.t.} \quad \sum_{j=1}^{n} d_j = \hat{d}$$

$$\mathbf{d} \geq 0.$$

Dynamic Pricing

That is, $\hat{r}(t, \hat{d})$ is the instantaneous maximum revenue rate given that the total demand rate (equivalently, the resource *consumption rate*) is constrained to be $\hat{d}$. It is easy to show that if $r(t, \mathbf{d})$ is jointly concave in $\mathbf{d}$, then $\hat{r}(t, \hat{d})$ will be concave in $\hat{d}$.

Using these new variables, we can then formulate (5.24) as

$$\max \quad \sum_{t=1}^{T} \hat{r}(t, \hat{d}(t)) \tag{5.26}$$

$$\text{s.t.} \quad \sum_{t=1}^{T} \hat{d}(t) \leq C$$

$$\hat{d}(t) \geq 0 \quad t = 1, \ldots, T.$$

Note that this is now a problem that is equivalent to a single-product pricing problem of the same form as (5.1) with a scalar demand rate $\hat{d}$ and revenue-rate functions $\hat{r}(t, \hat{d})$. Once we solve for the optimal demand rates $\hat{d}^*(t)$, we can then convert these into optimal vectors of demand rates $\mathbf{d}^*(t)$ by inserting $\hat{d}^*(t)$ into the optimization problem (5.25). Thus, the solution proceeds in two steps: first solve (5.26) to determine the optimal aggregate sales rate, and then solve (5.25) at each time t to disaggregate this optimal aggregate rate into a optimal vector of sales rates (equivalently prices) for each product.

This same action-space-reduction approach also works for stochastic versions of this problem, of the types examined in Section 5.2.2. To illustrate, consider the dynamic program (5.11) for the continuous, additive-uncertainty-demand model, but now suppose there are n products. The n-product version of (5.11) yields the dynamic program

$$V_t(x) = \max_{\mathbf{d} \geq 0} E\left[R(t, \mathbf{d}, \boldsymbol{\xi}(t)) + V_{t+1}(x - \sum_{j=1}^{n} D_j(t, d_j, \xi_j(t)))\right], \tag{5.27}$$

where $D_j(t, d_j, \xi_j(t)) = d_j + \xi_j(t)$ is the random demand for product j.

To reduce the action space, we again define the aggregate-demand rate $\hat{d} = \sum_{j=1}^{n} d_j$ and a maximized expected revenue rate using (5.25), where now $r(t, \mathbf{d}) = E[R(t, \mathbf{d}, \boldsymbol{\xi}_t)]$. Also, let

$$\hat{\xi}(t) = \sum_{j=1}^{n} \xi_j(t)$$

denote the aggregate noise term and

$$\hat{D}(t, \hat{d}, \hat{\xi}(t)) = \hat{d} + \hat{\xi}(t)$$

denote the aggregate (random) demand. When these transformed variables are substituted into the dynamic program (5.27), it reduces to the following equivalent single-product formulation

$$V_t(x) = \max_{\hat{d} \geq 0} \left\{ \hat{r}(t, \hat{d}) + G_{t+1}(x, \hat{d}) \right\},$$

where
$$G_{t+1}(x, \hat{d}) \equiv E[V_{t+1}(x - \hat{D}(t, \hat{d}, \hat{\xi}(t)))].$$

This has the same form as the single-product DP (5.11). Thus, the single-resource, n-product dynamic-pricing problem is really no more difficult to solve than the single-product problem.

This action-space reduction idea also extends to the general multiproduct ($m > 1$), multiresource problem (5.21) as well. In this case, one can show that the problem can be reduced to one with only m demand rates (one for each resource) rather than the original n rates (one for each product). Namely, let $\hat{\mathbf{d}} = (\hat{d}_1, \ldots, \hat{d}_m)$ define the maximized revenue rate at each time t

$$\hat{r}(t, \hat{d}) = \max \quad r(t, \mathbf{d})$$
$$\text{s.t.} \quad \mathbf{Ad} = \hat{\mathbf{d}}$$
$$\mathbf{d} \geq 0.$$

This maximized revenue-rate function and the new demand-rate variables $\hat{\mathbf{d}}$ are then used to reformulate the general problem (5.21) as

$$\max \quad \sum_{t=1}^{T} \hat{r}(t, \hat{\mathbf{d}}(t))$$
$$\text{s.t.} \quad \sum_{t=1}^{T} \hat{\mathbf{d}}(t) \leq \mathbf{C}$$
$$\hat{\mathbf{d}}(t) \geq 0 \quad t = 1, \ldots, T.$$

What these reductions show, in essence, is that the complexity of the multiproduct, multiresource dynamic-pricing problem is caused not by the number of products n but by the number of resources m, since ultimately m determines the dimensionality of both the state and action spaces.

5.5 Finite-Population Models and Price Skimming

We next consider what effect a finite-population assumption has on an optimal dynamic-pricing policy.[6] Recall that a finite-population model assumes that we sample customers without replacement from a finite number of potential customers. Thus, the history of demand (how many customers have purchased, how much they paid, and so on) affects the distribution of both the number and valuations of the remaining customers.

Because the finite-population assumption is more complex, we focus on deterministic models of this situation. However, we consider both a myopic and strategic customer version of the problem.

5.5.1 Myopic Customers

Recall that a myopic customer is assumed to purchase the first time the current price $p(t)$ drops below his valuation v. Combined with the finite-population assumption, this behavior can be exploited by the firm to achieve *price skimming*—a version of classical second-degree price discrimination.

Assume for simplicity that there is a finite population size N and that customers in this population have valuations v that are uniformly distributed on the interval $[0, \bar{v}]$. As an approximation, we assume that sales can occur in fractions, so the population can be regarded as continuous. The important point to note is that the fraction of customers who purchased until time t leave the population of customers for the remaining sale period.

As a result of the myopic-customer assumption, if the firm offers a price p, $N(1 - \bar{v}/p)$, customers will buy. And by the finite-population assumption, there will then be $N\bar{v}/p$ remaining customers, with valuations uniformly distributed on the interval $[0, p]$.

Now, consider a firm that sells a fixed capacity C of a product to this population over T time-periods. The firm is free to set different prices in each period. What is the optimal pricing strategy?

First, it is not hard to see that the optimal prices are decreasing over time, since (by the myopic-customer assumption) the only customers left at time t are those with values less than the minimum price offered in periods $1, \ldots, t-1$. (See Section 8.3 for further motivations for a

---

[6]Section 8.3.4 covers the economic aspects of a durable-goods monopolist under a finite-population, strategic-customer assumption. Here we concentrate on more operational results of dynamic pricing.

firm setting a decreasing schedule of prices.) Hence, the firm will sell nothing if it posts a price in period t that is higher than the minimum price offered in the past. This observation, applied inductively, shows that the optimal prices must decline over time. Moreover, note that if $p(t) \leq p(t-1)$ for all t, the revenue generated in period t is given by

$$p(t)\frac{N}{\bar{v}}(p(t-1) - p(t)),$$

where we define $p(0) = \bar{v}$. This is because $\frac{N}{\bar{v}}(p(t-1) - p(t))$ is the number of customers with valuations greater than $p(t)$ but less than the lowest previous price $p(t-1)$.

To see the effect the decreasing price schedule has on the optimal pricing policy, assume for simplicity that $C > N$, so the capacity constraint is never binding. In this case, the firm must solve

$$\max \sum_{t=1}^{T} \frac{N}{\bar{v}} p(t)(p(t-1) - p(t)) \quad (5.28a)$$
$$\text{s.t.} \quad p(t) \leq p(t-1), \quad t = 1, \ldots, T \quad (5.28b)$$
$$p(0) = \bar{v}, \quad (5.28c)$$
$$p(t) \geq 0. \quad (5.28d)$$

Note that the objective function is jointly concave in $p(t), t = 1, \ldots, T$. It is not hard to see that the constraints (5.28b) are redundant, since the objective function (5.28a) will penalize the use of a price $p(t) > p(t-1)$. Therefore, ignoring constraints (5.28b) and defining $p(T+1) = 0$, the first-order conditions imply the optimal unconstrained solution must satisfy

$$p(t) = \frac{p(t-1) - p(t+1)}{2}, \quad t = 1, \ldots, T.$$

One can easily verify that the solution

$$p^*(t) = \bar{v}\left(1 - \frac{t}{T+1}\right) \quad (5.29)$$

satisfies these first-order conditions. Since the optimization problem (5.28a–d) is strictly concave and (5.29) satisfies the inequality constraints $p(t) \leq p(t-1)$ for all t, it is in fact the unique optimal solution for (5.28a–d). This solution is illustrated in Figure 5.7(i).

The optimal pricing strategy effectively exploits the myopic behavior of customers to segment them into $T+1$ groups based on their valuations, and then price discriminates based on this segmentation. Specifically, as shown in Figure 5.7, segment t consists of those customer whose

(i) No capacity constraints

(ii) With capacity constraints

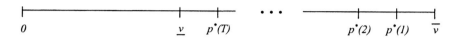

Figure 5.7. Optimal price-skimming solution for myopic customers: (i) no capacity constraints, (ii) with capacity constraints.

valuations are in the range $[p^*(t), p^*(t-1)]$, and these segments pay a declining price $p^*(t)$ given by (5.29). Segment $T+1$ has values in the range $[0, \overline{v}/(T+1)]$ and is not served at all.

There are several interesting observations about this solution. First, note we can write the optimal price in period t as

$$p^*(t) = \frac{p^*(t-1)}{2} + \frac{\overline{v}}{2}\left(1 - \frac{t}{T+1}\right).$$

The first term on the right, $p^*(t-1)/2$, is simply the single-period revenue-maximizing price, which follows from the fact that the remaining customers in period t have values uniformly distributed on $[0, p^*(t-1)]$. Therefore, the optimal price in period t is higher than the single-period revenue-maximizing price for period t (except in the last period $t = T$, where they are equal). Intuitively, this occurs because there is an additional benefit to the firm of raising its price in period t in the multiperiod setting; namely, it will have more customers to sell to in the future.

Second, note the price changes over time not because the distribution of valuations changes over time—as in the infinite-population model of demand—but because the firm seeks to price discriminate among the finite population of customers. For example, in an equivalent infinite-population model (essentially, the model of Section 5.2.1 with a linear-demand function), the distribution of values of customers is unaffected by past demand, and hence the distribution would still be uniform over $[0, \overline{v}]$ in each period. In this case, the optimal price to charge in each period would be a constant $\overline{v}/2$ rather than the declining price (5.29). Therefore, a finite population of customers creates an incentive to offer dynamically decreasing prices to achieve price discrimination, an incentive that is not present in infinite-population models.

Finally, note that if the number of periods T increases, the firm's revenues increase because one can show (after some algebra) that the optimal total revenue for T periods is

$$\sum_{t=1}^{T} p^*(t)\frac{N}{\bar{v}}(p^*(t) - p^*(t-1)) = \frac{N\bar{v}}{2}\left(\frac{T}{T+1}\right).$$

Indeed, as T tends to infinity, the firm achieves perfect price discrimination and captures the entire consumer surplus $N\bar{v}/2 = \int_0^{\bar{v}}(N/\bar{v})dv$; each customer ends up paying a price arbitrarily close to his valuation. In particular, a continuous-time model of this problem can achieve perfect price discrimination because the firm can continuously lower prices from $\bar{v}$ down to zero over the interval $[0, T]$. A number $dp(N/\bar{v})$ of customers with values $[p, p + dp]$ will buy when the price is p, so the firm achieves a revenue of $\int_0^{\bar{v}} p(N/\bar{v})dp = N\bar{v}/2$, which is the entire consumer surplus.

The introduction of a binding capacity constraint does little to change this basic story. Indeed, the solution (5.29) will not be feasible if $C < NT/(T+1)$. However, in this case, one can show that the optimal price is simply modified so that only those customers above a lower limit $\underline{v}$ are segmented, where

$$N(\bar{v} - \underline{v})\left(\frac{T}{T+1}\right) = C.$$

The optimal price in this case becomes

$$p^*(t) = \underline{v} + (\bar{v} - \underline{v})\left(1 - \frac{t}{T+1}\right),$$

and customers with valuations less than $\underline{v} + (\bar{v} - \underline{v})/(T+1)$ are not served. This solution is illustrated in Figure 5.7(ii).

5.5.2 Strategic Customers

One might question why customers would behave myopically when faced with a price-skimming strategy. Indeed, knowing that prices will decline over time, rational customers could do better (increase their net utility) by deviating from myopic behavior and delaying purchase until the price is much lower than their valuation. Such behavior is quite plausible and is a valid criticism of the myopic-customer model, but it complicates the analysis of the firm's optimal-pricing policy considerably. Most significantly, it turns the pricing problem into a game between the firm and its customers, in which we must analyze the equilibrium using game-theoretic tools.

Strategic customer behavior is, in fact, a central feature of the theory of optimal mechanism design discussed in Chapter 6 on auctions.

Both auction and list-price mechanisms are analyzed in Chapter 6, and we provide much of the analysis and insight about pricing with strategic customer behavior there (see also Section 8.3.4). Here we focus on the more limited topic of the effect of strategic customers on the price-skimming strategy alone. Further, to keep things simple, throughout this section we consider only the case where the firm has no capacity constraint $(C > N)$.

To proceed, one first has to make assumptions about whether the firm can credibly commit to a schedule of prices over time or whether the firm must follow a subgame-perfect equilibrium-pricing strategy. (See Appendix F for a discussion of subgame perfection.) In our case, requiring a subgame-perfect equilibrium means that the strategy for the firm at each time t has to be an equilibrium for the residual revenue-maximization game over the horizon $t, t+1, \ldots, T$, given whatever state the firm and customers were in period t.

For example, if the firm can commit to a price schedule, then a rational customer will simply look at the schedule of prices and (assuming no discounting of utility) decide to purchase in the period with the lowest price, and only customers with valuations above this lowest price will decide to purchase. So effectively, it is only the lowest price among the T periods that matters to customers. Given this fact (and ignoring capacity constraints), the firm will then set this minimum price as the single-period revenue-maximizing price, which, in the case where customer valuations are uniformly distributed on $[0, \bar{v}]$, is just $\bar{v}/2$. The firm will then set arbitrary but higher prices in the other periods. Which period the firm chooses for the minimum price doesn't matter unless revenues are discounted, in which case the firm would prefer collecting revenues sooner rather than later and would choose period 1. The total revenue the firm receives is then $N\bar{v}^2/4$, which is just the product of the price $\bar{v}/2$ and the number of customers willing to pay that price, $N\bar{v}/2$. One can formalize this reasoning and show that this is indeed the equilibrium strategy in the case where the firm has to commit to a price schedule.

Note that the fact that customers are rational has eliminated the ability of the firm to price discriminate; the firm is forced to offer a single uniform price to all customers. Moreover, the firm's revenue is strictly worse under this model. This is to be expected, the firm ought to do worse when customers are "smarter."

However, the single-period strategy outlined above is not always subgame-perfect. To see why, suppose this lowest price $\bar{v}/2$ occurs in period 1. Then in period 2, there will be a population of customers with values less than $\bar{v}/2$ who have not purchased. If the firm has any remaining supply after period t, it would rather sell the remaining stock at some positive price than let it go unsold. Thus, it has an incentive to lower the price in period 2 to capture some of the remaining customers. However, rational customers realize the firm faces this temptation after period 1 and, anticipating the price drop, do not purchase in period 1, so offering the lowest price in period 1 cannot be a subgame-perfect equilibrium.

Besanko and Winston [47] analyze the subgame-perfect pricing strategy. The equilibrium is for the firm to lower prices over time, similar to the price-skimming strategy of Section 5.5.1. In the case where revenues are not discounted, this equilibrium results in the firm setting a declining sequence of prices, where the price in the last period T is simply the single-period optimal price $\bar{v}/2$; all customers buy only in the last period. This case is essentially equivalent to the case where the firm can commit to a schedule of prices, with the exception that the firm is forced to offer the lowest price only in the last period.

The situation is somewhat more interesting if revenues and customer utility are discounted at the same rate. In this case, the subgame-perfect equilibrium has customers with high values buying in the early periods and those with lower values buying in later periods, again, as in the price skimming case of Section 5.5.1. However, unlike the price-skimming case, the equilibrium price in each period is *lower* than the single-period revenue-maximizing price for the customers remaining in that period. In particular, in period 1 the equilibrium price is less than $\bar{v}/2$, and the equilibrium price declines in subsequent periods. Thus, the firm is strictly worse off than when it can commit to a price schedule. This is because when the firm can commit to its price schedule, it can force all customers to purchase in period 1 by simply offering very high prices in periods $t > 1$ while setting a price of exactly $\bar{v}/2$ in period 1. All customers will then buy in period 1 at a price of $\bar{v}/2$.

Besanko and Winston [47] show that with strategic customers, the firm is always better off with fewer periods; that is, the firm's equilibrium revenue is decreasing in the number of periods. This is because the inability of the firm to commit to prices in later periods hurts it, and the more periods, the more often the firm falls victim to the temptation to lower prices. That is, it discounts early and often. This is to be contrasted with the case of myopic customers, where the firm's revenues are increasing in the number of periods. Thus, although the strategy looks like price skimming, rational customers create a qualitatively different situation for the firm than do myopic customers.

5.6 Promotions Optimization

In this section we look at normative models for retail and trade promotions. We first discuss promotions in general and how they differ from the sorts of dynamic-pricing problems considered thus far. We then look at two specific models of promotion optimization.

5.6.1 An Overview of Promotions

As mentioned in this chapter's introduction, promotions are short-run, temporary price changes that are frequently applied to replenishable and consumable goods (such as CPG products). Promotions are run either by the manufacturer (trade promotions) or by retailers (retail promotions or consumer promotions). Manufacturer may also give a discount directly to the end customer in the form of coupons and rebates. While manufacturers are interested in increasing sales or profits for their brand, retailers are interested in overall sales or profits for an entire product category.

A promotion generally increases sales to both the retailer and manufacturer, but there are a variety of factors at work behind the increase. Customers may increase their consumption of the product due to two fundamental effects: higher household inventories lead to fewer stockouts and therefore an increase in consumption; and higher inventories give customers greater flexibility in consuming the product because they don't have to worry about replacing the inventory at higher prices. For instance, Wansink and Deshpande [553] and Chandon and Wansink [104] show that larger household inventory causes faster usage rates if product-usage occasions are flexible (snack foods), products need refrigeration, or products occupy a prominent place in the pantry (for empirical evidence of this based on scanner data, see Ailawadi and Neslin [5]). Some other reasons promotions cause an increase in demand include customers switching from nondiscounted brands to the discounted brands and customers (or retailers for trade deals) stockpiling to take advantage of the low price (forward buying).

Not surprisingly a dominating factor behind the demand increase is the type of product. For example, products such as yogurt and potato chips tend to see an increase because of increased consumption, while for products such as tomato ketchup, diapers, and toilet paper, the sales increase is primarily because of brand switching or stockpiling.

Promotions, in the framework of RM, can be thought of as either (1) a manufacturer using price to dispose excess inventory, (2) a manufacturer trying to gain market share to induce customers to try out its products, (3) retailers experimenting with price to find optimal price points, (4) separating price-sensitive customers, who are willing to use coupons or who wait for deals, (5) retailers trying to increase store traffic, as customers once inside the store are likely to purchase other, nonpromotional items, or (6) a tactic for store brands or small firms to compete against the large advertising budgets of the established brands.

5.6.1.1 Types of Promotions

As mentioned, the main dichotomy in promotions is between retail promotions and trade promotions. Many promotional events are in fact closely coordinated between the manufacturer and the retailer. For instance, if the retailer runs an advertised promotion, the manufacturer may agree to bear a share of the advertising cost, or the trade deal may involve running an in-store display supplied by the manufacturer.

Retail promotions can be advertised or unadvertised (in-store promotions), often coordinated with temporary in-store displays. The price promotion part may take the form of a simple percentage off, coupons, or a "multibuy" (discount for multiple items packaged together), or an extra free (such as three for the price of two; or 15% more free). The latter two types are usually manufacturer-driven, as packaging may have to be changed.

Trade promotions traditionally are in the form of *off-invoice* as a percentage off the amount ordered during the promotion period. Surprisingly, many off-invoice promotions do not require the retailers pass the discount on to the customer, so they may just purchase more during the promotion period and sell it at regular price. The manufacturer would simply see a drop in orders once the promotion period is over.

More effective for the manufacturer is the use of mail-in coupons (direct discount to the final customer), or *scan-back deals*, in which the manufacturer reimburses retailers a certain amount for each unit sold, so the discount is on units sold to end customers rather than on units purchased by the retailer. Scan-back deals eliminate forward buying by the retailer and aligns the retailer's objectives with the manufacturer's.

5.6.1.2 Empirical Findings

The promotions literature is rich in empirical work—based mostly on scanner POS data—that analyzes the effects of promotions on sales and profits in different categories. The common trends that emerge from this research, summarized by Blattberg, Briesch, and Fox [74] as *empirical*

Table 5.8. Empirical generalizations on promotions.[a]

Finding	Supporting Literature
1. Temporary retail price reductions substantially increase sales.	Woodside and Waddle [580] Moriarty [390] Blattberg and Wisniewski [77]
2. Higher-market-share brands are less deal elastic.	Bolton [84] Bemmaor and Mouchoux [45] Vilcassim and Jain [535]
3. The frequency of the deals changes the consumer reference price.	Lattin and Bucklin [329] Kalwani et al. [281] Kalwani and Yim [282] Mayhew and Winer [367]
4. The greater the frequency of sales, the lower the height of the deal spike.	Bolton [84] Raju [434]
5. Cross-promotional effects are asymmetric, and promoting higher-quality brands impacts weaker brands disproportionately.	Blattberg and Wisniewski [77, 78], Krishnamurthi and Raj [315, 315] Cooper [128], Walters [549]
6. Retailers pass-through less than 100% of trade deals.	Chevalier and Curhan [114], Curhan and Kopp [138], Walters [549], Blattberg and Neslin [76]
7. Display and feature advertising have strong effects on item sales.	Woodside and Waddle [580], Blattberg and Wisniewski [77], Kumar and Leone [316]
8. Advertised promotions can result in increased store traffic.	Walters and Rinne [547], Kumar and Leone [316], Walters and MacKenzie [548], Grover and Srinivasan [225]
9. Promotions affect sales in complementary and competitive categories.	Walters and Rinne [547], Walters and MacKenzie [546], Mulhern [396], Walters [549], Mulhern and Leone [395]

[a] Source: Blattberg, Briesch and Fox [74].

generalizations,[7] are valuable both for the practitioner as well as the academic researcher. Table 5.8 gives the main findings. In addition to the findings in Table 5.8, Blattberg, Briesch, and Fox [74] report some conflicting findings with respect to the following four questions:

---

[7] Blattberg, Briesch and Fox [74] define an empirical generalization as follows: (1) the topic being studied is well-defined; (2) there are at least three articles by at least three different authors in which empirical research has been conducted in the specific area, and (3) the empirical evidence is consistent.

- Does the majority of promotional volume come from switchers rather than from customers increasing their consumption or category volume growth? The most likely explanation for the variation in the findings here may be the differences in the nature of the products; one can well imagine promotions causing consumption increase for yogurt, but not say, for toilet paper or ketchup.

- Do promotional elasticities exceed long-run price elasticities? That is, because of the temporary nature of a promotion, does it cause a greater increase in demand than if the firm were to permanently lower its price?

- Is the trough after the deal due to customers' accelerating their purchases and stockpiling, creating a drop in the normal sales after a promotion? Somewhat surprisingly, there is no consensus whether this happens.

- Is there is a negative long-term effect to promotions? Are promotions detrimental to long-term brand equity? The findings have been mixed, with some studies discovering a long-term negative effect, and some finding both a positive and negative impact due to promotions.

5.6.2 Retailer Promotions

We next examine two normative models of promotion optimization. In the first model, due to Greenleaf [221], a monopolist retailer is assumed to maximize profits from promoting a particular brand. (A "brand" is a particular-size of a given product.) Customers are assume to have a reference price (see Appendix E), assumed to be an exponentially smoothed average of past prices, as follows:

$$\bar{p}(t) = \alpha \bar{p}(t-1) + (1-\alpha)p(t-1) + \xi_t, \qquad (5.30)$$

where $0 \leq \alpha < 1$ is the smoothing parameter,[8] and ξ_t is a 0-centered random variable representing the error term.

Demand is assumed to be composed of two separable factors, a base demand $q(p(t))$ and a reference price factor $g(\bar{p}(t), p(t))$, as follows:

$$d(t) = q(p(t)) + g(\bar{p}(t), p(t)), \qquad (5.31)$$

where

$$g(\bar{p}(t), p(t)) = \begin{cases} \delta(\bar{p}(t) - p(t)) & \text{if } \bar{p}(t) > p(t) \\ \gamma(\bar{p}(t) - p(t)) & \text{otherwise.} \end{cases} \qquad (5.32)$$

---

[8] Based on scanner data, Greenleaf [221] finds $\alpha = 0.925$ for peanut butter, and Hardie, Johnson, and Fader [237] find $\alpha = 0.83$ for orange juice. It is also common in promotions models to assume *a priori* $\alpha = 0$—that is, the reference price is the previous period's price.

The parameters $\delta, \gamma > 0$ model customers' asymmetric price sensitivity to loss ($\bar{p}(t) \leq p(t)$) or gain ($\bar{p}(t) > p(t)$) perception: if customers value gains more than losses, $\delta > \gamma$, and if they are loss averse, $\delta \leq \gamma$. (Again, see Appendix E for a discussion of consumer-choice theory based on valuations of losses and gains.)

The reference-price dynamics given by (5.30) capture the effect of current promotions on future profits; frequent and deep promotions will reduce customers' reference price $\bar{p}(t)$ for the brand, and as a result they will start perceiving the normal price as a loss. So even though promotions generate short-run profits, it is in the retailer's long-run interest not to run promotions too frequently.

The retailer is assumed to maximize its discounted profits over an infinite planning horizon. The retailer's discount factor is β, $0 < \beta < 1$, and the marginal cost of production is c. This results in the following dynamic program:

$$\max_{p(t)} \sum_{t=0}^{\infty} \beta^t [(p(t) - c)d(t)], \tag{5.33}$$

with a state evolution equation given by (5.30).

Greenleaf [221], using simulations, shows that the optimal policy for the retailer obtained by solving (5.33) can be cyclical, oscillating between periods of high prices and periods of low prices. Kopalle, Rao, and Assunção [310], using analytical and numerical techniques, derive a number of interesting structural properties of (5.33). Specifically, they show that if the entire customer population is loss averse ($\delta \leq \gamma$), a constant-price policy is optimal. On the other hand, if customers value gains more than losses ($\delta > \gamma$), then a cyclical policy of hi-lo pricing is optimal for the retailer. In other words, the asymmetry in customer valuations for gains and losses can be sufficient motivation to run promotions. Moreover, they numerically show for this case that the difference between the high and low price increases as the gain coefficient δ increases for a fixed level of loss coefficient γ, and the high price increases as the memory parameter α in (5.30) decreases.

As we mentioned, retailers are more interested in category profits than in profits from promoting a particular brand. For a retailer managing n brands in a category, the objective therefore is to manage the n prices over time, $\mathbf{p}(t) = (p_1(t), \ldots, p_n(t))$. This requires solving the following optimization problem:

$$\max_{\mathbf{p}(t), t=1,2,\ldots} \sum_{t=1}^{\infty} \sum_{j=1}^{n} \beta^t [f_j(\mathbf{p}(t)) + (p_j(t) - c)g_j(\bar{p}_j(t), p_j(t))], \tag{5.34}$$

where $f_j(\mathbf{p}(t))$ is an aggregate profit function, dependent on the prices of all the brands but excluding reference price effects, which are captured by the second term, and $g_j(\bar{p}_j(t), p_j(t))$ is of the same form as (5.32). The state as before is the reference price, and the state equation is

$$\bar{\mathbf{p}}(t) = \alpha \bar{\mathbf{p}}(t-1) + (1-\alpha)\mathbf{p}(t).$$

Kopalle, Rao, and Assunção [310] analyze (5.34) and show once again that when $\delta_i > \gamma_i$ for all brands, a cyclical pricing policy of hi-lo pricing is optimal, and moreover, the cycles are *in phase*—that is, all the brands are priced high, or all the brands are discounted together. The reason is that hi-lo prices in phase minimize the cross-price effect, at the same time allowing the retailer to take advantage of the reference-price effect.

5.6.3 Trade-Promotion Models

As we discussed, a manufacturer offers rebates to retailers to promote its own brand. The cooperation might take many forms (such as joint advertising and store displays), and the contracts are varied (such as scan-backs and sale guarantees).

On the one hand, models for optimizing the manufacturer's promotions tend to be simpler than the retailer's problem, as the manufacturer is concerned with only one brand. But on the other hand, one has to model retailer pass-thru behavior. (Recall that pass-thru is the percentage of the discount the retailer passes on to the end consumer.) This requires modeling the vertical competition between manufacturer and retailer. In contrast to the previous section, however, one typically ignores reference-price effects because the discount is offered to the retailer rather than to the end customers.

The most widely used model for representing demand as a function of deal price and displays is the SCAN*PRO model of Section 9.6.4. Kopalle, Mela, and Marsh [309] analyze a Stackelberg game between a manufacturer and retailer, where the demand is given by the SCAN*PRO functional form. Silva-Russo, Bucklin, and Morrison [470] give a simpler mixed integer programming formulation (see also Tellis and Zufryden [507]), where the manufacturer assumes that retailers are passive, but they model retailers' pass-thru percentages. They report an implementation of the model at a large CPG manufacturer. The formulation does not by itself give insight into the optimal structure or policies for the manufacturer, but it is reasonably practical and captures the main concerns of the manufacturer in the formulation of its constraints.

5.7 Notes and Sources

The book by Nagle [400] provides a good general-management overview of pricing decisions. Elmaghraby and Keskinocak [177] provide a nice current survey on research in the area of dynamic pricing. As for the connection between pricing- and capacity-allocation decisions, see Walczak and Brumelle [543].

Smith and Achabal [480] study a continuous-time version of the problem with inventory-depletion effect as in Section 5.2.1.5. They also study the problem of selecting the optimal initial inventory and report summary results of tests of the model at several major retailers. Heching, Gallego, and van Ryzin [247] provide revenue estimates based on a regression test of this same type of deterministic model on data from an apparel retailer.

Gallego and van Ryzin [198] analyzed a continuous-time, time-homogeneous version of the stochastic model of Section 5.2.2, providing monotonicity properties of the optimal price, an exact solution in the exponential demand case, and proving the asymptotic optimality of the deterministic policy as in Section 5.2.2.3. Bitran and Mondschein [73] analyze a discrete-time model of the problem essentially the same as that presented in Section 5.2.2 and test in on apparel retail data. Zhao and Zheng [589] analyze the continuous-time model with a time-varying demand function and provide an alternative proof of monotonicity of the marginal values $\Delta V_t(x)$; they also provide results on the monotonicity of optimal prices over time. See also Kincaid and Darling [304] and Stadje [484]. Das Varmand and Vettas [145] analyze the problem of selling a finite supply over an infinite horizon with discounted revenues, where the discounting provides an incentive to sell items sooner rather than later and there is no hard deadline on the sales season.

Stochastic models with discrete price changes are analyzed in the continuous-time case in a series of papers by Feng and Gallego [185, 186] and Feng and Xiao [188, 189]. The problems differ in terms of whether there are two prices or more than two prices, whether the price changes are reversible or one-way changes. Feng and Gallego [186] extend the analysis also to the interesting case where demand is Markovian and may depend on the current inventory level—for example, as in the classical Bass model of new-product diffusion. The notion of the *maximum concave envelope* of prices is due to Feng and Xiao [188]. See also You [586] for a discrete-time analysis of the problem.

There is an extensive literature on production-pricing problems. Eliashberg and Steinberg [175] provide of review of joint pricing and production models. Single-period, convex-cost problems under demand uncertainty are analyzed by Karlin and Carr [293], Mills [386], and the early paper of Whitin [564]. The literature on single-period pricing under demand uncertainty (the price-dependent newsvendor problem) is surveyed by Petruzzi and Dada [417]. Multiperiod, convex cost models are analyzed by Hempenius [249], Thowsen [509], and Zabel [509].

Rajan, Rakesh, and Steinberg [433] analyze a deterministic model of dynamic pricing within an inventory replenishment cycle, where the motivation for dynamic pricing is the deterioration in the product as well as its declining market value with age (for example, pricing perishable foods).

The optimality of the greedy allocation algorithm for the deterministic production-pricing problem with capacity constraints was shown by Chann, Simchi-Levi, and Swann [105]. The optimality of the base-stock, posted-price policy discussed in Section 5.3.2 was proved by Federgruen and Heching [183]. The fixed-cost version of this problem was recently analyzed by Chen and Simchi-Levi [113].

Multiproduct, multiresource dynamic-pricing problems were analyzed in Gallego and van Ryzin [199], including bounds on the relationship between the stochastic and deterministic versions of the problem. The action-space-reduction approach is a recent result due to Maglaras and Meissner [354]. A related network pricing we have omitted is congestion pricing for communications service; see for example Pashalidis [413].

Stokey [490] analyzes a model of intertemporal price discrimination similar to that presented in Section 5.5.1. See also Kalish [279]. Stokey [491] analyzes a price-skimming model with rational customers under the assumption that the firm can commit to a price schedule. The material in Section 5.5.2 on the subgame-perfect pricing equilibrium for a firm faced with strategic customers is from Besanko and Winston [47].

The artificial-intelligence community also has recently become interested in dynamic pricing, using autonomous software agents. The approach is simulation based, with experiments using various strategies for the players. Although relevant, the approach is beyond the scope of this book, though the interested reader can refer to Morris, Ree, and Maes [393] and Morris and Maes [392].

The literature on promotions is rich in empirical work, which we have summarized, somewhat tersely, in Table 5.8. The material in Section 5.6.1.2 is entirely from Blattberg, Briesch, and Fox [74]. For more empirical generalization articles, see Bell, Chiang, and Padmanabhan [34] and Sethuraman and Srinivasan [458].

The standard reference on promotions is the book by Blattberg and Neslin [76]. There is a large body of work that tries to understand the interactions between the retailer and the manufacturer using game theory, which we do not have the opportunity to cover here—see Lal and

Villas-Boas [322, 323], Lal [325, 324], Rao, Arjuni, and Murthi [436], Gerstner and Hess [211] and Bell, Iyer, and Padmanabhan [35].

APPENDIX 5.A: Proof of Monotonicity Results
Proof of Proposition 5.1

Since there are multiple parts, we restate the proposition:
  If Assumptions 5.1 and 5.2 hold, then for all t:
(i) $G_t(x,d)$ is jointly concave in x and d.
(ii) $V_t(x)$ is concave in x.
(iii) $\frac{\partial}{\partial d} G_t(x,d)$ is increasing in x and decreasing in d.
  We first need a preliminary result:

LEMMA 5-5.A.1 *If Assumption 5.2 holds, then the truncated revenue function $r^+(t,d,x)$ is jointly concave in x and d on the for $x \geq 0$ and $d \in \Omega_d(t)$.*

Proof
By definition,

$$r^+(t,d,x) = p(t,d) E\left[\min\{D(t,d,\xi_t), x\}\right] = E\left[\min\{p(t,d)D(t,d,\xi_t), p(t,d)x\}\right].$$

By Assumption 5.2, the term $p(t,d)D(t,d,\xi_t)$ is concave in d, and $p(t,d)$ is concave as well. Also note if $x \geq 0$, then $p(t,d)x$ will also be jointly concave in x and d. This follows because the Hessian

$$\begin{vmatrix} xp''(t,d) & p'(t,d) \\ p'(t,d) & 0 \end{vmatrix}$$

is negative definite, since both $xp''(t,d) \leq 0$ and the determinant $-(p'(t,d))^2 \leq 0$. Therefore, $\min\{p(t,d)D(t,d,\xi_t), p(t,d)x\}$ is jointly concave because it is the minimum of two concave functions. Finally, taking expectations preserves concavity, hence $r^+(t,d,x)$ is jointly concave. QED

We are now ready to prove Proposition 5.1. Parts (i) and (ii) are related by induction. Indeed, we first show that if $G_{t+1}(x,d)$ is jointly concave in x,d and Assumptions 5.2 holds (so by Lemma 5-5.A.1 $r^+(t,d,x)$ is jointly concave in d and x), then $V_t(x)$ is concave in x. To do so, consider any two values nonnegative values x_1 and x_2 and any real α satisfying $0 \leq \alpha \leq 1$. For notational convenience define the convex combination $\bar{x} = \alpha x_1 + (1-\alpha) x_2$, and let d_i^* denote the value that maximizes $r(t,d) + G_{t+1}(x_i,d)$, for $i=1,2$ and define $\bar{d} = \alpha d_1^* + (1-\alpha) d_2^*$. Then

$$\begin{aligned} V_t(\bar{x}) &= \max_{d \in \Omega_d(t)} \{r(t,d) + G_{t+1}(\bar{x},d)\} \\ &\geq r^+(t,\bar{d},\bar{x}) + G_{t+1}(\bar{x},\bar{d}) \\ &\geq \alpha(r^+(t,d_1^*,x_1) + G_{t+1}(x_1,d_1^*)) + (1-\alpha)(r^+(t,d_2^*,x_2) + G_{t+1}(x_2,d_2^*)) \\ &= \alpha V_t(x_1) + (1-\alpha) V_t(x_2), \end{aligned}$$

where the last inequality follows from the joint concavity of r^+ and G_{t+1}. So $V_t(x)$ is concave in x provided r is concave (Assumption 7.2), and G_{t+1} is jointly concave.
  Likewise, we show that G_{t+1} is jointly concave if $V_{t+1}(x)$ is concave and Assumption 5.1 holds. To see this, consider any four nonnegative x_1, x_2, d_1, d_2 and any real

α satisfying $0 \leq \alpha \leq 1$. Define $\bar{x} = \alpha x_1 + (1-\alpha)x_2$ and $\bar{d} = \alpha d_1 + (1-\alpha)d_2$. Then

$$\begin{aligned}
G_{t+1}(\bar{x},\bar{d}) &= E[V_{t+1}(\bar{x} - D(t,\bar{d},\xi_t)]\\
&\geq E[V_{t+1}(\bar{x} - (\alpha D(t,d_1,\xi_t) + (1-\alpha)D(t,d_2,\xi_t))]\\
&\geq E[\alpha V_{t+1}(x_1 - D(t,d_1,\xi_t)) + (1-\alpha)V_{t+1}(x_2 - D(t,d_2,\xi_t))]\\
&= \alpha G_{t+1}(x_1,d_1) + (1-\alpha)G_{t+1}(x_2,d_2),
\end{aligned}$$

where the first inequality follows from Assumption 5.1, $V_{t+1}(x)$ is increasing in x, and the second inequality follows from the fact that $V_{t+1}(x)$ in concave.

Parts (i) and (ii) of Proposition 5.1 now follow from these two results using an induction argument and the fact that for all x, $V_0(x) = 0$, which is concave.

Finally, to show part (iii), note that the fact that $\frac{\partial}{\partial d}G_t(x,d)$ is decreasing in d follows from the concavity of G_t in d (part (i)). To show it is increasing in x, take a nonnegative α and d note that the difference

$$V_{t+1}(x - D(t,d+\alpha,\xi_t)) - V_{t+1}(x - D(t,d,\xi_t))$$

is oppositive since $D(t,d+\alpha,\xi_t) \geq D(t,d,\xi_t)$ by Assumption 5.1 and therefore is increasing in x by the concavity of $V_{t+1}(\cdot)$. Therefore, taking expectations above we then have that the difference

$$G_{t+1}(x, d+\alpha) - G_{t+1}(x,d)$$

is increasing in x as well. Since

$$\frac{\partial}{\partial d}G_t(x,d) = \lim_{\alpha \to 0} \frac{1}{\alpha}(G_{t+1}(x,d+\alpha) - G_{t+1}(x,d)),$$

it therefore follows that $\frac{\partial}{\partial d}G_t(x,d)$ is increasing in x. QED

Proof of Proposition 5.2

The proof here is essentially identical to Proposition 2-2.A.4 result for the discrete-choice single-resource model in Appendix 2.A.

We first show that $\Delta V_t(x)$ is decreasing in x. The proof is by induction on t. First, this is trivially true for $t = T+1$ by the boundary conditions $V_{T+1}(x) = 0$ for all x. Assume it is true for period $t+1$, and consider period t. Let d_i^* denote the optimal solution to (5.12) for inventory level $x+i$; that is, it is an optimal solution in the recursion

$$V_t(x) = \max_{d \in \Omega_d(t)}\{r(t,d) - d\Delta V_{t+1}(x+i)]\} + V_{t+1}(x)$$

and note that since $\Delta V_t(x+i) = V_t(x+i) - V_t(x+i-1)$, we can write

$$\begin{aligned}
\Delta V_t(x+2) - \Delta V_t(x+1) &= \Delta V_{t+1}(x+2) - \Delta V_{t+1}(x+1)\\
&\quad + (r(t,d_2^*) - d_2^*\Delta V_{t+1}(x+2))\\
&\quad - (r(t,d_1^*) - d_1^*\Delta V_{t+1}(x+1))\\
&\quad - (r(t,d_1^*) - d_1^*\Delta V_{t+1}(x+1))\\
&\quad + (r(t,d_0^*) - d_0^*\Delta V_{t+1}(x))
\end{aligned}$$

From the optimality of d_1^*, the following inequalities hold:

$$r(t,d_1^*) - d_1^*\Delta V_{t+1}(x+1) \geq r(t,d_2^*) - d_2^*\Delta V_{t+1}(x+1)$$

APPENDIX 5.A: Proof of Monotonicity Results

and
$$r(t, d_1^*) - d_1^* \Delta V_{t+1}(x+1) \geq r(t, d_0^*) - d_0^* \Delta V_{t+1}(x+1).$$
Substituting into (5.A.0) we obtain

$$\begin{aligned}
\Delta V_t(x+2) - \Delta V_t(x+1) &\leq \Delta V_{t+1}(x+2) - \Delta V_{t+1}(x+1) \\
&\quad + (r(t, d_2^*) - d_2^* \Delta V_{t+1}(x+2)) \\
&\quad - (r(t, d_2^*) - d_2^* \Delta V_{t+1}(x+1)) \\
&\quad - (r(t, d_0^*) - d_0^* \Delta V_{t+1}(x+1)) \\
&\quad + (r(t, d_0^*) - d_0^* \Delta V_{t+1}(x)).
\end{aligned}$$

Rearranging and canceling terms yields

$$\begin{aligned}
\Delta V_t(x+2) - \Delta V_t(x+1) &\leq (1 - d_2^*)(\Delta V_{t+1}(x+2) - \Delta V_{t+1}(x+1)) \\
&\quad + d_0^*(\Delta V_{t+1}(x+1) - \Delta V_{t+1}(x)).
\end{aligned}$$

By induction, $\Delta V_{t+1}(x+2) - \Delta V_{t+1}(x+1) \leq 0$ and $\Delta V_{t+1}(x+1) - \Delta V_{t+1}(x) \leq 0$ and since d values are at most one (expected demand in a period is at most one in the discrete Poisson case), $1 - d_2^* \geq 0$ and $d_0^* \geq 0$. Therefore, $\Delta V_t(x+2) - \Delta V_t(x+1) \leq 0$. (Note the concavity of $r(\cdot)$ is not required for this part of the proof.)

To show monotonicity in t, using the same notation note that

$$\begin{aligned}
\Delta V_t(x+1) &- \Delta V_{t+1}(x+1) \quad &(5.A.1) \\
&= (r(t, d_1^*) - d_1^* \Delta V_{t+1}(x+1)) - (r(t, d_0^*) - d_0^* \Delta V_{t+1}(x)) \\
&\geq (r(t, d_1^*) - d_1^* \Delta V_{t+1}(x+1)) - (r(t, d_0^*) - d_0^* \Delta V_{t+1}(x+1)) \\
&= (r(t, d_1^*) - r(t, d_0^*)) - \Delta V_{t+1}(x+1)(d_1^* - d_0^*), \quad &(5.A.2)
\end{aligned}$$

where the first inequality above follows by the fact that $\Delta V_{t+1}(x+1) \leq \Delta V_{t+1}(x)$. Now by the concavity of $r(t, d)$ we have that

$$(r(t, d_1^*) - r(t, d_0^*)) \leq \frac{\partial}{\partial d} r(t, d_1^*)(d_1^* - d_0^*).$$

But the first-order conditions imply $\frac{\partial}{\partial d} r(t, d_1^*) = \Delta V_{t+1}(x+1)$, so substituting above we have that
$$r(t, d_1^*) - r(t, d_0^*) \geq \Delta V_{t+1}(x+1)(d_1^* - d_0^*).$$
Substituting into (5.A.2) implies $\Delta V_t(x+1) - \Delta V_{t+1}(x+1) \geq 0.$ QED

Chapter 6

AUCTIONS

6.1 Introduction and Industry Overview

Auctions provide an alternative means of dynamically adjusting prices to match market conditions. An auction is simply a set of rules (called a *mechanism*) for specifying how information is revealed among customers and the firm, how goods are awarded to customers, and what payments are made from customers to the firm based on the revealed information. They differ from a dynamic posted-price mechanism in that typically customers are the ones who offer a price they are willing to pay—their *bid*—and the firm then decides which bids to accept. However, there are some auction formats that rather resemble posted pricing, in that the firm names a price, and customers simply indicate their willingness to buy at the offered price. As we show below, the prices in an auction depend both on the number of customers bidding and their valuations for items—and, not surprisingly, the more customers there are, or the more each customer values the items, the higher the prices generated by an auction. In this way, auction prices effectively "adapt" to market conditions, and hence they are often viewed as *price-discovery mechanisms*.

Auctions are important both practically and theoretically. On a practical level, auctions are encountered in many markets, including those for treasury bonds, livestock, used cars, electricity, foreign exchange, real estate, art and rare collectibles, fish, fresh flowers, industrial procurement, public-works contracts, and the sale of natural-resource rights (such as offshore oil and gas leases, logging rights, radio spectrum licenses, and so on). More recently, auctions have gained popularity with the success of e-commerce auction sites such as eBay.

From a revenue management perspective, in particular, auctions have some appealing features. First, they hold out the potential of achieving near-perfect, first-degree price discrimination, and although customers still retain some "information rent" that prevents a firm from capturing the entire consumer surplus, the revenue benefits over using a single price—or even second- and third-degree price discrimination—can be significant. Second, auctions have the potential to directly uncover these near-optimal prices without the need to estimate customers' demand functions or willingness to pay, though this statement again must be qualified somewhat as we explain shortly. Nevertheless, it is fair to say that most auction mechanisms generally require less information about customers than do alternative price-discrimination mechanisms.

On a theoretical level, auctions are important because they provide a rich framework for studying pricing mechanisms in settings where customers act strategically. Indeed, as we show in this chapter, auction theory can often be used to design *optimal mechanisms*—that is, mechanisms that maximize revenues among essentially all possible pricing mechanisms, under certain assumptions of course. In other cases, the theory provides convenient ways to compare the revenues produced by different pricing mechanisms. Also, the theory is based on a strategic (rational) consumer model, which adds to the realism of auction models relative to the (mainly) myopic models studied in Chapter 5.

We first look at some common examples of auctions in practice. Then, in Section 6.2, we describe the classical auction models and theory. Next, we look at dynamic auctions, both in the setting of selling a fixed capacity as in Chapter 2 and in a replenishment setting, where the firm orders and auctions over an infinite horizon as in the inventory-pricing problem of Section 5.3.2. Finally, we consider network auctions and discuss their relationship to the network RM problems of Chapter 3.

6.1.1   An Overview of Auctions in Practice

Auctions are used in a wide range of markets, including industrial, financial, and consumer markets. We briefly survey next each of these markets in turn.

6.1.1.1   Traditional Auction Houses

Traditional auction houses—the two largest being Christie's and Sotheby's—provide auctions for selling art, antiques, jewelry, wine, and other rare, high-value collectibles. Both have been doing so for a very long time indeed; Sotheby's was founded in 1744 and Christie's in 1766. Christie's is the market leader with sales of $2.3 billion in 2000. Both use a variation of an ascending, open-price (English) auction. (See Section 6.1.2 below for definitions of these auction types.) As of January 2000, Sotheby's started offering online auctions. These traditional auction houses generally limit themselves to high-value items, and their clientele are largely wealthy individuals and institutional collectors.

6.1.1.2   Financial-Market Auctions

Auctions have been used for many years in financial markets. Most government bonds and bills are sold at auctions, which are conducted at regular intervals to finance national debts. Investors (both institutional and individual) bid for the minimum interest rate they are willing to receive. The selling agency then sorts the bids and the bonds or bills are awarded to the lowest bidders until the desired amount of the issue is reached.

Auctions are also used by securities exchanges for trading stocks, bonds and foreign exchange. Typically, these are *double auctions* in which *bid* offers are made by customers and *ask* offers are made by sellers. A queue of bid and ask offers is maintained and trade takes place when the highest bid offer in queue exceeds the lowest ask offer in queue. (The rules for the price paid and how this matching takes place are usually specific to each exchange.)

6.1.1.3   Government Auctions

Governments use auctions for the sale of many public assets, including public lands, public industries (privatization sales), and natural-resource rights. A prominent example is the radio spectrum auctions for third-generation (3G) cellular phone service in Europe and the United States These spectrum auctions involved complex combinatorial features, in which communications companies bid for combinations of geographical areas to achieve coverage in a given market area. The sale prices produced by some of these auctions were staggering, and indeed the resulting debts incurred to finance these purchases have left many of the winning companies in a precarious financial position.

6.1.1.4   Industrial-Procurement Auctions

Auctions are also used in many industries for procurement of materials, services, and general subcontracting of production. Typically, this occurs through a request-for-quote (RFQ) process in which a buying firm details its requirements for a certain input, and selling firms submit price quotes to supply the input. Factors other than price, such as quality levels and delivery schedules, are typically important in the final selection as well.

Online versions of procurement auctions have also increased in the past several years. In the auto industry, the exchange Covisint was formed in early 2000 as a joint venture by Daimler-Chrysler, Ford Motor Company, and General Motors with technology provided by Commerce One and Oracle. The goal of the exchange is to facilitate integration and collaboration among suppliers and automakers, with the aim of lowering costs and facilitating more efficient business practices. The Covisint exchange supports a range of auction formats for procurement. FreeMarkets, which has been in operation since 1995, combines software products with market-making services that help facilitate real-time procurement auctions over the Internet. The company reports sales transactions of over $35 billion to date on their reverse-price auction systems and services. Many manufacturers also host their own private online auctions for procurement.

6.1.1.5 Consumer Online Auctions

Online consumer auctions have become popular, largely due to the success of eBay. eBay provides a platform for users to conduct auctions to buy and sell a wide range of items—a sort of Sotheby's OR Christie's for the common man.

An immense variety of items are sold on eBay—new, used, and collectibles, by both individuals and small businesses. It is by some measures, the most popular shopping site on the Internet as of this writing.[1] In 2001, eBay transacted more than $9.3 billion in gross merchandise sales. Most significantly, the company has proven that the Internet can be used to facilitate communication and trade among geographically dispersed individual buyers and sellers, allowing for the sort of real-time auction mechanisms that in the past required the physical presence of market participants.

Priceline.com provides a different online auction mechanism. It is based on what they term a "buyer-driven conditional purchase offer" mechanism,[2] in which customers declare what they are willing to pay for products and supplying firms accept or reject these offers. In return, consumers agree to varying degrees of flexibility in the brand and product features they receive for their offered price. This mechanism has proved quite popular as a channel for selling surplus airline seats and is gaining popularity for products such as discount phone service and home mortgages.

---

[1] For example, in 2000 eBay was the shopping site with the highest number of total user minutes according to Media Metrix.
[2] Priceline.com has been granted a United States patent for this invention.

Priceline is attractive to sellers in large part because the mechanism does not divulge the identity of the seller until after the purchase offer is accepted. (Customers bid on generic products and features, not specific brands.) This creates less of a pricing risk for a firm because it can discount without fear that its discounted prices will become widely known to other customers and to competitors. This feature produces *brand shielding* and such selling formats are often referred to as *opaque channels* in industry terminology. However, as we show in Section 6.3.3 below, under certain assumptions, this mechanism theoretically offers no benefit over list prices. (Priceline.com is discussed further in Chapter 10.)

6.1.2 Types of Auctions

There are a variety of mechanisms one can use to conduct an auction. For simplicity we focus first on the case of a firm auctioning a single indivisible good to a group of N customers. We then consider several variations of these simple, single-unit auctions.

6.1.2.1 Standard Auction Types

There are four common types of auctions for selling a single object:

- **Open ascending (English) auction** In an open ascending auction, the firm announces a progressively increasing sequence of prices. Customers indicate (say by raising their hand or showing a number) their willingness to buy an item at the announced price. The firm increases the price until only one customer is left willing to buy at the announced price. This is the mechanism commonly used to sell art and valuables at major auction houses such as Christie's and Sotheby's.

- **Open descending (Dutch) auction** In an open descending auction, the firm announces a progressively decreasing sequence of prices. The first customer to indicate willingness to buy at the announced price wins the item and pays the current price. The Aalsmeer and Naaldwijk flower markets in Holland have long used this type of auction, which explains the name.

- **Sealed-bid, first-price auction** In the sealed-bid, first-price auction, customers submit sealed bids to the firm. The customer submitting the highest bid wins the auction and pays the amount of his bid. This form of auction is used (in its minimization form) for awarding many government contracts.

- **Sealed-bid, second-price (Vickrey) auction** In the sealed-bid, second-price auction, customers again submit sealed bids, and the customer submitting the highest bid wins the auction. However, the amount the winner pays is equal to the second-highest bid submitted. While this auction form has certain desirable theoretical properties, as shown by Vickrey [533], it is somewhat less common in practice.[3]

These basic auction types can be varied: for example, one may impose a reserve price or minimum bid increments. Moreover, there are other, less standard, auction types that are encountered in practice as well, such as the *uniform price auction* used in many financial markets. The above four types, however, are the most common.

An auction is called a *reverse auction* if customers are competing to sell to the auctioneer by submitting cost (or willingness to sell) bids rather than price (or willingness to buy) bids, such as in a procurement auction. Reverse auctions are essentially equivalent to regular auctions if we put a "minus sign" on the rewards (one involves maximization of price while the other involves minimization of cost), and hence we do not address them separately here.

6.1.2.2 Multiunit Auctions

Multiunit versions of the above auction types can also be defined in the natural way. For example, suppose the firm has C homogeneous items to sell and each customer wants only at most one item. Then in the C-unit open ascending auction, the firm announces increasing prices, and customers indicate their willingness to pay the offered prices. The price is increased until only C customers remain and each is awarded an item at the prevailing price. In an open descending auction, the price declines until a customer indicates willingness to pay the announced price. The customer is awarded a unit at that price, and the firm continues to decrease the price until a second customer is willing to pay the announced price, and so on until all C units have been awarded.

In a sealed-bid, first-price auction, the C highest bids are accepted, and each pays his bid; in a sealed-bid, second-price auction, the C highest bidders are awarded the item and each pays the $(C+1)^{\text{st}}$ highest bid.

Again, more complex multiunit auctions exist in practice. For example, customers may bid for multiple units. In a sealed-bid, first-price auction, this is accomplished by having customers submit a *demand schedule*—a list of quantities and prices they are willing to pay for each marginal unit they buy. The firm then awards items to the C highest marginal values, which may involve awarding multiple units to a sin-

---

[3]Though Lucking-Reily [351] points out that the Vickrey auction is more commonly used than most people realize.

gle customer. In this chapter, we only consider the simple, single-unit demand version of multiunit auctions.

6.1.2.3 Combinatorial Auctions

Another complexity in many procurement auctions is that a customer may require several products simultaneously. For example, to complete production of a product, a manufacturer may need both metal and plastic resin, or to provide cell phone service in a particular region, a communications company may need licenses in several contiguous regions. Such problems create dependencies, in which customers are willing to pay more for certain combinations of items than the sum of what they would be willing to pay for each item alone.

In such cases, one can construct auctions where the customers submit bids for various combinations of items rather than individual bids for each item alone, and the firm must then decide on which combinations to award based on these bids. Such problems may require solving complex, combinatorial optimization problems to simply determine the winners of the auction. Understanding the customers' behavior in the face of such complex auctions is quite difficult. We examine one such combinatorial auction in Section 6.5 below, in which customers bid for "products" that require a subset of "resources" and the firm has to allocate a finite supply of these resources to the customers based on their bids. This problem closely matches the network problems of Chapter 3.

6.2 Independent Private-Value Theory

In this section, we present the basic theory of auctions for the so-called *independent private-value* model, which is the most widely studied in the literature. In addition, we focus here on the revenue-generating properties of auctions and largely ignore welfare and allocative efficiency properties. Readers interested in these properties and other extensions of the basic theory are referred to survey papers by Klemperer [305], Matthews [366], McAfee [369], and Milgrom [381].

6.2.1 Independent Private-Value Model and Assumptions

Consider an auction in which we are selling one or more homogeneous objects to N potential customers. Each customer desires at most one of the objects. Customer i values an object at v_i. The valuations v_i are private information to the individual customers, but it is common knowledge that v_i's are i.i.d. with a distribution F. We assume that

F is strictly increasing with a continuous density function $f(\cdot)$ and has bounded support on the interval $[0, \bar{v}]$, so $F(0) = 0$ and $F(\bar{v}) = 1$.

Note that the assumption that customers have i.i.d. valuations and all know F is not equivalent to saying all customers are the same. Indeed, because customers valuations are draws from a distribution, some customers will have high valuations, and some will have low ones; F merely describes the distribution of valuations in the customer population. In addition, customers know their valuation; thus, a customer with a high (low) valuation will *know* that his valuation is higher (or lower) than average and will bid accordingly. The assumption of i.i.d. valuations and symmetry is more precisely a statement about the views the participants hold about the market. It is equivalent to saying that all customers and the firm have the *same belief* about the likely valuations of other customers and that there is no discernable difference among customers *a priori*.

6.2.2 An Informal Analysis of Sealed-Bid, First- and Second-Price Auctions

First, to build some initial intuition we start with a somewhat informal analysis of the sealed-bid, first- and second-price auctions. A formal equilibrium analysis is then provided in Section 6.2.3.

A key feature of auction models is that they assume customers are rational; that is, they bid so as to maximize their surplus (the value of the item minus the price they pay). Hence, for each mechanism we need to analyze customers' bids as a function of their valuations—called their *bidding strategy*. When formulating his bidding strategy, a rational customer will take into account the bidding strategies of the other customers. Our auction analysis therefore relies on the concept of an *equilibrium set of strategies*; that is, a set of strategies such that each customer has no incentive to change his strategy provided the other customers do not change their strategies (Nash equilibrium in game-theory terminology; see Appendix F).

We are also interested in the revenues produced by a given auction. These revenues depend on the strategic, equilibrium response of customers. So changes in the mechanism will lead to changes in the equilibrium bidding strategies of customers, which in turn will affect the revenues the firm generates. Thus, a "good" mechanism induces a more profitable equilibrium, and this makes the revenue analysis of auctions qualitatively different from the analyses we have seen in the previous chapters.

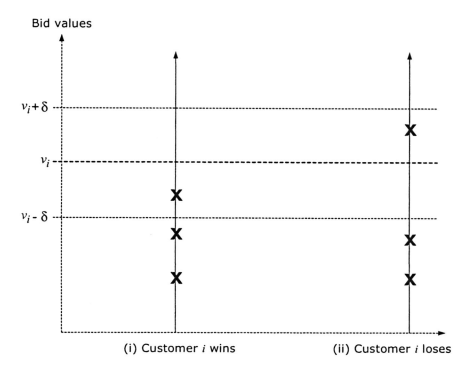

Figure 6.1. Perturbing the bid v_i in a second-price auction.

6.2.2.1 Equilibrium Strategies for a Second-Price Auction

Consider first a single-unit, sealed-bid, second-price auction with N customers. Recall that in this case each of the N customers submits a bid, the firm awards the item to the customer with the highest bid, and the winner pays the value of the second-highest bid. Let $\mathbf{b} = (b_1, \ldots, b_N)$ denote the vector of bids submitted by the N customers and let $b_{[i]}$ denote the ith reverse-order statistic: that is, $b_{[1]} \geq b_{[2]} \cdots \geq b_{[N]}$. Hence, $b_{[2]}$ denotes the value of the second-highest bid (the winner's payment). A bidding strategy for customer i specifies the bid customer i will submit as a function of his valuation v_i and is denoted $b_i(v_i)$. A bidding strategy that is an equilibrium strategy is denoted $b_i^*(v_i)$ (to denote that it is an optimal response to the strategies of other customers).

How would a rational customer bid in this type of auction? The answer, it turns out, is surprisingly simple. Each customer i cannot do better than to simply bid his own valuation v_i; that is, the strategy of bidding $b_i^*(v_i) = v_i$ is optimal for all customers i.

To see this, note that the amount the winner pays in a second-price auction is not affected by his bid since he pays an amount equal to the second-highest bid. In other words, a customer's bid affects whether he wins but not how much he pays if he wins. Now suppose customer i bids v_i. Consider the two possible cases—customer i wins or customer i loses—and see whether customer i can do better by changing his bid v_i in either case. The situation is shown in Figure 6.1.

First, consider case (i) on the left of Figure 6.1, where customer i wins the auction by bidding v_i. In this case, customer i has a surplus of $v_i - b_{[2]} \geq 0$. Now if he increases his bid to $v_i + \delta$, it has no effect because he is still the highest bidder and still pays an amount equal to the second-highest bid. So customer i cannot do better by increasing his bid. If customer i is a winner and decreases his bid to $v_i - \delta$, there is again no change in his surplus as long as he remains the highest bidder. However, if—as shown on the left-hand side of Figure 6.1—he lowers his bid enough to become the second-highest bidder, then he is no longer the winner and his surplus is zero. Since his surplus was positive beforehand, this is not an improvement either. Thus, customer i cannot do better than bidding v_i in case (i).

Now consider case (ii) on the right of Figure 6.1, in which customer i bids v_i and loses. Customer i's surplus in this case is zero because he does not get the item and pays nothing. Note also in this case, the highest bid is strictly greater than v_i; that is, $b_{[1]} > v_i$. Now if he decreases his bid to $v_i - \delta$, he remains one of the losers, and his surplus is still zero. If he increases his bid to $v_i + \delta$, again there is no change unless he increases his bid enough to become the new highest bidder. But in this case, he must pay an amount equal to the previous high-bidder's bid, which is strictly greater than his own valuation v_i (else he would have been the high bidder originally). So his surplus in this case is negative, he is worse off. Hence, he cannot do better than bidding v_i in case (ii) either. Therefore, in both cases (i) and (ii), bidding v_i is an optimal decision for customer i.

Note that this strategy is optimal regardless of the bids placed by other customers. Indeed, our analysis did not make any assumptions on the strategies used by other customers; the strategy $b_i^*(v_i) = v_i$ is optimal for *any realization* of competing bids. Such a strategy is called a *dominant strategy*, and the set of such strategies $b_i^*(v_i) = v_i, i = 1, \ldots, N$ is called a *dominant-strategy equilibrium* for the auction. (See Appendix F.)

A dominant-strategy equilibrium is a robust equilibrium. It applies under very general conditions; essentially, we need assume only that customers have private valuations (the valuation v_i that customer i has for the item is not influenced by the valuations of other customers) and customers are rational so that they recognize the benefit of this strategy. We need little else beyond these two assumptions. For example, customers can have different distributions of valuations, have different information about the distributions, and may be risk-averse. None of these change the equilibrium under the second-price mechanism because of the strong dominance of the bidding strategy.

Under this equilibrium, the firm earns a revenue equal to the second reverse-order statistic of the distribution $F(v)$, a quantity that is not difficult to evaluate (at least numerically). The following example illustrates both the equilibrium and the revenue calculation.

Example 6.1 There are N customers with valuations uniformly distributed on $[0,1]$, so $F(v) = v$ on this interval. Under the second-price auction, it is a dominant-strategy equilibrium for each customer to adopt the strategy $b^*(v_i) = v_i$.

The expected revenue earned by the firm is just $E[v_{[2]}]$—the value of the second highest bid. It is not hard to show for the uniform distribution that

$$E[v_{[2]}] = \frac{N-1}{N+1}.$$

Thus, the firm's average revenue increases with the number of bidders N.

6.2.2.2 Equilibrium Strategies for a First-Price Auction

Consider next a first-price auction, in which the highest bidder wins the item and pays his bid. We consider only symmetric bidding strategies in this case. That is, we assume each customer i uses the same bidding strategy $b(v)$ and therefore bids an amount $b_i = b(v_i)$. This is a reasonable assumption given that customer valuations are symmetric (have valuations independently drawn from the same distribution $F(v)$). We also assume that a customer's bid is increasing in his valuation (customers with higher valuations bid more), so the bid strategy $b(v)$ is increasing in v. This assumption is verified afterward. As before, an equilibrium strategy is denoted $b^*(v)$.

Again, we are looking for an equilibrium bid function such that if all customers are using the strategy $b^*(v)$, then no customer is able to improve his expected surplus by bidding anything other than $b^*(v)$. We then use the first-order conditions for this equilibrium to derive a differential equation for the bid function $b^*(v)$.

To begin, note that a customer i with valuation v_i will win the item if he is the highest bidder; that is, if $b^*(v_i) > b^*(v_j)$, for all $j \neq i$. So customer i's probability of winning is[4]

$$\begin{aligned} P(b^*(v_i)) &= \Pi_{j \neq i} P(b^*(v_i) > b^*(v_j)) \\ &= \Pi_{j \neq i} P(v_i > v_j) \end{aligned}$$

---

[4] Equation (6.1) contains a minor abuse of notation: the $P(\cdot)$ on the left-hand side represents a function of the bids (albeit a probability), while the right-hand side $P(\cdot)$ stands, as throughout the book, for probability of an event.

$$= F^{N-1}(v_i), \qquad (6.1)$$

where the second inequality follows from the assumption that strategy $b^*(v)$ is strictly increasing in v. Since the argument is generic to any customer i, we henceforth drop the subscript and consider an arbitrary customer with valuation v.

Now suppose our customer could improve his expected surplus by adopting the strategy of a customer with valuation $\tilde{v}$ different from v. Specifically, the customer would bid $b^*(\tilde{v})$ and thus win with probability $P(\tilde{v})$ but would still value the item at v. In this case, his expected surplus would be

$$S(b^*(\tilde{v}), v) = P(\tilde{v})(v - b^*(\tilde{v})). \qquad (6.2)$$

If the strategy $b^*(v)$ is truly an equilibrium, this surplus should be maximized at $\tilde{v} = v$ (otherwise, $b^*(v)$ would not be the customer's optimal bid). Therefore, applying the first-order optimality conditions, we can differentiate (6.2) with respect to $\tilde{v}$, set the result to zero at $\tilde{v} = v$, and obtain the following differential equation for $b^*(v)$:

$$b^{*\prime}(v) = \frac{P'(v)}{P(v)}(v - b^*(v)). \qquad (6.3)$$

The solution to this differential equation is somewhat tedious to derive, but one can verify that it is[5]

$$b^*(v) = v - \frac{\int_0^v P(s)ds}{P(v)}, \qquad (6.4)$$

where $P(v) = F^{N-1}(v)$ is the probability of winning given by (6.1).[6]

Note that the equilibrium strategy (6.4) is continuous and increasing in v and increasing in N (higher-valuation customers bid more; and the more customers there are, the higher a given customer bids). One can also show that it is the unique symmetric equilibrium for this problem (Riley and Samuelson [441]).

Note from (6.4) that $b^*(v) < v$, so customers in a first-price auction will bid strictly less than their valuation. Hence, unlike in the second-price auction, they *shade* their true valuations when bidding. This is to be expected because customers are required to pay what they bid, so they must shade their bids to make a positive surplus from winning.

---

[5]To check this, just use the fact that $b^{*\prime}(v) = [\int_0^v P(s)ds][P(v)]^{-2}P'(v)$, and substitute (6.4) into the right-hand side of (6.3).
[6]A boundary condition of $b^*(0) = 0$ is required as well; see Appendix 6.A.

Auctions 253

Finally, the revenue to the firm is the expected value of the highest bidder's bid because the winner pays his valuation. So the firm's expected revenue is $E[b^*(v_{[1]})]$. Again, the mean of this order statistic is not difficult to compute numerically or by simulation.

To illustrate, consider again the example of uniformly distributed valuations:

Example 6.2 There are N customers with valuations uniformly distributed on $[0,1]$, so $F(v) = v$ on this interval. In this case, $P(v) = v^{N-1}$ and the equilibrium bidding strategy is

$$\begin{aligned} b^*(v) &= v - \frac{\int_0^v P(s)\,ds}{P(v)} \\ &= v - \frac{\int_0^v s^{N-1}\,ds}{v^{N-1}} \\ &= v\left(1 - \frac{1}{N}\right). \end{aligned}$$

So each customer bids a fraction $1 - \frac{1}{N}$ of his valuation; hence, customers with higher valuations bid more, and the more customers N, the closer each bids to his actual valuation.

Since the highest bidder $v_{[1]}$ wins, the expected revenue to the firm is then $E[v_{[1]}](1 - \frac{1}{N})$. It is not hard to show for the uniform distribution that $E[v_{[1]}] = \frac{N}{N+1}$. Therefore, the firm's expected revenue is

$$E[v_{[1]}]\left(1 - \frac{1}{N}\right) = \frac{N}{N+1}\left(1 - \frac{1}{N}\right) = \frac{N-1}{N+1}.$$

Note that the expected revenue for the firm is the same in this example and in Example 6.2 for the second-price auction. In other words, the firm generates the same expected revenue regardless of which auction it runs. As we show below, this is not a coincidence; rather, it is a consequence of general conditions that guarantee that these two auctions are always revenue equivalent under the private-value model.

6.2.2.3 Strategic Equivalence of Open and Sealed-Bid Auctions

In the private-value model, the open descending (Dutch) auction is strategically equivalent to the sealed-bid, first-price auction in the sense that the equilibrium strategies for the two mechanisms are the same. That is, if $b^*(\cdot)$ is a symmetric equilibrium in a sealed-bid, first-price auction, then it is also a symmetric equilibrium in a open descending auction, and vice versa. This is true because in an open descending auction, each customer (knowing his valuation v) calculates his expected surplus at each price b, given that there are no other customers willing to buy at b. He then determines the value $b^*(v)$ at which this surplus is maximized and bids when the price drops to $b^*(v)$. But this is exactly the same calculation the customer must make when submitting a bid in a sealed-bid, first-price auction. Hence, the equilibrium strategies are the same.

Likewise, an open ascending auction can be shown to be strategically equivalent to a sealed-bid, second-price (Vickrey) auction under the independent private-value model. In an open ascending auction, it is always optimal for a customer to stay in the bidding as long as the announced price b is below his valuation v—and to drop out once the price exceeds v. But this is equivalent to the strategy of bidding $b^*(v) = v$ in a second-price auction, since in both cases if the customer wins, he ends up paying the valuation of the second-highest customer. And as we showed in Section 6.2.2.1, $b^*(v) = v$ is a dominant-strategy equilibrium in the Vickrey auction. Hence, the two auctions are strategically equivalent.

Because of this equivalence, we henceforth refer to these two cases as simply the *first-price* and *second-price* auctions—without specifying whether the mechanism is the open- or sealed-bid version.

6.2.3 Formal Game-Theoretic Analysis

We now formalize and generalize the analysis of bidding equilibria for a general auction mechanism. Formally, a *bidding strategy* for customer i is a function $b_i(v_i)$ that specifies the bid that customer i will submit conditional on his valuation v_i. We let $\mathbf{v} = (v_1, \ldots, v_N)$ denote the vector of valuations and $\mathbf{b}(\mathbf{v}) = (b_1(v_1), \ldots, b_N(v_N))$ denote the vector of bidding strategies used by the N customers. We let $\mathbf{v}_{-i} = (v_1, \ldots, v_{i-1}, v_{i+1}, \ldots, v_N)$; that is, the vector $\mathbf{v}$ without the ith component. Similarly, let

$$\mathbf{b}_{-i}(\mathbf{v}_{-i}) = (b_1(v_1), \ldots, b_{i-1}(v_{i-1}), b_{i+1}(v_{i+1}), \ldots, b_N(v_N))$$

denote the bid strategies for all customers other than i.

An *auction mechanism* is specified by a pair of mappings $\tilde{\mathbf{y}} : \Re_+^N \to \{0,1\}^N$ that defines the allocations of the goods and $\tilde{\mathbf{p}} : \Re_+^N \to \Re_+^N$ that defines payments made by the customers (equivalently, revenue received by the firm) as a function of their bids. The firm chooses the auction mechanism before the auction is conducted and announces it to all customers, so the mechanism too is common knowledge.

Suppose customer i chooses a strategy $b_i(v_i)$. Then $\tilde{y}_i(b_i(v_i), \mathbf{b}_{-i}(\mathbf{v}_{-i}))$ is the allocation of goods to customer i, which is equal to 1 if he is awarded a unit, and 0 otherwise. Given his bid $b_i(v_i)$, the probability that customer i is awarded a unit is given by

$$P_i(b_i(v_i)) = E_{\mathbf{v}_{-i}}[\tilde{y}_i(b_i(v_i), \mathbf{b}_{-i}(\mathbf{v}_{-i}))].$$

Similarly, $\tilde{p}_i(b_i(v_i), \mathbf{b}_{-i}(\mathbf{v}_{-i}))$ is the payment made by customer i given the bid vector $\mathbf{b}(\mathbf{v})$, and his expected payment is

$$R_i(b_i(v_i)) = E_{\mathbf{v}_{-i}}[\tilde{p}_i(b_i(v_i), \mathbf{b}_{-i}(\mathbf{v}_{-i}))].$$

Note the expected payment is the expected revenue received by the firm. When the number of players N is random, each player computes his optimal action by conditioning both on the valuations of the other players and the total number of players in the game.

Customers are assumed to be rational and attempt to maximize their expected net utility (the value of the item less the price paid to the firm). Therefore, customer i chooses his strategy $b_i(v_i)$ to maximize his expected surplus

$$S_i(b_i(v_i), v_i) = v_i P_i(b_i(v_i)) - R_i(b_i(v_i)). \tag{6.5}$$

For example, in the case of the single-unit, first-price auction, the item is awarded to the highest bidder who pays the auctioneer the value of his bid; all other bidders pay nothing. Then if $b_i = b_{[1]}$ (i is the highest bidder and wins the item), then $\tilde{y}_i(b_i(v_i), \mathbf{b}_{-i}(\mathbf{v}_{-i})) = 1$, and $\tilde{p}_i(b_i(v_i), \mathbf{b}_{-i}(\mathbf{v}_{-i})) = b_i(v_i)$, and if $b_i < b_{[1]}$ (i is not the winning bidder), $\tilde{y}_i(b_i(v_i), \mathbf{b}_{-i}(\mathbf{v}_{-i})) = 0$, and $\tilde{p}_i(b_i(v_i), \mathbf{b}_{-i}(\mathbf{v}_{-i})) = 0$. So the expected net utility is simply $P_i(b_i(v_i))(v_i - b_i(v_i))$.

We assume that customers choose their strategies without collusion. In this case, they play a noncooperative game of incomplete information. An appropriate solution concept in this context is that of the Bayesian equilibrium of Harsanyi [241], an extension of the ordinary Nash equilibrium [402]. Specifically, a vector of strategies $(b_1^*(\cdot), \ldots, b_N^*(\cdot))$ is an *equilibrium strategy* if, for all i, customer i's best response is to maintain his strategy $b_i^*(\cdot)$ provided all other customers maintain their strategies $\mathbf{b}_{-i}^*(\cdot)$. Formally,

$$S_i(b_i^*(v_i), v_i) \geq S_i(b, v_i) \quad \forall b \in [0, \bar{v}], \; \forall i = 1, \ldots, N.$$

In other words, no customer has an incentive to change his strategy if all other customers maintain their strategies. We further restrict ourselves and consider only *symmetric* equilibria; that is, strategies for which the equilibrium strategy $b_i^*(\cdot) = b^*(\cdot)$ is the same for all i. As mentioned, this assumption is reasonable given that customer valuations are assumed symmetric; however, it is a restriction nevertheless, and one cannot rule out the fact that asymmetric bidding strategies may exist. Henceforth, we let $b^*(\cdot)$ denote such a symmetric equilibrium strategy.

6.2.3.1 Direct-Revelation Mechanisms

The analysis of equilibrium bidding strategies is greatly simplified by considering what are called *direct-revelation mechanisms*. Essentially, a direct-revelation mechanism is one in which a customer's equilibrium strategy is to bid his true valuation v. For any mechanism that has an equilibrium it turns out, we can always find an equivalent direct-revelation mechanism.

To see this, note that if $b^*(v)$ is a symmetric equilibrium for some given auction mechanism, then the firm can always define an alternative mechanism (the direct-revelation mechanism) in which customers submit bids, the firm inserts these bids into the function $b^*(\cdot)$, and the resulting values are treated as bids under the rules of the original auction mechanism. The situation is illustrated in Figure 6.2. Since $b^*(\cdot)$ is an equilibrium strategy, it follows that under the direct-revelation mechanism it is an optimal strategy for every customer i to bid his valuation v_i, since otherwise it would contradict the fact that $b^*(\cdot)$ is an equilibrium strategy. Conversely, if there does not exist a direct-revelation mechanism defined by some $b^*(\cdot)$ in which bidding v is an equilibrium, then there cannot be any equilibrium bidding strategy under the original mechanism, otherwise the corresponding equilibrium $b^*(\cdot)$ would define such a direct-revelation mechanism.

In this way, we can reduce the equilibrium analysis of any mechanism to an analysis of the corresponding direct-revelation mechanism, in which case we can view the allocation and payments as being directly a function of the customers' valuations—denoted $y_i(v_i, \mathbf{v}_{-i})$ and $p_i(v_i, \mathbf{v}_{-i})$, respectively (because the optimal strategy is for customers to bid their valuations). This approach is illustrated in Figure 6.2.

Let $R_i(v)$ denote customer i's expected payment ($R_i(v_i) = E_{\mathbf{v}_{-i}}[p_i(v_i, \mathbf{v}_{-i})]$) under a direct-revelation mechanism. The equilibrium can be analyzed by noting that the expected surplus in the direct-revelation mechanism, defined by

$$S_i(v_i) = v_i P_i(v_i) - R_i(v_i),$$

must satisfy

$$S_i(v_i) \geq S_i(\tilde{v}) + P_i(\tilde{v})(v_i - \tilde{v}) \quad \forall \tilde{v} \in [0, \bar{v}]$$

for all customers i. In other words, for each customer i, revealing his true valuation v_i is no worse than pretending to have another valuation $\tilde{v}$. This condition is called the *incentive compatibility constraint* because it requires that it be in customer i's self-interest to truthfully reveal his valuation.

Original Mechanism

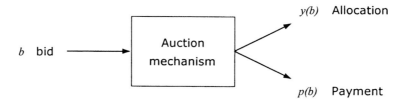

Direct-revelation mechanism

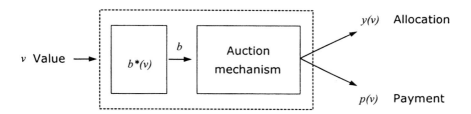

Figure 6.2. Illustration of the direct-revelation mechanism.

6.2.4 Revenue Equivalence

How much revenue is generated for the firm by a given auction mechanism? At first, answering this question would appear to be a hopeless task because each auction mechanism leads to different equilibrium bidding strategies and equilibrium payments. Finding and evaluating these various equilibria for a reasonable range of mechanisms (or ideally all possible mechanisms) is a daunting task. However, it turns out that the expected revenue generated by a private-value auction can be reduced to an analysis only of the resulting allocations $y_i(v_i, \mathbf{v}_{-i})$—without explicitly solving for the equilibrium bidding strategies. The only conditions required are that the functions $y_i(v_i, \mathbf{v}_{-i})$ are increasing in v_i (so that higher valuations lead to a higher probability of allocation)[7] and customers with valuation zero have zero expected surplus in equilibrium.[8] Specifically, we have:

---

[7]Verifying that the allocations are increasing in the valuations v_i may require analyzing monotonicity properties of the equilibrium strategy—that higher-valuation customers v bid higher in equilibrium.
[8]The requirement that customers with valuations zero have zero expected surplus is called a *participation constraint*; intuitively, it means we cannot force a customer to participate in an auction in which his expected surplus is negative.

THEOREM 6.1 *[Revenue Equivalence Theorem]* Consider the private-value model, in which there are C items and N customers with i.i.d. valuation independently drawn from a continuously differentiable, strictly increasing distribution F on $[0, \bar{v}]$. Consider any mechanism in which (i) the allocation to customer i, $y_i(v_i, \mathbf{v}_{-i})$, is increasing in v_i for all i, and (ii) customers with valuations of zero have zero expected surplus. Then the expected revenue for the firm is given by

$$E\left[\sum_{i=1}^{N} J(v_i) y_i(v_i, \mathbf{v}_{-i})\right], \tag{6.6}$$

where

$$J(v) = v - \frac{1 - F(v)}{f(v)}.$$

A proof of the this theorem is given in Appendix 6.A. Moreover, this revenue equivalence holds for the customers as well; a customer's expected payment is the same under all mechanisms satisfying the above conditions. However, the equivalence in both cases is only in expectation; the payments on a sample-path basis may be quite different under different mechanisms.

Note $J(v)$ is precisely the marginal revenue function (7.14) encountered in our analysis of revenue-maximizing prices in Chapter 7. This is not a coincidence; Bulow and Roberts [95] show the revenue function (6.6) can in fact be interpreted as a variant of third-degree monopoly price discrimination among the N customers (see Section 8.3.3.2).

In auction theory, $J(v_i)$ is sometimes referred to as customer i's *virtual value* because (6.6) implies that the firm can hope to collect only $J(v_i) < v_i$ from customer i (in expectation) and not his entire valuation v_i. The difference $v_i - J(v_i)$ is referred to as the *information rent* of customer i because it is the surplus that customer i retains due to his private information about his own valuation v_i.

As a result of Theorem 6.1, note that any two mechanisms that produce the same allocation for every realization of $v_1, \ldots, v_N$ (the same customers are awarded units under each mechanism) produce the same expected revenue for the firm. This is true despite the fact that the bidding strategies and payments may be very different under each mechanism. For this reason, Theorem 6.1 is referred to as the *revenue equivalence theorem*.

To illustrate, consider a standard first-price auction on a single unit. We saw in Section 6.2.2.2 that the equilibrium bidding strategy was strictly increasing in the value v and that the item is awarded to the highest bidder. In a second-price auction, the customer with the highest

Auctions 259

valuation also wins that auction. Thus, by Theorem 6.1, the expected revenue to the firm must be the same in each case. We illustrate this result with a continuation of our previous example:

Example 6.3 Suppose there are $N = 2$ customers with valuations uniformly distributed on $[0, 1]$. Let $v_{\max} = \max\{v_1, v_2\}$ and $v_{\min} = \min\{v_1, v_2\}$. From Example 6.2 we know that in a first-price auction, the customers will bid $v(1 - \frac{1}{N}) = v/2$ in equilibrium. The highest bidder will win, and the firm's expected revenue is

$$\frac{1}{2} E[v_{\max}] = (1/2)(2/3) = 1/3.$$

Now consider a second-price auction of Example 6.1. The highest bidder wins but pays the price of the second-highest bid, and each customer bids his valuation in equilibrium. The firm's expected revenue is

$$E[v_{\min}] = 1/3.$$

Hence, the two expected revenues are equal. Moreover, note that since $F(v) = v$, $J(v) = 2v - 1$, and therefore since in both auctions $y_i(v_1, v_2) = 1$ if and only if $v_i = v_{\max}$, we have

$$\begin{aligned} E[J(v_1)y_1(v_1, v_2) + J(v_2)y_2(v_1, v_2)] &= E[2v_{\max} - 1] \\ &= 2(2/3) - 1 \\ &= 1/3 \end{aligned}$$

as well. Finally, for $N > 2$ customers, the customer with valuation $v_{\max}$ wins, and the same analysis shows that

$$\begin{aligned} E\left[\sum_{i=1}^{N} J(v_i)y_i(v)\right] &= E[2\max\{v_1, \ldots, v_N\} - 1] \\ &= 2\left(\frac{N}{N+1}\right) - 1 \\ &= \frac{N-1}{N+1}, \end{aligned}$$

which is precisely the expected revenue found in Examples 6.1 and 6.2.

Similarly, in a standard C-unit auction, one can show that both the first-price and second-price auctions award the goods to the customers with the C highest valuations. Thus, the allocation $y(v)$ is the same for each v, and hence the two mechanisms generate the same expected revenue for the firm.

6.2.5 Optimal Auction Design

The revenue expression (6.6) can be used to design an optimal mechanism by simply choosing the allocation rule $\mathbf{y}^*(\mathbf{v})$ that maximizes

$$\sum_{i=1}^{N} J(v_i) y_i(v_i, \mathbf{v}_{-i}) \qquad (6.7)$$

subject to any constraints one might have on the allocation.

Toward this end, it is useful to make the same regularity Assumption 7.2 on the distribution function F that we impose in dynamic pricing problems: namely, that $J(v)$ is strictly increasing in v. Note that

$$J(v) = v - \frac{1}{\rho(v)},$$

where $\rho(v)$ is the hazard rate of the distribution F. The marginal revenue $J(v)$ satisfies this monotone condition as long as the hazard rate $\rho(v)$ is increasing—or not decreasing too quickly with v.[9]

To illustrate, consider designing an optimal C-unit auction using (6.6). Note that with C units to allocate and given a realization of $\mathbf{v}$, we want to maximize (6.7) subject to the constraint that

$$\sum_{i=1}^{N} y_i(v_i, \mathbf{v}_{-i}) \leq C,$$

and $y_i(\cdot) \in \{0, 1\}$ for all i. It is easy to see what the optimal allocation is by inspection. Indeed, define

$$v^* = \max\{v : J(v) = 0\} \qquad (6.8)$$

(and by convention, $v^* = \infty$ if $J(v) < 0, \; \forall v$). Then since $J(\cdot)$ is assumed to be increasing, it follows that it is never optimal to allocate a unit to a customer with valuation $v_i < v^*$ because awarding units to such customers results in a negative contribution to the sum (6.7). Among the remaining customers with $v_i \geq v^*$, we want to award units to those with the highest valuations v_i. Thus, the optimal allocation is to award units to the C highest-valuation customers v_i above v^*, and if there are less than C customers with $v_i \geq v^*$, to award units only to those customers and discard the remaining units.

How can we achieve such an allocation? One possibility is to introduce a *reserve price* into the standard first- or second-price C-unit auction mechanism. A reserve price is a lower bound on bids that the firm sets before the auction; only bids above the announced reserve price are considered.

To illustrate, consider a first-price auction; it is easy to see that if we set a reserve price of v^*, customers with valuations $v_i < v^*$ will not submit bids. One can show that the remaining customers with valuations

---

[9]More precisely, it is satisfied when $\rho'(v)/\rho^2(v) > -1$ for all $v \geq 0$. One can show that this condition is satisfied by many standard distributions, including the uniform, normal, logistic, exponential, and extreme value (double exponential) distributions [25].

$v_i \geq v^*$ will submit bids according to an increasing equilibrium strategy similar to (6.4). Indeed, the resulting symmetric equilibrium strategy is now

$$b^*(v) = v - \frac{\int_{v^*}^v F^{N-1}(s)ds}{F^{N-1}(v)}. \tag{6.9}$$

The C units are awarded to the C highest bids above the reserve price, and the resulting allocation is exactly the same as the optimal allocation. Hence, the first-price mechanism with reserve price v^* is optimal. A similar argument holds for the second-price auction, in which case one can show that it is optimal to post a reserve price of v^*, where winners pay the minimum of the $(C+1)^{\text{st}}$ highest bid above v^*, or v^* if there are fewer than $C+1$ bids above v^*. We therefore have the following theorem:

THEOREM 6.2 *Under the private-value model, the standard C-unit first-price and second-price auctions with reserve price v^* (given by (6.8)) are optimal for the firm.*

Hence, with a properly chosen reserve price, the standard first-price and second-price auctions are revenue maximizing among all possible pricing mechanisms. This is a rather remarkable result; under the private-value model assumptions, a firm simply cannot do better than to sell using one of these two auction formats. Again, we illustrate this result by returning to our uniform-distribution example:

Example 6.4 Suppose there are N customers with valuations uniformly distributed on $[0, 1]$. Since $F(v) = v$ implies $J(v) = 2v - 1$, the optimal reserve price $v^* = 1/2$ since this satisfies $J(v^*) = 0$. From (6.9), the customers with valuations over $1/2$ therefore bid

$$\begin{aligned} b^*(v) &= v - \frac{\int_{1/2}^v s^{N-1}ds}{v^{N-1}} \\ &= v(1 - \frac{1}{N}) + \frac{(1/2)^N}{Nv^{N-1}}, \end{aligned}$$

which is strictly greater than the bid of $v(1-1/N)$ submitted by these same customers in the first-price auction without reserve prices. Also, since the item is allocated to the highest-value customer and the distribution of the highest valuation is $F^N(v) = v^N$, the firm's optimal expected revenue is

$$\begin{aligned} E[J(\max\{v_1, \ldots, v_N\})] &= \int_{1/2}^1 (2v-1)Nv^{N-1}dv \\ &= \frac{N - 1 + (1/2)^N}{N+1}, \end{aligned}$$

which is again larger than the expected revenue of $(N-1)/(N+1)$ generated when no reserve prices are used but approaches the no-reserve-price revenue when N is large.

The fact that a reserve price primarily benefits the firm most when there are few customers is intuitive. In essence, the reserve price serves to create "extra competition" for customers—forcing them to bid higher in a first-price auction or pay more if they win in a second-price auction—than they otherwise would without a reserve price. However, with lots of competition from other customers, the need for the firm to introduce this extra incentive is less important, as the customers themselves create sufficient competition.

6.2.6 Relationship to List Pricing

How is the optimal auction mechanism related to a traditional list-price mechanism? There are several close connections worth examining.

First, note that a list-price mechanism qualifies as one of the possible allocation and payment mechanisms studied above for the C-unit auction. In particular, using a fixed list price p, customers indicate their willingness to pay p (just as in an ascending auction). If there are C or fewer customers willing to pay p, each receives a unit and pays the fixed amount p; if there are more than C customers willing to pay p, the C units are randomly rationed to these customers. This produces an allocation and payment rule just as in the standard auction types. In the list price case, it is easy to see that it is a dominant strategy for a customer to "bid" (indicate his willingness to pay p) if his valuation v is more than p. Thus, a dominant, symmetric equilibrium strategy exists in which all customers with valuations greater than p attempt to buy.

Given this observation, it is easy to compute the firm's expected revenue. Let $N(v)$ denote the number of customers with valuations greater than v. Then the expected revenue to the firm as a function of p is

$$r(p) = pE[\min\{N(p), C\}]. \tag{6.10}$$

Another way of deriving this revenue is to use the expression (6.6) for the firm's equilibrium revenue and to note that for the list-price mechanism

$$E\left[\sum_{i=1}^{N} J(v_i)y_i(v_i, v_{-i})\right] = E[\min\{N(p), C\}]E[J(v)|v > p],$$

since $E[\min\{N(p), C\}]$ units are allocated, and each customer i to which we allocate a unit ($y_i = 1$) has a valuation $v_i > p$, so $E[J(v)|v > p]$ is the expected value of the corresponding term $J(v_i)$. Using the fact that $J(v) = v - (1 - F(v))/f(v)$, it is then a simple exercise to show that $E[J(v)|v > p] = p$, which gives us the same expression as (6.10).

A direct optimization of (6.10) does not lead to a clean expression, but several special cases are simple and provide useful insight. We look at these next.

6.2.6.1 Capacity Is Unconstrained

The first case is where the number of customers $N \leq C$, so there are always fewer customers than there are units. In this case, $E[\min\{N(p), C\}] = E[N(p)] = N(1 - F(p))$ and the revenue for a list price of p is $r(p) = Np(1 - F(p))$. Differentiating and setting the result equal to zero, we find that the optimal price p^* satisfies

$$p^* f(p^*) - (1 - F(p^*)) = 0.$$

But since $f(p) > 0$ (F is strictly increasing), rearranging this is equivalent to $J(p^*) = 0$, which is the condition for the revenue-maximizing price and is also the condition for determining the optimal reserve price. Thus, the optimal price when $N \leq C$ is the same as the optimal reserve price—that is, $p^* = v^*$. Moreover, the revenue under this optimal price can be written

$$E[N(v^*)]E[J(v)|v > v^*] = E\left[\sum_{i=1}^{N} J(v_i)\mathbf{1}(v_i > v^*)\right]$$

$$= E\left[\max_{y_i \in \{0,1\}} \{\sum_{i=1}^{N} J(v_i)y_i\}\right],$$

where $\mathbf{1}(v_i > v^*)$ is the indicator function of the event $v_i > v^*$. But the expression $E\left[\max_{y_i \in \{0,1\}} \{\sum_{i=1}^{N} J(v_i)y_i\}\right]$ above is simply the optimal auction revenue when $J(v)$ is strictly increasing and $N \leq C$. Thus we have

PROPOSITION 6.1 *If capacity is not constrained in a C-unit, private-value auction ($N \leq C$), then using a fixed list price p^* satisfying $J(p^*) = 0$ is an optimal mechanism for the firm.*

6.2.6.2 Large Capacity and Sales Volumes

Another case in which list pricing is provably good is when both the number of customers and the number of units for sale is large. Specifically, let θ be a positive integer, and consider a problem with θC units and θN customers for some $N > C > 0$. If we set a fixed price of p, then the number of customers willing to purchase at this price is denoted $N_\theta(p)$ with mean $N\theta(1 - F(p))$. Moreover, by the law of large numbers, as $\theta \to \infty$,

$$\frac{N_\theta(p)}{\theta} \to N(1 - F(p)) \quad (a.s.),$$

and the firm's revenue satisfies

$$p\frac{\min\{N_\theta(p), \theta C\}}{\theta} = p\min\{\frac{N_\theta(p)}{\theta}, C\} \qquad (6.11)$$

$$\to p\min\{N(1-F(p)), C\} \quad (a.s.).$$

Just as in the capacity-constrained pricing problem of Section 5.2.1.2, the asymptotically optimal price is given by

$$p^* = \max\{p^0, \bar{p}\},$$

where p^0 is the revenue-maximizing price, determined by $J(p^0) = 0$, and $\bar{p}$ is the run-out price, determined by equating the expected number of customers willing to pay $\bar{p}$ to the supply C, so $N(1-F(\bar{p})) = C$. When $p^* = \bar{p}$, the expected revenue is $C\bar{p}$, and when $p^* = p^0$, the expected revenue is $Np^0(1-F(p^0))$.

Similarly, one can analyze the scaled optimal auction revenue. Note that the scaled expected optimal auction revenue can be written

$$\frac{1}{\theta} E \left[\sum_{i=1}^{\min\{\theta C, N_\theta(v^*)\}} J(v_{[i]}) \right],$$

where $v_{[i]}$ denotes the ith largest valuations $(v_{[1]} \geq v_{[2]} \geq \cdots \geq v_{[N]})$.

First, consider the case $N(1-F(v^*)) > C$. Then as $\theta \to \infty$, with probability one $N_\theta(v^*) > \theta C$. So the above becomes

$$\begin{aligned}
\frac{1}{\theta} E\left[\sum_{i=1}^{\theta C} J(v_{[i]})\right] &\to N \int_{F^{-1}(1-\frac{C}{N})}^{\bar{v}} J(v) f(v) dv \\
&= N \left[\int_{\bar{p}}^{\bar{v}} v f(v) dv - \int_{\bar{p}}^{\bar{v}} (1-F(v)) dv \right] \\
&= N\bar{p}(1-F(\bar{p})) \\
&= C\bar{p},
\end{aligned}$$

which is exactly the asymptotic fixed-price revenue given by (6.11) when $p = \bar{p}$.

In the alternative case where $N(1-F(v^*)) < C$, as $\theta \to \infty$, with probability one $N_\theta(v^*) < \theta C$, and similar reasoning shows that the

$$\frac{1}{\theta} E[\sum_{i=1}^{\theta C} J(v_{[i]})] \to Np^0(1-F(p^0)),$$

which is again exactly the asymptotic fixed-price revenue given by (6.11) when $p = p^0$. These arguments can be formalized to show

PROPOSITION 6.2 *If the number of customers and the number of units for sale in the private-value auction model are, respectively, θN and*

θC *for some integers* θ, *and* $N > C > 0$. *Then as* $\theta \to \infty$, *a listprice mechanism is asymptotically optimal, in the sense that the ratio of the optimal expected list price revenue to the optimal expected auction revenue tends to one.*

As a result, in high-sales-volume settings, using a fixed price will be near optimal. This implies auction benefits are something of a "smallnumbers" phenomenon, which is consistent with the auctions one encounters in practice.

6.2.6.3 Dynamic Pricing

Another close connection between auctions and list-price mechanisms is obtained by considering a dynamic pricing policy as a particular allocation and payment mechanism in a private-value auction model. Making this connection yields several important insights.

For example, consider the problem of selling a single unit to a population of N strategic consumers. As in the price-skimming model of Section 5.5.2, the private-value model considers the N customers to have i.i.d. private valuations v_i for the item. The Dutch-auction mechanism calls for the firm to continuously reduce the price over time until a customer decides to bid at the offered price. The customer then pays this offered price. However, this is precisely what happens in a (continuoustime) dynamic pricing policy as well, so a descending dynamic price can effectively achieve the Dutch-auction outcome. By simply adding an optimal reserve price v^*, below which we will not lower the price, such a dynamic pricing mechanism becomes optimal.

More generally, by the revenue equivalence theorem, *any* dynamic pricing policy that results in the C highest-valuation customers, with valuations in excess of v^*, receiving the units, will be revenue-maximizing for the firm and thus produce the same expected revenue as the optimal auction.

For example, Bulow and Klemperer [94] analyze the C-unit, privatevalue model under a dynamic pricing mechanism. In their mechanism, the firm uses a list price that is lowered continuously until one or more customers offers to buy at the current price. If the number of customers willing to buy at the current price is less than the remaining supply, these customers are awarded the items at this price, and the firm continues to lower the price. If the number of customers willing to pay the current price exceeds the remaining supply, the firm does not sell the items; it instead increases the price discontinuously and then tries again to lower the price. Since a customer's probability of getting an item is higher when he attempts to buy early (if he attempts to buy and fails, he can always try again later, so his probability of obtaining the item cannot decrease by attempting to buy early), it is not hard to show that customers with the highest valuations are the ones that attempt to buy first. Therefore, the firm allocates the items first to the customers with the highest valuation. As a result, Bulow and Klemperer [94] argue that by revenue equivalence, if the firm does not lower the price below v^*, this dynamic pricing mechanism is optimal. Similar arguments hold for many other dynamic pricing policies as well.

This shows there is a rather close connection between optimal auction theory and dynamic pricing theory with strategic consumers. Indeed, using the revenue equivalence theorem, the seemingly difficult task of analyzing the customer equilibrium produced by a dynamic pricing strategy is greatly simplified, and it shows in fact that a range of pricing mechanisms are optimal.

6.2.7 Departures from the Independent Private-Value Model

Many of our conclusions thus far depend to a greater or lesser extent on the assumptions of the independent, private-value model. What happens when these assumptions are relaxed? In this section, we look briefly at a few cases that are especially relevant for RM. Each has implications for the types of auctions that are optimal for the firm.

6.2.7.1 The Common-Value Model

The private-value model assumes that each customer's valuation is independent of the valuations of other customers. Thus, if a customer learns the value that another customer places on the item, it has no impact on his valuation. Such an assumption is reasonable if the item is going to be used for personal enjoyment or consumption. However, in other cases the item may have a common commercial value, may be resold at some future point in time, or may be of uncertain quality, so the valuations others have on the item could reveal useful information about the value of the item to a given customer.

A canonical example of such a setting is selling an offshore oil lease. The value of the lease to a customer is dependent on two key factors: the volume of oil it contains and the cost of extracting that oil. Typically, there is a high degree of uncertainty about both these factors. Because of differences in survey data or technological expertise, different customers may have independent information on the value of a given lease, and so on. As a result, knowing how another customer values the lease may change your assessment of its profitability.

A simple model of such a setting is the following: consider auctioning a single item that has a *common value* a, which is the same for all customers. However, the value a is uncertain. All customers have the same prior knowledge of a, embodied as a distribution over values of a. This distribution is common knowledge. A value of a is drawn from this distribution, and then each customer i receives a (noisy) signal t_i of the form

$$t_i = a + \xi_i,$$

where $\xi_i, i = 1, \ldots, N$ are i.i.d. random-noise terms with mean zero. The distribution of ξ_i is also common knowledge.

Note that given only the signal t_i, customer i's expected value for the item is

$$E[a|t_i] = E[t_i - \xi_i|t_i] = t_i,$$

with variance given by the variance of ξ_i. However, if one were to aggregate the customers' signals by averaging them, the estimate would be $a + \frac{1}{N}\sum_{i=1}^{N} \xi_i$, which provides a much better (lower variance) information on the value a than do the signals t_i alone.[10] More generally, customer i's estimate of a may be altered by information he receives about the signals of other customers. This sort of behavior significantly affects the auction outcomes.

For example, one phenomenon that arises in this setting is the so-called *winner's curse*. To illustrate the idea, consider a sealed-bid, second-price auction. Suppose customer i were to bid his expected valuation t_i for the item, as in the private-value case. The customer might (incorrectly) reason that bidding his own expected valuation t_i is a dominant strategy because bidding more than t_i increases his chance of winning only in cases where his expected surplus is negative, and bidding less than t_i decreases his chances of winning only in cases where his expected surplus is positive. The reasoning is false, however, because customer i's expected valuation conditional on winning the auction is less than his unconditional expected valuation t_i. Indeed,

$$E[a|t_i = \max\{t_1, \ldots, t_N\}] = t_i - E[\max\{\xi_1, \ldots, \xi_N\}] < t_i,$$

since $E[\max\{\xi_1, \ldots, \xi_N\}] > E[\xi_i] = 0$ (provided that ξ_i has nonzero probability of exceeding its mean zero).

Intuitively, winning should indicate to customer i that his noise term ξ_i is the largest and therefore his initial estimate t_i is upwardly biased. Therefore, if he were to bid his unconditional estimate t_i, winning the

---

[10] For example, the variance of the aggregate signal $\frac{1}{N}\sum_{i=1}^{N} \xi_i$ is a factor $1/\sqrt{N}$ smaller than the variance of ξ.

auction would indeed be bad news. It would indicate his expected surplus was negative; hence the *winner's curse*. To overcome this "curse," a rational customer must adjust his bid downward, considering the fact that it is the expected valuation of the item conditioned on having the highest signal that matters in determining his winnings.

The tendency of customers to reduce their bids to avoid the winner's curse changes the revenue equivalence of the basic auction types. In particular, while the sealed-bid auction conveys no information to customers, the Dutch (open descending price) and English (open ascending price) auctions provides them some information because they can observe how many customers are still willing—or not willing—to buy at the current price, when each drops out, and so on. The information about other customers' valuations tends to reduce the negative impact of the winner's curse.

For example, when an item has a common-value component, one can show the firm is better off using an English (open ascending price) auction than a sealed-bid, second-price auction—auctions that are strategically equivalent under the private-value model. Moreover, one can show that if the firm has its own signal (some private information) positively correlated with the item's value (like past price data of similar items or an appraisal), it benefits by sharing that information with the customers. This is because customers will tend to increase their estimate of the item's value as a result and bid more aggressively. Reserve prices also benefit the firm, but unlike in the private-value case, the optimal reserve price may vary with the type of auction and the number of customers.

6.2.7.2 Risk Aversion

Another factor affecting the results of the independent, private-value model is the assumption that both the firm and customer are risk-neutral. (See Appendix E for a discussion of risk preferences.) While the assumption of risk neutrality for a firm is often reasonable (for example, when the firm is a large, participating in many auctions over time), the assumption of risk neutrality for individual consumers is typically less realistic. However, it is easy to determine the relative performance of the standard auction types under risk aversion.

First, consider the case where the firm is risk-neutral and the customers are risk-averse. By revenue equivalence, note that a customer i's expected payment conditioned on his valuation v_i being the highest is the same under the first- and second-price auctions, since this expected payment is simply the expected revenue to the firm. However, the customer who wins a first-price auction pays a certain amount $b^*(v_i)$, while the same customer in a second-price auction will pay an uncertain amount with the same mean—namely, the valuation of the second-highest customer conditioned on the fact that v_i is the highest valuation. Thus, a risk-averse customer will prefer the first-price auction to the second-price auction. Given this preference, in the first-price auction risk-averse customers will tend to increase their bids above the risk-neutral equilibrium bid $b^*(v)$. (Bidding one's own valuation is still a dominant strategy under risk aversion in the second-price auction, so the bidding strategy in this case is not affected.) The higher resulting equilibrium bids in the first-price auction mean that the firm's expected revenue is higher as well, so the firm prefers this auction format.

Now consider the opposite case, where the firm is risk-averse and the customers are risk-neutral. By the same reasoning as above, the firm's revenue in the first-price auction conditioned on the winning value being v is certain while in the second-price auction it is uncertain. Therefore, unconditioning on v, the revenue in the second-price auction is more variable as well—also with the same mean as in the first-price auction. Thus, a risk-averse firm will also prefer the first-price auction.

The fact that the firm prefers the first-price auction in both cases (and is no worse in the second-price auction if all parties are risk-neutral) has been offered as one explanation for the relative popularity of first-price auctions over second-price auctions in practice.

6.2.7.3 Asymmetry Among Customers

Yet another departure from the private-value model is to relax the assumption of symmetry. The simplest case is to assume that there are two types of customers, types 1 and 2, with different valuations for the item drawn from different distributions, denoted $F_1(v)$ and $F_2(v)$, with corresponding marginal revenue (virtual value) functions $J_1(v)$ and $J_2(v)$. For example, type 1 customers may be experienced customers, and type 2 customers may be novice customers, or type 1 customers may be individuals while type 2 are industrial customers.

To see what can happen in this case, assume the first N_1 customers are of type 1 and the next N_2 are of type 2, and assume the marginal revenue functions are both increasing. The optimal allocation for the firm is obtained by maximizing

$$E\left[\sum_{i=1}^{N_1} J_1(v_i) y_i(v_i, \mathbf{v}_{-i}) + \sum_{i=N_1+1}^{N_1+N_2} J_2(v_i) y_i(v_i, \mathbf{v}_{-i})\right],$$

subject to the constraint that the total allocation is one

$$\sum_{i=1}^{N_1} y_i(v_i, \mathbf{v}_{-i}) + \sum_{i=N_1+1}^{N_1+N_2} y_i(v_i, \mathbf{v}_{-i}) = 1.$$

As before, it is optimal to allocate the item to the customer with the highest marginal value. However, note since $J_1(v)$ and $J_2(v)$ may differ, the customer with the highest marginal value is not necessarily the one with the highest valuation v.

This has important consequences for the optimal auction. For example, it means that it can be optimal for the firm to set different reserve prices for different types of customers, and the firm may systematically favor one class of customers over another in awarding the item. Indeed, one can show that in certain cases, it is optimal for the firm to favor the type of customers that tend to value the item less. The rationale for this is that by favoring these low-value types, the firm encourages the high-value types to bid even higher. The resulting higher equilibrium bids it receives from the high-value types more than compensates the firm for the loss he occasionally takes in favoring the low-value types.

In other words, it is optimal for the firm to *discriminate* among customers in the offering terms for the auction. This behavior is similar to the classical third-degree price-discrimination policy of offering different prices to different customer groups based on their different willingness to pay. (See Section 8.3.3.1.)

6.2.7.4 Collusion

The private-value model assumes the firm defines a game among the customers, intended to extract the highest prices possible from them. A key assumption in this game is that customers do not cooperate. Yet in practice, there is the possibility of *collusion* among customers, in which a coalition of customers (popularly called a *bidding ring*) cooperates and agrees to submit bids that are designed to reduce the price paid by the winner. Such collusion has been reported, for example, in the awarding of some government contracts.

There are several practical devices to reduce the likelihood of collusion among customers, most of which involve reducing the ability of customers to communicate among themselves. For example, one technique is to keep the identity of all customers secret, so customers cannot identify each other and form a bidding ring. (Though this may fail if the number of potential customers is so small that most customers know, *a priori*, the pool of likely participants—such as major suppliers in a procurement auction). Another technique is to reduce the amount of information relayed about bids to the minimum necessary to conduct the auction. For example, the firm might report only the highest current bid in an ascending auction, not the number of bids received, the time bids were received, or the history of bid values. This prevents customers from using such data to "signal" their intentions to each other.[11]

Because collusion can take so many forms, it is difficult to make general recommendations on the firm's "optimal response" to collusion. Nevertheless, to give some sense of the effect that it has consider a case where all N customers in the private-value model can collude perfectly. That is, they can get together and agree to submit bids, make payments, and allocate the item among themselves to maximize the surplus they receive as a group. In this case, the group of N customers effectively acts as a single "big customer" with valuation $\hat{v} = \max\{v_1, \ldots, v_N\}$, with distribution

$$P(\hat{v} \leq v) = F^N(v),$$

where $F(\cdot)$ is the distribution of the valuations for the N customers with density $f(\cdot)$. The marginal value for this distribution is

$$J_N(v) = v - \frac{1 - F^N(v)}{NF^{N-1}(v)f(v)}.$$

Of course, when faced with a single customer, the optimal auction is still the usual one: conduct a first or second-price auction with a reserve price set according to the marginal value of the single customer. So assuming $J_N(v)$ is increasing, the firm should set a reserve price v^* satisfying

$$J_N(v^*) = 0.$$

In this case, the bidding ring will be forced (yet willing) to pay v^* when its maximum valuation, $\hat{v}$, is at least this large. Also, one can show that this optimal reserve price is higher than the noncooperative optimal reserve price and that it increases with the number of customers N in the coalition. Thus, the possibility of collusion creates an incentive to use higher reserve prices than when customers do not collude. Indeed, the desire to thwart collusion is one of the main motivations for using reserve prices in practice.

---

[11] For example, at a keynote address to the Institute for Mathematics and its Applications (IMA) in December 2000, Robert Weber reported an instance in which bidders in a auction used the least significant digits in their bid amounts as a signaling mechanism. To overcome this, the auctioneer imposed larger minimum bid increments, thus reducing—or at least raising the cost of—this sort of signaling.

6.3 Optimal Dynamic Single-Resource Capacity Auctions

We next consider a dynamic auction problem that is in essence the auction equivalent of the single-resource problem of Chapter 2. In contrast to the traditional auction problem, in this case the firm receives bids from T groups of customers who are separated over time. In particular, in each period t, we assume that a new set of customers arrives and bids for the remaining capacity. The firm must determine winners in period t before observing the bids (or even the number of customers) in future periods. This dynamic feature parallels the traditional RM model, in which the firm must determine the capacity to sell in a given period before observing demand in future periods.

Such separation of customers over time is common in RM practice, a canonical example being the airline industry. Leisure travelers typically make travel plans months in advance of departure because they frequently must coordinate their vacation travel with other arrangements, like reserving resort accommodations, taking time off work or finding child care, and so on. In contrast, business travelers may not even know of their need to travel until a few days in advance of departure. As a result, if an airline were to conduct a single auction months in advance of departure, they would likely lose many business travelers; if they conducted a single auction a week before departure, they would likely lose many leisure travelers. This creates an incentive for them to conduct auctions at multiple points in time.

Other industries face similar situations, in which customers' needs are realized at different points of time (the need to buy a gift for a birthday, for example)—or are based on other contingent events (a new order to a manufacturer triggering a need for new supplies) that effectively separate customers in time. In such situations, a firm attempting to use a single auction at a single point in time would find itself eliminating many potential customers. By conducting multiple auctions over time, it can reach a larger pool of customers.

We next look at the optimal auction-design problem for this dynamic auction setting. We also compare the optimal auction to a traditional RM mechanism based on using dynamic list prices and capacity controls.

6.3.1 Formulation

A firm has an initial capacity of C units of a good that it wants to sell over a finite time horizon T. It does this by conducting a sequence of auctions indexed by $t = 1, \ldots, T$.

Customers are separated in time. In period t, N_t risk-neutral potential customers arrive. N_t is a nonnegative, discrete-valued random variable distributed according to a known p.m.f. $g(\cdot)$ with support $\{0,\ldots,M\}$ for some $M > 0$ and strictly positive first moment.

The assumptions parallel the private-value model: Each customer wishes to purchase at most one unit and has a reservation value v_i^t, $1 \leq i \leq N_t$. When the context is clear, we omit the time index and write v_i. Reservation values are private information, i.i.d. samples from a distribution $F(\cdot)$, which, as in the private-value model, is assumed strictly increasing with a continuous density function $f(\cdot)$ on the support $[0,\bar{v}]$, with $F(0) = 0$ and $F(\bar{v}) = 1$. To simplify notation and subsequent analysis, we assume that the distribution functions g and F do not depend on the time t but the extension to time-dependent distributions is straightforward.

The distributions F and g are assumed common knowledge to the firm and all potential customers (although this assumption can be relaxed for the second-price mechanism below). In addition, customer i knows his own (private) valuation v_i. Without loss of generality we assume that the unit salvage value for the firm at time $t = 0$ is $v_0 = 0$.

The firm's problem is to design an auction mechanism that maximizes its expected revenue. To do so, it must solve for an optimal allocation $y(v)$ in each period, given the values of N_t, v in each period and knowing only the probabilistic information (distributions) of these values in future periods.

Define the value function $V_t(x)$ as the maximum expected revenue obtainable from periods $t, t+1, \ldots, T$ given that there are x units in period t. Using (6.6) for the expected revenue in each period, the Bellman equation for $V_t(x)$ in terms of the allocation variables $y(v)$ can be written

$$V_t(x) = E_{N_t,v}\left[\max\left\{\sum_{i=1}^{N_t} J(v_i)\, y_i + V_{t+1}(x-k) : y_i \in \{0,1\},\right.\right.$$
$$\left.\left. k = \sum_{i=1}^{N_t} y_i,\ k \leq x\right\}\right], \quad (6.12)$$

where k is the total number of units awarded in period t. The boundary conditions are

$$V_{T+1}(x) = 0, \quad x = 1,\ldots,C, \quad (6.13)$$

where C denotes the initial capacity. An allocation $y(\cdot)$ that achieves the maximum above given x, t and v will be an optimal dynamic allocation policy. (See Appendix D.)

6.3.2 Optimal Dynamic Allocations and Mechanisms

We first analyze the theoretical properties of the dynamic program (6.12)–(6.13). From this structure, one can show that variants of the classic first- and second-price auctions are optimal for this problem.

6.3.2.1 Optimal Allocations

As in the traditional single-resource RM model, the solution of the dynamic program (6.12)–(6.13) hinges on the monotonicity of the marginal values $\Delta V_t(x) = V_t(x) - V_t(x-1)$. Indeed, one can show the following [542]:

PROPOSITION 6.3 $\Delta V_t(x)$ is decreasing in x for any fixed t and is decreasing in t for any fixed x.

These are quite natural economic properties. At any point in time, the marginal benefit of each additional unit declines because the future number of customers is limited; therefore, the chance of selling the marginal unit—and the expected revenue if we sell it—decreases. Similarly, for any given remaining quantity x, the marginal benefit of an additional unit decreases with t because as time progresses, the number of future customers declines; therefore, the chance of selling the marginal unit—and the expected revenue if we sell it—goes down.

Proposition 6.3 simplifies the optimal allocation. To see this, note that since $J(\cdot)$ is assumed to be increasing, if the firm decides to award k units, it is optimal to allocate them to the highest $J(v_i)$'s (that is, to the highest v_i's). Therefore, define

$$R(k) \equiv \begin{cases} 0 & \text{if } k = 0 \\ \sum_{i=1}^{\min\{k, N_t\}} J(v_{[i]}) & \text{if } k > 0, \end{cases} \quad (6.14)$$

and note that

$$R(k) = \max\left\{ \sum_{i=1}^{N_t} J(v_i) y_i : y_i \in \{0, 1\}, \sum_i y_i = \min\{k, N_t\} \right\}.$$

Also, define $\Delta R(i) \equiv R(i) - R(i-1)$. Then the formulation (6.12) can be rewritten in terms of k as follows:

$$V_t(x) = E_{N_t, v}\left[\max_{0 \leq k \leq x}\left\{ \sum_{i=1}^{k} [\Delta R(i) - \Delta V_{t+1}(x - i + 1)] \right\} \right] + V_{t+1}(x), \quad (6.15)$$

where the sum is defined to be 0 if $k = 0$. Let k^* be the optimal solution above (the optimal number of bids to accept) at time t in state x.

Auctions

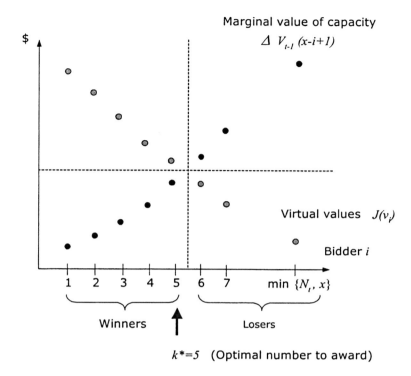

Figure 6.3. Illustration of optimal allocations in the dynamic auction model.

Let N_t denote a realization of the random variable N_t and v be a realization of customers' types. The following proposition characterizes the optimal allocation and follows from (6.15) and Proposition 6.3:

PROPOSITION 6.4 *For any realization (N_t, v), the optimal number of units to allocate in state (x,t) is given by*

$$k^* = \max\{1 \leq k \leq \min\{x, N_t\} : \Delta R(k) > \Delta V_{t+1}(x - k + 1)\}$$

if $R(1) > \Delta V_{t+1}(x)$ and by $k^* = 0$ otherwise. Moreover, it is optimal to award these k^* units to those customers with the highest valuations v_i.

This shows how the firm should run the auction—provided it can infer the valuations v_i of the customers. In particular, note that $\Delta R(i) = J(v_{[i]})$ for $i = 1, \ldots, \min\{x, N_t\}$, so the decision rule in Proposition 6.4 about the optimal number of bids to accept is simply based on sorting the values v_i and progressively awarding items to the highest-value customers until $J(v_{[i]})$ drops below the marginal opportunity cost $\Delta V_{t+1}(x - i+1)$. The situation is illustrated in Figure 6.3. Thus, given the customer valuations v_i and the value function $V_{t+1}(x)$, the optimal allocation rule is simple.

6.3.2.2 Optimal Mechanisms

We next demonstrate that appropriately modified versions of two standard procedures—the first- and the second-price auctions—achieve the optimal dynamic allocation.

Second-price auction In a straightforward application of the second-price mechanism in the dynamic auction setting, it is no longer optimal for customers to bid their valuation. The following informal reasoning shows why. Suppose it is optimal to bid truthfully under the second-price mechanism and let

$$\hat{v}_i \equiv J^{-1}(\Delta V_{t+1}(x - i + 1)), \quad i \geq 1. \tag{6.16}$$

The thresholds $\hat{v}_i$ are directly computable from the solution of (6.23) described in the previous section, which uses common knowledge information, and are in principle known to all customers and the firm. Following Theorem 6.4, the firm will accept bid $v_{[i]}$ as long as $v_{[i]} > \hat{v}_i$. Now suppose the firm decides to award k units. That means $v_{[i]} > \hat{v}_i$, $i = 1, \ldots, k$; and $v_{[i]} \leq \hat{v}_i$, $i = k+1, \ldots, N_t$. However, if the first loser, $v_{[k+1]}$, had bid $\hat{v}_{k+1} + \varepsilon$ instead (which in fact verifies $v_{[k+1]} < \hat{v}_{k+1} + \varepsilon$), the firm would include him among the winners and award $k+1$ units, and the customer would pay only $v_{[k+2]}$ and make a positive profit. Hence, customers have some incentive to bid above their own valuations (a pure second-price mechanism fails to elicit truthful bids).

However, the following modification to the second-price mechanism avoids this pitfall. In each period t, the firm first computes the thresholds $\hat{v}_i$ using the current capacity x. Given the vector of submitted bids **b**, the firm will award k units, where

$$k = \max\{i \geq 1 : b_{[i]} > \hat{v}_i\}, \tag{6.17}$$

and $k = 0$ if $b_{[1]} \leq \hat{v}_1$; and all winners will pay

$$b_{[k+1]}^{(2\text{nd})} = \max\{b_{[k+1]}, \hat{v}_k\}, \tag{6.18}$$

where $b_{[k+1]}$ is the $(k+1)^{\text{st}}$ highest bid and $\hat{v}_k$ is the threshold to award the kth unit. Ties between bids are broken by randomization. For simplicity we refer to (6.17)-(6.18) as the *modified second-price* mechanism. One can then show the following result [542]:

PROPOSITION 6.5 *For the modified second-price auction with allocation and payments given by (6.17)-(6.18), a customer's dominant strategy*

Auctions 277

is to bid his own valuation. Moreover, under this dominant-strategy equilibrium, the modified second-price mechanism is optimal.

First-Price auction In a first-price auction, items are awarded to the highest bidders, and winners pay their bids. This type of mechanism may be more natural in many applications.

To establish that the first-price auction achieves the same expected revenue as the second-price mechanism described above, one needs to show that (1) items are again awarded according to the optimal allocation rule derived in the previous section and (2) customers with zero value have zero expected surplus. To do this, it suffices to show that there exists a symmetric equilibrium bidding strategy $b^*(\cdot)$ that is strictly increasing in the customer's valuation. In this case, the firm can use this bid function to invert a bid and infer the customer's valuation, which it can then use to correctly compute the number of items to award.

The main result for this case is the following [542]:

PROPOSITION 6.6 *Under the first-price auction, there exists a symmetric, strictly increasing, bidding strategy equilibrium $b^*(v_i)$. The strategy b^* depends on the current values of t and x as given by*

$$\hat{b}^*(v_i) = v_i - \frac{\int_{\hat{v}_1}^{v_i} P(v)\,dv}{P(v_i)} \quad \text{and} \quad b^*(v_i) \equiv \lim_{\varepsilon \downarrow 0} \hat{b}^*(v_i - \varepsilon), \quad (6.19)$$

where $P(v)$ is the probability that a customer with valuation v is among the winners,

$$P(v) = \begin{cases} 0 & \text{if } k^* = 0, \\ \sum_{n=1}^{M} \left\{ \sum_{k=0}^{k^*-1} \binom{n-1}{k} [1 - F(v)]^k [F(v)]^{n-1-k} \right\} g(n) & \text{if } k^* \geq 1, \end{cases}$$
(6.20)

$k^* = \max\{0 \leq i \leq \min\{x, N_t\} : v_i > \hat{v}_i\}$, *and by convention $\hat{v}_0 < 0$. Moreover, under this symmetric equilibrium, the first-price auction is optimal.*

Note that (6.19) shows—since winners are required to pay what they bid—that under a first-price mechanism customers shade their valuations to make some positive surplus. Since $b^*(\cdot)$ is strictly increasing, the units are sold to the players with the highest valuations. Moreover, once the firm observes bids $b_1^*, \ldots, b_{N_t}^*$, it can calculate the valuations $v_1, \ldots, v_{N_t}$ through the well-defined inverse bidding function $b^{*-1}(\cdot)$.

An important practical observation from this result is that the optimal first-price mechanism is not greedy, in the sense that it does not maximize the sum of *observable* revenue in the current period plus the expected revenue to go, because the firm compares the values of $J(v_i)$'s with the marginal value $\Delta V_t(\cdot)$ rather than with the bids themselves. As a result, the firm may (1) accept bids below the marginal value when $J(v_{[k]}) > \Delta V_{t+1}(x - k + 1) \geq b^*(v_{[k]})$ and (2) reject bids that are above marginal value when $b^*(v_{[k]}) > \Delta V_{t+1}(x - k + 1) \geq J(v_{[k]})$. Numerical experiments show that both cases may occur.

This behavior is somewhat counterintuitive because at first blush is seems that any bid that exceeds the marginal value of capacity ought to be worth accepting. However, such reasoning neglects the effect that the acceptance policy has on the bidding strategy of the customers. If the firm accepts all bids that are *ex post* profitable, then customers end up bidding *less* in equilibrium than they do when the firm follows the optimal acceptance strategy. The net result is to lower the firm's total revenue. In short, the firm has to occasionally refuse profitable bids to induce the customers to bid more aggressively—and in equilibrium it benefits by taking these short-run losses. This is simply an extension of the rationale for using reserve prices in a standard auction.

6.3.3 Comparisons with Traditional RM

We next compare the optimal auction mechanisms with a variation of a traditional quantity-based RM mechanism as in Chapter 2. The firm sets a list price at the beginning of each period and calculates a threshold on the number of units it is willing to award at the list price. Both the price and the capacity limit are optimized. We call this mechanism the *dynamic list price, capacity-controlled mechanism* (DLPCC). Note that unlike in a traditional RM mechanism, in DLPCC prices are set optimally rather than being given exogenously.

Customers who are interested in acquiring one unit at that list price submit *acceptances* (an offer to buy). If the number of acceptances exceeds the capacity limit set by the firm, the units are randomly rationed to the customers. It is easy to see in this case that a dominant strategy for customers is to submit an acceptance if and only if their valuations exceed the firm's reserve price.

6.3.3.1 Theoretical Comparisons

One can show that the DLPCC mechanism is, in fact, optimal in several cases. Indeed, we have [542]

PROPOSITION 6.7 *The DLPCC is optimal if the following cases:*
(i) There is at most one customer per period ($N_t \leq 1$ (w.p.1)).
(ii) There are more units to sell than there are potential customers

$(\sum_t N_t < C \ (w.p.1))$.
(iii) Asymptotically as the number of customers and units to sell grows large $(C, N_t \uparrow \infty)$.

That is, unless customers can be aggregated in time, the number of customers and objects is not too large, and there is some scarcity, there is no advantage to using a bidding mechanism over simple list pricing. These results are analogous to those in Section 6.2.6 for the single-period auction.

6.3.3.2 Numerical Comparisons

We next consider some numerical examples that illustrate the conditions under which an optimal pricing mechanism significantly outperforms DLPCC. In the examples that follow, the dynamic program associated with the optimal mechanism is solved using simulation, and customers' valuations are assumed to be uniformly distributed.

The first experiment shows how the revenue changes as the *concentration of customers*, defined as the number of customers per period, is varied. The firm starts with $C = 16$ units, and the total number of customers in all periods is constant at 64. The number of periods varies from 1 to 64, so that the number of customers per period varies. That is, the example runs from 64 customers in one period (high concentration of customers) to one customer in each of 64 consecutive periods (low concentration of customers).

The results are given in Table 6.1. Observe that the optimal revenue increases as the concentration of customers increases. This is intuitive, since as the firm observes more customers' valuations per period, it is making allocation decisions with reduced uncertainty about future bid values. Moreover, an increase in concentration increases direct bidding competition amongst customers. The gap reaches over 6% in the extreme case of a single period with 64 customers, which is significant. The second experiment compares the suboptimality gaps of the DLPCC mechanism under various levels of capacity and demand. The number of periods is kept constant at $T = 5$. The number of customers per period is fixed at $N_t = 10, 30, 50$ and 100; and for each of these, three choices of capacity—$C = 0.1 \, T \, N_t$, $C = 0.3 \, T \, N_t$ and $C = 0.5 \, T \, N_t$—are used. Results are shown in Table 6.2. The gaps for DLPCC tend to decrease from left to right (which corresponds to increasing the capacity to demand ratio) and from top to bottom (which corresponds to increasing proportional number of customers per period and number of units in stock) in each table. Note that the gaps of 2% or more occur only in the

Table 6.1. Dynamic auction revenues for different concentrations of customers.

Customers Per Period	Number of Periods	Optimal Revenue Mean	95% CI	DLPCC Revenue Mean	Gap
1	64	11.410	(11.390, 11.430)	11.412	0.16%
2	32	11.434	(11.420, 11.448)	11.401	0.41%
4	16	11.480	(11.466, 11.495)	11.383	0.98%
8	8	11.534	(11.511, 11.556)	11.348	1.79%
16	4	11.621	(11.602, 11.639)	11.292	2.99%
32	2	11.722	(11.704, 11.740)	11.201	4.59%
64	1	11.796	(11.780, 11.812)	11.060	6.36%

Table 6.2. DLPCC suboptimality gaps relative to a dynamic auction for different demand to capacity ratios.

N_t	$C = 0.1\,T\,N_t$	$C = 0.3\,T\,N_t$	$C = 0.5\,T\,N_t$
10	2.37%	2.32%	0.58%
30	1.77%	1.77%	0.38%
50	1.43%	1.43%	0.21%
100	1.06%	1.13%	0.14%

case where the number of customers is moderate (such as 10) and the capacity is constrained ($C = 0.1\,TN_t$).

Other numerical experiments of [542] show that the relative benefit of the dynamic auction increase as the variance in the customer's reservation value v_i increases and as the variance in the number of customers N_t increases. Hence, variability in the demand environment appears to favor the use of a dynamic auction mechanism.

6.4 Optimal Dynamic Auctions with Replenishment

We next consider an infinite-horizon auction problem with replenishment, which is essentially the auction equivalent of the dynamic pricing and inventory problem of Section 5.3.2. A firm orders, stores, and then sells units of a homogeneous good over an infinite time horizon. The firm starts a period with an initial (integral) inventory x, and it reorders at a unit cost c at the end of the period. Replenishment orders arrive instantly, and we do not allow backlogging. In each period, a convex,

Auctions 281

strictly increasing holding cost of $h(x)$ is charged on the starting inventory level x.[12]

The firm sells its goods through a sequence of auctions indexed by $t \geq 1$. The problem is assumed to be stationary, so the statistics of demand are the same for all periods t. Private-value assumptions apply. In each period, N risk-neutral customers arrive. N is a nonnegative, discrete-valued random variable distributed according to a known probability mass function $g(\cdot)$ with support $[0, M]$ for some $M > 0$ and with a strictly positive first moment. Each customer requires one unit and has a private valuation v_i, $1 \leq i \leq N$, i.i.d. with a distribution $F(\cdot)$, which is strictly increasing with a continuous-density function $f(\cdot)$ on the support $[0, \bar{v}]$. We assume that the marginal value $J(\cdot)$ derived from $F(\cdot)$ is strictly increasing.

As in the single-resource capacity auction case, we use $\mathbf{v}$ both for the random vector of valuations (from the firm's perspective) and for a particular realization. The distribution functions g and F are constant through time t and are assumed common knowledge to the firm and all potential customers. We assume that both the number of customers N and their valuations $\mathbf{v}$ are independent from one period to the next. Thus, each period is an independent draw of N and $\mathbf{v}$.

The firm's problem is to design an auction mechanism and find a replenishment policy that maximizes its expected discounted profit. As before, we analyze this by first finding an optimal allocation and then finding mechanisms that achieve the optimal allocation.

6.4.1 Dynamic Programming Formulation

We analyze this problem using a dynamic programming formulation in terms of allocation variables $y(\mathbf{v})$. Define the value function $V(x)$ as the maximum expected discounted profit given an initial inventory $x = 0, 1, \ldots$, which satisfies the Bellman's equation:

$$V(x) = E\left[\max_{\substack{y \in \{0,1\}^N \\ q \in Z_+}} \left\{ \sum_{i=1}^{N} J(v_i) y_i + \alpha V(x - k + q) - h(x - k + q) \right. \right.$$

$$\left. \left. - cq : \sum_{i=1}^{N} y_i = k,\ k \leq x, \right\} \right] \quad (6.21)$$

---

[12]One can also analyze the case where holding cost is charged on the ending inventory level. The results are qualitatively the same as long as the holding cost $h(x)$ is linear.

where $0 < \alpha < 1$ is the discount factor, k is the total number of units awarded, and q is the replenishment order for the next period. Note from first principles that the state space can be bounded by M because at most M customers will arrive in any period, and since we can reorder at the end of every period, there is no need to stock more than M. Our objective is finding an optimal stationary policy consisting of an allocation $\mathbf{y}(\cdot)$ and a replenishment order $q(\cdot)$, that achieves $V(x)$.

Assuming $J(\cdot)$ is monotone increasing (Assumption 7.2), it again follows that if the firm allocates k units, it is optimal to allocate them to the highest $J(v_i)$'s (to the highest v_i's). So, as before, define

$$R(k) \equiv \begin{cases} 0 & \text{if } k = 0 \\ \sum_{i=1}^{\min\{k,N\}} J(v_{[i]}) & \text{if } k > 0, \end{cases} \quad (6.22)$$

and note that $R(k)$ is a random function that solves

$$R(k) = \max\{\sum_{i=1}^{N} J(v_i) y_i : 0 \leq y_i \leq 1, \sum_i y_i = \min\{k, N\}\}.$$

Therefore, we can rewrite (6.21) in terms of k as follows:

$$V(x) = E\left[\max_{\substack{0 \leq k \leq x \\ q \in \mathcal{Z}_+}} \{R(k) + \alpha V(x - k + q) - h(x - k + q) - cq\}\right],$$

$$x = 0, 1, \ldots, M. \quad (6.23)$$

Note that above we are assuming that excess stock can be eliminated without cost (*free disposal*) when $N < k \leq x$. This assumption is not essential for the analysis, but it helps to simplify the notation.

6.4.2 Optimal Auction and Replenishment Policy

We next characterize the optimal auction and replenishment policy for this problem. The first statement is presented in algorithmic form [527]:

PROPOSITION 6.8 *Consider the inventory-pricing problem described in (6.23). Define the optimal base-stock level by*

$$z^* = \max\{z \in \mathcal{Z}_+ : \alpha \Delta V(z) - \Delta h(z) - c > 0\}.$$

Then the optimal stationary policy is to allocate units to customers and replenish stock as follows:

STEP 1 (Allocate units):
FOR $k = 1, 2, \ldots, \min\{x, N\}$, allocate the kth unit if either:
(i) $x - k \geq z^*$ and $J(v_{[k]}) > \alpha \Delta V(x - k + 1) - \Delta h(x - k + 1)$
(ii) $x - k < z^*$ and $J(v_{[k]}) > c$
ELSE GOTO STEP 2.

STEP 2 (Replenish stock):
IF $x - k < z^*$, then order up to z^*, i.e., $q = z^* + k - x$; ELSE order nothing $(q = 0)$.

The policy says that while the current inventory is above the optimal base-stock level z^* (case (i)), then we will award the kth unit if the benefit from accepting the kth bid (its virtual value $J(v_{[k]})$) exceeds the profit of keeping the kth unit for the next period less the marginal holding cost for keeping it. The kth unit is not replenished in this case. Once the inventory reaches the optimal level z^* (case (ii)), the firm awards a unit as long as the benefit from accepting a bid exceeds the cost of replacing the unit awarded; each such unit is replenished. This policy is illustrated in Figure 6.4.

An interesting result of this allocation policy is that when the inventory is less than the optimal base-stock level z^*, the firm can achieve the optimal allocation by simply running a standard first-price or second-price auction in each period with a fixed reserve price

$$\hat{v} \doteq J^{-1}(c). \tag{6.24}$$

Indeed, the following characterization of the optimal policy in this case [527]:

PROPOSITION 6.9 *Once the inventory reaches z^* units, the optimal policy in all subsequent periods is to (i) run a standard first- or second-price z^*-unit auction with fixed reserve price $\hat{v}$ and then (ii) at the end of each period, order up to the optimal base-stock level z^*.*

Since the problem is over an infinite horizon and the optimal policy calls only for ordering when the inventory drops below z^*, the firm eventually reaches a point where the above simple auction and replenishment policy are optimal for all remaining time. That is, z^* is the unique recurrent state in the resulting Markov chain that governs the evolution of the inventory over time under the optimal policy.

This result is significant on several levels. First, it shows that the classical first-price and second-price mechanisms remain optimal in the

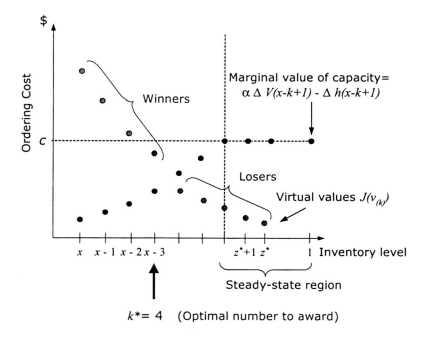

Figure 6.4. Illustration of optimal allocations in the dynamic-auction model with replenishment.

dynamic-inventory setting. These are both familiar auction mechanisms, which are easy for customers to understand and easy for firms to implement. The inventory-replenishment policy is also a familiar and simple base-stock policy. This combination makes the optimal policy quite practical. On a theoretical level, the result is as simple as one could hope for in this setting. Finally, it is convenient as well from a computational perspective because it reduces the optimal policy to a search over the single parameter z^*, as we show next.

6.4.3 Average-Profit Criterion

Consider maximizing the long-run average profit. One can show that the optimal policy for the α-discounted problem is in fact Blackwell optimal;[13] that is, it is simultaneously optimal for all discounted problems with discount factors $\alpha \in (\bar{\alpha}, 1)$, for some $0 < \bar{\alpha} < 1$. As a result, one can show (see [527]) that the optimal average-profit policy will again be to run a standard first-price or second-price auction in each period with

---
[13] See Bertsekas [57, Section 4.2, Definition 1.1]) for a formal definition of Blackwell optimality.

reserve price $\hat{v} = J^{-1}(c)$ and then order up to a fixed base-stock level z^* at the end of each period.

Indeed, because of this fact, one can develop a simple procedure for finding the optimal base-stock level z^*. Let

$$\Pi(z) \equiv E\left[\max_{0 \leq k \leq \min\{z,N\}} \{R(k) - ck\}\right] - h(z) \qquad (6.25)$$

be the average profit when following a policy of reordering up to a fixed base-stock level z. We know that such a policy will be optimal for some z^*; therefore, we simply need to search for a value z that maximizes $\Pi(z)$. In fact, one can verify that the profit function $\Pi(z)$ is concave in z and that $z \leq M$. $\Pi(z)$ can be evaluated by simulation and in special cases by closed-form expressions. Taking advantage of the concavity of $\Pi(z)$, a binary search over the range for z^* therefore gives an overall algorithm complexity of $O(N \log M)$. Henceforth, we denote the optimal objective value $\Pi^* \equiv \Pi(z^*)$.

6.4.4 Comparison with a List-Price Mechanism

We next consider how the optimal auction policy compares with a traditional, fixed-price policy. Specifically, we consider the base-stock, list price policy of Section 5.3.2, in which the firm sets a fixed list price p in each period and then replenishes by ordering up to a fixed base-stock level z. To be consistent, we assume we incur the holding cost at the beginning of the period, and we assume customers who are interested in acquiring one unit at the posted price submit acceptances. If the number of acceptances exceeds the current inventory of the firm, we randomly ration the units to the customers. It is easy to see that under this pricing mechanism, a dominant strategy for customers is to submit an acceptance if and only if their own valuations are higher than the list price.

We compare the profits earned under the optimal mechanism with those under the base-stock, list-price mechanism for an optimal choice of p and z. We give theoretical comparisons first, followed by a numerical comparison of the two policies.

6.4.4.1 Theoretical Comparisons

We restrict ourselves to the average-cost case, where the optimal profit is given by optimizing (6.25) over z, though similar results can be developed for the discounted case. One can show the following [527]:

PROPOSITION 6.10 *The base-stock, list price policy is optimal when (i) The number of customers is at most one—$N \leq 1$ with probability* one.

(ii) The number of customers is large, $N \to \infty$, and the holding cost is linear, $h(z) = a + hz$.
(iii) The holding cost is zero, $h(x) = 0$.

Part (i) shows that if the firm is receiving isolated bids (as in some consumer online auctions, such as Priceline.com's mechanism), there is no inherent advantage to using auctions over list pricing. Some aggregation of customers is needed to gain a strict advantage through an auction mechanism. Intuitively, this is because one needs to generate some bidding competition among customers to realize a benefit from an auction. With at most one customer bidding, there is no competition. Part (ii) is analogous to the finite-horizon problem. As the number of customers in each period becomes large, the fraction with valuations above any given price p converges to a deterministic function of p, and hence the ratio of the auction and the fixed-price revenues tends to one. The intuitive reason for part (iii) is that with no holding cost, the firm will stock the maximum inventory M at the start of each period under both the optimal auction and list price policies. As a result, there is no rationing of product, and thus customers do not face any bidding competition. Without bidding competition, the auction produces the same profits as the base-stock, list price policy.

6.4.4.2 Numerical Comparisons

We next present the results of some numerical simulations from [527] with the average-profit criterion. The following base case is used as a starting point. The ordering cost is normalized at $c = 1$; customers' valuations are assumed uniform of width $\Delta = 0.5$ centered at c (that is, customers' valuations are centered at the cost, with Δ representing the dispersion in valuations); there are a constant $N = 50$ customers per period; and the holding cost is linear of the form $h(z) = c \delta z$, where $\delta = 1\%$ is the one-period interest rate.

The individual parameters of this base case are varied to see the effect on the absolute and relative performance of each policy. Along with expected profit, a *fill rate* is computed for each policy, defined as the expected number of customers who are awarded an item divided by the expected number who attempt to purchase (those with valuations above the reserve price in the auction or those with valuations above the fixed price in the base-stock, list price case).[14] The fill rate gives a measure of

---

[14] Formally, if $N(v)$ denotes the number of customers with valuations greater than v, then the fill rate is the ratio $E[\min\{N(\hat{v}), z^*\}]/E[N(\hat{v})]$ in the auction case and $E[\min\{N(p^*), z^*_{LP}\}]/E[N(p^*)]$ in the list price case, where p^* is the optimal list price.

the scarcity of inventory relative to demand and is a traditional service measure in inventory problems.

The first experiment shows how the profit is affected by the number of customers in each period. The number of customers N is assumed constant, but N is varied from 1 to 1,000. All other parameters are the same as in the base case. The results are summarized in Table 6.3. As

Table 6.3. Dynamic auction and replenishment profits for different numbers of customers.

| N Customers | Base-Stock, Auction | | | Base-Stock, List Price | | | Profit |
per Period	Profit	z^*	Fill Rate	Profit	z^*_{LP}	Fill Rate	Gap
1	0.02	1	100.00%	0.02	1	100.00%	0.00%
5	0.13	2	90.39%	0.12	3	98.74%	3.20%
10	0.27	4	95.93%	0.26	4	96.62%	2.50%
50	1.40	14	95.03%	1.38	16	98.88%	1.62%
100	2.83	26	94.90%	2.80	30	99.32%	1.31%
1000	28.72	242	95.86%	28.54	259	99.80%	0.62%

one would expect, the profits and inventory levels increase in both policies as the number of customers increases. Also, as shown theoretically in Proposition 6.10, the base-stock, list-price mechanism is optimal in the limiting case of just one customer per period. In the other extreme, as N gets large, again the base-stock, list price profit approaches the optimal auction profit, as predicted by the asymptotic result of Proposition 6.10. The biggest benefit from the auction occurs at a moderate value of five customers per period, where it achieves a 3.2% increase in profits over list pricing.

Note that the fill rate and inventory level are also higher in the base-stock, list price case. This suggests that the auction policy deliberately introduces some scarcity in the available goods to create more bidding competition among the customers.

The next experiment shows the effect of varying the interest rate δ—or equivalently varying the holding cost rate since $h(z) = c\,\delta\,z$ (with $c = 1$ in our case). Typically, this interest rate represents a cost of capital plus a rate of depreciation in the product's value over time. Table 6.4 shows the results. The small difference in the expected profits for the lowest interest rate confirms the result of Proposition 6.10—low holding cost leads to high inventory levels, which reduces the bidding competition and hence the benefit of the auction. As the interest rate rises, the auction performs relatively better, achieving a large 21.67% improvement over list pricing when the interest rate reaches 10%. This is simply the reverse effect: a high holding cost means the firm is unwilling to stock much

Table 6.4. Dynamic auction and replenishment profits for different holding costs.

Interest Rate δ	Base-Stock, Auction			Base-Stock, List Price			Profit Gap
	Profit	z^*	Fill Rate	Profit	z^*_{LP}	Fill Rate	
0.01%	1.560	21	99.97%	1.560	23	100.00%	0.01%
0.10%	1.543	18	99.58%	1.541	20	99.92%	0.14%
1.00%	1.404	14	95.03%	1.381	16	98.88%	1.62%
5.00%	0.932	10	77.36%	0.845	11	93.10%	9.37%
10.00%	0.502	7	55.77%	0.393	7	82.41%	21.67%

inventory. Since the number of customers per period is unchanged, the number of customers per unit of inventory increases; more competition among customers is created and hence the auction mechanism performs relatively better.

It is worth pointing out, however, that there are few practical situations where interest rates of over 1% per period are observed, especially if one is considering auctions that are held relatively frequently (such as weekly). Rates this high are observed for products such as personal computers, which become obsolete quickly, but for most goods, weekly rates of less than 1% are the norm. This suggests that either the product has to suffer rapid depreciation or selling events have to be relatively infrequent (such as monthly or semiannual periods, not weekly) for the firm to realize a significant benefit from using auctions over list pricing.

Finally, as in the finite-horizon problem without replenishment, numerical experiments show that variability in the valuations v_i or variability in the number of customers N increases the relative benefits of the auction policy.

6.5 Network Auctions

We next consider an auction mechanism for a network RM problem of the type studied in Chapter 3, which is based on Cooper and Menich [129]. Customers in this case bid for products (combinations of resources), and the firm awards resources based on these product-level bids. Such auctions are also relevant to procurement settings, where customers bid for a mix of inputs required to produce a given product. Customers desire the entire *bill of materials* (the complete set of resources) and a firm, with stockpiles of the various resources, must solicit bids and award the resources given a collection of *package* bids.

We next look at such an auction based on a network version of the Vickrey (second-price, sealed-bid) auction (a so-called Vickrey-Clarke-Groves mechanism [533, 120, 226]). We describe the basic mechanism and the resulting equilibrium bidding strategies and then explore the

Auctions   289

connections between this problem and traditional, network-capacity-control problems.

6.5.1 Problem Definition and Mechanism

The problem definition and notation are similar to those in Chapter 3, but are slightly modified to be consistent with the auction notation of this chapter. There are N customers, each with a private valuation v_j for one unit of product j, which requires one or more of m resources. We define $a_{ij} = 1$ if the product required by customer j uses resource i and $a_{ij} = 0$ otherwise. The incidence matrix is defined by $\mathbf{A} = [a_{ij}]$. The vector of current remaining capacities of the m resources is $\mathbf{x} = (x_1, \ldots, x_m)$. Because the mechanism is based on a generalized Vickrey auction, minimal assumptions about customers and their valuations are needed, as we show below. In fact, we require only that customers are rational and that their valuations are finite.

The mechanism is defined as follows. As in the classical auction setting, let $\mathbf{y} = (y_1, \ldots, y_N)$ denote the allocation vector, $\mathbf{p} = (p_1, \ldots, p_N)$ denote the payment vector, and $\mathbf{b}(\mathbf{v}) = (b_1(v_1), \ldots, b_N(v_N))$ denote the vector of bidding strategies. Customers submit a sealed bid $b_j(v_j)$ for their desired product j. The firm collects all N bids $\mathbf{b} = (b_1, \ldots, b_N)$ and then solves the following integer program:

$$z^*(\mathbf{b}) = \max \quad \mathbf{b}^\top \mathbf{y} \qquad (6.26)$$
$$\text{s.t.} \quad \mathbf{A}\mathbf{y} \leq \mathbf{x}$$
$$y_j \in \{0,1\}, \quad j = 1, \ldots, N.$$

Let $\mathbf{y}^*(\mathbf{b})$ denote an optimal solution to this integer program. The set of winning customers is denoted $\mathcal{W}(\mathbf{b}) = \{j : y_j^* = 1\}$.

It is important to note that the optimal value of this integer program is *not* the revenue earned by the firm; rather, it is solved simply as a means of determining winners and losers in the auction. The revenue to the firm will be determined by the vector of payments $\mathbf{p}(\mathbf{b})$ that are requested from the winning customers, which we look at next.

Note that the surplus of customer j is

$$S_j(\mathbf{b}, v_j) = v_j y_j^*(\mathbf{b}) - p_j(\mathbf{b}). \qquad (6.27)$$

Let $\mathbf{e}_j$ denote the jth unit vector and note that $\mathbf{b} - b_j \mathbf{e}_j$ is the vector of bids with the jth component replaced by zero—that is, the vector of bids without the bid of customer j. Then the scheme calls for the winning customers to pay

$$p_j(\mathbf{b}) = y_j^*(\mathbf{b}) \left[b_j - (z^*(\mathbf{b}) - z^*(\mathbf{b} - b_j \mathbf{e}_j)) \right]. \qquad (6.28)$$

Note that the term $z^*(\mathbf{b}) - z^*(\mathbf{b} - b_j \mathbf{e}_j)$ is simply the network benefit of having customer j's bid. And also clearly $z^*(\mathbf{b}) - z^*(\mathbf{b} - b_j \mathbf{e}_j) \leq b_j$, since when adding customer j's bid of b_j the optimal value of the problem (6.26) cannot increase by more than b_j. If $z^*(\mathbf{b}) - z^*(\mathbf{b} - b_j \mathbf{e}_j) < b_j$, it is because other winning bids were displaced to include the bid of customer j in the optimal solution. Hence, $b_j - (z^*(\mathbf{b}) - z^*(\mathbf{b} - b_j \mathbf{e}_j))$ represents the *displacement cost* produced by including customer j in the winning set, and hence in this scheme a customer pays his displacement cost. As we show below, this displacement-cost interpretation of the payment has a close connection to the bid-price values from the network problems studied in Chapter 3.

6.5.2 Equilibrium Analysis

We next analyze the equilibrium produced by this mechanism. An important relation is obtained by rewriting (6.28) as

$$p_j(\mathbf{b}) = z^*(\mathbf{b} - b_j \mathbf{e}_j) - \sum_{j' \neq j} b_{j'} y_{j'}^*(b). \tag{6.29}$$

This holds because

$$y_j^*(\mathbf{b})(-z^*(\mathbf{b}) + z^*(\mathbf{b} - b_j \mathbf{e}_j)) = -z^*(\mathbf{b}) + z^*(\mathbf{b} - b_j \mathbf{e}_j),$$

which is trivially true when $y_j^*(\mathbf{b}) = 1$; when $y_j^*(\mathbf{b}) = 0$, then it is true because in this case $z^*(\mathbf{b}) = z^*(\mathbf{b} - b_j \mathbf{e}_j))$.

Therefore, substituting (6.29) into (6.27), we find that the customer j's net utility can be written as

$$S_j(\mathbf{b}, v_j) = v_j y_j^*(\mathbf{b}) + \sum_{j' \neq j} b_{j'} y_{j'}^*(\mathbf{b}) - z^*(\mathbf{b} - b_j \mathbf{e}_j).$$

This shows that customer j's payoff does not depend on his bid b_j but only on whether his bid places him in the winning set $\mathcal{W}(\mathbf{b}) = \{j : y^*(j) = 1\}$. Thus, as in a second-price auction, one can show that if a customer bids less than his valuation, it reduces his chances of winning only in cases where he would have a positive net surplus, and bidding more than his valuation increases his chance of winning only in cases where his net surplus is negative. Indeed, one can prove

PROPOSITION 6.11 *Under a sealed-bid mechanism with allocations determined by an optimal solution $y^*(\mathbf{b})$ to (6.26) and payments determined by (6.28), then $b_j^*(v_j) = v_j$ is a dominant strategy for all customers, and hence $b^*(v) = v$ is a dominant-strategy equilibrium.*

As a result of this fact, we can assume $\mathbf{b} = \mathbf{v}$, and the equilibrium revenue collected by the firm is therefore given by

$$\sum_{j=1}^{N} p_j^*(\mathbf{v}) = \sum_{j=1}^{N} y_j^*(\mathbf{v}) v_j - z^*(\mathbf{v}) + z^*(\mathbf{v} - \mathbf{e}_j v_j)$$

$$= \sum_{j=1}^{N} z^*(\mathbf{v} - \mathbf{e}_j v_j) - (N-1) z^*(\mathbf{v}).$$

Note that this revenue is less than $z^*(\mathbf{v})$ since $p_j(\mathbf{v}) \leq v_j$ and $p_j(\mathbf{v}) = 0$ when $y_j^*(v) = 0$. So the revenue collected by the firm is less than the optimal solution to the integer program (6.26) as claimed.

6.5.3 Relationship to Traditional Auctions

We next show that this mechanism is indeed the network generalization of the second-price mechanism in a traditional C-unit auction. To see this, note we can formulate the C-unit auction as an instance of the network auction with $m = 1$ and $x_1 = C$. In this case, solving the integer program (6.26) is trivial. We simply award the C items to the C customers with the highest bids b_j, which is the same as the classical second-price allocation. Also, note that the optimal value is

$$z^*(\mathbf{b}) = \sum_{j'=1}^{C} b_{[j']},$$

where $b_{[j]}$ denotes the jth highest bid. As a result, by (6.28) each winner pays

$$p_j(\mathbf{b}) = y_j^*(\mathbf{b}) [b_j - z^*(\mathbf{b}) + z^*(\mathbf{b} - \mathbf{e}_j b_j)]$$

$$= b_j - \sum_{j'=1}^{C} b_{[j']} + (\sum_{j'=1}^{C} b_{[j']} - b_j + b_{[C+1]})$$

$$= b_{[C+1]},$$

which is just the usual second-price auction payment with no reserve price. Thus, the allocations and payments reduce to those of the classical C-unit second-price auction in the $m = 1$ case.

6.5.4 Relationship to Traditional Network RM

This network-auction mechanism also has an interesting connection to bid prices in traditional network RM. We proceed informally here to illustrate the ideas, but the connections can be made rigorous.

Consider the linear programming relaxation of (6.26), which is

$$z^*(\mathbf{b}) = \max \ \mathbf{b}^\top \mathbf{y} \qquad (6.30)$$
$$\text{s.t.} \ \mathbf{Ay} \leq \mathbf{x}$$
$$0 \leq y_j \leq 1, \ j = 1, \ldots, N.$$

Note that this is the exactly same form as the deterministic linear programming (DLP) model of Section 3.3.1, interpreting the demands for product j to be one for all j. Let $y^*(\mathbf{b})$ denote the optimal solution of (6.30). As in the DLP model, let $\boldsymbol{\pi} = (\pi_1, \ldots, \pi_m)$ denote a vector of optimal dual variables for the capacity constraints $\mathbf{Ay} \leq \mathbf{x}$.

Note that if we remove customer j from the problem, then the reduction in revenue in this relaxed problem is zero if $y_j^*(\mathbf{b}) = 0$, while if $y_j^*(\mathbf{b}) = 1$, it is approximately given by[15]

$$z^*(\mathbf{b}) - z^*(\mathbf{b} - \mathbf{e}_j b_j) \approx b_j - \sum_{i \in A_j} \pi_i,$$

since removing customer j eliminates his bid b_j but frees up a unit of capacity on each resource i used by product j, and $\boldsymbol{\pi}$ gives the marginal benefit of this freed-up capacity. So the right-hand side above is approximately the net benefit of having customer j in the problem.

As a result, the amount a winning customer j pays from (6.28) is approximately

$$p_j(\mathbf{b}) = b_j - (z^*(\mathbf{b}) - z^*(\mathbf{b} - \mathbf{e}_j b_j))$$
$$\approx \sum_{i \in A_j} \pi_i.$$

Thus, roughly speaking, winning customers pay the bid prices of the set of resources required by the product they are bidding for.[16] Of course, the actual bidding mechanism uses an integer program rather than a linear program, but the connection is still close.

For example, if we allow continuous allocations in the auction (customers can receive a fractional quantity of the product they bid on and are willing to accept any quantity between 0 and 1), the two problems

---

[15] Here, we are ignoring the possibility that the dual is degenerate, and we are assuming the allowable decrease in the right-hand side of the constraints $\mathbf{Ay} \leq \mathbf{x}$ is at least one, so that $\mathbf{A}_j^\top \boldsymbol{\pi}$ measures the change in the optimal objective function when the capacity is reduced by the vector A_j.

[16] Despite the close connection between bid prices and the price paid by customers in this network Vickrey auction, the use of the term bid price is purely a coincidence; the two problems were not connected in the literature or in practice until quite recently.

coincide exactly. In this case, by linear programming duality, one can say that a customer j gets a positive allocation only if his bid b_j is at least as large as the sum of the bid prices, $\sum_{i \in A_j} \pi_i$. The payment of these customers with positive allocations is also given by the sum of the bid prices.

6.5.5 Revenue Maximization and Reserve Prices

While the network mechanism outlined above has a well-defined, dominant-strategy equilibrium, it is not revenue maximizing for the firm. To see this, suppose that $z^*(\mathbf{b}) - z^*(\mathbf{b} - \mathbf{e}_j b_j) = b_j$ for any customer j. This would occur, for example, if the resources required by the product requested by customer j are not capacity constrained, so that including j as one of the winners would not displace any other winners. In this case, the payment according to (6.28) is simply $p_j(b_j) = 0$. Therefore, any customers requesting unconstrained resources would win and pay nothing. However, clearly, the firm would increase its revenue by charging these customers something positive.

Just as in the classical auction, reserve prices can be used to increase the revenue in the network case. However, there is no theory showing how to construct optimal reserve prices in this case. Still, one can heuristically consider a scheme whereby the firm imposes reserve prices, denoted $\hat{\pi}_i$, on each resource i and requires each customer j to submit bids that exceed the sum of the reserve prices of the resources requested—that is, $b_j \geq \mathbf{A}_j^\top \hat{\pi}$. It is still a dominant strategy for customer j to bid his valuation $b_j = v_j$, provided it exceed $\mathbf{A}_j^\top \hat{\pi}$; otherwise, his dominant strategy is not to bid at all.

Numerical results show that one can increase revenue significantly by using such reserve prices. For example, Table 6.5 shows the simulated revenues for an example with two resources and three customer types. Type 1 customers require only resource 1, type 2 customers require only resource 2, and type 3 customers require both resources 1 and 2. The number of customers of each type is an independent Poisson random variable. Customers of types 1 and 2 have valuations with a mean of 100 and variance of 10; customers of type 3 have a valuation with mean 200 and variance of 20. There are 20 units of capacity for each of the two resources. Two demand scenarios are tested—a high-demand scenario in which the mean number of customers of each type is 15 and a low-demand scenario in which the mean number of customers of each type is 5. Symmetric reserve prices are used for each of the two resources, but they are varied.

Table 6.5 shows the effect of the reserve prices on the average revenues in the two scenarios. Note that in the high-demand scenario (Poisson-

15), the reserve price has a minimal effect on the average revenue for low reserve prices, with a maximum occurring at $70. However, revenues decrease significantly once the reserve price approaches $100, the mean valuation that customers have for each resource. In contrast, in the low-demand scenario (Poisson-5), the reserve price significantly increases revenues, achieving a maximum with a reserve price of $80. Again, revenues fall when we increase the reserve price beyond this point. This behavior is consistent with a traditional C-unit auction, where reserve prices affect only the revenue when there are fewer than C customers willing to bid above the reserve price.

Table 6.5. Network auction simulation results: average revenues as a function of reserve price.[a]

Resource Reserve Price	Average Revenue (Poisson): 15 Customers per Product	Average Revenue (Poisson): 5 Customers per Product
0	3652.7	3.6
10	3645.6	101.0
20	3701.3	799.0
30	3697.6	618.0
40	3680.6	798.0
50	3722.2	1018.7
60	3713.8	1209.7
70	3737.8	1408.8
80	3725.7	1566.9
90	3703.3	1468.5
100	2936.6	970.2
110	1019.7	374.2
120	46.7	16.2

[a] *Note:* Data as reported by Cooper and Menich [129].

6.6 Notes and Sources

The formal study of auctions stems from the seminal work of Vickrey [533], who derived the equilibrium strategies and the revenue equivalence of standard first and second-price auctions. The extensive two-volume collection edited by Klemperer [306] provides an excellent source for much of the literature on auction theory; see also Klemperer's [305] excellent survey article contained therein. Other survey articles on the private-value model are Matthews [366], McAfee [369], and Milgrom [381].

The analysis of optimal auction mechanisms for the private-value model, as described in Section 6.2.5, stems from the seminal paper of

Myerson [398]. Maskin and Riley [364] extended Myerson's optimal auction analysis to multiunit auctions. The optimal discriminatory auction discussed in Section 6.2.7.3 is addressed in more detail in the survey of McAfee and McMillan [369] but was again originally due to Myerson [398].

The dynamic RM auction model in Section 6.3 is from Vulcano, van Ryzin, and Maglaras [542]. The infinite-horizon version with replenishment discussed in Section 6.4 is from van Ryzin and Vulcano [527].

The problem and results in Section 6.5 on network auctions are from Cooper and Menich [129]. For a an in-depth survey of other combinatorial auctions, see de Vries and Vohra [156].

APPENDIX 6.A: Proof of the Revenue-Equivalence Theorem

This surprisingly simple proof of the revenue-equivalence Theorem 6.1 is from Klemperer [305]:

Proof

Consider any symmetric equilibrium. Let $P(v)$ denote the probability that a customer winning under this equilibrium given his valuation is v (a type v customer), and let $S(v)$ denote the expected surplus of a customer with valuation v, defined by

$$S(v) = vP(v) - R(v),$$

where $R(v)$ is the expected payment. Since we are assuming an equilibrium, we must have that

$$S(v) \geq S(\tilde{v}) + (v - \tilde{v})P(\tilde{v}), \quad \forall \tilde{v}.$$

This follows because $P(\tilde{v})$ is the probability a customer wins if they were to follow the strategy of a customer with valuation $\tilde{v}$ instead of v. And if a customer with a valuation v wins by doing so, they value the item an amount $v - \tilde{v}$ different from a type $\tilde{v}$ customer. Hence, the right-hand side above is the expected surplus for a customer of valuation v if he follows the strategy of a type $\tilde{v}$ customer. However, as we are in equilibrium, a type v customer's surplus cannot be improved by deviating from the equilibrium strategy v.

Considering that a type v customer would not want to mimic a type $v + dv$ customer, we then have

$$S(v) \geq S(v + dv) + (-dv)P(v + dv),$$

and similarly since a type $v + dv$ customer would not want to mimic a type v customer,

$$S(v + dv) \geq S(v) + (dv)P(v).$$

Combining these two inequalities and we have that

$$P(v + dv) \geq \frac{S(v + dv) - S(v)}{dv} \geq P(v).$$

Since by assumption the allocation $y_i(v_i, \mathbf{v}_{-i})$ is increasing in v_i for all $\mathbf{v}_{-i}$, then $P(v)$ is increasing in v (since $P(v) = P(y_i(v, \mathbf{v}_{-i}) = 1)$), so the above inequalities are always feasible. Letting $dv \to 0$, shows

$$\frac{d}{dv} S(v) = P(v),$$

where upon integrating, we obtain

$$\begin{aligned} S(v) &= S(0) + \int_0^{\bar{v}} P(x)dx \\ &= \int_0^{\bar{v}} P(x)dx, \end{aligned} \quad (6.A.1)$$

where the last equality follows by the assumption $S(0) = 0$ (i.e., customers with valuation zero have zero expected surplus).

Next, note that the expected payment, $R(v) = vP(v) - S(v)$, is equal to the expected revenue received by the firm. This means the firm's expected revenue from type v customer is

$$\int_0^{\bar{v}} (vP(v) - S(v))f(v)dv. \quad (6.A.2)$$

To evaluate this, note that by (6.A.1), we have

$$\begin{aligned} \int_0^{\bar{v}} S(v)f(v)dv &= \int_0^{\bar{v}} f(v)\left[\int_0^v P(x)dx\right] dv \\ &= \int_0^{\bar{v}} (1 - F(v))P(v)dv, \end{aligned} \quad (6.A.3)$$

where the last equality is obtained by integrating by parts, since

$$\begin{aligned} \int_0^{\bar{v}} f(v)\left[\int_0^v P(x)dx\right] dv &= \left[F(v)\int_0^{\bar{v}} P(x)dx\right]\Big|_0^{\bar{v}} \\ &\quad - \int_0^{\bar{v}} F(v)P(v)dv \\ &= \int_0^{\bar{v}} (1 - F(v))P(v)dv \end{aligned}$$

Substituting (6.A.3) into (6.A.2), the firm's expected revenue from each customer i is

$$\int_0^{\bar{v}} \left(v - \frac{1 - F(v)}{f(v)}\right) P(v)f(v)dv = E_{v_i}[J(v_i)P(v_i)],$$

where recall that $J(v) = v - \frac{1-F(v)}{f(v)}$ (which is always well defined since, by assumption, F is strictly increasing, so we always have that $f(v) > 0$). Summing over all N customers, the firm's total expected revenue is

$$\sum_{i=1}^{N} E_{v_i}[J(v_i)P(v_i)].$$

APPENDIX 6.A: Proof of the Revenue-Equivalence Theorem

Finally, noting that the allocation variable $y_i(v_i, \mathbf{v}_{-i}) = 1$, if customer i is awarded an item and is zero otherwise, we have that $P(v_i) = E_{\mathbf{v}_{-i}}[y_i(v_i, \mathbf{v}_{-i})|v_i]$. Hence, the above expected revenue can be written

$$E_\mathbf{v}\left[\sum_{i=1}^{N} J(v_i) y_i(v_i, \mathbf{v}_{-i})\right].$$

QED

PART III

COMMON ELEMENTS

Chapter 7

CUSTOMER-BEHAVIOR AND MARKET-RESPONSE MODELS

This chapter reviews the basic theory of consumer choice, aggregate demand, and the operational, market-response models that are used in both quantity- and price-based revenue management. Because demand results from many individuals making choice decisions—choices to buy one firm's products over another, to wait or not to buy at all, to buy more or fewer units—we begin by looking at models of individual-choice behavior. When added up, these individual purchase decisions determine aggregate demand, so we next discuss aggregate-demand functions and their properties. Our treatment of the theory is somewhat abbreviated, aimed more at developing an intuitive and practical understanding of the concepts. The Notes and Sources section at the end of the chapter provides references that offer more extensive treatment of consumer behavior theory. Appendix E at the end of the book provides a basic reference on consumer theory, including utility theory, reservation prices, and risk preferences.

7.1 The Independent-Demand Model

Before delving into more complex models of demand, we first briefly review the *independent-demand model*, which is the basis of much of the material in Chapters 2 and 3 on quantity-based RM. This model is rather simple: it assumes that demand for each product is an independent stochastic process, not influenced by the firm's availability controls. Further, as we have seen in Section 2.2, static models of quantity-based RM also assume that the demand for products arrives in a specified order over the booking period, with demand for the lower-priced products appearing first. Thus, the independent model does not endogenize customer behavior, neither choice behavior nor purchase-timing behavior.

While it is easy to criticize the simplistic nature of this model, one can make a few theoretical and practical arguments in support of it. As discussed in Chapter 2, in standard quantity-based RM practice the customer is faced with a menu of possible products differentiated by prices and restrictions. As a result, if the firm offers n products, customers are approximately segmented into n separated populations (one for each product) according to their preference for the different product restrictions and prices. If customers are sufficiently well segmented by the restrictions (in the sense that most of the customers who are eligible to purchase one product are not eligible to buy another), then the independent-demand assumption is not unreasonable. However, this argument is admittedly weakened by the fact that (at least in the airline case) most restrictions are progressively relaxed as the fares get higher. So a customer who is eligible for one fare class is normally eligible for all classes with higher fares. We must then assume that customers are unwilling to purchase these higher fares.

Second, the independent-demand model is reasonable if the market is competitive and products are commodities—defined as products in which the identity of the supplier is of little importance to customers. In such cases, firms are price takers and can control only the quantity they sell; customers, in turn, base their choices only on price and are willing to buy from any firm offering the market price. (See Section 8.2.) Hence, if a given commodity product is not available at one firm—in particular, if its availability is closed by RM system controls—then demand for that product effectively disappears because customers will purchase the product from a competitor rather than switching to alternative products.

Third, the model reflects the current airline and hotel industry practice of separating pricing and capacity-control decisions, reflecting the different scope of the two in these industries—pricing decisions are made infrequently, while capacity-control is done in real-time; prices are set at a market level that includes a large number of flight departures (for airlines) and for an entire season (for a hotel) while capacity control is exerted on individual flights and dates. The implicit assumption in traditional quantity-based RM is that when prices for the products change, the change in the demand being observed will influence the forecasts and this changed forecasts, together with the new price of each product, will lead to changes in the capacity controls on each flight. This sort of quasi-static view of the price-demand relationship lies at the heart of current RM practice, and indeed the success of traditional RM methodology points to the practical utility of the overall approach.

Finally, the independent-demand model considerably simplifies the RM forecasting and optimization tasks. Forecasting can use historical demand data in standard time series forecasting methods, and we can solve stochastic optimization models based on the independent-demand model (at least approximately) relatively efficiently.

Yet despite these arguments in support of the independent-demand model, the fact that it ignores consumer behavior is conceptually unsatisfying and, more important, limits the full potential of RM methods. To counter its simplifications, a number of ad-hoc methods, such as the sell-up model discussed in Section 2.6, have been proposed. The discrete-choice model of Section 2.6 is a more recent alternative that overcomes the limitations of the independent-demand model. This latter model has more in common with the customer-choice-behavior view of demand that is the focus of this chapter.

7.2 Models of Individual Customer Choice

We next look at the basic approaches for modeling individual customer purchase decisions. In Chapter 9, we discuss methods for estimating the parameters of these models.

7.2.1 Reservation-Price Models

The simplest practical models of customer choice directly model customers' reservation prices for particular items. Each customer is assumed to follow a simple decision rule: if his reservation price (or valuation) v equals or exceeds the offered price p, the customer purchases the product; otherwise, he will not purchase the product. Moreover, he buys at most one unit of the product.

A customer's reservation price is specific to each individual and typically is private information unknown to the firm. However, based on management judgment, historical observed purchase behavior or other observable characteristics of the individual (such as place, time, and channel of purchase), the seller can attempt to model the distribution of the reservation prices across a population of customers and estimate at least the parameters of the distribution. This leads to a problem of finding a distribution $F(\cdot)$ such that the probability that a customer's reservation price is below p is given by $F(p) = P(v \leq p)$.

Often, however, the distribution of reservation prices is modeled indirectly by assuming an aggregate-demand function, as we discuss in Section 7.3 below. Hence, we postpone further discussion of reservation-price modeling until that point.

7.2.2 Random-Utility Models

Random-utility models are based on a probabilistic model of individual customer utility. (See Appendix E for a formal discussion of utility theory.) They are useful for several reasons. First, probabilistic models can be used to represent heterogeneity of preference among a population of customers. Second, they can model uncertainty in choice outcomes due to the inability of the firm to observe all the relevant variables affecting a given customer's choice (other alternatives, their prices, the customer's wealth, and so on). Third, they can model situations where customers exhibit *variety-seeking behavior* and deliberately alter their choices over time (movie or meal choice, for example). Finally, probabilistic choice can model customers whose behavior is inherently unpredictable—that is, customers who behave in a way that is inconsistent with well-defined preferences and at best, exhibit only some probabilistic tendency to prefer one alternative to another. Luce [349] developed a model of this type of random-choice behavior based purely on a set of axioms on choice probabilities, analogous to the axioms used to define classical deterministic utility functions. (See Appendix E.)[1]

For all these reasons, it is often reasonable to assume that a firm has only probabilistic information on the utility function of any given customer, and this can be modeled by assuming that customers' utilities for alternatives are themselves random variables. Specifically, let the n alternatives be denoted $j = 1, \ldots, n$. A customer has a utility for alternative j, denoted U_j. Without loss of generality we can decompose this utility into two parts, a *representative* component u_j that is deterministic and a mean-zero *random* component, ξ_j. Therefore,

$$U_j = u_j + \xi_j, \tag{7.1}$$

and the probability that an individual selects alternative j from a subset S of alternatives is given by[2]

$$P_j(S) = P(U_j \geq \max\{U_i : i \in S\}). \tag{7.2}$$

In other words, the probability that j has the highest utility among all the alternatives in the set S.

The representative component u_j is often modeled as a function of various observable attributes of alternative j. A common assumption is

---

[1] The distinction between models based on randomized preferences and those based on random-choice behavior is important primarily to behavioral theorists. A seminal work in this area is Block and Marschak [79]. However, for most RM problems what matters most is the demand process produced by a given model.

[2] That customers choose based on maximizing utility is itself an assumption. See Appendix E for a discussion of utility maximization as a model of customer choice.

the *linear-in-attributes* model

$$u_j = \beta^\top x_j, \tag{7.3}$$

where β is a vector of parameters and x_j is a vector of attribute values for alternative j, which could include factors such as price, measures of quality and indicator variables for product features. Variables describing characteristics of the customer (segment variables) can also be included in x_j.[3]

This formulation defines a general class of random-utility models, which vary according to the assumptions on the joint distribution of the utilities $U_1, \ldots, U_n$. Random-utility models are no more restrictive in terms of modeling behavior than are classical utility models; essentially, all we need assume is that customers have well-defined preferences so that utility maximization is an accurate model of their choice behavior. (Theorem E.3 in Appendix E.) However, as a practical matter, certain assumptions on the random utilities lead to much simpler models than others. We look at a few of these special cases next.

7.2.2.1 Binary Probit

If there are only two alternatives to choose from (such as buying or not buying a product) and the error terms $\xi_j, j = 1, 2$, are independent, normally distributed random variables with mean zero and identical variances σ^2, then the probability that alternative 1 is chosen is given by

$$P(\xi_2 - \xi_1 \leq u_1 - u_2) = \Phi(\frac{u_1 - u_2}{\sqrt{2}\sigma}), \tag{7.4}$$

where $\Phi(\cdot)$ denotes the standard normal distribution. This model is known as the *binary-probit* model. While the normal distribution is an appealing model of disturbances in utility (it can be viewed as the sum of a large number of random disturbances), the resulting probabilities do not have a closed-form solution. This has led researchers to seek other, more analytically tractable, models.

7.2.2.2 Binary Logit

The binary-logit model applies also to a situation with exactly two choices, similar to the binary-probit case, but is simpler to analyze. The assumption made here is that the error term $\xi = \xi_1 - \xi_2$ has a *logistic*

---

[3]Specifically, the utility can also depend on observable customer characteristics, so for customer i the utility of alternative j is u_{ij}. For simplicity, we ignore customer-specific characteristics here, but they can be incorporated into all the models that follow.

distribution—that is,

$$F(x) = \frac{1}{1+e^{-\frac{x}{\mu}}},$$

where $\mu > 0$ is a scale parameter and $-\infty < x < \infty$. Here ξ has a mean zero and variance $\frac{\mu^2 \pi^2}{3}$. The logistic distribution provides a good approximation to the normal distribution, though it has "fatter tails." The probability that alternative 1 is chosen is given by

$$P(\xi_2 - \xi_1 \leq u_1 - u_2) = \frac{e^{\frac{u_1}{\mu}}}{e^{\frac{u_1}{\mu}} + e^{\frac{u_2}{\mu}}}. \tag{7.5}$$

7.2.2.3 Multinomial Logit

The multinomial-logit model (MNL) is a generalization of the binary-logit model to n alternatives. It is derived by assuming that the ξ_j are i.i.d. random variables with a Gumbel (or double-exponential) distribution with cumulative density function

$$F(x) = P(\xi_j \leq x) = e^{-e^{-(\frac{x}{\mu}+\gamma)}},$$

where γ is Euler's constant $(= 0.5772\ldots)$ and μ is a scale parameter. The mean and variance of ξ_j are

$$E[\xi_j] = 0, \quad Var[\xi_j] = \frac{\mu^2 \pi^2}{6}.$$

The Gumbel distribution has some useful analytical properties, the most important of which is that the distribution of the maximum of n independent Gumbel random variables with the same scale parameter μ is also a Gumbel random variable. If two random variables ξ_1 and ξ_2 are Gumbel distributed with mean 0 and scale parameter μ, then $\xi = \xi_1 - \xi_2$ has a logistic distribution with mean 0 and variance, $\frac{\mu^2 \pi^2}{3}$, leading to the binary-logit model.

For the MNL model, the probability that an alternative j is chosen from a set $S \subseteq \mathcal{N} = \{1, 2, \ldots, n\}$ that contains j is given by

$$P_j(S) = \frac{e^{\frac{u_j}{\mu}}}{\sum_{i \in S} e^{\frac{u_i}{\mu}}}. \tag{7.6}$$

If $\{u_j : j \in S\}$ has a unique maximum and $\mu \to 0$, then the variance of the $\xi_j, j = 1, \ldots, n$ tends to zero and the MNL reduces to a deterministic model—namely

$$\lim_{\mu \to 0} P_j(S) = \begin{cases} 1 & \text{if } u_j = \max_{i \in S}\{u_i\} \\ 0 & \text{otherwise.} \end{cases}$$

Conversely, if $\mu \to \infty$, then the variance of the $\xi_j, j = 1, \ldots, n$ tends to infinity and the systematic component of utility u_j becomes negligible. In this case,

$$\lim_{\mu \to \infty} P_j(S) = \frac{1}{|S|}, \quad j \in S,$$

which corresponds to a uniform random choice of the alternatives in S. Hence, the MNL can model behavior ranging from deterministic utility maximization to purely random choice.

The MNL has been widely used as a model of customer choice. However, it possesses a somewhat restrictive property known as the *independence from irrelevant alternatives* (IIA) property—namely, for any two sets $S \subseteq \mathcal{N}$, $T \subseteq \mathcal{N}$ and any two alternatives $i, j \in S \cap T$, the choice probabilities satisfy

$$\frac{P_i(S)}{P_j(S)} = \frac{P_i(T)}{P_j(T)}. \tag{7.7}$$

Equation (7.7) says that the relative likelihood of choosing i and j is independent of the choice set containing these alternatives. This property is not realistic, however, if the choice set contains alternatives that can be grouped such that alternatives within a group are more similar than alternatives outside the group because adding a new alternative reduces the probability of choosing similar alternatives more than dissimilar alternatives. A famous example illustrating this point is the "blue-bus/red-bus paradox," (Debreu [150]):

Example 7.1 An individual has to travel and can use one of two modes of transportation: a car or a bus. Suppose the individual selects them with equal probability. Let the set $S = \{\text{car, bus}\}$. Then

$$P_{\text{car}}(S) = P_{\text{bus}}(S) = \frac{1}{2}.$$

Suppose now that another bus is introduced that is identical to the current bus in all respects except color: one is blue and one is red. Let the set T denote $\{\text{car, blue bus, red bus}\}$. Then the MNL predicts

$$P_{\text{car}}(T) = P_{\text{blue bus}}(T) = P_{\text{red bus}}(T) = \frac{1}{3}.$$

However, as bus color is likely an irrelevant characteristic in this choice situation, it is more realistic to assume that the choice of bus or car is still equally likely, in which case we should have

$$P_{\text{car}}(T) = \frac{1}{2}$$

$$P_{\text{blue bus}}(T) = P_{\text{red bus}}(T) = \frac{1}{4}.$$

As a result of IIA, the MNL model must be used with caution. It should be restricted to choice sets that contain alternatives that are, in some sense, "equally dissimilar." Example 9.18 provides one empirical test for the IIA property.

Despite this deficiency, the MNL model is widely used in marketing. (See Guadagni and Little's [227] work on determining brand share in the presence of marketing variables such as advertising and promotion.) It has also seen considerable application in estimating travel demand. (See Ben-Akiva and Lerman [48].) The popularity of MNL stems from the fact that it is analytically tractable, relatively accurate (if applied correctly), and can be estimated easily using standard statistical techniques. (See Example 9.6.)

Variations of the MNL have been introduced to avoid the IIA problem, the most prevalent of which is the nested MNL [49]. Our next section looks at some generalizations of the MNL that avoid the IIA property.

7.2.3 Customer Heterogeneity and Segmentation

RM often relies on the premise that different customers are willing to pay different amounts for a product. For example, demand functions arise from heterogeneity in the reservation prices of customers. In many situations, this level of modeling of heterogeneity is sufficient or is the only practical approach.

Yet a more accurate representation of demand is achievable if customers can be segmented into groups with similar preferences and price responses. This entails classifying customers into K segments, where each segment has its own choice model. If done properly, each of these segment-level models predicts the behavior of the segment better than a common choice model. In the extreme case, one could potentially define a different segment for each customer. However, a model of heterogeneity has to find the right balance between estimability and accuracy; each segment should not be so narrowly defined or so small as to make estimation impossible, yet it should be sufficiently small that customers within a segment have relatively homogeneous price and marketing variable responses. The aim is to maximize between-group variation but minimize within-group variation with respect to market responses. (Many of the techniques used to identify and segment customers are based on cluster analysis.) We next look at a few common approaches along these lines.

7.2.3.1 Finite-Mixture Logit Models

In the basic MNL model with linear-in-attribute utilities, the coefficients β in (7.3) are assumed to be the same for all customers. This may not be an appropriate assumption if there are different segments with different preferences. Moreover, as we've seen, the assumption leads to the IIA property, which may not be reasonable in certain contexts. If we can identify each customer as belonging to a segment, then it is an easy matter to simply fit a separate MNL model to the data from each segment. However, a more sophisticated modeling approach is needed if segment membership is not observable.

Assume that customers within each segment follow a MNL model with identical parameters and that customers have a certain probability of belonging to a segment (called a *latent segment*), which has to be estimated along with the MNL parameters for each segment. This results in the so-called *finite-mixture logit models*.

Assume that there are L latent segments and that the probability that a customer belongs to segment l is given by

$$q_l = \frac{e^{\nu_l}}{\sum_{i=1}^{L} e^{\nu_i}}, \quad l = 1, \ldots, L.$$

All customers in segment l are assumed to have utilities determined by an identical vector of coefficients β_l. Then the probability of choosing alternative j in this finite-mixture logit model is given by

$$P_j(S) = \sum_{l=1}^{L} q_l \frac{e^{\beta_l^\top x_j}}{\sum_{i \in S} e^{\beta_l^\top x_i}}, \quad j \in S.$$

One then tries to estimate the coefficients of the model (β and q_l, $l = 1, \ldots, L$) using, for example, maximum-likelihood methods. This model often provides better estimates of choice behavior than the standard MNL model, at the expense of a more complicated estimation procedure.

7.2.3.2 Random-Coefficients Logit Models

Another approach to modeling heterogeneity is to assume that each customer has a distinct set of coefficients β that are drawn from a distribution—usually assumed normal for analytical convenience—over the population of potential customers. This leads to what is called the *random-coefficients logit model*. The coefficients may also be correlated, both within themselves as well as with the error term, though we focus here on the simpler case where the coefficients are mutually independent.

Here again the utility of alternative j is given, similar to the MNL model, as

$$U_j = \beta^\top x_j + \xi_j, \quad j = 1, \ldots, n.$$

However, β is now considered a vector of random coefficients, each element of which is assumed to be independent of both the other coefficients in β and the error term ξ_j. Furthermore, the components of β

are assumed to be normally distributed with a vector of means **b** and a vector of standard deviations $\boldsymbol{\sigma}$. The components of the random vector $\boldsymbol{\beta}$ corresponding to characteristic m, denoted β_m, can be decomposed into

$$\beta_m = b_m + \sigma_m \zeta_m,$$

where ζ_m, $m = 1, \ldots, M$ is a collection of i.i.d. standard normal random variables.

It is convenient to express the utility as a systematic part and a mean-zero error term as before. To this end, define the composite random-error term

$$\nu_j \;=\; [\sum_{m=1}^{M} x_{mj}\sigma_m \zeta_m] + \xi_j, \quad j = 1, \ldots, n. \qquad (7.8)$$

Then a customer's random utility is given by

$$U_j(\boldsymbol{\nu}) = \mathbf{b}^\top \mathbf{x}_j + \nu_j, \quad j = 1, \ldots, n, \qquad (7.9)$$

where ν_j is given by (7.8). Hence, the key difference between the standard MNL and the random-coefficient logit is that the error terms ν_j are no longer independent across the alternatives (and somewhat less important, they are no longer Gumbel distributed). The following example illustrates the idea:

Example 7.2 Suppose that there are three alternatives $j = 1, 2, 3$ with two characteristics each $m = 1, 2$ and that the values of the characteristics are given as in Table 7.1.

If the parameter means are estimated as $b_1 = 1, b_2 = 1$, then the logit model would have a customer choosing one of the three products with an equal probability. In contrast, the random-coefficients logit model would have customers with a high preference for characteristic 1 (ζ_m high) consider alternatives 1 and 2 as closer substitutes than alternative 3. Customer preferences and product characteristics interact via (7.8).

Note also that the IIA property of standard logit is partially mitigated in this model. A customer with a high preference for characteristic $m = 1$ will choose alternative 2 with high probability if the choice set $\{2, 3\}$ is offered and will choose 1 or 2 with equal probability if the choice set $\{1, 2, 3\}$ is offered.

7.3 Models of Aggregate Demand

Even with transaction-level data, it is often easier to model and estimate aggregate demand rather than individual customer-choice decisions. Figure 7.1 illustrates how the heterogenous reservation prices of individual demand translate into a price versus quantity relationship for aggregate demand. Depending on the model, this aggregate demand

Table 7.1. Attribute weights x_m^j for attributes $m = 1, 2$ in alternative $j = 1, 2, 3$.

	Alternatives (j)		
m	1	2	3
1	200	200	-100
2	-100	-100	200

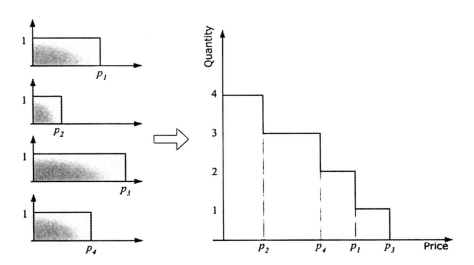

Figure 7.1. Individual demand with different reservation prices and the aggregate demand.

could be defined at the product, firm, or market level. If defined at the product or firm level, interactions with demand for other products (cross-elasticities) and dependence on historical demand or product attributes may have to be incorporated in its specification. In this section, we look at some commonly used aggregate-demand models.

7.3.1 Demand Functions and Their Properties

For the case of a single product, let p and $d(p)$ denote, respectively, the (scalar) price and the corresponding demand at that price. Also let Ω_p denote the set of feasible prices (the domain) of the demand function. For most demand functions of interest, $\Omega_p = [0, +\infty)$ but some functions (such as the linear-demand function) are not well defined for all nonnegative prices.

7.3.1.1 Regularity

It is often convenient to make the following regularity assumptions about the demand function:

ASSUMPTION 7.1 (REGULARITY: SCALAR CASE)
(i) The demand function is continuously differentiable on Ω_p.
(ii) The demand function is strictly decreasing, $d'(p) < 0$, on Ω_p.
(iii) The demand function is bounded above and below:

$$0 \leq d(p) < \infty, \quad \forall p \in \Omega_p.$$

(iv) The demand tends to zero for sufficiently high prices—namely,

$$\inf_{p \in \Omega_p} d(p) = 0.$$

(v) The revenue function $pd(p)$ is finite for all $p \in \Omega_p$ and has a finite maximizer $p^0 \in \Omega_p$ that is interior to the set Ω_p.

These are not restrictive assumptions in most cases and simply help avoid some technical complications in both analysis and numerical optimization. For example, consider a linear demand model (defined formally in Section 7.3.3.1)

$$d(p) = a - bp, \quad p \in \Omega_p = [0, a/b]. \tag{7.10}$$

This is trivially differentiable on Ω_p, is strictly decreasing if $b > 0$, is nonnegative and bounded for all $p \in \Omega_p$, tends to zero for $p \to a/b$ and the revenue $ap - bp^2$ and has a finite maximizer $p^0 = \frac{a}{2b}$.

7.3.1.2 Market-Share and Reservation-Price Distribution

It is sometimes convenient to express the demand function in the form

$$d(p) = N(1 - F(p)), \tag{7.11}$$

where $F(p)$ is a cumulative distribution function and N is interpreted as the *market size*. $1 - F(p)$ is then interpreted as the fraction of the market that is willing to buy at price p; equivalently, $F(p)$ is the distribution of reservation prices v in the customer population. The derivative of $F(p)$ is denoted $f(p) = \frac{\partial}{\partial p} F(p)$.

For example, consider again the linear-demand function (7.10). This can be written in the form (7.11) if we define $N = a$ and $F(p) = pb/a$. Since $F(\cdot)$ is the probability distribution of a customer's reservation price v, reservation prices are uniformly distributed in the linear-demand-function case.

7.3.1.3 Elasticity of Demand

The *price elasticity* of demand is the relative change in demand produced by a relative change in price. It is defined by

$$\epsilon(p) \equiv \frac{p}{d}\frac{\partial d}{\partial p} = \frac{\partial \ln(d)}{\partial \ln(p)}.$$

Note that elasticity is defined at a particular price p.

To illustrate, for the linear-demand function (7.10), $\frac{\partial d}{\partial p} = -b$, so the elasticity is

$$\frac{p}{d}\frac{\partial d}{\partial p} = -\frac{bp}{a - bp}.$$

Products can be categorized based on the magnitude of their elasticities. A product with $|\epsilon(p)| > 1$ is said to be *elastic*, while one with a elasticity value $|\epsilon(p)| < 1$ is said to be *inelastic*. If $|\epsilon(p)| = \infty$, demand for the product is said to be *perfectly elastic*, while if $|\epsilon(p)| = 0$, demand is said to be *perfectly inelastic*. Table 7.2 shows a sample of estimated elasticities for common consumer products. While many factors affect elasticity, these estimates give some sense of the relative magnitudes of elasticities.

7.3.1.4 Inverse Demand

The *inverse-demand function*, denoted $p(d)$, is the largest value of p which generates a demand equal to d—that is,

$$p(d) \equiv \max_{p \in \Omega_p}\{p : d(p) = d\}.$$

Given an inverse-demand function, one can view demand rather than price as the decision variable, since every choice of a demand d implies a unique choice of price $p(d)$. This is useful, as it is often easier analytically and computationally to work with demand rather than price as the decision variables in optimization problems.

The inverse may not be well-defined, however—for example, for values of d corresponding to points at which the demand function $d(p)$ has a jump discontinuity. Also, there may be not be a price p that produces any given value of demand d (for example, if demand remains bounded as p tends to zero yet d is large). Since not all values of d may be obtainable, we let Ω_d denote the set of achievable demand values. This set plays a role analogous to Ω_p for the demand function.

Under the regularity Assumption 7.1, the demand function is strictly decreasing and continuously differentiable on Ω_p, so the inverse-demand function is always well defined and continuously differentiable on the set $\Omega_d = \{x : x = d(p), p \in \Omega_p\}$. Indeed, under Assumption 7.1, the

Table 7.2. Estimated elasticities (absolute values) for common products.[a]

Product	$|\epsilon(p)|$
Inelastic	
Salt	0.1
Matches	0.1
Toothpicks	0.1
Airline travel, short-run	0.1
Gasoline, short-run	0.2
Gasoline, long-run	0.7
Residential natural gas, short-run	0.1
Residential natural gas, long-run	0.5
Coffee	0.25
Fish (cod) consumed at home	0.5
Tobacco products, short-run	0.45
Legal services, short-run	0.4
Physician services	0.6
Taxi, short-run	0.6
Automobiles, long-run	0.2
Approximately unit elasticity	
Movies	0.9
Housing, owner occupied, long-run	1.2
Shellfish, consumed at home	0.9
Oysters, consumed at home	1.1
Private education	1.1
Tires, short-run	0.9
Tires, long-run	1.2
Radio and television receivers	1.2
Elastic	
Restaurant meals	2.3
Foreign travel, long-run	4
Airline travel, long-run	2.4
Fresh green peas	2.8
Automobiles, short-run	1.2–1.5
Chevrolet automobiles	4
Fresh tomatoes	4.6

[a] *Source:* Reported by Gwartney and Stroup [232], collected from various econometric studies.

demand function is continuous, decreasing, and bounded and tends to zero for sufficiently high prices. One can also verify that the domain of the inverse-demand function is always an interval of the form $\Omega_d = [0, \bar{d}]$ for some upper bound $\bar{d}$.

Equation (7.11) expressed in terms of the reservation-price distribution, $F(p)$, the inverse-demand function is defined by

$$p(d) = F^{-1}(1 - d/N),$$

where $F^{-1}(\cdot)$ is the inverse of $F(\cdot)$.
To illustrate, the inverse of the linear-demand function (7.10) is

$$p(d) = \frac{1}{b}(a - d),$$

and the set of feasible demand rates is $\Omega_d = [0, a]$.

7.3.1.5 Revenue Function

The *revenue function*, denoted $r(d)$, is defined by

$$r(d) \equiv dp(d).$$

This is the revenue generated when using the price p and is of fundamental importance in dynamic-pricing problems. For example, the linear-demand function (7.10) has a revenue function

$$r(d) = \frac{d}{b}(a - d).$$

For most dynamic-pricing problems, we require that this revenue function be concave, as in the linear example above. This condition leads to well-behaved optimization problems.

7.3.1.6 Marginal Revenue

Another important quantity in pricing analysis is the rate of change of revenue with quantity—the *marginal revenue*—which is denoted $J(d)$. It is defined by

$$\begin{aligned} J(d) &\equiv \frac{\partial}{\partial d} r(d) \\ &= p(d) + dp'(d). \end{aligned} \quad (7.12)$$

It is frequently useful to express this marginal revenue as a function of price rather than quantity. At the slight risk of confusion over notation, we replace d by $d(p)$ above and define the marginal revenue as a function of price by[4]

$$J(p) \equiv J(d(p)) = p + d(p)\frac{1}{d'(p)}. \quad (7.13)$$

---

[4] By the inverse-function theorem, $p'(d) = 1/d'(p)$.

Note that $J(p)$ above is still the marginal revenue with respect to quantity—$\frac{\partial}{\partial d} r(d)$—but expressed as function of price rather than quantity; in particular, it is not the marginal revenue with respect to price.[5]

Expressing marginal revenue in terms of the reservation-price distribution $F(p)$, we have that

$$J(p) = p - \frac{1}{\rho(p)}, \tag{7.14}$$

where $\rho(p) \equiv f(p)/(1 - F(p))$ is the *hazard rate* of the distribution $F(p)$.[6] The marginal revenue function plays an important role in pricing problems. It is also central to the design of revenue-maximizing auctions, where it is referred to as the *virtual utility*, for reasons that are discussed in Chapter 6.

To illustrate, consider the marginal revenue of the linear-demand function of (7.10) as a function of d,

$$J(d) = \frac{\partial}{\partial d}\left[\frac{d}{b}(p-d)\right] = \frac{1}{b}(a - 2d).$$

Substituting $d(p) = a - bp$ for d above we obtain the marginal revenue as a function of price

$$J(p) = \frac{1}{b}(a - 2(a - bp)) = 2p - \frac{a}{b}.$$

It is frequently useful to make the following assumption about the marginal revenue:

ASSUMPTION 7.2 (MONOTONE MARGINAL REVENUE) *The marginal revenue $J(d)$ defined by (7.11) is strictly decreasing in the demand d. Equivalently, the marginal revenue $J(p)$ defined by (7.13) is strictly increasing in the price p.*

---

[5] The relationship between the marginal revenue with respect to price and quantity is as follows: since $r = pd$, then $\frac{\partial r}{\partial d} = d\frac{\partial p}{\partial d} + p$ and $\frac{\partial r}{\partial p} = p\frac{\partial d}{\partial p} + d$. Therefore, $\frac{\partial r}{\partial p} = \frac{\partial d}{\partial p}(p + d\frac{\partial p}{\partial d}) = \frac{\partial d}{\partial p}\frac{\partial r}{\partial d}$. (This also follows from the chain rule.)

[6] To see this, note that $\frac{\partial}{\partial x} F^{-1}(x) = 1/f(x)$, so

$$\begin{aligned} J(p) &= p + d(p)\frac{1}{d'(p)} \\ &= p - d(p)\frac{1}{Nf(p))} \\ &= p - \frac{1 - F(p)}{f(p)}, \end{aligned}$$

where the first equality follows from (7.13) and the next two from (7.11).

Note that this condition guarantees that the revenue function $r(d)$ is a concave function of the demand d, which again is a useful property in optimization models because it guarantees that first-order conditions are sufficient for determining an optimal price. This property is satisfied, for example, by the linear-demand function.

Slightly weaker conditions than those of Assumption 7.2 will also ensure that pricing-optimization problems are well behaved. In particular, if the revenue function is strictly unimodal,[7] this is often sufficient to ensure that there is a unique optimal price. (This is true for simple unconstrained pricing problems, for example.) Recall that $f(p)$ denotes the reservation-price density (derivative of $F(p)$) and $\rho(p) = f(p)/(1 - F(p))$ denotes the hazard rate. Then the following sufficient conditions on the reservation-price distribution ensure strict unimodality of the revenue function (see Ziya et al. [591]):

PROPOSITION 7.1 *Suppose that the reservation-price distribution $F(p)$ is twice differentiable and strictly increasing on its domain $\Omega_p = [p_1, p_2]$ ($F(p_1) = 0$ and $F(p_2) = 1$). Suppose further that $F(\cdot)$ satisfies any one of the following conditions:*

(i) $2\rho(p) > -\frac{f'(p)}{f(p)}$ for all $p \in \Omega_p$.

(ii) $\frac{2}{p} > -\frac{f'(p)}{f(p)}$ for all $p \in \Omega_p$.

(iii) $\rho(p) + \frac{1}{p} > -\frac{f'(p)}{f(p)}$ for all $p \in \Omega_p$

Then the revenue functions $r(p) = pd(p) = pN(1 - F(p))$ is strictly unimodal on Ω_p (equivalently, the revenue function $r(d) = p(d)d$ is strictly unimodal on $\Omega_d = [N(1 - F(p_1)), N(1 - F(p_2))]$.

Ziya et al. [591] show there are demand functions that satisfy one condition but not the others, so the three conditions are distinct.

Another desirable property of the marginal revenue function is that it spans the range $[0, +\infty)$ as p ranges over Ω_p (equivalently, d ranges over Ω_d). This is because in optimization problems, the first-order conditions typically involve equating marginal revenue to a nonnegative value (such as a cost or a Lagrange multiplier). If the marginal revenue spans the range $[0, +\infty)$, then the solutions of the first-order conditions are always in Ω_p (or Ω_d), and therefore, the explicit price (or demand) constraints can be safely ignored. We formalize this property in the following assumption:

---

[7] A function $f(x)$ defined on the domain $[a, b]$ is said to be a *unimodal function* if there exists an $x^* \in [a, b]$ such that $f(x)$ is strictly increasing on $[a, x^*]$ and $f(x)$ is strictly decreasing on $[x^*, b]$.

ASSUMPTION 7.3 *The range of the marginal revenue defined by (7.11) and (7.13) spans* $[0, +\infty)$. *That is, for every* $x \in [0, +\infty]$, $\exists d \in \Omega_d$ *such that* $J(d) = x$; *equivalently,* $\exists p \in \Omega_p$ *such that* $J(p) = x$.

Note that the linear-demand function does not satisfy this condition because the marginal revenue is $J(d) = (1/b)(a - 2d)$ and $\Omega_d = [0, a]$, so the marginal revenue ranges over $[-a/b, a/b]$. Other common demand functions, however, do satisfy this assumption, as described below.

7.3.1.7 Revenue-Maximizing Price

Under Assumption 7.2, the revenue is maximized at the point where the marginal revenue becomes zero. Assumption 7.1, part (v), requires that the maximizer is an interior point of the domain Ω_p, in which case the *revenue-maximizing price* p^0 is determined by the first-order condition

$$J(p^0) = 0.$$

Similarly, the revenue-maximizing demand, denoted d^0, is defined by

$$J(d^0) = 0.$$

They are related by

$$d^0 = d(p^0).$$

For example, for the linear-demand function we have $J(p) = 2p - a/b$ so $p^0 = \frac{a}{2b}$, an interior point of the set $\Omega_p = [0, a/b]$. The revenue-maximizing demand is $d^0 = a/2$.

Note from (7.13) that since $J(p) = p(1 + \frac{d}{p}\frac{\partial p}{\partial d})$, $\frac{\partial d}{\partial p} < 0$ (from Assumption 7.1, part (ii)), and $\frac{\partial d}{\partial p}\frac{p}{d} = \epsilon(p)$ is the price elasticity, we have

$$J(p) = p\left(1 - \frac{1}{|\epsilon(p)|}\right). \quad (7.15)$$

Thus, marginal revenue is increasing if demand is elastic at p (that is, if $|\epsilon(p)| > 1$), and marginal revenue is decreasing if demand is inelastic at p (that is, if $|\epsilon(p)| < 1$). At the critical value $|\epsilon(p^0)| = 1$, marginal revenue is zero and revenues are maximized.

If $J(p)$ is not monotone but one of the conditions of Proposition 7.1 is satisfied, then p^0 is a price such that $r(p^0)$ is increasing for $p < p^0$ and is decreasing for $p > p^0$; moreover, $p^0 = \inf\{p : |\epsilon(p)| \geq 1\}$.

Figure 7.2 illustrates the idea. Here, the revenue function $r(d)$ for the linear-demand function is plotted above, while the marginal-revenue function $J(d)$ is plotted below. Moving to the right corresponds to increasing the demand d and decreasing the price p. The inelastic-demand region is to the right of p^0, and the elastic region is to the left of p^0.

Customer-Behavior and Market-Response Models 319

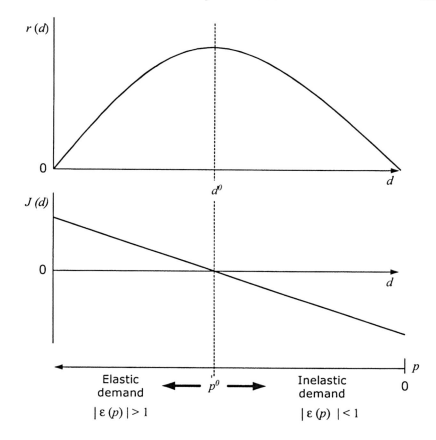

Figure 7.2. Revenue and marginal-revenue curves.

Starting at the far right with a price of zero, the demand is very inelastic; large relative changes in price (for example, doubling the price from δ to 2δ) result in small relative changes in demand. As a result, raising the price increases revenues. To the left, at very high price levels, relatively small decreases in price result in large increases in demand. Consequently, decreasing price improves revenues. The optimal price p^0 is the boundary of these two regions.

If there is a cost for providing the product—either a direct cost or opportunity cost—it is always optimal to price in the elastic region. To see this, let $c(d)$ denote the cost, so that $r(d) - c(d)$ is the firm's profit. Then the optimal price will occur at a point where $J(d) = r'(d) = c'(d)$. Assuming cost is strictly increasing in quantity, $c'(d) > 0$, the optimal price will be at a point where marginal revenue is positive—in the elastic region. Thus, it is almost never optimal to price in the inelastic region.[8]

7.3.2 Multiproduct-Demand Functions

In the case where there are $n > 1$ products, let p_j denote the price of product j and $\mathbf{p} = (p_1, \ldots, p_n)$ denote the vector of all n prices. The demand for product j as a function of $\mathbf{p}$ is denoted $d_j(\mathbf{p})$, and $\mathbf{d}(\mathbf{p}) = (d_1(\mathbf{p}), \ldots, d_n(\mathbf{p}))$ denotes the vector of demands for all n products. Again, Ω_p will denote the domain of the demand function. We also use the notation $\mathbf{p}_{-j} = (p_1, \ldots, p_{j-1}, p_{j+1}, \ldots, p_n)$ to denote all prices other than p_j.

Paralleling the single-product case, the following regularity assumptions for the multiproduct-demand function help ensure the resulting optimization models are well behaved:

ASSUMPTION 7.4 (REGULARITY: n-PRODUCT CASE) *For* $j = 1, \ldots, n$:
(i) $d_j(\mathbf{p})$ demand is strictly decreasing in p_j for all $\mathbf{p} \in \Omega_p$.
(ii) The demand function is continuously differentiable on Ω_p.
(iii) The demand function is bounded above and below: $0 \leq d_j(\mathbf{p}) < +\infty$, $\forall \mathbf{p} \in \Omega_p$.
(iv) The demand function tends to zero in its own price for sufficiently high prices—that is, for all $\mathbf{p}_{-j}$, $\inf_{p_j \in \Omega_p} d_j(p_j, \mathbf{p}_{-j}) = 0$.
(v) The revenue function $\mathbf{p}^\top \mathbf{d}(\mathbf{p})$ is bounded for all $\mathbf{p} \in \Omega_p$ and has a finite maximizer $\mathbf{p}^0$ that is interior to Ω_p.

As in the scalar case, we let $\mathbf{p}(\mathbf{d})$ denote the inverse-demand distribution; it gives the vector of prices that induces the vector of demands $\mathbf{d}$. In the multiproduct case, this inverse is more difficult to define generally, and in most cases we simply assume it exists. (For the common demand functions of Section 7.3.3, the inverse can be defined either explicitly or implicitly.) Likewise, we denote by Ω_d the domain of the inverse-demand function, the set of achievable demand vectors $\mathbf{d}$.

The revenue function is defined by

$$r(\mathbf{d}) = \mathbf{d}^\top \mathbf{p}(\mathbf{d}),$$

which again represents the total revenue generated from using the vector of demands $\mathbf{d}$—or equivalently, the vector of prices $\mathbf{p}(\mathbf{d})$. Paralleling

---

[8]The only exception is if the firm *benefits* from disposing of products—that is, if it has a negative cost. For example, this could occur if there is a holding cost incurred for keeping units rather than selling them. In such cases, it may be optimal to price in the inelastic region.

Assumptions 7.2 and 7.3, in the multiproduct case it is often convenient to make the following assumption:

ASSUMPTION 7.5 *The multiproduct revenue function satisfies*
(i) $r(\mathbf{d})$ is jointly concave on Ω_d.
(ii) For every $\mathbf{x} \in \Re_+^n$, there exists a $\mathbf{d} \in \Omega_d$ such that $\nabla_{\mathbf{d}} r(\mathbf{d}) = \mathbf{x}$.

Again, these assumptions help simplify the resulting pricing optimization problems and, and though more difficult to check than in the single-product case, are satisfied by several common demand functions.

The *cross-price elasticity* of demand is the relative change in demand for product i produced by a relative change in the price of product j. It is defined by

$$\epsilon_{ij}(p) = \frac{p_j}{d_i}\frac{\partial d_i}{\partial p_j} = \frac{\partial \ln(d_i)}{\partial \ln(p_j)}.$$

If the sign of the elasticity is positive, then products i and j are said to be *substitutes*; if the sign is negative, the products are said to be *complements*. Intuitively, substitutes are products that represent distinct alternatives filling the same basic need (such as Coke and Pepsi), whereas complements are products that are consumed in combination to meet the same basic need (such as hamburgers and buns).

7.3.3 Common Demand Functions

The demand function of a product can depend on variables other than its price (such as product attributes or, marketing variables such as advertising, brand name, competitor's prices and past market share), and modeling demand as a function of all relevant variables makes a model more realistic and accurate. The variables can either be current or lagged, when past-period variables affect demand. Here we focus on demand functions that depend only on current prices. A few other market-response functions that include nonprice variables are discussed in Section 9.6.4.

Table 7.3 summarizes the most common demand functions and their properties, and Figure 7.3 shows graphs of a few of these. All these functions satisfy the regularity conditions in Assumptions 7.1 and 7.4, the exception being the constant-elasticity-demand function, which does not satisfy part (v) of either assumption as explained below.

Table 7.3. Common demand functions.

	$d(p)$	$p(d)$	$r(d)$	$J(d)$	$\|\epsilon(p)\|$	p^0
Linear	$a - bp$	$\frac{1}{b}(a-d)$	$\frac{d}{b}(a-d)$	$\frac{1}{b}(a-2d)$	$\frac{pb}{a-bp}$	$\frac{a}{2b}$
Log-linear (exponential)	ae^{-bp}	$\frac{1}{b}\ln(\frac{a}{d})$	$\frac{d}{b}\ln(\frac{a}{d})$	$\frac{1}{b}(\ln(\frac{a}{d})-1)$	pbe^{-1}	$\frac{e}{b}$
Constant elasticity	ap^{-b}	$(\frac{a}{d})^{1/b}$	$a^{1/b}d^{1-1/b}$	$(1-\frac{1}{b})(\frac{a}{d})^{1/b}$	b	$\begin{cases} 0 & b>1 \\ +\infty & b<1 \\ \text{all } p \geq 0 & b=1 \end{cases}$
Logit	$\frac{e^{-bp}}{1+e^{-bp}}$	$\frac{1}{b}\ln(\frac{1}{d}-1)$	$\frac{d}{b}\ln(\frac{1}{d}-1)$	$\frac{1}{b}\left(\ln(\frac{1}{d}-1)-\frac{1}{d-1}\right)$	$\frac{p}{b(1+e^{-bp})}$	No closed-form formula

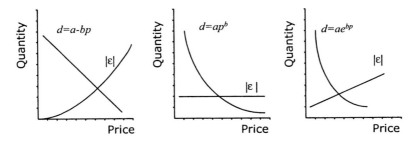

Figure 7.3. Some common aggregate demand (price-response) functions: (i) linear-demand function (ii) constant-elasticity-demand function (iii) exponential (log-linear) demand function.

7.3.3.1 Linear Demand

We have already seen the case of a linear-demand function in the scalar case. To summarize, it is

$$d(p) = a - bp,$$

where $a \geq 0$ and $b \geq 0$ are scalar parameters. The inverse-demand function is

$$p(d) = \frac{1}{b}(a - d).$$

The linear model is popular because of its simple functional form. It is also easy to estimate from data using linear-regression techniques. However, it produces negative demand values when $p > a/b$, which can cause numerical difficulties when solving optimization problems. Moreover, as mentioned, it does not satisfy Assumption 7.3. Hence, one must typically retain the price constraint set $\Omega_p = [0, a/b]$ when using the linear model in optimization problems.

In the multiproduct case, the linear model is

$$\mathbf{d(p) = a + Bp},$$

where $\mathbf{a} = (a_1, \ldots, a_n)$ is vector of coefficients and $\mathbf{B} = [b_{ij}]$ is a matrix of price sensitivity coefficients with $b_{ii} \leq 0$ for all i and the sign of b_{ij}, $i \neq j$ depending on whether the products are complements ($b_{ij} < 0$) or substitutes ($b_{ij} > 0$). If $\mathbf{B}$ is nonsingular, then the inverse-demand function exists and is given by

$$\mathbf{p(d) = B^{-1}(d - a)}.$$

One sufficient condition for $\mathbf{B}^{-1}$ to exist is that the *row* coefficients satisfy[9]

$$b_{ii} < 0 \quad \text{and} \quad |b_{ii}| > \sum_{j \neq i} |b_{ij}|, \quad i = 1, \ldots, n. \tag{7.16}$$

Roughly, this says that demand for each product i is more sensitive to a change in its own price than it is to a simultaneous change in the prices of all other products. An alternative sufficient condition for $\mathbf{B}^{-1}$ to exist is that the *column* coefficients satisfy

$$b_{jj} < 0 \quad \text{and} \quad |b_{jj}| > \sum_{i \neq j} |b_{ij}|, \quad j = 1, \ldots, n. \tag{7.17}$$

Equation (7.17) says that changes in the price of product j impacts the demand for product j more than it does the total demand for all other products combined. In the case of substitutes ($b_{ij} > 0, i \neq j$), this is equivalent to saying there is an aggregate market expansion or contraction effect when prices change (for example, the total market demand strictly decreases when the price of product j increases, and demand for product j is not simply reallocated one for one to substitute products).

7.3.3.2 Log-Linear (Exponential) Demand

The log-linear—or exponential—demand function in the scalar case is defined by

$$d(p) = e^{a-bp},$$

where $a \geq 0$ and $b \geq 0$ are scalar parameters. This function is defined for all nonnegative prices, so $\Omega_p = [0, +\infty)$. The inverse-demand function is

$$p(d) = \frac{1}{b}(a - \ln(d)).$$

The log-linear-demand function is popular in econometric studies and has several desirable theoretical and practical properties. First, unlike the linear model, demand is always nonnegative so one can treat price (or quantity) as unconstrained in optimization problems. Second, by taking the log of demand, we recover a linear form, so it is also well suited to estimation using linear regression. However, demand values of zero are not defined when taking logarithms, which is problematic in settings where sales are infrequent.

The multidimensional log-linear form is

$$d_j(\mathbf{p}) = e^{a_j + \mathbf{B}_j^\top \mathbf{p}}, \quad j = 1, \ldots, n,$$

---

[9] As noted by Maglaras and Meissner [354] from conditions in Horn and Johnson [258].

Customer-Behavior and Market-Response Models   325

where a_j is a scalar coefficient and $\mathbf{B}_j = (b_{j1}, \ldots, b_{jn})$ is a vector of price-sensitivity coefficients. Letting $\mathbf{a} = (a_1, \ldots, a_n)$ and $\mathbf{B} = [b_{ij}]$, as in the linear model, and taking the logarithm, we have

$$\ln(\mathbf{d}(\mathbf{p})) = (\ln(d_1(\mathbf{p})), \ldots, \ln(d_n(\mathbf{p}))) = \mathbf{a} + \mathbf{B}\mathbf{p},$$

so again the log-linear model can be estimated easily from data using linear regression provided the data is not too sparse.

The inverse-demand function can be obtained as in the linear case if $\mathbf{B}$ is nonsingular, in which case

$$\mathbf{p}(\mathbf{d}) = \mathbf{B}^{-1}(\ln(\mathbf{d}) - \mathbf{a}),$$

and one can again use the sufficient conditions (7.16) or (7.17) to check that $\mathbf{B}^{-1}$ exists.

7.3.3.3 Constant-Elasticity Demand

The constant-elasticity-demand function in the single-product case is of the form

$$d(p) = a p^{-b},$$

where $a > 0$ and $b \geq 0$ are constants. The function is defined for all nonnegative p, so $\Omega_p = [0, +\infty)$. Since $\partial d / \partial p = -a b p^{-(b+1)}$, the elasticity is

$$\epsilon(p) = \frac{p}{d} \frac{\partial d}{\partial p} = -b,$$

a constant for all values p (hence the name). The inverse-demand function is

$$p(d) = \left(\frac{a}{d}\right)^{1/b}.$$

Note that because elasticity is constant, from (7.15) the marginal revenue will always be positive or will always be negative for all values of p (unless by chance $|\epsilon(p)| = 1$, in which case it is zero for all values of p). Thus, this function usually violates Assumption 7.1, part (iv), because either the marginal revenue is always positive so $p_0 = +\infty$ or the marginal revenue is always negative, so $p_0 = 0$, both extreme points of the set Ω_p (unless, again the elasticity is exactly one, in which case all values of p are revenue maximizing). From this standpoint, it is a somewhat ill behaved demand model in pricing-optimization problems, though in cases where revenue functions are combined with cost functions this behavior is less problematic.

The multiproduct constant elasticity model is

$$d_i(\mathbf{p}) = a_i p_1^{b_{i1}} p_2^{b_{i2}} \ldots p_n^{b_{in}}, \quad i = 1, \ldots, n,$$

where the matrix of coefficients $\mathbf{B} = [b_{ij}]$ defines the cross (and own) price elasticities among the products, since

$$\epsilon_{ij}(\mathbf{p}) = \frac{\partial d_i/d_i}{\partial p_j/p_j} = b_{ij}.$$

Note that the inverse-demand function $\mathbf{p}(\mathbf{d})$ exists if the matrix $\mathbf{B}$ is invertible, since $(\log(d_1(\mathbf{p})), \ldots, \log(d_n(\mathbf{p}))) = \mathbf{a} + \mathbf{B}p$ (here $\mathbf{a} = (a_1, \ldots, a_n)$) and $\log(\cdot)$ is a strictly increasing function.

7.3.3.4 Logit Demand

The logit demand function is based on the MNL model of Section 7.2.2.3. Recall that in the MNL the utility of each alternative $j = 1, \ldots, n$, is assumed to be of the form

$$U_j = u_j + \xi_j,$$

where u_j is the mean utility of choice j and ξ_j is an i.i.d., random-noise term with a Gumbel distribution with mean zero and scale parameter one. For the logit-demand function, we also include a no-purchase alternative (indexed by zero) with utility

$$U_0 = u_0 + \xi_0,$$

where ξ_0 is an independent Gumbel random variable with mean zero and scale parameter 1. Since utility is ordinal, without loss of generality we can assume $u_0 = 0$. The choice probabilities are then given by (7.6) with the no-purchase alternative having a value $e^{u_0} = 1$.

As mentioned, it is common to model u_j as a linear function of several known attributes including price. Assuming the representative component of utility u_j is linear in price and interpreting the choice probabilities as fractions of a population of customers of size N lead to the class of logit-demand functions.

For example, in the scalar case, we assume $u_1 = -bp$, and this gives rise to a demand function of the form

$$d(p) = N \frac{e^{-bp}}{1 + e^{-bp}},$$

where N is the market size, $1 - F(p) = \frac{e^{-bp}}{1+e^{-bp}}$ is the probability that a customer buys at price p, and b is a coefficient of the price sensitivity. The function is defined for all nonnegative p, so $\Omega_p = [0, +\infty)$. There is no closed-form expression for the inverse-demand function, but it is easy to see that $d(p)$ is strictly decreasing in p, so the inverse exists and is well defined.

In the multiple-product case, the demand function is given by

$$d_j(\mathbf{p}) = N \frac{e^{-b_j p_j}}{1 + \sum_{i=1}^{n} e^{-b_i p_i}}, \quad j = 1, \ldots, n,$$

where again $\mathbf{b} = (b_1, \ldots, b_n)$ is a vector of coefficients and

$$P_j(\mathbf{p}) = \frac{e^{-b_j p_j}}{1 + \sum_{i=1}^{n} e^{-b_i p_i}}$$

is the MNL probability that a customer chooses product j as a function of the vector of prices $\mathbf{p}$.

One potential problem with the MNL demand model is that it inherits the IIA property (7.7). This causes problems if groups of products share attributes that strongly affect the choice outcome. To illustrate what can go wrong, consider the cross-price elasticity of alternative i with respect to the price of alternative j, $\epsilon_{ij}(\mathbf{p})$. This is given by

$$\begin{aligned}
\epsilon_{ij}(\mathbf{p}) &= \frac{\partial \ln d_i(p)}{\partial \ln p_j} \\
&= -p_j b_j \frac{e^{-b_j p_j}}{1 + \sum_{k=1}^{n} e^{-b_k p_k}}. \quad (7.18)
\end{aligned}$$

Notice that this cross-price elasticity is not dependent on i, and therefore cross-elasticity is the same for all alternatives i other than j.

The implications of this constant cross-price elasticity can be illustrated by an example of automobile market shares.[10] Consider a pair of subcompact cars and an expensive luxury car. If we lower the price of one of the subcompact cars by 10%, then (7.18) says that the percentage change in the demand for the other subcompact car will be the same as the percentage change in the demand for the luxury car (if the other subcompact car demand drops by 20%, then the luxury car demand will also drop by 20%). Such behavior is not very realistic. This IIA behavior stems fundamentally from the i.i.d. assumption on the random-noise terms ξ's of the MNL model. (See Berry [53] for a discussion, and a possible way around these restrictions on cross-price elasticities.)

7.3.4   Stochastic-Demand Functions

A deterministic demand function $d(p)$ can be used to define a stochastic model of demand in a variety of ways. In the stochastic case, we let

---

[10] If the population is homogeneous, the choice probabilities represent market share, and the MNL can be used to estimate market shares.

$D(p,\xi_t)$ denote the random demand as a function of the price p and a random-noise term ξ_t. The three most common random-demand models are discussed below.

7.3.4.1 Additive Uncertainty

In the additive model, the demand is a continuous random variable of the form

$$D(p,\xi) = d(p) + \xi,$$

where ξ is a zero-mean random variable that does not depend on the price. In this case, the mean demand is $d(p)$, and the noise term ξ shifts the demand randomly about this mean.

Note that this additive disturbance has the property that the elasticity of demand depends on ξ. This follows since

$$\epsilon(p,\xi) = \frac{p}{D(p,\xi)}\frac{\partial D(p,\xi)}{\partial p} = \frac{\epsilon(p)}{1+\xi/d(p)},$$

where $\epsilon(p) = \frac{p}{d(p)}\frac{\partial d(p)}{\partial p}$ is the deterministic elasticity. So if a realization of ξ is less than zero, the elasticity of demand in the stochastic model is greater than the deterministic elasticity, and if the realization of ξ is greater than zero, it is smaller.

One potential problem with the additive uncertainty model is that demand could be negative if $d(p)$ is small and the variance of ξ is large. For this reason, the additive model should be used with caution in applications where the coefficients of variation for the demand uncertainty is high.

7.3.4.2 Multiplicative Uncertainty

In the multiplicative model, the demand is again a continuous random variable but of the form

$$D(p,\xi) = \xi d(p),$$

where ξ is a nonnegative random variable with mean one that does not depend on the price p. In this case, the mean demand is again $d(p)$, and the noise term ξ simply scales the mean demand by a random factor. For the multiplicative model, the elasticity of demand for any given realization of ξ is the same as the deterministic elasticity, since

$$\epsilon(p,\xi) = \frac{p}{d(p,\xi)}\frac{\partial d(p,\xi)}{\partial p} = \epsilon(p),$$

where again $\epsilon(p)$ is the deterministic elasticity. Thus, the random-noise term does not affect the elasticity of demand; it affects only the magnitude of demand.

Note also that one can also combine the multiplicative and additive uncertainty models, leading to a demand function of the form

$$D(p,\xi) = \xi_1 + \xi_2 d(p),$$

where ξ_1 is a zero-mean random variable and ξ_2 is a nonnegative, unit-mean random variable.

7.3.4.3 Poisson and Bernoulli Uncertainty

Poisson and Bernoulli models of uncertainty are used in the dynamic models of demand discussed in Chapter 5. In the Bernoulli model, $d(p)$ is simply a probability of an arrival in a given period. So $d(p)$ is the probability that demand is one in a period, and $1 - d(p)$ is the probability demand is zero. As a result, the mean demand in a period is again $d(p)$, and we can represent the demand as a random function

$$D(p,\xi) = \begin{cases} 1 & \xi \leq d(p) \\ 0 & \xi > d(p) \end{cases},$$

where ξ is a uniform $[0,1]$ random variable.

For example, consider a situation in which the buyer in the period has a reservation price v that is a random variable with distribution $F(\cdot)$. If the firm offers a price of p, they will sell a unit if $v \geq p$, which occurs with probability $1 - F(p)$. This corresponds to setting $d(p) = 1 - F(p)$ above.

In the Poisson model, time is continuous, and $d(p)$ is treated as a stochastic intensity or rate. That is, the probability that we get a unit of demand in an interval of time from t to $t + \delta$ is $\delta d(p) + o(\delta)$ and the probability that we see no demand is $1 - \delta d(p) + o(\delta)$ (all other events have probability $o(\delta)$).

The Poisson and Bernoulli models are useful for several reasons. First, they translate a deterministic demand function directly into a stochastic model, without the need to estimate additional parameters (such as variance). They also are discrete-demand models—as opposed to the continuous demand of the additive and multiplicative models—and more closely match the discreteness of demand in many RM applications. At the same time, the Poisson and Bernoulli models assume a specific coefficient of variation, which may or may not match the observed variability. The additive and multiplicative models, in contrast, allow for different levels of variability in the model, as the complete distribution of the noise term can be specified.

7.3.4.4 Stochastic Regularity

As in the deterministic case, it is useful analytically to make some regularity assumptions about the stochastic demand functions. In particular:

ASSUMPTION 7.6 (STOCHASTIC-DEMAND-FUNCTION REGULARITY)
The variance of demand is uniformly bounded, $E[|D(p,\xi_t)|^2] \leq K < +\infty$ for $p \geq 0$.

This condition is not very restrictive and is required only to ensure that stochastic optimization problems are well behaved.

7.3.5 Rationing Rules

A final demand-modeling issue concerns how capacity is allocated to customers in cases where demand exceeds supply. For example, suppose capacity is 100 units and the firm commits to a fixed price of $10 per unit before knowing the demand realization. If the demand at this price turns out to be 120, then what assumptions do we make about which customers get the capacity and which do not? Do we assume that the capacity is allocated to customers with the highest valuations (thereby increasing the customer surplus), or should we assume that it is allocated randomly—for example, on a first-come, first-serve basis? The rules used for allocating capacity to customers when demand exceeds capacity are called *rationing rules* in economics.

There are two classical rationing rules: (1) The *efficient-rationing rule* (also called *parallel rationing*), in which it is assumed that units are allocated to customers with the highest valuations, and (2) the *proportional-rationing rule* (also called *randomized rationing*), in which it is assumed that capacity is allocated randomly, so the allocation is independent of the customers' valuations. While the former is more efficient from a consumer surplus standpoint, it is difficult to achieve in most posted-price settings (though some types of auctions implement it very well).

In quantity-based RM applications the most natural assumption is the proportional-rationing rule because when a given product is open, it is normally purchased on a first-come first-served basis. Therefore, provided there is no correlation between valuations and order of arrival, the inventory is sold independent of valuations.

7.4 Notes and Sources

Kreps [313] provides a comprehensive and readable treatment of the classical rational theory of consumer choice, including preference relations, utility theory, and choice under uncertainty. See also the microeconomics text of Mas-Collel et al. [365].

Random-utility models originated with the early work of the mathematical psychologist Thurston [511, 510] and were later formalized by economists, most notably Manski [358] and McFadden [372, 373]. (See also the edited volume by Manski and McFadden [357].) The limitations of the MNL as a model for transportation demand are discussed in detail by Oum [412]. The Gumbel distribution, which plays a central role in the MNL, is one of the distributions of extremes examined in Gumbel [229].

Kamakura and Russell [286], Chintagunta [117], and Allenby, Arora and Ginter [8] are some marketing-science papers that use the finite-mixture logit models. The finite-mixture and random-coefficient models are said to be heterogeneous in preferences; that is, customers use the *same* choice model but have different preferences (for example, use different coefficients) within that choice model. Another source of heterogeneity—called *structural heterogeneity*—is when customers in different segments use fundamentally different decision processes in making their purchase decisions. Such structural heterogeneity is studied in Kannan and Wright [287] and Kamakura, Kim, and Lee [285]. Finally, Dirichlet distributions have been used to model heterogeneity in brand-choice behavior (Fader and Lattin [179]; Jain, Bass, and Chen [266]).

An excellent comprehensive text on both the theory and application of discrete-choice models for demand estimation is Ben-Akiva and Lerman [48]. See also the book by Anderson et al. [16] for another good text on discrete-choice theory and economic-modeling applications of the theory.

Chapter 8

THE ECONOMICS OF RM

8.1 Introduction

Many topics traditionally studied by economists are central to understanding revenue management. After all, using price as a means for balancing supply and demand and achieving an efficient allocation of goods is one of the central themes of economics. Economic theories on rationing, free entry, price discrimination, monopoly pricing, pricing under capacity restrictions, oligopoly pricing, multiproduct pricing, differentiated products pricing, among others, provide explanations and predictions of price formation under various market conditions. This body of work is more theoretical than operational in nature but provides fundamental insights into RM—insights that are often lost in the morass of operational details surrounding the main-stream RM methodology that has been our focus thus far.

While economic theory has much to say about RM practice, at the same time RM is an anomaly of sorts, insofar that the practice often deviates from classical economic predictions in important respects. For example, on the one hand, products such as airline seats are widely considered to be commodities and the markets free (for example, the service is not highly differentiated, competition is fierce, prices are transparent, efficient electronic markets exist in the form of central reservation systems, and so on); yet prices exhibit wide dispersion and are far from being equal to marginal cost even under intense competition. What can explain this? Is it that airline markets are not as competitive as one might imagine, and the practice of RM is just a manifestation of monopoly price discrimination? Alternatively, can it be explained by other factors, such as high fixed costs or uncertain and variable demand? How can airlines and hotels sustain the sale of the same physical product at multiple prices in a competitive market?

Indeed, many papers in the economics literature point to airline RM as an example of price discrimination—a monopoly practice—while others cite the deregulation of the airline industry and the subsequent drop in prices (on average) as a model of the benefits of a competitive free market. Which is it? Similarly, while retail markdown pricing might look like monopoly price skimming, retailing is notoriously competitive, especially in this modern era of transparent online prices. So should we use the monopoly or competitive explanation of pricing in this industry?

Unfortunately, the answer to such questions is often "all of the above," in the sense that in most real-world RM contexts there are many economic forces at work operating at different levels and different time scales. Economic theory, in contrast, tends to isolate and study one effect at a time. This fact is important to keep in mind when reading this chapter.

Still, one can attempt to organize RM practice in a hierarchy, similar to what was discussed in Chapter 1. At a high level, RM can be considered to be a three-stage "game." In the first stage, firms make structural design decisions. They may design their products based on customer preferences, on existing competitive products, or in anticipation of new products that might be offered by rivals. For instance, a retailer making stocking decisions prior to a season decides what assortments, styles, and colors to stock. An airline picks its routes and schedules and may create discount products with restrictions as a result of the entry (or threat of entry) of a low-cost carrier. A hotel picks its location, the quality and size of its rooms, the level of service it offers, and so forth.

In the next stage, firms may commit to either a quantity or price decision. A retailer might commit to an order quantity before the start of a sales season; a hotel commits to the number of rooms it provides; an airline commits to a flight schedule and an assignment of aircraft to each route, and so on. Firms may commit to prices as well. For example, as mentioned in earlier chapters, airlines typically commit to market-level prices for a wide range of flight departures. (This fact is the basis for the exogenous-price assumptions in the classic single-resource and network RM problems.) This is done both for advertising purposes (say, to advertise only one fare for all flights in a market), for competitive reasons (say, to match a marketwide fare offered by a competitor), and for administrative reasons—such as reducing the complexity and cost of managing pricing decisions.

Once these precommitment decisions are made, firms may have some recourse decision that allows them to adjust to changes in market conditions. Again, this might be a price or quantity recourse decision, and the scope and flexibility of the recourse decisions vary from industry to industry. For example, the ability to change prices often depends on the channels used: price changes may be rapid and costless in an electronic distribution channel (such as a central reservation systems or website) but may be much less flexible in physical channels like retail stores. Also, prices may have to be fixed if they are advertised heavily or have to be published in a physical catalog. If prices are rigid, firms may be able to adjust their quantity decisions—how much to sell, to whom, and through which channels. Again, this is the situation in the classical quantity-based RM industries like airlines and hotels. The ability to use quantity as a recourse decision is highly dependent on the type of product and the technology and costs of production and distribution. Airlines and hotels benefit from the fact that their "products" (different fare classes and rate categories) are all supplied by the same physical inventory; hence, they have a high degree of flexibility in reallocating capacity. In contrast, an automaker cannot use subcompact cars to satisfy demand for luxury vehicles and therefore has much less quantity flexibility. As another example, while retailers can in theory reallocate stock from one store to another as a recourse, often they find the handling and transportation costs prohibitive.

At each of these three stages of the game—structural design, precommitment and recourse—economic forces come into play. And the nature of the market often changes dramatically from one stage to the next; firms may face a highly competitive decision at the precommitment stage but enjoy near monopoly power once they reach the recourse stage—or vice versa. Although analyzing a complete model of RM— from product design to the final recourse decisions—is currently beyond the field's technical scope, as a start one can construct simple models that isolate one or two stages of the game and impose one or two relevant variations at a time. Taking this point of view, there is indeed a rich body of economics work that becomes relevant to RM.

In particular, economic analysis allows us to address the following resource-level questions: How would a monopoly set the multiple prices in quantity-based RM? How do they compare with single prices? Is there equilibrium in capacity, allocations, and prices for two competing firms practicing RM? Why do firms fix prices and manipulate allocations in quantity-based RM?

We can also address the following industry-structure and competition questions: Why does an industry have products with such similar restrictions for the sale of its products? Is there an equilibrium in the *types* of restrictions? Is RM the best sale mechanism for a monopolist? What about in an oligopoly? Why do we see "price wars" or "fare sales"?

Finally, economic theory provides insights into the following welfare and regulatory questions: Does RM provide the optimum number of products (variety) for customers (as would a welfare maximizer)? Can RM be sustained under perfect competition? Is dynamic pricing conducive to tactical collusion? Does RM increase overall welfare? Is it beneficial to the consumer (increasing total consumer surplus)?

This remainder of the chapter looks at these questions and is organized as follows. The next three sections (Sections 8.2– 8.4) look at economic models of RM and pricing by market conditions: perfect competition, monopoly, and oligopoly. The strategic variables in each case could be price or quantity—or both. For instance, capacity might be chosen first, followed by price competition. For each market condition, we start with the basic economic models and then specialize to models that incorporate additional features relevant to RM industries.

For the sake of the reader not familiar with standard economics models of competition, we begin with the basics, though our coverage here is brief. Also, some basic notions of game theory are assumed in the discussion that follows. Appendix F provides a primer on game theory, and the Notes and Sources section at the end of this chapter has additional references.

8.2 Perfect Competition

Perfect competition represents an extreme form of market competition in which the decisions of individual firms are severely constrained by market forces. It should be considered more as an abstraction of market conditions than a literal model of real world markets, though some centralized financial and commodity markets do come close to achieving perfect competition. Here we briefly review the assumptions of the competitive-market model and then examine some variations of the model relevant to RM.

8.2.1 Perfectly Competitive Markets

A market under perfect competition has two main characteristics. The first is that the goods produced by all firms are *commodities*— defined formally as goods for which customers exhibit no preference over the source of supply. In other words, customers simply do not care from whom they buy a commodity. A bushel of wheat, a gallon of gasoline, a megawatt of electricity or a share of IBM stock—all are goods typically considered to be commodities. Note that simply saying "goods are identical" is not sufficient to define a commodity. For example, you may prefer gas station A to gas station B simply because station A is located on your side of the road and station B is not. So factors like location, timing of availability (such as winter fruit from the Southern Hemisphere), terms of trade, and costs of switching suppliers may create preferences for suppliers even if the goods themselves are identical. Most often in real world markets, perfect competition arises only when goods are traded in a centralized (physical or virtual) location and the terms of trade are standardized. Were you to buy shares of IBM stock on the New York Stock Exchange, you would have no idea who previously owned these shares—and would most likely not care to know either. Such a situation defines a true commodity market.

If a firm is selling a commodity, it cannot control the price at which it sells. This follows because if a firm were to demand a price even marginally higher than the market price, then customers would simply buy from an alternative (and perfectly equivalent) source, and the firm's demand would drop to zero. Conversely, a firm has no incentive to sell at less than the market price either because customers are willing to pay it the same market price as other (again perfectly equivalent) sources. Thus, in a perfectly competitive market firms are simply *price takers*, able to sell at the market price but unable to directly control this price. This feature of competitive markets is also referred to as the *law of one price*.

The second key characteristic of a perfectly competitive market is that there are a large number of firms and each supplies only a negligible fraction of the total supply. The key assumption here is that the individual quantity decisions of a firm have no impact on the market price (each firm is only a drop in the bucket of total supply, as it were). For example, selling your 100 shares of IBM will have a negligible impact on the price of IBM because upward of 10 million shares might be traded in a single day. This is to be contrasted with the case of oligopolistic commodity markets (see the discussion of Cournot competition in Section 8.4.1.1), where there are several large firms that—while not directly controlling the market price—can influence it because their volume represents a significant fraction of the total market (for example, how much oil Saudi Arabia decides to sell on the world oil market influences the price of oil, even if Saudi companies cannot demand a price higher than the market price).

8.2.2 Firm-Level Decisions Under Perfect Competition

Under perfect competition, firms can sell as much as they want at the market price[1], and the only factors limiting a firm's output are their own capacity constraints or costs of production. More precisely, if p is the market price, x is the quantity supplied, and $c(x)$ is the firm's cost function (either direct or opportunity costs or both)—which we assume for simplicity is increasing and convex—then a firm in a competitive market chooses its quantity x by solving

$$\max_{x \geq 0}\{px - c(x)\},$$

which leads (assuming an interior solution) to the first-order condition

$$p = c'(x).$$

Hence, the market price is equal to the marginal cost of production for each firm supplying the market. Note that the firm's decisions are entirely supply-driven in this case—based only on their technology and costs of production rather than any demand-side considerations. Therefore, under the competitive model it would seem there is little scope for the sorts of demand decisions that lie at the heart of RM. Yet as we show below, adding some small modifications to the competitive model leads to interesting insights about RM practices.

If marginal costs are constant ($c(x) = cx$), the first-order profit condition reduces to $p = c$; in other words, the market price just covers the marginal cost of production, and hence firms earn zero profit. This fact leads to the so-called *zero-profit equilibrium*, in which a zero-profit condition for all firms is assumed as essentially the "definition" of a competitive market. The zero-profit equilibrium is a convenient assumption, obviating the need to perform a complex game-theoretic analysis of a large number of competing firms, and in places we assume it in this section. But this assumption in not completely innocuous and arguably oversimplifies the effects of competition.

8.2.3 Precommitment and Demand Uncertainty

One important modification of the perfect-competition model occurs when firms precommit to capacity and price in a market where aggregate demand is uncertain. We have already mentioned several examples of

---

[1] Here, the phrase "selling as much as they want" should be interpreted in light of the assumption that the maximum quantity any firm would choose to supply is still a negligible fraction of the total market supply.

RM industries where both precommitment and demand uncertainty are central features of the industry.

To see the effect that these factors have, consider a commodity market. Let k be the marginal cost of adding one more unit of capacity and c be the marginal cost of serving an additional customer. Further, suppose that a firm precommits to both the capacity it provides, x, and the price it charges, p. Finally, assume that aggregate market demand, D, is uncertain. However, the market is competitive, so each firm earns zero profit in equilibrium. We show that under these conditions, the law of one price breaks down, and different firms may offer different prices (price dispersion)—or a single firm may sell the same commodity at multiple prices, a situation reminiscent of traditional quantity-based RM.

To take a simple concrete case, suppose the total market demand D is d with probability $\frac{1}{2}$ (the low-demand scenario) and $2d$ with probability $\frac{1}{2}$ (the high-demand scenario). Then consider the following perfectly competitive equilibrium: the total market capacity is $2d$, with d units priced at a low price $p_2 = c+k$ and the remaining d units priced at a high price $p_1 = c + 2k$. Note that because customers (as in any commodity market) have no preference for the source of supply, they first buy from the firm with the lowest price (those pricing at $p_2 = c + k$). Since $P(D \geq d) = 1$, these low-price firms always sell out and earn a per-unit profit of

$$P(D \geq d)(p_2 - c) - k = P(D \geq d)k - k = 0.$$

On the other hand, if demand is high ($D = 2d$), a total of d customers will be forced to buy from the high-price firms at a price of $p_1 = c+2k$. (We assume for simplicity in this example that all customers are willing to buy at this price.) However, these suppliers also earn an expected per-unit profit of zero, since

$$P(D \geq 2d)(p_1 - c) - k = P(D \geq 2d)(2k) - k = 0.$$

In effect, the high-price firms "specialize" in supplying the market in the high-demand scenario. Their high prices compensate them for not selling as often, yet on average they earn the same zero profit as their low-price competitors. Note that this equilibrium is also beneficial to customers in the sense that it meets demand under all scenarios.

This same outcome can be interpreted in a way that even more closely resembles quantity-based RM; namely, each individual firm could price half of its output at p_1 and the other half at p_2. The firm would then sell its low-price allocation under both scenarios but would sell the high-price allocation only under the high-demand scenario.

This example of price dispersion in zero-profit equilibrium was first pointed out by Prescott [428] and is referred to as the *Prescott equilibrium*. To generalize the result, let $F(p)$ denote the fraction of customers with reservation values less than p (the reservation-price distribution), and let N denote the market size, so $N(1-F(p))$ is the aggregate-demand function. Suppose N is a discrete random variable, taking on m discrete levels $n_1 \geq n_2 \geq \cdots \geq n_m$, with $P(N = n_i) = f_i$, $i = 1, \ldots, m$. Then Prescott [428] shows that the unique competitive equilibrium involves price dispersion and is given by the set of m prices

$$p_i = c + \frac{k}{\sum_{j=1}^{i} f_j}, \quad i = 1, \ldots, m \quad (8.1)$$

and corresponding capacities

$$x_i = n_i(1 - F(p_i))\left[1 - \sum_{j=i+1}^{m} \frac{x_j}{n_i(1 - F(p_j))}\right], \quad i = 1, \ldots, m. \quad (8.2)$$

Again, this could be interpreted as each firm specializing and offering one of the m prices or as individual firms offering a menu of m price levels and limiting the quantity they sell at each price point. Dana [142] extends this equilibrium to the case where the market size has a continuous distribution, in which case the equilibrium then has a continuum of prices.

The main point is that by simply introducing two features into the classical competitive model—capacity and price precommitment and aggregate-demand uncertainty—the law of one price breaks down, and price dispersion (either among firms or within a single firm) is the unique competitive equilibrium. Moreover, as Dana [142] points out, this equilibrium provides one possible explanation for classical airline RM, as it clearly reflects the important characteristics of multiple prices and capacity controls on the quantity sold at each price. The important feature to emphasize here is that the prices and capacity controls arise because of precommitment and demand uncertainty alone; they are not used to achieve other objectives like price discrimination or revenue maximization per se.[2]

---

[2]Though a firm may still price to break even, which may require maximizing its profits to avoid a loss.

8.2.4 Peak-Load Pricing Under Perfect Competition

RM is strongly associated with service industries. Because services are not storable, the capacity that a firm provides in one period cannot be used to supply demand in future periods, and the manufacturer's classic strategy of building inventory to meet future demand is not an option. Consequently, any capacity level the firm chooses may turn out to be excessive during low-demand periods and inadequate during high-demand periods.

Peak-load pricing is one solution to this dilemma. The basic idea is to try to level out demand by pricing differently in peak and off-peak periods, thereby achieving more efficient capacity utilization. Peak-load pricing has been extensively studied by economists (see Crew et al. [136]), both in the context of regulated industries (electric utilities, postal service, telecom) and in degregulated contexts. It provides one economic explanation for classical quantity-based RM practices and also explains differential pricing in other contexts. We first look at the case where peak periods can be identified a priori and then discuss the case where the peak period is uncertain.

8.2.5 Identifiable Peak Periods

Here we consider a simple model of peak-load pricing, due to Bergstrom and MacKie-Mason [51], in which the peak period is known. There are two periods: period 1 is the peak period, and period 2 is the off-peak period (a Friday versus Saturday flight, weekday versus weekend hotel night, prime-time versus late-night TV slot, and so on). A firm must first select a capacity x for use in both periods, and the demand served in each period t, denoted d_t, cannot exceed this capacity. The unit capacity cost is k, and the unit variable cost for servicing customers is c. Bergstrom and MacKie-Mason [51] assume that customers have a utility for consumption $U(d_1, d_2)$ that is homothetic (a utility function that can be written as $U(d) = g(h(d))$, where $g(\cdot)$ is a monotonically increasing function and $h(d)$ is homogeneous of degree 1), twice differentiable and strictly concave.

Denote the marginal rate of substitution between peak and nonpeak consumption by

$$MRS\left(\frac{d_1}{d_2}\right) \equiv \frac{\partial U(d_1, d_2)/\partial d_1}{\partial U(d_1, d_2)/\partial d_2}.$$

Assume that the firm is in a competitive market and is constrained to operate at zero profit.[3]

Under the assumptions on the utility, given any prices p_1, p_2 on the peak and off-peak periods, respectively, demand will be determined by the same ratio of prices. That is, the demand function satisfies

$$MRS\left(\frac{d_1}{d_2}\right) = \frac{p_1}{p_2}.$$

Define the function $\mathcal{X}(\rho)$ implicitly by $MRS(\mathcal{X}(\rho)) = \rho$, so that $\mathcal{X}(\rho)$ is the ratio of demand for the peak period to demand for the non-peak period corresponding to a price ratio of $\rho = p_1/p_2$. Because period 1 is the peak period, we assume $\mathcal{X}(1) > 1$; that is, demand for period 1 is higher than demand for period 2 when both are priced the same. For any ρ such that $\mathcal{X}(\rho) \geq 1$, there exists a unique set of equilibrium prices p_1, p_2 and demands d_1, d_2 that makes the peak demand equal to capacity.

The following simple example from Bergstrom and MacKie-Mason [51] illustrates this equilibrium:

Example 8.1 (PEAK-LOAD PRICING WITH SUBSTITUTION) Suppose the marginal cost c is zero and customers' utility is linear of the form

$$U(d_1, d_2) = 2d_1 + d_2,$$

so customers value a unit from the peak period twice as much as a unit from the off-peak period. If the firm used only a single price, all customers would buy in the peak period, and therefore by the zero-profit constraint the firm would have to charge a price of $p = k$ to recover its capacity cost (the price would be set equal to the marginal cost of capacity).

Now consider the effect of allowing the firm to charge different prices in each period. In this case, $MRS\left(\frac{d_1}{d_2}\right) = 2$, so the ratio of prices is also $p_1/p_2 = 2$. At this ratio of prices, customers are indifferent between the peak and off-peak periods because although they value the peak twice as much, its price is twice as high. Hence, they split their consumption evenly over both periods. In this case, the firm sells all its output in both periods, and therefore by the zero profit condition it must have $p_1 + p_2 = k$ to recover its capacity cost. Combining this condition with the fact that $p_1/p_2 = 2$ implies that $p_1 = 2k/3$ and $p_2 = k/3$; both prices are strictly less than the marginal cost of capacity.

Notice in this example that by using peak-load pricing, the firm uses strictly lower prices in both the peak and nonpeak periods compared with uniform pricing. Thus, the differential pricing benefits customers.

---

[3] Bergstrom and MacKie-Mason [51] allow the slightly more general assumption that the firm is constrained to operate at a fixed return on capital, which could be due to either regulation or competition.

However, this is not always the case. Other utility functions lead to peak-period prices that are higher than those charged under uniform pricing (though the off-peak price is always lower).

As for the impact on capacity, peak-load pricing is commonly thought to reduce the amount of capacity a firm provides because customers are encouraged to shift their consumption from peak to off-peak periods. However, as in Example 8.1 above, peak-load pricing may lower prices in all periods, which can stimulate overall demand relative to uniform pricing. If the demand stimulation effect of these price reductions is strong enough, the capacity provided by the firm may actually increase to meet the increased demand. Bergstrom and MacKie-Mason [51] show that whether capacity increases or decreases under peak-load pricing depends in a simple way on the price elasticity of demand.

8.2.6 Uncertainty over the Timing of Peak Loads

The above analysis assumes the firm knows which periods are peak and which are off-peak. However, a firm may face uncertainty about the periods customers prefer. For example, a heat wave may cause a surge in the demand for electricity on random days or a surprise play-off victory by a football team may cause a jump in demand for flights to, and hotels in, the winning team's home town. With random peaks like this, it is not possible to set peak prices *ex ante* as in the previous section. So how should a firm in this situation respond? Here we look at a model from Dana [143] that addresses this question, as the results closely resemble classical quantity-based RM. For concreteness, we place the work in the context of airlines, but the model applies more generally.

Consider a market with two flight times, morning and afternoon, which are served using the same aircraft (same capacity); sales occur over two periods. The cost to supply a single unit of capacity, which can be used at both flight times, is $2k$ (or k per-unit-per-flight), and the variable cost to serve each customer is c (incurred only when the unit is sold). Firms precommit to both their prices and the capacity they provide at each price, as in the Prescott model.

There are N customers who want to fly on these flights, but one departure time (the peak time) is more popular than the other. N_1 will prefer the peak flight, and N_2 will prefer the off-peak flight, with $N_1 > N_2$ and $N_1 + N_2 = N$. The customers' preferred flight is (in the aggregate) uncertain, however; specifically, we assume that with probability $1/2$, N_1 customers prefer the morning flight and N_2 prefer the afternoon flight, while with probability $1/2$ the reverse is true: N_1 prefer the afternoon flight, and N_2 prefer the morning flight. The flight capacity is $N/2$.

Customers have identical utility v for travel but have heterogeneous disutility w (*waiting cost*) for traveling at their less preferred time. w is assumed to be an i.i.d. random variable with a continuous distribution function $F(w)$ defined on $[0, \bar{w}]$. The waiting costs are assumed uncorrelated with the preferred flight time. We call customers with waiting costs $w < k$ *leisure customers* (L) and those with waiting costs $w \geq k$ *business customers* (B). Let $N_L = NF(k)$ and $N_B = N(1 - F(k))$ denote, respectively, the number of leisure and business customers.[4] Customers arrive in random order to make their purchases (proportional rationing, Section 7.3.5).

Dana [143] shows that if $\bar{w} > k$, so some customers are always willing to pay the capacity cost to fly at their preferred time, there exists a unique competitive equilibrium for this model in which x_1 units are offered at each flight time at a high price $p_1 = c + 2k$ and x_2 units are offered at each flight time at a low price $p_2 = c + k$, where

$$x_1 = (N_1 - N_2)\left(\frac{1 - F(k)}{1 + F(k)}\right)$$

and

$$x_2 = N_2 + (N_1 - N_2)\left(\frac{F(k)}{1 + F(k)}\right).$$

Arriving customers buy their preferred flight as long as the low price is available. Once the low-price seats sell out, however, leisure customers will buy their less preferred flight at the low price because their waiting cost w is less than the price premium of k for the high-price units on their preferred flight. Business customers, however, would rather pay the higher price on their preferred flight because their waiting cost w is greater than the price differential k. In this way, customers with low waiting costs (leisure customers) are shifted to the off-peak flight, while customers with high waiting costs (business customers) are accommodated on the peak flight. Moreover, note that this shifting of demand occurs without firms' prior knowledge of which flight time is preferred by customers.

The model also has interesting implications as to which customers benefit from the differential pricing equilibrium in the case when $N_2 = 0$ (everyone prefers the peak flight). In this case, the expected price paid

---

[4]This sort of law-of-large-numbers assumption concerning individual uncertainty is typical of many economics models; for example, it is the reason that in Chapter 5 we express a deterministic demand function as $d(p) = N(1 - F(p))$, where N is the market size and $F(p)$ is the distribution of individual's reservation prices. It is reasonable if the population is large.

by business customers is given by

$$E[p] = \left(\frac{N_B + N_L}{N_B + 2N_L}\right)(c + 2k) + \left(\frac{N_L}{N_B + 2N_L}\right)(c + k),$$

which is decreasing in the number of leisure customers N_L and is strictly less than $c + 2k$, which would be the equilibrium price if business customers were to be served separately.[5] Thus, business travelers pay less than they would if they were to be served without leisure customers, and the more leisure customers, the lower the average price paid by business customers. Leisure customers pay a price of $c + k$ but are sometimes forced to fly at their less preferred time, and as a result one can show that leisure customers have an expected utility that is decreasing in the number of business customers N_B (the more business customers, the worse things are for leisure customers because they are forced to fly at less preferred times more often).

For these reasons, this model suggests that in quantity-based RM, it may be that it is leisure customers who subsidize business customers—and not the other way around—despite the fact that business customers are the ones paying more on average. The following remark of a Marriott Hotels manager reinforces this point of view:

> The fact of the matter is, if it weren't for incremental leisure guests, business guests would have to pay a higher price for their rooms in order for the hotel to meet financial obligations. I'd like to offer all our guests a $79 room, but in order to cover the costs of the hotel and ensure returns to our investors we must differentiate. The bottom line is this: either we accommodate both guests, one paying $79 and one paying $125, or we ask the business guest to pay $145.[6]

8.2.7 Advance Purchases in Competitive Markets

A common feature of quantity-based RM is the use of reservations and advance-purchase discounts. But why do such discounts exist, and what economic purpose do reservations serve? Here we look at an explanation based on the competitive model proposed by Dana [141]. The model can be viewed as an extension to Prescott's model of Section 8.2.3 that adds several essential features of advance-purchase markets. We provide a simplified example of Dana's model, described in the airline context.

---

[5]This follows since business travelers are willing to pay more than k, the capacity cost per customer if loads were balanced, to fly on their preferred flight; hence, the business-customers-only equilibrium has all business customers traveling on the peak flight at a cost per customer of $c + 2k$.
[6]This quote is from Richard Hank in *Yield Management: Strategies and Tactics*, Educational Institute of the American Hotel and Motel Association, 1990 (Dana [143]).

The key feature of the model is that there are two types of customers—those who are certain early on of their need to travel but who have low valuations (such as leisure travelers) and those with a high degree of uncertainty early on about their need to travel but with high valuations given the need to travel (such as business travelers). Let B denote the business customers and L denote the leisure customers, with willingness to pay (given the need to travel) of v_B and v_L, respectively, where $v_B > v_L$. Let N_B and N_L denote the number of type B and L customers, respectively.

Sales take place in two periods. Period 1 is an advance-purchase period where the product can be purchased for later consumption in period 2. In period 2 there is a spot market where the product can also be purchased for subsequent consumption. Customers are served in random order in each period (proportional rationing, Section 7.3.5).

The demand of low-valuation customers is certain; that is, all N_L customers are sure they want to travel in period 2. However, the demand of high-valuation customers is subject to two types of uncertainty. First, the aggregate number of business customers is uncertain. We assume there is a population of N_B total business travelers and m aggregate-demand states d_i, each occurring with probability f_i, $i = 1, \ldots, m$, where $\sum_{i=1}^{m} f_i = 1$. In demand state i, d_i out of the N_B business customers want to travel. We assume demand states are ordered so that $N_B \geq d_1 \geq d_2 \geq \cdots \geq d_m$.

Second, we assume each type B customer has individual uncertainty about the need to travel.[7] Let $\theta_i = d_i/N_B$ denote the conditional probability that a type B customer wants to travel in aggregate demand state i. The number of customers is assumed large, however, so exactly d_i customers will want to purchase in period 2 in aggregate-demand state i. The unconditional probability that a type B customer wants to travel, denoted π_B, is then given by

$$\pi_B = \sum_{i=1}^{m} f_i \theta_i.$$

On the supply side, the production costs are the same as in the Prescott model; there is a marginal cost of capacity k, which is incurred whether service is provided or not, and a marginal cost of service c, which is incurred only if service is provided. Further, firms must precommit to both their capacities and prices in the two periods.

---

[7] Dana [141]'s full model allows for individual uncertainty over the need to travel among both types of customers as well as aggregate uncertainty over the number of customers of each type. He also allows for parallel as well as proportional rationing.

8.2.7.1 Equilibrium Without Advance Purchase

We first consider the equilibrium of this model without an advance-purchase market, in which all customers are served in a single spot market. This situation is simply the Prescott equilibrium of Section 8.2.3. Specifically, there will be m prices offered in the spot market given by the Prescott prices (8.1). We illustrate this outcome with an example:

Example 8.2 (Dana [141]) Suppose there are $N_L = 100$ leisure customers with valuations $v_L = \$8$ and $N_B = 1,000$ business customers with valuations $v_B = \$20$. The aggregate number of business customers takes on three states, $d_1 = 300, d_2 = 100, d_3 = 0$, each with probability $1/3$ ($f_i = 1/3$ for all i). The capacity cost is $k = \$6$, and the variable cost of service is $c = 0$. The resulting equilibrium prices, capacities, and sales are shown in Table 8.1.

The prices in Table 8.1 follow from (8.1). In demand state 3, there are 100 total customers, no business customers ($d_3 = 0$), and $N_L = 100$ leisure customers, all of whom are willing to purchase at the \$6 price. In demand states 1 and 2, only business customers will purchase at the higher prices because they value travel at $v_B = \$20$, which is greater than p_1 and p_2, while leisure customers value travel only at $v_L = \$8$, which is less than both p_1 and p_2. However, by proportional rationing, a fraction $\frac{d_i}{d_i + N_L}$ of the low-price (\$6) units sell to business customers. Thus, in state i, the residual number of business customers remaining after the low-price units sell out is

$$d_i - \left(\frac{d_i}{d_i + N_L}\right) 100.$$

In demand state 2 ($d_2 = 100$), this residual demand is equal to 50; hence, the market provides an additional 50 units at the price $p_2 = \$9$. In demand state 1 ($d_i = 300$), the residual business demand is 225, of which 50 units are sold at $p_2 = \$9$; hence, the market provides an additional 175 units at the highest price $p_1 = \$18$ to satisfy demand in this last state. Note that for all units, however, the expected price is equal to the capacity cost of \$6.

Table 8.1. Prices and capacities for Example 8.2 without an advance-purchase market.

State (i)	d_i	Prices (p_i)	Capacities (x_i)	Total Sales
1	300	$p_1 = c + k/(1/3) = \$18$	175 @ \$18	325
2	100	$p_2 = c + k/(2/3) = \$9$	50 @ \$9	150
3	0	$p_3 = c + k/1 = \$6$	100 @ \$6	100

8.2.7.2 Equilibrium with Advance Purchase

Now consider the same model with an advance-purchase market. We assume that resale is prohibited, so customers buy for their own use and not for speculation. Further, we assume that a firm that sells an advance-purchase ticket must provide a unit of capacity and then service each such purchase. Let p_0 denote the price in the advance-purchase market.

Then the zero-profit condition implies that the advance-purchase price must be $p_0 = c + k$. The prices in the second period will then follow a Prescott-like equilibrium and will in general be no less than $c+k$ because of the aggregate-demand uncertainty.

Now consider the effect on leisure (type L) customers. Since they face an expected price greater than $c + k$ in period 2, the leisure customers (who are certain to want to travel) prefer purchasing in the advance-purchase market. If a business customers (type B) purchases in advance, their expected surplus is $\pi_B v_B - p_0$, which they compare with the expected surplus they get if they purchase in period 2. This surplus is

$$\sum_{i=1}^{m} f_i \theta_i \left(v_B - \sum_{k=i}^{m} \left(\frac{x_k}{d_i}\right) p_k \right),$$

provided the prices $p_i \leq v_B$ for all i.[8] (The term inside the parentheses is simply the type B value minus the expected price paid in demand state i; this is multiplied by the probability a type B will want to travel and the demand state is i.) If the business customers choose period 2, then the advance-purchase mechanism segments the market. The result is easiest to see by continuing Example 8.2:

Example 8.3 Consider the data given in Example 8.2, but suppose there exists an advance-purchase market. Table 8.2 shows the equilibrium prices and capacities that result.

The table is explained as follows. Note that 100 units are provided in the advance-purchase market; the leisure (type L) customers buy these 100 units. The equilibrium prices in period 2 are also shown and are identical to those in Example 8.2. The equilibrium sales follow assuming only business (type B) customers purchase in period 2. Since there are $\pi_B N_B$ of these customers who want to purchase, in state 1, 100 units are sold at p_2 and the remaining 200 type B customer purchase at the highest price $p_1 = \$18$; in state 2, 100 units are sold at a price $p_2 = \$9$; in state 3, no units are sold.

To verify that type B customer will indeed purchase in period 2, note they have an unconditional probability

$$\pi_B = (1/3)(0) + (1/3)(100/1,000) + (1/3)(300/1,000) = 2/15$$

of needing to travel. Hence, their expected surplus in the advance-purchase market is

$$\pi_B v_B - p_0 = (2/15)20 - 6 \approx -3.33,$$

while their expected surplus if they purchase in period 2 is

$$(1/3)(100/1,000)(20 - 9) + (1/3)(300/1,000)(20 - (100/300)9 - (200/300)18) \approx 0.87.$$

Since this later surplus is clearly greater (and positive), the type B customers prefer to purchase in period 2 as claimed.

Table 8.2. Prices and capacities for Example 8.3 with an advance-purchase market.

State (i)	d_i	Prices (p_i)	Capacities (x_i)	Total Sales
Adv. purch.		$p_0 = c + k = \$6$	100 @ \$6	100
1	300	$p_1 = c + k/(1/3) = \$18$	200 @ \$18	400
2	100	$p_2 = c + k/(2/3) = \$9$	100 @ \$9	200
3	0	$p_3 = c + k/1 = \$6$	0 @ \$6	100

This example is not unique. Indeed, Dana [141] shows the market is always segmented provided $\pi_B < 1$. This is true regardless of the values v_B and v_L and holds for all distributions of aggregate uncertainty; type B customers always buy in period 2, while type L customers always prefer buying in the advance-purchase market. In particular, it is not necessary to have valuations correlated with individual demand uncertainty as in our examples. This segmentation outcome is also the unique competitive equilibrium.

Finally, note that comparing Examples 8.2 and 8.3, the leisure (type L) customers significantly benefit from the advance-purchase market; they are never rationed out of the market (as they are in the spot-market case), and they pay lower prices.

8.3 Monopoly Pricing

A *monopoly* arises when a single firm (the *monopolist*) becomes the exclusive provider of a given product or service. It is the polar opposite of the competitive-market model because a monopoly firm dominates its market and can dictate the terms of trade, including both prices and quantities. A monopoly firm has considerable market power and— unless regulated—attempts to extract the maximum possible profit from its market.

Like the model of perfect competition, a perfect monopoly is essentially a theoretical abstraction, as often, viable substitutes exist that reduce the market power of a monopolist. To take an airline example, if the price of a short-haul flight is too high, potential customers may choose to drive or take the train instead. Such a carrier may find itself without much market power despite the fact that it is nominally a monopolist. On the other hand, the same carrier in a long-haul market would likely ignore the competition from automobile and train modes of transport. Therefore, the extent of monopoly power often depends on the availability and quality of substitutes. Alternatively, one can view

---

[8] If the equilibrium price p_i exceeds the customers' valuation v_B, then customers will not buy, and their surplus in state i is zero. Thus, the corresponding terms in the sum will be zero.

substitute products as making the monopolist's aggregate demand more price elastic.

Even if a monopolist has significant market power, the possibility of entry by competitors may reduce its pricing power. That is, the monopolist may have to price low enough to discourage potential competitors from entering its market. For instance, going back to the airline example, if a monopoly carrier made disproportionately high profits, its success would likely tempt other carriers to enter its market and provide competing service. Anticipating the competition and the lower profits that would result, the monopolist might conclude it is better off in the end to price less aggressively. Indeed, if the threat of entry is severe—the so-called case of *free entry*—the monopolist might even be forced to price *as if* it were in a competitive market. This phenomenon is called *contestability*—or *monopolistic competition*—and has had considerable influence on the economic debate surrounding market regulation.

In this section, we begin by examining pricing of a single product in a perfect monopoly. Next, we look into reasons for a monopolist to use multiple prices and practice price discrimination—a commonly cited explanation for RM practices.

8.3.1   Single-Price Monopoly

The monopolist is assumed to know the aggregate (deterministic) demand as a function of price, $d(p)$. Let $p(d)$ be the inverse-demand function and $c(x)$ the cost of producing x items. We assume that $c(\cdot), p(\cdot)$ and $d(\cdot)$ are twice differentiable functions, $c(\cdot)$ is an increasing function, and $d(\cdot)$ is a decreasing function. Then the monopolist's problem is to set a price $p \geq 0$ to maximize its profit, defined as

$$V(p) = pd(p) - c(d(p)).$$

Taking the derivative of $V(\cdot)$ with respect p and setting the result to zero, we obtain the following necessary condition for the optimal price p^*:

$$J(p^*) = \frac{\partial c(d(p^*))}{\partial d}, \qquad (8.3)$$

where recall from Chapter 7 that $J(p) = p + d(p)/d'(p)$ is the marginal revenue defined by (7.14). The firm will produce (and sell) the quantity at p^*, $x^* = d(p^*)$. The conditions (8.3) says that at the optimal price, marginal revenue is equal to marginal cost.[9] We call p^* the *optimal*

---

[9]Note that under perfect competition the price is equal to marginal cost, whereas for a monopoly the price will exceed the marginal cost, and hence the quantity produced will be lower. The resulting welfare loss is called the *deadweight loss of monopoly*.

The Economics of RM 351

unconstrained price for a monopolist selling a single product at a single price. The following example illustrates this calculation:

Example 8.4 (LINEAR-DEMAND FUNCTION) Consider the demand function $d(p) = a - p$, where a is a constant. Let c be a constant marginal cost ($c < a$). Then the profit function for the monopolist is given by

$$V(p) = p(a-p) - c(a-p).$$

The single price p^* that maximizes $V(\cdot)$ is $p^* = (a+c)/2 > c$ (differentiating $V(\cdot)$ with respect to p and setting the result to zero), and the optimal quantity is

$$d^* = (a-c)/2.$$

We can also rewrite (8.3) in terms of demand elasticity at p^*, $\epsilon(p^*) = \frac{p^*}{x^*}\frac{\partial d(p^*)}{\partial d}$, to obtain

$$\frac{p^* - c'(x^*)}{p^*} = \frac{1}{|\epsilon_p(p^*)|}.$$

The left-hand side can be interpreted as the profit margin (ratio of the price in excess of the marginal cost to price), while the right-hand side is the inverse of the magnitude of demand elasticity. Hence, the more elastic the demand, the lower the monopolist's profit margin.

8.3.2 Monopoly with Capacity Constraints

If the monopolist faces capacity constraints, it may affect its optimal price. In particular, this will occur if the aggregate demand at the optimal unconstrained price exceeds the capacity, in which case the monopolist will want to set a price higher than the optimal unconstrained price. The higher price will ensure that demand is within the capacity constraint.

This intuitive fact is formalized as follows: Let C denote the capacity of the monopolist. The profit optimization problem for the monopolist is

$$\max \quad pd(p) - c(d(p)) \quad (8.4)$$
$$\text{s.t.} \quad 0 \leq d(p) \leq C.$$

The Kuhn-Tucker necessary conditions for (8.4) imply that a nonnegative Lagrangian multiplier $\pi^* \geq 0$ exists such that p^* and π^* satisfy

$$J(p^*) - \frac{\partial c(d(p^*))}{\partial d} - \pi^* \frac{\partial d(p^*)}{\partial p} = 0 \quad (8.5)$$
$$\pi^*(d(p^*) - C) = 0.$$

If $d(p^*) < C$, then $\pi^* = 0$ and the optimal price will be the same as the optimal unconstrained price. If $d(p^*) = C$, the optimal capacity-constrained price will in general be higher than the unconstrained price. This is the same phenomenon we observed in Section 5.2.1.2, where the optimal constrained price is simply the maximum of the unconstrained optimal price and the stock-clearing price (the price at which demand is exactly equal to capacity).

8.3.3 Multiple-Price Monopoly and Price Discrimination

While the monopolist can earn a positive profit by maximizing its selling price, it can earn even more profit by selling its product at multiple prices. Many of these multiprice strategies have been offered as explanations of RM. Here we look at the variety of reasons a monopolist might want to use multiple prices, beginning with price discrimination.

8.3.3.1 Price Discrimination

Price discrimination is said to occur when two or more similar goods are sold at prices that are in different ratios to marginal costs (Stigler [489]). This definition, though somewhat technical, allows us to precisely compare prices of products that may differ in some respects; however, it also opens up ambiguities about when to consider two products similar goods. Two hotel rooms for the same day and for the same length of stay may appear to be clearly identical. Yet if the time of purchase is considered an attribute of the products, then a hotel room sold two days in advance and one sold two months in advance may not be identical. (See Phlips [420] for a discussion of many such examples.) The essential question, however, is whether differences in price can be explained by differences in costs alone or by other factors, like attempts to extract consumer surplus.

Classification of Price Discrimination Price discrimination can be classified into three types, depending on the degree of discriminating power (this classification first given by Pigou [421]):[10]

- *First-degree (or perfect) price discrimination* involves charging each customer the maximum amount he is willing to pay. To do so, the firm must have information on each customer's willingness to pay and be able to vary price by customer and by unit. First-degree price

---

[10]Though the definitions here are more modern and do not correspond exactly to Pigou's original taxonomy.

discrimination is essentially a theoretical abstraction, as it is nearly impossible to elicit each customer's individual reservation price, except in situations where the firm knows customers extremely well, and even then, it may require a protracted negotiating process.

- *Second-degree price discrimination* occurs when the firm discriminates by offering a menu of possible *purchase contracts* (terms of trade) and customers decide which contracts to purchase. Examples include quantity discounts, two-part tariffs (fixed plus variable fees), and bundle pricing. The key characteristic is that customers *self-select* the purchase contract that they like best.

  The advance-purchase and Saturday-stay restrictions of traditional airline RM can be considered a form of second-degree price discrimination. The restrictions attempt to segment the customer population, exploiting correlations between price sensitivity and product preferences. They are designed so that customers self-select the product (and pay the price) targeted for their segment. Hotels follow a similar practice; they may set a weekly rate targeted at vacation guests and a daily rate targeted at business guests (this is a form of nonlinear tariff). Notice that self-selection is bound to be rather porous, in the sense that it is difficult to make sure that customers from one segment will not buy contracts targeted to other segments.

- *Third-degree price discrimination* occurs when the firm divides the customers into different groups based on some identifiable characteristics (called a *sorting mechanism*) and then sets a separate price for each group. It is assumed that the firm can identify, at the time of purchase, the characteristics that identify the segment of the customer. All the members of a group pay the same amount, but they are prohibited (or are in someway discouraged) from purchasing at the price set for other groups. Prices that differ by geographic region, discounts for senior citizens, tour-group rates and conference rates offered by hotels and cruise-lines are some examples.

In some industries like broadcasting and media, a combination of first-degree and second-degree price discrimination occurs when a bundle is priced individually based on the customer relationship, customer budgets, as well as the quantity purchased. Indeed, in practice the various forms of price discrimination are more commonly used in combination than in isolation.

Conditions for Price Discrimination Several conditions are required to implement price discrimination:

- *Variance in customer preferences* If all customers are exactly identical in terms of price sensitivity and nonprice preferences, there is little scope for price discrimination. Moreover, in the case of second-degree discrimination it must be possible to find self-selection mechanisms that are correlated with differences in willingness to pay, so it is in a customer's interest to buy at the prices targeted for their group (called *incentive-compatible mechanisms*). In the case of third-degree price discrimination, the firm has to identify observable characteristics that are correlated with willingness to pay (the *sorting mechanism*).

- *No resale* To implement price discrimination, the product should also be either impossible to resell after purchase or contractually prohibited from resale, else arbitrage is possible. That is, someone (or a firm) could buy at the lower prices and resell the good to those customers who face higher prices, undercutting the firm's ability to price discriminate. Such nontransferability is implemented in various ways in RM situations. Airlines and hotels prohibit transferability contractually and add high penalties on change of itinerary or dates. In industries such as broadcasting or energy, private contracts prohibit transferability. Personal services such as health care are also not transferable. Rules against scalping prevent resale in ticket markets. And so on.

- *Monopoly power* Firms should have some degree of monopoly power to sustain a structure of price discrimination. As we mentioned before, under perfect competition in a commodity market, firms have no power to set prices, and the law of one price applies. Nevertheless, the monopoly power need not be absolute, in the sense that there is a single seller without any threat of entry. When products are differentiated in some way or sold in dispersed markets, it is possible to maintain price discrimination even with limited market power, and a competitive equilibrium with multiple prices can still exist. The degree to which such discrimination can be practiced decreases as the number of competitors increases or the product differentiation decreases.

8.3.3.2 Optimal First- and Third-Degree Price Discrimination

How does a monopolist price if it can price discriminate? We begin with the case of first- and third-degree price discrimination, which are the easiest to analyze.

The Economics of RM 355

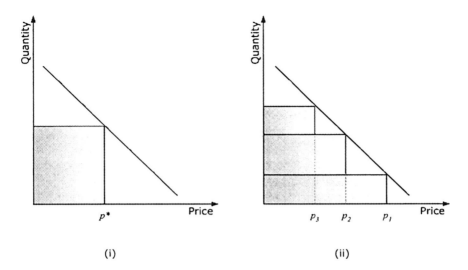

Figure 8.1. Revenue from selling a product at a (i) a single price and at (ii) three different prices to different customers.

Graphical Analysis Consider a simple linear-demand curve as shown in Figure 8.1. For simplicity, assume that marginal costs are zero. The shaded area in Figure 8.1(i) represents the maximum revenue obtained by selling at a single price, and p^* denotes the revenue-maximizing price.

Now consider what happens if the monopolist sells at multiple prices. To make the situation concrete, suppose three prices are offered, $p_1 \geq p_2 \geq p_3$. Further, suppose the monopolist can exercise *perfect price discrimination*, in the sense that those customers with reservation prices above p_1 buy at price p_1, those with reservation prices in the interval $[p_1, p_2]$ buy at p_2, and those with reservation prices in the interval $[p_2, p_3]$ buy at p_3.[11] Then the shaded area in Figure 8.1(ii) represents the revenue that is obtained.

As is evident by comparing the shaded areas in Figure 8.1, the revenue that can be obtained by discriminating and selling at three prices is significantly greater than the revenue obtained by selling at a uniform price. And as is intuitively clear from the picture, the more prices used, the greater the revenue generated by the monopolist.

Indeed, as the number of prices tends to infinity, we achieve first-degree price discrimination, since in the limit every customer will end

---

[11] A monopolist who can implement this "perfect price discrimination" is also called the *Pacman monopolist* by von der Fehr and Kühn [540] because the monopolist "eats its way down" the demand curve with this pricing.

up paying exactly his reservation price for the product. The monopolist then captures the entire consumer surplus (the entire area under the demand curve in Figure 8.1).

While this graphical argument is commonly used to explain price discrimination, it sidesteps the issue of how such a perfectly discriminating outcome might actually be achieved. One possibility is that the monopolist offers the product for sale in high-to-low price order (first p_1, then p_2, then p_3) and customers behave myopically, buying the first time the price drops below their reservation price.[12] Another possibility is that there is some observable attribute (such as height) that is perfectly correlated with a customer's reservation price and the monopolist price discriminates based on this observable attribute (for example, taller people pay more). As both examples suggest, however, such perfect price discrimination is a bit far-fetched in practice.

General Analysis A more general (and realistic) case of third-degree price discrimination is to assume that there exists some sorting mechanism that allows the monopolist to divide customer into n segments. For simplicity, assume a constant marginal cost of c. For example, in the airline case we might have $n = 2$, where the sorting is based on whether a customer is staying a Saturday night or not. Let $d_i(p_i)$ be the demand function from segment i when it is offered the price p_i. The aggregate demand function $d(p)$ is then

$$d(p) = \sum_{i=1}^{n} d_i(p).$$

That is, if the uniform price p is offered to all n groups, the total demand from all groups is exactly the aggregate demand.

The monopolist's optimal prices are then obtained by solving

$$\max_{\{p_1,\ldots,p_n\}} \sum_{i=1}^{n} (p_i - c) d_i(p_i) = \sum_{i=1}^{n} \max_{p_i} (p_i - c) d_i(p_i).$$

As the problem is separable by groups, the optimal prices are simply determined by applying (8.3) to each group, yielding

$$J_i(p_i^*) = c,$$

where $J_i(p_i) = p_i + d_i(p_i)/d_i'(p_i)$ is the marginal revenue of group i.

Assuming the monopolist can prevent customers from switching to products not intended for him, then *any* sorting mechanism will result

---

[12] See Chapter 5 for a discussion of this myopic-behavior assumption.

in (weakly) higher profits, since

$$\max_p (p-c)d(p) = \max_p \sum_{i=1}^n (p-c)d_i(p) \leq \sum_{i=1}^n \max_{p_i}(p_i - c)d_i(p_i).$$

However, if there is no correlation between the segments and willingness to pay, then the discrimination will simply produce n "copies" of the original demand function; $d_i(p) = \alpha_i d(p_i)$ for all i, where $\sum_{i=1}^n \alpha_i = 1$. In this case, the inequality above is tight and there is no benefit from discrimination. If there are correlations between price sensitivity and segment (that is, the demand elasticity varies from segment to segment), there will be a strict benefit to discrimination.

Of course, all this hinges on the ability of the monopolist to maintain the separation of segments. Indeed, if the separation breaks down, the monopolist may be worse off using price discrimination, as the following example illustrates:

Example 8.5 Demand is given by $d(p) = a - p$ as in Example 8.4. The marginal cost is $c < a$.

Optimistic monopolist The monopolist believes it can perfectly segment customers and set two prices for the product $p_1 > p_2$ and that all customers with valuations greater than or equal to p_1 will buy at p_1 and those with valuations between p_1 and p_2 will buy at p_2. Then its problem is

$$\max_{p_1 \geq p_2 \geq 0} V(p_1, p_2) = (p_1 - c)(a - p_1) + (p_2 - c)(a - p_2 - (a - p_1)).$$

One can verify that the optimal solution to this problem is to set $p_1^* = (2a+c)/3$ and $p_2^* = (a+2c)/3$.

Worst-case monopolist The worst that can happen is that the sorting mechanism cannot be maintained and *all* customers buy at the lowest price. In this case, the monopolist's profit from $p_1^* = (2a+c)/3$ and $p_2^* = (a+2c)/3$ is

$$V(p_1, p_2) = (p_2^* - c)(a - p_2^*).$$

In short, the profit function and outcome are the same as those of the single-price monopolist. However, since the price p_2^* is strictly less than the optimal single-price of $p^* = (a+c)/2$, this will result in lower profits. Hence, there is a danger that if the monopolist has optimistic beliefs about its ability to sort customers yet the sorting mechanism is easy to bypass, then price discrimination will generate lower profits than the single-price strategy.

The issue of creating incentives for customer to "stay in their designated segment" is the essence of second-degree price discrimination, which we look at next.

8.3.3.3 Optimal Second-Degree Price Discrimination

Now suppose the monopolist cannot discriminate based on any observable characteristics. In other words, customers reveal their "types" only through their purchase behavior, and they are not *ex ante* identifiable. In this case, the monopolist is forced to design a mechanism such that customers self-select the designated product.

To make things simple, suppose the monopolist identifies a purchase condition—a contract—for which customers have a utility or disutility (the Saturday stay) that divides customers into n groups of size $N_i, i = 1, \ldots, n$. Assume further that all customers in group i have the same reservation price for contract j, denoted v_{ij}. We assume $v_{ii} \geq v_{ij}$ for all j; that is, customers in group i prefer contract i. Assume for simplicity the marginal cost is $c = 0$.

If the monopolist sets a price of p_i for contract i, then a customer in group i receives a utility $v_{ii} - p_i$, from purchasing it. The customer will then self-select contract i if it provides positive utility,

$$v_{ii} - p_i \geq 0, \tag{8.6}$$

and it has the highest utility among all n contracts; that is, if

$$v_{ii} - p_i \geq v_{ij} - p_j, \quad j = 1, \ldots, n. \tag{8.7}$$

Equation (8.6) is called the *participation constraint* because it ensures customers have nonnegative surplus from participating (purchasing). Equation (8.7) is called the *incentive compatibility* constraint and ensures that customers in group i have an incentive to in fact select contract i. The monopolist then must solve the problem

$$\max_{\{p_1,\ldots,p_n\}} (p_i - c) N_i,$$

subject to satisfying (8.6) and (8.7) for all groups i.

Note that if the monopolist could identify a customer's group and charge them based on their identity (third-degree discrimination), then it could eliminate constraint (8.7). In this case, it would charge each group i a price just sufficient to satisfy the participation constraint (8.6)— namely, their maximum willingness to pay $p_i = v_{ii}$. However, because of the need to induce customers to self-select and satisfy the incentive compatibility constraint (8.7), the monopolist may be forced to charge some groups strictly less than their maximum willingness to pay v_{ii}. The monopolist will therefore in general make less profit under second-degree price discrimination because of the need to create self-selection incentives. The following example illustrates this idea:

Example 8.6 An airline offers an unrestricted full-price fare p_1 and a Saturday-stay discounted fare p_2. Business travelers want to return before the weekend and have a valuation $v_{B1} = 700$ for returning midweek and a value $v_{B2} = 200$ for staying over on Saturday. Leisure customers have no time preference and value both times at 100, so their values are $v_{L1} = v_{L2} = 100$. Suppose there is one business customer, two leisure customers, and no capacity constraints.

If the airline could identify the business and leisure passengers and charge them accordingly (third-degree discrimination), it would be optimal to charge the business customer \$700 for his desired flight and the two leisure customers \$100. However, if it cannot identify customers, then the airline must ensure that the business customer prefers the full-fare ticket. This will happen only if $v_{B1} - p_1 \geq v_{B2} - p_2$. This constrains the monopolist's price. In fact, an optimal set of prices to charge in this case is $p_1 = 600$ and $p_2 = 100$. This ensures that the business customer's surplus $v_{B1} - p_1 = 700 - 600 = 100$ from the full-fare is no less than his surplus from staying a Saturday night and buying the discounted fare, $v_{B2} - p_2 = 200 - 100 = 100$. Hence, we see the monopolist is forced to lower its full-price ticket by \$100 to induce self-selection.

A multiprice policy need not be motivated by price discrimination alone. A monopolist might use multiple prices because of uncertainty in aggregate demand or because it fetches revenue with more consistency, among other reasons, as we discuss next.

8.3.3.4 Capacity Constraints

Capacity constraints combined with aggregate uncertainty in demand can create an incentive to use multiple prices, as illustrated in the following example:

Example 8.7 (Wilson [568]) Suppose the capacity is 100 and there are two types of customers: there are 50 customers who are willing to pay \$12 and an additional 100 customers willing to pay only \$10. Customers arrive at random order (with a proportional rationing), and there is no way to distinguish the high-valuation customers from the low-valuation ones.

Charging a single price of \$12 fetches \$600, and charging a single price of \$10 fetches \$1,000. However, if the first 75 units are sold at \$10 and the remaining 25 at \$12, then on average, one-third of the first 75 units (or 25 units) are sold to the high-valuation customers (at less than the price they are willing to pay, \$10), and the remaining 50 units are sold to the low-valuation customers. The last 25 units will be sold only to the remaining high-valuation customers at \$12. So the expected revenue will be \$1,050.[13]

Example 8.7 works because total demand exceeds capacity, and therefore multiple prices serve to efficiently ration capacity to high-value customers. The next case shows multiple prices will lead to more revenue even when total demand is no more than capacity.

---

[13]The calculation, admittedly somewhat crude, goes like this. Of the first 75 seats, 25 are sold to the \$12 customers by proportional rationing. So 25 of the \$12 customers remain to buy the last 25. So the revenue is \$750 from the first 75 units and \$300 from the last 25 units.

8.3.3.5 Precommitment and Demand Uncertainty

Just as in the Prescott equilibrium in the competitive case (Section 8.2.3), the need to commit to prices combined with aggregate demand uncertainty create an incentive for a monopolist to use multiple prices. The following example illustrate this idea:

Example 8.8 (Dana [142]) A monopolist must commit to both its capacity C and its prices prior to the realization of demand. The capacity cost is $k = \$2$, and the variable cost to serve is zero. Demand is high with a probability of 0.5 and low with probability 0.5. In the low-demand state, there are 50 customers with a reservation value of \$10; in the high-demand state, there are 100 customers with a reservation value of \$12.

If the monopolist sets a single price of \$10, it is optimal to provide $C = 100$ units of capacity, and the monopolist's expected profit is $\$750 - \$200 = \$550$. If the monopolist sets a price of \$12, it is optimal to provide $C = 50$ units of capacity and the expected profit is $\$600 - \$100 = \$500$.

However, by using multiple price and capacity controls, the monopolist can generate more profit. Specifically, the monopolist could provide capacity of $C = 100$, sell the first 50 units at $p_2 = \$10$, and sell the remaining 50 at a price of $p_1 = \$12$. Therefore, if demand is low, its revenue will be \$500, and if demand is high, its revenue will be \$1,100, giving an expected profit of $\$800 - \$200 = \$600$; hence, setting multiple prices and limiting sales at the lower price generate more revenue.

Note that in this example the quantities supplied as the low and high price are exactly the Prescott quantities from Section 8.2.3; however, the prices are different. Namely, the Prescott prices (8.1) in this example are $p_2 = k = \$2$ and $p_1 = 2k = \$4$. The difference here is that the monopolist sets prices based on the maximum willingness to pay in each scenario rather than on the need to simply cover capacity costs, as is required in the zero-profit competitive equilibrium.

8.3.3.6 Risk Reduction

Even if there is no increase in expected revenue, a monopolist might still prefer a multiprice policy in the face of aggregate demand uncertainty because it lowers the *variance* of its revenue (for example, the monopolist may be risk-averse), as shown by the following example:

Example 8.9 Suppose that the firm has a resource with two units of capacity and there are three customers each demanding one unit. There is uncertainty in the valuations of the customers: the firm knows only that the customer will buy with probability 1 if it prices at \$62.5 and with probability 0.5 if it prices at \$100.

Then the best single-price policy can be verified to be \$100, with a total expected revenue of \$1,100/8 (the firm will sell zero units with a probability 1/8, exactly one unit with a probability 3/8 and two units with a probability 1/2). A multiprice policy of selling the first seat at \$62.5 but the second at \$100 fetches the same expected revenue of \$1,100/8 (the firm will sell no units with a probability zero, exactly one unit at \$62.5 with a probability 1/4 and 2 units, one at \$62.5 and the other at \$100

The Economics of RM

with a probability 3/4). However, the variance of the revenue obtained is reduced from 4,843.75 to 1,875, as seen from the calculations in Table 8.3 for the single-price policy and in Table 8.4 for the multiple-price policy.

Table 8.3. Revenue and variance calculations for Example 8.9 with a single-price policy.

Units Sold	Prob.	Rev.	Prob. × Rev.	Prob. × (Rev. - Avg)^2
0	0.125	0	0	2,363.2
1	0.375	100	37.5	527.3
2	0.5	200	100	1,953.1
Sum	1		137.5	4,843.75

Table 8.4. Revenue and variance calculations for Example 8.9 with a multiple-price policy.

Units Sold	Prob.	Rev.	Prob. × Rev.	Prob. × (Rev. - Avg)^2
0	0	0	0	0
1	0.25	62.5	15.625	1,406.25
2	0.75	162.5	121.875	468.75
Sum	1		137.5	1875

In this example, the firm can cut its uncertainty in the number of units sold by selling the first few at a low price more consistently and then taking the risk of having unused capacity on only the final few high-priced units.

8.3.3.7 Fairness

An oft-forgotten consequence of selling at a fixed price is the inherent unfairness of the distribution of goods (see Section 7.3.5 on models of rationing). Customers who happen to arrive first get served with a higher probability, regardless of their valuations. A monopolist might be motivated to enforce fairness for the sake of long-run customer satisfaction. Multiple prices may be seen as one way to rectify this unevenness in allocation.

Example 8.10 Consider again a set of customers with private valuations v for a single unit of a good. The firm has one unit for sale, and there are exactly two customers in the market who arrive sequentially, in period 1 and period 2 respectively. The firm knows only that the valuations v are normally distributed with a mean of $200 and standard deviation of $30.

The optimal single price for the firm can be verified to be equal to $179.56 as follows. Let $F(200, 30)$ be a normal distribution representing customer valuations with mean 200 and standard deviation 30. For the single price case, we need to solve $\max_p p(1 - F(p)) + pF(p)(1 - F(p))$, which yields $p^* = 179.56$. The optimal single price of \$179.56 fetches a revenue of \$168.53. The first customer then gets the good with probability 0.752, and the second with probability 0.186, even though their valuations follow the same distribution.

What if the monopolist wants both customers to have the same probability of obtaining the good? Let p_1 be the price charged in period 1 and p_2 for period 2. If we impose the restriction that customers have equal probabilities of getting the item, we need to solve $\max_{p_1,p_2} p_1(1 - F(p_1)) + p_2 F(p_1)(1 - F(p_2))$, subject to $1 - F(p_1) = F(p_1)(1 - F(p_2))$. This yields $p_1^* = 201.75, p_2^* = 159.53$ for a revenue of \$172 and both the first and the second period customer will take home the good with a probability of 0.47.

While the firm gives up some revenue as well as increases its risk of not selling the item, it decreasing the variance in customer surplus, and one can argue that this might have a long-term benefit to the firm. Conceivably other fairness criteria can also be used—equal *ex post* expected payments, equal expected surplus, are some possibilities.

8.3.3.8 Demand Learning

The price and demand in a given period provides information to the monopolist that it can potentially use to adjust the price in the subsequent period. This learning can lead to different prices in different periods as illustrated in the following two-period model:

Example 8.11 (Lazear [332]) A firm has one unit of a good remaining and two periods in which to sell it. All customers have a uniform valuation v, unknown to the firm. The firm has a prior distribution of v as uniform between 0 and 1.

If the monopolist prices at a price p_1 in the first period and the unit does not sell, it knows that $v < p_1$. The firm's posterior distribution of v is then uniformly distributed between 0 and p_1. It can then use this updated information to set its price p_2 in the second period.

Let $F_1(\cdot)$ be the prior c.d.f. in period 1 (and p.d.f. $f_1(\cdot)$) and $F_2(v)$ the posterior c.d.f. in period 2. Then the monopolist's problem is to solve

$$\max_{p_1,p_2} p_1(1 - F_1(p_1)) + p_2(1 - F_2(p_2))F_1(p_1). \tag{8.8}$$

The optimality condition for (8.8) is

$$p_2 = \frac{F_1(p_1) - F_1(p_2)}{f_1(p_2)},$$

which implies p_1 has to be greater than p_2. For the case of $F_1(\cdot)$ uniformly distributed, $p_2 = p_1/2$. The solution to (8.8) turns out to be $p_1 = 2/3$, $p_2 = 1/3$.

8.3.3.9 Changing Customer Valuations over Time

Finally, multiple prices could be used just because customer valuations change over time. A fashion good is not as fashionable after a while,

The Economics of RM

and customers might value it less. The following example illustrates this phenomenon:

Example 8.12 (Lazear [332]) Customers value a good at v during the first period, and at v/α during the second ($\alpha > 1$). The firm estimates the distribution of v as uniform between 0 and 1. Then its pricing problem is

$$\max_{p_1,p_2} p_1(1 - F_1(p_1)) + p_2(1 - F_2(\alpha p_2))F_1(p_1). \qquad (8.9)$$

Solving the optimality conditions result in optimal prices:

$$p_1 = \frac{2\alpha}{4\alpha - 1}$$
$$p_2 = \frac{p_1}{2\alpha}.$$

8.3.3.10 The Effects of Discounting

We next illustrate the effect of discounting as discussed in Lazear [332]. Consider the two-period learning model of Example 8.11. Let's say the firm now has a discount rate r (a discount *factor* of $\beta = 1/(1+r)$), so a sale at price p_2 in period 2 has a present value of $p_2/(1+r)$. Then its objective function changes from (8.8) to

$$\max_{p_1,p_2} p_1(1 - F_1(p_1)) + \frac{p_2}{1+r}(1 - F_2(p_2))F_1(p_1). \qquad (8.10)$$

The optimal solution (with a prior distribution of v uniformly distributed between 0 and 1) changes to

$$p_1 = \frac{2(1+r)}{4(1+r) - 1}$$
$$p_2 = p_1/2.$$

Note the difference between this and the solution to (8.9).

The argument is that given *any* price in period 1, the firm should set the monopoly price in period 2, based on the posterior distribution. This would give $p_2 = p_1/2$ for the uniform case. Now substitute this in (8.10), and differentiate to get p_1.

Notice that the policy in the second period does not change from Example 8.11. For large values of r, the firm prefers to sell in period 1 than in period 2, so it prices lower than in Example 8.11 (for example, $r = 1$, $p_1 = 4/7 < 2/3$) during the first period.

8.3.4 Strategic Customer Behavior

Customer are said to be *strategic* if they optimally adapt their behavior in response to changes in a firm's pricing policies or sales mechanism. We have already considered strategic behavior in Section 5.5.2

in the context of price skimming, and in Chapter 6 in the context of strategic bidding in auctions. Here we reexamine the effect of strategic behavior—in particular, strategic purchase-timing behavior—on the optimal pricing decisions of a monopolist.

8.3.4.1 The Coase Problem Without Discounting

Questions concerning a monopolists's ability to charge different prices over time were raised as far back as Coase [124]. The *Coase conjecture*[14], applied to a *durable good*[15] with an infinite life, goes like this. Consider a monopolist selling a durable good with a constant marginal cost of production. As in our graphical example of third-degree price discrimination in Section 8.3.3.2, a monopolist would ideally like to price its good high at first to sell to high-valuation customers. Once this high-value demand is exhausted (the durable-goods assumption), the monopolist then lowers the price to sell to the remaining customers with lower valuations, and so on. This results in a decreasing sequence of prices. However, rational customers will anticipate the lower future prices and refuse to buy until the price drops to the minimum price, so the monopolist can sell only at the lowest price. Hence, the firm is forced to sell at a uniform price.

Moreover, the monopolist may be forced to set its uniform price at marginal cost. This occurs, as in the price skimming model of Section 5.5.2, if the monopolist cannot credibly *commit* to its pricing strategy. Any fixed price higher than marginal cost is not subgame perfect because once the firm sells to the high-valuation customers, it is then optimal for it to subsequently lower the price (provided the initial price is above the marginal cost). Anticipating the firm's temptation to lower prices, rational customers will not purchase at the initial high price. Pricing at marginal cost, therefore, is the only strategy that is subgame-perfect. This gives the Coase conjecture: a durable-goods monopolist is forced to price at marginal cost, recovering the competitive outcome.

As McAfee and Wiseman [370] point out, intuitively this result occurs because the monopolist is, in a sense, forced to "compete with itself" in future periods; purchasing in the future is a perfect substi-

---

[14]The Coase Conjecture, verified under certain conditions in the models of Stokey [491], Bulow [93], and Gul, Sonnenschein, and Wilson [228] for high discount factors, is still referred to as a conjecture, and is distinct from the Coase theorem.

[15]By *durable good*, we mean here a good that does not need to be replaced in the foreseeable future, usually taken to be three years (automobiles, major household appliances, personal computers, and so on), so that once customers buy, they do not need to buy again in subsequent periods. This is in contrast to nondurable goods (groceries, gasoline, electricity), which need to be purchased frequently over time.

tute for purchasing now (assuming either arbitrarily fast price decreases or no discounting of utility over time, or both). A qualitative recommendation that results from the Coase conjecture is that it is better for the monopolist to lease at a uniform price rather than sell outright because then a change in its lease price affects both current and future customers equally. The firm can then credibly commit to offering the optimal monopoly price. (Wilson [569] offers this as an explanation for IBM's traditional emphasis on leasing mainframe computers.)

While airline seats, retail fashion goods, and hotel rooms may seem the antithesis of a durable good, the differences are not as great as they appear at first blush. For one, a customer who purchases a unit of a perishable resource (either a fashion item, an airline seat, or a hotel room) is unlikely to buy another unit during the relevant life cycle (sale period) of the product, so the product effectively has an *infinite life* in this sense. There may be a limited number of potential customers as well, so as for the durable good, customers are exhausted over time. If the product is perishable and capacity is limited, most customers are aware that the RM monopolist has considerable difficulty committing to its price.[16] Most important, the RM monopolist is asking the same questions as the durable-goods monopolist. A retailer may wonder if its markdown pricing policy will induce some high-valuation customers to delay their purchase until the markdowns occur. Similarly, an airline or hotel manager might wonder if business customers are buying advance-purchase discounts, resulting in a dilution of revenue. For all these reasons, the durable-goods monopoly is a relevant model for understanding RM practice.

8.3.4.2 The Coase Problem with Discounting

The above discussion ignores customers' discount factors. We next describe the Coase problem and results with discounting. Our model and notation follow von der Fehr and Kühn [540].

Assume that initially there is a group of customers who are willing to buy during the subsequent periods but that customers (the buyers, B) discount their valuations by a rate $r_B > 0$, and the firm (the seller, S) discounts its revenues at a rate of $r_S > 0$. Let $\beta_B = 1/(1 + r_B)$ and $\beta_S = 1/(1 + r_S)$ be the corresponding discount factors. The monopolist sells to a set of customers $\mathcal{N}$ who demand exactly one unit of the good, which has a marginal cost of zero to produce. $\mathcal{N}$ can either be finite

---

[16] A good industry example of this behavior is seen in the cruise-line industry, where last minute discounts are pervasive. As a result, most industry insiders feel customers are now "trained" to wait until the last minute to purchase.

($\mathcal{N} = \{1, 2, \ldots, N\}$) or a continuum ($\mathcal{N} = [0, 1]$). Customers' valuations are given by $v : \mathcal{N} \to (0, \bar{v}]$, and we assume that the customers are indexed by nondecreasing valuations. The utility for a customer with value v who purchases in period t at price p_t is given by $\beta_B^t(v - p_t)$. The firm can choose its prices from a set P, which can be either a discrete price grid or a continuum $(0, \infty)$. The number of periods T can be either finite or infinite.

The firm sets a price p_t in period t based on past history. Given a price p_t in period t and the history of all prices up until time t, customers decide whether to purchase or wait for the next period. The firm can potentially condition its strategy based on the purchase behavior of each single customer. Both customers and the firm are assumed to be risk-neutral utility maximizers. The following result then holds:

THEOREM 8.1 *([26]; proof in [540]) If $\mathcal{N}$ is finite and the price-set $P = \Re_+$, then for every $\beta_B < 1$, there exists a $\bar{\beta}_S \in (0, 1)$ such that for all $\beta_S \in (\bar{\beta}_S, 1)$, the unique subgame-perfect equilibrium in prices is perfectly discriminating.*

Note that this is solely an existence result. Essentially, Theorem 8.1 says that if the firm is sufficiently patient, it will (credibly) threaten to wait as long as it takes for the high-valuation customer to buy. Only then will it lower the price. Since there is a finite number of customers, each customer has a significant impact on the firm's profit, again making the threat credible. Customers faced with a firm like this are better off purchasing immediately as soon as the price drops below their valuations; that is, it is optimal for customer to behave *myopically*. Although the number of periods can be infinite, this is not strictly necessary, provided the firm can condition prices on which customers purchased so far, as the following example from Bagnoli et al. [26] shows:

Example 8.13 The market has two customers with reservation values v_1, v_2, and there are two periods of sale. Assume $v_1 > 2v_2 > 0$. Assume also that customers purchase at any price that gives them a nonnegative utility (as opposed to the utility maximizers of Theorem 8.1). In period 2, both customers are better off buying if the price p_2 is below their valuations. The firm knowing this sets $p_2 = v_1$ if customer 1 has not yet purchased the good, and $p_2 = v_2$ if customer 1 already purchased in period 1. (Note that this policy requires the firm to know who bought in the first period.) Given that the second stage is played this way, we can verify that the firm obtains the maximum payoff by setting $p_1 = v_1$ (by calculating the payoff for the two options v_1 and v_2). The firm will discriminate perfectly, and its discounted payoff is then $v_1 + \beta_S v_2$.

For the next result, assume that $\mathcal{N} = [0, 1]$, and let the marginal cost c be a real number but not an integer. (This is an odd—yet not that restrictive—assumption, but if c is integer, von der Fehr and Kühn [540] show that multiple equilibria can exist.) P is a set of uniformly spaced grid of prices, $P = c, c+1, c+2, \ldots, p_H$. We then have

THEOREM 8.2 *([228], [21]; proof in [540])* *There exists a $\bar{\beta}_B \in (0,1)$ such that for all $\beta_B \in (\bar{\beta}_B, 1)$ and all $\beta_S < 1$, there exists a unique subgame-perfect equilibrium with the Coase property; that is, p_1 is the competitive price $\lfloor c \rfloor + 1$, and all customers with valuations greater than $\lfloor c \rfloor + 1$ buy in the first period.*

For this result to hold two conditions are important. First, there is a continuum of customers, so no customer by himself is important enough to the firm, and therefore the firm's threat not to reduce prices until a measure of high-value customers buys ceases to be credible. Second, the firm can cut the price only in discrete amounts (here by 1, but more generally by an amount k as in [540]), so if the customers' discount factor is sufficiently close to 1, they can wait for a price cut. (If prices are continuous, no matter how patient the customers, the firm can cut in small enough increments to force the customers to purchase immediately.)

While the opposing conclusions of Theorems 8.1 and 8.2 might appear confusing at first, the important point is the dependence of the results on the relevant factors, such as the number of customers, the relative value of the discount factors, the firm's knowledge of customer valuations, and the information the firm gains after each sale about who bought and who didn't.

These relationships provide insights into RM. For one, they suggest a firm's ability to discriminate intertemporally depends on who is more patient, the firm or the customer. If the firm is patient, it can credibly sustain price discrimination; conversely, if consumers are patient, it cannot. This is key issue when selling a perishable product; if customers know that the good is perishable or has a fixed deadline for sale, they know that the discount factor for the firm is low. On the other hand, customers are aware that there is a fixed capacity, and there is a chance that demand will exceed capacity and they will be rationed out of the market. The longer customers wait, the smaller their chances of getting the good, so their discount factor is also low (unless of course they know that demand is certain to be below capacity). However, it is likely that only the firm knows the information about aggregate demand (say from the historical demand). This suggests it is strategically important for the firm not to reveal information on its demand state—such as remaining capacity, historical demand, or current rate of sale. For their part, customers have an incentive to conceal information about their valuations, even *ex post*, because otherwise the firm can identify who purchased and who did not and change its pricing based on the history of valuations. This latter point serves as something of theoretical justification for the often wary public reaction to the concept of *personalized pricing*. (See Johnson [269].)

Finally, we have also assumed that all customers *want* to purchase the items at time zero and discount their valuations in each subsequent period. For the monopolist, this leads at best to a price-skimming policy or at worst to a constant-price policy. A more general assumption would be that customers' desired purchase times are randomly distributed during the horizon and that they discount both having to buy earlier or later than they would like.

8.3.4.3 Quasi-Myopic Behavior

Somewhat in between the myopic- and strategic-customer model is a model where the firm sets its prices so that it is in the best interests of the strategic customer to *be* myopic (myopic behavior is incentive-compatible). This, of course, gives the firm more leeway and leads to a revenue that, while not as good as the perfect-discrimination price, is not as bad as the Coasian prediction. One such model is found in Harris and Raviv [239], which we describe as given in Wilson [569] (pp.242–244).

Example 8.14 (MYOPIC INCENTIVE CONSTRAINED PRICING) A monopolist faces N potential customers whose valuations are distributed over the discrete set of values $v_1 > v_2 > \cdots > v_m$ with a known, discrete distribution. Each customer demands one unit of the product. The firm has an inventory of C units and assume $N > C$. The firm sets a declining price schedule $p_1 \geq p_2 \geq \cdots \geq p_k$, with price p_j valid in period j. The probability of the inventory is still available in period j is given by r_j. It is assumed that the customer is able to calculate this number. A customer with a valuation v that is planning on purchasing in period j ($v \geq p_j$) will have an expected surplus of $(v - p_j)r_j$. Period j is targeted at the customer with valuation v_j. The firm sets the prices such that each customer is indifferent between purchasing at the intended time or delaying to a later period. That is, the prices satisfy the indifference constraint

$$r_{k+1} \equiv 0,$$
$$p_{k+1} \equiv \infty,$$
$$(v_j - p_j)r_j = (v_j - p_{j+1})r_{j+1}, \quad j = k, \ldots, 1.$$

This recursive constraint, in fact, determines the prices set by the monopolist:

$$p_k = v_k,$$
$$p_j = v_j + (p_{j+1} - v_j)\frac{r_{j+1}}{r_j}, \quad j = k-1, \ldots, 1.$$

The monopolist's decision is essentially to determine k, the lowest-valuation customer that it is willing to sell to. (Note that only the lowest-valuation customers that can buy are left with zero surplus.)

8.3.5 Optimal Mechanism Design for a Monopolist

Thus far we have assumed that the selling mechanism of the monopolist is given. But what if the monopolist has the freedom to chose among all possible mechanisms? Which is the optimal one to use?

In a sense, this question was answered in Section 6.2.5 on optimal auctions. There we saw that under the assumptions of the private-value auction model, a first- or second-price auction with an appropriately chosen reserve price is optimal among all possible selling mechanisms, including simple posted prices, price discrimination, and dynamic pricing. (See the discussion in Section 6.2.6.) Here we consider a variation of this same optimal-mechanism design result, due to Harris and Raviv [239], in which customers take on a discrete number of types. (This contrasts with the private-value auction model, in which customers take on a continuum of types.) We show that an alternative pricing mechanism is optimal in this case, though it is closely related to an auction mechanism and is essentially a variation of the optimal auction results.

There are N potential customers indexed by i, and the firm has a finite capacity C. The firm and all customers are risk-neutral. The marginal cost of serving a customer is zero. Customer i is willing to pay an amount v_i to acquire the product. Customers' valuations are private information, and the firm and customers other than i know only that the distribution of v_i is equally likely to be one of m values $w_1 > w_2 > \cdots > w_m > 0$. (So the situation is one of asymmetric information.) For simplicity, assume the values $w_1, \ldots, w_m$ are on an equally spaced grid so that

$$w_1 = \alpha$$
$$w_j = w_1 - (j-1)\delta, \quad j = 2, \ldots, m,$$

where $\alpha > 0$ and $\delta > 0$ are given constants. Customers' valuations are independent draws from this discrete distribution. Let $\mathbf{v} = [v_1, \ldots, v_N]$, with $\mathbf{v}_{-i}$ denoting as usual the vector without the ith component.

As in the optimal auction analysis of Section 6.2.3, the monopolist's problem can be reduced to a search for an optimal direct-revelation mechanism, in which the customers truthfully reveal their valuations and the monopolist's problem is to find only the optimal direct-revelation mechanism. Formally, a direct-revelation mechanism is defined by an allocation rule (a collection of functions $y_i(\mathbf{v})$ that describe the number of units allocated to each customer $i = 1, \ldots, N$ as a function of the values $\mathbf{v}$) and a payment rule (another collection of functions $r_i(\mathbf{v})$, $i = 1, \ldots, N$ that describe what each customer i is required to pay in return for his allocation). Customers are asked to report their values v, and then they receive their allocation and make payments according to the announced allocation and payment functions. One can show it is sufficient to consider only symmetric allocation and payment rules, in which case the allocation to customer i is only a function of his value v_i and the values $\mathbf{v}_{-i}$ of the other customers. That is, $y_i(\mathbf{v}) = y(v_i, \mathbf{v}_{-i})$ and $r_i(\mathbf{v}) = r(v_i, \mathbf{v}_{-i})$ for all i. So in what follows, we consider only symmetric mechanisms.

For the mechanism to be a direct-revelation mechanism, each customer i should have an incentive to report his true value v_i rather than some other value $\tilde{v}$. This is enforced by the following incentive compatibility constraint:

$$S(v_i) \geq S(\tilde{v}) + P(\tilde{v})(v_i - \tilde{v}) \quad \forall v, \tilde{v} \in \{w_1, \ldots, w_m\}, \quad (8.11)$$

where $S(\cdot)$ is a customer's expected surplus, given by

$$S(v_i) = v_i P(v_i) - R(v_i),$$

where $P(v_i) = E_{\mathbf{v}_{-i}}[r(v, \mathbf{v}_{-i})]$ denotes the probability of obtaining the good and $R(v_i) = E_{\mathbf{v}_{-i}}[r_i(v_i, \mathbf{v}_{-i})]$ denotes the expected payment of customer i.

Harris and Raviv [239] analyze this problem and show that, as in the optimal-auction problem, the above optimization problem can be reduced to a function of the allocation variables alone; namely,

$$\max \; E\left[\sum_{i=1}^{N} \hat{J}(v_i) y(v_i, \mathbf{v}_{-i})\right]$$

$$\text{s.t.} \; \sum_{i=1}^{N} y(v_i, \mathbf{v}_{-i}) \leq C$$

$$y(v_i, \mathbf{v}_{-i}) \geq 0, \quad \forall i,$$

where

$$\hat{J}(v) = w_j - (j-1)\delta \quad \text{if} \; v = w_j.$$

The function $\hat{J}(v)$ is analogous to the virtual value we saw in Chapter 6; note that it is strictly less than the customer's value v because, as in an auction, customers retain some "information rents" due to the asymmetry of information. Like the virtual value in the auction, $\hat{J}(v)$ also has the interpretation as the monopolist's marginal-revenue function. To see why, let d_j denote the expected demand at price w_j, and observe that

$$d_j = \frac{j}{m},$$

$d_{j-1} < d_j$ (increasing price decreases demand). The expected revenue at price w_j is then
$$R(d_j) = d_j w_j.$$
Therefore, the marginal revenue obtained by increasing demand from d_{j-1} to d_j is
$$\begin{aligned} \frac{R(d_j) - R(d_{j-1})}{d_j - d_{j-1}} &= jw_j - (j-1)w_{j-1} \\ &= w_j - (j-1)\delta \\ &= \hat{J}(w_j). \end{aligned}$$

And again as in the auction-design problem, the optimal allocation $y^*(v_i, \mathbf{v}_{-i})$ above is straightforward to implement: rank the customers by their virtual values $\hat{J}(v)$, and allocate the items to those customers with the highest virtual values $\hat{J}(v)$, stopping when either all C units are exhausted or the virtual value of the customers drops below zero.

Harris and Raviv show that the following *priority-pricing* scheme achieves this optimal allocation: the firm announces a schedule of prices $p_1 \geq p_2 \geq \cdots \geq p_h$, and each customer self-selects the price he is willing to pay. However, customers selecting higher prices have higher priority because the firm allocates items to customers in order from highest to lowest price. Formally, let h be a cut-off index defined such that

$$\begin{aligned} \hat{J}(w_j) &\geq 0 \quad \text{for } j \geq h \\ \hat{J}(w_j) &< 0 \quad \text{for } j < h. \end{aligned} \quad (8.12)$$

Then we have the following result:

THEOREM 8.3 ([239]) (i) *If $C < N$, the optimal scheme is a priority-pricing scheme with priority prices $p_1^*, \ldots, p_h^*$ (h is defined by (8.12)) given by*

$$\begin{aligned} p_h^* &= w_h \\ p_j^* &= w_j - \left(\frac{\delta}{z_j^*}\right) \sum_{i=j+1}^{h} z_i^*, \quad j = h-1, h-2 \ldots, 1, \end{aligned}$$

where z_j^* is the probability of purchase for a customer with valuation w_j under the optimal allocation:
$$z_j^* = P^*(w_j) = E_{\mathbf{v}_{-1}}[y^*(w_j, \mathbf{v}_{-1})], \quad j = 1, \ldots, m.$$

(ii) If $C \geq N$, the optimal marketing scheme is to set a single price equal to w_h, where h, again, is as given by (8.12). This is the smallest

reservation value such that the marginal expected value is greater than or equal to zero (the marginal cost).

One can interpret Theorem 8.3 in terms of our earlier optimal auction results of Chapter 6 as follows. The priority price p_j^* is equivalent to the optimal bid that a customer of type w_j would make in a first-price auction, and p_h^* is equivalent to the optimal reserve price. And as in a first-price auction, customers are sorted and awarded units according to their bid values. Here, because types are discrete, the firm can precompute the optimal bids and offer them as posted priority prices.[17] Customers will pick the priority price that corresponds to their optimal first-price bid. In this sense, the priority pricing mechanism is essentially equivalent to an optimal first-price auction mechanism.

8.3.6 Advance Purchases and Peak-Load Pricing Under Monopoly

As mentioned in Sections 8.2.7 and 8.2.4, advance-purchase markets and peak-load pricing are common features of many quantity-based RM industries. Here we look at the monopoly analysis of Gale and Holmes [196] of these two phenomena, who use mechanism design techniques to show that advance-purchase discounts achieve the revenue of the best possible sale mechanism (and is superior to peak-load pricing). We present their explanation in terms of an airline example, although similar reasoning can be given for other industries where identical (substitutable) inventory is sold for use in different periods.

The model is similar in spirit to the one discussed in Section 8.2.6. Consider a monopoly that has two flights for a future day of departure, one flight departing at a time of high demand and the other at a time of low demand. The capacity of each flight is C. The total number of customers N exceeds C but is less than $2C$. The time before departure is divided into two periods, an early period (call it period 1) and a late period nearer the time of departure (call it period 2). We assume that there is a continuum of customers and that the capacity and the population size are normalized so that the total population size is one ($N = 1$). All the customers are identical and have a reservation price of r.

The customer purchase model is essentially identical to that of Section 8.2.6; during period 1, customers are uncertain which one of the flight times they prefer. They will realize this only during period 2.

---

[17]Indeed, note the similarity between the formula for the optimal price p_j^* above and the optimal bidding strategy (6.4) of Section 6.2.2.2 for the first-price auction.

However, throughout they know their private waiting cost w (even during period 1) for taking their less preferred flight. The waiting cost w varies from individual to individual and (from the perspective of the firm) is an i.i.d. random draw from a distribution $F(w)$ with density $f(w)$ over $[0, \bar{w}]$. A customer then, is willing to pay $v - w$ for his less preferred flight. Of the total number of customers, a fraction $\alpha > 0.5$ will realize in period 2 that they prefer the first flight. So there is congestion on this flight, but there is enough capacity to satisfy all potential demand on the second flight—that is, $0.5 < C < \alpha$. The monopolist is aware that the first of the two flights is going to be the peak-demand flight.

The monopolist would like to design a mechanism that maximizes its profits. However, since the customers' cost information w is private, the mechanism should be such that it is in the customers' best interests to truthfully reveal their costs.

There are a number of different pricing options: using a single-price for both flights, a peak price for the first flight and a lower price for the second flight, or an advance-purchase discount (APD) in period 1. In fact, there are an infinite number of possible sales mechanisms. Although one could use the theory of mechanism design to find the best among all possible forms of pricing policies, here we focus on comparing peak-load pricing with advance-purchase discounts.

In peak-load pricing the peak flight is sold at a price of $v - \hat{w}_1$ and the off-peak flight at a lower price of $v - \hat{w}_2$. (A single-price policy will lead to equal or lower revenue than a peak-load pricing policy as the monopolist has the option of setting $\hat{w}_1 = \hat{w}_2$.)

Let w_C be the unique solution satisfying

$$\alpha[1 - F(w_C)] = C.$$

Then at a price of v for flight 1 and $v - w_C$ for flight 2, the monopolist is assured of filling flight 1. Indeed, it is not hard to show that $\hat{w}_1 = 0$ (a price of v for flight 1) is always optimal and that it is optimal to have $\hat{w}_2 < w_C$. So the optimal $\hat{w}_2$ maximizes the total revenue $V_{peak}(w)$, defined by

$$V_{peak}(w) = \underbrace{vC}_{\text{Flight 1}} + \underbrace{(v-w)\{(1-\alpha) + \alpha F(w)\}}_{\text{Flight 2}}, \quad 0 \leq w \leq w_C.$$

The first-order conditions imply

$$(v - w)f(w) - F(w) = \frac{1}{\alpha} - 1 > 0, \qquad (8.13)$$

and $\hat{w}_2$ is the solution to this equation.

If, instead, the monopolist prices both flights at a cost of v during period 1 and offers an advance-purchase discount of $\alpha\hat{w}$ during period 1 for the low-demand flight, any customer with a time cost below $\hat{w}$ will find it beneficial to buy the low-demand flight early; if he waits, he has to pay v, but if he buys in advance, he has to pay only $v - \alpha\hat{w}$ (with a probability α that it won't be his desired flight and he will incur an expected time cost of $\alpha\hat{w}$). So it is incentive compatible for the customer to follow the monopolist's plan.

The monopolist will set $\hat{w}$ to maximize its profit function under the advance-purchase discount scheme which is given by,

$$V_{APD}(w) = \underbrace{v(C + (1-\alpha)[1 - F(w)])}_{\text{Regular Sales}} + \underbrace{(v - \alpha w)F(w)}_{\text{Advance Purchases}}, \quad (8.14)$$

$$0 \leq w \leq w_C.$$

Differentiating $V_{APD}(\cdot)$, we get $\hat{w}$ as the solution to

$$(v - w)f(w) - F(w) = 0. \quad (8.15)$$

We assume $(v - w) - \frac{F(w)}{f(w)}$ is increasing in w so this solution is uniquely defined.[18] Comparing (8.15) with (8.13), $0 \leq \hat{w}_2 \leq \hat{w}$, as shown in Appendix 8.A, the revenue from giving advance-purchase discounts exceeds that from peak-load pricing—that is, $V_{APD}(\hat{w}) \geq V_{peak}(\hat{w}_2)$.

The intuition behind this result is that in a peak-load pricing scheme, the monopolist has to give a discount to *all* customers who fly on the low-demand flight—even those customers with high valuations who prefer the low-demand flight. In contrast, the advance-purchase mechanism—because it is a self-selection mechanism—gives discounts only to customers with low waiting costs.

Note also that the discounts are offered in period 1, before customers know their preferences. This uncertainty plays a role: the monopolist can offer a discount of $\alpha\hat{w}$, instead of $\hat{w}$ in the earlier period, because there is some chance customers will end up on their preferred flight. The lower discount that results makes this a profitable policy. This advantage of advance-purchase selling is reiterated by Shugan and Xie [467], who note that in the advance-purchase market a firm sells to customers who are uncertain about their time preferences, and this puts the firm in a better position during the consumption period when customers (privately) know their time preferences. In this sense, the information asym-

---

[18]This assumption is equivalent to the familiar monotone marginal-revenue assumption, Assumption 7.2, but expressed in terms of distribution of the waiting cost w rather than in terms of the distribution of the utility $v - w$.

metry between customers and the firm is lower in the advance-purchase market, which benefits the firm.

From a social-welfare point of view also, the advance-purchase discount mechanism turns out to be superior, as the overall sum of customer and firm surplus is maximized. Those with high costs for taking the less desirable flight do not take it, and those with lower costs take the low-demand flight. Here, advance-purchase restrictions serve as the optimal mechanism for allocating resources, while being compatible with customers' interests in revealing their private information.

8.4 Price and Capacity Competition in an Oligopoly

An oligopoly is a market in which a limited number of firms compete to supply the same (or similar) good. As in a monopoly, individual firms are sufficiently large so that their actions (prices and quantities) affect market demand. Yet, firms do not operate in isolation as in a monopoly, and the actions of one firm affect the demand of its competitors as well. This creates a strategic interaction among the firms' decisions, which is the distinguishing feature of oligopolistic competition. Oligopoly is arguably the most interesting market condition for studying RM because it is the prevailing competitive situation in many RM industries.

The two classic models of oligopoly are the Cournot model (competition in quantities [132]) and the Bertrand model (competition in prices [54]). The standard equilibrium concept for an oligopoly is the (pure-strategy) Nash equilibrium. (See Appendix F for basic definitions and concepts related to Nash equilibria.) The equilibria need not be unique.

The goal of oligopoly analysis is to derive Nash equilibria, analyze their properties, and study the resulting implications for individual firms and the market as a whole. When multiple Nash equilibria exist, the predictive power of the model is diminished because it is not clear which of the multiple equilibria will emerge in an industry.[19] Therefore, proving there is a unique equilibrium—if this is the case—is important. Sometimes, no pure-strategy equilibrium exists at all. In such cases, a *mixed-strategy equilibrium* may exist, in which firms randomize over their strategies, though such equilibria typically give less insight into market outcomes.

---

[19]Ways to predict the "most-likely" equilibria when multiple equilibria exist is an active topic of research in game theory. See Samulson [453].

We begin by looking at static (one-shot) oligopoly models where firms sell a homogeneous product. We then look at dynamic models with repeated interactions between firms. Finally, we consider models where firms sell differentiated products.

8.4.1 Static Models

Static models of oligopoly are one-stage models of competition in which firms make their decisions simultaneously. We look first at competition in quantities (Cournot competition) then at competition in prices (Bertrand competition).

Do firms practicing RM compete on price or quantity? It's difficult to say. On the one hand, as capacity is fixed in many RM contexts, this would suggest that price is the relevant strategic variable. So Bertrand competition might seem more appropriate. On the other hand, the main decision variables in quantity-based RM are capacity allocations, which are quantity variables. This would seem to imply that Cournot models are more relevant for quantity-based RM. Yet, neither the Bertrand nor the Cournot model is completely adequate for describing quantity-based RM. In Section 8.4.3, we model dynamic RM interactions in price and allocations in a duopoly (two-firm) market, which comes closer to fully describing RM competition. Still, for broad questions on long-term market behavior, strategic-capacity investment, and price competition, Cournot and Bertrand models are relevant to RM.

8.4.1.1 Static Cournot Model

Cournot's paper [132] is one of the earliest models of oligopolistic competition. He studied a market of n firms that choose their production quantity simultaneously. The market then determines the price that clears total output.

Let $\mathbf{x} = (x_1, \ldots, x_n)$ denote the output of the n firms, and $X = x_1 + \cdots + x_n$ denote the aggregate supply. The price at which the market clears the supply X is given by an inverse-demand function $p(X)$. Note that the exact mechanism by which this supply is sold is unspecified in the model, though one can imagine a fictional auctioneer who sells the total supply by searching for a price that exactly clears the market. (Indeed, the absence of a realistic price-setting mechanism is a persistent criticism of the Cournot model.) Nevertheless, the model captures many important aspects of strategic interaction and as a result has become a mainstay of the industrial organization literature.

For simplicity we assume a constant marginal cost c for each firm. Then firm i maximizes its profit

$$\max_{x_i} V_i(\mathbf{x}) = p(X)x_i - cx_i. \tag{8.16}$$

This leads to the first-order condition (called the *Cournot pricing formula*):

$$p(X) + x_i p'(X) = c, \ i = 1, \ldots, n.$$

The Cournot equilibrium is given by the simultaneous solution of these n first-order conditions, and it exists under quite general conditions. For instance, a sufficient condition is that the profit function $V(\cdot)$ be quasiconcave in x_i and the strategy spaces compact. The *markup* of firm i is given by $p(X) - c = -x_i p'(X) > 0$, if $x_i > 0$. The size of the markup indicates the degree of market power that firms have.

Let $\alpha_i \equiv x_i/X$ denote the market share of firm i and $\epsilon(X) = p(X)/Xp'(X)$ denote the market-demand elasticity (with respect to quantity) at X. Rewriting the Cournot pricing formula in terms of these quantities,

$$\frac{p(X) - c}{p(X)} = \frac{\alpha_i}{|\epsilon(X)|}.$$

Thus, a firm's markup is directly proportional to its market share and is lower if the demand elasticity is higher. All firms that produce make positive profits. However, if all the firms have the same marginal costs, as the number of firms increases, Cournot profits approach zero; equivalently, the equilibrium approaches that of perfect competition. We illustrate this outcome for the case of linear demand and constant marginal costs in the following example:

Example 8.15 (COURNOT WITH LINEAR-DEMAND FUNCTIONS) Consider a market with n firms with a constant marginal cost c of production and linear inverse-demand function $p(d) = a - d$, $a > c$. So firm i's payoff function is

$$V_i(\mathbf{x}) = x_i(a - \sum_{j=1}^{n} x_j) - cx_i. \tag{8.17}$$

Differentiating (8.17) with respect to x_i and setting it to 0 yields

$$a - \sum_{j=1}^{n} x_j - c - x_i = 0. \tag{8.18}$$

The symmetric equilibrium $x^* = x_1^* = \cdots = x_n^*$ is given by

$$x^* = \frac{a-c}{n+1}, \tag{8.19}$$

and the market-clearing price at the Cournot equilibrium by

$$p^* = a - nx^* = c + \frac{a-c}{n+1}. \tag{8.20}$$

So the equilibrium price is strictly above marginal cost, but as $n \to \infty$, $p^* \to c$, the competitive price.

For general (nonquasiconcave) payoff functions, an equilibrium need not exist.

8.4.1.2 Cournot with Uncertain Demand and Exogenous Price

We next look at a variation of the Cournot model with demand uncertainty that more closely matches quantity-based RM situations. It is the oligopoly analogue of the well-known *newsvendor problem* in operation management, and one can think of it as a stylized model of n firms competing in capacity allocations under demand uncertainty.

Consider first the following monopoly newsvendor model. A firm faces a stochastic demand D for its good and must produce before observing the value of D. D is a continuous random variable with known distribution $F(\cdot)$. There is a variable cost c for producing units, and each unit sells at a fixed price p. The newsvendor problem is a good model for situations where prices are exogenous, and capacity decisions have to be made *ex ante* and cannot be changed after observing high or low demand states.

The (monopolist) newsvendor problem is to order a quantity x to maximize the firm's expected profits (assuming salvage value of 0)

$$\max_x \quad V(x) = pE[\min(D, x)] - cx. \tag{8.21}$$

This is a well studied problem (see Winston [571], pp.907–909) and serves as the basic template for the overbooking problems of Chapter 4. From the first-order conditions for (8.21), one can show that the optimal output quantity x^* satisfies

$$F(x^*) = 1 - c/p.$$

This result is analogous to Littlewood's rule (2.2) from Chapter 2.

The oligopoly version of the newsvendor problem can be considered as Cournot competition with uncertain demand and an exogenously determined price. The exogenous price can be thought of being set competitively at a aggregate level, while capacity allocations take place at a lower, operational level (for example, airlines matching fares in broad O-D markets but tactically competing in terms of allocating capacity on individual flight departures).

The Economics of RM

Each firm chooses a quantity x_i before demand is realized. Customers choose among the available firms, and if a firm is out of stock, customers may switch and purchase from a competing firm. (Customers are willing to substitute among firms.) Hence, the demand that firm i sees depends on the quantity decisions of its competitors, $\mathbf{x}_{-i}$. Specifically, firm i's payoff function is given by

$$V(x_i, \mathbf{x}_{-i}) = pE[\min(R_i(x_i, \mathbf{x}_{-i}), x_i)] - cx_i, \quad (8.22)$$

where $R_i(x_i, \mathbf{x}_{-i})$ is the *effective demand* function (a random variable) for i that is a function of D and the quantities $\mathbf{x}$, supplied by all firms. Because price is not guaranteed to adjust to the supply, as in the Cournot model, there is a possibility of the market supplying too much or too little capacity for a given demand realization.

The effective demand may be defined by various reallocation rules or through discrete-choice models of customer behavior. (See Lippman and McCardle [346] and Mahajan and van Ryzin [355].) The only assumptions required, however, are (1) that the effective demand for firm i is stochastically decreasing in $\mathbf{x}_{-i}$; that is, the more capacity its competitors provide, the lower the demand seen by firm i; and (2) the distribution of the effective demand is continuous on a bounded interval $[0, u_i]$ for all $\mathbf{x}_{-i}$.

The payoff function (8.22) for each firm i is concave in the firm's own quantity x_i. Hence, by a standard result in game theory for games with quasiconcave payoff functions (see Appendix F), it follows

THEOREM 8.4 *([346]) There exists a pure-strategy equilibrium in inventory levels $(x_1^*, x_2^*, \ldots, x_n^*)$ in the n-firm competitive newsvendor game.*

A quantity of interest is the total capacity provided by the firms and how it compares with that of the monopoly case. Consider for illustration purposes the following duopoly setting:

Example 8.16 Consider a duopoly newsvendor competition, and let D_i denote the "native demand" for firm i. That is, these are the customers whose first preference is for firm i. However, customers will buy from the competing firm if their preferred firm has no remaining capacity. The effective demand for firms 1 and 2 is then given by

$$R_1(\mathbf{x}) = D_1 + (D_2 - x_2)^+ \quad \text{and} \quad R_2(\mathbf{x}) = D_2 + (D_1 - x_1)^+,$$

where $\mathbf{x} = (x_1, x_2)$ are the capacities of the two firms. The aggregate demand is $D = D_1 + D_2$, and it is assumed to have a continuous, strictly increasing c.d.f. $F(\cdot)$. Each firm has a variable cost c for its capacity.

Recall that for the monopoly case, the optimal quantity x^* is determined by $F(x^*) = 1 - c/r$. The equilibrium duopoly quantities, $\mathbf{x}^* = (x_1^*, x_2^*)$, in contrast,

satisfy
$$P(R_i(\mathbf{x}^*) \leq x_i^*) = 1 - c/r, \quad i = 1, 2. \tag{8.23}$$

Now suppose $x_1^* + x_2^* < x^*$; that is, the total equilibrium capacity is less than the monopoly capacity. Then

$$\begin{aligned} c/r &= P(D > x^*) \\ &< P(D > x_1^* + x_2^*) \\ &= P(R_1(\mathbf{x}^*) > x_1^* \cap R_2(\mathbf{x}^*) > x_2^*) \\ &\leq P(R_1(\mathbf{x}^*) > x_1^*), \end{aligned}$$

a contradiction. So we must have $x_1^* + x_2^* \geq x^*$; the total duopoly capacity is therefore at least as large as the monopoly capacity.

This example suggests that if firms compete in allocations, they will tend to use allocations that are higher than the monopoly allocations. Lippman and McCardle [346] show that this result holds in general for certain reallocation rules, and Mahajan and van Ryzin [355] show that it holds when customers choose sequentially according to a general discrete-choice model. However, Netessine and Rudi [405] provide examples where competitive quantities can be lower than the monopoly quantities, at least for one of the firms.

An interesting case in point is when "competing goods" are in fact really substitute goods offered by the same firm. For example, they could represent different departure times offered by the same airline. The model here suggests that if firms are not aware of customer substitution behavior, they may in effect be "competing with themselves" in setting allocations for their substitute goods. This can lead to distortions in the capacity allocations and lower profits compared with the case where allocations are coordinated.

8.4.1.3 RM Duopoly Games

Consider the case of a duopoly in which both firms have a fixed capacity but sell in two classes, denoted H (high) and L (low). Each firm offers identical prices $p_L < p_H$. Customers substitute among airlines within the same class, and the strategic decisions are how much of the capacity to allocate to class L. This problem has been studied by Netessine and Shumsky [404], and we review their model here.

Demand for each firm and class is modeled as random variables D_{ki} for $k = L, H$, and $i = 1, 2$, with demand reallocation occurring as follows. If class H is closed for firm 1, all residual H demand for 1 goes to class H of firm 2 and vice versa. Similarly, if class L is closed for firm 1, all residual L demand for firm 1 goes to class L of firm 2 and vice versa. Note there is no buy-up; a L customer is assumed to never purchase a

The Economics of RM

H fare, and a H customer never purchases a L fare. We also assume that all L demand appears before H demand. This corresponds to a duopoly version of the independent-class assumption of Chapter 2 and Appendix E.

We next argue that a pure-strategy Nash equilibrium should exist for this game, using a bipartite graph framework and Littlewood's rule (Section 2.2.1). Let the strategy space for firm 1 be the high-fare protection level, y_1, over the set of integers between $[0, C]$, and for firm 2, let its strategy space be $C - y_2$ defined over $[0, C]$, where y_2 is firm 2's high-fare protection level.

By Littlewood's rule, the optimal protection level for H is independent of the demand from L customers. As L demand does not buy up, this two-class RM game can be defined in terms of the protection levels for H demand only.

Form a bipartite graph with C nodes on one side representing firm 1's strategy and C nodes on the other side, with node j representing firm 2's strategy of protecting $C - j$ units (Figure 8.2). Call this the *equilibrium graph*. Arc (j, i) represents firm 1's best response (order i units) if firm 2 protects $C - j$ units. The following proposition is proved in Appendix 8.A:

PROPOSITION 8.1 *Let $l_1 < k_1$ and $l_2 > k_2$. Then the equilibrium graph for the duopoly RM game cannot have the two best-response arcs for firm 1 of the form (k_2, k_1) and (l_2, l_1). Similarly for firm 2.*

If the graph has no crossing arcs, as in Proposition 8.1, then there has to exist an equilibrium pair of arcs, as seen by simply following a sequence of best-response arcs: we either have an equilibrium pair of arcs, or we have to double back and create a crossing pair of arcs (Talluri [503]). Thus, the game has an equilibrium.

Littlewood's rule assumes that the demands for the two classes are not correlated. However what if demand is correlated across the two firms (but still without buy-up from L to H)? That is, the L demand for firm 1 is correlated with the L demand for firm 2, and similarly, the H demand. Notice that in this case, each firm will still set its best-response using Littlewood's rule: For a fixed protection level of the other firm, the L and H demand each firm sees is uncorrelated. So with this type of correlation in demand, the protection level set for B is still independent of the demand forecast of L, and by the above argument there is still a guaranteed equilibrium.

Netessine and Shumsky [404] analyze this game for the more complicated case where there is a correlation between the L demand and H demand, and also for the case of "vertical competition" with two

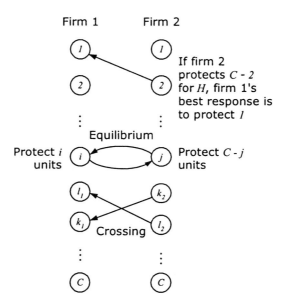

Figure 8.2. The equilibrium bipartite graph.

airlines serving different markets and operating connecting itineraries. For the case of correlation between L and H demand, they show that an equilibrium need not always exist. The best response functions for their example without an equilibrium (under their continuous model) are shown in Figure 8.3.

The correlations between the L and H demands can cause a firm's objective function (say firm 1) to be multimodal (for a fixed allocation of the other airline). As the other firm increases its booking limit for the lower class, the optimal solution for firm 1 can jump from one of the peaks of the multimodal function to another peak, causing the jump seen in Figure 8.3. In addition, one of the main results of [404] are conditions on the correlation matrix for existence of pure-strategy equilibrium.

8.4.1.4 Static Bertrand Models

The second fundamental model of oligopoly competition is the Bertrand model of price competition. The main assumptions in this model are that firms produce an identical commodity and that, as in the case of perfect competition, all customers buy only from firms offering the lowest price. Firms compete on price, and each firm produces a quantity sufficient to satisfy *all* the demand it faces at its offered price. Firms choose their prices noncooperatively and simultaneously.

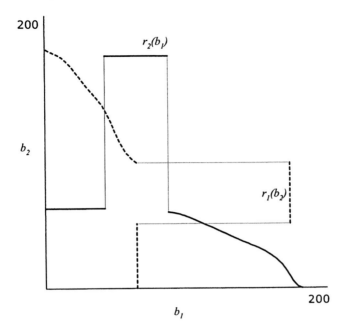

Figure 8.3. Best responses as a function of booking limits (both firms have a capacity of 200). The best responses increase very slightly initially and then again after the jump ([405]). The correlations in the demands are as follows: $\text{Corr}(D_{L_1}, D_{H_1}) = \text{Corr}(D_{L_2}, D_{H_2}) = -0.9$ and $\text{Corr}(D_{L_1}, D_{L_1}) = \text{Corr}(D_{H_2}, D_{H_2}) = -0$.

We present the model for a duopoly, but the analysis extends easily to the n firm case.

Assume there are two firms, denoted 1 and 2. Also, for simplicity assume each firm has the same marginal cost c. As earlier, let the market-demand function be denoted by $d(p)$. The demand for firm 1 at price p_1 is given by the demand function $d_1(\cdot)$:

$$d_1(p_1, p_2) = \begin{cases} d(p_1) & \text{if } p_1 < p_2 \\ d(p_1)/2 & \text{if } p_1 = p_2 \\ 0 & \text{if } p_1 > p_2. \end{cases} \quad (8.24)$$

(Similarly for firm 2.) Note the inherent discontinuity in the demand function. If firm 2 prices are even a tiny amount below 1's price, firm 1 will lose all its demand. Firm 1's profit function is then given by

$$V_1 = (p_1 - c)d_1(p_1, p_2).$$

Let the Nash equilibrium in prices be denoted (p_1^*, p_2^*). The equilibrium in the Bertrand model is in fact very clear-cut: the unique equilibrium is for both firms to price at the marginal cost c, identical to the perfectly competitive prediction. The explanation is as follows. From firm 1's perspective, if firm 2 prices at p_2, firm 1 has to price at least at p_2 or else it receives no demand at all. If $p_2 > c$, it is in firm 1's interest to undercut firm 2 slightly and price at $p_1 = p_2 - \delta$ for some small $\delta > 0$. It then receives all the demand and makes a positive profit. But the symmetric argument holds for firm 2, which in turn has an incentive to slightly undercut firm 1. This downward sequence of prices continues until both firms' prices reach marginal cost, which is the only equilibrium.

Thus, under Bertrand's model, firms make zero profits, even when faced with just one competitor. This result suggests that if the only differences between firms are their prices, the competitive outcome will be quite bleak indeed. While stylized, the intuition behind the Bertrand model is compelling and helps explain why firms often strive so hard to differentiate themselves and avoid head-to-head price competition as much as possible.

The situation is different, however, if one firm's marginal cost is lower than the others. For example, if $c_1 < c_2$ (firm 1 is the "low-cost leader"), then by the same reasoning as above the unique equilibrium is $p_1 = p_2 = c_2$. In this case, the low-cost firm makes a positive profit since $p_1 > c_1$, but the high-cost firm makes zero profit. It makes sense then in this situation for the low-cost firm to emphasize price as the main point of differentiation, since it benefits from pure price competition. A real world example of this situation is the low-cost airlines, which—because of their lower cost structures—benefit greatly by emphasizing price competition. Walmart is another well-known example of a firm with a clear low-cost, low-price strategy of this sort.

The extreme prediction of the Bertrand model, however, demands that we examine the main assumptions of the model more closely:

- Products are homogeneous and identical.

- There are no capacity constraints on a firm's production.

- Firms have constant marginal costs of production.

- There is one and only one opportunity for strategic interaction.

- The demand model is deterministic.

Relaxing any one of these assumptions leads to more realistic predictions, albeit at the expense of more complicated analysis. In particular, for RM industries products are usually differentiated, capacity restrictions are important, demand is uncertain, and firms interact frequently over time. Although we don't know of static models of oligopoly competition that incorporate all these features, there are a number of economic models in the literature that relax one or the other and derive alternative equilibria predictions. Repeated interactions are the subject of Section 8.4.2, and Section 8.4.3 covers competition between firms with differentiated products. In the next section, we look at the effects of capacity constraints on the Bertrand model.

8.4.1.5 Bertrand-Edgeworth

Consider Bertrand's model in which firms have capacity constraints. Demand in excess of the capacity constraints is reallocated to other firms according to a rationing rule. In this case, multiple prices can exist in the market. For instance, firm 1 can price higher than firm 2 and still survive by receiving all the residual demand that exceeds firm 2's capacity (assuming that the demand is high enough to cover its costs). In general, when there are decreasing returns to scale (the marginal cost increases as capacity increases), the model is called the Bertrand-Edgeworth model of price competition. Edgeworth showed that an equilibrium need not exist in such a case. (However, mixed-strategy equilibria exist as shown below.) The market will never settle down, and prices will cycle between high and low values—a pricing pattern known as the *Edgeworth cycle*. This is illustrated in the following example:

Example 8.17 An extreme instance of increasing marginal cost occurs when there are rigid capacity constraints (marginal costs above the capacity constraint are infinite). In the presence of capacity constraints, we have to specify a rationing rule. We assume the randomized rationing rule (Section 7.3.5), where customers arrive at random and purchase the lowest-priced product that provides them with a positive surplus.

Suppose there are two firms, each with a capacity of 1 and a marginal cost of $1. The demand consists of three customers, two of whom have a willingness to pay of $1 and the third with a willingness to pay of $3. Note if one firm prices at $1, the other firm has an incentive to price at $3 since it will then capture the $3 customer with probability 1/2, generating a positive expected profit of $3(1/2) - $1 = $0.5. However, if one firm prices at any price $p > \$1$, then the other firm can increase its expected profit by pricing at $p - \epsilon$ to take away the $3 customer from its competitor. This means that each firm has an incentive to undercut its competitor if the competitor prices above marginal cost, and yet each firm also has an incentive to raise its price if its competitor prices at marginal cost. Firms therefore cycle continuously between high and low prices (the Edgeworth cycle).

This example shows that capacity constraints provide an explanation for unstable pricing. For example, the volatility in airline pricing—and the on-and-off price wars observed in that industry—are arguably examples of this type of Bertrand-Edgeworth phenomenon.

Although there may be no pure-strategy equilibrium for the Bertrand model with capacity restrictions, one can show that there exists a mixed-strategy equilibrium. Assume a linear inverse-demand curve $p(d) = a - d$, $a > c$. Each of the n firms faces a strict capacity limit of C and sets prices $p_1, \ldots, p_n$ simultaneously and independently.

THEOREM 8.5 *([341], [314], [88]) Assume the efficient-rationing rule (customers with the highest valuations are served first (7.3.5)). Then*
(i) If $C \geq (a-c)/(n-1)$, an equilibrium is given by $p_i = c$, $i = 1, \ldots, n$, and firms make zero profits.
(ii) If $C \leq (a-c)/(n+1)$, the equilibrium is given by $p_i = a - nC$, $i = 1, \ldots, n$.
(iii) If $(a-c)/(n+1) < C < (a-c)/(n-1)$, a mixed-strategy equilibrium exists and in any such equilibrium, expected profit is given by $(a - (n-1)C - c)^2/4$.

There may be many other factors besides capacity constraints that lead to the nonexistence of pure-strategy equilibria in a Bertrand style competition or lead to equilibria with positive profits. Indeed, consider a duopoly where there is a fixed cost F of entry, in addition to the marginal cost of production, so that $c(x) = F + cx$ for $x > 0$ and $c(x) = 0$ otherwise. Then according to the assumptions of Bertrand competition, each firm if it entered would price at the marginal cost c, but then both firms would end up making negative profits, so no firm supplies the market (assuming both firms make simultaneous entry decisions). There are a number of such variations of the basic Bertrand model. (See Vives [539] and Tirole [513].)

8.4.1.6 Cournot *Then* Bertrand-Edgeworth

Here we consider a two-stage oligopoly model in which firms choose capacities in the first stage, and then in the second stage—knowing each others' capacities—they choose prices. This is a reasonable abstraction, for example, of competing retailers ordering quantities at the beginning of the season and then pricing the products to clear their stocks; or airlines making short-term schedule and capacity decisions in a market and then pricing their products afterward.

Consider a duopoly with a linear-demand function ($p(d) = a - d$) and identical marginal costs of capacity c, as given in Example 8.15 and Theorem 8.5.[20] We assume $a > c$. Once capacity is chosen in stage 1,

---

[20] We limit our discussion to a linear-demand model for simplicity. The results of Kreps and Scheinkman [314] apply to more general, differentiable, and concave demand functions.

there are no further costs; that is, the marginal cost to serve demand is zero. Kreps and Scheinkman [314] show that in this two-stage game under efficient rationing (Section 7.3.5), there is an equilibrium in which firms choose exactly the Cournot capacities in the first stage. Thus, the severe zero-profit outcome of Bertrand price competition is considerably tempered when firms must precommit capacity. (However, Davidson and Deneckere [147] show that this result is sensitive to the form of rationing, specifically, it holds only for efficient rationing.)

To show this outcome, we use Theorem 8.5, extended to a duopoly with asymmetric capacities C_1 and C_2. To do so, let $R_i(x_j), i \neq j$ denote the optimal quantity response of firm i to firm j's capacity choice x_j, assuming there is no cost to capacity. That is,

$$R_i(x_j) = \arg\max_{x \geq 0} \{x(a - x - x_j)\}. \tag{8.25}$$

Kreps and Scheinkman [314] show the following crucial result

THEOREM 8.6 ([314]) (i) A pure-strategy equilibrium exists only if $C_1 \leq R_1(C_2)$ and $C_2 \leq R_2(C_1)$, and the prices at this equilibrium are given by $p_1 = p_2 = (a - C_1 - C_2)$.
(ii) If $C_1 > R_1(C_2)$ or $C_2 > R_2(C_1)$, there exists a mixed-strategy equilibrium, and the highest-capacity firm (that is firm 1) makes a profit in the second stage equal to $R_1(C_2)(a - R_1(C_2) - C_2)$.

We can use this result to show that the Cournot symmetric equilibrium, denoted x^*, is an equilibrium of the two-stage game. For our linear-demand function, $x^* = (a - c)/3$ (equation (8.19)), and $R_i(x^*) = (2a + c)/6$. So, $R_1(x^*) \geq x^*$, $R_2(x^*) \geq x^*$.

To show that $C_1 = x^*$, $C_2 = x^*$ is an equilibrium, assume that firm 2 chooses $C_2 = x^*$. Then if firm 1 chooses $C_1 \leq R_1(x^*)$,

$$R_2(C_1) = (a - C_1)/2 \geq (a - x^*)/2 = (2a + c)/6 \geq x^*.$$

So by Theorem 8.6(i), it has a pure-strategy equilibrium price $(a - C_1 - x^*)$ and its revenue, $C_1(a - C_1 - x^* - c) \leq x^*(a - 2x^* - c)$. So $C_1 = x^*$.

If $C_1 > R_1(x^*)$, by Theorem 8.6(ii) (as $C_1 > R_1(x^*) \geq x^* = C_2$, firm 1 is the higher-capacity firm), the revenue in the second stage and cost in the first stage sum to

$$R_1(x^*)(a - R_1(x^*) - x^*) - cR_1(x^*).$$

But it follows from (8.19) that

$$R_1(x^*)(a - R_1(x^*) - x^*) - cR_1(x^*)$$
$$= \frac{2a + c}{6}(a - \frac{2a + c}{6} - \frac{a - c}{3}) - c\frac{2a + c}{6}$$

$$\leq \frac{a-c}{3}(a - \frac{a-c}{3} - c)$$
$$= x^*(a - 2x^* - c).$$

So $C_1 = x^*$, $C_2 = x^*$ is indeed an equilibrium in capacity choices as claimed. Kreps and Scheinkman [314], in fact, show that this Cournot equilibrium in the first stage is unique.

8.4.2 Dynamic Models

While static models of competition provide important insights on RM competition, in real life, competitors often interact with each other repeatedly over time. For example, retailers order and price their products every sales season and repeatedly compete with each other in each season; airlines periodically adjust schedules fleet assignments, and prices over time, competing with each other in each period. Such repeated competitive interaction can lead to very different equilibrium outcomes.

One important consequence of repeated interaction is the possibility of collusion. Of course, overt collusion in setting prices or quantities is nearly always prohibited by law (although there are some significant exceptions, such as the OPEC oil cartel). Yet under repeated oligopolistic competition, another form of collusion, called *tacit collusion*, is a possible equilibrium outcome. Even though firms do not communicate with each other or form explicit agreements to collude, it can be in each firm's interest in equilibrium to set prices that approach collusive levels. In this section, we look at models and conditions to sustain or break such tacit collusion.

Chamberlin [103] was one of the first economists to point out that repeated interaction between oligopolists can facilitate collusion. With repeated interactions firms can threaten retaliation (price wars) and thus ensure that competitors do not deviate from collusive prices. Stigler [488] modeled factors, such as the credibility of threats, the speed with which competitive moves are detected, and the speed of information transmission, and studied how they affect collusion. While a number of static models (kinked-demand theory and conjectural variations theory) have been developed to address collusion, modern industrial organization theory relies on repeated games to model competitive interactions in an oligopoly.

Models of finitely repeated pricing games, somewhat surprisingly, turn out to say little more than what their static counterparts do. The subgame-perfect equilibrium (see Appendix F) of a T-period Bertrand game turns out to be the Bertrand equilibrium (that is, marginal-cost pricing) repeated T times. The argument for this is relatively simple and follows by backward induction: In the last period, the static equilibrium is the unique marginal-cost equilibrium. Given this outcome for the last period, the equilibrium for the next-to-the-last period is again the marginal-cost equilibrium, and so on. This is disappointing, because with a repeated game one would like to uncover new equilibrium possibilities.

In infinite games the situation changes; threats, retaliation, and punishment become credible as firms can punish a deviating firm with a price war for a sufficient number of periods and then revert back to collusive pricing. The basic requirement for this to be implementable is that each oligopolist be able to observe defections from collusive pricing quickly. When such detection lags are short and price changes are costless and frequent, a collusive outcome is a possible—albeit nonunique—equilibrium. On the other hand when detection is difficult, the collusiveness breaks down, and firms revert to marginal-cost pricing even in supergames.

Thus secret price cuts and private deals undermine collusive pricing. Most RM industries have a mixture of public and private prices. Airlines post their fares on reservation systems (public tariffs) and at the same time make numerous private arrangements with corporations or individual travel agents (negotiated tariffs). A similar situation holds for hotels, cruise lines, and rental-car companies. It is not unusual to see identical pricing by firms for products posted publicly but large variations for prices negotiated privately. Retail prices by and large are public information, but at the same time, it is also difficult to gather price information.[21]

8.4.2.1 Dynamic Finite Bertrand-Edgeworth

As we mentioned, the finite-stage version of the Bertrand pricing game leads to an equilibrium where firms follow the static Bertrand equilibrium (Section 8.4.1.4) in each stage. This need not be the case if firms have capacity constraints, however. We consider here a finite-stage duopoly version of the Bertrand-Edgeworth model of Section 8.4.1.5, due to Dudey [164].

Consider two firms with capacities C_1 and C_2 that sell a homogeneous good to N customers (N is assumed to be known to both firms) at prices p_1 and p_2, respectively. Each customer has a reservation price v and chooses to buy from the low-price firm, and if both prices are the same, will choose one of the two firms with equal probability. If $p_1 > v$

---

[21] Notable exceptions occur in the U.S. pharmaceutical industry, which maintains a database of prices, and e-commerce sites—however, even the latter make it difficult for robots to gather prices automatically.

and $p_2 > v$, the customer does not purchase from either firm. Firms can change the price dynamically and quote a price for each customer separately. If a firm sells out its capacity, its price is assumed to be set at $v+1$. Assume the marginal cost of a sale is c, where $c < v$. Assume the discount factor is one for both firms.

Let period t correspond to customer t, $t = 1, \ldots, N$. At any given period t, there are $d(t) = N - t$ remaining customers. Let $x_1(t)$ and $x_2(t)$ represent remaining capacities for the firms at the beginning of period t (so $x_1(1) = C_1$ and $x_2(1) = C_2$). Let $p_1(t)$ and $p_2(t)$ represent prices quoted by the firms for the customer in period t. Firms are assumed to know t, N, v and the history of prices prior to t, but not necessarily $x_1(t)$ and $x_2(t)$. However, knowing the price history, t, N, and v, firms can accurately infer $x_1(t)$ and $x_2(t)$.

These informational assumptions of the model are rather restrictive, as in practice firms usually do not know either v or N (and hence cannot infer t or the remaining capacities of the other firm). However, this model is a reasonable stylized representation of RM dynamics under competition.

The equilibrium analysis breaks up into different cases:

Case 1: ($C_1 < N \leq C_2$) Firm 1 cannot supply the entire market, but firm 2 can. In this case, let firm 1's strategy be

$$p_1(t) = \begin{cases} v & \text{if } 0 < x_1(t) < d(t) \leq x_2(t) \\ c & \text{if } d(t) \leq \min(x_1(t), x_2(t)) \\ v+1 & \text{if } x_1(t) = 0 \end{cases} \quad (8.26)$$

and firm 2's strategy

$$p_2(t) = \begin{cases} v+1 & \text{if } 0 < x_1(t) < d(t) \leq x_2(t) \\ c & \text{if } d(t) \leq \min(x_1(t), x_2(t)) \\ v & \text{if } x_1(t) = 0. \end{cases} \quad (8.27)$$

These strategies are shown in [164] to be a *unique* equilibrium to the finite-stage game. Note that if $x_1(t) = d(t)$ at any time t, the result is a dynamic Bertrand game from t onward, and both firms make zero profits. So it turns out that firm 2 is better off letting the smaller-capacity rival sell out and then move to monopoly price. To see why, consider a point when, say, firm 1 has a capacity of $C_1 = 100$ units, there are 101 remaining customers and firm 2 has a capacity of $C_2 > 101$. Firms have no reason to price at values other than v, $v+1$, or c. Given firm 1's strategy, if firm 2 prices at v until $x_1(t) = d(t)$ (until at least one customer buys from firm 2 first), its expected profit[22] is

$$\frac{v-c}{2} + \frac{v-c}{4} + \frac{v-c}{8} + \cdots + \frac{v-c}{2^{100}} < v - c. \qquad (8.28)$$

However, if it follows the strategy (8.27), waiting for firm 1 to sell out, its profit is $(v-c)$. Firm 1 earns $(v-c)C_1$, and firm 2 earns $(v-c)(N-C_1)$. Depending on N and C_1, firm 1, the one with the smaller capacity, can make higher profits.

Case 2: $(C_1 + C_2 \leq N)$ Both firms set their prices equal to v unless their capacity is 0, in which case they set it to $v + 1$. Firm 1 makes a profit of $(v-c)C_1$, and firm 2 a profit of $(v-c)C_2$.

In general, if at any period t, $x_1(t) + x_2(t) \leq d(t)$ and $\min(x_1(t), x_2(t)) \geq 0$, both firms set their prices equal to v.

Case 3: $(C_1 \geq N; C_2 \geq N)$ Both firms set their prices equal to c unless their capacity is 0, in which case they set it to $v + 1$. Both firms make 0 profits.

In general if at any period t, $\min(x_1(t), x_2(t)) \geq d(t)$, both firms set their prices equal to c.

Case 4: $(N/2 < C_1 = C_2 < N)$ Both firms set their prices equal to $c + (v-c)(N - 2C_1 + 1)$, or in general, if at any stage $d(t)/2 < x_1(t) = x_2(t) < d(t)$, both firms set their period t prices equal to $c + (v-c)(d(t) - 2x_1(t) + 1)$.

This is the strangest case because firms can price below cost. Say $C_1 = C_2 = 100$ and $N = 101$. If customer 101 purchases from one of the firms, then the subsequent equilibrium yields $(v-c)99$ for the other firm and $(v-c)$ for the firm that made the sale to customer 101. If instead, both firms set their prices equal to $c - 98(v-c)$, then both firms earn $(v-c)$. This is an equilibrium because, given the other firm sets $c - 98(v-c)$, raising the price above $c - 98(v-c)$ will fetch $(v-c)$, and lowering it below $c - 98(v-c)$ will fetch less than $(v-c)$.

So even though the static Bertrand-Edgeworth need not have a pure-strategy equilibrium, the finite-stage version can have unique equilibrium strategies in a duopoly. While the existence of the equilibrium makes this model appealing, the conclusions are rather odd indeed (especially, Case 4). There is also a disproportionate sharing of profits, with the firm with the smaller capacity always coming out best. One is therefore led to suspect the assumptions of the model.

---

[22] Calculated as follows: The first term represents the next customer buying from firm 2, after which, they revert to Bertrand. This makes firm 2 a profit of $(v-c)/2$. The second term represents the next customer buys from firm 1, and the one after that buys from firm 2, in which case firm 2 makes $(v-c)/4$, and so on.

392 THE THEORY AND PRACTICE OF REVENUE MANAGEMENT

8.4.2.2 Dynamic Infinite Bertrand

We next consider infinitely repeated Bertrand competition. It turns out that almost any price up to the monopoly price can be sustained in equilibrium, and there are an infinite number of possible equilibria. Nevertheless, it is a good place to start understanding the supergame framework.

Consider two firms competing in a repeated Bertrand game over an infinite horizon. In each period there is exactly one customer who purchases based on the Bertrand formula (8.24). Future profits are discounted by a factor β. In period t, firms simultaneously set prices $p_1(t)$ and $p_2(t)$, and collect a payoff

$$V_i(p_1(t), p_2(t)) = (p_i(t) - c)d_i(p_1(t), p_2(t)).$$

Firms maximize

$$\lim_{T \to \infty} \sum_{t=0}^{T} \beta^t V_i(p_1(t), p_2(t)).$$

The one-shot Bertrand equilibrium, where both firms price at marginal cost c, can be shown to be a (stationary) equilibrium to this infinite game.

However, there are many other possible equilibria based on the concepts of credible threats and trigger strategies. Let p^* be the monopoly price (maximizing $(p-c)d(p)$). A *trigger strategy* for a firm is to price at p^* if the other firm has priced at p^* for all preceding periods and to price at c otherwise. So each firm *threatens* the other firm with marginal-cost pricing forever if the other firm deviates in any period from the (collusive) price p^*.

A trigger-price strategy is an equilibrium as long as $\beta > 1/2$. To see this, note that if a firm deviates by Δ from p^* in period t, it receives a profit of $p^* - \Delta - c$ in that period and zero from then on. On the other hand, if it sticks to the trigger strategy, it receives a profit of

$$\frac{1}{2}(p^* - c)(1 + \beta + \beta^2 + \cdots) = \frac{1}{2(1-\beta)}(p^* - c) \geq p^* - \Delta - c,$$

as long as $\beta \geq 1/2$. In fact, one can substitute any p between c and p^* and reason the same way to show that it too can be sustained in equilibrium.

8.4.2.3 Cournot Followed by Dynamic Infinite Bertrand-Edgeworth

The two-stage Kreps and Scheinkman model of Section 8.4.1.6 showed that an equilibrium for the first-stage Cournot game in which the payoff is given by a second stage of Bertrand-Edgeworth competition is the same as the one-shot Cournot capacity prediction of Section 8.4.1.1, and both firms set the market-clearing price in the second stage. Do the results change if the second-stage game is an infinite-period Bertrand-Edgeworth game? That is, firms, after they choose their capacities, have *many* opportunities to adjust their prices. This question was explored by Benoit and Krishna [46], and we present their results below.

First of all, one equilibrium to this Cournot and infinite Bertrand-Edgeworth game turns out to be setting Cournot capacities in the first stage and the Cournot prices for every period during the price competition phase. Call this the *stationary Cournot equilibrium*. But—and this is generally no surprise in a supergame—there are many other equilibria for this Cournot and infinite Bertrand-Edgeworth game.

THEOREM 8.7 *([46], [88]) For all discount factors $\beta \geq 1/2$, and for all choice of capacities (x, x) chosen by the firms, there exists a price $\bar{p}$ such that $p_1(t) = p_2(t) = \bar{p}, \forall t$ is a subgame-perfect equilibrium for the second-stage infinite-horizon Bertrand-Edgeworth game.*

All equilibria other than the stationary Cournot equilibrium share some common properties. *All equilibria for the Cournot and infinite Bertrand-Edgeworth game, except the stationary Cournot equilibrium, will have the firms choosing excess capacity in the first stage*—excess capacity in the sense that firms do not use this capacity in the second stage but build it solely to threaten their rivals with a price drop if they deviate from the collusive prices. We show this informally by the following arguments:

If $p_1 = p_2$, firms will share the demand equally, and if $p_1 < p_2$, firm 1 will first get a demand of

$$\tilde{d}_1(p_1, p_2, x_1, x_2) = \min(x_1, d(p_1)),$$

and firm 2 will get a residual demand of

$$\tilde{d}_2(p_1, p_2, x_1, x_2) = \min(x_2, \max(0, d(p_2) - \tilde{d}_1)).$$

If $p_1 > p_2$ firm 2 gets the demand first and then firm 1 will get the residual demand. *Excess capacity* is defined as when either firm chooses capacity $x_i > \tilde{d}_i(p_1, p_2, x_1, x_2)$.

Define firm 1's minmax revenue as the least amount firm 2 can hold firm 1's revenue down to. That is,

$$\begin{aligned} v_1(x_1, x_2) &= \min_{p_2} \max_{p_1} p_1 \tilde{d}_1(p_1, p_2, x_1, x_2) \\ &= \max_{p_1} p_1 \tilde{d}_1(p_1, 0, x_1, x_2), \end{aligned}$$

and vice versa for firm 2.

Suppose that both firms choose $x_i = \tilde{d}_i(p_1, p_2, x_1, x_2)$ in the first stage in a stationary perfect-equilibrium path, and that (x_1, x_2) are not the Cournot equilibrium capacities (say, because x_1 is not a best-response for firm 1 in the first stage Cournot game). Then there exists some capacity y such that

$$p(y + x_2)y - cy > p(x_1 + x_2)x_1 - cx_1.$$

Consider

$$v_1(y, x_2) = \max_{p_1} p_1 \min[y, \max(0, d(p_1) - x_2)] \geq p(y + x_2)y, \quad (8.29)$$

as $d(p(y + x_2)) = y + x_2$. So firm 1 would prefer to use capacity y and get its minmax revenue for the second-stage price game than use x_1. Similarly for firm 2. Hence (x_1, x_2) must be the Cournot equilibrium.

If exactly one firm—say, firm 1—has excess capacity; that is,

$$x_1 > \tilde{d}_1(p_1, p_2, x_1, x_2).$$

Then (shortening $\tilde{d}_1(p_1, p_2, x_1, x_2)$ to $\tilde{d}_1$),

$$p(\tilde{d}_1 + x_2)\tilde{d}_1 - c\tilde{d}_1 > p(\tilde{d}_1 + x_2)\tilde{d}_1 - cx_1$$

and using the same argument as of (8.29)

$$v_1(\tilde{d}_1, x_2) \geq p(\tilde{d}_1 + x_2)\tilde{d}_1.$$

So firm 1 would use capacity $\tilde{d}_1(p_1, p_2, x_1, x_2)$ rather than x_1 in the stationary perfect-equilibrium path, and we can say:

PROPOSITION 8.2 *If (x_1, x_2) and $(p_1(t), p_2(t)) = (p_1, p_2)$ is a stationary perfect equilibrium path of this (two-stage, infinite) game, then either (i) (x_1, x_2) is a Cournot equilibrium or (ii) for all t, both firms have excess capacity.*

Benoit and Krishna [46] furthermore prove the following theorem for nonstationary equilibrium paths:

THEOREM 8.8 *([46]) If (x_1, x_2) and $(p_1(t), p_2(t))$, $t = 1, \ldots$, form a subgame-perfect equilibrium path of the Cournot and infinite Bertrand-Edgeworth game, either*
(i) (x_1, x_2) is a Cournot equilibrium and $(p_1(t), p_2(t))$, Cournot prices, for all t or
(ii) for an infinite number of periods t,

$$x_1 + x_2 > \tilde{d}_1(p_1(t), p_2(t), x_1, x_2) + \tilde{d}_2(p_1(t), p_2(t), x_1, x_2).$$

The intuition is that firms cannot avoid excess capacity; if they both choose small capacities, each has an incentive to expand because its capacity-constrained competitor is unable to retaliate in the price-competition phase. Yet if both choose large capacities, it is unprofitable for each to fully utilize their capacities.

8.4.3 Product Differentiation

Thus far we have assumed that firms produce an identical, perfectly substitutable product (a commodity). Yet this is rarely the case in real world markets. Customer tastes and preferences differ, and it is in the firms' interest to differentiate their products, both to exploit customers' willingness to pay for different features and to mitigate the effects of competition. RM industries are characterized by their rich diversity of products and features. The features may differ physically (fashion apparel, airline schedules and service, hotel location, network ratings) or may differ in their terms of trade. For such differentiated products markets, what does economic theory say about competitive outcomes?

The effect is significant. Even in a Bertrand-style pricing game, equilibrium prices can exceed marginal cost. The premium is directly proportional to a measure of the customer's taste for diversity and also the number of firms in the market.

8.4.3.1 Static Model

Consider a static-pricing game of a duopoly with a product each. For illustration, we derive the equilibrium price assuming customer purchase behavior follows a multinomial-logit (MNL) model.

Let N denote the number of customers, p_1 denote the price of firm 1 and p_2 that of firm 2. The products are differentiated by a set of attributes, and customers preference for these attributes may make them choose a product from a firm even if the product is priced higher than the other.

The customer's taste for such diversity is modeled by a factor $\mu > 0$. The probability that a customer will buy product i is given by

$$\gamma_i(p_1, p_2) = \frac{e^{-p_i/\mu}}{e^{-p_1/\mu} + e^{-p_2/\mu}}.$$

When μ tends to zero, it means customers are indifferent to the product characteristics and buy based purely on price, whereas when μ is very high, they are quite insensitive to price and buy each product with equal probability.

Let c be the marginal cost of production and K the fixed cost of production. Then the profit function for firm i is given by

$$V_i(p_1, p_2) = (p_i - c)N\gamma_i(p_1, p_2) - K.$$

This function turns out to be strictly quasiconcave (under the logit assumption of our model; see [16], p.222). Differentiating the profit functions of firms 1 and 2 with respect to p_1 and p_2 and setting them to zero simultaneously gives the symmetric equilibrium for the firms as

$$p^* = c + 2\mu.$$

The result can be extended to a n-firm oligopoly to give

$$p^* = c + \frac{n\mu}{n-1}.$$

Thus, the equilibrium price is above marginal cost, and the lower the price sensitivity, the higher the premium above marginal cost—quite a different outcome than the one obtained in the commodity Bertrand competition case. This outcome suggests why firms strive so hard to differentiate their products and services; it is one way to avoid the zero-profit outcome of head-to-head price competition. (For variations and analysis on optimal product diversity, see Anderson, de Palma, and Thisse [16].)

8.4.3.2 Dynamic RM Competition in Allocations and Prices

In Section 8.4.2.1, we have seen a dynamic model of price competition in a duopoly where the firms have fixed capacities: The price competition in each period was modeled as a Bertrand game—firms can observe each others' remaining capacities, and the buyers are assumed to always purchase from one of the firms. In this section, we relax these assumptions the following way. Firms sell differentiated products and make available an offer set as in quantity-based RM. Customers consider the offer sets from each firm and make a purchase decision based on a choice rule. Customers have a no-purchase option, and they decide not to purchase with a certain positive probability. In contrast to the findings of the model in Section 8.4.2.1, an equilibrium need not always exist under this model, and the no-purchase option plays a crucial role on whether it does or not.

Model The model, in fact, can be considered the duopoly version of the dynamic discrete-choice model of quantity-based RM of Section 2.6.2 (with elements from the differentiated products competition model of

Section 8.4.3.1). Let's recall some of the elements of this model. Time is discrete, and there are T periods until the time of service. The resources of both firms are consumed simultaneously at time T. Bookings occur during the intervals 0 to T with at most one arrival during each period. A customer arrives in a period, observes the available choices of the two firms, and then, based on the prices and attributes of the fare products, either decides to buy one of the products of one of the firms or decides not to purchase any of the available products. Let 0 represent this no-purchase alternative.

We assume a MNL customer choice model with a no-purchase alternative. In each period there is a probability λ of a customer arrival. If a customer does not purchase in a period, he does not reappear in a later period. Since the choice model is MNL, a customer's probability of choosing an available product j of firm i, when firm 1 offers the set A_1 and firm 2 offers the set A_2 can be represented for convenience by

$$\gamma_j^i(A_1, A_2) = \frac{w_j^i}{\sum_{j \in A_1} w_j^1 + \sum_{j \in A_2} w_j^2 + w_0},$$

where the weights w_j^1 and w_j^2 are formed by the prices p_j^1 and p_j^2 and possibly other attributes of the products and firms, and w_0 is the weight of the no-purchase alternative.

Firms have capacities of x_1 and x_2 units and n_1 and n_2 RM products respectively that share the capacities. Firms can observe each others' remaining capacities and choose offer sets simultaneously in each period. They fix the prices of the products $p_1^1, p_2^1, \ldots, p_{n_1}^1$ and $p_1^2, p_2^2, \ldots, p_{n_2}^2$ initially and keep them fixed throughout the booking horizon. At the beginning of each period, each firm makes available a subset of its fare products. So even though prices are fixed, the firms can influence the prices that a consumer sees in each period by changing the offer set. Although specific to a choice rule with the the logit functional form, this model captures both customer choice and the dynamics of availability controls in quantity-based RM.

Equilibrium Analysis The state space is the firms' remaining capacities, x_1 and x_2. The reaction function of a firm is based on its own remaining capacity, its competitor's remaining capacity, and its competitor's current offer set.

As we saw in Section 2.6.2, under the MNL customer-choice model in the monopoly case a firm's optimal offer set has the nested-by-revenue order property, which means that each firm needs to consider only complete sets. That is, for firm 1, sets of the form $C_{k_1} = \{1, 2, \ldots, k_1\}$, for $k_1 = 1, 2, \ldots, n_1$ and for firm 2, sets of the form $C_{k_2} = \{1, 2, \ldots, k_2\}$, for

$k_2 = 1, 2, \ldots, n_2$ are optimal. This property carries over to the duopoly case, so we assume from now on that the strategy spaces for the firms are these complete sets. We represent the collection of all complete sets as $\mathcal{N}_1$ for firm 1 and $\mathcal{N}_2$ for firm 2. We will say that $C_{k_1} < C_{k_2}$ if $k_1 < k_2$, or, as the sets are nested, $C_{k_1} \subset C_{k_2}$. If $C_{k_1} < C_{k_2}$, we denote the set difference as $C_{k_2} - C_{k_1}$.

The value function for firm 1 at time t, given firm 2 offers C^2, is given by the following:

$$V_t^1(x_1, x_2 | C^2) = \max_{C^1 \subseteq \mathcal{N}_1} \{\lambda \sum_{j \in C^1} \gamma_j^1(C^1, C^2) p_j^1 \tag{8.30}$$
$$+ (1 - \lambda + \lambda \gamma_0(C^1, C^2)) V_{t+1}^1(x_1, x_2)$$
$$+ \lambda \sum_{j \in C^1} \gamma_j^1(C^1, C^2) V_{t+1}^1(x_1 - 1, x_2)$$
$$+ \lambda \sum_{j \in C^2} \gamma_j^2(C^1, C^2) V_{t+1}^1(x_1, x_2 - 1) \},$$

where $V_{t+1}(\cdot, \cdot)$ is the equilibrium revenue to go from period $t+1$ onward.

There are two things to note about (8.30): (1) it is defined at time t only if there is an equilibrium from t until T, and (2) if there are multiple equilibria, we assume (exogenously) that one equilibrium is chosen so that the value function is again uniquely defined. The value function for firm 2 is defined similarly. We let $V_t^1(x_1, x_2 | C^1, C^2)$ denote firm 1's revenue if it uses C^1 to react to firm 2's C^2.

While equation (8.30) looks complicated, in words it simply says that given that firm 2 chooses offer set C^2, firm 1's revenue is the current period's revenue plus the expected revenue to go in the next period, which depends on the new capacity state for each firm in the next period. To analyze this recursion, let us rewrite equation (8.30), using the fact that

$$\sum_{j \in C^2} \gamma_j^2(C^1, C^2) = 1 - \gamma_0(C^1, C^2) - \sum_{j \in C^1} \gamma_j^1(C^1, C^2),$$

as

$$V_t^1(x_1, x_2 | C^2) = \tag{8.31}$$
$$\max_{C^1 \subseteq \mathcal{N}_1} \{\lambda \sum_{j \in C^1} \gamma_j^1(C^1, C^2)[p_j^1 + V_{t+1}^1(x_1 - 1, x_2) - V_{t+1}^1(x_1, x_2 - 1)]$$
$$+ (1 - \lambda) V_{t+1}^1(x_1, x_2)$$
$$+ \lambda \gamma_0(C^1, C^2)[V_{t+1}^1(x_1, x_2) - V_{t+1}^1(x_1, x_2 - 1)]\}$$
$$+ \lambda V_{t+1}^1(x_1, x_2 - 1).$$

Let $\Delta = V_{t+1}^1(x_1, x_2 - 1) - V_{t+1}^1(x_1 - 1, x_2)$ and $\delta = V_{t+1}^1(x_1, x_2 - 1) - V_{t+1}^1(x_1, x_2)$. Let $g(\cdot|x_1, x_2)$ represent the term

$$g(C^1|x_1, x_2) = \sum_{j \in C^1} w_j^1 [p_j^1 - \Delta] - w_0 \delta.$$

For simplicity we just write $g(C^1|x_1, x_2)$ as $g(C^1)$, whenever there is no room for confusion. Notice that $g(\cdot)$ for firm 1 is independent of the strategies of firm 2, and vice versa for firm 2.

The following relationships are intuitive and not hard to show rigorously:

$$V_{t+1}^1(x_1, x_2) \leq V_{t+1}^1(x_1, x_2 - 1),$$
$$V_{t+1}^1(x_1 - 1, x_2) \leq V_{t+1}^1(x_1, x_2),$$

and

$$V_{t+1}^1(x_1 - 1, x_2) \leq V_{t+1}^1(x_1, x_2 - 1).$$

Note that $\Delta - \delta = V_{t+1}^1(x_1, x_2) - V_{t+1}^1(x_1 - 1, x_2) \geq 0$.

As prices p_j^i are decreasing in j, as firm i offers larger sets, the function $g(C^i)$ increases monotonically first and then decreases monotonically (that is, it is unimodal). The function $g(C^i)$ can also be negative. It could be the case that $g(\cdot)$ is negative for all strategies of firm i. Figure 8.4 shows the two possibilities for firm i, where $\bar{G}^i$ is the set that has the maximum value of g. We call the case where $g(\bar{G}^i) \geq 0$ as Case I, and Case II when $g(\bar{G}^i) < 0$.

PROPOSITION 8.3 *[503] When both firms are in Case I, or both are in Case II, there exists an equilibrium in offer sets.*

The difficulty is if one firm is in Case I and the other in Case II. Consider the following example:

Example 8.18 Suppose firm 1 and firm 2 have the data given in the following table:

	Firm 1 $i = 1$	Firm 2 $i = 2$
Δ	100	100
δ	50	10
w_1^i	1	1
w_2^i	1	10
w_0	10	10
p_1^i	200	300
p_2^i	67	108.18

Then the payoff and best-response arcs are as given in Figure 8.5, and it can be seen that there is a cycle and no equilibrium, even for the MNL choice function.

400   THE THEORY AND PRACTICE OF REVENUE MANAGEMENT

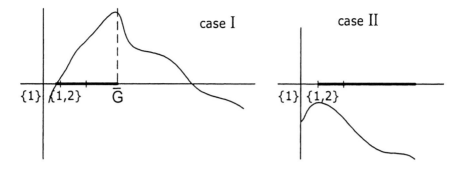

Figure 8.4. The two cases for the function $g(\cdot)$. The best responses will lie in the regions marked with bold lines ([503]).

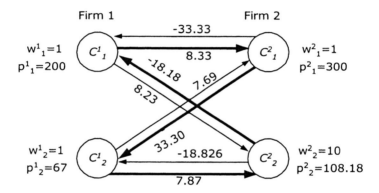

Figure 8.5. Payoff values for Example 8.18. Best responses are in thick lines.

So even though the single-period problem always has an equilibrium (as $\Delta = \delta = 0$), the competitive game over a finite number of periods, may end up having no equilibrium. Notice that this is not a repeated game: the dynamic program has strong intertemporal relationships, and the parameters of the game change over time.

One can give some conditions on the choice model parameters that guarantees existence of an equilibrium. Define for $C_k > C_l$,

$$q^i_{C_k - C_l} = \frac{\sum_{j \in C^i_k - C^i_l} w^i_j p^i_j}{\sum_{j \in C^i_k - C^i_l} w^i_j},$$

which represents the weighted-average price for the products from k to l.

The Economics of RM

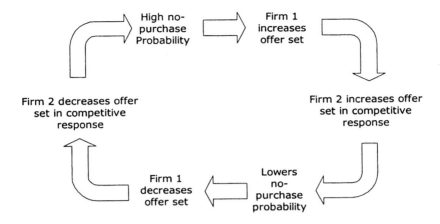

Figure 8.6. No-purchase probabilities causing a best-response cycle.

PROPOSITION 8.4 *If for $i = 1, 2$ and for any C_k^i, C_l^i the parameters of the choice model satisfy*

$$\sum_{j \in C_l^i} p_j^i w_j^i > q_{C_k - C_l}^i (w_0 + \sum_{j \in C_l^i} w_j^i), \tag{8.32}$$

then there exists a pure-strategy Nash equilibrium in the subsets that each firm offers at every time interval t.

The condition for Proposition 8.4 holds, for instance, when $w_0 = 0$. That is, if a period has a customer arrival and available inventory, that period will have a sale.

In the finite Bertrand-Edgeworth game of Section 8.4.2.1 an equilibrium always exists. In that model too firms can observe each other's capacities, but the customer in each period will always purchase from one of the two firms (equivalent to $w_0 = 0$) if there is available inventory. So the no-purchase option plays a crucial role indeed; its relative magnitude determines whether we have an equilibrium in the game or not. Figure 8.6 shows the intuition behind why modeling the no-purchase option introduces the instability into the game under certain circumstances.

While Example 8.18 is somewhat discouraging, it is not indicative of a nonequilibrium in many interesting cases, such as the following: if the two firms have equal capacities, identical products, and customers, then will we see equilibrium in the offer sets? Example 8.18 is not applicable, and it is possible that one can derive conditions on Δ and δ to show the existence of an equilibrium. Indeed, Example 8.18 is not even conclusive for the general case; it does not show that the Δ and δ used actually arise in a dynamic game. So there are still many gaps in our understanding of quantity-based RM competition, even for a duopoly.

The results of this section also extend to a choice model where customers choose the lowest-available product in the offer sets. If there is a single segment (as we assumed here) of customers, then this is essentially equivalent to firms offering a single price in each time-period—that is, dynamic pricing. Customers then have three options: no-purchase or purchase at the price offered by firm 1 or from firm 2. One needs to impose a few conditions on the parameters of the choice model, however: the weights are decreasing functions of the price, and $w_j p_j$ is unimodal. For instance, the logit model with weights of the form used in Section 8.4.3.1 satisfy this property.

8.5 Notes and Sources

A good rigorous treatment of the foundations of microeconomics can be found in Mas-Colell, Whinston, and Green [365]. Economic theory specifically related to prices and price formation can be found in Simon [471] and Stigler [489]. Good references to the study of industrial organization are Tirole [513] and Shy [469].

Advances in proving the existence of Cournot equilibrium for general functions have been made by McManus [378], Roberts and Sonnenshein [444], Bamon and Fraysee [27], Novshek [408] and Amir [13]. Vives [539] covers oligopoly pricing using results on submodular functions. In addition, consult the survey of Shapiro [463] on models of oligopoly. Stigler [488] is the classic article on tacit collusion and punishment strategies that spurred the application of repeated-games framework to firm interactions in an oligopoly.

Price discrimination, classification of discrimination policies, economic reasoning, along with a large number of examples can be found in Phlips [420] and in the survey of Varian [531]. In multiproduct firms, Clemens [121] is one of the earliest articles that talks about price discrimination.

Many of the monopoly models and examples we discuss in the chapter are standard, or have their sources cited in the text. The mechanism design principles used in the Harris and Raviv [239] model and the Gale and Holmes model on advance-purchase discounts come from Myerson [398] and Harris and Townsend [240] and are also quite standard now.

The literature on intertemporal price discrimination and the Coase conjecture is extensive, and besides the references given within the chapter, the reader can consult Stokey [490, 491].

In addition to the pricing methods mentioned in the text, there are many randomized pricing strategies that allocate the sale of the good with a certain probability and the price varies according to the probability. In other words, the dimension of differentiation is the reliability of the service. See, for example, Tschirhart and Jen [517].

The relationship between auctions and posted-price selling is a fascinating topic that can use further investigation. Some useful references on this topic are Wang [551, 552] and Ziegler and Lazear [590].

Economists have also studied retail store price dispersion and tacit collusion and price extensively. Here we mention some of the classical papers. Sales Salop and Stiglitz [452] explain price dispersion in retail stores due to customers' uncertainty of retail prices. Sobel [482] studies the timing of sales in an oligopoly to attract low-valuation customers. Models of price wars are given in Green and Porter [222] and empirical evidence in Porter [427]. Eaton and Engers [170] study collusive price formation among firms selling differentiated products (the ones covered in this chapter were for homogeneous products).

The standard reference on discrete-choice models of product differentiation is Anderson, de Palma, and Thisse [16]. (See also Anderson and de Palma [17].) An interesting topic that we have ignored in this chapter is how firms choose the features of their product (product-design competition). The survey article of Lancaster [328] is a good introduction to the economics of product variety. Optimal provision of products in an oligopoly has been studied by Anderson, de Palma, and Nesterov [15]. Shaked and Sutton [459] study a three-stage game of entry (first stage), choice of product quality (second stage) and then price (third stage) to understand how firms differentiate their products.

The bipartite graph construction and related equilibrium proofs in the chapter appendix are from Talluri [503].

The problem of peak-load pricing dates back to the work of Steiner [486] and Boiteux [82]. Crew et al. [136] provide a survey of the literature on peak-load pricing. See also Bergstrom and MacKie-Mason [51] for a concise analysis of the problem.

In terms of literature on advance-purchase discounts, Shugan and Xie [467] provides a nice analysis of the topic, including conditions under which advance selling are profitable and resulting managerial implications. Dana [141] analyzes advance-purchase discounts under perfect competition. Shugan and Desiraju [466] also analyze the economics of advance purchases combined with capacity rationing and overbooking.

There is considerable literature analyzing trade and retail promotions and the interactions between manufacturers, retailers, and customers in a game-theoretic framework, that we have not covered. A few references on this topic are Lal [325, 324], Lal and Villas-Boas [322, 323], Bell, Iyer, and Padmanabhan [35], Rao, Arjuni, and Murthi [436], and Gerstner and Hess [211].

APPENDIX 8.A: Proofs

Proof that $V_{APD}(\hat{w}) \geq V_{peak}(\hat{w}_2)$
Proof

From optimality of $\hat{w}_2$,
$$V_{peak}(\hat{w}_2) \geq V_{peak}(\hat{w}),$$
which can be expanded as
$$rC + (r - \hat{w}_2)[(1 - \alpha) + \alpha F(\hat{w}_2)] \geq \\ rC + (r - \hat{w})[(1 - \alpha) + \alpha F(\hat{w})],$$
and optimality of $\hat{w}$ implies
$$V_{APD}(\hat{w}) \geq V_{APD}(\hat{w}_2),$$
which can be expanded as
$$r\{C + (1 - \alpha)[1 - F(\hat{w})]\} + (r - \alpha\hat{w})F(\hat{w}) \geq \qquad (8.A.1) \\ r\{C + (1 - \alpha)[1 - F(\hat{w}_2)]\} + (r - \alpha\hat{w}_2)F(\hat{w}_2).$$

After some algebraic manipulation, the right-hand side of (8.A.1) can be written as
$$\begin{aligned}
r\{C + (1 - \alpha)[1 - F(\hat{w})]\} &+ (r - \alpha\hat{w})F(\hat{w}) \geq \qquad (8.A.2) \\
r\{C + (1 - \alpha)[1 - F(\hat{w}_2)]\} &+ (r - \hat{w}_2)\alpha F(\hat{w}_2) \\
+ (1 - \alpha)(r - \hat{w}_2) &+ (1 - \alpha)rF(\hat{w}_2) - (1 - \alpha)(r - \hat{w}_2) \\
= \{rC + (r - \hat{w}_2)&[(1 - \alpha) + \alpha F(\hat{w}_2)]\} \\
+ (1 - \alpha)\hat{w}_2 & \\
\geq V_{peak}(\hat{w}_2). &
\end{aligned}$$

So $V_{APD}(\hat{w}) \geq V_{peak}(\hat{w}_2)$ as claimed. QED

Proof of Proposition 8.1
Proof
$l_2 > k_2$ implies $C - l_2 < C - k_2$. If firm 2 chooses k_2, firm 1's best response was k_1. If firm 2 chooses l_2 (that is, protects less for its H demand), then firm 1 should see more of a spillover of firm 2 H demand (firm 1's demand is stochastically more than if firm 2 chooses k_2). So by Littlewood's rule, its protection level should increase, or $l_1 \geq k_1$. QED

Proof of Proposition 8.3
Proof
Suppose there exist two such arcs. So if firm 2 offers the complete set k_2, then firm

APPENDIX 8.A: Proofs

1's best response is k_1, and if firm 2 offers the complete set l_2, then firm 1's best response is l_1. Because the customer's choice rule is the MNL, this means that

$$\frac{\sum_{j \leq k_1} w_j^1 p_j^1}{\sum_{j \leq k_1} w_j^1 + \sum_{j \leq k_2} w_j^2 + w_0} \geq \frac{\sum_{j \leq l_1} w_j^1 p_j^1}{\sum_{j \leq l_1} w_j^1 + \sum_{j \leq k_2} w_j^2 + w_0}. \quad (8.A.3)$$

Similarly, when firm 2 offers l_2,

$$\frac{\sum_{j \leq l_1} w_j^1 p_j^1}{\sum_{j \leq l_1} w_j^1 + \sum_{j \leq l_2} w_j^2 + w_0} \geq \frac{\sum_{j \leq k_1} w_j^1 p_j^1}{\sum_{j \leq k_1} w_j^1 + \sum_{j \leq l_2} w_j^2 + w_0}. \quad (8.A.4)$$

Now note that since $k_1 > l_1$, the denominator of the left-hand side of (8.A.3) is greater than that of the right-hand side and that this implies $\sum_{j \leq k_1} w_j^1 p_j^1 \geq \sum_{j \leq l_1} w_j^1 p_j^1$. Now if $x/y \geq z/w$ and $x \geq z$, then $x/(y+v) \geq z/(w+v)$ for any constant $v > 0$. Since $l_2 > k_2$,

$$\frac{\sum_{j \leq k_1} w_j^1 p_j^1}{\sum_{j \leq k_1} w_j^1 + \sum_{j \leq l_2} w_j^2 + w_0} \geq \frac{\sum_{j \leq l_1} w_j^1 p_j^1}{\sum_{j \leq l_1} w_j^1 + \sum_{j \leq l_2} w_j^2 + w_0}. \quad (8.A.5)$$

But this contradicts the fact that if firm 2 offers l_2, the largest best-response set for firm 1 is l_1: it is, in fact, $k_1 > l_1$. $\quad QED$

Chapter 9

ESTIMATION AND FORECASTING

9.1 Introduction

A revenue management system requires forecasts of quantities such as demand, price sensitivity, and cancellation probabilities, and its performance depends critically on the quality of these forecasts. Indeed, some industry estimates suggest that a 20% reduction of forecast error can translate into a 1% incremental increase in revenue generated from the RM system (Poelt [424]). While it is difficult to generalize from such figures, there is little doubt that good forecasting is vitally important for RM. In practice, forecasting is a high-profile task of RM, consuming the vast majority of development, maintenance, and implementation time.

The term *forecast* may conjure up the notion of a single number, such as the demand for a specific day on a specific flight in the future or demand for a particular item at a retail store (a so-called *point estimate*). A certain amount of misunderstanding about RM forecasting is not uncommon among nontechnical analysts and managers, who are accustomed to thinking of forecasts as a single number. However, a point estimate is almost never accurate; a forecast is more complicated than a single number and needs to be understood in statistical terms that account for the inherent uncertainty in predicting future outcomes.

In this chapter, we examine forecasting for RM. We start with an overview of the role of forecasting in RM—surveying the available data sources, forecasting strategies and methodologies, and factors involved in actually operationalizing a RM forecasting system. The remainder of the chapter describes estimation and forecasting methods in more depth.

While our discussion is centered on RM forecasting, most of the techniques we describe are not particular to RM and as such can safely be described as *standard*. Indeed, there are many excellent textbooks devoted to estimation and forecasting, and it is not our intention to approach them in scope and depth. Rather, this chapter is intended as a primer on the subject—sufficient in coverage to give a good sense of the range of methods and issues involved in RM forecasting but not providing an in-depth reference on any one method. We do, however, give enough detail to understand and implement at least a basic version of each method. To implement and maintain a high-quality, production-level RM forecasting system, one needs to know more about the nuances of each forecasting method, and the reader in this situation is encouraged to consult a specialized source for such information. (The Notes and Sources section at the end of the chapter contains references to a number of books dedicated to estimation and forecasting.)

9.1.1 The Forecasting Module of RM Systems

RM forecasting presents many challenges to a system designer. For one, a significant amount of programming work is involved in collecting and manipulating the data to convert it into the required data feeds for the forecasting module. Large volumes of transactional data have to be gathered from multiple sources, either in real time or on an overnight batch basis. The database design is an important issue because in many large-scale implementations, an immense number of records have to be retrieved, updated, and added within a small time window. Data backup procedures take up further time. All these data and systems issues must be addressed prior to actual forecasting itself.

Figure 9.1 shows a schematic of the process flow of a typical quantity-based RM system and where the forecasting module resides in the process. The outputs of the forecasting module are fed to the optimization module, which produces RM controls such as markdown prices, booking limits, bid prices, and overbooking limits. In the stage between forecasting and optimization, most RM systems also give analysts monitoring and overriding ability over the forecasts. These so-called *user influences* are used to either increase or decrease the forecasts at different levels of aggregation before they are used in optimization. Indeed, in most quantity-based RM systems, analysts are not permitted to change capacity controls directly but can change them only indirectly by manipulating the forecast inputs. This practice is based on the belief (widespread among RM practitioners) that knowledge of markets or special conditions can sometimes make human analysts better than algorithms at forecasting demand, but rarely, if ever, can human analysts set RM controls better than optimization algorithms.

Estimation and Forecasting 409

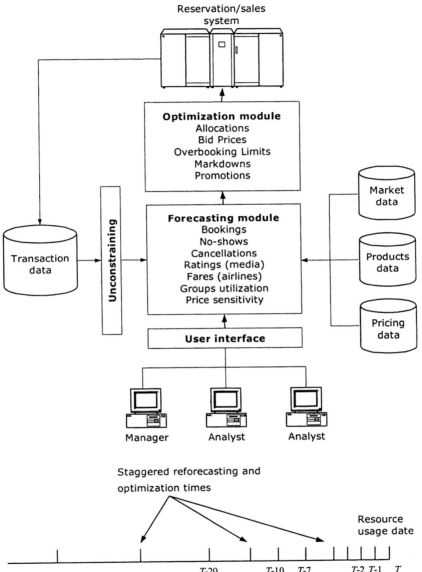

Figure 9.1. Forecasting module in a RM system. Periodic reforecasting and optimization timeline.

In most RM systems, forecasting is automated, transactional, and data-driven—as opposed to qualitative (such as expert opinion) or survey-based. This is primarily due to the sheer volume of forecasts that have to be made and the tight processing-time requirements. These practical constraints limit the choice of forecasting algorithms. They also limit the types of data that can reasonably be collected and the amount of time a user can spend calibrating and verifying forecasts. Often, certain forecasting procedures, even if they give superior forecasts, may not be a viable option because they take too long to run, require data that is too expensive to collect (say, using surveys), or require too much expert, manual effort to calibrate.

For quantity-based RM, most systems use time-series methods, which use historical data to project the future. For price-based RM systems (retail RM, for example), one is usually interested in forecasting demand as a function of marketing variables such as price or promotion. As a result, causal forecast models, which use explanatory variables such as prices, weather or economic indicators, play a bigger role in price-based RM.

9.1.2 What Forecasts Are Required?

RM forecasting requirements are driven by the input requirements of the optimization module. As we saw in previous chapters, most optimization models use stochastic models of demand and hence require an estimate of the complete probability distribution or at least parameter estimates (e.g., means and variances) for an assumed distribution. Moreover, forecasting aggregate demand is just one of a host of quantities that need to be estimated in a RM system. Many other features of the demand—how it evolves over time, what percentage cancel, how it responds to a promotion—are also important in making good control decisions, and need forecasting.

Quantity-Based RM Forecasts Quantity-based RM industries like airlines and hotels have a wide variety of forecasting requirements. For example, in addition to the demand data, characterizing the way reservations for different customer types arrive during the booking period is important for some optimization models. Thus, so-called *booking-curve* or *booking-profile forecasting* is usually an important task.

Cancellation and no-show probabilities usually have to be estimated as well. Cancellation probabilities tend to be a function of time. (A customer who books early may have a higher probability of cancelling than one who books later.) Therefore, forecasting a cancellation *curve* over time may be more appropriate, giving better information to the optimization module. No-shows occur at the time of service; hence, assuming that a customer shows up with a certain probability is often an appropriate model of no-shows (see Chapter 4). Both no-show and cancellation-rate forecasts have to be calculated for future customers as well as customers who have already booked. For existing reservations, however, additional sources of information, such as the customer's own past history of cancelling, whether the reservation has been paid and ticketed, and characteristics of how the reservation was made (channel, time, etc.) can be used, increasing the data-gathering requirements.

Revenue values are also critical inputs to optimization modules. Often, these values change over time or are uncertain, so the prices at which the products will be sold in the future may have to be forecasted. Predicting revenue values can be a major challenge, especially when prices change rapidly and competitive forces drive pricing.

Optimization models may also require estimates of cross-selling and up-selling probabilities. The buy-up factors discussed in Section 2.6.1 may have to be estimated from historical data. Spill and recapture are two other quantities that are sometimes required in setting (or at least managing) RM controls. *Spill* refers to the amount of demand that is lost by closing down a class or because a compartment is sold out, while *recapture* is the amount of this spilled demand that is recaptured by the firm's substitute products. The discrete-choice model of Section 2.6.2, requires estimates of the parameters of a choice model, sometimes by channel of distribution or by customer segment.

Price-Based RM Forecasts For price-based RM, somewhat different forecasts are required. One common requirement is an estimate of the parameters of a demand function—or at least an estimate of the price sensitivity at the current price. Cross-price elasticity estimates may also be required when there are significant substitution effects (say, for category pricing in a retail store), which vastly increases the scale of the forecasting task. In addition, forecasts of demand at the current price, the size of the potential customer population, stockout and low-inventory effects, and switching behavior may be required. Such estimates require looking at the historical price-demand relationship of the product or similar products or at intertemporal panel tracking data. Some retailers have also tried intelligent experimentation in real time to estimate how consumers will respond to various price changes (*live price testing*).

In summary, the forecasting requirements in even a modestly large RM system are daunting, indeed. It is little wonder, then, that developing a good forecasting system is so vitally important for a successful RM implementation.

9.1.3 Data Sources

Data is the life-blood of a forecasting system. Therefore, identifying which sources of data are available and how they can best be used is an important first step in developing a forecasting system.

Most RM systems in practice rely primarily on historical sales data to construct forecasts. While this leads to highly efficient systems for data collection, forecast calibration, and automated forecasting, relying on historical data has its weaknesses. For example, in industries where products change frequently—when an airline offers service to a new city or at a new time for instance—there is often little historical data on which to base forecasts. Similarly, in media RM, forecasts for rating of new programs must be constructed despite the fact that their ratings often have little relationship to the ratings observed for past programs. Fashion apparel retailers, for instance, have to estimate sales of new styles that may be only vaguely similar to the styles sold in the past. In addition, even if the product stays constant, major changes in the economy, competing technologies, or industry structure may render past data of little use in predicting the future.

In short, if no explicit relationships with external data sources are tracked, the forecasting system will be "blind" to outside events. The same is true with respect to changes in competitors' products and prices. In cases where such external data is ignored, it is common practice in RM to rely on analysts to monitor outside events and compensate by adjusting forecasts appropriately through so-called *user influences*.

9.1.3.1 Sales-Transaction Data Sources

The main sources of data in most RM systems are transactional databases—for example, reservation and property management systems (PMSs), CRM and ERP databases, and retail inventory and scanner databases. Further descriptions of these data sources are given in Chapters 10 and 11. These sources may be centralized, independent entities shared by other firms in the industry (such as GDSs of the airline industry selling MIDT data), a centralized facility within a company that interfaces with several local systems (a retail chain's point-of-sale (POS) system linking all its stores), a local reservation system (a hotel PMS), or customer-oriented databases with information on individual customers and their purchase history (customer-relationship management (CRM) systems and PNR databases).

For quantity-based RM, the most widely used transactional data source is the reservations database. Reservation databases typically store customer data in two formats: either as an aggregate number

Estimation and Forecasting

of bookings in a class (total bookings) or as information about each individual booking, called a *customer booking record (passenger name record (PNR)* in the airline industry). Forecasts may be based on either the aggregate bookings or individual customer booking records. The aggregate bookings data contain information only on the total number booked in each fare class, while the individual booking records contain much more specific information on each customer—such as their name, address, booking time, number of units booked, amount paid, frequent flyer or other loyalty program number, booking class, cancellation time, capacity used (length of stay and room number for a hotel, car type and duration for a rental-car company, or itinerary for an airline), ancillary spending (dining expenses, telephone calls). The customer record may also contain links to other customer records (for example, for a group booking) that may be useful for forecasting.

For retail RM, factory-shipment data, store-level scanner data, consumer-panel data, regional demographic data, and advertising and promotions data are the primary data used. Industry-wide aggregate scanner data (sold by firms such as Information Resources, Inc. and A.C. Nielsen). Warehouse-shipment data can be obtained from Selling Areas Marketing, Inc. (SAMI), which provides sales, average price, and distribution information for the U.S.

Panel data, obtained from tracking purchases of a group of panelists over time, provides valuable information on cross-sectional and intertemporal purchase behavior. Such data are widely used in retail and media industries. A panel member's purchase data is also linked to promotions, availability, displays, advertising, couponing, and markdowns through the time of purchase, allowing for precise inferences on preferences and marketing influences. Many marketing research companies provide such panel-data services.

9.1.3.2 Controls-Data Sources

In addition to sales information, databases often store information on the controlling process itself. Examples of this kind of data include records of when a class is closed for further bookings, snapshots of bid prices used in the control, overbooking authorizations, past prices, and promotion activities. Such information is of great use in correcting for unobservable no-purchase decisions by potential customers (Section 9.4).

Industrywide database systems (such as a GDSs in the travel industry) may also yield additional information for forecasting—for example, the availability of competitor bookings, prices of competing products, and market share. Many airline GDS companies make this information available on a weekly or monthly basis on tapes called *market infor-*

mation data tapes (MIDT). Although few airlines at present use this information in their RM systems, it is useful for estimating competitive market share, and for longer-term, strategic planning and analysis.

For markdown pricing and other price-based RM applications, the control decisions are past history of prices and promotions. Most retail POS systems store this information routinely. For in-store displays or bundle pricing, the POS data has to be merged with a marketing database. Inventory data is also provided by many retail POS systems, and this data is useful for correcting for stockouts and broken-assortments effects (missing color-size combinations).

9.1.3.3   Auxiliary Data Sources

A few auxiliary data sources also play a big role in RM forecasting in some industries. For instance, currency exchange-rate and tax information is necessary to keep track of revenue value for sales in different countries. In the airline industry, the schedules and possible connections (provided by firms such as OAG) are required to determine what markets are being served. If a ticket is sold across multiple airlines, the various prorating agreements affect the ultimate revenue value of each product sold. A revenue accounting system keeps track of such agreements and calculates the net revenue.

In broadcasting, ratings, customer location, and demographic information is required. A causal forecasting method may take into account information on the state of the economy, employment, income and savings rates, among other factors. A rental-car firm can use advance travel bookings to predict its own demand at airport locations. Information on ad-hoc events (special events) like conferences, sports events, concerts, holidays, is also crucial in improving the accuracy of forecasts. Many forecasting systems allow the users to manually enter information on such events.

Many retail RM systems also use weather data, which is supplied by several independent vendors via daily automated feeds. Short-term weather forecasts guide discounting and stocking decisions (for example, a snow-storm could suggest high demand for snow shovels). Weather data also plays an important role in energy forecasting for electric power generators and distributors.

Macroeconomic data (such as GNP growth rates and housing starts) is rarely used in automated, tactical forecasting but frequently plays a role in aggregate forecasts of factors such as competitors' costs, industry demand and market share, and broad consumer preferences. Statistics on cost of labor are published by the Bureau of Labor Statistics (BLS) in the United States in a monthly publication called *Employment and*

Earnings, which provides average hourly earnings for workers by product category. BLS also provides monthly producer price indexes on raw materials.

For products sold through distributors, important data is not always available. For example, an automobile manufacturer may not know the final price paid by a customer because dealers have no obligation to report this information back to the manufacturer. Similarly, trade promotions may lead to increased shipments for the manufacturer, but the distributor may simply stock up during the trade deal and sell at a normal price, reducing the impact of the promotion. Lack of such information is one of the impediments for many firms contemplating RM.

9.1.3.4 Partial-Bookings Data

In most quantity-based RM applications, bookings occur over an extended period of time, yet the product or service is provided on a very frequent basis, often daily. For example, an airline may sell seats on a flight that operates every day, but bookings can occur over a period of 12 months prior to departure; hotels take reservations for rooms for each day, yet bookings are made many days or weeks in advance. In such situations, there are often large quantities of so-called *partial-booking data* in the reservation system. While incomplete, such data is quite useful for forecasting.

Figure 9.2 shows an example of partial-bookings data, indicating the number of bookings observed each day for capacity in the past as well as the future. The y-axis represents the date of service (such as the departure date in the airline case or the check-in date for a hotel), and the x-axis represents the number of days prior to the date of the service.

One way to use these partial histories of bookings is to forecast the *increments* of demand for each booking day, rather than forecasting the total demand to come. Thus, for example, data on the demand received on the 12th day prior to service can be used to predict demand on the 12th day prior to service in the future, even though the data may be from a booking history that is incomplete. Such methods are discussed in more detail in Section 9.3.9.

9.1.4 Design Decisions

After the data sources are identified, one has to make a number of design decisions regarding the forecasting strategy and methodology. Here we look at the main design decisions in qualitative terms.

Number of Days Prior to the Usage of the Resource									Resource-Usage Date
-8	-7	-6	-5	-4	-3	-2	-1	0	
6	9	20	24	33	41	54	57	70	10-Jun
8	14	20	23	39	50	55	59	61	11-Jun
1	3	3	3	6	12	14	20	28	12-Jun
6	6	10	11	13	19	22	24		13-Jun
3	11	19	25	30	31	33			14-Jun
1	1	3	10	16	20				15-Jun
0	1	2	8	13					16-Jun
1	12	24	30						17-Jun

Cumulative Bookings

Figure 9.2. Wedge shape of cumulative current-bookings data for a sequence of usage dates on each of the days prior to usage.

9.1.4.1 Parametric or Nonparametric Forecasts?

As mentioned in Section 9.1.2, in most RM forecasting we are interested in estimating a probability distribution of future demand—or in estimating demand as a function of price variables or product attributes. Such estimates can be made in one of two ways. The first is to assume a specific functional form and then estimate the parameters of this functional form. This approach is called *parametric estimation*. Alternatively, distributions or functions can be estimated directly based on observed historical data, without assuming any *a priori* functional form. This approach is called *nonparametric estimation*. Choosing between a parametric or nonparametric approach to forecasting is a basic design decision.

While nonparametric methods are in a sense more general, they are not necessarily a better choice. Nonparametric estimates suffer from two serious drawbacks: First, because they do not use a functional form to "fill in" for missing values, they often require much more information than is available in many RM applications to obtain reasonable estimates of a distribution or demand function. Second, even with sufficient data, nonparametric estimates may not be as good at predicting the future, even if they fit the historical data well. Parametric models are better able to "smooth out" the noise inherent in raw data, which often results in a more robust forecast. Indeed, we know of no RM systems that use purely nonparametric methods to estimate demand, though several use nonparametric methods in selected places. Neural networks, which are sometimes viewed as *semiparametric* methods, have been reported in several RM applications, and these we cover later in this chapter.

Estimation and Forecasting 417

Parametric methods usually are much more modest in their data requirements, have the advantage of providing estimates of demand that extend beyond the range of the observed data (allow for extrapolation), and are generally more robust to errors and noise in the data. The disadvantage of parametric techniques is that some properties of the distribution must be assumed—for example, that it is symmetric about the mean, has certain coefficients of variation, or has certain *tail behavior* (the characteristics of the demand distribution for extreme values of demand). Thus, parametric methods can suffer in terms of overall forecasting accuracy if the actual demand distribution deviates significantly from these assumptions (called *specification errors*). Because they are more widely used in RM, we focus on parametric methods in this chapter.

9.1.4.2 Levels of Aggregation

Forecasts can be made at different levels of aggregation as well, and how to aggregate data and forecasts is another important design decision. To give an example, airlines price their products by fare-basis codes with a large number of fares-basis codes sold within each booking class. Capacity control, however, is usually performed at the booking-class level. How, then, should forecasting be handled? Should the demand be forecast for each fare product (basis code) or each booking class? That is, should we aggregate all the fare products in a booking class and forecast at the level of the booking class? Or should we forecast at the fare-product level and aggregate these forecasts into a forecast for booking-class demand?

Another level-of-aggregation design decision comes up in network RM (Chapter 3), where the optimization system requires a forecast of demand for each multiresource product in the network. In principle, the forecasts should be at the level of the network products (O&Ds or lengths of stay) as this is the level required by network-optimization models. However, many reservation systems do not collect data at this level of detail. In the airline case, for example, reliable data may exist only for individual flight legs, and we may have to heuristically disaggregate leg-level forecasts into O&D, product-level forecasts.

Ultimately, however, we need to produce the forecasts required by the optimization module. Continuing the airline example, this would imply generating forecasts at the fare-product level if we were using a bid-price control, but perhaps at the booking-class level if we were using a resource level, booking-class-based control. On the other hand, the requirements of the optimization module can often be manipulated. For example, one can simply forecast at the booking-class level and then assume that all the demand in this booking class has the same (say, the average) fare. In the network case, the RM system might be using a simple single-resource heuristic to approximate the network RM problem, in which case an aggregate forecast for each resource independently may do just fine. Thus, the forecasting and optimization design decisions are intimately related. Indeed, in practice it is hard to change one without affecting the other.

In retail RM as well, the level of aggregation in forecasting is largely governed by the data and optimization requirements. Store-level pricing requires store-level estimates of demand and price sensitivity for each product, whereas a model that optimizes prices set on a chainwide basis may not require this same level of detail. If household purchase data (panel data) is available or if experiments or surveys can be conducted, then one can forecast based on models of individual purchase behavior and combine these to determine an aggregate demand function. However, if only aggregate POS sales data can be obtained, one has to estimate the aggregate demand function directly.

9.1.4.3 Bottom-Up versus Top-Down Strategies

Broadly speaking, there are two different strategies for aggregating forecasting. In a *bottom-up forecasting strategy*, forecasting is performed at a detailed level to generate *subforecasts*. The end forecast is then constructed by aggregating these detailed subforecasts. In a *top-down forecasting strategy*, forecasts are made at a high level of aggregation—a *superforecast*—and then the end forecast is constructed by disaggregating these superforecasts down to the level of detail required. The following are examples of bottom-up and top-down forecasting strategies.

Example 9.1 (BOTTOM-UP FORECASTING) An airline is interested in getting forecasts of load factors (occupancy) for each flight in each compartment for an upcoming season. The airline stores data of each past customer, itinerary, and fare class. This itinerary-level data is first used to make forecasts for the number of customers expected to book for each itinerary and fare-class combination for every day of the season. Next, these detailed forecasts are added together to produce an aggregate forecast for the seasonal load-factors.

Example 9.2 (TOP-DOWN FORECASTING) A hotel is interested in a forecast of the number of people expected to book for each future date in each room-category and length-of-stay combination. The hotel first forecasts the total number of guests who will book to arrive on each day in each rate category (the superforecast). Then it forecasts the fraction of guests that stay for a specified length of time (a length-of-stay distribution). Finally, it combines these two components to arrive at an end forecast of expected number who will start their stay on a specific date and stay a certain number of days by multiplying the forecast for the aggregate number of guests on a specific date by the estimate of the fraction that will stay for a given length of time.

Which strategy is most appropriate is not always clear-cut. It depends on the data that is available and accessible to an automated system on a daily basis, the outputs required, and the types of forecasts already being made and available for use. Moreover, the "right" answer in most cases is that *both* strategies are required because certain phenomena can be estimated only at a low level of aggregation, while others can be estimated only at a high level of aggregation.

For example, it is clear that in an airline network if one wants an estimate of demand for each itinerary and fare-class combination, then aggregate booking-class or flight-leg data will not be sufficient; data for each passenger itinerary is required. At the same time, such passenger-level data is sparse, with often only a handful of bookings occurring for any given combination in a year. Hence, aggregate phenomena such as daily or weekly seasonalities, holiday effects, or upward or downward trends in total demand are—for all practical purposes—unobservable at the disaggregate level; one must look at aggregate booking data over many itinerary and fare-class combinations to observe such effects. Even with good passenger-level data, one may therefore need to aggregate data and perform aggregate forecasts to identify important "large-scale" phenomena. As a result, hybrid combinations of bottom-up and top-down approaches are the norm in practice.

9.2 Estimation Methods

Estimation is the problem of finding model parameters that best describe a given set of observed data. Forecasting, in contrast, involves predicting future, unobserved values. Thus, estimation is generally *descriptive* (characterizing what *has been* observed), while forecasting is *predictive* (characterizing what *will be* observed). Roughly, in the RM context, estimation is the calibration of a forecasting model's parameters (hence it also is called *forecast calibration*) and is done relatively infrequently; while forecasting is the use of the estimated model to predict future values, and is performed frequently on an operational basis.

For example, an estimate of price sensitivity based on past sales data may be used in a forecast of future demand. Similarly, many forecasting methods are based on estimating the parameters of a dynamic model from historical demand data, which is subsequently used to predict future values of demand. Yet this distinction between estimation and forecasting is not always very sharp. In some methods, such as the Kalman filtering, estimation and forecasting work in lock-step, one after another.

Here we examine methodology for parameter estimation and discuss some of the theoretical and practical issues that arise.

9.2.1 Estimators and Their Properties

An estimator represents, in essence, a formalized "guess" about the parameters of the underlying distribution from which a sample (the observed data) is assumed to be drawn. Estimators can take on many forms and can be based on different criteria for a "best" guess. We use demand estimation as our example, but the ideas apply more generally.

9.2.1.1 Nonparametric Estimators

Let the random variable Z_k denote the kth observation of demand, and let $F(z) = P(Z_k \leq z)$ denote the distribution of Z_k. Nonparametric estimation methods do not make any assumptions on the underlying distribution $F(\cdot)$. For example, we could estimate $F(z)$ by simply computing the fraction of observations in the sample that are less than or equal to z for each value of z. This empirical distribution then forms a nonparametric estimate of the true distribution $F(z)$.

Nonparametric estimates of this type have the advantage of not requiring any assumptions on the form of the distribution. However, as mentioned earlier, they typically require more data to produce accurate estimates and do not allow one to extrapolate beyond the observed data easily. For example, if there were no observations less than 10 in a data set, then the empirical distribution would estimate that $P(Z_k \leq z) = 0$ for all values of z less than 10.

9.2.1.2 Parametric Estimators

For a parametric estimator, we assume that the underlying distribution of Z_k is of the form

$$P(Z_k \leq z | \boldsymbol{\beta}, \mathbf{y}_k) = F(z, \boldsymbol{\beta}, \mathbf{y}_k), \qquad (9.1)$$

where $\mathbf{y}_k = (y_{k1}, \ldots, y_{kM})$ is a vector of M explanatory (independent) variables (time, indicators of holiday events, prices, lagged observations of Z_k itself, and so on) and $\boldsymbol{\beta} = (\beta_1, \ldots, \beta_M)$ is a M-dimensional vector of parameters. For ease of exposition, we assume that the dimension of $\boldsymbol{\beta}$ and $\mathbf{y}_k$ are the same, though this is not necessary.

Assume we have a sequence of N independent observations $z_1, \ldots, z_N$, with values y_{km}, $k = 1, \ldots, N$, $m = 1, \ldots, M$ for the explanatory variables, alternatively represented by vectors $\mathbf{y}_1, \ldots, \mathbf{y}_N$ or by a $N \times M$ matrix $\mathbf{Y} = [y_{km}]_{k=1,\ldots,N,\ m=1,\ldots,M}$. The estimation problem, then, is to determine the unknown parameters $\boldsymbol{\beta}$ using only the sample of the N observations $\{Z_k\}$ (the data) and the values of the explanatory variables $\mathbf{Y}$ corresponding to each observation (characteristics of the observed data).

Estimation and Forecasting 421

It is usually convenient to express the relationship between the demand and the explanatory variables by a simple functional form consisting of a systematic (deterministic) component and an additive noise component:[1]

$$Z_k = \zeta(\boldsymbol{\beta}, \mathbf{y}_k) + \xi_k, \qquad (9.2)$$

where the randomness comes from the error term ξ_k. Many of the regression and time-series forecasting models of this chapter can be viewed as manifestations of (9.2). The following is a common example of (9.2):

Example 9.3 (LINEAR MODEL) Consider the linear model of demand

$$Z_k = \boldsymbol{\beta}^\top \mathbf{y}_k + \xi_k, \quad k = 1, \ldots, N, \qquad (9.3)$$

where ξ_k are i.i.d. $N(0, \sigma^2)$ random variables, independent also of the explanatory variables $\mathbf{y}$. Z is often referred to as the *dependent* variable and the vector $\mathbf{y}$ as the *independent* variables. The distribution of Z in terms of (9.1) is then

$$F_{Z_k}(z|\boldsymbol{\beta}, \mathbf{y}_k) = P(Z_k \leq z|\boldsymbol{\beta}, \mathbf{y}_k) = \Phi(\frac{z - \boldsymbol{\beta}^\top \mathbf{y}_k}{\sigma}),$$

where $\Phi(\cdot)$ is the standard normal distribution.

9.2.1.3 Properties of Estimators

If the N observations, $\mathbf{z}_N = (z_1, \ldots, z_N)$, are considered independent realizations of $\mathbf{Z}_N = (Z_1, \ldots, Z_N)$, then the estimator based on these observations is a function of N i.i.d. random variables, $\hat{\boldsymbol{\beta}}(\mathbf{Z}_N)$, and is therefore itself a random variable. What properties would we like this (random) estimator to have?

Bias For one, it would be desirable if the expected value of the estimator equaled the actual value of the parameters—that is, if

$$E[\hat{\boldsymbol{\beta}}(\mathbf{Z}_N)] = \boldsymbol{\beta}.$$

If this property holds, the estimator is said to be an *unbiased estimator*; otherwise, it is a *biased estimator*. The estimator of the mth parameter, $\hat{\beta}_m$, is said to have a *positive bias* if its expected value exceeds β_m, and a *negative bias* if its expected value is less than β_m.

If the estimator is unbiased only for large samples of data—that is, it satisfies

$$\lim_{N \to \infty} E[\hat{\boldsymbol{\beta}}(\mathbf{Z}_N)] = \boldsymbol{\beta},$$

then it is called an *asymptotically unbiased estimator*. All unbiased estimators are, of course, also asymptotically unbiased.

---
[1] We drop the notation conditioning on $\boldsymbol{\beta}$ and $\mathbf{y}$, $(\cdot|\boldsymbol{\beta}, \mathbf{y})$, when it is obvious from the context.

Efficiency An estimator $\hat{\beta}(\mathbf{Z})$ is said to an *efficient estimator* if it is unbiased and the random variable $\hat{\beta}(\mathbf{Z})$ has the smallest variance among all unbiased estimators. Efficiency is desirable because it implies the variability of the estimator is as low as possible given the available data. The Cramer-Rao bound[2] provides a lower bound on the variance of *any* estimator, which can be used to prove an estimator is efficient. In particular, if an estimator achieves the Cramer-Rao bound, then we are guaranteed that it is efficient. An estimator can be inefficient for a finite sample but *asymptotically efficient* if it achieves the Cramer-Rao bound when the sample size is large.

Consistency An estimator is said to be *consistent* if for any $\delta > 0$,

$$\lim_{N \to \infty} P(|\hat{\beta}(\mathbf{Z}_N) - \beta| < \delta) = 1,$$

that is, if it converges in probability to the true value β as the sample size increases. Consistency assures us that with sufficiently large samples of data, the value of β can be estimated arbitrarily accurately.

Ideally, we would like our estimators to be unbiased, efficient, and consistent, but this is not always possible. We revisit these properties in Section 9.5.1.2 on specification errors.

9.2.2 Minimum Square Error (MSE) and Regression Estimators

One class of estimators is based on the *minimum mean-square error* (MSE) criterion—also referred to as *regression estimators*. MSE estimators are most naturally suited to the case where the forecast quantity has an additive noise term as in (9.2). Given a sequence of observations $z_1, \ldots, z_N$ and associated vectors of explanatory variable values $\mathbf{y}_1, \ldots, \mathbf{y}_N$, the MSE estimate of the vector β is the solution to

$$\min_{\beta} \sum_{k=1}^{N} [z_k - \zeta(\beta, \mathbf{y}_k)]^2, \tag{9.4}$$

where $\zeta(\beta, \mathbf{y}_k)$ is as defined in (9.2). The minimization problem (9.4) can be solved using standard nonlinear optimization methods such as conjugate-gradient or quasi-Newton. However, the problem is greatly simplified if the function $\zeta(\beta, \mathbf{y}_k)$ and the error terms have a specialized form, as shown next.

Ordinary Least-Squares (OLS) and Linear-Regression Estimators If the function ζ in (9.1), the error terms ξ_k in (9.2), and explana-

---
[2] See DeGroot [151], pp. 420–430 for a discussion of the Cramer-Rao bound.

tory variables $\mathbf{y}_k$ satisfy the assumptions listed in Table 9.1, the MSE estimates are also known as the *ordinary least-squares (OLS) estimators*—or *linear-regression estimators*. Specifically, suppose the observations Z_k are linear functions of M explanatory variables of the form,

$$Z_k = \boldsymbol{\beta}^\top \mathbf{y}_k + \xi_k.$$

Furthermore, suppose the explanatory variables $\mathbf{y}_k$ are uncorrelated and the error term ξ_k are independent, normal random variables that have means of zero and identical variances (homoscedasticity). Then the OLS estimators are the values $\boldsymbol{\beta}$ that solve

$$\min_{\boldsymbol{\beta}} \sum_{k=1}^{N} \left[z_k - \boldsymbol{\beta}^\top \mathbf{y}_k\right]^2.$$

We can write equation (9.2) in matrix form as

$$\mathbf{Z} = \mathbf{Y}\boldsymbol{\beta} + \boldsymbol{\xi}, \qquad (9.5)$$

where $\mathbf{Y} = [y_{km}]_{k=1,\ldots,N,\ m=1,\ldots,M}$, and $\mathbf{Z} = (Z_1,\ldots,Z_N)$. The MSE estimates for $\boldsymbol{\beta}$, given N observations $\mathbf{z} = (z_1,\ldots,z_N)$, are then

$$\hat{\boldsymbol{\beta}} = (\mathbf{Y}^\top \mathbf{Y})^{-1} (\mathbf{Y})^\top \mathbf{z}, \qquad (9.6)$$

assuming the matrix $\mathbf{Y}^\top \mathbf{Y}$ is invertible.

Example 9.4 Consider the following model of demand:

$$Z_k = \beta + \xi_k. \qquad (9.7)$$

This model has one scalar parameter β, which is constant over time, and is equivalent to having $M = 1$ and $\mathbf{Y} = (1,\ldots,1)$ in (9.5). Assume ξ_k is normally distributed with mean 0 and constant variance. Then if we have N observations, $z_1,\ldots,z_N$, the MSE estimate $\hat{\beta}$ based on this data solves

$$\min_{\beta} \sum_{k=1}^{N} [z_k - \beta]^2.$$

Applying (9.6) and noting that $\mathbf{Y}^\top \mathbf{Y} = N$ and $\mathbf{Y}^\top \mathbf{z} = \sum_{k=1}^{N} z_k$, we obtain

$$\hat{\beta} = \frac{1}{N} \left(\sum_{k=1}^{N} z_k \right),$$

which is simply the sample mean of the data.

Table 9.1. Assumptions of ordinary least-squares (OLS) estimation.

Assumption	Violation	Test	Fix
$\zeta(\beta, y_k)$ Linear	Nonlinear relationship	Specification tests (Section 9.5.1.2)	Transformations, Nonlinear Regression
Homoscedasticity (ξ's have constant variance across observations)	Heteroscedasticity (Variances of ξ's different for different observations)	White [563]	Transformations; GLS or MLE estimation
Errors ξ's are uncorrelated across observations	Serial correlation of observations	Durbin-Watson [166–168]; von Neumann ratio test [243]	GLS, MLE
Errors ξ's are normally distributed	Non-normal observed errors	Shapiro-Wilk	Transformations; MLE
Explanatory variables $\mathbf{y}$ are uncorrelated	Multicollinearity (some of the elements of $\mathbf{y}$ are strongly related)	Belsley-Kuh-Welsch test [44]	Drop some of the variables

From (9.6) it can be seen that the OLS estimator $\hat{\beta}$ is a linear function of the random observations $\mathbf{Z}$, which makes computing the estimates quite easy. In addition, the OLS estimators have several desirable properties: they are consistent, unbiased, and efficient under very general conditions. For these reasons, the MSE/linear-regression estimator is popular in practice.

Regression is widely used in price-based management for estimating price sensitivity, market shares, and the effects of various marketing variables (such as displays and promotions) on demand. Regression estimates are somewhat less common in quantity-based RM forecasting application such as airline and hotel RM because in these applications it is often difficult to obtain data on the exogenous explanatory variables as an automated data feed. When regression is used in quantity-based RM, typically the only explanatory variables in the model are the historical demand data itself (the explanatory variables are past demand observations). However, in such cases formal time-series models of the type discussed in Section 9.3.2 are usually preferred.

When any of the assumptions of the OLS regression in Table 9.1 is violated, one has to resort to more advanced regression techniques such as *generalized least squares* (GLS), *seemingly unrelated regressions* (SUR), and two-stage and three-stage least squares (2SLS, 3SLS) (see Greene [220]). A description of these methods is beyond the scope of this chapter.

9.2.3 Maximum-Likelihood (ML) Estimators

While regression is based on the least-squares criterion, *maximum-likelihood (ML) estimators* are based on finding the parameters that maximize the "likelihood" of observing the sample data, where *likelihood* is defined as the probability of the observations occurring. More precisely, given a probability-density function f_Z of the process generating Z_k, $k = 1, \ldots, N$, which is a function of a vector of parameters β and the observations of the explanatory variables, $\mathbf{y}_k$, the likelihood of observing value z_k as the kth observation is given by $f_Z(z_k|\beta, \mathbf{y}_k)$. The likelihood of observing the N observations $(z_1, \mathbf{y}_1), \ldots, (z_N, \mathbf{y}_N)$ is then

$$\mathcal{L} = \prod_{k=1}^{N} f_Z(z_k|\beta, \mathbf{y}_k). \tag{9.8}$$

The ML estimation problem is to find a β that maximizes the likelihood $\mathcal{L}$. It is more convenient to maximize the log-likelihood, $\ln \mathcal{L}$, because this converts the product of functions in (9.8) to a sum of functions. Since the log function is strictly increasing, maximizing the log-likelihood is equivalent to maximizing the likelihood. This gives the ML problem:

$$\max_{\beta} \sum_{k=1}^{N} \ln f_Z(z_k|\beta, \mathbf{y}_k).$$

In special cases, this problem can be solved in closed form. Otherwise, if the function $f_Z(\cdot)$ is a differentiable function, gradient-based optimization methods such as Newton's method can be used to solve it numerically.

ML estimators have good statistical properties under very general conditions; they can be shown to be consistent, asymptotically normal, and asymptotically efficient, achieving the Cramer-Rao lower bound on the variance of estimators for large sample sizes.

Example 9.5 (ESTIMATING THE MEAN OF A NORMAL DISTRIBUTION) Consider the following model of demand from Example 9.4:

$$Z_k = \beta + \xi_k. \tag{9.9}$$

Recall that the model assumes that the scalar parameter β is constant over time, and ξ_k is normally distributed with mean 0 and constant variance σ. Suppose we have N observations, $z_1, \ldots, z_N$. Then the ML estimator solves

$$\max_{\beta} \prod_{k=1}^{N} \frac{1}{\sigma\sqrt{2\pi}} e^{-(z_k-\beta)^2/2\sigma^2}.$$

Taking the log of the objective function yields

$$\max_{\beta} -\frac{N\ln(2\pi)}{2} - N\ln(\sigma) - \sum_{k=1}^{N} \frac{(z_k - \beta)^2}{2\sigma^2}.$$

Differentiating with respect to β and setting the result to zero, one can show that the ML estimator is

$$\hat{\beta} = \frac{1}{N} \sum_{k=1}^{N} z_k,$$

which is just the sample mean. Note despite the fact that the estimation criterion is different, this estimator is the same as the MSE estimator of Example 9.4.

Example 9.6 (ESTIMATING THE PARAMETERS OF MULTINOMIAL-LOGIT MODEL) The MNL discrete-choice model is described in Section 7.2.2.3. The data consists of a set of N customers and their choices made from a finite set S of alternatives. Associated with each alternative j is a vector $\mathbf{y}_j$ of explanatory variables (assume for simplicity there are no customer-specific characteristics). The probability that a customer selects alternative i is then given by (assuming that all customers face the same choice-set of products)

$$P_i(S) = \frac{e^{\beta^\top \mathbf{y}_i}}{\sum_{j \in S} e^{\beta^\top \mathbf{y}_j} + 1}, \tag{9.10}$$

where β is a vector of (unknown) parameters. Let $c(k)$ be the choice made by customer k. The likelihood function is then

$$\mathcal{L} = \prod_{k=1}^{N} \left[\frac{e^{\beta^\top y_{c(k)}}}{\sum_{j \in S} e^{\beta^\top y_j} + 1} \right].$$

The maximum-likelihood estimate $\hat{\beta}$ is then determined by solving

$$\max_{\beta} \ln \mathcal{L}. \tag{9.11}$$

While this maximum-likelihood problem cannot be solved in closed form, it has good computational properties. Namely, there are closed-form expressions for all first and second partial derivatives of the log-likelihood function, and it is jointly concave in most cases (McFadden [372]; Hausman and McFadden [244]). The ML estimator has also proved to be robust in practice. (See Ben-Akiva and Lerman [48] for further discussion and case examples.)

9.2.4 Method of Moments and Quantile Estimators

While MSE and ML estimators are the most prevalent, several other estimators are also used in practice. Two common ones are the *method of moments* and *quantile estimators*.

In the method of moments, one equates moments of the theoretical distribution to their equivalent empirical averages in the observed data. This yields a system of equations that can be solved to estimate the unknown parameters β. The following example illustrates the idea:

Example 9.7 (ESTIMATING THE PARAMETERS OF A NORMAL DISTRIBUTION) Suppose we want to estimate the parameters of a normal distribution. The sample mean and sample second moment are computed as follows:

$$\bar{z} = \frac{1}{N} \sum_{k=1}^{N} z_k$$

$$\overline{z^2} = \frac{1}{N} \sum_{k=1}^{N} z_k^2.$$

Equating these to the theoretical mean and second moment yields the system of equations

$$\bar{z} = \mu$$
$$\overline{z^2} = \sigma^2 + \mu^2.$$

Solving for μ and σ gives the estimates $\hat{\mu} = \bar{z}$ and $\hat{\sigma} = \sqrt{\overline{z^2} - (\bar{z})^2}$.

Alternatively, we can use quantile estimates based on the empirical distribution to estimate the parameters β of a distribution. For example, we might estimate the mean of a normal distribution by noting that as the normal distribution is symmetric, the mean and median are the same. Hence, we can estimate the mean by computing the median of a sequence of N observations. More generally, one can compute a number of quantiles of a data set and equate these to the theoretical quantiles of the parametric distribution. In general, if m parameters need to estimated, m different quantiles are needed to produce m equations in m unknowns (for a normal distribution, for example, one could equate the 0.25 and 0.75 quantiles of the data to the theoretical values to get two equations for the mean and variance). Quantile estimation techniques are sometimes preferred, as they tend to be less sensitive to outlier data than are MSE and ML estimators.

9.2.5 Endogeneity, Heterogeneity, and Competition

Table 9.1 lists the standard problems associated with classical regression—correlation of the error terms, collinearity, and so on—and techniques for dealing with violations of the assumptions. Such problems and their corrective measures are well known and can be found in many standard econometric books. In this section, we focus on a few nonstandard estimation problems that are of particular importance for RM applications—endogeneity, heterogeneity, and competition.

9.2.5.1 Endogeneity

The model (9.2) is said to suffer from endogeneity if the error term ξ is correlated with one of the explanatory variables in $\mathbf{y}$. This is a common problem in RM practice, both in aggregate-demand function estimation and in disaggregate, discrete-choice model estimation.

For example, products may have some unobservable or unmeasurable features—quality, style, reputation—and the selling firm typically prices its products accordingly. So if there are two firms in the market with similar products and one has higher nonquantifiable quality, we may observe that the firm with the higher-quality product has both a larger market share and a higher price. A naive estimation based on market shares that ignores the unobserved quality characteristics would lead to the odd conclusions that higher price leads to higher market share! Such effects are widespread in price-elasticity estimation because we can rarely observe all relevant product and firm characteristics and price is usually correlated with many of these unobservable characteristics.

Econometricians call this problem *endogeneity* or *simultaneity*. The technical definition is that the random-error term in (9.2) is correlated with one of the explanatory variables, $E[\mathbf{Y}^\top \boldsymbol{\xi}] \neq 0$, or equivalently (in the case of linear regression) these vectors are not orthogonal. So while $\boldsymbol{\xi}$ is supposed to represent all unobservable customer and product characteristics that influence demand for a given set of explanatory variables $(Z|\mathbf{y})$, some of the explanatory variables $\mathbf{y}$ also contain information on the unobservable attributes through their correlation with $\boldsymbol{\xi}$.

Econometric techniques to correct endogeneity fall under a class of methods called *instrumental-variables (IV) techniques*, attributed to Reiersøl [438] and Geary [202]. Two-stage and three-stage least-squares methods (2SLS and 3SLS) are some of the popular IV techniques. Instrumental variables are exogenous variables that are correlated with an explanatory variable but are uncorrelated with the error term $\boldsymbol{\xi}$. If there are such IVs, we can use them to "remove" the problematic correlation between the independent variables $\mathbf{y}$ and $\boldsymbol{\xi}$.

We illustrate the idea for the case of linear regression. In (9.5) suppose

$$E[\mathbf{Y}^\top \boldsymbol{\xi}] \neq 0.$$

However, suppose there exist M instrumental variables (we can use some of the y's to construct this vector of IVs) for each observation so that we have a $N \times M$ matrix $\mathbf{V}$ with the property that $E[\mathbf{V}^\top \boldsymbol{\xi}] = 0$, and $E[\mathbf{V}^\top \mathbf{Y}]$ is nonsingular. Then the IV estimator is

$$\hat{\beta}_{\mathrm{IV}} = [\mathbf{V}^\top \mathbf{Y}]^{-1} \mathbf{V}^\top \mathbf{Z}, \tag{9.12}$$

where $\mathbf{Z} = \mathbf{Y}\beta + \boldsymbol{\xi}$. The IV estimator is a consistent estimator of β, which can be shown by substituting $\mathbf{Z} = \mathbf{Y}\beta + \boldsymbol{\xi}$ in (9.12):

$$\begin{aligned} E[\hat{\beta}_{\mathrm{IV}}] &= E[[\mathbf{V}^\top \mathbf{Y}]^{-1} \mathbf{V}^\top (\mathbf{Y}\beta + \boldsymbol{\xi})] \\ &= \beta + E[[\mathbf{V}^\top \mathbf{Y}]^{-1} \mathbf{V}^\top \boldsymbol{\xi}]. \end{aligned}$$

For a given set of N observations $\mathbf{z} = (z_1, \ldots, z_N)$, the IV estimator can be calculated by the sample average

$$\hat{\beta}_{\mathrm{IV}} = \beta + [\frac{\mathbf{V}^\top \mathbf{Y}}{N}]^{-1} [\frac{\mathbf{V}^\top \boldsymbol{\xi}}{N}],$$

which converges by the weak law of large numbers to β w.p.1 as $\frac{\mathbf{V}^\top \mathbf{Y}}{N} \to E[\mathbf{V}^\top \mathbf{Y}]$ and $\frac{\mathbf{V}^\top \boldsymbol{\xi}}{N} \to 0$ w.p.1.

A regression with an IV transformation is called an *IV regression* (see Greene [220] and Woolrdige [581] for details and examples of IV methods) and a *generalized IV regression* if we use more than M IV variables. There are no mechanically generated IVs that work for all cases. It often requires considerable ingenuity to find good IVs and to argue that they can in fact serve to correct for endogeneity. This is what makes the technique rather difficult to apply, requiring the skills of an experienced econometrician.

In nonlinear problems, IV techniques become more difficult to apply. For example, one often encounters endogeneity when estimating discrete-choice demand models such as the MNL model from aggregate data (prices correlated with unobservable product characteristics). However, the problem is hard to correct because the aggregate demand is a nonlinear function of the utilities of each product and the endogeneity is present in the equation for the utilities. So using any IV technique for correcting for endogeneity becomes computationally challenging, as pointed out by Berry [53]. Berry [53] and Berry, Levinsohn, and Pakes [52] recommend that for the case of discrete-choice models in an oligopoly setting, one use measures of the firm's costs and the attributes of the products of the other firms as IVs. See also Besanko, Gupta, and Jain [63] for estimating a logit model in the presence of endogeneity due to competition.

9.2.5.2 Heterogeneity

Customer heterogeneity is important to understand in RM. In Section 7.2.3 we examined a few models of heterogeneity—namely, the finite-mixture logit model and the random-coefficients discrete-choice model. Here, we discuss how to estimate these models.

Estimation of the finite-mixture logit model is relatively straightforward. First, we must determine the number of segments. If there is no *a priori* knowledge of the number, we iterate the estimation procedure, increasing or decreasing the number of segments in each round, using suitable model-selection criteria (see Section 9.5.1) to decide on the optimal number of segments. For a given number of segments L, we find the parameters that maximize the log-likelihood function. For the finite-mixture logit model of Section 7.2.3.1, this would amount to maximizing the following likelihood function based on the purchase histories of N customers:

$$\mathcal{L} = \prod_{k=1}^{N} \sum_{l=1}^{L} \frac{e^{\nu_l}}{\sum_{i=1}^{L} e^{\nu_i}} \frac{e^{\beta^{l\top} \mathbf{y}_{c(k)}}}{\sum_{j=1}^{n} e^{\beta^{l\top} \mathbf{y}_j}},$$

where $c(k)$ is the choice made by customer k. The only difficulty, from an optimization point of view, is that taking logs on both sides does not convert the right-hand side into a sum of terms, so the maximization is somewhat more challenging than the estimation of standard logit models.

Estimation of the random-coefficient logit, likewise, uses maximum-likelihood estimation and is more difficult in general than the standard multinomial logit. Consider the model given in Section 7.2.3.2. Assuming the parameters follow a normal distribution, the likelihood function that needs to be maximized is given by

$$\mathcal{L} = \prod_{k=1}^{N} \int_{\beta} \frac{e^{\beta^\top y_{c(k)}}}{\sum_{j=1}^{n} e^{\beta^\top y_j}} f(\beta) d\beta, \qquad (9.13)$$

where f is the M-dimensional joint normal p.d.f. (with an identity covariance matrix if the taste parameters are independent). If the distributions of the parameters β are modeled as a joint normal distribution with a general covariance matrix structure, then evaluation of the integral is quite difficult in practice. However, the extreme value distribution has been integrated out in (9.13), and we do end up with a logit-like term inside the integrals.

One of the problems dealing with unobservable heterogeneity in the population is that we often have to assume a distribution of heterogeneity without having much evidence as to its specification. Many times, a distribution is chosen for analytical or computational convenience. Unfortunately, a situation can arise where two radically different distributions of heterogeneity equally support the aggregate demand observations. This was pointed out by Heckman and Singer [248], who illustrated this overparameterization with the following example:

Example 9.8 Consider an aggregate-demand function based on a heterogeneity parameter θ. The variance on the distribution of θ represents the degree of heterogeneity. Let the demand for a particular value of θ be given by the distribution

$$G_1(z|\theta) = 1 - e^{-z\theta}, \quad z \geq 0, \ \theta > 0,$$

and let θ be equal to a constant η with probability 1 (essentially saying the population is homogeneous). The aggregate-demand distribution then is $F_1(z) = 1 - e^{-z\eta}$.

Consider another possible specification where

$$G_2(z|\theta) = 1 - \int_{z(2\theta)^{-0.5}}^{\infty} \frac{2}{\sqrt{2\pi}} e^{-w^2/2} dw, \quad z \geq 0$$

and the distribution of θ given by $\eta^2 e^{-\eta^3 \theta}$. This also turns out to lead to an aggregate-demand distribution given by $1 - e^{-z\eta}$. So based only on aggregate demand data, it is impossible to identify which specification is correct.

Therefore, one should proceed with caution when inferring a functional form for unobserved heterogeneity from aggregate data.

Nonparametric methods avoid the problem of having to specify a distribution, and Jain, Vilcassim, and Chintagunta [267] follow this strategy. Assume that the coefficients of the MNL model β in (9.10) are randomly drawn from a discrete multivariate probability distribution $G(\Theta)$. That is, the kth customer is assumed to make his choice using β_k, whose components are drawn from $G(\Theta)$. $G(\cdot)$ is considered a discrete distribution with support vectors $\theta_1, \ldots, \theta_L$. They estimate the number of support vectors L, the location of the support vectors, and the probability mass θ_i associated with the ith support vector from observed data.

9.2.5.3 Competition

If one has access to information on prices and demand for an entire market (such as MIDT data for airlines and scanner-panel data sold by marketing research firms), it is possible to separately estimate competitive- and own-price effects. A common strategy in such cases is to assume a model of competition between the firms, derive the equilibrium conditions implied by this model, and then estimate the parameters subject to these equilibrium conditions. We illustrate this approach with an example:

Example 9.9 Assume a homogeneous population of customers who choose among n products according to the MNL choice rule. Then the theoretical share of product j is given as in Section 7.2.2.3,

$$P_j = \frac{e^{\boldsymbol{\beta}^\top \mathbf{y}_j}}{\sum_{i=1}^n e^{\boldsymbol{\beta}^\top \mathbf{y}_i}}, \quad (9.14)$$

where price is one of the explanatory variables in $\mathbf{y}_j$. One way to estimate the parameters $\boldsymbol{\beta}$ is by equating the observed market share to the theoretical prediction of equilibrium. It is convenient to take logs in doing this, which yields the following system of equations relating market shares to choice behavior:

$$\ln P_j = \boldsymbol{\beta}^\top \mathbf{y}_j - \ln(\sum_{i=1}^n e^{\boldsymbol{\beta}^\top \mathbf{y}_i}), \quad j = 1, \ldots, n. \quad (9.15)$$

Next assume that prices are formed by a Bertrand-style competition in prices (see Section 8.4.1.4). Let c_j be the constant marginal cost of production for product j. The profit function for product j is given by

$$V_j(p_j) = (p_j - c_j) N P_j, \quad (9.16)$$

where N is the size of the population. Let β_p be the coefficient of price in (9.14). Differentiating (9.16) with respect to p_j and setting it to zero, we get the first-order equilibrium conditions,

$$(p_j - c_j)\beta_p P_j(1 - P_j) + P_j = 0, \quad j = 1, \ldots, n. \quad (9.17)$$

The vector of parameters $\boldsymbol{\beta}$ is then estimated by attempting to fit a solution to (9.15) and (9.17) simultaneously. This can be done using, say, nonlinear least-squares estimation.

9.3 Forecasting Methods

We next turn to forecasting methods, which explicitly attempt to "predict" the future values of a sequence of data. For RM, we are mostly interested in forecasting demand (demand to come, as well as aggregate demand for the resource and at various levels of aggregation), though in many cases one also needs to forecast quantities such as market prices, length of stay (in hotel RM), cancellation and no-show rates, and so on. Indeed, the methods presented here, by and large, apply to a wide variety of forecasting tasks, though for purposes of illustration we focus on demand forecasting as our canonical application.

Forecasting is a vast topic, spanning a diverse range of fields including statistics, computer science, engineering, and economics. Over the years, a core set of forecasting methods have been developed and new improvements continue despite the maturity of the field. Some of these forecasting methods are based on rigorous mathematical and statistical foundations, while others are largely heuristic in nature.

Yet despite this long history and vast body of research on forecasting, there are few published reports that document the performance of various forecasting methods in RM applications. Presentations on forecasting by practitioners at industry conferences often suffer from the proprietary nature of the material, with key details either omitted or disguised. The same can be said of most presentations by RM system vendors. Nevertheless, one can still glean some useful insights into current practice from these sources.

For one, most forecasting algorithms in RM practice are variations of standard methods, and most are not particularly complicated or mathematically sophisticated. Also, many vendors use multiple algorithms, which allow users the option of choosing one or more methods, or, alternatively, the system may combine the forecasts from the various methods itself (see Section 9.3.11). Finally, the majority of forecasting effort in practice is directed at data-related tasks—collection, preprocessing and cleansing—rather than on forecasting methodology per se.

In terms of forecasting methods, the emphasis in RM systems is on speed, simplicity, and robustness, as a large number of forecasts have to be made and the time available for making them is limited. For example, if an airline has 50,000 itineraries in 10 fare classes that it reforecasts 40 times during a sales period (typical numbers for a medium-size airline), then they must forecast nearly *2 million* demand quantities every day! And this does not include forecasts of important auxiliary quantities such as cancellation and no-show rates. It is little wonder, then, that fast, simple methods are preferred in RM systems.

Forecasting is normally performed overnight in a batch process and then fed to the optimization modules, so the time window for completing all control operations ranges from six to eight hours at most. Forecasting model calibration (estimation), in turn, can only be done off line and infrequently.

Robustness of the forecasts is also important in practice for these same reasons. If a large number of forecasts are off widely and the system starts generating exceptions, analysts may be overwhelmed by the amount of manual intervention required. Hence, performance—in terms of forecast accuracy under "normal" data conditions—while always a desirable criteria, has to be balanced against these "real world" speed constraints and robustness considerations. We next provide an overview of RM forecasting algorithms, starting with ad hoc and time-series methods and progressing to Bayesian, state-space (Kalman filter), and machine-learning (neural network) methods.

9.3.1 Ad-Hoc Forecasting Methods

The first class of methods we look at are known as ad-hoc forecasting methods because their reasoning is largely heuristic in nature. The term *ad hoc* is somewhat misleading, however, as many of these methods turn out to have good theoretical properties despite their heuristic origins. They are also sometimes referred to as *structural* forecasting methods because they proceed by assuming a compositional structure on the data, breaking up and composing the series into hypothesized patterns (see Figure 9.3). These include the following three types of components:

- **Level** The typical or "average" value of the data, though in ad-hoc methods the level is not defined as a statistical average in any formal sense.

- **Trend** A predictable increase or decrease in the data values over time. Most often these are modeled as linear increases or decreases, but other functions may be used.

- **Seasonality** A periodic or repeating pattern in the data values over time—for example, as produced by day-of-week or time-of-year effects.

Ad-hoc forecasting methods are intuitive, are simple to program, and maintain and perform well in practice. For these reasons, they are prevalent in RM practice.

A common strategy of ad-hoc forecasting methods is to try to "smooth" the data or average-out the noise components to estimate the level, trend, and seasonality components in the data. These estimates of

Estimation and Forecasting 435

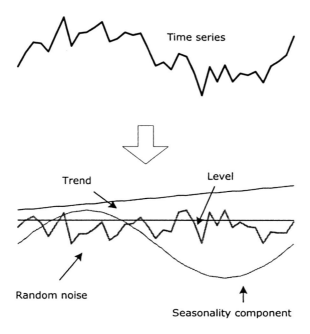

Figure 9.3. Time series of data and its components.

the smoothed series are then used to forecast future values, as we show next.

9.3.1.1 M-Period Moving Average

Let t represent the current time, and suppose we want to forecast values at time $t+k$ in the future, $\hat{Z}_{t+k}$, called the *k-period ahead forecast*. Let $z_1, \ldots, z_t$ denote the observed demand data, and $\hat{Z}_{t+1}, \ldots, \hat{Z}_{t+K}$ denote the forecasts. To forecast one period ahead, one simple approach is to use the average of the past M observations. That is, the forecast for period $t+1$ is given by

$$\hat{Z}_{t+1} = \frac{z_t + z_{t-1} \ldots + z_{t-M+1}}{M}, \qquad (9.18)$$

called the *simple M-period moving-average forecast*. M is called the *span* of the moving average. The formula for the k-period ahead forecast is given by

$$\hat{Z}_{t+k} = \hat{Z}_{t+1}, \quad k = 2, \ldots, K.$$

A different way of writing (9.18) is

$$\hat{Z}_{t+1} = \hat{Z}_t + \frac{z_t - z_{t-M}}{M},$$

which is computationally faster. If t is less than M (that is, in the initial stages of forecasting), one can use $M = t$.

The moving-average method is very simple and fast, but its motivation is largely heuristic. The idea is simply that the most recent observations serve as better predictors for the future than do older data. Therefore, instead of taking the forecast as the average of *all* the data, we average only the M most recent data observations.

The moving-average forecast responds more quickly to underlying shifts in the demand process if the span M is small, but a small span results in a more volatile forecast (one that is more sensitive to noise in the data). In practice, M may range from 3 to 15, but the value depends heavily on the data characteristics and the units used for the time intervals.

When the data exhibits an upward or downward trend, the moving average method will systematically underforecast or overforecast. To handle such cases, variations such as double or triple moving average have been developed, but for such data one of the exponential smoothing methods given next is usually preferred.

9.3.1.2 Exponential Smoothing

Exponential-smoothing methods are among the most popular forecasting methods used in RM practice because they are simple and robust and generally have good forecast accuracy. We look at three variations of exponential smoothing. First, however, we formally define the following component estimates of the forecast:

A_t = the estimate of the level (average) for period t,
T_t = the estimate of the trend for period t, and
S_t = the estimate of the seasonality factor for period t.

See Figure 9.3 for an illustration of these components.

Simple Exponential Smoothing This simplest version of exponential smoothing is defined by a single parameter, $0 < \alpha < 1$, called the *smoothing constant for the level*. The forecast for time-period $t + 1$ is given by

$$\hat{Z}_{t+1} = A_t = \alpha z_t + (1 - \alpha)\hat{Z}_t. \qquad (9.19)$$

The k-period ahead forecast is then simply

$$\hat{Z}_{t+k} = \hat{Z}_{t+1}, \ k = 1, \ldots, K.$$

Estimation and Forecasting

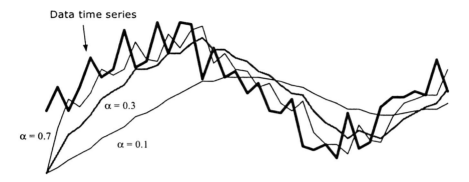

Figure 9.4. Exponential smoothing with different smoothing parameters α.

The choice of α is a design decision and is usually calibrated prior to starting the forecasting system. Smaller values of α smooth the forecasts more, leading to more stability, while larger values of α make the forecast more responsive to recent changes in level but also more susceptible to noise. In practice, α is typically set between 0.05 and 0.3 in RM applications. In addition, more advanced adaptive variations of the smoothing methods attempt to automatically optimize the value of α based on its observed performance.

Some motivation for the exponential smoothing method can be obtained by expanding the recursive formula (9.19), substituting repeatedly for $\hat{Z}_{(\cdot)}$:

$$\begin{aligned}
\hat{Z}_{t+1} &= \alpha z_t + (1-\alpha)\hat{Z}_t \\
&= \alpha z_t + (1-\alpha)(\alpha z_{t-1} + (1-\alpha)\hat{Z}_{t-1}) \\
&= \alpha z_t + \alpha(1-\alpha)z_{t-1} + (1-\alpha)^2 \hat{Z}_{t-1} \\
&= \vdots \\
&= \alpha \sum_{j=0}^{\infty}(1-\alpha)^j z_{t-j}.
\end{aligned} \quad (9.20)$$

Thus, we see the forecast for period $t+1$ is a weighted combination of all previous observations with the weights "exponentially" decreasing at a rate of $(1-\alpha)$. High values of α make the decrease rapid, and the forecasts will be more responsive to recent observations, while low values of α will spread the weights over a longer period, and the forecasts will react more slowly to changes in demand. Figure 9.4 illustrates the role of the smoothing parameter on a sample time series of data.

The smoothing parameters play a similar (albeit more complicated) role in the next two models.

Exponential Smoothing with Linear Trend Let $0 < \alpha < 1$ and $0 < \beta < 1$ be two parameters representing the smoothing factors for the underlying level and trend, respectively. Then the forecast for time-period $t+1$, $\hat{Z}_{t+1}$ is given by the following formulas:

$$\hat{Z}_{t+1} = A_t + T_t, \qquad (9.21a)$$
$$A_t = \alpha z_t + (1-\alpha)(\hat{Z}_t + T_t) \qquad (9.21b)$$
$$T_t = \beta(\hat{Z}_t - \hat{Z}_{t-1}) + (1-\beta)T_{t-1}. \qquad (9.21c)$$

The k-period ahead forecast is given by

$$\hat{Z}_{t+k} = A_t + kT_t, \quad k = 1, \ldots, K.$$

Note T_t is the estimate of the trend factor in each period and is smoothed using β.

Exponential Smoothing with Trend and Seasonality (Holt-Winter's Method) This method is applicable to data series that exhibit seasonal variations (for example, monthly, quarterly, or half-yearly variations). Let $0 < \alpha < 1$, $0 < \beta < 1$, and $0 < \gamma < 1$ be three parameters used to control the smoothing on the underlying level, trend, and seasonality, respectively. Let L represent the periodicity of the seasonality—that is, the number of periods after which the seasons repeat. L depends on the length of the periods and the seasonality—for instance, if we are constructing quarterly forecasts and the seasonality is by quarter, $L = 4$, or if we are constructing monthly forecasts and the seasonality is by month, $L = 12$. Then the forecast for time-period $t+k$ is given by the formula,

$$\hat{Z}_{t+k} = (A_t + kT_t)S_{t+k-L}, \quad k = 1, \ldots, K, \qquad (9.22)$$

and the three components of this forecast are updated as follows:

$$A_t = \alpha\left(\frac{z_t}{S_{t-L}}\right) + (1-\alpha)(\hat{Z}_t + T_t) \qquad (9.23a)$$
$$T_t = \beta(\hat{Z}_t - \hat{Z}_{t-1}) + (1-\beta)T_{t-1} \qquad (9.23b)$$
$$S_t = \gamma\left(\frac{z_t}{S_t}\right) + (1-\gamma)S_{t-L}. \qquad (9.23c)$$

In (9.23c), S_t is the new estimate of the seasonality factor for period t. These factors are updated once each season and are smoothed with the previous estimate of the seasonality factor, of L periods in the past, using γ. Equation (9.23a) "deseasonalizes" the data by replacing z_t by $\frac{z_t}{S_{t-L}}$ and then updates this deseasonalized data using the same procedure as in exponential smoothing with a linear trend.

The deseasonalized forecast is "reseasonalized" in (9.22) by multiplying by the estimated seasonality factor S_{t+k-L} to generate the forecast $\hat{Z}_{t+k}$. More than one seasonal factor can be incorporated into the model (such as both a day-of-week factor and a monthly factor) by simply keeping two multiplicative seasonal factors and updating them as in (9.23a) and (9.23c).

9.3.2 Time-Series Forecasting Methods

In contrast to ad-hoc forecasting methods, time-series methods are based on well-specified classes of models that describe the underlying time series of data. These models have relatively simple mathematical structures, yet the model classes are rich enough to represent a wide range of data characteristics. Since the models are well specified, it is possible to derive "optimal" (MSE or ML) forecasting methods for each one. In this way, the forecasting procedure is specifically tailored to the underlying data-generation model. This formal representation of the dynamics governing the time series and the rigorous development of optimal forecasting methods is what distinguishes time-series methods.

The collection of random variables $\{Z_t\}$ is called a *time series* if it represents successive observations taken over time. The values Z_t are assumed to be generated by a dynamic system, which may depend on past values Z_s for $s \leq t$ and a series of random disturbances $\{\xi_t\}$. At time t, we have observations of the past data values $z_t, z_{t-1}, \ldots$ and would like to forecast the future values of the time series—for example, forecasting the value k units in the future, or Z_{t+k}. We might be interested in a single point estimate, $\hat{Z}_{t+k}$, of this future value Z_{t+k} or an estimate of the parameters of its distribution.

A time-series forecasting process proceeds in two basic steps. First, we make a hypothesis about the specific type of process generating the time series of data. Various model-identification techniques can be employed to help determine which models best fit the data. Once the model is identified, we estimate its parameters. Finally, we apply the corresponding optimal forecasting method specific to that model.

One distinct advantage of time-series methods is that they explicitly model the correlations between successive data points and exploit any dependence to make better forecasts. However, it is up to the RM system designer to decide if such correlations exist (for example, whether there are "runs" in the data, where high-demand observations are often followed by other high-demand observations). Moreover, even when correlation exists, the designer must decide if it is worth building in this extra complexity to obtain better forecasts because these models require relatively large samples of data (usually at least 50 observations) to calibrate accurately.

In what follows, we present several time-series models and methods for forecasting and updating estimates of their parameters. But first, we introduce two important concepts central to time-series forecasting: stationarity and autocorrelation.

9.3.2.1 Stationary Time Series

Stationarity is an important property of a time series that greatly simplifies the forecasting task. Simply put, a time series is stationary if its statistical properties do not change over time. More formally, if $Z_t, \ldots, Z_{t+k}$ and $Z_{t+m}, \ldots, Z_{t+k+m}$ are two sets of k random variables from the series, then the series is said to be *stationary* if the joint distribution of these two sets of variables is the same for all choices of time t and all pairs of values k and m.

To understand why the stationarity assumption simplifies forecasting, consider the problem of estimating the first two moments (means, variances, and covariances) of a collection of N random variables from a nonstationary time series. Nonstationarity means that these N observations were generated by a random process whose joint distribution could be different at each time. Therefore, to estimate the first and second moments, we need to estimate N expected values, N variances, and $N(N-1)/2$ covariances—a total of $N^2/2 + 3N/2$ parameters. However, if the series is stationary, all the expected values and variances will be the same, as Z_t and Z_{t+k} have the same marginal distribution. Moreover, there are only $N-1$ distinct covariances because the joint distribution of Z_t and Z_{t+k} is the same as that of Z_{t+m} and Z_{t+k+m} (for all t, k and m), and hence their respective covariances are the same. Therefore, the number of parameters we need to estimate if the series is stationary is only $2 + (N-1)$, a much more manageable task. To simplify things even further, one often makes further structural assumptions that guarantee that a large number of the covariances are identically zero, making the estimation problem even simpler.

How serious is the assumption of stationarity? At first glance, it seems quite restrictive. In fact, many time series encountered in practice are clearly nonstationary. For example, any time-series data with a trend or seasonal pattern is not stationary (if the series shows an increasing trend, the underlying distributions of the successive random variables are certainly not identical). However, even if the time series itself is not stationary, transformations of the series—such as the difference between successive values—may be stationary. Indeed, time-series forecasting methods for nonstationary data typically involve transforming the data to obtain related stationary series; forecasts based on this transformed stationary series are then used construct a forecast for the original time series.

9.3.2.2 Autocorrelation

As we show below, entire classes of stationary time-series methods are specified through their covariance structure over time—that is, the covariance of Z_t and Z_{t+k} for all k. The autocorrelation function (ACF) and partial autocorrelation function (PACF) are the key tools to analyze this covariance structure. They serve as "signatures", as it were, of a time-series model, and by comparing these signatures to the "sample" signatures obtained from our data we can determine which models are most appropriate.

Specifically, the jth *autocovariance* function is defined as the covariance between Z_t and Z_{t+j}:

$$\gamma_j = \text{Cov}(Z_t, Z_{t+j}).$$

The autocovariance function measures the dispersion or variance of the process. However, two data series that are identical except for the scale of measurement will have different autocovariance functions. Therefore, it is better to deal with the *autocorrelation* function, defined as the autocovariance function divided by the variance

$$\rho_j = \frac{\gamma_j}{\gamma_0},$$

which is scale invariant.

Given a data series $z_1, \ldots, z_N$, the jth *sample autocovariance* function is given by

$$c_j = \frac{\sum_{t=1}^{N-j}(z_t - \bar{z})(z_{t+j} - \bar{z})}{N},$$

where $\bar{z}$ is the sample mean

$$\bar{z} = \frac{1}{N} \sum_{t=1}^{N} z_j.$$

The jth *sample autocorrelation function* is given by

$$r_j = \frac{c_j}{c_0}.$$

The partial autocorrelation function (PACF) is defined as

$$\text{Corr}(Z_t, Z_{t+j} | Z_{t+1}, \ldots, Z_{t+j-1})$$

and can be shown to be equal to the ratio of two determinants involving the autocorrelations (see Wei [560], pp.15–22). A sample PACF can be defined analogous to the sample ACF, but it is considerably more complex to compute. However, most statistical packages automatically compute and plot the sample ACFs and PACFs, so the complexity of the calculations is not a major concern. An example of a sample autocorrelation function and a partial autocorrelation function is shown in Figure 9.5.

9.3.3 Stationary Time-Series Models

We first consider stationary time-series models. To begin, define a *linear filter* as a stochastic process $\{Z_t\}$ that can be written as an infinite weighted sum of random variables as follows:

$$Z_t = \mu + \xi_t - \psi_1 \xi_{t-1} - \psi_2 \xi_{t-2} - \cdots, \quad (9.24)$$

(the minus sign on the ψ's is by convention), where ψ_t and μ are constant parameters and the random variables ξ_t (called *white-noise disturbances*) are assumed to be i.i.d. normally-distributed random variables with a mean of 0 and standard deviation σ_ξ for all t. The stochastic process $\{\xi_t\}$ is therefore a stationary process. We define μ to be the level of the series, which is assumed to be constant. If the sequence $\psi_1, \psi_2, \ldots$ is finite or is infinite and convergent, then one can show that the process $\{Z_t\}$ is stationary and μ is the mean of the series ($E[Z_t] = \mu$).

We can rewrite equation (9.24) to express Z_t in terms of $Z_{t-1}, Z_{t-2}, \ldots$, and ξ_t as follows:

- First, eliminate ξ_{t-1} from (9.24), and write ξ_t, ξ_{t-1} in terms of the remaining variables and parameters,

$$\xi_t = Z_t - \mu + \psi_1 \xi_{t-1} + \psi_2 \xi_{t-2} + \cdots \quad (9.25)$$

$$\xi_{t-1} = Z_{t-1} - \mu + \psi_1 \xi_{t-2} + \psi_2 \xi_{t-3} + \cdots . \quad (9.26)$$

- Substitute (9.26) in (9.25) to obtain

$$Z_t = \mu(1 + \psi_1) - \psi_1 Z_{t-1} + \xi_t + (-\psi_1^2 - \psi_2)\xi_{t-2} \quad (9.27)$$
$$+ (-\psi_1 \psi_2 - \psi_3)\xi_{t-3} + \cdots .$$

- Repeat this process to eliminate ξ_{t-2}, ξ_{t-3} and so on to obtain an equation where Z_t is expressed solely in terms of $Z_{t-1}, Z_{t-2}, \ldots$, and ξ_t:

$$Z_t = \delta + \xi_t + \theta_1 Z_{t-1} + \theta_2 Z_{t-2} + \cdots, \quad (9.28)$$

Estimation and Forecasting 443

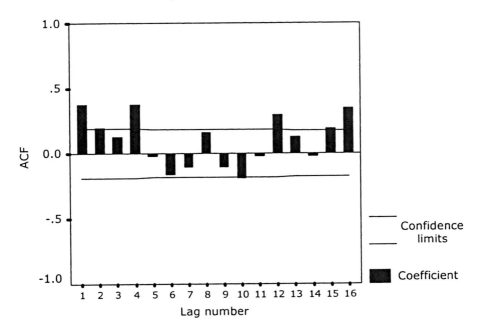

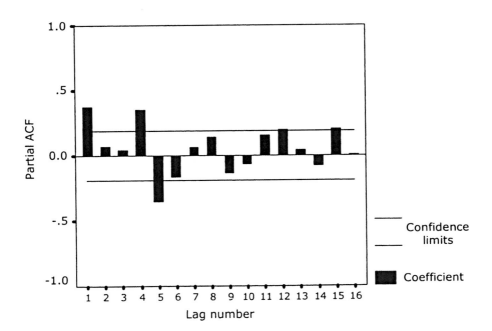

Figure 9.5. Plots of sample autocorrelation function and a partial autocorrelation function.

where δ and $\theta_1, \theta_2, \ldots$ are a new set of constants that depend on μ and $\psi_1, \psi_2, \ldots$.

The random variable Z_t can represent either a stationary or nonstationary process depending on the properties of the parameters θ_i (equivalently, ψ_i). Three important stationary time-series models arise from using (9.24) and (9.28):

- **Moving average process (MA(q))** This process requires that only a finite number of ψ's be nonzero in (9.24). A qth order MA process is given by

$$Z_t = \mu + \xi_t - \psi_1 \xi_{t-1} - \cdots - \psi_q \xi_{t-q}. \tag{9.29}$$

- **Autoregressive process (AR(p))** This process requires that only a finite number of θ's be nonzero in (9.28):

$$Z_t = \delta + \xi_t + \theta_1 Z_{t-1} + \cdots + \theta_p Z_{t-p}. \tag{9.30}$$

- **Autoregressive moving average process (ARMA(p,q))** This process is a combination of MA and AR process

$$\begin{aligned} Z_t &= \delta + \xi_t + \theta_1 Z_{t-1} + \cdots + \theta_p Z_{t-p} \\ &\quad - \psi_1 \xi_{t-1} + \cdots + \psi_q \xi_{t-q}. \end{aligned} \tag{9.31}$$

An AR process is stationary if the roots of the polynomial $1 + \theta_1 x + \cdots + \theta_p x^p$ are greater than one. An MA process is called *invertible* if all the roots of the polynomial $1 + \psi_1 x + \cdots + \psi_q x^q$ are greater than one. One can show that a finite-order stationary AR process can be expressed as an infinite-order MA process, and conversely, a finite-order invertible MA process can be written as an infinite-order AR process. This relationship is useful because if a fitted AR model contains a large number of parameters, it is possible that the corresponding MA model will have fewer parameters, and vice versa. An ARMA model, being a combination of an AR and an MA process can, in principle, reduce the number of parameters even further. Every ARMA(p,q) model has what is called a *pure MA representation*—that is, it can be written as the following infinite sum (see Wei [560], p.58 for a derivation):

$$Z_t = \mu + \xi_t - \psi_1 \xi_{t-1} - \cdots - \psi_q \xi_{t-q} - \cdots.$$

In most practical applications of these models, p and q rarely exceed 2. The means and covariances for ARMA series with small values of p and q are given in Table 9.2. Recall γ_k denotes the covariance

Estimation and Forecasting 445

Table 9.2. Means and covariances of some stationary time-series processes.

Process	Z_t	Mean	Autocovariances (γ_k)	
AR(1)	$\delta + \theta_1 Z_{t-1} + \xi_t$	$\frac{\delta}{1-\theta_1}$	$\theta_1^k \frac{\sigma_\xi^2}{1-\theta_1^2}$	
AR(2)	$\delta + \theta_1 Z_{t-1}$ $+\theta_2 Z_{t-2} + \xi_t$	$\frac{\delta}{1-\theta_1-\theta_2}$	$\theta_1 \gamma_{k-1} + \theta_2 \gamma_{k-2} + \sigma_\xi^2$ $\theta_1 \gamma_{k-1} + \theta_2 \gamma_{k-2}$	$k=0$ $k>0$
MA(1)	$\mu - \psi_1 \xi_{t-1} + \xi_t$	μ	$(1+\psi_1^2)\sigma_\xi^2$ $-\psi_1 \sigma_\xi^2$ 0	$k=0$ $k=1$ $k \geq 2$
MA(2)	$\mu - \psi_1 \xi_{t-1}$ $-\psi_2 \xi_{t-2} + \xi_t$	μ	$(1+\psi_1^2+\psi_2^2)\sigma_\xi^2$ $-\psi_1(1-\psi_2)\sigma_\xi^2$ $-\psi_2 \sigma_\xi^2$ 0	$k=0$ $k=1$ $k=2$ $k \geq 3$
ARMA(1,1)	$\delta + \theta_1 Z_{t-1}$ $-\psi_1 \xi_{t-1} + \xi_t$	$\frac{\delta}{1-\theta_1}$	$\frac{1+\psi_1^2-2\theta_1\psi_1}{1-\theta_1^2}\sigma_\xi^2$ $\frac{(\theta_1-\psi_1)(1-\theta_1\psi_1)}{1-\theta_1^2}\sigma_\xi^2$ $\theta_1 \gamma_{k-1}$	$k=0$ $k=1$ $k \geq 2$

$E[Z_t Z_{t+k}] = E[Z_t Z_{t-k}]$. Note that for stationary processes, the covariances are independent of t, with γ_0 representing the variance. In some cases (as for AR(2)), the covariances do not have a closed-form formula but can be derived as solutions to a set of equations (see Wei [560] for derivations). The AR and MA processes have distinctive ACF and PACFs. Figure 9.6 shows some typical theoretical ACF and PACFs. The forms of these ACF and PACFs provide important clues as to which model is most appropriate for the observed data. Such model-identification issues are discussed in Section 9.3.5.

Once we decide that a set of time-series data is an MA(q) or an AR(p) process, we can proceed to identify the parameters of the model by using ML or MSE criteria. We can then use the models for forecasting in a relatively straightforward manner, as shown in the following example.

Example 9.10 We illustrate the forecasting process on the following data set

$\{z_t\}_{t=1,\ldots,24} =$
$\{25.11, 17.23, 17.87, 17.80, 17.49, 17.99, 18.59, 19.08,$
$19.55, 19.50, 20.74, 21.32, 20.76, 21.10, 21.03, 21.75,$
$21.17, 19.01, 18.95, 17.75, 17.22, 16.52, 17.35, 16.61\}.$

Assume the data comes from an AR(2) process,

$$Z_t = \delta + \xi_t + \theta_1 Z_{t-1} + \theta_2 Z_{t-2}. \tag{9.32}$$

The forecasting process proceeds as follows:

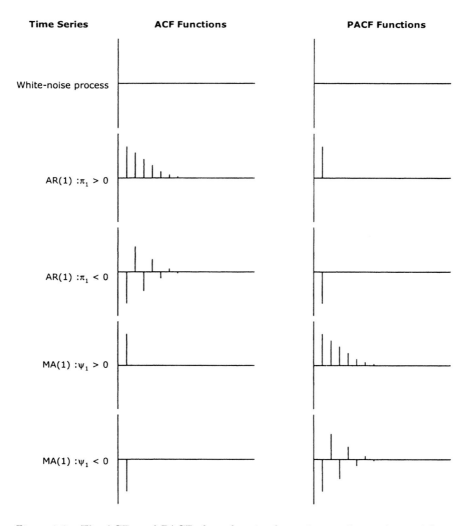

Figure 9.6. The ACFs and PACFs for a few simple stationary time-series models.

Parameter estimation We first estimate the parameters δ, θ_1, and θ_2 in (9.32) by MSE estimation. This we do by solving the following optimization problem (note that an AR(2) process requires at least two initial points, so we begin with data

Estimation and Forecasting

point 3):[3]

$$\min_{\delta,\theta_1,\theta_2} \sum_{k=3}^{t}(z_k - (\delta + \theta_1 z_{k-1} + \theta_2 z_{k-2}))^2 \qquad (9.33)$$

Let $\hat{\delta}$, $\hat{\theta}_1$, and $\hat{\theta}_2$ denote the parameters that minimize the mean-square error on these data. For an AR(2) process, (9.33) has a closed-form solution, but in general numerical optimization is required to find the minimum. (Most statistical software packages solve this optimization problem automatically.) The parameters that minimize the mean-square error for the data set (9.10) turn out to be $\hat{\delta} = 1.5$, $\hat{\theta}_1 = 0.891$, and $\hat{\theta}_2 = 0.0285$.

Forecast For an AR(2) process, the one-step forecast depends on the two previous observations. In general, the k-period forecasts $\hat{Z}_{t+k}$ for $k = 1, 2, 3, \ldots$ are then given by (assume $t > 2$)

$$\begin{aligned} k &= 1: & \hat{Z}_{t+1} &= 1.5 + 0.891 z_t + 0.0285 z_{t-1} \\ k &= 2: & \hat{Z}_{t+2} &= 1.5 + 0.891 \hat{Z}_{t+1} + 0.0285 z_t \\ k &\geq 3: & \hat{Z}_{t+l} &= 1.5 + 0.891 \hat{Z}_{t+k-1} + 0.0285 \hat{Z}_{t+k-2}. \end{aligned} \qquad (9.34)$$

The results of the forecast are given in Table 9.3.

9.3.4 Nonstationary Time-Series Models

As mentioned, most time-series data encountered in practice are nonstationary. In such cases, stationary time-series models may not fit the data well and can produce poor forecasts. Techniques to deal with nonstationary data try to make the data stationary by a suitable transformation, so that one can then apply a stationary time-series model to the transformed data. The resulting stationary forecasts are then transformed back to their original nonstationary form. Differencing successive points in the time series is one such technique.

Time series that are stationary after successive differencing are called *homogenous* nonstationary series. This means that after differencing the series is adequately represented by an ARMA model of the form (9.31). Other transformations, such as taking the logarithm of the series, can make a series stationary if the relative or percentage changes are stationary rather than the differences. For ease of exposition, however, we focus only on differencing in this section.

Given a time series Z_t, define a new time series W_t as $W_t = Z_t - Z_{t-1}$. W_t is called the first-difference of the Z_t series. As we mentioned, there is often good reason to suspect that even if Z_t is not stationary, W_t might be. A series with a linear trend, for instance, has constant differences and

---

[3]Solving the minimization problem (9.33) could be computationally quite intensive, especially if it has to be re-solved after each observation. A practical alternative is to estimate parameters only periodically—say, after every 50 observations.

Table 9.3. Results of the AR(2) forecasting example.

Period	Data	Forecast	Error	MAD
1	25.11			
2	17.23			
3	17.87	17.57	0.30	0.30
4	17.80	17.91	-0.11	0.21
5	17.49	17.87	-0.37	0.26
6	17.99	17.59	0.40	0.30
7	18.59	18.03	0.56	0.35
8	19.08	18.57	0.51	0.38
9	19.55	19.03	0.51	0.40
10	19.50	19.46	0.05	0.35
11	20.74	19.43	1.31	0.46
12	21.32	20.53	0.79	0.49
13	20.76	21.08	-0.32	0.48
14	21.10	20.61	0.50	0.48
15	21.03	20.89	0.14	0.45
16	21.75	20.84	0.92	0.48
17	21.17	21.48	-0.31	0.47
18	19.01	20.98	-1.97	0.57
19	18.95	19.04	-0.09	0.54
20	17.75	18.92	-1.17	0.57
21	17.22	17.86	-0.64	0.58
22	16.52	17.35	-0.82	0.59
23	17.35	16.71	0.64	0.59
24	16.61	17.43	-0.82	0.60

its first-difference series would be stationary. If W_t still is not stationary, we can construct a new series that is the differences of W_t and examine if it is stationary, and so on.

An *autoregressive integrated moving-average process*, ARIMA(p,d,q), is one whose d^{th} differenced series is an ARMA(p,q) process. As for the case of ARMA models, the parameters p, d, q are usually small (less than or equal to 2) in real-world forecasting models.

How do we decide how many differences to take or whether to difference at all? The ACF is helpful in this regard. If the series is nonstationary, the sample ACF shows high values for many periods, whereas if the series is stationary, it damps down to zero quickly, often within four or five periods. We can then difference the data and analyze the resulting ACF to see if the results indicate stationarity. If not, then more differencing may be needed.

The ARIMA(p,d,q) model is designed for homogeneous, nonstationary time series. For example, when there is a trend (linear or nonlinear), then successive differencing of ARIMA converts the series to a stationary series. If, however, the data has a seasonal pattern in addition to a trend, a more involved procedure is required. One option is to consider the series as a product of two stationary series—one that represents the seasonal component and another that represents a stationary time series. We can then difference the seasonal component by the period of seasonality, and the other component can be treated as a stationary time series. However, model identification, parameter estimation, and forecasting are considerably more complicated for this sort of model and are beyond the scope of this chapter.

Finally, we note there is a heuristic relationship between ARIMA process and the simple exponential smoothing method (9.3.1.2). To see this, consider the following ARIMA$(0, 1, 1)$ series:

$$Z_{t+1} - Z_t = \xi_{t+1} - \psi \xi_t \tag{9.35}$$

and

$$\xi_t = Z_t - Z_{t-1} + \psi \xi_{t-1}. \tag{9.36}$$

Substituting successively for $\xi_t, \xi_{t-1}, \ldots$ in the form (9.36) into (9.35), we obtain

$$Z_{t+1} = \xi_{t+1} + (1 - \psi) \sum_{j=0}^{\infty} \psi^j Z_{t-j}.$$

Note the similarity with the simple exponential smoothing method (9.20), where $\alpha = (1 - \psi)$. Box and Jenkins [85] and Harvey [243] derive many connections like this between ad-hoc models and ARIMA models.

9.3.5 Box-Jenkins Identification Process

Determining the model that best represents a given time series is more of an art than a science. Often many different models must be tried before one can narrow down the choice of a "best" model. However, the Box-Jenkins method provides a framework to formalize the model-selection process. It recommends an iterative methodology of choosing the model, validating it, and modifying it to identify the best possible time-series model. Here, we briefly review this methodology.

The first step in the process is *identification*. In this step, the sample ACF and PACF functions are plotted to tentatively identify the most likely candidate for a model. These correlograms are then compared with the correlograms of a standard process such as MA(q), AR(p), or ARMA(p, q) for small values of p and q. For instance, if the sample ACF stops after q spikes, an MA(q) model would be appropriate; if the sample PACF stops after p spikes, an AR(p) model would be appropriate; if neither looks like the right model but the correlograms still decline exponentially toward zero, an ARMA model would be more suitable.

The next step is an *estimation* step in which the model parameters are estimated from the data. Usually, these are least-squares or maximum-likelihood estimates.

The final step is the *diagnostic* step, which verifies that the chosen model and parameters indeed fit the data well. We do this by taking the ACF of the residual series (actual data values subtracted from the model prediction data) and performing various statistical tests (such as the Box-Pierce test) to see if it represents white noise. If the model performs poorly on these tests, the model is rejected, and another model is tested.

Once the model has been selected, we can then use it to generate forecasts as illustrated in Example 9.10. In practice, once a system is operational, the model itself is rarely altered. In contrast, nonparametric or semiparametric methods, such as neural-network methods, adapt the model automatically based on recently observed data. Indeed, the substantial amount of manual work and statistical skills required to implement the Box-Jenkins methodology are its main disadvantages in practice, especially in a RM context where one needs a highly automated forecasting system with minimum manual intervention. As a result, time-series methods have not found much favor in current RM practice. But their performance, when sufficiently tuned and calibrated, can be significantly better than the simpler ad-hoc forecasting methods of Section 9.3.1. So even if they are not used operationally, time-series methods play an important role as reference methods when evaluating simpler forecasting methods.

9.3.6 Bayesian Forecasting Methods

Bayesian methods are a large class of forecasting methods that use the Bayes formula to merge a prior belief about forecast values with information obtained from observed data. The methods are especially useful when there is no historical data, a common occurrence when new products are introduced. For example, an airline may start flying on a new route and have no historical demand information on the route. Fashion apparel products often change every season, and hence demand may be unrelated to the historical sales of past products. Similarly, a TV broadcaster has no historical demand information on demand for a new series. Nevertheless, in each of these cases forecasters may have some subjective beliefs about demand, based on human judgment or alternative data sources (such as test marketing and focus groups). Bayesian methods provide a rigorous and systematic way of specifying such prior beliefs and then updating them as demand data is observed. Hence, they make it possible to combine subjective knowledge with information obtained from data and observations.

9.3.6.1 Basic Bayesian Forecasting

As before, let $Z_1, Z_2, \ldots$ be a sequence of i.i.d. random variables representing a data-generation process. We assume Z_t has a density function $f(z|\theta)$ that is a function of a single, unknown parameter θ. For example, Z_t might have a Poisson distribution, and the parameter θ might be the mean λ. Since θ is unknown, it too is assumed to be a random variable with a probability density $g(\theta)$. This density, called the *prior*, represents our current belief about the value of the parameter θ. Roughly, if we are confident about the value of θ, then the density $g(\theta)$ would be tightly concentrated (have a low variance); conversely, if we are very unsure about the value of θ, then it would be more spread out (have a higher variance). A prior with a large variance is called a *diffuse* prior.

When new data is observed, we may change our belief about the parameter θ. The procedure for formalizing this updating is given by Bayes rule. Let $g_0(\theta)$ represent our initial ($t = 0$) prior distribution and z_1 denote our first observation. Then after observing demand, our *posterior distribution* of θ is given by

$$g_1(\theta) = \frac{g_0(\theta)f(z_1|\theta)}{\int_\theta g_0(\theta)f(z_1|\theta)d\theta}. \tag{9.37}$$

The Bayes estimator of θ is then the expected value of θ based on the posterior distribution (that is, once the information from the observed demand had been incorporated):

$$\theta^* = E[\theta] = \int_\theta \theta g_1(\theta)d\theta. \tag{9.38}$$

The estimator θ^* has several nice theoretical properties. In particular, one can show that it minimizes the variance of the forecast error.

The value θ^* is used in forecasting by setting $\hat{Z}_t = E[Z_t|\theta^*]$. Once the next data value z_2 is observed, we repeat the procedure to get $g_2(\theta)$, and so on. Thus, $g_t(\theta)$ represents our current (time t) belief about θ. (Note that it is a function of the history of observations, $z_1, \ldots, z_t$.)

What makes Bayes estimation practical is that for certain prior distributions of the parameters θ and certain corresponding sample distributions of the random variable Z, the posterior distributions of the parameters in (9.37) have the same distributional form as the prior, and their parameters are given by closed-form updating formulas. A pair of distributions that has this property is said to be a *conjugate family of prior distributions*. We list below some well-known pairs of conjugate families of prior distributions (see DeGroot [151] for derivations):

- **Beta-binomial** $Z_1, Z_2, \ldots, Z_N$ are 0-1 random variables from a Bernoulli distribution with $P(Z_t = 1) = \theta$, and θ has a beta distribution with parameters α, β. After observing $z_1, z_2, \ldots, z_N$, θ has a beta distribution with parameters $\alpha + \sum_{k=1}^{N} z_k$ and $\beta + N - \sum_{k=1}^{N} z_k$.

- **Poisson-gamma** $Z_1, Z_2, \ldots, Z_N$ have a Poisson distribution with mean λ, and λ has a Gamma distribution with parameters α, β. After observing $z_1, z_2, \ldots, z_N$, λ has a gamma distribution with parameter $\alpha + \sum_{k=1}^{N} z_k$ and $\beta + N$.

- **Normal-normal** $Z_1, Z_2, \ldots, Z_N$ have a normal distribution with a known variance σ^2 but an unknown mean μ ($\theta = \mu$), and suppose μ has a normal distribution with mean η and variance v^2. The posterior distribution of μ is a normal distribution with mean

$$\frac{\sigma^2 \eta + v^2 \sum_{k=1}^{N} z_k}{\sigma^2 + Nv^2} \tag{9.39}$$

and variance

$$\frac{\sigma^2 v^2}{\sigma^2 + Nv^2}. \tag{9.40}$$

The following example illustrates the use of these formulas for forecasting:

Example 9.11 (BAYESIAN FORECASTING) Consider the following time series:

$$Z_t = \mu + \xi_t, \tag{9.41}$$

where ξ_t is normally distributed with a mean of 0 and a known variance σ^2—that is, the random variables $Z_1, Z_2, \ldots$ are assumed to be from a normal distribution $N(\mu, \sigma^2)$.

Suppose our prior distribution on μ is modeled as being normal with mean η_0 and variance v_0^2. The value η_0 can be thought of as representing our "best guess" of μ and the value v_0 as representing our degree of confidence in this guess.

After an observation z_1 is made, our estimate on the distribution of μ is updated using the update formulas in (9.39) and (9.40).

$$\eta_1 = \frac{\sigma^2 \eta_0 + v_0^2 z_1}{\sigma^2 + v_0^2} \tag{9.42a}$$

$$v_1^2 = \frac{\sigma^2 v_0^2}{\sigma^2 + v_0^2}. \tag{9.42b}$$

After the next observation z_2 is made, they are again updated as follows:

$$\eta_2 = \frac{\sigma^2 \eta_1 + v_1^2 z_2}{\sigma^2 + v_1^2}$$
$$= \frac{\sigma^2 \eta_0 + v_0^2 (z_1 + z_2)}{\sigma^2 + 2v_0^2},$$
$$v_2^2 = \frac{\sigma^2 v_1^2}{\sigma^2 + v_1^2}$$
$$= \frac{\sigma^2 v_0^2}{\sigma^2 + 2v_0^2},$$

and so forth. After each observation k, the revised forecast of μ is given by

$$E[\mu|z_1, \ldots, z_k] = \eta_k.$$

Notice the ease with which new forecasts can be computed in Example 9.11. The method is also parsimonious with data: only the current estimates need to be stored and updated; all the previous information is contained in the current estimates. However, for distributions that are not conjugate, the updating formulas get complicated, and the Bayesian method loses its attractive properties.

9.3.6.2 Hierarchical and Empirical Bayes Methods

Hierarchical Bayes methods are an appealing way to combine sales data from multiple locations or sources. For example, a manufacturer might be forecasting the sales of its brand across multiple retail chains, a retailer might combine the demand data for a product from multiple stores locations, or an airline might combine data from multiple flights serving a given market.

The method works as follows: Let n be the number of sources and $Z_1, \ldots, Z_n$ represent the random variables of demand at each source. Let $\mathbf{z}_1, \ldots, \mathbf{z}_n$ denote N-vectors of observations of demand at each source ($\mathbf{z}_k$ is assumed to be a vector of N i.i.d. realizations of the random variable Z_k). Let $\theta_1, \ldots, \theta_n$ be the parameters of the distributions of $Z_1, \ldots, Z_n$, respectively, with densities $f_{\theta_k}(\cdot)$. We assume for simplicity that the θ_k's are scalars.

How should we combine these observations? The answer depends on how the parameters θ_k are related. If the parameters $\theta_1, \ldots, \theta_n$ are completely unrelated, we can estimate each independently. If they are all the same, $\theta_1 = \cdots = \theta_n$, we can simply pool all the data together to forecast a single number. However, neither assumption may be satisfactory in a given practical forecasting situation. That is, the sources may be related but not necessarily identical. Hierarchical Bayes methods address this intermediate case. They posit the parameters $\theta_1, \ldots, \theta_n$ as realizations of a common (across the n sources) prior distribution of θ and use the information from "all other" data to obtain a prior for the parameter of each specific source, which is then updated in a Bayesian manner using that source's data.

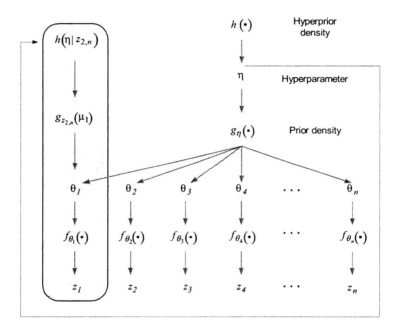

Figure 9.7. The hierarchical Bayes model (adapted from [172]) showing the Bayesian estimation procedure for θ_1 using "other" data, $\mathbf{z}_2, \ldots, \mathbf{z}_n$.

Figure 9.7 shows the hierarchical Bayes model for forecasting θ_1. First, $\theta_1, \ldots, \theta_n$ are assumed to be i.i.d. realizations of a density $g_\eta(\theta)$, where η is a *hyperparameter* from a *hyperprior* density $h(\cdot)$. Both η as well as $h(\cdot)$ are unknown. Then we estimate θ_1 in this framework using not just $\mathbf{z}_1$ but also the other data $\mathbf{z}_{-1} \equiv \{\mathbf{z}_2, \ldots, \mathbf{z}_n\}$. Let $\boldsymbol{\theta} = (\theta_1, \ldots, \theta_n)$ and $\boldsymbol{\theta}_{-1} = (\theta_2, \ldots, \theta_n)$. Using the other data, we perform a Bayesian update on the hyperparameter η to obtain the posterior distribution of

Estimation and Forecasting

$h(\cdot)$, $h(\eta|\mathbf{z}_{-1})$:[4]

$$h(\eta|\mathbf{z}_{2,n}) \propto h(\eta) \int \prod_{k=2}^{n} g_\eta(\theta_k) \prod_{k=2}^{n} f_{\theta_k}(\mathbf{z}_k) d\theta. \qquad (9.43)$$

The integral term in (9.43) is the probability of observing $\mathbf{z}_2, \ldots, \mathbf{z}_n$ for a given η. We can use this to obtain a prior density for θ_1, using Bayes rule:

$$g_{\mathbf{z}_{-1}}(\theta_1) \propto \int h(\eta|\mathbf{z}_{-1}) g_\eta(\theta_1) d\eta. \qquad (9.44)$$

From this prior density, we calculate the posterior density of θ_1 based on the data set $\mathbf{z}_1$:

$$p_{\mathbf{z}_{2,n}}(\theta_1|\mathbf{z}_1) \propto g_{\mathbf{z}_{-1}}(\theta_1) f_{\theta_1}(\mathbf{z}_1), \qquad (9.45a)$$
$$\propto g_{\mathbf{z}_{-1}}(\theta_1) L_1(\theta_1), \qquad (9.45b)$$

where $L_1(\theta_1)$ is the likelihood function of θ_1 given $\mathbf{z}_1$. We can interpret $L_1(\theta_1)$ as the information on θ_1 obtained solely from $\mathbf{z}_1$, while $g_{\mathbf{z}_{-1}}(\theta_1)$ is the "correction" based on the information from the other data $\mathbf{z}_{-1}$. Notice that throughout, we do not need to estimate or know the value of η: it is integrated out in (9.44). However, we do need to know the form of the function $h(\cdot)$ to calculate (9.44). This hyperprior density is somewhat removed from the actual data and hence is difficult to interpret or assign a priori.

One way of avoiding specifying the hyperprior density $h(\cdot)$ is to use what is called an *empirical Bayes approximation* to $g_{\mathbf{z}_{-1}}(\theta_1)$. The empirical Bayes approximation proceeds as follows. Suppose we represent the likelihood (with respect to η) given the other data $\mathbf{z}_{-1}$ as

$$L_{\mathbf{z}_{-1}}(\eta) = \int \prod_{k=2}^{n} g_\eta(\theta_k) \prod_{k=2}^{n} f_{\theta_k}(\mathbf{z}_k) d\theta$$

---

[4]To avoid excessive notation, we do not write down the normalizing factor and just represent the density as being proportional ($\propto$) to the right-hand side. So the use of Bayes rule in (9.43) should be read as

$$h(\eta|\mathbf{z}_{2,n}) = \frac{h(\eta) \int \prod_{k=2}^{n} g_\eta(\theta_k) \prod_{k=2}^{n} f_{\theta_k}(\mathbf{z}_k) d\theta}{\int h(\eta) \int \prod_{k=2}^{n} g_\eta(\theta_k) \prod_{k=2}^{n} f_{\theta_k}(\mathbf{z}_k) d\theta d\eta}$$
$$\propto h(\eta) \int \prod_{k=2}^{n} g_\eta(\theta_k) \prod_{k=2}^{n} f_{\theta_k}(\mathbf{z}_k) d\theta.$$

The density can easily be recovered by dividing by the integral. We do likewise for all subsequent applications of Bayes rule in this section.

and let $\hat{\eta}$ denote the ML estimate

$$\hat{\eta} = \arg\max_{\eta} L_{\mathbf{z}_{-1}}(\eta). \tag{9.46}$$

Then instead of using the exact density $g_{\mathbf{z}_{-1}}(\theta_1)$ of (9.44) in (9.45b), we use the approximation $g_{\hat{\eta}}(\theta_1)$ to obtain the *MLE posterior density* of θ_1

$$p_{\hat{\eta}}(\theta_1|\mathbf{z}_1) \propto g_{\hat{\eta}}(\theta_1)L_1(\theta_1).$$

Of course, (9.46) should be easy to solve or else this approximation will be difficult to implement. One can even substitute a MSE estimator or a method-of-moments estimator (Sections 9.2.2 and 9.2.4) instead of the ML estimate of (9.46) if these make the computations more tractable. So one has to choose the densities of $g_\eta(\cdot)$ and $f_\theta(\cdot)$ judiciously for computational convenience. But in the end, the advantage of this method of empirical approximation is that it does not require an estimate of $h(\cdot)$.

We illustrate the hierarchical Bayes model with a retail RM example:

Example 9.12 (SHRINKAGE ESTIMATION OF RETAIL PRICE AND PROMOTIONAL ELASTICITIES ([75])) A manufacturer sells a product through multiple (n) chains (collection of stores). Periodically the manufacturer offers promotions and wants to gauge the effect of the promotions on sales. The model of sales during a promotional campaign is the following:

$$\begin{aligned}SL_t = {} & \theta_1 + \theta_2 PR_t + \theta_3 DD_{t-1} + \theta_5 AD_t \\ & + \theta_6 DP_t + \theta_7 FL_t + \theta_8 CD_t + \theta_9 SL_{t-1} + \xi_t,\end{aligned} \tag{9.47}$$

where
- SL_t = logarithm of sales in period t
- PR_t = relative price in period t (regular price divided by an average of competitive regular prices)
- DD_t = deal discount in period t (normal shelf price minus actual divided by normal shelf price)
- AD_t = feature advertising in period t (proportion of stores in chain using the ad)
- DP_t = display in period t (proportion of stores in chain displaying the brand)
- FL_t = 0-1 indicator variable, 1 if period t is the final period of a multi-week deal, and 0 otherwise
- CD_t = maximum deal discount for competing brands in chain in period t

The data consists of T periods of sales data from the n chains. Let z_t^i represent the log sales of chain-brand i at time t, and y_{mt}^i, the mth covariate (explanatory variable in (9.47), $m = 1, \ldots, M$ ($M = 7$)) value for period t for chain i. The regression models for the log sales for the n chains are given by

$$Z_t^i = \sum_{m=1}^{M} \theta_m^i y_{tm}^i + \xi_t^i, \quad i = 1, \ldots, n;\ t = 1, \ldots, T,$$

Estimation and Forecasting   457

where the ξ_t^i are assumed to have a normal distribution with mean zero and a common variance σ^2, and to be independent. Let $\boldsymbol{\theta}^i = (\theta_1^i, \ldots, \theta_M^i)$, and $\boldsymbol{\xi}^i = (\xi_1^i, \ldots, \xi_{p_i}^i)$. We assume σ^2 is known,[5] and let $\boldsymbol{\sigma}^2$ denote $\sigma^2 \mathbf{I}$.

This is a straightforward regression problem if the n stores are estimated separately. Let $\mathbf{Z}^i = (Z_1^i, \ldots, Z_T^i)$, and $\mathbf{Y}^i$ the matrix whose elements are y_{tm}^i, $t = 1, \ldots, T$; $m = 1, \ldots, M$. Then the regression equation for store i in matrix form is

$$\mathbf{Z}^i = \mathbf{Y}^i \boldsymbol{\theta}^i + \boldsymbol{\xi}^i.$$

The MSE estimates for $\boldsymbol{\theta}^i$ are given by (same as (9.6))

$$\hat{\boldsymbol{\theta}}^i = ((\mathbf{Y}^i)^\top \mathbf{Y}^i)^{-1} (\mathbf{Y}^i)^\top \mathbf{Z}^i.$$

However, estimating by chain reduces the size of the data sets and often leads to odd predictions with wrong signs on the coefficients or similar calibration problems.

We can build instead a hierarchical model assuming that each parameter θ_m^i comes from a prior normal distribution

$$\theta_m^i \sim N(\mu_m, \varsigma_m), \quad i = 1, \ldots, n; \ m = 1, \ldots, M.$$

$\eta_m = \{\mu_m, \varsigma_m\}$, $m = 1, \ldots, M$, are the hyperparameters with an unknown distribution $h(\cdot)$, generating the parameters θ_m^i, $m = 1, \ldots, M$; $i = 1, \ldots, n$. Let $\boldsymbol{\Sigma} = \text{diag}[\varsigma_1, \ldots, \varsigma_M]$, and $\boldsymbol{\mu} = (\mu_1, \ldots, \mu_M)$.

If we knew $\eta_m = \{\mu_m, \varsigma_m\}$, $m = 1, \ldots, M$, and we had a prior $\hat{\boldsymbol{\theta}}^i$ of the parameters, then we could have estimated the mean of the posterior distribution of $\boldsymbol{\theta}^i$ by Bayes theorem as (using a vector version of the formulas in Example 9.11)

$$\mathbf{Q}_i^{-1}(\sigma^{-2}(\mathbf{Y}^i)^\top \mathbf{Y}^i \hat{\boldsymbol{\theta}}^i + \boldsymbol{\Sigma}^{-1} \boldsymbol{\mu}), \tag{9.48}$$

where

$$\mathbf{Q}_i = \sigma^{-2}(\mathbf{Y}^i)^\top \mathbf{Y}^i + \boldsymbol{\Sigma}^{-1}.$$

The new updated mean of (9.48) is in a sense a convex combination of the prior mean $\hat{\boldsymbol{\theta}}$ and the actual (unknown) mean $\boldsymbol{\mu}$. The mean is "shrunk" toward the hyperparameter $\boldsymbol{\mu}$ by the *shrinkage factor* $\mathbf{Q}_i^{-1} \sigma^{-2} (\mathbf{Y}^i)^\top \mathbf{Y}^i$.

The estimate (9.48) is unusable however as we do not know $\eta_m = \{\mu_m, \varsigma_m\}$, $m = 1, \ldots, M$ (i.e., $\boldsymbol{\mu}$ and $\boldsymbol{\Sigma}$). If we had estimates of the hyperparameters from "other" data however, we can use them instead in (9.48) for any given chain i. So the estimates of $\boldsymbol{\theta}^i$ would be a convex combination of a hyperprior estimate of μ's from data other than from chain i and the data of chain i. This is the idea behind the hierarchical Bayes method.

In practice, obtaining the ML estimates of η from the other data may be too difficult. But this does not prevent us from using any reasonable estimate that we can obtain based on the other data. Blattberg and George [75] give a variety of alternatives for the hyperparameter estimates for this regression problem. Also, here

---

[5]For any given set of estimators of $\boldsymbol{\theta}^i$, $\hat{\boldsymbol{\theta}}^i$, a good estimate of σ^2 is

$$\hat{\sigma}^2 = \frac{\sum_{i=1}^n \sum_{t=1}^T (z_t^i - \sum_{m=1}^M \hat{\theta}_m^i y_{tm}^i)^2}{\sum_{i=1}^n T - nM + 2}.$$

we have constrained the θ_m^i's to have identical variances ς_m. See [75] for alternative constraints with different interpretations. Blattberg and George [75] also consider weekly sales data for a national brand and show that hierarchical Bayes methods improve predictive performance.

9.3.7 State-Space Models and Kalman Filtering

Like time-series methods, state-space methods assume the time series $\{Z_t\}$ is driven by an underlying dynamic system. The system is defined by a "state" together with a system of equations for describing how the state and observable outputs (say, the time-series data) evolve over time as function of possibly random inputs. The future behavior of the system can be completely described by the present state and future inputs, a feature known as a *Markovian representation* of the system. However, the current state most often is not directly observable and must be estimated based on observed data. The following example illustrates a simple case of such a system:

Example 9.13 Consider a series being generated by the following model:

$$Z_t = \mu_t + \xi_t, \quad \xi_t \sim N(0, \sigma_\xi^2) \tag{9.49a}$$
$$\mu_t = \phi\mu_{t-1} + \eta_t, \quad \eta_t \sim N(0, \sigma_\eta^2), \tag{9.49b}$$

where μ_t is the underlying mean of data Z_t. Here the mean μ_t (a scalar) is the state of the system, which we cannot observe directly. The mean evolves according to the state equation (9.49b), which is a linear function of the past state (μ_{t-1}) and a process noise term η_t. The observable output Z_t is described by the observation equation (9.49b) and is equal to the mean μ_t plus a measurement noise term ξ_t.

For a time series generated by (9.49a)–(9.49b), a forecasting method might proceed as follows: (i) keep a current estimate of the underlying state $\hat{\mu}_t$, (ii) forecast $\hat{Z}_{t+1} = \hat{\mu}_t$, (iii) after observing the data at time $t+1$, update our current estimate of state to $\hat{\mu}_{t+1}$ and repeat. (Details of how this can be done are discussed below.)

One can view many forecasting models in a state-space framework. For example, in simple exponential smoothing equation (9.19), the level factor A_t can be interpreted as the unobservable state, while Bayesian forecasting methods can be viewed as an attempt to estimate an unobservable "state" (the unknown parameters of the distribution). More generally, if we define the "state" at time t as consisting of the complete history of observations and actions up to time t, then this state would contain all the information relevant for forecasting. Thus, at an abstract level, all forecasting models can be cast in a state-space model framework. However, such an abstract description is of little practical value because the dimension of the state increases without bound over time. Hence, for the state-space approach to be useful, we need a more compact (finite-dimensional) representation of the state, as in

Example 9.13. In this section, we focus on the best known state-space forecasting method: the Kalman filter.

9.3.7.1 The Kalman Filter Formulation

The Kalman filter is based on a finite-dimensional system of linear state and observation equations and zero-centered Gaussian (normally distributed) noise terms. Under these conditions, the Kalman filter provides an efficient algorithm for estimating the state and for forecasting.

Formally, let the n dimensional real vector $\mathbf{y}_t$ represent the state at time t. The state is assumed to evolve according to a linear system equation:

$$\mathbf{y}_t = \mathbf{A}\mathbf{y}_{t-1} + \boldsymbol{\nu}_t, \qquad (9.50)$$

where $\boldsymbol{\nu}_t$ is a n-vector of random variables, called the *process noise*, and $\mathbf{A}$ is a known $n \times n$ matrix of parameters. We assume $\boldsymbol{\nu}_t$ is a Gaussian (*white-noise*) process—a set of i.i.d. random variables from a normal distribution $N(\mathbf{0}, \mathbf{Q})$, where $\mathbf{Q}$ is a known $n \times n$ matrix called the *process-noise covariance* matrix.

There is a m-dimensional vector $\mathbf{z}_t$ of observations,[6] which is related to the state by the following observation equation:

$$\mathbf{z}_t = \mathbf{H}\mathbf{y}_t + \boldsymbol{\xi}_t,$$

where $\mathbf{H}$ is a known $m \times n$ matrix of parameters, and $\boldsymbol{\xi}_t$ is a m-vector of i.i.d. random variables, called the *measurement noise*, that we assume has a normal distribution $N(\mathbf{0}, \mathbf{R})$, with a known $m \times m$ *measurement noise covariance* matrix $\mathbf{R}$. While we assume the matrices $\mathbf{A}, \mathbf{H}, \mathbf{Q}$, and $\mathbf{R}$ are known, in practice they are usually estimated from data as discussed later.[7] To illustrate this formulation, we give an example of the AR(2) model in state-space form:

Example 9.14 Consider the AR(2) process described in Section 9.3.2, where

$$\mathbf{z}_t = \delta + \xi_t + \theta_1 z_{t-1} + \theta_2 z_{t-2}. \qquad (9.51)$$

---

[6] Note that the observation is a vector here, in contrast to the scalar observations of previous sections. We also use **z** and **y** to represent the *random variables* generating z and y, instead of **Z** and **Y** as in the rest of this chapter, to avoid confusion with our matrix notation convention.
[7] Here we have also assumed that the matrices $\mathbf{A}, \mathbf{H}, \mathbf{Q}, \mathbf{R}$ are constant across time. However, the theory and the Kalman filter forecasting equations hold even when this data changes over time. The Gaussian distribution assumption on the error terms is also not strictly necessary, although it is commonly assumed in most applications.

We can rewrite equation (9.51) as a system of state-space equations, as a combination of a state-evolution equation,

$$\underbrace{\begin{bmatrix} z_t \\ z_{t-1} \\ \delta \end{bmatrix}}_{\mathbf{y}_t} = \underbrace{\begin{bmatrix} \theta_1 & \theta_2 & 1 \\ 1 & 0 & 0 \\ 0 & 0 & 1 \end{bmatrix}}_{\mathbf{A}} \underbrace{\begin{bmatrix} z_{t-1} \\ z_{t-2} \\ \delta \end{bmatrix}}_{\mathbf{y}_{t-1}} + \underbrace{\begin{bmatrix} \xi_t \\ 0 \\ 0 \end{bmatrix}}_{\boldsymbol{\nu}_t},$$

and as a measurement equation,

$$z_t = \underbrace{\begin{bmatrix} 1 & 0 & 0 \end{bmatrix}}_{\mathbf{H}} \underbrace{\begin{bmatrix} z_t \\ z_{t-1} \\ \delta \end{bmatrix}}_{\mathbf{y}_t} + \underbrace{0}_{\xi_t}.$$

In a similar fashion, the general ARMA(p,q) model can also be formulated in a Kalman-filter framework (see Wei [560], p.385), as can many of the other time-series models of Section 9.3.2 (see Harvey [243]).

In a forecasting context, the state can be viewed as the (unobservable) parameters of the true underlying demand-generation process. Each observation gives additional information of the parameters, and this information can be used to update our current estimate of the state via the state-evolution equation. With the updated state, a forecast for period $t+1$ can be made using the prediction equation for period $t+1$, substituting the state obtained for period t. The Kalman filter provides an efficient recursive algorithm for performing these operations.

9.3.7.2 The Kalman Filter Forecasting Algorithm

We first state the Kalman filter forecasting algorithm, and then explain the intuition behind it and some of its formal properties.

The algorithm proceeds as follows. Let the subscript indexing $(\cdot)_{t|t-1}$ denote the value of the variable at time t based on all the information up to time $t-1$ (before the observation in period t). At each time t, we keep an estimate of the underlying state $\hat{\mathbf{y}}_{t|t-1}$ that encapsulates all the information gained from past observations. After time t, we get a new observation $\mathbf{z}_t$ and update our estimate of state to $\hat{\mathbf{y}}_{t|t}$ using $\hat{\mathbf{y}}_{t|t-1}$ and $\mathbf{z}_t$ (by (9.50)). We then make a forecast for time $t+1$, $\hat{\mathbf{z}}_{t+1} = \mathbf{H}\hat{\mathbf{y}}_{t+1|t}$, with $\hat{\mathbf{y}}_{t+1|t} = \mathbf{A}\hat{\mathbf{y}}_{t|t}$.

Let $\mathbf{e}_{t|t-1} = \mathbf{y}_t - \hat{\mathbf{y}}_{t|t-1}$ and $\mathbf{e}_{t|t} = \mathbf{y}_t - \hat{\mathbf{y}}_{t|t}$ represent, respectively, errors from the true state before and after the state estimates have been updated. Let $\mathbf{P}_{t|t-1} = E[\mathbf{e}_{t|t-1}\mathbf{e}_{t|t-1}^\top]$ and $\mathbf{P}_{t|t} = E[\mathbf{e}_{t|t}\mathbf{e}_{t|t}^\top]$ represent, respectively, the error covariance matrices. The algorithm is as follows:

Initialization: Let time $t = 0$. Assume initial values of $\mathbf{P}_{0|0}$ (say $\mathbf{I}$) and the initial state $\hat{\mathbf{y}}_{0|0}$.

Forecasting step: At time t, project the error, state, and forecast:

$$\hat{\mathbf{y}}_{t+1|t} = \mathbf{A}\hat{\mathbf{y}}_{t|t}$$
$$\mathbf{P}_{t+1|t} = \mathbf{A}\mathbf{P}_{t|t}\mathbf{A}^\top + \mathbf{Q}$$
$$\hat{\mathbf{z}}_{t+1} = \mathbf{H}\hat{\mathbf{y}}_{t+1|t}.$$

Measurement updating step: After observing $\mathbf{z}_{t+1}$, update

$$\hat{\mathbf{y}}_{t+1|t+1} = \hat{\mathbf{y}}_{t+1|t} + \mathbf{K}_{t+1}(\mathbf{z}_{t+1} - \hat{\mathbf{z}}_{t+1}), \qquad (9.52)$$

where the matrix $\mathbf{K}_{t+1}$ is given by

$$\mathbf{K}_{t+1} = \mathbf{P}_{t+1|t}\mathbf{H}^\top (\mathbf{H}\mathbf{P}_{t+1|t}\mathbf{H}^\top + \mathbf{R})^{-1}.$$

Update the error covariance

$$\mathbf{P}_{t+1|t+1} = (\mathbf{I} - \mathbf{K}_{t+1}\mathbf{H})\mathbf{P}_{t+1|t}.$$

The matrix $\mathbf{K}_t$ is known as the *Kalman gain*. The crucial step is (9.52), which calculates the *a posteriori* estimate of the state *after* observing the measurement in period $t + 1$ from the *a priori* estimate (*before* observing the measurement in period $t + 1$). If the disturbances are normal, the distribution of the initial state will be normal, and the mean and variance of the *a priori* estimate of the state are given by $\hat{\mathbf{y}}_{t+1|t}$ and $\mathbf{P}_{t+1|t}$. The conjugate distribution of a normal distribution is again normal and after observing the measurement $\mathbf{z}_{t+1}$, the *a posteriori* distribution of $\mathbf{y}_{t+1|t+1}|\mathbf{z}_{t+1}$ is also normal with mean given by (9.52). This mean-state vector also turns out to be the minimum mean-square estimate of $\mathbf{y}_{t+1|t+1}$ given all the information up to time $t + 1$. Even when the disturbances are not normal, the Kalman filter equations can be shown to be the best *linear* estimator, in the sense of minimizing the mean-square error among all linear updates of the form $\hat{\mathbf{y}}_{t+1|t+1} = \hat{\mathbf{y}}_{t+1|t} + \mathbf{K}(\mathbf{z}_{t+1} - \hat{\mathbf{z}}_{t+1})$; that is, the Kalman gain is the matrix $\mathbf{K}$ that minimizes $(\mathbf{z}_{t+1} - \mathbf{H}\hat{\mathbf{y}}_{t+1|t+1})^\top (\mathbf{z}_{t+1} - \mathbf{H}\hat{\mathbf{y}}_{t+1|t+1})$.

An attractive property of the Kalman filter is the recursive nature of the algorithm. At each step, we need only to maintain the current estimate of the state and the estimate of the covariance matrix. As new observations come in, we can then easily update these two quantities.

462 THE THEORY AND PRACTICE OF REVENUE MANAGEMENT

Moreover, updating these estimates by the Kalman filter equations is computationally very efficient, which is one of the most appealing features of the algorithm. The following is a simple example of the operation of the Kalman filter:

Example 9.15 (FORECASTING USING THE KALMAN FILTER) Let the state evolution equations for a 1-dimensional state be given by

$$y_t = y_{t-1} + \nu_t$$

and the measurement be given by the process

$$z_t = y_t + \xi_t,$$

where $\nu_t \sim N(0, Q)$ and $\xi_t \sim N(0, R)$. Then the state update equations of the Kalman Filter are

$$\hat{y}_{t+1|t} = \hat{y}_{t|t}$$
$$P_{t+1|t} = P_{t|t} + Q$$
$$\hat{z}_{t+1} = \hat{y}_{t+1|t},$$

and the measurement equations to update the state and measurement are

$$\hat{y}_{t+1|t+1} = \hat{y}_{t+1|t} + K_{t+1}(z_{t+1} - \hat{z}_{t+1}), \tag{9.53}$$

where K_{t+1} is given by

$$K_{t+1} = \frac{P_{t+1|t}}{(P_{t+1|t} + R)}.$$

Update the error covariance by

$$P_{t+1|t+1} = (1 - K_{t+1})P_{t+1|t}.$$

To start off the forecasting process, at $t = 0$, we need to assign some values to $\hat{y}_{0|0}$ and $P_{0|0}$. Rather arbitrarily let's set $\hat{y}_{0|0} = 1$. As with Bayesian methods, the quantity $P_{0|0}$ should reflect our degree of certainty about our estimate of the state $\hat{y}_{0|0}$. A value of $P_{0|0} = 0$ would imply that we are completely sure of our initial estimate; more often, we choose some value $P_{0|0} \neq 0$. The precise value is not critical—the Kalman filter algorithm is quite robust this way—but the more uncertain we are of our estimate, the higher this value should be (something like $P_{0|0} = 2$ would generally suffice for this case).

Notice the similarity between (9.53), which can be rewritten in terms of z and $\hat{z}$'s as

$$\hat{z}_{t+1} = K_t z_t + (1 - K_t)\hat{z}_t$$

and the simple exponential smoothing formula (9.19), repeated here:

$$\hat{z}_{t+1} = \alpha z_t + (1 - \alpha)\hat{z}_t.$$

Estimation and Forecasting

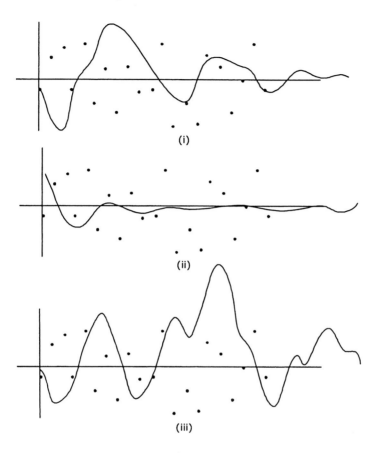

Figure 9.8. Kalman filter smoothing. (i) A data stream generated with a measurement error ξ around the mean and the Kalman filter uses a value $R = Var(\xi)$ (ii) Kalman filter uses a value of $R > Var(\xi)$ (the filter is smoothed too much), (iii) the Kalman filter uses a value $R < Var(\xi)$, and the forecasts follow the noise too quickly.

Indeed, the Kalman gain K_t can be considered as an adaptive smoothing factor that changes over time based on the observed data. As $t \to \infty$, one can also show the Kalman gain converges to a constant matrix K, which means, after many observations, the Kalman filter will converge to the simple exponential smoothing formula (9.19). However, the Kalman gains are in fact the "optimal" weighting factors, in the sense that for linear state and measurement processes, they minimize the mean-square error.

9.3.7.3 Estimating the Matrices A, H, Q, and R

Lastly, we address the question of estimating the matrices $\mathbf{A}, \mathbf{H}, \mathbf{Q}$, and $\mathbf{R}$. Although the Kalman filter equations are easy to apply if these matrices are known, in practice it is highly unlikely that we know their exact values. For instance, in the state-space formulation of the AR(2) process (9.51), we do not know the components θ_1 and θ_2 of the matrix $\mathbf{A}$. However, these parameters can be estimated by maximum-likelihood methods based on an initial set of observations (see Harvey [242] and Harvey [243], p.91). The values used for $\mathbf{Q}$ and $\mathbf{R}$ will also affect the behavior of the algorithm. If the values we choose for $\mathbf{Q}$ and $\mathbf{R}$ are much higher than the true variance in the process and measurement error terms, then the forecasts tend to be very reactive to noise, and if they are much smaller than the actual variances, the forecasts are much smoother (see Figure 9.8). Again, these variances can also be estimated by maximum-likelihood methods.

9.3.8 Machine-Learning (Neural-Network) Methods

All the forecasting methods we have discussed thus far follow the same underlying strategy: posit a functional form for the relationship between the observed data and various factors (such as noise terms, time, past observations, and causal factors) and then estimate the parameters of this function using historical data. In contrast, *machine-learning*—or specifically, *neural-network*—methods do not make a functional assumption *a priori*; rather, they use interactions in a network-processing architecture to automatically identify the underlying function that best describes the demand process. The methods are based on artificial intelligence approaches that mimic the way the human brain learns from experience. In theory, with the appropriate architecture and training procedure, neural networks are capable of approximating any nonlinear functional form after a sufficient degree of "learning" on samples generated by that function.

Neural networks have found wide applicability in pattern recognition, classification, reconstruction, biology, computer game playing, and time series forecasting. Business applications have been reported in market analysis, bond rating, credit-risk evaluation, and financial series forecasting. Some RM vendors and airlines have implemented neural-network forecasting methods as well [496].

Neural-network forecasting encompasses a large class of architectures and algorithms, and the literature is extensive. Here we only describe the workings of a simple neural network with the most basic of training

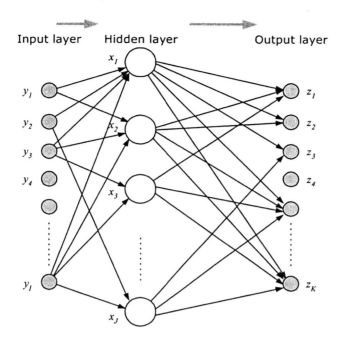

Figure 9.9. A three-layer feed-forward neural network.

algorithms. However, this introduction should provide a good sense of the overall approach.

9.3.8.1 An Overview of Neural Networks

A neural network consists of an underlying directed graph and a set of additional quantities defined on the graph. In an important class of neural networks, the nodes of the network are arranged in consecutive layers, and the arcs are directed from one layer to the next, left to right as shown in Figure 9.9. Such networks are called *feed-forward* networks or *perceptrons* and form the most important class of neural networks used for forecasting. We limit our discussion here to feed-forward networks.

The first layer is called the *input layer* and the last is called the *output layer*, with the layers in the middle being the *hidden layers*. Most networks in practice have at most one or two hidden layers. A network with a single hidden layer has been shown to be able to approximate most nonlinear functional forms [397]. The training data is "fed" to the input layer, and the forecasts are "read" from the output layer. Typically, in demand-forecasting applications, each node in the input layer corresponds to an explanatory variables (analogous to the y's in the linear-regression equation (9.3)), and each node in the output layer corresponds to a future forecast. For example, if we want to use the 20 most recent historical observations to make forecasts for the next three periods, the network would have 20 input nodes (one for each historical observation) and three output nodes (one for each forecast), with a certain number of hidden nodes in between.

More generally, a neural-network architecture is defined by a graph $G = (\mathcal{N}, \mathcal{A})$, where $\mathcal{N}$ is a set of nodes and $\mathcal{A}$ is a set of directed arcs. The following quantities are defined on the network:

- A *state variable*, s_j, associated with each node $j \in \mathcal{N}$. Typically, state is binary (every node is either active (state 1) or inactive (state 0)) or it is continuous, usually taking on values between 0 and 1. The state can change for each set of inputs or in an online forecasting application after every new observation. Thus, states are said to *evolve* over discrete units of time t, $t = 0, 1, 2, \ldots$, and we represent the state of node j at time (observation) t as $s_j(t)$.

- A *weight*, w_{ij}, associated with each directed arc $(i, j) \in \mathcal{A}$.

- An *activation threshold value* ν_j associated with each node $j \in \mathcal{N}$. Typically, the activation threshold value serves as a threshold for making the node active or inactive. For example, if the sum of the weights of incoming arcs exceeds ν_j, then consider node j active and inactive otherwise.

- An *activation function* (or *transfer function*), which determines the state of node j as a function of the states of other nodes i with arcs into j (with arcs of the form (i, j)), the arc weights w_{ij}, and the activation threshold ν_j: $f_j(\{s_i, w_{ij} : (i,j) \in \mathcal{A}\}, \nu_j)$. The activation functions can be different for each layer (or even each node). Typically, the activation functions act on the *sum* of the weights of arcs from *active* nodes coming into node j, in which case, the activation threshold for j can be represented as $f_j(\sum_{\{(i,j) \in \mathcal{A}\}} w_{ij} s_i - \nu_j)$. Activation functions serve to make the nodes active or inactive.

Some examples of transfer functions f include the following:

- A *linear function*, where $f(h) = h$.
- The *Heavyside step function*, which is a simple threshold value comparison between $\sum_{\{(i,j) \in \mathcal{A}\}} w_{ij} s_i$ and ν_j:

$$f(h) = \begin{cases} 1 \text{ (active)} & \text{if } h \geq 0 \\ 0 \text{ (inactive)} & \text{otherwise.} \end{cases}$$

Estimation and Forecasting 467

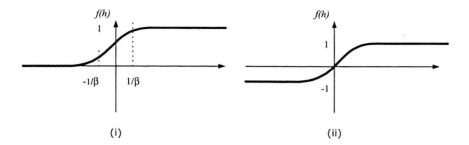

Figure 9.10. The (i) Fermi function and the (ii) tanh activation functions.

- The *logistic sigmoid functions* (Figure 9.10), which are a class of monotonic, differentiable functions $f(h)$ with

$$\lim_{h \to -\infty} f(h) = 0,$$

and

$$\lim_{h \to \infty} f(h) = 1.$$

An example of a logistic sigmoid function is the *Fermi function*:

$$f(h) = \frac{1}{1 + e^{-2\beta h}}. \qquad (9.54)$$

- The tanh function:[8]

$$f(h) = \tanh(\beta h) = \frac{1 - e^{-2\beta h}}{1 + e^{-2\beta h}}. \qquad (9.55)$$

The value of the transfer function is taken to be the *state* of the node. The state is binary (0 or 1) for the Heavyside step function and continuous for the linear function (from $-\infty$ to ∞) and the logistic sigmoid functions (between 0 and 1).

9.3.8.2 Training and Forecasting

Calibration of a neural network is called *training the network*. A set of training data is used to calibrate the weights and the values of the threshold functions. Once these parameters are determined, the network can be used for forecasting. Thus, the three main steps are defining the

---

[8]The tanh function can be shown to be equivalent to the Fermi function after a linear transformation of the inputs and outputs (see Bishop [69], p.127). However, the tanh function has been found to give faster training convergence and is generally preferred.

network, training, and forecasting. We illustrate these steps on the three-layer network of Figure 9.9.

Defining the Network The input is a set of I values of independent variables associated with each observation, represented by I input nodes, and the output is a forecast for K future periods, represented by K output nodes. The inputs could consist of all variables that would influence the demand. For instance, if the forecast is for demand in a particular market for an airline, the input variables, in addition to historical demand in that market, could consist of variables such as schedule frequency, capacity, time in market or economic indicators. Assume there are J nodes in the hidden layer. We index arcs from the input layer to the hidden layer as (i, j) and arcs from the hidden layer to the output layer as (j, k).

We next need to define the transfer functions. We use the tanh function (9.55) as the activation function $f(h)$ for the nodes of the hidden layer and a linear function $\tilde{f}(h) = h$ as the activation function for the nodes of the input and output layers. These functions are defined by the activation thresholds ν_i, the arc weights w_{ij}, w_{jk}, and the parameter β of the tanh function.

Let y_i represent the state of input node i, and x_j the state of node j of the hidden layer, and $\hat{Z}_k$ the state of node k of the output layer. The inputs to the hidden layer are formed by a weighted combination of values of the states of the input layer

$$h_j = (\sum_{i=1}^{I} w_{ij} y_i) - \nu_j,$$

and the state of the hidden node j is therefore

$$x_j = f(h_j).$$

The inputs to the output layer in turn are a weighted combination of the states of the hidden layer and the activation thresholds of the output nodes:

$$h_k = (\sum_{j=1}^{J} w_{jk} x_j) - \nu_k.$$

The state of the output node k is then $\tilde{f}(h_k) = h_k$. This completes the definition of the network.

Training Once a network topology is chosen, we have to determine values for the arc weights and node activation thresholds. This training is, in all respects but terminology, equivalent to estimating the parameters of any other forecasting model from historical data—except that we are not working with a simple functional form for the demand generating process but rather from a complicated network of interacting functions.

One of the first, and still quite popular, methods of training is the *error back-propagation method*. The method uses a squared error criterion and prescribes an iterative procedure to update the weights to minimize the squared error. Appendix 9.A gives an application of this algorithm to the three-layer network of Figure 9.9.

Forecasting Once training is complete, we have a set of values for the parameters of the network, ν_i, w_{ij}, w_{jk}, and the parameters β of the tanh function. Since the state of the input nodes y_i is equal to $\tilde{f}_i(h)$, and since we chose $\tilde{f}$ to be the linear function, the input state is simply the input to node m. Again, the inputs to the hidden layer are a weighted combination of values of the states of the input layer

$$h_j = (\sum_{i=1}^{I} w_{ij} y_i) - \nu_j,$$

so the state of the hidden node j is computed as

$$x_j = f(h_j).$$

The inputs to the output layer are again a weighted combination of the states of the hidden layer and the activation thresholds of the output nodes:

$$h_k = \sum_{j=1}^{J} w_{jk} x_j - \nu_k.$$

The final forecast is then given by state of the output nodes:

$$\hat{Z}_k = \tilde{f}(h_k) = h_k.$$

9.3.8.3 More Advanced Neural Networks

The network architecture and training algorithms described thus far form the most basic neural-network methodology. But other variations of this method are available. Even for the simple method presented here, we have not delved into procedures to choose the number of hidden layers or the number of nodes in each layer, or the best choice of the transfer functions. For example, there are many procedures to automatically prune or grow the network topology based on the observed data and the network's predictive performance.

As far as training goes, we have described only one of the earliest and the most basic of training algorithms. A significant amount of the neural network literature is devoted to improving training, in terms of speeding up the convergence or ensuring the convergence is to the right parameters (global convergence), and avoiding overfitting (Section 9.5.1.4). The interested reader should consult a textbook on neural networks before deciding among these various options.

9.3.9 Pick-up Forecasting Methods

Pick-up forecasting methods exploit some unique characteristics of reservation data in quantity-based RM, where the period between repeated service offerings is shorter than the period over which reservations are made (for example, an airline offers a daily flight between two cities but accepts reservations for these flights up to 90 days prior to departure). They are best viewed as a forecasting strategy—specifying a method for disaggregating and aggregating reservations data—rather than a class of fundamentally new forecasting algorithms.

As we mentioned in Section 9.1.3.4, reservations data has a "wedge-shaped" form, in which one has a partial and evolving picture of demand over time. Figure 9.11 shows this evolution of demand in matrix and graphical form for resources sold on consecutive dates. Rather than relying only on complete booking histories for forecasting, pick-up methods exploit both the complete and partial-bookings data to make better forecasts. The main idea is to forecast incremental bookings (booking obtained over short intervals of time prior to service) and then aggregate these increments to obtain a forecast of total demand to come.

We illustrate this idea with the *additive pick-up method*. Suppose for the data in Figure 9.11, we want to forecast for 13-June when we have one day remaining. The historical observed bookings on the day of departure are 8, 2, and 13 (for the service dates 12 June, 11 June and 10 June respectively). From this data $\{8, 2, 13\}$, we make an incremental forecast for zero-day prior for 13 June (bookings expected on 13 June) as, say, the mean value of 7.6. Similarly, for the forecast for 14 June, we first construct two incremental forecasts, one for zero-day prior and the other for one-day prior; sticking to our averaging method, this yields incremental forecasts of 7.6 and 3.75, respectively. Then the forecast of demand to come for 14 June is the sum of these two increments or $7.6 + 3.75 = 11.35$, and so on, for the other dates in the future.

Formally, the k-day ahead forecast of demand to come is given by

$$\hat{Z}(t+k) = \sum_{i=0}^{k} \hat{Z}_{[i]}(t+k),$$

Estimation and Forecasting

Number of Days Prior to the Usage of the Resource									Resource-Usage Date
-8	-7	-6	-5	-4	-3	-2	-1	0	
6	3	11	4	9	8	13	3	13	10-Jun
8	6	6	3	16	11	5	4	2	11-Jun
1	2	0	0	3	6	2	6	8	12-Jun
6	0	4	1	2	6	3	2		13-Jun
3	8	8	6	5	1	2			14-Jun
1	0	2	7	6	4				15-Jun
0	1	1	6	5					16-Jun
1	11	12	6						17-Jun

Incremental bookings

Number of Days Prior to the Usage of the Resource									Resource-Usage Date
-8	-7	-6	-5	-4	-3	-2	-1	0	
6	9	20	24	33	41	54	57	70	10-Jun
8	14	20	23	39	50	55	59	61	11-Jun
1	3	3	3	6	12	14	20	28	12-Jun
6	6	10	11	13	19	22	24		13-Jun
3	11	19	25	30	31	33			14-Jun
1	1	3	10	16	20				15-Jun
0	1	2	8	13					16-Jun
1	12	24	30						17-Jun

Cumulative bookings

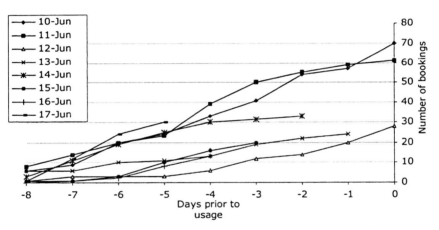

Figure 9.11. Incremental bookings, cumulative bookings, and cumulative booking curves for eight consecutive days of a resource. 12 June is the current date with full historical bookings. 13 June till 17 June have partial booking information.

where $\hat{Z}_{[i]}(\cdot)$ represents the incremental bookings forecast i days prior to the time of service. The forecasts $\hat{Z}_{[i]}(t + k)$ are constructed using the available historical i-day-prior incremental bookings. In principle, any time-series method can be used to make these incremental forecasts.

In the *multiplicative pick-up method*, the forecast is performed on data normalized as a fraction of current bookings. So if k days prior to the resource usage date there are 100 total bookings on hand and on $(k-1)$ days prior 10 bookings were observed, then the incremental increase is 10% or 0.1. The incremental bookings data is first converted to such fractions. In our example in Figure 9.11, to make a forecast for 13 June, we convert the zero-day prior incremental bookings into $\{8/14, 2/55, 13/54\}$ (14, 55, and 54 are the total bookings on hand for 12 June, 11 June, and 10 June, respectively). Similarly, the one-day prior fractions for 14 June are $\{2/22, 6/14, 4/55, 3/54\}$. We can take the average of these fractions to obtain the forecast of the pick-up fraction zero-day prior and one-day prior. This would be 0.284 for zero-day prior and 0.162 for one-day prior, the average multiplicative "pick-up" over current bookings. A forecast of demand to come for 14 June would be $0.284 \times (33 + 0.162 \times 33) + 0.162 \times 33 = 16.23$. This is higher than given by the additive pick-up method, reflecting the underlying assumption of the multiplicative method that future bookings are positively correlated with current bookings. Other aggregation strategies and variations are possible.

Again, the advantage of pick-up methods is that they use all the available bookings information. Moreover, as partial bookings are recent data, using this data can make the forecast more responsive to shifts in demand. While the idea is simple and mostly heuristic, pick-up methods are widely used in quantity-based RM and reported to perform well.

9.3.10 Other Methods

Several other methods of forecasting have been reported in RM. The *Delphi method* is a formal procedure for extracting analyst and managers' opinion on expected demand. It is used primarily in cases where there is no historical information, where there is an unexpected demand shock, or in some cases when RM is done manually. Fuzzy logic (Ting and Tzeng [512]) and expert systems (Basgall [29]) have been proposed as the basis for a second level of automation in RM forecasting. These systems attempt to replicate the rules used by human analysts when monitoring and overriding a RM system. Chaos-theoretical models for forecasting market response have been proposed by Mulhern and Caprara [394], although we are not aware of widespread use of these techniques in RM. Another forecasting method proposed for RM is based on fitting historical booking to a set of cumulative booking curves. The current bookings on hand are extrapolated using these curves to give the forecast. This approach is similar in spirit to the multiplicative pick-up method discussed above.

9.3.11 Combining Forecast Methods

With computing power and storage becoming cheaper by the day, an increasingly feasible forecasting strategy is to simultaneously use several forecasting methods and pick the "best" one. Of course, identifying which method is best becomes another forecasting exercise in itself, and there have been many proposals for such a model-picking strategy.

Moreover, it may not even be necessary to identify the best-performing method: a linear combination of the forecasts with an appropriate set of weights can turn out to be consistently superior to any one of the constituent methods. This idea was proposed in an article by Bates and Granger [30] and subsequently much investigated by forecasting researchers. The intuition behind this result is that if the errors produced by two forecasting methods are negatively correlated, then combining them will reduce the overall forecast error.

So what is the best set of weights for such a linear combination? This can be determined by finding weights that minimize the mean-squared error of the combined forecast. Although it is difficult to obtain such weights analytically, various heuristics have been proposed (see the Notes and Sources of this chapter for references). The weights themselves can adapt to fresh data and be updated from period to period.

We give one set of weights proposed by Bates and Granger [30] to combine forecasts from two different models. Let MSE_i be the mean-squared error of model $i, i = 1, 2$. Let ρ be the coefficient of correlation between the errors in the forecasts of the two models. Then define the weights as α and $(1 - \alpha)$, where α is given by:

$$\alpha = \frac{MSE_2 - \rho\sqrt{MSE_1}\sqrt{MSE_2}}{MSE_1 - MSE_2 - 2 - \rho\sqrt{MSE_1}\sqrt{MSE_2}}.$$

Then the combined forecast is given by

$$\hat{Z} = \alpha\hat{Z}_1 + (1-\alpha)\hat{Z}_2.$$

Another combination scheme, this time using adaptive weights that vary over time, is to set $\alpha = \alpha(t)$, at time t, where

$$\alpha(t) = \sum_{i=1}^{t} \frac{MSE_2(t)}{MSE_1(t) + MSE_2(t)},$$

where $MSE_i(t)$ is the mean squared error of model i at time t. The interested reader should consult Montgomery et al. [388], Gupta and Winston [230], and Foster and Vohra [192] for other similar rules and their properties.

9.4 Data Incompleteness and Unconstraining

We next look at forecasting from data that is either missing or partially observable, a common situation in RM. Indeed, once a product is closed or capacity is sold out, we normally stop observing demand at that point because most reservation systems record only actual bookings and not "attempted bookings." Ignoring this censoring can cause a significant bias in the forecasts. For instance, consider a product that had been closed consistently in the past. Its observed demand would be uniformly zero, and a forecast based on this data would forecast demand as zero. However, if the optimization system had opened this product, a positive demand might have been observed.

Incompleteness can occur in price-based RM when sales (and no-sales) are not directly observable. This can make it difficult to obtain complete information on customer purchase behavior. For example, if a customer decides not to purchase because some alternative is not available in the retail store, this information frequently goes unrecorded. Ignoring these lost sales can lead to a bias in the forecasts if the data is not corrected to account for the missing information.

Of course, companies that sell directly through their own call centers or websites have the potential, in theory, to capture attempted reservations or no-purchase outcomes. However, in our experience few actually do. And given the significant role that third-party reservations systems and distribution channels play in many RM industries, the problem of incomplete data remains an important one in RM forecasting.

Fortunately, there are several good methods available for correcting for incomplete data, which we discuss here. Our description of these methods is focused primarily on quantity-based RM because this is where the incomplete-data problem is most acute. However, the techniques are also used for estimating parameters in price-based RM, such as when correcting for stock-outs or unobservable heterogeneity in retail RM.

9.4.1 Expectation-Maximization (EM) Method

The *expectation-maximization (EM) method* is the most widely used method for correcting for constrained data in quantity-based RM. While the algorithm can be described in generic form, it is easiest to understand it by looking at specific examples. Because of its importance, we give two such examples below, one for the independent fare class model and the other for the discrete choice demand model.

9.4.1.1 Unconstraining in an Independent Booking Class Model Using EM

Consider the independent-demand model of Section 2.2, in which the demand for each product is assumed to be independent of the demand for other products. Since most current quantity-based RM implementations assume independent demand for products, the method described here (or variations of it) is very prevalent in practice.

Suppose we have $M + N$ observations of bookings for a given product, $z_1, \ldots, z_{M+N}$, of which M observations are constrained because the product was closed. We ignore the time-series aspect of the observations and treat $z_1, \ldots, z_{M+N}$ as an unordered set of observations generated by an i.i.d. process. Specifically, if the time-series data has trend or seasonality, the EM algorithm cannot be applied as shown below. (Combining unconstraining with time-series forecasting is more complicated. See McGill [376].) Our goal is to find the parameters of an underlying demand distribution for these observations.

Assume that the underlying demand distribution is normal with mean μ and standard deviation σ. (The same unconstraining procedures can be applied—albeit with different formulas—for many other distribution as well.) We further assume that *all* the observations come from a common distribution and that the observations are constrained at random, i.e., they appear randomly in the sample.[9] Since we are treating the observations as unordered, assume $z_1, \ldots, z_M$ are constrained (*right censored*) at booking limits $b_1, \ldots, b_M$, so that $z_1 = b_1, \ldots, z_M = b_M$. The remaining N observations are unconstrained.

If the data were not constrained, then it would be easy to construct the complete-data likelihood function. Namely,

$$L(\mu, \sigma, M+N) = \prod_{i=1}^{M+N} \frac{1}{\sqrt{2\pi}\sigma} e^{\frac{-(z_i-\mu)^2}{2\sigma^2}}, \quad (9.56)$$

with the complete-data log-likelihood function given by

$$\ln L(\mu, \sigma, M+N) = -\frac{M+N}{2}\ln 2\pi - (M+N)\ln\sigma - \frac{\sum_{i=1}^{M+N}(z_i-\mu)^2}{2\sigma^2}. \quad (9.57)$$

The μ and σ that maximizes $\ln L(\cdot)$ in (9.57) are given by the closed-form solution

$$\hat{\mu} = \frac{1}{M+N}\sum_{i=1}^{M+N} z_i$$

$$\hat{\sigma}^2 = \frac{1}{M+N}\sum_{i=1}^{M+N}(z_i-\hat{\mu})^2.$$

---

[9] In the RM context this assumption implies that there is no correlation among demand on days when the product is sold out. Strictly speaking, this assumption rarely holds in RM practice, but it is common to ignore this correlation possibility as the alternative statistical methods are considerably more complicated.

(See Example 9.5.) However, we do not know the true values of the M constrained observations $z_1, \ldots, z_M$ and therefore cannot use this procedure directly.

The EM method uses this complete-data likelihood function in an iterative algorithm with an alternating *E-step* and *M-step* (hence the name). The E-step replaces the censored data by estimates of their uncensored values using the current estimates of the mean and standard deviation. The M-step then maximizes the complete-data log-likelihood function based on this updated data to obtain new estimates of the mean and standard deviation. The procedure is then repeated until the parameter estimates converge. The advantage of this approach is that it is much easier to estimate the complete-data likelihood than it is to estimate the incomplete-data log-likelihood function. Hence, even though we have to solve the complete-data likelihood problem many times, the overall algorithm is still very efficient.

Specifically, for our normal distribution example, let $\mu^{(k)}, \sigma^{(k)}$ represent the estimates of the parameters of the normal distribution after the kth iteration of the algorithm. The steps of the EM algorithm for our time series follow:

STEP 0 (Initialize): Initialize μ and σ to be $\mu^{(0)}$ and $\sigma^{(0)}$. Good candidates for these starting values are the sample mean and sample standard deviation of all the unconstrained observations.

Let $\delta > 0$ be a small number, to be used as a stopping criterion.

$$\mu^{(0)} = \frac{\sum_{i=M+1}^{M+N} z_i}{N}$$

$$\sigma^{(0)} = \sqrt{\frac{\sum_{i=M+1}^{M+N}(z_i - \mu^{(0)})^2}{N}}.$$

STEP 1 (E-step): Calculate the expected value of the censored data in the log-likelihood function assuming that they come from a normal distribution X with parameters $(\mu^{(k-1)}, \sigma^{(k-1)})$. That is, for $i = 1, \ldots, M$ calculate

$$\hat{Z}_i^{(k-1)} \doteq E[X|X \geq b_i, X \sim N(\mu^{(k-1)}, \sigma^{(k-1)})]$$

and

$$(\hat{Z}_i^2)^{(k-1)} \doteq E[X^2|X \geq b_i, X \sim N(\mu^{(k-1)}, \sigma^{(k-1)}].$$

The formulas for these conditional expectations are somewhat complex but involve simply evaluating two integrals.

Estimation and Forecasting 477

Next, for each censored observation $i = 1, \ldots, M$, replace z_i by $\hat{Z}_i^{(k-1)}$ and z_i^2 by $(\hat{Z}_i^2)^{(k-1)}$ to form the complete-data log-likelihood function $Q(\mu, \sigma)$ as in (9.57). Note in this way we are simply replacing the constrained values in the log-likelihood function by their expected values given the current estimates of the mean and standard deviation.

STEP 2 (M-step): Maximize $Q(\mu, \sigma)$ with respect to μ and σ to obtain $\mu^{(k)}, \sigma^{(k)}$, yielding

$$\mu^{(k)} = \frac{1}{M+N} \left[\sum_{i=1}^{M} \hat{Z}_i^{(k-1)} + \sum_{i=M+1}^{M+N} z_i \right]$$

and

$$\sigma^{(k)} = \frac{1}{M+N} \left[\sum_{i=1}^{M} \left((\hat{Z}_i^2)^{(k-1)} - 2\hat{Z}_i \mu^{(k-1)} + (\mu^{(k-1)})^2 \right) \right.$$
$$\left. + \sum_{i=M+1}^{M+N} \left(z_i - \mu^{(k-1)} \right)^2 \right].$$

STEP 3 (Convergence test): IF $\|\mu^{(k)} - \mu^{(k-1)}\| < \delta$ and $\|\sigma^{(k)} - \sigma^{(k-1)}\| < \delta$, THEN STOP;
ELSE, $k \leftarrow k + 1$, GOTO STEP 1.

If the expected log-likelihood is continuous in the parameters (μ and σ in our case), a result by Wu [582] shows that if the sequence of EM estimates converges, the limiting value will be a stationary point of the incomplete log-likelihood function. Whether the sequence diverges—or converges to something other than the global maximum—is more difficult to determine and depends on the characteristics of the data set. In practice, however, the EM method has proved to be very robust.

Once convergence has been achieved—say, in iteration K—the unconstrained values for $z_i, i = 1, \ldots, M$ can be taken as $E[X|X \geq b_i]$, where X is normally distributed with $\mu^{(K)}, \sigma^{(K)}$:

Example 9.16 Consider the data set of bookings in Table 9.4 from 11 Jan to 29 Jan. The data on 13 Jan, 16 Jan and 18 Jan is constrained at the booking limit 17, 22, and 15 respectively. Assume the data comes from a normal distribution. Based on the constrained data, the parameters of the normal distribution $(\mu^{(0)}, \sigma^{(0)}) = (22.526, 7.537)$.
Let C be the capacity constraint, D the demand, $z^{(k)}$ the unconstrained value at the kth iteration, and $\bar{C}^{(k)} = \frac{C - \mu^{(k)}}{\sigma^{(k)}}$. Then at the $(k+1)^{\text{st}}$ iteration, replace $\bar{z}^{(k)}$ by

$$E[\bar{z}^{(k)} | \bar{z}^{(k)} \geq C; D \sim N(\mu^{(k)}, \sigma^{(k)})]$$

Table 9.4. EM algorithm iterations on constrained data.

Date	Constrained Bookings	Iteration 1	Iteration 2	Iteration 3
11-Jan	22	22	22	22
12-Jan	15	15	15	15
13-Jan	17*	23.544*	24.216*	24.275*
14-Jan	33	33	33	33
15-Jan	16	16	16	16
16-Jan	22*	25.416*	25.699*	25.724*
17-Jan	22	22	22	22
18-Jan	15*	23.579*	23.963*	23.963*
19-Jan	22	22	22	22
20-Jan	17	17	17	17
21-Jan	23	23	23	23
22-Jan	19	19	19	19
23-Jan	31	31	31	31
24-Jan	17	17	17	17
25-Jan	30	30	30	30
26-Jan	23	23	23	23
27-Jan	31	31	31	31
28-Jan	12	12	12	12
29-Jan	41	41	41	41
Mean	22.526	23.502	23.572	23.577
S.D	7.537	7.178	7.185	7.186

given by the following formula for the normal distribution

$$z^{(k+1)} = C + \frac{\sigma}{\sqrt{2\pi}} e^{-0.5(\bar{C}^{(k)})^2} - \bar{C}^{(k)} P(D_{\sim N(\mu^{(k)},\sigma^{(k)})} \geq \bar{C}^{(k)}).$$

So at the first iteration, replace 17 by 23.544, 22 by 25.416, and 15 by and 23.579. At the second iteration, replace 23.544 by 24.216, and so on. As can be seen from Table 9.4, the algorithm quickly converges (in this case; convergence is much slower in general) to $\mu = 23.577, \sigma = 7.186$.

9.4.1.2 Unconstraining in a Discrete-Choice Dynamic Model Using EM

We next consider the problem of unconstraining under the dynamic discrete-choice model of Section 2.6.2. Recall that in this model there is an arrival probability λ in each period and consumers select among the available classes according to a discrete-choice model. The RM control problem is then to decide which products to make available at each point in time. We consider here a multinomial-logit model similar to Example 9.6, where the probability that an arriving customer purchases

Estimation and Forecasting 479

alternative i from a set S is given by

$$P_i(S) = \frac{e^{\beta^\top \mathbf{y}_i}}{\sum_{j \in S} e^{\beta^\top \mathbf{y}_j} + 1},$$

where $\mathbf{y}_j$ is a vector of attributes of alternative j and β is a vector of parameters. The no-purchase probability is

$$P_i(S) = \frac{1}{\sum_{j \in S} e^{\beta^\top \mathbf{y}_j} + 1}.$$

The difficulty here is estimating the parameters β and λ from purchase data. Specifically, if we have only purchase data, it is impossible to distinguish a period without an arrival, from a period in which there was an arrival but the arriving customer did not purchase. With this incompleteness in the data, the complete-data maximum-likelihood estimation procedure of Example 9.6 cannot be used.

However, we can again apply the EM algorithm to correct for the missing data. The broad strategy is the same as the one for the normal distribution case in Section 9.4.1: start with arbitrary initial estimates of the parameters $\hat{\beta}$ and the arrival rate $\hat{\lambda}$. Then use these estimates to compute the conditional expected value of $\mathcal{L}$, $E[\mathcal{L}|\hat{\beta}, \hat{\lambda}]$ (the expectation step). Maximize the resulting expected log-likelihood function to generate new estimates $\hat{\beta}$ and $\hat{\lambda}$ (the maximization step), and repeat till the procedure converges.

Suppose there are T periods. Let P denote the set of periods in which customers purchase and $\bar{P}$ denote period in which there are no purchase transactions. Let $a_t = 1$ if there is an arrival in period t and $a_t = 0$ if there is no arrival. Let $j(t)$ denote the choice made by an arrival in period t. We can then write the complete log-likelihood function as

$$\mathcal{L} = \sum_{t \in P} \left[\ln(\lambda) + \beta^\top \mathbf{y}_{j(t)} - \ln(\sum_{j \in S} e^{\beta^\top \mathbf{y}_j} + 1) \right]$$
$$+ \sum_{t \in \bar{P}} \left[a_t \left(\ln(\lambda) - \ln(\sum_{j \in S} e^{\beta^\top \mathbf{y}_j} + 1) \right) + (1 - a_t) \ln(1 - \lambda) \right].$$
(9.58)

The unknown data are the values a_t, $t \in \bar{P}$ in the second sum. However, given estimates $\hat{\beta}$ and $\hat{\lambda}$, we can determine their expected values (denoted $\hat{a}_t$) easily via Bayes rule:

$$\hat{a}_t \doteq E[a_t | t \in \bar{P}, \hat{\beta}, \hat{\lambda}]$$

$$P(a_t = 1 | t \in \bar{P}, \hat{\beta}, \hat{\lambda})$$
$$= \frac{P(t \in \bar{P} | a_t = 1, \hat{\beta}, \hat{\lambda}) P(a_t = 1 | \hat{\beta}, \hat{\lambda})}{P(t \in \bar{P} | \hat{\beta}, \hat{\lambda})}$$
$$= \frac{\hat{\lambda} P_0(S | \hat{\beta})}{\hat{\lambda} P_0(S | \hat{\beta}) + (1 - \hat{\lambda})}, \quad (9.59)$$

where
$$P_0(S | \hat{\beta}) = \frac{1}{\sum_{j \in S} e^{\hat{\beta}^\top \mathbf{y}_j} + 1}$$

is the no-purchase probability for an arrival in period t given $\hat{\beta}$.

Substituting $\hat{a}_t$ into (9.58) we obtain the expected log-likelihood

$$\begin{aligned}
E[\mathcal{L} | \hat{\beta}, \hat{\lambda}] &= \sum_{t \in P} \left[\beta^\top \mathbf{y}_{j(t)} - \ln(\sum_{j \in S} e^{\beta^\top \mathbf{y}_j} + 1) \right] \\
&\quad - \sum_{t \in \bar{P}} \hat{a}_t \ln(\sum_{j \in S} e^{\beta^\top \mathbf{y}_j} + 1) \\
&\quad + \sum_{t \in P} \ln(\lambda) \\
&\quad + \sum_{t \in \bar{P}} (\hat{a}_t \ln(\lambda) + (1 - \hat{a}_t) \ln(1 - \lambda)). \quad (9.60)
\end{aligned}$$

As in the case of the complete log-likelihood function, this function is separable in β and λ. Maximizing with respect to λ we obtain the updated estimate

$$\lambda^* = \frac{|P| + \sum_{t \in \bar{P}} \hat{a}_t}{|P| + |\bar{P}|}. \quad (9.61)$$

This is intuitive; our estimate of lambda is the number of observed arrivals $|P|$, plus the estimated number of arrivals from unobservable periods $\sum_{t \in \bar{P}} \hat{a}_t$, divided by the total number of periods $|P| + |\bar{P}| = |T|$. We can then maximize the first two sums in (9.60) to obtain the updated estimate β^*. Note that this expression is of the same functional form as the complete data case (9.11). The entire procedure is then repeated.

Summarizing the algorithm:

STEP 0 (Initialize): $\hat{\beta}^{(0)}$ and $\hat{\lambda}^{(0)}$, k=0.

STEP 1 (E-step): For $t \in \bar{P}$, use the current estimates $\hat{\beta}^{(k)}$ and $\hat{\lambda}^{(k)}$ to compute $\hat{a}_t^{(k)}$ from (9.59).

STEP 2 (M-step): Compute $\lambda^{(k+1)}$ using (9.61).
Compute $\beta^{(k+1)}$ by solving

$$\max_{\beta} \left\{ \sum_{t \in P} \left(\beta^\top \mathbf{y}_{j(t)} - \ln(\sum_{j \in S} e^{\beta^\top \mathbf{y}_j} + 1) \right) - \sum_{t \in \bar{P}} \hat{a}_t^{(k)} \ln(\sum_{j \in S} e^{\beta^\top \mathbf{y}_j} + 1) \right\}$$

STEP 3 (Convergence test): IF $\|(\hat{\lambda}^{(k+1)}, \hat{\beta}^{(k+1)}) - (\lambda^{(k)}, \beta^{(k)})\| < \delta$, THEN STOP;
ELSE $k \leftarrow k + 1$ AND GOTO STEP 1.

One interesting fact is that there can be multiple pairs (β, λ) that produce the same probabilities of sales. In this case, the EM and logit estimates will find only one such pair. To take a trivial case, suppose there is only $n = 1$ fare product and that y_1 and β are scalars. The probability that we observe a sale if this fare product is open is then

$$p = \lambda \frac{e^{\beta y_1}}{e^{\beta y_1} + 1}.$$

It is clear that there are a continuum of values (β, λ) that will produce the same value p. However, the maximum-likelihood estimate will identify only one such pair. This difficulty is not a fault of the EM or logit method per se; it is a reflection of the fact that—as in this simple example—there may be more than one model that produces the same purchase probabilities. In such cases, it is simply not possible to uniquely identify the model from observed data; there is, in effect, a degree of freedom that we cannot resolve.

9.4.2 Gibbs Sampling

While the EM algorithm is the most popular and widely used method for unconstraining in RM applications, there are alternative statistical methods to deal with constrained data. We briefly describe one technique here, called *Gibbs sampling*, which is part of a broader set of methods called *Markov-chain Monte Carlo (MCMC)* methods. Although not widely used in forecasting for quantity-based RM, they have found application in price-based RM (Allenby and Rossi [10], Allenby, Arora, and Ginter [8]), econometrics (Chib and Greenberg [115]), and missing-data problems (Schafer [456]).

MCMC methods simulate a (typically intractable) target distribution $f(\cdot)$ of a (multidimensional) random variable $\mathbf{Z}$ by repeatedly simulating a sequence $\{Z^{(1)}, \ldots, Z^{(k)}, \ldots\}$, where each element in the sequence depends on the previously generated element, and the limiting distribution of $Z^{(k)}$ as $k \to \infty$ is the target distribution $f(\cdot)$. By generating enough of these sequences, we can reconstruct the entire distribution $f(\cdot)$.

We first describe the Gibbs sampling method in general and then apply it to the censored normal example in Section 9.4.1.1. Let a random vector Z be partitioned into J subvectors

$$Z = (Z_1, Z_2, \ldots, Z_J).$$

Let $f(Z)$ be the joint distribution of Z—that is, the target distribution. The Gibbs algorithm is applicable whenever $f(Z)$ is unknown, intractable, or difficult to sample from, but all the distributions $f(Z_i|Z_{-i})$, for $i = 1, \ldots, J$, (Z_{-i} is the vector X but without the ith block) have known distributions that are easy to sample from.

Let $Z^{(k)} = (Z_1^{(k)}, Z_2^{(k)}, \ldots, Z_J^{(k)})$ be the generated sample at the kth iteration.

Gibbs algorithm: Repeat the following steps till convergence (the criteria for which are discussed later):

- Generate $Z_1^{(k+1)}$ from $f(Z_1|Z_2^{(k)}, \ldots, Z_J^{(k)})$
- Generate $Z_2^{(k+1)}$ from $f(Z_2|Z_1^{(k+1)}, Z_3^{(k)}, \ldots, Z_J^{(k)})$
- $\vdots$
- Generate $Z_J^{(k+1)}$ from $f(Z_J|Z_1^{(k+1)}, Z_2^{(k+1)}, \ldots, Z_{J-1}^{(k+1)})$

The stationary distribution of the sequence $\{Z^{(k)}\}_{k=0,1,2,\ldots}$, under relatively mild conditions, can be shown to converge to the joint distribution $f(Z)$.

The use of Gibbs sampling for parameter estimation usually proceeds in a Bayesian framework, in which we assume a prior distribution on the parameters, and—from a practical point of view—choose a conjugate family of distributions for the parameters.

To illustrate, let's see how to apply Gibbs sampling method to estimate the unconstrained mean and variance of a sample from a censored normal distribution with unknown mean and standard deviation, μ, σ.

Assume as in the previous section that we have a sequence of $M + N$ independent observations $\{z_i\}$, where the first M observations are constrained at $b_i, i = 1, \ldots, M$. Our problem is to estimate μ and σ. It is convenient to assume that $\mu|\sigma$ (μ given σ) has a "prior" normal distribution and σ, a "prior" inverted chi-square distribution,[10] denoted by χ^{-2}. This particular choice of distributions ensures the posterior distributions of $\mu|\sigma$ and σ are normal and inverted chi-square again.

The vector $\mathbf{Z}$ is then assumed to consist of two blocks—the first, of the unknown parameters $[\mu, \sigma]$, and the second, the vector of censored observations $(Z_1, \ldots, Z_M)$. The Gibbs algorithm begins with initial values for these two subvectors. For instance, as we did in the case of the EM application, take (μ, σ) initially equal to the sample mean and standard deviation of $\{z_i\}_{i=M+1,\ldots,N}$ and set the vector $(z_1^{(0)}, \ldots, z_M^{(0)})$ equal to the vector of censored values $(b_1, \ldots, b_M)$.

At the $(k+1)^{\text{st}}$ step, generate

$$(Z_1^{(k+1)}, \ldots, Z_M^{(k+1)}) \sim N(\mu^{(k)}, \sigma^{(k)} \mid Z_1 \geq b_1, \ldots, Z_M \geq b_M)$$

as M independent draws.

Next generate new values for

$$(\mu^{(k+1)}, \sigma^{(k+1)})$$

from a normal and inverted chi-square distribution, respectively, as follows:

$$\mu^{(k+1)} \sim N(\bar{y}, \sigma^{(k)})$$
$$\sigma^{(k+1)} \sim (M+N-1)\bar{S}^2 \chi^{-2}_{M+N-1},$$

where $\bar{y}$ and $\bar{S}$ are the sample mean and standard deviation of the M generated values and the N unconstrained values:

$$[z_1^{(k+1)}, \ldots, z_M^{(k+1)}, z_{M+1}, \ldots, z_{M+N}].$$

This procedure is repeated until the distributions of μ, σ, $z_1, \ldots, z_M$ reach stationarity. However, testing for stationarity of a distribution can be problematic (Section 9.3.2.1), so in practice a number of heuristic termination criteria are used [456]. The resulting expected value of μ and σ can then be used as our parameter estimates.

9.4.3 Kaplan-Meir Product-Limit Estimator

The Kaplan-Meir product-limit (PL) estimator ([289]) is another approach to censored-data estimation. Its origins lie in survival analysis (with continuous distributions), but here we present it in terms of censored demand observations. It is a nonparametric method, the output of

---
[10] A random variable Y has an inverted chi-square distribution if Y^{-1} has a chi-square distribution.

which is an estimate of the complete distribution (as in Gibbs sampling) rather than the parameters of an assumed distribution.

As before, assume we have $M+N$ observations $z_1, \ldots, z_{M+N}$, with the first M being constrained (right-censored) at the values $b_1, \ldots, b_M$. So the observations are of the form $z_i = \min(Z_i, b_i)$, where b_i is the booking-limit (called the *limits of observation*; the event $Z_i > b_i$ is called a *loss*). As earlier, Z_i is considered independent of b_i. The *survival function* of Z_i is defined as $G(z) = P(Z_i > z)$, and an estimate of it is equivalent to an estimate of the distribution of Z.

The PL estimate $\hat{G}(z)$ of the survival function is then given as follows. List and label the $M+N$ observations in order of increasing magnitude, so that $0 \leq z_{(1)} \leq z_{(2)} \leq \cdots \leq z_{(M+N)}$. For a particular value z, let $S_z = \{r | z_{(r)} \leq z, z_{(r)} < b_{(r)}\}$. That is, S_z is the set of indices in the ordered list that are not constrained by the booking limits and have values less than z. Then

$$\hat{G}(z) = \prod_{r \in S_z} \frac{N-r}{N-r+1}, \tag{9.62}$$

where each term above is an estimate of the conditional probability that the demand exceeds x_r given that it exceeds x_{r-1}. The main idea behind Kaplan-Meir estimate is best explained via a simple example:

Example 9.17 Suppose that we have four observations with bookings $\{5, 10^*, 11, 18\}$, where the superscript * signifies a constrained observation. Suppose we are interested in the probability that $Z \geq 15$. If we ignore the constrained observation (that is, base our estimate on the unconstrained reduced sample), we get an estimate of $1/3$ (one of the three unconstrained values exceeds 15).

However, we can also view $P(Z \geq 15)$ as equal to $P(Z \geq 15 | Z \geq 10) P(Z \geq 10)$. Then we estimate $P(Z \geq 10) = 3/4$ (based on the full sample) and $P(Z \geq 15 | Z \geq 10) = 1/2$ (based on the sample of last two observations), and we obtain $P(Z \geq 15) = 3/8$. So the estimate of $P(Z \geq 10)$ helps in obtaining a better estimate of $P(Z \geq 15)$.

Kaplan and Meier show that the estimator $\hat{P}(z)$ in (9.62) gives the distribution that maximizes the likelihood of the observations. The curve given by (9.62) is remarkably easy to compute and makes no parametric assumptions. However, it can be inefficient (Miller [384]) and difficult to compare by eye (Efron [173]), and it is also difficult to compute confidence intervals for a Kaplan-Meier estimator.

9.4.4 Plotting Procedures

A hybrid parametric/nonparametric approach to censored data is based on simply fitting a parametric distribution to an nonparametric survivor function estimate $\hat{G}(z)$, such as derived using the Kaplan-Meir estimator. Such methods are called *plotting procedures*, because they correspond to plotting an empirical distribution and then inferring parameters from this plotted distribution.

To take a simple case, if the distribution is assumed to be exponential, so that $P(Z > z) = e^{-\lambda z}$, then we have that

$$\ln(G(z)) = -\lambda z.$$

Hence, if we plot the empirical function $\ln(\hat{G}(z))$, it should roughly be linear, with slope equal to $-\lambda$. One could estimate this slope via linear regression, for example. In the case of a normal distribution with mean μ and standard deviation σ, the distribution is

$$F(z) = \Phi(\frac{z-\mu}{\sigma}),$$

where $\Phi(z)$ is the standard normal distribution. Hence,

$$\Phi^{-1}(1 - S(z)) = \frac{z-\mu}{\sigma},$$

where $\Phi^{-1}(\cdot)$ is the inverse of the standard normal distribution. Therefore, by plotting $\Phi^{-1}(1-\hat{S}(z))$ we should expect to see roughly a straight line with slope $1/\sigma$ and intercept $-\mu/\sigma$. Again, values for the slope and intercept can be determined using linear regression.

While somewhat less rigorous in a strict statistical sense than other censored-data methods, plotting procedures can be attractive in practice because they are simple and intuitive.

9.4.5 Projection-Detruncation Method

The *projection-detruncation method* is similar in spirit to the EM algorithm. It has been used in the PODS simulations for quantity-based RM and its origin is credited to Hopperstad ([256, 42, 587]).

The variation over the EM method of Section 9.4.1.1 is that in the kth E-step of the algorithm, instead of replacing the constrained values by an estimate of the conditional mean

$$\hat{Z}_i^{(k-1)} = E[X | (\mu^{(k-1)}, \sigma^{(k-1)}), X \geq b_i],$$

it replaces the values by the solution $\hat{Z}_i^{(k-1)}$ of the following equation

$$\int_{\hat{Z}_i^{(k-1)}}^{\infty} f(x|(\mu^{(k-1)}, \sigma^{(k-1)}))dx = \tau \int_{b_i}^{\infty} f(x|(\mu^{(k-1)}, \sigma^{(k-1)}))dx,$$

(9.63)

where τ is a fixed constant throughout the algorithm. While there is no formal theoretical justification of (9.63) or a proof of convergence, the heuristic interpretation is as follows. Note that (9.63) can be written

$$P\left(Z_i > \hat{Z}_i^{(k-1)} \middle| Z_i > b_i, \mu^{(k-1)}, \sigma^{(k-1)}\right) = \tau.$$

So $\hat{Z}_i^{(k-1)}$ corresponds to selecting a fixed fractile of the conditional distribution given the current parameter estimates $\mu^{(k-1)}, \sigma^{(k-1)}$. For example, selecting $\tau = 1/2$ would correspond to estimating $\hat{Z}_i^{(k-1)}$ as the *median* of the conditional distribution, whereas the EM method uses the *mean* of the conditional distribution. Hence, by using a small τ value the constrained observations are unconstrained more aggressively than may be the case in the EM method. Whether this leads to more accurate estimation of the mean or a faster convergence than the EM algorithm is not known, however. Zeni [587] gives an example comparing the estimates of the two methods for $\tau = 0.15$, and the estimate of the mean of the projection-detruncation method is nearly 10 percent higher than that given by the EM algorithm, though one can arguably attribute this to the choice of τ.

9.5 Error Tracking and System Control

As mentioned, all forecasts are subject to some degree of error. Hence, understanding and responding correctly to forecast errors are important tasks in practice. Here we review the main methods for error tracking and system control.

A forecaster needs to consider several types of errors. The difference between the observed data and a model fit to this data is called the *estimation error*. Such error could be due to many factors: natural randomness in the demand process, unobservable characteristics of the products or demand, misspecifications, unrealistic model assumptions such as independence of the variables or error terms. We group all such errors—errors in the estimation of the parameters of the model or the specification of the model—as estimation errors.

Forecasting error, on the other hand, is the difference between a model's predictions for a *future* observation and the subsequent observation. The difference between forecast and estimation errors is a matter of timing. Large estimation errors might compel us to refine the model or "fix" it in some way *now* because we are aware of the errors. Forecasting errors, on the other hand, are unknown at the time of the model specification and are realized only over time. There is also a dynamic, online aspect to forecasting error and system control that is distinct from the one-shot nature of estimation.

It is natural to suppose that a model that fits historical data well that, say has low estimation errors, will also generalize well and give low forecast errors. This, however, is not the case. As we show in Section 9.5.1.4, it is not uncommon to fit a model to give near-zero estimation errors based on observed data, but then find that it has atrocious predictive power. Indeed, forecasting can be said to be the art of understanding estimation errors (their sources and reasons) and then selecting and training a model properly for optimum prediction power.

9.5.1 Estimation Errors

We first look at issues involved in analyzing estimation errors—in particular, bias, specification error, model-selection criteria, and overfitting.

9.5.1.1 Bias Detection and Correction

Bias in the parameter estimates of a model is called *estimation bias*. This could arise because of the lack of a good estimator, incomplete data, or nonconvergence of the estimation procedures. A bias in the parameter estimates of a model leads to a bias in the forecasts, and in general, it is desirable to eliminate it. If the cause of the bias were known, we would, of course, fix the bias by eliminating the cause, but this is not always possible—for lack of development time, investigation time, or data, and so on. If this is the case, a simple and general method for correcting for parameter bias is the so-called *jackknife estimator* (Quenouille [431]; Tukey [519]), which we describe next.

Suppose θ is a parameter and $\hat{\theta}$ an estimator of the parameter based on an i.i.d. sample $Z_1, \ldots, Z_N$. Suppose that $\hat{\theta}$ is a biased estimator of the following form

$$E[\hat{\theta}] = \theta + a_1/N + O(1/N^2).$$

That is, a order $1/N$ term and a second-order error term. The jackknife estimator is calculated as follows. Let $\hat{\theta}_{-i}$ be the estimator $\hat{\theta}$ applied to the sample with the ith observation removed. Define

$$\tilde{\theta}_i = N\hat{\theta} - (N-1)\hat{\theta}_{-i}, \quad i = 1, \ldots, N.$$

Define the (first-order) jackknife estimator as

$$\tilde{\theta} = \frac{\sum_{i=1}^{N} \tilde{\theta}_i}{N} = N\hat{\theta} - (N-1)\frac{\sum_{i=1}^{N} \hat{\theta}_i}{N},$$

which has the rather nice property that

$$E[\tilde{\theta}] = \theta + O(1/N^2).$$

Higher-order jackknife estimators can be defined that eliminate higher-order biases. Besides bias correction, the jackknife is a valuable tool for interval estimation and has connections to bootstrap methods (Miller [385]; Davison and Hinkley [148]).

While bias is usually undesirable, biased estimators may occasionally be beneficial if they lead to lower variance (more efficient) estimates. To give an example, if some of the explanatory variables in a linear regression are correlated (multicollinearity), the coefficients of the regression will have a high variance. A method for reducing this variance is *ridge regression*, which minimizes an objective consisting of the sum of the variance of the parameter estimates and the bias squared, so a small amount of bias is deliberately accepted (Judge et al. [273]).

9.5.1.2 Specification Errors

Specification errors are errors resulting from flawed model assumptions; that is, errors arising from a model that does not reflect the underlying data-generating process. In short, how can we be certain that the function ζ used (9.2) is indeed the "right" function to use, both explaining observed values of Z as well as providing good predictive power for future observations? Managerial judgment, visual inspection, data analysis, and statistical tests all play a role in answering this question.

Specification tests are designed to test whether a given model and its corresponding assumptions are correct. Failure to pass such a test could mean one of the following: the functional form is inadequate to represent the data-generating process; the functional form is correct, but the wrong set of independent variables have been used in the model; both the functional form and variable choice are correct, but the error term distribution is misspecified; or assumptions on the error term of the model (such as homoscedasticity or independence of errors) are violated.

There are several tests to check for misspecification (see also Section 9.2.2). The simplest ones are graphical, such as plotting values of the empirical distribution against the fitted distribution to look for a straight-line relationship, or Q-Q plots, in which the quantiles of the theoretical distribution are plotted on the x-axis and the ordered fractions of the observed values on the y-axis (a good fit is when all the values are along the diagonal). Testing an empirical distribution against a given theoretical distribution can be done using statistical procedures such as the Kolmogorov-Smirnov test. We refer the reader to DeGroot [151], pp. 554–559 for details on such tests.

Coefficient of Determination for Regressions The statistic most widely used in regressions to measure goodness of fit is the *coefficient of* *determination* (R^2), defined as follows for N observations:

$$R^2 = 1 - \frac{\sum_{j=1}^{N}(z_j - \hat{z}_j)^2}{\sum_{j=1}^{N}(z_j - \bar{z})^2}, \tag{9.64}$$

where z_j are the observations, $\hat{z}_j$ is the estimate for observation j based on the estimated parameters, and $\bar{z}$, is the mean of N observations. The R^2 value varies between 0 and 1 and signifies the percentage of the total variation in the dependent observations explained by the regression relationship. Thus, a high value of R^2 is desirable. Most commercial statistical programs (SAS, SPSS, R, S, IMSL, MINITAB, Statistica, and so on) compute this statistic automatically.

However, the choice of functional form is important, and one should not rely on quantitative measures alone. A forecaster's business intuition about the relationships and causal variables ought to play as big a role as formal statistical tests. A good R^2 value or a good visual fit does not imply a regression has good explanatory power, as we discuss below in Section 9.5.1.4 on overfitting.

The statistics of regression is concerned with many more issues than just estimating parameters and calculating R^2 values. Statistical tests exist for determining which of the independent variables is redundant, their degree of importance in determining the independent variable, their goodness of fit to the functional form, the appropriateness of the functional form and the assumptions on the errors, and so forth. For example, if the parameter estimates are assumed to be normal, then a t-test can be used to determine if the estimate is within a given interval about the true parameter value with a certain level of confidence. Similarly, a F-test can be used to test if some of the parameters are effectively redundant (values close to zero) and can be eliminated. The details of such tests are beyond the scope of this chapter, but these tests are standard and described in most statistics or econometrics texts (Kvanli et al. [318]; Judge et al. [273]; Draper and Smith [161]; Guttman [231]; Neter and Wasserman [403]).

Tests Against an Alternate Specification One form of a specification test is to test a null hypothesis that a given specification is correct against an alternate (usually more general) specification hypothesis. Depending on the type of null hypothesis, there are three classical specification tests one can use: *likelihood ratio* (LR), *Wald*, and the *Lagrange multiplier* (LM) tests. We describe only the LR test here.

Let β denote the vector of model parameters. Let the null hypothesis H_0 be that $\beta \in \Omega_0$ and the alternate hypothesis be that $\beta \in \Omega$, where typically $\Omega_0 \subset \Omega$. Then the likelihood of the observed data is as defined in (9.8). The likelihood ratio is

$$\theta = \frac{\sup_{\beta \in \Omega_0} \mathcal{L}(\beta)}{\sup_{\beta \in \Omega} \mathcal{L}(\beta)}. \tag{9.65}$$

If this ratio is small, the null hypothesis is rejected. That is, there is a significant loss of likelihood by restricting the parameter set to Ω_0. One attractive feature of the likelihood ratio is that the statistic

$$LR = -2 \ln \theta \tag{9.66}$$

is asymptotically χ^2 distributed, and this fact can be used for hypothesis testing.

Tests for Misspecification In contrast to the tests in the previous section, a test for misspecification does not specify a single alternate hypothesis. Instead, the null hypothesis is that the specification is correct and the alternate hypothesis is that there is a misspecification. Naturally, this is appealing as we are testing against a large number of alternative specifications using a single test. We describe next, informally, a general misspecification test strategy due to Hausman (Hausman [245]; also attributed to Durbin [169] and Wu [583]). We illustrate it by applying it to testing the IIA property in a discrete-choice model (Section 7.2.2.3).

To describe the idea behind the Hausman test, consider a specification test as in the previous section. The null hypothesis H_0 is that a given specification is true; the alternate hypothesis H_1 is that another specification is true. Let $\hat{\beta}_0$ be a consistent and asymptotically efficient estimator achieving the Cramer-Rao bound on the variance of the parameters (Section 9.2.1.3) of the specification under H_0. (In most cases, there would exist such an estimator if the null hypothesis were true; for instance, the maximum-likelihood estimators are consistent and asymptotically efficient [220] under some mild regularity conditions.) If instead H_1 were true, then $\hat{\beta}_0$ will be biased and inconsistent under H_1 (provided H_0 and H_1 are sufficiently different and assuming that the specification of H_1 uses the same vector of parameters as that of H_0). Let $\hat{\beta}_1$ be some other estimator for the specification of H_0—consistent but asymptotically inefficient under H_0, but consistent under H_1 also. If such estimators exist, then one can construct a test statistic out of the difference $\hat{q} = \hat{\beta}_0 - \hat{\beta}_1$, as this difference should be approximately centered at zero.

Hence, to test for misspecification when there is no alternate specification, one can proceed by choosing two distinct estimators for the null hypothesis specification—one efficient and one not efficient but more robust (consistent even under a mispecification) than the first one. Then, if the model is correctly specified, the difference between the estimators will very likely have a mean away from zero. To apply the statistic, the variance of $\hat{q}$, $V(\hat{q})$ has to be calculated, which fortunately turns out to be equal to the difference of the variances of $\hat{\beta}_0$ and $\hat{\beta}_1$. The test statistic used is $\hat{q}^\top V(\hat{q})^{-1}\hat{q}$, which can be shown to have an asymptotically χ^2 distribution (Hausman [245]; MacKinnon [352]). With no misspecification, $\hat{q}$ will tend to 0 w.p.1.

This specification test strategy, called the *Hausman-type test*, is quite general and has found many applications in econometrics. We illustrate the test by an example relevant to RM and price-response estimation.

Example 9.18 (HAUSMAN-MCFADDEN SPECIFICATION TEST FOR THE MNL DISCRETE-CHOICE MODEL ([244])) Given a set of observations of choices among n alternatives made by a population of N individuals, we would like to know if the MNL model is the correct specification for the choice process. Assume that the no-purchase choices are also observed.

Recall that the MNL model is characterized by the IIA property (Section 7.2.2.3): the ratio of the probabilities of choosing any two alternatives is independent of the attributes or the availability of a third alternative. Let $\mathcal{N} = \{1,\ldots,n\}$ be the set of alternatives, the probability of choosing alternative i is given by (7.6)

$$P_i(\mathcal{N}) = \frac{e^{\boldsymbol{\beta}^\top \mathbf{y}_k}}{\sum_{j\in\mathcal{N}} e^{\boldsymbol{\beta}^\top \mathbf{y}_j}},$$

where $\mathbf{y}_j$ is the M-vector of attributes and relevant characteristics of the decision maker for alternative j, $j = 1,\ldots,n$, and $\boldsymbol{\beta}$ is a M-vector of parameters to be estimated (assumed to be jointly normal with a covariance matrix $\boldsymbol{\Sigma}$).

If S a subset of the alternatives, $S \subset \mathcal{N}$, then if the IIA property holds, for $i \in S$,

$$P_i(\mathcal{N}) = P_i(S) P_S(\mathcal{N}), \qquad (9.67)$$

where

$$P_S(\mathcal{N}) = \sum_{j\in S} P_j(\mathcal{N}).$$

If the IIA property fails to hold, there has to be a set S where (9.67) fails to hold. So if we restrict our population to customers who purchased only in S, we obtain an estimate $\hat{\beta}_S$ based only on this data, with its covariance matrix estimated by $\hat{\Sigma}_S$. Let $\hat{\beta}_\mathcal{N}$, $\hat{\Sigma}_\mathcal{N}$ be the corresponding estimates for the full choice set.

Note that there may be some elements of the M-vector of parameters that may not be identifiable from data restricted to purchases in S (for instance, alternative-specific variables where the alternatives are not in S). If such is the case, we have to restrict ourselves to a subvector corresponding to explanatory variables that vary within S, but for simplicity, assume that this subvector coincides with the full M-vector of explanatory variables.

The Hausmann specification test is based on the difference $\hat{q} = \hat{\beta}_\mathcal{N} - \hat{\beta}_S$. If the IIA property holds, the two estimates $\hat{\beta}_\mathcal{N}$ and $\hat{\beta}_S$ should coincide, and $\hat{q}$ will be a consistent estimator of 0. Then if $\mathbf{V}(\hat{q})$ is the variance-covariance matrix of $\hat{q}$ ($\mathbf{V}(\hat{q}) = \hat{\mathbf{\Sigma}}_\mathcal{N} - \hat{\mathbf{\Sigma}}_S$), the test statistic

$$\hat{q}^\top \mathbf{V}(\hat{q})^{-1} \hat{q}$$

is asymptotically χ^2 distributed with degrees of freedom given by the rank of $\mathbf{V}(\hat{q})$.

The null hypothesis can then be accepted or rejected with a specified degree of confidence. In principle, this has to be tested for all possible subsets S of $\mathcal{N}$. Also, there is no guarantee that the variance-covariance matrix $\mathbf{V}(\hat{q})$ is invertible. Hausmann and McFadden report that the test is not very powerful unless deviations from MNL are substantial.

9.5.1.3 Model Selection

Model selection is one of the most subtle tasks in estimation. There are no clear-cut rules; intuition, judgment, experience, and repeated testing are required to find a model that generalizes well and has good predictive power. We have already seen one iterative process for choosing a model—the Box-Jenkins methodology of Section 9.3.5 for time-series models. In this section we present additional statistical guidelines, less elaborate than Box-Jenkins, for selecting a model.

Formally, these are decision rules for selecting one of K possible models $M_1, \ldots, M_K$. The models can be time-series models or regression models or others, each with a set of parameters that we assume are estimated by a maximum-likelihood procedure. Let $\mathcal{L}^*(\boldsymbol{\beta}_k)$ represent the maximum-likelihood of model M_k based on the N observations $z = \{z_1, \ldots, z_N\}$, where $\boldsymbol{\beta}_k$ is the parameter vector of model M_k of dimension m_k.

Selection Criteria The simplest way to select a model is to rank the models according to some goodness-of-fit criterion and choose the highest-ranking one. Various decision rules have been proposed to do this, the two classical ones being the following:

- **Bayes information criterion (BIC)** The BIC of model k is defined as
$$BIC_k = -\ln \mathcal{L}^*(\boldsymbol{\beta}_k) + \frac{1}{2} m_k \ln N,$$
and the best model is the one with the smallest BIC.

- **Akaike information criterion (AIC)** The AIC of model k is defined as
$$AIC_k = -\ln \mathcal{L}^*(\boldsymbol{\beta}_k) + m_k,$$
and the best model is the one with the smallest AIC.

A number of competing criteria have been proposed, including the Fisher information criterion (FIC) [442, 559], cross-validation (CV) [492,

12], final prediction error (FPE) [464], generalized information criterion (GIC) [435].

Bayesian Selection Both the BIC and AIC have theoretical roots in the Bayesian model-selection methodology, which we describe next. Let $f_k(z|\beta_k)$ be the density function of model M_k. Let $g(\beta_k)$ be the prior distribution of the parameters of model k.

Given the data, which model is most likely? By Bayes formula,

$$P(M_k|z) = P(z|M_k)\frac{P(M_k)}{P(z)},$$

where $P(z|M_k)$ is the likelihood function for model M_k with the prior $g(\beta_k)$. Consider the posterior odds of a model M_i over M_k:

$$\underbrace{\frac{P(M_i|z)}{P(M_k|z)}}_{\text{Posterior odds}} = \underbrace{\frac{P(z|M_i)}{P(z|M_k)}}_{\text{Bayes Factor } B_{ik}} \times \underbrace{\frac{P(M_i)}{P(M_k)}}_{\text{Prior odds}}.$$

The Bayes factor indicates whether model M_i is preferred to model M_k; if B_{ik} is > 1, then M_i is preferred.

Varying i and summing over $i = 1,\ldots,K$, we get the posterior probability of model M_k as

$$P(M_k|z) = \left(\sum_{i=1}^{K} \frac{P(M_i)}{P(M_k)} B_{ik}\right)^{-1}.$$

Computing the Bayes factors can be difficult in practice, as calculating $P(z|M_k)$ involves multiple integration over the prior density $g(\beta_k)$:

$$P(z|M_k) = \int_{\beta_k} f(z_1,\ldots,z_n|\beta_k) g(\beta_k) d\beta_k.$$

One alternative is to use a holdout sample to get estimates of β_k and then use $P(z|M_k(\hat{\beta}_k))$ instead of computing the integral explicitly. The prior distribution $g(\cdot)$ is typically also calculated from a holdout sample.

Variable Selection Another task in model selection is deciding, within a given model class, which variables should be included. It is generally undesirable to include too many variables. Correlations among independent variables can lead to erroneous coefficient estimates, as in the phenomenon of multicollinearity in linear regression. Even if the explanatory variables are independent, the principle of *Occam's razor*[11]

---

[11]The Occam's razor principle of scientific investigation states that if E represents the evidence and $P(H|E)$ the probability of a specified hypothesis H given the evidence, if

$$P(H_1|E) = P(H_2|E) = \cdots = P(H_k|E),$$

prescribes that one should make do with as few variables as possible to achieve a given level of predictive power.

Formally, given a choice of M possible explanatory variables $Y_1, \ldots, Y_M$, which is the best subset to use? We can use the model-selection criteria discussed above (AIC, BIC, FIC), treating each subset choice as a different model. However, for large M this is computationally quite burdensome as there are $O(2^M)$ possible combinations of variables. A simpler methodology, often employed in practice, is to begin with an initial subset and then try adding one variable at a time—testing to see if it increases some measure of predictive power. Similarly, one can begin with a full set and remove one variable at a time, testing for loss of predictive power at each step. See Miller [382] for a comprehensive treatment of subset selection procedures.

More sophisticated search techniques for variable subset selection, based on hierarchical Bayes models and Gibbs sampling, have also been proposed (Mitchell and Beauchamp [387]; George and McCulloch [207]).

9.5.1.4   Overfitting

In this section we look at a common problem with fitting a model to training data—namely, *overfitting*. Rather than discuss it generally, we illustrate the problem of overfitting with an example.

Consider a set of data that is generated by the following formula (unknown to the forecaster):

$$Z_t = 0.2\cos(2\pi t/10) + 0.5\sin(2\pi t/10) + \xi_t. \tag{9.68}$$

If we perform a nonlinear regression on the first 10 points using a 10th degree polynomial of the form $\zeta(t) = \sum_{i=0}^{10} w_i (t/10)^i$, we obtain the fit shown in Figure 9.12. This on surface appears to fit the data well. A cubic polynomial fit to the same data set does not providing as exact a fit on the first 11 points. However, using the formula for the 10th degree polynomial for forecasting is disastrous; for instance, its projection for the 12th data point is -21.77, while the actual value is 0.66, and the accuracy of projections further in the future is even worse. The cubic polynomial, in contrast, has less forecast error. The 10th degree polynomial is an over-fit; it has too many degrees of freedom (in this case, 11 parameters for 11 data points!). We are in effect "fitting our model to noise" by using it. A model is said to *generalize* well if it performs well on data that it has *not* been trained on. In forecasting, we are looking

---

for hypotheses $H_1, H_2, \ldots, H_k$, then the simplest of $H_1, H_2, \ldots, H_k$ is to be preferred (Kotz and Johnson [311]).

Estimation and Forecasting

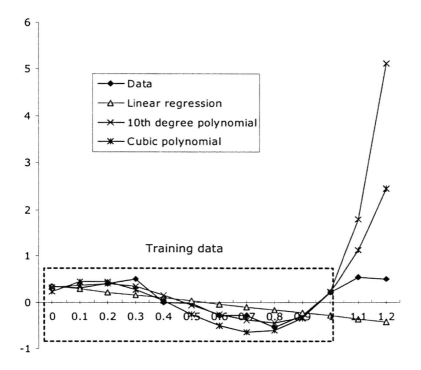

Figure 9.12. Overfitting example. Projection of the three polynomials fit to the first 10 points of equation (9.68) (x-axis plots $t/10$).

for good generalization properties rather than good explanations of past data.

Such overfitting problems come up during the model-selection phase for model-based methods and can be limited by considering only models that are "reasonable" from a subjective, business point of view, rather than trying blindly to find the best-fitting model based on past data. For neural networks, the problem is more subtle and difficult to detect. Because there is no explicit functional form that we choose—and because three-layer neural networks can approximate practically any function—the danger that we might overtrain and fit the network to noise is very high indeed. A good strategy to avoid overfitting is to keep a holdout sample and use the forecast errors on the holdout sample rather than on the training data to guide training.

9.5.2 Forecasting Errors and System Control

An analysis of the forecast errors is often as important as the forecast itself. Forecast error analysis is useful for the several reasons. First, the historical observed forecasting errors give a measure of the confidence one can have in the forecasting system or algorithm. Forecast errors can be used to estimate the variance in the underlying demand process and hence can be used to estimate second-order parameters of the distribution. Errors can also be used to filter out outlier data. Finally, errors can be used to track the forecast and signal unusual events or instability in the system. We look at each of these applications below.

9.5.2.1 Measures of Forecast Errors

Suppose we have been running our forecasting system for N periods and have already constructed N forecasts and made observations of the forecast relative to the actual values on these N periods. Then the forecast error for a particular period t is given by

$$e_t = z_t - \hat{Z}_t,$$

where z_t is the observed value and $\hat{Z}_t$ is the forecasted value for period t.

The following are some measures of forecast error that are used in practice:

- **Sum of forecast errors**:

$$E_N = \sum_{t=0}^{N} e_t.$$

- **Mean error**:

$$\bar{E}_N = \frac{E_N}{N}.$$

The mean error is an estimate of the forecast bias. If the forecasting system is unbiased, the mean bias should converge to 0 as N increases.

- **Smoothed error**: This is given by the following recursive formula:

$$E_N^\alpha = \alpha e_N + (1-\alpha) E_{N-1}^\alpha,$$

where $0 < \alpha < 1$ is a smoothing constant.

- **Mean absolute deviation (MAD)**:

$$MAD_N = \frac{\sum_{t=0}^{N} |e_t|}{N}.$$

- **Mean squared error (MSE):**

$$MSE_N = \frac{\sum_{t=0}^{N} e_t^2}{N}.$$

- **Mean absolute percentage error (MAPE):**

$$MAPE_N = \sum_{t=0}^{N} \frac{|e_t/z_t|}{N}.$$

The quantity $|e_t/z_t|$ is called the *relative error* and is not defined if z_t is 0; hence the MAPE calculation should omit such values.

- **Tracking signal (TS):**

$$TS_N = \frac{E_N}{MAD_N}.$$

It is strongly recommended that at least one of MAD, MSE, or MAPE and the TS be used to monitor a forecasting system. The primary role of MAD, MSE, and MAPE measures is to evaluate the performance of the forecasting system. Lower numbers mean better forecasts.

Among MAD, MSE and MAPE, the choice of which one to use depends strongly on the nature of the forecasts. MSE penalizes large errors for a single observation much more than MAD. Therefore, it is a better measure to detect if a few observations have large errors. If we are interested in overall performance, then MAD is generally a better choice. MAPE is useful for comparing performance across different time series, as the errors are measured relative to the data values.

9.5.2.2 Bias Detection and Correction

In addition to measuring forecast performance, a system should also monitor forecast bias. Tracking signal (TS) tests are used to monitor automated forecasts to see if the system is generating consistently biased forecasts. Typically, if the TS number exceeds a bound, an alert is generated for analysts to investigate. Most often in practice such bias is caused by a special, one-off event, but occasionally a recalibration may be required because of a fundamental change in the demand process.

There are two common tests for detecting a systematic bias in the forecast from observed errors. First, assume that the forecast is measured on a set of N observations. Let

$$\hat{T}_N = \frac{\sqrt{N}\bar{E}_N}{\sqrt{MSE_N}}.$$

Then if the forecast is unbiased, the statistic $\hat{T}$ has approximately a t-distribution with $N - g$ degrees of freedom, where g is the number of parameters in the model that are being estimated. (See Abraham and Ledolter [1], p.372.) For large N, $\hat{T}$ is approximately a standard normal (mean zero, variance one) random variable. For a given significance level, a statistical test can then be devised with the null hypothesis that the forecast is unbiased.

A second, more popular operational test for bias is to compare the absolute value of the tracking signal with a constant. (See Montgomery [388].) The forecasting system is declared biased if

$$|TS_N| > K_1.$$

The constant K_1 is usually set to be between 4 and 6. Similar tests exist using variations of the tracking signal formula, one with smoothed error in the numerator of the TS definition and a constant between 0.2 and 0.5 in the right-hand side of the bias test, and another where MSE is used instead of MAD in the denominator of the tracking signal formula and the constant in the bias test changed to be between 2 and 3.

If one knows that the forecasting system has a bias, then it would appear trivial to fix the bias—just multiply or add a correction factor. Or better still, recalibrate the system or modify the forecasting algorithms; for instance, the forecast bias could be because of a bias in the estimation of the parameters of the model (Section 9.2.1.3). But this assumes we have a precise idea of the magnitude of the bias and that it is more or less constant. As for recalibrating the model, this is often an expensive process and can involve a considerable amount of research and experimentation to come up with a better (unbiased) estimate.

9.5.2.3 Outlier Detection and Correction

Outliers are extreme values of data that are caused by corrupted records or special nonrecurring conditions in the demand process. Outlier data can severely disrupt a forecasting system. Smoothing methods—like the moving-average method—are especially susceptible to outliers because the presence of an unusual data point will distort the forecasts for several successive periods.

One technique to guard against outliers is to presmooth the data to make them more robust to the presence of outliers. The *moving-median smoothing* method in one example. Here the data is preprocessed by the following transformation:

$$\tilde{z}_t = \text{Median}(z_{t-1}, z_t, z_{t+1}),$$

and the forecast is trained as if $\tilde{z}_t$ were the real data sequence. This is an example of a nonlinear smoothing method. Many such nonlinear data smoothers exist. (See Tukey [520].) Care should be exercised, however, when the data has seasonality or other periodic effects. The data may first have to be deseasonalized before using such filters.

Another technique is to try to identify and remove outliers before feeding the data to the forecasting system. One such outlier identification test is to consider a data point t an outlier if

$$\left|\frac{e_t}{MAD_t}\right| > K_2,$$

where the value of K_2 is chosen to be between 5 and 6.

9.6 Industry Models of RM Estimation and Forecasting

In this section we give some examples of specific RM forecasting models. The models are intended to be representative of those used in a particular industry to forecast a particular quantity of interest: for example, no-shows, cancellations, and groups forecasting in the airline, rental-car, and hotel industries; ratings forecasting in the media industry; sales response functions in the retail industry; promotion effects forecasting for manufacturers; and load forecasting in the electricity and gas industries. Many variations of these models are possible, and the examples presented here are intended only as illustrations—not recommendations—of forecasting approaches.

9.6.1 Airline No-Show and Cancellations Forecasting

Forecasts of cancellation and show-up rates are key inputs to the overbooking module of an airline RM system. In addition to the statistical and operational techniques discussed in this chapter so far, this example also highlights the use of data-mining algorithms for forecasting.

The first problem in cancellation forecasting is coping with reservations data. If one uses only net-bookings data for forecasts—not uncommon in RM systems—new bookings may hide cancellations. For instance, if in a period there are 100 bookings on hand, and during the period 20 new bookings are realized, but 10 current bookings cancel, then net-bookings data may make it appear that there have been 10 new bookings and 0 cancellations. Cancellation forecasts based on such data will then be biased. Similarly, go-shows or walk-ups—that is, people who show up without reservations (distinguished from regular bookings by the fact that it is lumpy demand occurring at the time of service)—

500 THE THEORY AND PRACTICE OF REVENUE MANAGEMENT

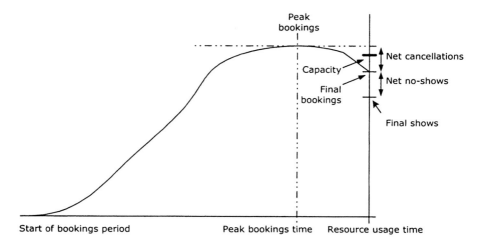

Figure 9.13. A booking curve showing net cancellations and net no-shows.

may also hide no-shows. So depending on the data being used and the requirements of the overbooking optimization, we may need to decide whether we are forecasting gross (actual) cancellations or net (observed) cancellations (Figure 9.13).

Both no-show and cancellation rates can be defined at different levels of aggregation, for the entire cabin or by fare class. Defining rates by fare class is more accurate as significant differences may exist between fare classes—for example, some may have penalties for cancelling while others may not. Cancellations can also be defined over different intervals of time, as incremental cancellations over a given period or total cancellations over the entire booking period.

Besides the level of aggregation, the cancellation rate and no-show rate can have different interpretations—(1) as the probability that a given individual booking will cancel or no-show or (2) as a fraction of the total number of bookings at a given point of time (either current time or some time in the future) that are likely to cancel or no-show. The second interpretation leads to the concept of a cancellation curve over the booking period. The cancellation rate may change over time as very early bookings tend to have higher cancellation rates than later ones (see Figure 9.14). A full cancellation curve is usually needed only in dynamic overbooking models.

For illustration, consider the binomial model of Section 4.2.1. If there are N current bookings, the cancellation rate p_C is the probability that a booking will cancel before the time of service. We define the no-show rate p_{NS} similarly. Both p_C and p_{NS} are assumed to be constant and

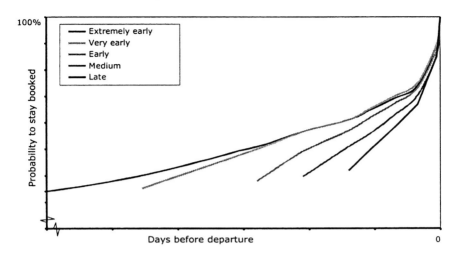

Figure 9.14. Cancellation probabilities as a function of booking time (of the customer for a set of airline data). Data from Westerhof [561].

equal for all bookings, irrespective of when they booked. They can change over time but in a Markovian sense; they will depend only on the current time but not on the history of the bookings. We will also assume that the probabilities are independent of the number of current bookings (this only determines the expected number of cancellations and no-shows by the Binomial distribution).

The simplest procedure for forecasting p_C is to construct a time series of historical fractions who cancelled at a given point in time and use a standard forecasting method on this time series of data. For the case of cancellations, the series is constructed by taking

Net cancellations = max{0, Current bookings-Final bookings}.

Final bookings are bookings just before no-shows and go-shows are accounted for. Net no-shows are calculated as final bookings less final consumed inventory. If there are customers who are denied service, then they would be subtracted from the net no-shows.

Net no-shows rate p_{NS} can be forecasted from the time series of historical no-show fractions. The fractions of no-shows and cancellations over time are calculated by dividing final consumed inventory and final bookings by the bookings on hand at each point of time. If the data is detailed enough to extract actual cancellations and no-shows (as opposed to net cancellations and no-shows), then these can be used to calculate a time-series forecast of the actual cancellation and no-show rates.

No-shows and cancellation forecasting can further be divided into two parts: forecasting the cancellation behavior of customers who already booked and those who will book in the future. For the former, it is possible to exploit correlations between customer cancellation probabilities and purchase characteristics like time of booking, source of booking, amount paid, cancellation penalties, and refund policies associated with the fare, to improve the forecasts.

Kalka and Weber [280], Feyen and Hüglin [190], and Westerhop [561] report airline no-show and cancellation forecasting for existing customers using data-mining and data-discovery tools on PNR data. Some of the attributes used are origin, destination, flight time, return trip, booking class, number of passengers traveling together, flight time, number of connections, and connection time. Feyen and Hüglin [190] use logistic regression on the attributes and the observed rates for prediction while Kalka and Weber [280] use induction trees. (See Quinlan [432].)

The methodology in Kalka and Weber [280] can be illustrated in Figure 9.15 for two attributes—flight time and booking class. The historical bookings and cancellations are mapped to the attribute space, and we partition the space by partitioning the ranges on the attributes. This is somewhat analogous to clustering points into groups, except that we are now interested in rules for partitioning each attribute dimension, rules that subsequently will be used to categorize new observations with its likelihood of cancellation. A cancellation probability is calculated for each box as the fraction of bookings in that box that cancel. For any new booking, its cancellation probability is derived by looking up the box it falls in and taking its corresponding cancellation probability. Data-mining tools use artificial intelligence rules-based techniques to partition the customer attribute space and construct an induction tree that gives a sequence of rules to be applied to classify observations.

9.6.2 Groups Demand and Utilization Forecasting

Bookings for units of five or more are usually classified as groups. Groups in RM can either be ad-hoc groups (one-shot groups such as school excursions or crews) or series groups (repeating groups—for example, bookings by a package-tour operator). (See Sections 10.1.2 and 10.2.1.) In this section we describe the forecasting tasks associated with groups.

Forecasting of group bookings demand is rarely done. This is because ad-hoc groups are such rare events that it makes it difficult to try to forecast demand from such sources. Series groups, on the other hand, are negotiated so far in advance that they make forecasting unnecessary.

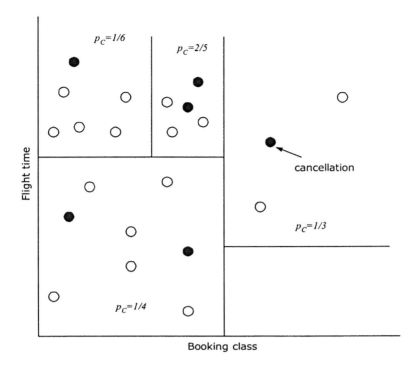

Figure 9.15. Induction tree on cancellations data.

However, group-utilization forecasting is an important task. Group utilization is defined as the percentage of a group reservation that will eventually show up. In principle, it is no different from cancellations and no-show forecasting. However, it is treated separately because groups may act as a unit, with strong correlation between the members of the unit. For instance, a group may cancel as a whole, in which case there is a sudden lumpy change in the available capacity. Because of its potential impact on availability and the higher risk involved in groups canceling as a whole, group utilization is usually tracked separately from regular cancellations and no-shows by dedicated analysts or managers.

Analysts also have better information about groups reservations than individual bookings because the reservation is usually made directly through personal contact. The source of a group reservation (such as a tour operator, cruise-line or agency) and type of group (such as a corporate meeting or convention) also helps in tracking historical usage rates. Group utilization forecasting differs if the group is an ad-hoc group or a series group. Ad-hoc groups are more likely to cancel as a whole (or not pay by the deadline), while series groups, being negotiated contracts for a long period, tend to survive till the usage date with only partial cancellations.

A forecast of group utilization is made on a historical series of utilization ratios constructed for groups with similar characteristics (same group type, group booking source, market, and so on). The forecast is usually made for each individual group reservation and updated constantly as new information and confirmations come in. Causal models are widely used to forecast group reservations because of the rich data available specific to each individual group. Group utilizations have been found to be correlated with group type (ad hoc or series), origin location of bookings, group size, penalty costs, historic cancellations, booking source, time of booking, and group travel purpose, among others. Bayesian models are also suitable because they allow analysts' beliefs on the group's utilization to be incorporated into the forecast.

9.6.3 Sell-Up and Recapture Forecasting

Sell-up and recapture are used in some RM models as discussed in Section 2.6.[12] The sell-up probability for a class is the probability that customers for that class will buy-up to at least one of the other higher classes (of the same resource) if their class is closed (this is called *differential sell-up rate* in Gorin [217]; the sell-up rate used in Belobaba and Weatherford [37] is only between the class and the next highest class). Recapture occurs when the customer buys an alternative resource (say, on a different date or time) if their requested class is closed.

There are several difficulties in estimating buy-up and recapture probabilities. For example, how do we tell if a customer is an "original" customer or a "recaptured" customer? Looking at transactional data alone, this is impossible to determine. It is common practice to pass this burden on to analysts, who are required to input buy-up and recapture probabilities for each market using their best judgment. Given the number of markets/resources/date combinations, often a single number is used for each market or for the firm as a whole.

Other approaches are based on data. Gorin [217] proposes the following formula for estimating the sell-up rate of a class for a resource: Let s_i represent total number bookings in class i (over a collection of sample historical data for the resource). Assume the classes are indexed with the lower index having a higher fare. Then Gorin [217] defines the

---

[12]Andersson [18] defines a customers who neither sells-up nor is recaptured but buys an alternative product from a competitor, as a *deviation*.

Estimation and Forecasting

sell-up rate p_s as

$$p_s = \frac{(s_{i-1}|i-1 \text{ closed and } i \text{ open}) - (s_{i-1}|i-1 \text{ and } i \text{ both open})}{(s_i|i \text{ open})}.$$

However, he also states that this estimate likely is biased, so it should be used with caution. Recapture effects are not considered by Gorin.

Consumer-choice models provide a more systematic approach to buy up and recapture estimation. An early attempt at such a model is by Maynes and Wood [368], who build an econometric model of demand for three latent market segments as a function of price, schedule attributes, and competitor prices and availability. The ratio of the forecasts of demand for a class on a resource and a lower fare class on the same resource provides the sell-up probability for the lower class. This approach can be extended to estimate recapture rate as well. However, these rates are calculated on a pairwise basis only, independent of what other options are available at that time.

Andersson [19] presents a richer model of consumer behavior based on utilities and discrete-choice theory. At any given point of time, a choice set S is defined as a set of competing resource/class combinations for class j on resource i. If j is closed on i, then define S_{-ij} as the set S without class j on resource i. The estimate of the recapture rate is then defined as follows:

$p(k,l|S_{-ij})$ = Probability that resource k, class l is chosen when i,j is closed but all other choices in S_{-ij} are open.

$p(k,l|S)$ = Probability that resource k, class l is chosen when all choices in S are open.

Then the recapture rate by combination k,l from i,j, denoted $p_{i,j}(k,l)$, is defined as

$$p_{i,j}(k,l) = p(k,l|S_{-ij}) - p(k,l|S).$$

The probabilities $p(k,l|S_{-ij})$ and $p(k,l|S)$ can be estimated using an appropriate discrete-choice model. Andersson [19] (see also Köhler [308]) reports a study at Scandinavian Airlines where the choice probabilities were estimated using a MNL model fit from both transactional data and passenger surveys. The passenger surveys were in the form of games (lasting around 10 minutes), presenting alternatives of price, departure time, restrictions, and airline brand name.

9.6.4 Retail Sales Forecasting

Retail RM requires an estimate of a demand function. Besides price, advertising, product features, past sales, economic conditions, store location, brand effects, weather, and competitor actions are some factors that strongly affect sales. Consequently, in retail marketing forecasts, in contrast to airline or hotel RM, causal models are widely used.

As discussed in Chapter 7, there are two basic approaches to demand function modeling. One way of incorporating the effects of marketing variables on sales is through models of individual consumer choice behavior. Then in a bottom-up forecasting fashion, these individual choices are aggregated to get total demand. Another approach—called *aggregate forecasting*—is to model aggregate demand directly as a function of price and other marketing variables. We focus on this latter approach here, as it is prevalent both in marketing theory and practice, and as we covered discrete-choice models earlier.

Let Z denote sales, and let the marketing variables be represented by $y_1, \ldots, y_M$ for multivariate models and by y for univariate models: Consider a basic sales response model of the form

$$Z = f(y_1, \ldots, y_M) + \xi.$$

The functional forms are usually designed such that either (1) absolute change in the marketing variables leads to an absolute change in sales or (2) percentage (relative) change in the marketing variables leads to an absolute change in sales. For instance, the function $Z = \beta y$ is of the former kind (since $\partial Z = \beta \partial y$), while the function $Z = \beta \ln y$ is the latter kind ($\partial Z = \beta \partial y / y$).

The function $f(\cdot)$ can be a *static* function of variables of the current period only, or a *dynamic* function capturing the effects of marketing variables in past periods (for example, advertising done in the past month has an effect on the sales of this month). Below are some examples of static sales response functions.

- **Semilogarithmic model:**

$$Z = \beta_0 + \beta_1 \ln y_1 + \cdots + \beta_M \ln y_M + \xi. \qquad (9.69)$$

Percentage sale in a marketing variable leads to an absolute change in the sales.

- **Multiplicative or power model:**

$$\ln Z = \beta_0 + y_1^{\beta_1} + \cdots + y_M^{\beta_M} + \xi. \qquad (9.70)$$

The β's have the interpretation of elasticities. A more general form of (9.70) is called the *interactive* model and is given by the sum of all possible products of the variables:

$$Z = \sum e^{\beta_0} y_1^{\beta_1} \cdots y_m^{\beta_m}. \qquad (9.71)$$

It is rarely used in this full form.

- **Exponential model:**

$$\ln Z = \ln Z_{\max} - \beta^T y + \xi. \tag{9.72}$$

Sales exhibit increasing returns to scale as the value of the marketing variable (say price) goes down to zero. $Z_{\max}$ represents the maximum possible sales.

- **Log-reciprocal or S-shaped model:**

$$\ln Z = \beta_0 - \frac{\beta_1}{y} + \xi, \; \beta_0 > 0. \tag{9.73}$$

This function possesses an inflection point at $y = \beta_1/2$. Sales show increasing marginal returns for y less than the inflection point and decreasing marginal returns from then on.

Other S-shaped curves are possible using *logistic* models such as the following *log-linear* and *double-log* models:

$$\ln(\frac{Z}{Z_{\max} - Z}) = \beta_0 + \sum_{j=1}^{M} \beta_j y_j + \xi \tag{9.74}$$

$$\ln(\frac{Z - Z_{\min}}{Z_{\max} - Z}) = \ln \beta_0 + \sum_{j=1}^{M} \beta_j \ln y_j + \xi. \tag{9.75}$$

- **Gutenberg model:**

$$Z = \beta_0 - \beta_1 \sinh[\beta_2(y - \bar{y})] + \xi. \tag{9.76}$$

$\bar{y}$ is a reference value for the marketing variable (for instance average competition price). The Gutenberg model is a complicated but flexible function. Simon [471] gives an application using this model.

Next we give some examples of dynamic sales response functions, in which the sales in a period is a function of variables of the past (lagged) periods, future (lead) customer actions or the current period:

- **Geometric distributed-lag model:** This is a dynamic model that relates the sales in period t to observed values in previous periods with exponentially decreasing weights:

$$Z_t = \beta_0 + \beta_1(1 - \alpha)\sum_{j=0}^{\infty} \alpha^j Z_{t-j} + \xi, \; 0 < \alpha < 1. \tag{9.77}$$

- **SCAN*PRO model:** This is a widely used store-sales model (proposed by Wittink et al. [572]) for determining the effect of promotions on sales. Denote, for brand i in store k in period t,

Z_{ikt}	Sales
W_t	Week-of-the-year indicators ($W_t = W_{t+52}$)
p_{ikt}	Discounted price
$\tilde{p}_{ikt}$	Nondiscounted price
F_{ikt}	0-1 indicator variable, for feature
D_{ikt}	0-1 indicator variable, for display
v_{ikt}	Inventory
ξ_{ikt}	Error term.

  Then the model is given by

  $$Z_{ikt} = e^{\beta_{0ikt}} \delta_{ikt}^{W_t} \beta_{2ik}^{F_i kt} \beta_{3ik}^{D_i kt} \beta_{4ik}^{v_i kt} e^{\xi_{ikt}}$$
  $$\prod_{m=1, m\neq i}^{M} \left(\frac{p_{mkt}}{\tilde{p}_{mkt}}\right)^{\beta_{1mkt}} \beta_{2imk}^{F_m kt} \beta_{3imk}^{D_m kt}. \quad (9.78)$$

  The δ's and β's for each period have to be estimated from data.

There are literally hundreds of models such as these studied by marketing scientists, with many empirically tested on real-world data. Once a model has been fixed, regression is the most common approach for estimating static models, while time-series methods (Section 9.3.2) are common for estimating dynamic models.

9.6.5 Media Forecasting

Forecasting for broadcast media presents some unique challenges. (Media RM is discussed in detail in Chapter 10.) Prices for advertising are quoted as cost per thousand impressions. For print and television firms, the circulation and ratings determine how much the firm can charge for their advertisement space. Internet media rates are based on page-views or click-through metrics. Market-research firms such as Nielsen, IRI, and Media Metrix are dedicated to measuring the size of the circulation (print), page-views (Internet) and audience (television, radio).

A broadcaster faces two main forecasting tasks. One is to forecast ratings for shows by day-of-week and season; the other is to forecast demand for advertising slots for these shows. Forecasting the latter is usually much easier than the former because the network has knowledge of its customers—their historical preferences and buying patterns, required demographics, and in many cases, even their advertising budget. A forecast of demand is first constructed by making an estimate of each customer's demand (often manual, based on last year's demand) and then making a tentative sales plan satisfying the customer's preferences based on the ratings forecasts (Bollapragada et al. [83]). In this section, we concentrate on ratings forecasts, which is a good example of a rather difficult causal forecasting problem.

Few TV or radio managers rely on formal ratings forecasting models for their own programming decisions; surveys, gut feeling, and innate programming intuition seem to be the dominant methodologies in practice. These forecasts, though often subjective, can be helpful for RM purposes as well, as they reflect managerial judgment (for instance, they can be used to form priors in a Bayesian framework).

Recently, several methods have been proposed based on formal models of consumer viewing behavior. Television viewing habits are conceptualized as a two-stage process. In the first stage, the individual decides whether or not to watch TV. This leads to a forecast of the total aggregate TV viewing population at any given time. Once a decision to watch TV is made, the individual chooses one of the available programs, which leads to show-level ratings. (See Gensch and Shaman [206].) This two-stage model suggests using a time-series model to predict aggregate viewership by time and day of week based on recent programming data, and then a discrete-choice model to predict ratings by show. Past viewership, viewing time, seasonality, and regional differences are good predictors of aggregate viewership, while the show characteristics, slot, show-promotion, lead-ins (the popularity of the program before) and lead-outs (and the program that runs after) influence the market share of a show.

For example, Reddy, Aronson, and Stam [437] building on the work of Horen [257], use a regression model to predict the ratings of TV shows running for multiple seasons and hence with some historical data. Shows are classified into homogeneous types, based on their characteristics (movie, news, afternoon talk show). The model is:

$$Z_t^i = \beta_0 + \sum_l \beta_l^A A_l^i + \sum_j \beta_j^S S_j^i + \sum_k \beta_k^D D_k^i + \qquad (9.79)$$
$$\sum_m \beta_m^T T_m^i + \sum_p \beta_p^R R_p^i + \sum_{u \neq i, (u,i) \in U} \beta_i^{INT} I_u^i + \xi_{it},$$

with the following variable definitions,

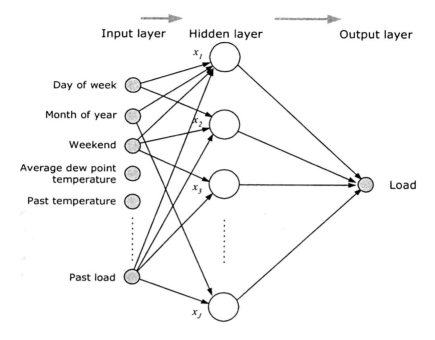

Figure 9.16. Neural network for gas-load forecasting.

Z_t^i Rating of show i in time slot t
A_l^i Measure of the relative perceived attractiveness of show i of type l (managerial rating from 0 to 10)
S_j^i 0-1 indicator variable, 1 if show i is of type j, 0 otherwise
D_k^i 0-1 indicator variable, 1 if show if show i is scheduled on day k, 0 otherwise
T_m^i 0-1 indicator variable, 1 if show i is in time slot m, 0 otherwise
R_p^i 0-1 indicator variable, 1 if show i is an hour-long show, 0 otherwise
I_u^i 0-1 indicator variable, 1 if show u (of type s) leads into show i (of type r), 0 otherwise
ξ_{it} Residual error term.

9.6.6 Gas-Load Forecasting

We next look briefly at a gas-load forecasting system using neural-network methods, reported to be implemented at Williams Gas (Lamb and Logue [327]). The model forecasts short-term (between one to five days ahead) demand for gas in a pipeline. The pipeline has thousands of meters drawing gas from it, each with variable demand. The factors that affect this demand were identified as

- Weather parameters (such as temperature, humidity, wind direction, and so on supplied by weather data vendors)

- Historic load
- Calendar (hour of the day, day of the week, holiday, month)
- Expected gas demand for each meter.
- Price (historical, current and competitor prices).

Figure 9.16 shows an example of a three-layer neural network using some of these potential inputs.

9.7 Notes and Sources

We have given many reference in the text of the chapter. Here we gather some general references on the topics of the chapter along with some additional pointers.

Regression-related topics can be found in any advanced statistic or econometric books. Here are some references: DeGroot [151] and Kvanli [318] for introductory statistics; Maddala [353], Greene [220], and Judge et al. [273] for econometrics. Some books devoted exclusively to regression are Draper and Smith [161], Guttman [231], and Neter and Wasserman [403].

Books on forecasting are available at all levels. We recommend Montgomery et al. [388] for a general introduction to operational forecasting and Harvey [242, 243] for a more advanced treatment of time-series analysis and Kalman filtering.

For books on neural networks, we recommend Bishop [69] for a very readable yet rigorous introduction to neural networks (albeit for pattern recognition) Our treatment follows also Müller, Reinhardt, and Strickland [397]. Some useful survey papers on the use of neural networks in forecasting are Poli and Jones [423], Cheng and Titterington [110], Zhang, Patuwo, and Hu [588], Hill, Marquez, O'Connor, and Remus [252], Hill, O'Connor, and Remus [253]. The application of neural networks to predict consumer choice can be found in West, Brockett, and Golden [562] and Dasgupta, Dispensa and Ghose [144].

For estimation of price-response functions and market-share models, see the following marketing science text books: Eliashberg and Lilien [174], Wedel and Kamakura [558], Hanssens, Parsons, and Schultz [235], Cooper and Nakanishi [127], Dasgupta, Dispensa, and Ghose [144], Hruska [259], West, Brockett, and Golden [562], Hill et al. [253], Zhang [588], and Lee et al. [335]. Kalyanam [283] proposes a Bayesian mixture model of pricing specifications when there is no consensus on the right model.

See Berry, Levinshohn, and Pakes [52], Berry [53], Besanko, Gupta, and Jain [63] and Chintagunta, Kadiyali, and Vilcassim [116] for estimation in a competitive market similar to the method described in Example 9.9. The problem of endogenity in estimation has received much recent attention in the marketing science literature spurred by the paper of Berry [53]. See also Chintagunta, Kadiyali and Vilcassim [116] and Villas-Boas and Winer [536] for further studies on endogeneity. Nevo [406] gives an excellent practical guide to estimating random-coefficient logit models of demand.

The Bayesian method of updating of parameters can be incorporated into many of the time-series methods of Section 9.3.2 also in a fairly straightforward manner. (See, for instance, Montgomery et al. [388].) The empirical Bayes method that we cover in this chapter is not the only possibility for handling hierarchical Bayes methods. See Lindley and Smith [345] and Blattberg and George [75] for alternatives. Hierarchical Bayes methods have also found application in modeling heterogeneity in preferences in discrete-choice models (Albert and Chib [6]; Allenby and Rossi [11]; Huber and Train [260]), and in conjoint analysis (Allenby and Ginter [9]; Lenk et al. [339]).

Literature on combining forecasts is also quite vast, given its promise of returning more than the sum of its parts. The standard references in this area are Newbold and Granger [407], Granger and Newbold [218], Makridakis and Winkler [356], Clemen and Winkler [122], Clemen [123], Gupta and Wilson [230], Schmittlein, Kimm, and Morrison [457], Morrison and Schmittlein [391], and Foster and Vohra [192]. See also Montgomery et al. ([388], p.192).

One of the few textbooks dedicated to the EM algorithm is McLachlan and Krishnan [377]. The book also contains many applications, convergence properties and lists a large number of EM references. Connections with the Gibbs method is mentioned, but the reader should refer to Schafer [456] dedicated to MCMC methods. For an introduction to Gibbs Sampling, see the article of Casella and George [101] and Gilks, Richardson and Spiegelhalter [212]. For Gibbs algorithm applied to missing-data problems, see Gelfand, Smith and Lee [203].

Both the origins of EM and Gibbs sampling (at least their ideas) can be traced far back, but Dempster et al. [152] and Geman and Geman [205] are credited with their invention and popularization.

The original paper of Kaplan and Meier [289] is still a good introduction to the Kaplan-Meier estimator. Many books on survival analysis (Miller [383]; Cox and Oakes [134]) also describe the method in detail. Logistic regression has been proposed as a parametric alternative to the Kaplan-Meier curve, with good properties and flexibility, and with all the advantages of a parametric form (Efron [173]).

The unconstraining methods described here are not the only alternatives though. It is possible to use the bootstrap, jackknife as well as regression with censored data to get estimates of the parameters of a censored sample. See Efron [171] and Davison and Hinckley [148] for an introduction to using bootstrap for unconstraining and for regression with censored data.

APPENDIX 9.A: Back-Propagation Algorithm for Neural-Network Training

We illustrate the back-propagation algorithm for training a neural network on our example of Figure 9.9.

Because we chose the linear function $\tilde{f}(h) = h$ as the activation function for input and output nodes, we represent, by a slight abuse of notation, the nth instance of an input and its corresponding output of the neural net as $\mathbf{y}_n = (y_{n1}, \ldots, y_{nI})$ and $\hat{\mathbf{Z}}_n = (\hat{Z}_{n1}, \ldots, \hat{Z}_{nK})$ respectively. As always, let $\mathbf{z}^n$ be the actual observation at the nth instance. Assume we are given a set of N training data instances (a set of N input-output pairs $(\mathbf{y}_n, \hat{\mathbf{Z}}_n)$, $n = 1, \ldots, N$ that we will use to determine weights of the neural network).

For training instance n, the state of node j in the hidden layer is then $x_{nj} = f(h_{nj})$, where $h_{nj} = \sum_{i=1}^{I} w_{ij} y_{ni} - \nu_j$, and the state of node k of the output layer is $\hat{Z}_{nk} = \tilde{f}(h_{nk}) = h_{nk}$, where $h_{nk} = \sum_{j=1}^{J} w_{jk} x_{nj} - \nu_k$.

For the given set of transfer functions $f(\cdot), \tilde{f}(\cdot)$, our objective is to choose the weights w_{ij}, w_{jk} and the activation threshold values ν_j, ν_k such that they minimize the squared deviation between the output values of the network and the actual observations:

$$S[w_{ij}, \nu_j, w_{jk}, \nu_k] = \frac{1}{2} \sum_{n=1}^{N} \sum_{k=1}^{K} [z_{nk} - \hat{Z}_{nk}]^2.$$

The error back-propagation method performs this minimization iteratively in two stages (for a three-layer network), the first stage corresponding to the output layer and the second stage to the hidden layer. At each stage, the weights and threshold values are updated in the spirit of the steepest-descent algorithm of nonlinear optimization (see Bertsekas [59]), as follows:

STEP 0: Choose an initial set of values for the w's and ν's. Choose a step-size δ (which can either be fixed or chosen according to a step-size selection rule; see Bertsekas [59]).

STEP 1: Update the w's for arcs between the hidden layer and the output layer as follows:

$$w_{jk} \leftarrow w_{jk} + \Delta w_{jk},$$

where

$$\Delta w_{jk} = -\delta \frac{\partial S}{\partial w_{jk}} = \delta \sum_{n=1}^{N} [z_{nk} - \tilde{f}(h_{nk})] \tilde{f}'(h_{nk}) \frac{\partial h_{nk}}{\partial w_{jk}}.$$

Update the ν's of the output layer as follows:
$$\nu_k \leftarrow \nu_k + \Delta\nu_k,$$
where
$$\Delta\nu_k = -\delta\frac{\partial S}{\partial \nu_k} = \delta\sum_{n=1}^{N}[z_{nk} - \tilde{f}(h_{nk})]\tilde{f}'(h_{nk})\frac{\partial h_{nk}}{\partial \nu_k}.$$

STEP 2: Update the w's for arcs between the input layer and the hidden layer as follows:
$$w_{ij} \leftarrow w_{ij} + \Delta w_{ij},$$
where (applying differentiation using the chain rule)
$$\begin{aligned}\Delta w_{ij} &= -\delta\frac{\partial S}{\partial w_{ij}} \\ &= \delta\sum_{n=1}^{N}[\sum_{k=1}^{K}(z_{nk} - \tilde{f}(h_{nk}))\tilde{f}'(h_{nk})]w_{jk}f'(h_{nj})\frac{\partial h_{ni}}{\partial w_{ij}}.\end{aligned}$$

Update the ν's of the hidden layer as follows:
$$\nu_j \leftarrow \nu_j + \Delta\nu_j,$$
where
$$\begin{aligned}\Delta\nu_j &= -\delta\frac{\partial S}{\partial \nu_j} \\ &= \delta\sum_{n=1}^{N}[\sum_{k=1}^{K}(z_{nk} - \tilde{f}(h_{nk}))\tilde{f}'(h_{nk})]w_{jk}f'(h_{nj})\frac{\partial h_{nj}}{\partial \nu_j}.\end{aligned}$$

STEP 3: If convergence criterion is not met, GOTO STEP 1.

Chapter 10

INDUSTRY PROFILES

Implementing revenue management requires an understanding of real world market conditions. Regulations, technology standards, consumer behavior, product characteristics, pricing policies, and industry distribution practices are all important factors that affect the way RM is practiced. This chapter explores these institutional factors affecting RM in industries that are both mature and emerging users of RM. For each industry we begin by describing its products, consumers, and pricing practices. We then summarize the current state of RM in the industry and the key issues affecting its RM practices. Our progression is from industries in which RM is a mature practice to those in which it a relatively new or emerging practice.

A word of caution is in order here, however. Industry practices can change rapidly as new technologies and business models emerge, and such changes can fundamentally alter the way RM is practiced. Consequently, this chapter represents at best a snapshot of current RM practice. Continuing innovations in business models and technologies will no doubt keep RM an evolving discipline for many years to come.

10.1 Airlines

As the earliest and largest user of RM, the airline industry deserves special attention. Hence, we begin our industry discussion with an in-depth look at RM practices in the airline industry.

10.1.1 History

As mentioned in Chapter 1, RM has its origins in the rise of capacity-controlled discount fares after the deregulation of the U.S. airline industry. Before deregulation, the only service options offered by commercial airlines were first-class and coach-class service. Fares on a route were identical for all carriers and set by the Civil Aeronautics Board (or by the International Air Transport Association (IATA) on international flights) based on standard costs.

The first innovation in fare structures occurred in international markets with the development of APEX—*advance-purchase excursion*—fares. APEX fares offered travelers the option of buying a coach-class seat at a discount but were restricted to round-trip travel and required an advance purchase and a minimum stay.

The period after deregulation in the United States was characterized by successive innovations in creating discounted products. As discussed in Chapter 1, American Airlines introduced "Super Saver" fares in 1975. These fares had a seven-day advance-purchase requirement and minimum-stay conditions and required round-trip travel. Advance-purchase restrictions were lengthened progressively over the years, culminating in 1985 with American Airlines' introduction of "Ultimate Super Saver" fares that required a 30-day advance purchase. In 1987, Texas Air Corporation introduced a "Max Saver" fares that had the further restriction of being non-refundable. The practice of limiting refundability (or imposing cancellation or change-of-itinerary fees) became an industrywide practice shortly thereafter. In the mid-1980s, airlines introduced Saturday-night stay requirements to further prevent business travelers from buying discounted products.

Today, most airlines offer discounts based on a relatively stable set of restrictions, typically a combination of advance-purchase restrictions of 7, 14, 21, and 30 days, the requirement to stay a Saturday night, nonrefundability, and penalties for changes in the itinerary after purchase. The low-cost carriers, which concentrate primarily on the leisure market, use primarily advance-purchase discounts and change penalties.

10.1.2 Customers, Products, and Pricing

Airlines serve a wide range of customers, both individual travelers as well as groups. The classic segmentation of individual travelers is between business and leisure customers.

Those traveling for business purposes have strong time preferences. They thus tend to value schedule convenience and booking/cancellation flexibility and are considered relatively price-insensitive, because, in most cases, their travel expenses are paid by their employers or charged to clients.

Leisure travelers, on the other hand, tend to be more sensitive to price because they are paying from their own pockets. However, because they are traveling for discretionary purposes, they tend to have more flexibility in their travel dates and will modify their schedule to find a good deal. They are also willing, and even prefer sometimes, to precommit to travel many days ahead of departure.

The two segments differ also in their travel-time preference, with business travelers preferring to leave on weekdays and return by the weekend, and leisure travelers preferring to depart at the end of the week and stay over a weekend. Leisure travel peaks around major holidays, while business travel drops at these points in time.

Of course, this is at best a crude description of the behavior of the many customer segments an airline serves. Some leisure travelers with high disposable incomes are more sensitive to schedule convenience and in-flight amenities than they are to price. Many business travelers are as price-sensitive as leisure customers (for example, those who are self-employed). Business travelers sometimes travel over the weekend, and leisure travelers often need to travel midweek. Some business travelers can easily commit to an advance reservation with no refund, while some leisure customers decide to travel at the last minute. Therefore, in general, there are many variations in preferences for schedules, routings, and in-flight service among travelers, and there are many differences in their ability to plan and commit to their travel plans.

Besides individual travelers, airlines also serve various wholesale travel groups. For example, cruise lines will often book blocks of seats connecting to their various sailings for packaged holidays. Tour operators also purchase blocks of seats to offer as part of combined air-hotel packages.

Airline sales can also be classified by sales channel. Travel agencies have long been the dominant sale channel, but Internet sales (either own-website sales or third-party travel sites) are rapidly growing in importance. Consolidators and wholesalers are two other significant sales channels in many airline markets.

All these differences in sales channels, customer types, and behavior affect RM and are targeted by various airline products, as explained below.

10.1.2.1 Itineraries and Combinability Rules

Airline products are itineraries (seats on a routing on a date and time in the future) on its network of flights. An itinerary may involve multiple connections.[1] Because of the many connection possibilities, an airline with 500 flights a day may offer hundreds of thousands of possi-

---

[1] Most full-service carriers have significant network traffic. The newer low-cost carriers deal exclusively in point-to-point service and do not offer connections.

ble itineraries for sale. Added to this complexity, airlines offer different compartments of service (first, business, coach), and within each compartment, multiple fare products with different rules and restrictions—all offered for travel up to a year in the future. Prices are set at the level of itinerary, date of travel, fare product, and point of sale—requiring hundreds of thousands of products to be priced regularly.

Because of the immense number of combinations involved, airlines post prices for only a fraction of their itineraries. They then define rules on how these simpler fares can be combined (so-called *combinability rules*). Another pricing method—called *constructed fares*—involves specifying a base fare between regions (such as Spain to North America) and then defining add-on charges based on specific origins and destinations (for example, Spain-Chicago, +30 Euros). It is the responsibility of the reservation system and the travel agent to follow these combinability and fare-construction rules, interpret them properly, and charge customers the correct fare.

10.1.2.2 Interlining

An itinerary can also consist of flights involving several airlines (called *interlining*). The price for interline flights depends on the agreements between the two carriers. The revenue is then split by a mutual proration agreement or, lacking such an agreement, by IATA-specified guidelines. The prorated revenue settlement is most often done by agencies dedicated to this task, which introduces significant delays in even accounting for revenue.

10.1.2.3 Pricing Itineraries

Even if all flights are on a single airline, pricing an itinerary is complicated by the fact that there are often many different ways to do so. Figure 10.1 shows an example of an itinerary and some ways of pricing it as a combination of one-ways, round-trips, open-jaws and one-ways with stopovers, and so on. A pricing solution is made up of these components. For each one of these components, in turn, there are many possible fare products. Each fare product has its own rules and restrictions attached to it. If a passenger qualifies for all the restrictions *and* all the combinability rules are satisfied *and* the airline reservation system indicates that the booking classes for the desired fares on *all* the legs are open, then the agent can book the itinerary. The rules indicate conditions on the amount of time available for purchase, the cancellation penalties, and so forth, and footnotes indicate exceptions and clauses. This rule and footnote interpretation is complicated, with sequential sets of logical clauses specified at one or more levels (such as flight level, geographic

Industry Profiles

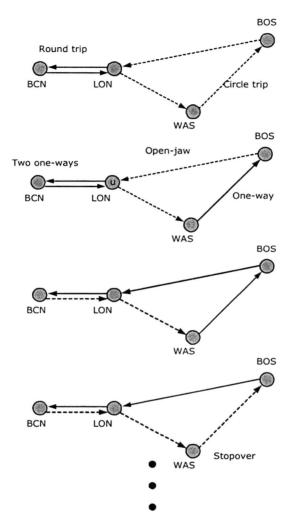

Figure 10.1. Different ways of pricing an air travel itinerary.

level and periods of application). Indeed, the industry at present time has more than thirty categories of rules, restrictions, and footnotes. As a result, an itinerary such as the one shown in Figure 10.1 can have several thousand possibilities for pricing.

A travel agent or GDS searches among the valid combinations (usually to find the cheapest). However, because of the large number of possible combinations involved and the difficulty in interpreting the combinability and footnote rules, even automated systems sometimes miss the lowest possible price. This is one reason that one travel agent can sometimes produce a lower price than another. Web-based price search engines (e.g., Expedia, Orbitz, Travelocity) also show differences in their ability to search these combinations correctly and find the best price, as their performance depends on the quality of the rules interpretation code and the number of combinations priced.

10.1.2.4 Managing Prices

Prices are distributed through two agencies, ATPCO and SITA.[2] ATPCO (Airline Tariff Publishing Company) is an industry organization that aggregates fares and rules from a large number of airlines and distributes them to GDSs. Anyone can subscribe to its service for a fee. Distribution is electronic with different frequency of downloads that can be restricted by filters to reduce the volume of data. A change service sends new and changed fares only, by market and region.

Airline pricing departments price their fare products by monitoring competitors' prices and their own traffic data. Airline analysts closely monitor competitor price changes. Pricing departments respond to competitors' price moves very quickly, often filing a response on the same day.

Fares available for such public distribution are called *public fares*. However, in addition to public fares, airlines offer a large number of *private fares*. Private fares are discounts or special rates given to important travel agencies, wholesalers, and corporations. They are not revealed to the public (or competition) and are available only by corporate agreements or qualified agents. In some international markets, private fares constitute 90% of sales, though in Europe and the United States, they are a smaller fraction of total sales. Private fares are sometimes sold via GDSs (a special code is required to access them), though many are off line paper agreements. Field sales agents monitor these private fares and send intelligence back to the airline when possible, though pricing analysts usually have only a vague idea of the private fares offered by competitors.

10.1.2.5 Priceline.com and Internet-Only Fares

The Internet has opened up a new sales channel for airline tickets with its own fare structure and sale practices. Many airlines use their websites to make last minute fares available at very low prices, a tactic

---

[2]Schedule information is distributed mainly by another organization called the Official Airline Guide (OAG).

little used previously. These fares are not advertised ahead of time and are usually not sold through any other channel.

Intermediaries, such as Priceline.com and Hotwire.com, have also created new channels and pricing mechanisms in the U.S. airline industry, and similar sites have emerged in other countries. Priceline.com defines its process as a "buyer-driven commerce," and it has many similarities with reverse auctions (see Section 6.1.2). A customer specifies an itinerary and a price they are willing to pay. The request is a commitment by the customer to buy at the offered price; if an airline accepts the offer, the flight is booked and the customer is charged. However, customers must accept considerable uncertainty over the details of their itinerary, including not knowing the airline they will fly, the number of connections it will have, or the exact time of arrival and departure. Moreover, they cannot change or cancel the booking once it is made. Customers can also buy only economy, round-trip tickets and cannot use frequent-flyer points.

Once a request is made, the Priceline.com system searches for an airline willing to sell below that price and sends an accept/reject decision back to the customer within 15 minutes. Priceline.com keeps the margin between the customer-quoted price and the airline price.

The Priceline.com mechanism is designed to appeal to price-conscious customers who are flexible about their travel times and routings. The deliberate uncertainty introduced into the transaction (in airline, arrival time, routing) makes it unattractive for most business travelers (and many leisure travelers as well). The fact that the airline's identity is uncertain also offers *brand shielding*, allowing airlines to discount without making the discounts widely known, thus limiting the threat to their main channels of distribution. (For these reasons, such fares are called *opaque fares* in the industry.)

Several airlines have also started similar sale procedures on their own websites, combining them with regular sales and last minute, Internet-only offers and posted-price offers.

10.1.3 RM Practice

Since the practice of RM in the airline industry is quite mature, several standard practices have developed. Here we summarize and discuss these practices.

10.1.3.1 Fare Classes and Fare Basis

RM systems for airlines book reservations in fare classes (or *booking classes*). Each compartment (first, business, and coach) has a number of fare classes—typically eight or more for coach, one or two for business, and one or two for first. These booking classes are represented by letters; some industry-standard booking classes are F for first-class, J and C for business class, Y for full-fare coach with no restrictions, and M, B, K, H, Q, Z, and others for the discounted fare classes in coach. The exact codes used for the classes vary from airline to airline.

Each fare class is used to book tickets sold under different *fare codes* or *fare-basis codes*. Each of these fare-basis codes (with names such as "QXE30") have specific fares associated with them, and the main requirements for booking under that fare basis code is encoded (somewhat cryptically) into the name of the fare code. For example, the 30 in QXE30 represents a 30-day advance-purchase requirement. Table 10.1 gives an example of fare codes and their mappings to booking classes and restrictions. The primary reason for grouping fare-basis codes into

Table 10.1. An example of airline fare codes, classes and their restrictions.

Fare Code	Fare	Booking Class	Main Restrictions
QDBQ	137	Q	Senior fare, Sat. night stay, 1-month max. stay
WDRK4DLZ	86	W	Youth fare, 3-day AP, 4-day min, 1-month max. stay
SDFKSLMC	110	S	Sat. night stay, 1-month max stay
KDAP	167	K	4-day min., 1-month max. stay
BDBO	198	B	Senior Fare, 6-month max. stay
BDZZ	198	B	Youth, 6-month max. stay
BDAP	216	B	3-month max. stay
YD	260	Y	

fare classes is that many reservation systems can accommodate only a relatively small number of fare classes (typically five to eight) per cabin. Thus, grouping fare codes into fare classes allows an airline the flexibility to post a wide range of fares yet control their availability through a smaller number of fare classes.

10.1.3.2 Booking Processes and Availability

Airlines start selling seats on flights up to a year before departure. However, since flight schedules are usually not finalized until three months prior to the departure date, most bookings that are made very early on are tentative, consisting mostly of group bookings by tour operators. Most regular fare class bookings come in during the last two or three months before departure. Typically leisure passengers book earlier than business passengers and the restrictions imposed on the cheaper fare classes try to exploit this preference of business passengers to book late.

A typical booking process proceeds as follows. An airline posts availability in each fare class to the reservation systems stating the availability of seats in each fare class. This is done using codes such as Flight 314: Y4 M4 B0 ..., a notation that indicates up to four seats are available in Y class, four in M class, and zero in B (put another way, Y and M are open, B is closed). When a customer requests an itinerary, a travel agent retrieves this availability information from the GDS. If a fare class is open, the travel agent is allowed to make a booking in that fare class. Within a fare class the agent quotes a fare for the itinerary based on one of the fare codes. The fare code is then recorded under the passenger name record (PNR).

Booking data are also grouped based on fare classes rather than fare codes for forecasting purposes in most RM systems. This aggregation often makes it difficult to precisely estimate demand and revenues from historical data.

10.1.3.3   Global Distribution Systems (GDSs)

A *global distribution systems* (GDS) provides centralized control and distribution of bookings. There are a number of GDSs currently in operation worldwide (see Table 11.5). The operation of these GDSs is governed by regulations intended to prevent the host airline or airlines from biasing the display to their advantage, though there has been considerable controversy surrounding "display bias" over the years.

GDSs communicate with the host reservation system of each airline to periodically obtain availability information. A travel agency subscribes to a GDS and makes bookings through it. The GDS in turn sends messages to the host reservation system of each airline involved in the itinerary of a given booking. Airlines are charged a fee for each one of these GDS booking transactions.

When the host reservation system of an airline closes a booking class, it sends a message to all GDSs indicating that a particular class on a particular flight is closed. The GDSs in turn display the new availabilities to travel agents' queries. The communication requirements between the travel agent, GDS, and the host reservation system are very demanding, usually requiring that the connect and transaction be completed in a second or less. A few million transactions are processed by the GDSs each day.

For competitive purposes, so as not to reveal their inventory decisions to competing airlines, airlines do not reveal their complete availability information to GDSs. For example, an airline may post an availability of Y4 to a GDS; meaning four seats are authorized for sale in Y class, even though there may be 50 seats remaining on the flight.

10.2 Hotels

The hotel industry is another industry in which RM is well established. Hotels are categorized as business, extended-stay, resorts, or a mix of business and leisure and also by size (large, small) and location (airport, urban, central business district or CBD, highway, beach). Hotels may be managed by independent owners, as part of a chain that is managed directly by employees of a single corporation, or as part of a franchise. Some hotel companies manage only individual properties, while large hotel chains can own thousands of properties under multiple brand names. Chains sometimes manage a property without taking ownership. This diversity in the types and operations of hotels means RM practices in the industry also vary quite a bit.

10.2.1 Customers, Products, and Pricing

Like airlines, hotels have both individual and group customer segments. Free individual travelers (*FIT*s), are guests who book their own rooms, whether for business or leisure. Some FIT segments include corporate, long-stay guests (those who stay greater than one week), individual vacation packages marketed by the hotel itself (such as honeymoon or golf), weekend packages, and walk-in customers. In addition to FITs, hotels receive demand for single rooms from travel packages sold by travel agencies or airlines. The groups segment is made up of tour groups, conference groups, incentive groups (such as salesforce reward parties), ad-hoc groups (an excursion group), and recurring groups (airline crew, cruise-line).

However, despite some of the similarity with airline customer types, the segmentation mechanisms used in hotel RM are somewhat different from those used by the airlines. For example, advance-purchase discounts, a prominent segmentation mechanism of airlines, are not that commonly used by hotels. The equivalent of the Saturday-night-stay restriction of airlines—intended to restrict access for business customers—is the weekend rate, applicable only for stays on Friday and Saturday nights.

10.2.1.1 Room Revenues

Rooms are the primary source of revenue in most hotels, but hotels also generate significant revenues from secondary sources such as food and beverage sales, function space, activities (golf, ski, entertainment), and gambling (in the case of casinos). For this reason, the value of a customer to the hotel may be hard to determine exactly. For example, a customer's restaurant spending for food and beverages is uncertain at the time of booking. Table 10.2 summarizes the typical revenue sources for a hotel. Despite their importance to customer profitability, these additional sources of revenue are often not accounted for in hotel RM systems, though some RM systems do work with the average *net* revenue for each rate product.

10.2.1.2 Room Types

The rooms of a hotel are usually classified into several *room types* with up to 40 room types, each with a potentially different rate. Some examples of room types are presidential suites, suites, deluxe rooms, business-floor rooms, standard rooms, executive rooms, lower-floor rooms, preferential rooms and room with a view. Other classifications include smoking or nonsmoking rooms, and single or double bed, with small differences in prices between these classifications.

Even though there can be many different room types, they are normally grouped together into three or four categories for capacity-control purposes. For example, the classification may be reduced to suites, business rooms, and standard rooms, equivalent to airline compartments. There is normally a gradation of rates in these categories and a large difference in the average rate between categories. The rates for each room type are again grouped together into only a few classes for RM purposes.

10.2.1.3 Room Rates

Rates start off with what is called a *rack rate*—or the *published rate*—which is the highest rate for a given room type (equivalent to Y for the coach class in airlines). Rates go down as a percentage off the rack rate. The rates are usually referred to as 90%, or 80% (off rack).

A customer can qualify for a particular rate based on his affiliation (company, government, diplomats), membership (automobile clubs such as AAA), or individually negotiated discounts. Travel agencies negotiate discounted rates, called wholesaler's rates, which can be lower than corporate rates. Even these *wholesalers' rates* vary significantly from one vendor to another. It is not unusual for a large hotel to end up with up to 150 rates. The rates are usually adjusted only once or twice a year. Hotels typically aggregate both the rates and the customer types, leading to about 10 to 12 *rate bands* (or classes) for inventory control.

Pricing for a multiresource inventory (multinight stay request) for hotels is almost universally taken as the sum of the daily rates. This is in contrast to airline pricing where the fares of a multileg itinerary have little to do with the prices on the individual segments. This difference arises because a multinight stay does not constitute a different "market" per se, whereas in the airline industry different itineraries serve quite distinct geographical markets with potentially very different levels of competition. The only exception to this simple way of pricing multiday stays is when a hotel offers one or two nights free for longer stays (akin to a volume discount).

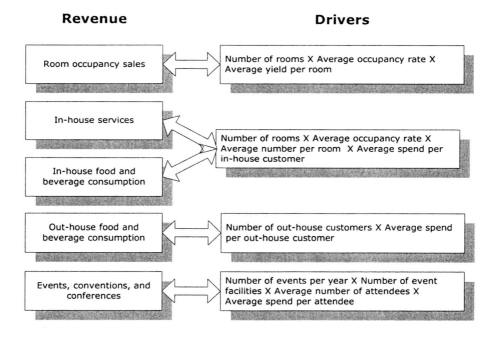

Figure 10.2. The main revenue sources and revenue drivers for a hotel.

10.2.2 RM Practice

As mentioned, hotel RM practices tend to exhibit greater variation than airline RM practices, mainly due to the more fragmented nature of the industry. Still, one can identify common elements.

10.2.2.1 Booking Process

In a typical large hotel, approximately 60 to 80% of bookings are made directly with the hotel, either locally, through the Internet, or through a centralized call center. The remaining bookings come from GDSs.[3]

Customer bookings arrive anywhere from one to 18 months in advance. However, a large part of the reservation activity—at least corporate and transient bookings—happens during the last few days preceding the room-usage date. Reservations may or may not be guaranteed by credit card, though this practice varies by region and country. Not surprisingly, cancellation rates are quite different when reservations are guaranteed by a credit card versus when they are not guaranteed.

Hotels typically follow one of two distinct policies with respect to rate quotations. One policy is for the reservation agent to quote the "best-available rate"; the second is to use a "top-down" quotation policy, where the quote starts with the higher rates and the agent in essence "bargains down" the rate depending on the customer response, by offering different room types or packages. This latter policy is partly due to the richer variety of the inventory in hotel products and can be considered a type of information discovery mechanism between the agent and customer. However, the practice of negotiating rates, which is a rather unpleasant experience to many customers, is declining with the increased use of Internet booking engines where the norm is to offer a menu of available rooms and packages and customers self-select the class and room-type combination.

Corporate bookings in many hotels follow a somewhat different process. If the request is turned away because that corresponding class is closed, then reservation agents typically attempt to sell a higher room category, rather than quote a higher rate for the requested category. This is normally because of contractual restrictions between the hotel and the corporation or wholesaler. Some corporate contracts guarantee last-room availability, which ensures the customer can book a room if one is physically available.

10.2.2.2 Property Management Systems (PMS)

Hotel reservations are controlled by a *property management system (PMS)* that automates the flow of the vast information required in the operation of a hotel.

---

[3]The Hotel Electronic Distribution Network Association (HEDNA) reported that in 1999 the GDSs delivered over 43 million bookings for hotels, with a value in excess of $12.5 billion [96].

Table 10.2. Features of a hotel property management system (PMS).

Reservations Options
Individual, group, shared, and multirate reservations
Add, change, cancel bookings including multiple legs
Standard, package, negotiated, and group rates
Multicurrency rate displays
Reservation agent opportunity messages
Online automatic up-sell messages
Geographic and regional property search
Property detail
User definable products available by rate code
Unlimited rate availability strategies and restrictions
Multiple rate codes for automatic entitlement rate display
Wait-list capability; automatic wait-list queue
Query system for reservation action items

Rates and Inventory Control
Multilevel inventory control by room class and room type
Maximum occupancy by room type
Rate plans
Property rates in multiple currencies
Rates confirmed in multiple currencies
Support for packages
Negotiated rates
Tax calculations
Available rate codes based on arrival date, number of nights etc.
Supports generic room types
Room and rate management
Transaction activity by agent
Online written, e-mailed, or faxed confirmations
Capture brochure requests
Revenue and forecasting reports
Security controls
User-definable work menus

Customer profiles
Multiple address capability for each profile
Member or club numbers
Profile merging
Client ID number automatically or manually assigned
Profiles include preferences for products, room types.
Preferences may be used as defaults when assigning rooms
Ability to over-ride overbooking parameters
Notation of special commission for client, agent, or source
Membership tracking
Profile relationship linking
Credit-card numbers
History statistics
Groups and Blocks

Room and rate restrictions
Open
Closed
Closed to arrival
Minimum length of stay
Maximum length of stay
Maximum number of persons
Seasonal close
Stay through
Advance booking
Open/close room types by rate code, property, room class, etc.
Administration

Group rooms control
Room inventory allotments with cutoff date or days
Online customer surveys
Tour series (recurring group stays)
Global distribution system interface
Interfaces to GDSs

The PMS records many transactions, like meal and beverage sales, in real time, and many others, like arrival updates and billing, in batch mode at night. In addition, the PMS controls many other functions such as accounting, billing, employee records, security, and supplies inventory and ordering. The features of a modern PMS are given in Table 10.2.

The PMS is usually linked to external GDSs or, for a hotel chain, to a hotel's own corporate reservation system. Most of the principal GDSs, as well as many travel and tourism Internet sites, list hotels with live (seamless) connections either to the hotel-chain reservation system or directly to the hotel PMSs. The GDSs list room types and attributes (bed type) and typically store only the rack rate and one or two discounts below the rack rate. Therefore, GDS bookings come only in a relatively limited number of classes.

The PMS communicates with a GDS through a *consolidating system* (belonging to so-called *switch companies*). The two leading consolidating systems belong to Pegasus (formerly THISCO—The Hotel Industry Switch Company) and WizCom (owned by Cendent Corp. and previously owned by Avis Rental Car Company). These consolidating systems broadcast the PMS availability message to the major GDSs and also consolidate the GDSs' booking messages. As a result, the communication between PMSs and GDSs is not instantaneous, but the systems are moving toward a query-reply mode.[4] Hotels have to pay a fee for each booking through a GDS (approximately $4.50), the switch company (approximately $0.45), and also a commission to the hotel chain/representative (if applicable). In addition, frequent-flier costs (about $8 to $9 per booking) and travel-agent fees (about 10% of revenues), reduce the revenue received for a room.

10.2.2.3 Overbooking and Cancellations

No-show rates in hotels range from 7% to 20% depending on the rate category. Cancellations and no-show depend on time of booking (later bookings tend not to no-show), credit card guarantees, whether the room is being shared, and so on. A cancellation happens not only when the customer calls to cancel but also if the customer decides to check out early. This means that the future capacity of the hotel is often uncertain.

Overbooking is widely practiced in the hotel industry. The hotel equivalent of an airline denied-boarding is when a customer is "walked" to another hotel. In general, hotels are conservative in overbooking, and walking a customer is a relatively rare event. Sometimes hotels walk a

---

[4]See the discussion of seamless availability in Section 11.2.3.2. Many GDSs today have this capability.

"less valuable" customer (a one-night stay guest) even when a room is available, to avoid walking a "more valuable" customer (long-stay guest) who is slated to arrive later. Aggregate supply and demand in the locality is taken into consideration when setting overbooking limits. For example, if there is a convention or festival in the city and all the surrounding hotels are likely to be full, a lower overbooking limit is normally set.

Overbooking of a different type occurs in resorts and leisure hotels that work with tour operators. The hotels sign an agreement with the tour operator to guarantee a minimum number of rooms (*allotments*), but the tour operator is usually not obliged to fill the rooms. The hotel can sign for more allotments than it has rooms, hoping that by pooling the allotments across tour operators the final show demand will be less than its capacity.

10.2.2.4 Capacity Controls

Hotel capacity controls follow the traditional nested allocation and bid price[5] schemes of airlines, with a few important differences. For one, control is often based on the length of stay. A *minimum length-of-stay control* is often used to accept only stays over a certain duration. The rationale for this type of control is that during high-demand period, the hotel does not want short-stay customers occupying rooms for a small number of days and displacing demand of longer-stay customers. A *maximum length-of-stay control* is the opposite; it sets an upper bound on the duration of stay, so lower-revenue, long-stay guests do not displace higher-revenue, short-stay guests. A *closed-to-arrival control* restricts bookings that start on a selected date. An *open-for-day use* means only bookings with zero nights (only day use) will be offered (in a specified rate code).

These length-of-stay controls are somewhat redundant if a hotel RM system uses a bid price system; nevertheless, many PMS systems still offer them to support incumbent hotel pricing structures and management practices. (See Table 10.2.)

Most hotel RM systems make intraday forecasts and optimize much more frequently than do airline systems. It is not unusual to see optimization being run every hour as the date of usage nears.

---

[5] With bid price implementations dominating. In the hotel industry, bid prices are sometimes referred to as *hurdle rates*.

10.3 Rental Car

RM practices in the rental-car industry have similarities to both the airline and hotel RM. However, again there are differences worth noting.

10.3.1 Customers, Products, and Pricing

There are six major rental-car companies in the U.S.: Hertz, Avis, National, Budget, Alamo, and Dollar, with nearly 95% market share between them. The business is somewhat similar to that of hotel chains: some own all their properties, some work on a franchise basis, and a few, especially at remote locations, just take bookings and subcontract out actual car rentals to a local company.

A significant percentage of their business comes from airport locations, so it is no surprise that the customer segments for rental cars closely mirror those of airlines. The deregulation of the airline industry also affected rental-car companies significantly, increasing the volume of business and changing the mix of business and leisure customers.

The product of a rental-car company is a combination of car type (there can be up to 20 car types), insurance options, pickup and drop-off location, advance purchase restrictions, and length of rental. Many corporations and travel agencies negotiate special rates with car-rental companies, which are accessed through special discount codes. These contracts are usually for a fixed-per-day price across all or most locations for a given period. Segmentation also occurs in the channels of distribution, with special discounts for booking directly on the company's website.

There are some subtle differences in customer booking and rental behavior compared with airlines and hotels. A customer who shows up at a hotel or for an airline flight at the last minute (called a *walk-up*) is usually willing to pay a high price for the service because the limitations imposed by airline schedules or hotel locations restrict their alternatives. Such walk-ups, therefore, pay near full price in those industries.

On the other hand, most car-rental counters are located at airports, clustered together; so walk-up customers have many alternatives to choose from with little search costs for shopping around. Therefore, the prices quoted to such walk-up customers are heavily influenced by the local availability of cars and competition. In periods of low demand, prices may actually be lower on the day of the rental than they are when booked in advance. On the other hand, during peak-demand periods, carefully setting aside inventory for walk-ups can have revenue benefits, just as in the airline industry. Such last minute pricing for day-zero customers is typically the responsibility of local field managers, and RM systems have to account for the resulting uncertainty in day-zero prices.

10.3.2   RM Practice

One significant feature of car-rental RM is the nature of capacity. Capacity is much more flexible than it is in either airline or hotel RM. For example, a rental-car company may operate more than one location in a city or a geographical area (for example, a downtown and an airport location). Inventory at each one of the locations can be pooled (*intrapool*), allowing greater flexibility in adjusting capacity to meet demand. Even if there is only one location in a given area, capacity can usually be increased or decreased by *interpool* moves, by moving cars from nearby cities, and also by controlling the sale of older vehicles and turn-backs to manufacturers. Small adjustments to the fleet size at a location can therefore be made on a weekly basis if need be.

Available capacity is also affected by customers who rent at one location and drop off at another (*migratory inventory*, Carroll and Grimes [100]), creating a network imbalance, or by customers who return the car earlier or later than their planned return date (akin to a hotel guest who understays or overstays). This means the capacity itself is often uncertain.

Free upgrading, in which a customer is given a car of higher rental value for no extra charge, is also an important factor in rental-car RM. Indeed, when demand for a lower car type exceeds the available inventory and the forecasted demand for a higher category car type is low, car-rental companies often plan to give free upgrades. This practice is analogous to planned overbooking over multiple compartments by airlines, where economy passengers who cannot be accommodated in the coach compartment get free upgrades to business-class (see Section 4.5). However, the practice is more prevalent in car-rental RM because there are many more inventory types and the capacities are more evenly balanced across the different car types.

Business customers typically select mid- and full-size vehicles, have insurance protection and gas included in the rate, and rent and return during the week. Leisure customers drive smaller cars or vans and on average rent longer than business customers. As a result, RM product-segmentation restrictions include Saturday-night stay, minimum length of stay, and weekend rates to stimulate midweek rentals. Many rates have blackout periods where they are not available (such as holiday weekends).

Carroll and Grimes [100] describe an implementation of a RM system at Hertz. Although at the core, the system calculates marginal values

Industry Profiles 533

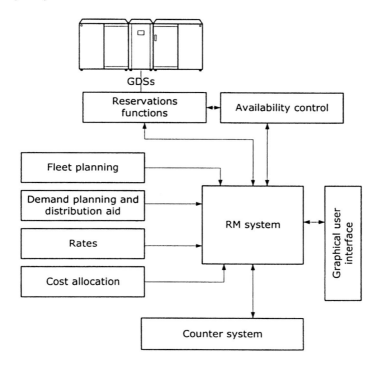

Figure 10.3. A rental car RM system implementation.

and uses a bid-price control to make accept/reject decisions for bookings, it differs from airline RM systems in some significant ways. Most notably, RM is very closely integrated with capacity planning—how many cars to purchase, where to deploy them, what products to offer and sell. Figure 10.3 shows the relationship between the RM system and the capacity management systems.

10.4 Retailing

RM in retailing is a relatively new but growing practice. Apparel and grocery retailers have to deal with highly perishable and seasonal products. High-tech retailers (PCs, consumer electronics) have similar problems, as their inventory loses value rapidly due to technological obsolescence. These characteristics mean that tactical demand management is important economically for retailers. Recently, a number of specialized software firms have entered the market for providing RM systems to retailers, and several major retailers have adopted or are testing these systems. Retail RM differs from the industries we've discussed thus far in that dynamic pricing, in the form of discounts, markdowns,

Table 10.3. World's top 10 retailers, store types, and their revenues for the year 2002.[a]

	Country	Name	Format	Retail Sales (US$ mill)	Income (Loss) (US$ mill)
1	U.S.	Wal-Mart	Discount, hypermarket, supermarket, superstore, warehouse	217,799	6,671
2	France	Carrefour	Convenience, discount, hypermarket, supermarket	61,565	1,069
3	Netherlands	Ahold	Cash and carry, convenience, discount, drug, hypermarket, specialty, supermarket	57,976	1,207
4	U.S.	Home Depot	DIY, specialty	53,553	3,044
5	U.S.	Kroger	Convenience, discount, specialty, supermarket, warehouse	50,098	1,043
6	Germany	Metro	Cash and carry, department, DIY, hypermarket, specialty, superstore	43,357	398
7	U.S.	Target	DIY, specialty	39,455	1,368
8	U.S.	Albertson's	Drug, supermarket, warehouse	37,931	501
9	U.S.	Kmart	DIY, specialty	36,151	(2,418)
10	U.S.	Sears	DIY, specialty	35,843	735

[a] *Source:* Deloitte, Touche, and Tohmatsu, "2003 Global Powers of Retailing," Stores, January 2003.

and promotions—rather than capacity controls—are used to manage demand.

10.4.1 Customers, Products, and Pricing

Characterizing retailing practices is difficult because retailers sell very different products using a variety of different formats and channels. Broadly, retailers can be classified as either selling durable or nondurable goods. Durable-good sales constitutes between 35 to 45% of total retail sales in the U.S. [485]. Table 10.3 gives a list of the top 10 retailers of the world and their revenues for the year 2002. The revenue of the largest retailer, Walmart, exceeds $200 billion, which gives an idea of the importance of even small incremental gains from RM. Table 10.3 also shows the different categories of stores for each of the firms. Among the

Industry Profiles 535

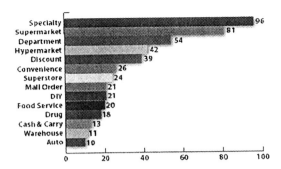

Figure 10.4. Store type breakdown for the top 200 retailers (*Source:* Deloitte, Touche, and Tohmatsu, "2003 Global Powers of Retailing," *Stores*, January 2003).

top 200 retailers of the world, the breakdown by store type is shown in Figure 10.4.

Some retailers—such as a grocery stores, department stores, and e-commerce sites like Amazon.com—sell an assortment of products from different manufacturers and suppliers. Others sell only own-brand or private-label merchandize—such as direct-to-consumer firms like Gateway and Dell and private-label retailers like The Gap or Eddie Bauer. There are specialty stores that carry a deep selection of one type of product, and mass merchants and department stores that sell a tremendous range of products. Retailers may sell through physical stores, catalogues, and online—or some combination of all three. Competition is generally considered intense in the retailing industry because consumers typically have many alternatives and can buy through many different retail channels. Profit margins in the retail industry rarely exceed 3%, as can be seen from Table 10.3.

However, customer do vary in terms of their geographical location, their preference for different channels, the importance they place on customer service and the in-store experience, their preference to buy early rather than late in the season, and their willingness to spend time searching products and prices. As a result, time of purchase, sales channel, and location tend to be the major segmentation mechanisms used in retail RM.

Regardless of the format, most retailers have to manage prices for thousands of *stock-keeping units (SKUs)* and control a large number of in-store and out-of-store promotional campaigns. Moreover, prices can vary based on the channel of distribution and by geographical region or country. There are also several ways of implementing price changes—

including coupons, promotions, markdowns, and tie-ins. Thus, the number of prices that need to be managed can be very large indeed.

The challenges of RM vary depending on the type of retail environment. As a sample of these differences, we next look at RM practices in apparel, grocery, and Internet retailing.

10.4.1.1 Apparel Retailing

Apparel has short life cycles and is usually progressively marked down, and at the end of the sales season taken off shelves and sold at clearance prices or through discount outlets. Figure 10.5 shows the magnitude of markdowns over the last 20 years. Indeed in retail management, a firm's markdown dollar budget, defined as the original list price minus the final sale price, is a closely tracked number.

The duration of the apparel sales season may be anywhere from a few months to a year. Most apparel is manufactured overseas (for U.S. and Europe) and has to be ordered well ahead of the sales season. The production and ordering cycle is often too long to reorder during a season, so retailers must precommit to the quantity stocked of each item.[6]

Forecasting is an important and difficult task in apparel retailing. Items have to be ordered by size, color, and style. For a retailing chain, inventory also has to be allocated by store and the retailer may occasionally need to redistribute inventory. As apparel items are often new and unique every season, there may be little historical data available for forecasting. Hence, the judgment of store buyers plays an important role and some RM systems use Bayesian forecasting techniques to merge a buyer's prior beliefs with observed in-season sales data (see Section 9.3.6).

The initial price of items in apparel retailing is generally determined manually (at least for "designer merchandize" [506]) because of the judgment required to evaluate brands, quality, and design attractiveness. In addition, price targets are often part of the initial product planning for an item. Once prices are set, RM systems are used to manage the timing and depth of markdowns based on sales trends (at the store or regional levels), inventory levels, forecasts and managerial targets, and business rules. Promotional events also influence the markdown strategy. RM systems also typically provide "what-if" analysis that allows managers to estimate the impact of markdowns. In addition to markdown decision support, the RM system may provide forecasts, initial buy recom-

---

[6]The exception is the use of "quick response" supply strategies, which attempt to provide within-season replenishment of apparel, using both fast logistics and domestic manufacturers.

Industry Profiles

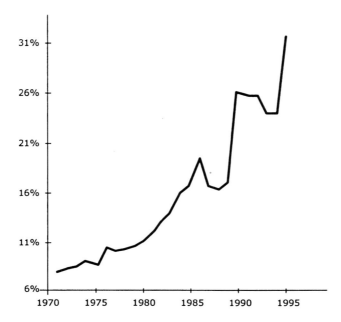

Figure 10.5. Growth of department-store markdowns as a percentage of store sales (*Source:* Merchandizing and Operational Results of Department and Speciality Stores, National Retail Federation).

Table 10.4. U.S. apparel sales by channel.[a]

Category	1999 Sales (US$ mill)	2000 Sales (US$ mill)	2000 Market Share %
Brick-and-mortar	167,346	169,256	92.9
Catalog	9,428	7,177	3.9
Online/Internet	2,904	5,873	3.2
Total apparel	179,678	182,306	100

[a] *Source:* NPD Group Inc.

mendations, store-level allocations, size allocations, and replenishment recommendations.

10.4.1.2 Grocery

Grocery retailing differs from apparel in several respects. First, most goods are consumables that are purchased repeatedly by consumers over time. In addition, goods are replenished frequently. As a result, inventory constraints and seasonalities are less of an issue in grocery retail RM. However, there are significant substitution effects in grocery re-

Table 10.5. U.S. apparel sales by category.[a]

Segment	Volume (US$ millions)	% Change vs. 1999	%
Men's	63,300	-0.3	34.7
Women's	96,588	2.1	53
Boys'	7,653	-2.2	4.2
Girls'	7,114	1.1	3.9
Infants' and toddlers'	7,651	13.1	4.2
Total	182,306	1.5	100

[a] *Source:* NPD Group Inc.

tailing. Consumers have choices of many brands, and brand-switching behavior is common. Consumers also stockpile, buying many units of a product when the price is low, which reduces their future purchases. Therefore, grocers have to consider the value of consumers' purchases over time and plan to attract both current and repeat purchases. Consumers also tend to be more price-sensitive with repeat-purchase products, and in general, there is less differentiation of products and ambience among grocers. All this leads to more direct and intense price competition. Margins in the grocery business are also very low.

Other factors link the RM decisions for related products in grocery retailing. Stores are often more concerned with the profits they generate on a *category* of products than on the profits they generate on any given brand or given product within a category. This has led many stores to adopt a *category management* approach to pricing and inventory decisions.

Additionally, consumers often shop a grocery store for a bundle of goods. As a result, stores may stock and price certain products more for their effect on consumers' store-choice decisions rather than for the profits they generate directly. Also, consumer perceptions on whether the retailer has high or low prices are formed largely through repeat-purchase items (such as milk or bread). Low prices (even at or below cost) on these items can drive traffic to a store and benefit overall sales and customer satisfaction. However, a low-price "loss-leader" strategy is normally applied only to a small subset of products. These so-called *market-basket effects* play an important role in RM in the grocery sector.

Grocery stores also carry a tremendous variety of products. An average grocery store may stock nearly 40,000 SKUs. The product mix consists of both perishables (meat, vegetables and diary products), nonperishables (consumer packaged-goods or dry goods; toilet paper, canned tomatoes, pet food), and frozen and chilled goods, each requiring a different pricing strategy.

Because of the large assortment of products, *menu costs*—the costs of changing prices—is an important consideration for grocery retailers. Electronic devices that are connected to a database and can change prices automatically (electronic shelf labels or ESLs) reduce these costs but are not yet widely deployed. In the vast majority of cases, grocers implement price changes by having clerks manually retag products with new price labels. This process is both expensive and time-consuming. Consequently, prices for many items are relatively stable and change only for competitive reasons or for manufacturer promotions or discounts.

The initial price for fruits and vegetables tends to follow a market price determined by overall supply and demand in wholesale markets. Retail prices are then marked down as the expiration date nears. For nonperishables, the price is influenced strongly by the manufacturer's wholesale pricing. Volume discounts (truckload pricing) is common and may influence a retailers ordering and pricing strategy. Trade promotions at the wholesale level are also a common practice and retailers may or may not pass these trade promotions on in the form of lower retail prices.

Promotions are very common in the grocery industry, and retail RM systems help grocery stores plan and execute promotions optimally. RM models recommend which product in which category to promote, the size of the products to promote, the best advertising strategy (amongst display, local ad or regional ad) and the optimal discounting to meet the store's objectives. RM software may also recommend baseline prices for products, depending on the store positioning (for example, quality products, convenience store, warehouse food market), store location, and local demographics.

10.4.1.3 Internet Retailing

The RM issues faced by Internet retailers are somewhat distinct from those of bricks-and-mortar retailers. For one, an unprecedented level of customization and customer profiling can be performed while the customer is shopping online. The Internet also makes prices more transparent to consumer, which makes price comparison easier; "shopbots" and price-comparison sites gather information from multiple vendors, allowing consumers to compare prices easily and in real time. At the same time, prices can be changed by online retailers at near-zero cost.

These features have led to two early predictions on how the Internet would affect retail pricing practices: first, that Internet prices would be significantly lower than in traditional brick-and-mortar store prices, and prices would vary little from one retailer to another; and second, that retailers would move to real-time dynamic pricing, customizing price based on supply and demand and their knowledge of the individual customer. However, neither of these predictions has proved to be true. Internet prices show significant variations among retailers—as much, if not more, than the price variations observed among bricks-and-mortar retailers—with the largest retailers not necessarily selling at the lowest price. A survey by McKinsey [126] showed that Internet shoppers rate price as relatively low in importance in their purchase decisions; brand, product information, and customer service rate consistently ahead of price. The same McKinsey survey [126] finds only 8% of Internet shoppers are bargain hunters, who would buy only at the cheapest e-retailer.

To quote Eli Katz, president of e-Commerce for Fragrance Counter and Cosmetics Counter [505], a retailer that does not discount heavily:

> There's always going to be customers who look for price, and there's always going to be customers who look for service, just like some people shop at Saks and some shop at Walmart. That will hold true on the Web as well.

The second prediction, that firms would practice dynamic pricing, has also only partially borne out. To quote Ken Harris, CIO of Gap Inc. [250]:

> I wouldn't rule it [dynamic pricing] out. Right now, I don't think it's quite ready. It's important that consumers understand how pricing is determined and that they feel it's fair.

Stephen Hamlin, VP of operations for iQVC, the Internet arm of the QVC Shopping Network [250] echoes this sentiment:

> One thing we really believe in is unilateral prices. If it's on TV for one price, it's on the website for that same price; if it changes on TV, it changes on the Web.

Moreover, there has also been some high-profile consumer backlash against dynamic pricing online. The most famous such incident occurred in 2000 when Amazon.com was discovered to be offering different prices for the same DVDs to different customers based on their profiles.[7] An incident with one consumer was widely reported in the news media, and Amazon quickly stopped the practice in the wake of this publicity. (See Section 11.5 for further discussion in a RM implementation context.)

While explicit price manipulation has not been well received by consumers, other marketing tactics are practiced very efficiently on the Web. Tools such as *collaborative filtering* (a data-mining technology that infers customer tastes based on "similar" customer purchases) create product

---

[7] "On the Web, Price Tags Blue," *Washington Post*, September 27, 2000.

recommendations and presentations in real-time based on customer information and current shopping behavior. Volume discounts are offered to encourage customers to buy larger quantities or bundles are created to increase their buy size. Online promotions and coupons are tailored to separate the price-sensitive customer and induce a customer to make a purchase. Such tactics are largely rules-based in current e-CRM applications, though it is likely that models and algorithms will play a more significant role in optimizing these decisions in the future.

Another new use of the Internet for e-tailers is online price testing. Selling at a single price provides very little information about the price sensitivity of customers. Because of the low menu costs for e-tailing, experiments are conducted online to gauge price sensitivity. A few price-optimization vendors provide this capability in their products.

10.4.2 RM Practice

There has been significant commercial activity in retail RM in recent years, with several new technology vendors offering software systems targeted at different segments of the retail industry. These retail RM applications focus on improving gross margins by optimizing base prices, markdowns, and promotions.

As mentioned, a key difference between retail RM and traditional airline and hotel RM is that it is price-based rather than quantity-based—that is, market-response models are utilized for dynamic pricing. This requires an ability to estimate the demand effects of short-term price changes.

Another important difference is that historical data is often inadequate for making good demand forecasts, especially for seasonal, fashion and high-tech products. Thus, there is more emphasis on forecasting demand for an item based on its in-season sales and the sale of "comparable" items. Experimenting with price changes at a sample of locations is also a common technique for gauging price sensitivity.

10.4.2.1 Features of a Retail RM System

Besides the price-optimization functions, retail RM applications provide important operational and productivity benefits, such as

- Automating routine price changes by location and channel,

- Monitoring profit or sales targets for items and categories,

- Tracking the performance of promotions and advertising campaigns,

- Maintaining consistent pricing and rounding rules,

- Automating price matching based on competitor prices,

- Supporting price-sensitivity experiments, and

- Generating periodic reports and statistics to track pricing performance.

Retail RM systems may also include other nonprice decision support tools for initial store allocation and assortment decisions, and reordering and replenishment of products. Indeed, the current trend is for retail RM to be combined within a SCM system under the label *retail analytics*.

10.4.2.2 Data Sources

The data available for retail RM is very rich indeed. Point-of-sale (POS) scanner data provides detailed and complete transaction data in an electronic format and is available almost instantaneously to retailers (and manufacturers). E-commerce channels add click-stream data, making it possible to monitor shopper browsing behavior and estimate customer responses.

In addition, *panel data* tracking total household purchases for a sample of consumers, available from market-research firms like IRI and Nielsen, allows companies to track and estimate brand-switching and market-share information over time. Other firms, such as AWIS Weather Services and Meteorlogix, supply forecasts of weather by micro-region that can be incorporated into sales-forecasting models for traffic and weather related items. Demographic data (age, income level, housing costs) by micro-region is also sold by many firms.

Finally, most large retailers have inventory-control systems, ERP systems, and SCM systems that link to supplier inventory systems, accounting systems (for cost and labor rates), financials, and inventory-management systems.

Retail RM systems use some or all of these data sources to perform demand forecasting and pricing optimization. They download current data from a store's POS and ERP/SCM systems and combine this with the store's historical data to calibrate forecasting models of demand. The models often require some input from buyers and analysts—for example, to identify a past product that is "similar to" a new product the store is introducing.

10.5 Media and Broadcasting

As with retailing, RM in the media and broadcasting industry is a relatively new practice. Indeed, to date, only a few networks and television stations are reported to practice RM. These include the Seven

Network in Australia, NBC in the U.S. [83], and CBC of Canada [184]. Still, the advertising market is large and has many of the characteristics conducive to the practice of RM. The sale of advertising time in broadcasting, though superficially similar to the sale of airline seats or hotel rooms in the sense that it too is a sale of a perishable commodity, has features and practices that make it quite distinct.

10.5.1   Customers, Products, and Pricing

Advertising time is normally bought by ad agencies on behalf of clients. The station either sells the space directly or through national sales representatives working on behalf of the television station or network. Advertising agencies also sometimes buy blocks of time in anticipation of demand from clients. Advertisers can be local, regional, or national and are additionally classified by industry (beverage product category, automotive or local dealers, and so on).

Advertisers also vary in their time sensitivity. Some are advertising for specific promotional events, so timing and placement of ads are critical, and they may require this space on very short notice. For others, their ads are for general brand awareness or for public information campaigns, so the exact timing of ads is less important and moreover can be planned well ahead of time.

10.5.1.1   Advertising Product

The advertising product is classified based on demographics (such as, adults 18+ or females 13–18) and also the time of the spot (prime-time, late-night). Table 10.6 shows how inventory is specified in television, radio, and print media.

Demographic differences in a show's viewership mean that the value of a particular slot for a particular advertiser can vary, so segmentation of buyers based on demographics is common. The demographics-based requirements of advertisers also give broadcasters the flexibility to substitute times and programs. For example, if it turns out that a particular time slot is oversold, a customer will usually accept a different time slot with similar demographics, though many advertisers have strong preferences for specific programs.

10.5.1.2   Sales Process

Advertising space is typically sold in both an upfront (long-term) and scatter (short-term) markets—and also on an opportunistic basis (called *remnant* space for print media). The television market is further divided into national, cable, and local (spot) markets.

Table 10.6. Inventory definitions in television, radio, and print media.

Media	Inventory Description
Television	Day-part (time part into which a broadcast day is divided) (e.g., daytime, early fringe, access, prime, evening news, late fringe) Audience composition (demographics), (e.g., women 25–54 Ratings (e.g., guaranteed 53.5) Preemptability
Radio	Day-part (e.g., morning drive, mid day, afternoon drive, night Audience composition Ratings
Print	Total pages, issue dates, positioning, separation, cancellation options

Table 10.7. A sample advertising purchase plan.

Qtr.	Purchased	Firm	Option	Option Date
3Q	$3 million	$3 million	$0 million	–
4Q	$10 million	$6 million	$4 million	1 Aug.
1Q	$4 million	$4 million	$0 million	–
2Q	$6 million	$4 million	$2 million	10 Jan.

Up to 90% of national sales are upfront. Upfront sales are made at least a year in advance of airing, usually during the June and July period. Scatter sales are made each quarter, roughly three to six weeks in advance of airing. Opportunistic sales (similar to airline last minute sales) are the sale of distressed inventory just before airing. The prices for upfront sales tend to be 20% to 40% less than for scatter sales; prices for opportunistic sales can be as low as 50% below those of scatter sales.

Customers can purchase inventory as well as options on additional inventory, with a deadline for exercising the option. The combination is called a *purchase plan*. Table 10.7 shows a sample purchase plan.

10.5.1.3 Prices and Ratings

The sale price of an advertising slot is based on its *gross rating points* (GRP)—the percentage of a demographic group in a market viewing a program at a point in time. The GRP is measured after the show goes on air, based on a sample of viewers that keep a meter or a diary record. Therefore, at the time a slot is sold, the exact GRP of the slot is uncertain, and only estimates can be provided. Yet buyers pay for the actual GRP, typically on a *cost per thousand impressions* (CPM) basis. Consequently, if the GRP turns out to be less than the promised

GRP, the station may have to compensate the advertiser—usually by means of free additional time (called *make-goods* in the industry). On the other hand, if the GRP turns out to be more than that specified in the contract, the buyers usually do not have to pay more, and hence the station or network has lost a revenue opportunity. Ratings for new television shows in particular are highly uncertain.

CPM is usually guaranteed for upfront sales, and the advertisers specify the target CPM, demographics, frequency, and any other restrictions that they may have. In the scatter market, ratings are not normally guaranteed, and prices can vary depending on the viewership.

10.5.1.4 Preemption

Preemption of scheduled advertisements is an accepted practice in the broadcast industry. That is, even if the station sells a particular time slot to an advertiser, if a higher offer is subsequently received from another advertiser the station may preempt the original advertiser, offering either to substitute another slot or return their money. This preemption is common in large media markets but less prevalent in smaller markets, where maintaining customer relations takes precedence. Hybrid practices, in which only a certain number of slots are sold as preemptable slots, also exist in the industry.

10.5.1.5 Print Media

The sale of advertising in the print media is similar to broadcasting. The prices are based on readership and demographics, on the size of the advertisement, and on the number of issues that the advertisement will be run. Magazines tend to be much more willing than broadcasters to negotiate prices because their capacity is more flexible. Despite the similarities between print and broadcast media, we know of no reported implementations of RM in the print media, though it remains a promising area for applying RM principles.

10.5.2 RM Practice

Broadcasters use a wide variety of pricing structures in an attempt to enhance revenues. For example, the price for a slot may be based on the loyalty of the advertisers (the frequency rate card), whether the advertiser is a local or a national advertiser (the national and regional rate cards), the ratings (the grid rate card), and finally on how rapidly the inventory is being sold. These structures evolved partly to segment customers based on location and access to information, partly to differentiate the product based on demographics, and are also partly rough-cut attempts at RM (for example, the prices may be raised based on remaining inventory levels). Taking a cue from airlines, some stations are attempting to further segment their customers based on time of purchase and penalties for cancellations.

The actual prices for each category of inventory are determined by competition and historical rates. In addition, many stations raise prices after a certain percentage of inventory is sold or offer last-minute discounts if time is unsold. Again, as mentioned above, it is not unusual to see (at least in major U.S markets) a system of preemption. However, advertisers typically get a lower price if they agree to allow their ads to be preempted. Controlling the use of preemptable and nonpreemptable slots is an important challenge for RM in the industry.

Another important factor in the sale of airtime is that it is rarely sold in units of a single time slot. Rather, a package deal is normally negotiated between the agency and station representatives. For example, the advertising agency may have a target GRP and demographic in mind for its client for a campaign over a certain period of time, and the agency's goal is to buy a package from a range of television stations, programs and time slots to meet that target. This makes it important for the broadcaster to know the value of each of its time slots to negotiate effectively. Therefore, estimates of the marginal opportunity costs of time slots (their bid prices) are useful information to a broadcaster.

This package-deal nature of sales introduces network effects into the evaluation and negotiation process. Just as an itinerary for an airline is a collection of resources, a package for a television or radio station is the sale of capacity for a collection of its shows. A station may therefore decide to accept an entire package even if it is losing money on certain time slots in the package, provided the overall net revenue contribution of the deal is positive.

Finally, a significant obstacle to RM in broadcasting is that sales practices in the industry vary widely, often depending on the traditions of each local market. This makes it difficult to construct a common RM model and system that is appropriate for a large number of media firms.

10.6 Natural-Gas Storage and Transmission

The natural-gas industry in the United States has undergone a process of deregulation since the mid-1980s. These structural reforms have led to a number of innovations, including experimentation with RM techniques.

Table 10.8. An example of a pipeline delivery contract.

Rate schedule	FT-A
Service requester name	XYZ PAPER PRODUCTS CO
Service requester	6139711
Service requester proprietary	70860
Shipper affiliated indicator	N
Agent name	ABC ENERGY RESOURCES, L.P.
Agent affiliated indicator	N
Contract number	13
Contract effective date	9/1/93
Contract effective thru date	4/30/04 23:59
Rollover period	
Maximum daily quantity	3,556
Negotiated rate indicator	N
Footnote	

Location Name	Location	Location Prop	Receipt Delivery	Location Zone	Location Segment	Quantity
AAA PLANT DEHYDRATION	38535	10144	R	0	SU	3556
BBB STORAGE INJECTION	125643	60018	D	4	S2	3556

10.6.1 Customers, Products, and Pricing

In 1985, *Federal Energy Regulatory Commission* (FERC) of the United States issued Order 436, which required pipelines to provide open access to their facilities, allowing consumers to contract separately for purchases of gas and for transportation services. This encouraged better balancing of supplies of gas among producers and consumers. The *Natural Gas Wellhead Decontrol Act of 1989* required the removal of all price controls on wellhead sales by 1993, allowing natural-gas prices to be freely set in the market. Similar deregulation is in progress in a number of other parts of the world as well.

Because of these reforms, the gas industry has gradually moved away from long-term minimum-purchase contracts between the pipelines and producers toward short-term contracts and spot-markets for buying and selling gas. This has tremendously increased the volatility of both prices and demand for pipeline services.

The gas-industry distribution structure consists of local distribution companies (LDCs), pipeline companies, retail marketing companies, and wholesale marketing companies. The customer base of these firms is diverse, ranging from large industrial users to individual homeowners.

Table 10.9. Sample base-line delivery tariffs for interruptible and noninterruptible delivery. The high rates for the non-interruptible service are used only on rare occasions. Normally the rates are calculated from traded natural gas (locational) futures prices on NYMEX (New York Mercantile Exchange) and other commodity exchanges (e.g., The Intercontinental Exchange).

Interruptible Base Transportation Rates (IT)
(Per Dekatherm)

Receipt Zone	Delivery Zone							
	0	L	1	2	3	4	5	6
0	$0.185		$0.319	$0.425	$0.484	$0.553	$0.626	$0.746
L		$0.157						
1	$0.325		$0.259	$0.368	$0.425	$0.495	$0.568	$0.688
2	$0.425		$0.368	$0.177	$0.235	$0.315	$0.377	$0.497
3	$0.484		$0.425	$0.235	$0.144	$0.306	$0.367	$0.487
4	$0.564		$0.506	$0.315	$0.306	$0.169	$0.197	$0.317
5	$0.626		$0.568	$0.377	$0.367	$0.197	$0.176	$0.278
6	$0.746		$0.688	$0.497	$0.487	$0.317	$0.278	$0.208

Base reservation Firm Transportation Rates (FT-A)
(Per Dekatherm)

Receipt Zone	Delivery Zone							
	0	L	1	2	3	4	5	6
0	$3.10		$6.45	$9.06	$10.53	$12.22	$14.09	$16.59
L		$2.71						
1	$6.66		$4.92	$7.62	$9.08	$10.77	$12.64	$15.15
2	$9.06		$7.62	$2.86	$4.32	$6.32	$7.89	$10.39
3	$10.53		$9.08	$4.32	$2.05	$6.08	$7.64	$10.14
4	$12.53		$11.08	$6.32	$6.08	$2.71	$3.38	$5.89
5	$14.09		$12.64	$7.89	$7.64	$3.38	$2.85	$4.93
6	$16.59		$15.15	$10.39	$10.14	$5.89	$4.93	$3.16

LDCs form the end stage of the gas supply chain, delivering gas to customers. They typically have to purchase in spot markets (at least for excess requirements), where prices are volatile, and sell at relatively fixed prices to their consumers. Futures contracts, swaps, and options are extensively used to manage the resulting risk. Retail marketing companies, which do not own physical distribution facilities, handle only the marketing and billing functions and contract with LDCs for delivery.

Wholesalers are one level up in the natural-gas supply chain. They buy from gas producers and deliver to LDCs and large industrial accounts. Their orders are based on forecasts of the demand from large customers as well as aggregate demand from LDCs and retail marketers.

Table 10.10. Sample natural gas transportation and storage products.[a]

Product Code	Description
FT-A	Gas-pipeline's firm transportation service that provides the customer with the highest priority and most reliable transportation service. Customers have the right to nominate at primary, secondary, or tertiary priorities based on zone and leg entitlements. An FT-A shipper may also utilize authorized overrun on receiving advanced approval by the pipeline. Authorized overrun allows firm shippers to schedule volumes in excess of their contractual limits for an additional charge. These nominations are scheduled at a higher priority than interruptible transportation.
FT-A (Zone 0L)	A lower-priced firm transportation service restricted to transportation within Zone 0L, a subset of gas pipeline's Zone 01.
FT-BH	A lower-priced firm back-haul transportation service restricted to pure displacement nominations from qualified receipt points. Nominations at secondary points are allowed provided they are within the path of the primary route and there is no forward-haul component. Deliveries on laterals constitute forward haul.
EDS/ERS	Extended delivery or receipt service enables FT-A shippers to nominate deliveries in zones downstream or receipts in zones upstream of their FT-A contractual rights for an incremental charge plus additional fuel.
IT	Gas pipeline's interruptible transportation service.
FS	Firm storage service from facilities located in the production area or the market area. This service has a space charge, injection limit, injection charge, withdrawal limit, and withdrawal charge associated with it.
IS	Interruptible storage service from facilities located in the production area or the market area. This service provides customers flexibility in supply and market options.
SA	Gas pipeline's pooling service that provides shippers with free supply aggregation throughout the system. Pool-to-pool transactions and imbalance trades are free of charge provided the pooling areas are the same. Between different pooling areas, these transactions require transportation and an associated charge.
LMS	Gas pipeline's load-management service that provides balancing to customers at receipt or delivery points, including pipeline customers customers Gas Pipeline Company cashes out imbalances on a monthly basis. Cost of imbalance, assuming no penalty situation, is established based on pipeline position, location, and type of imbalance, and the arithmetic average price of gas published in Natural Gas Week's Gas Price Report.

[a] Source: Tennessee Gas Pipeline website.

Wholesalers purchase delivery services from pipeline companies, who route gas from the source (wellhead) to the end market. As part of deregulation in the U.S., pipeline companies were required to unbundle purchase and delivery of natural gas (FERC Order 636, April 1992). Thus, pipeline companies generally are restricted only to transmitting or storing gas.

Besides these traditional firms in the industry, new intermediaries have entered as a result of deregulation. Some of these firms are simply asset management companies who hedge risks and make profits through trading energy-related contracts. Some are market makers who create packages and contracts from different pipeline vendors and suppliers and become "virtual suppliers" themselves. Internet market places, such as Intercontinental Exchange (ICE) and (the now defunct) EnronOnline (EOL), were started to facilitate trading of these new instruments.

10.6.2 RM Practice

For pipeline companies, in particular, the unbundling of transportation and purchases of gas has increased their dependence on transportation revenue, which is now estimated to constitute nearly 93% of their total revenue. These changes and the increased uncertainty in demand has made revenue managing pipeline capacity sales all the more critical.

Pipelines are essentially involved in the sale of space—the capacity for transmission or storage. Indeed, pipeline RM is somewhat similar to airline RM in that demand is for transport over a network with many interconnection points and routings. Pipelines have to price their space based on future demand forecasts as well as available capacity. Demand is realized in the form of forward contracts (essentially reservations) with various forms of options (analogous to airline cancellations and no-shows).

To give an example of a RM problem in the industry, consider the pipeline in Figure 10.6 connecting cities A, B, C, and D. For a certain date the residual capacities of the network are as shown in the figure. Consider the following bids for future capacity (Dth=Dekatherms):

A-B 2,000 Dth at 0.15$/Dth

A-C 5,000 Dth at 0.20$/Dth

B-C 3,000 Dth at 0.12$/Dth

Selling A-C would fetch the highest revenue, but it would exhaust capacity on B-C and prevent future sales. If the bids are indivisible, the firm may be better off rejecting the A-C bid. If bids are divisible, a

Industry Profiles 551

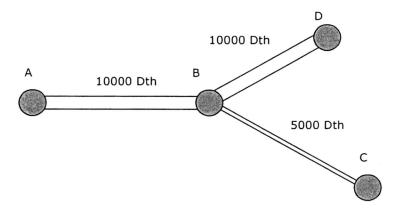

Figure 10.6. Residual capacity (in dekatherms) on a gas pipeline network for a specific date.

maximum-revenue network-optimization problem gives the optimal allocation. For a future date, a RM system would make a forecast of all future bids (volume and price, by origin and destination) for capacity on this network and use one of the network methods of Chapter 3 to determine which bids should be accepted now for this future date.

In addition to the capacity optimization illustrated above, current RM systems for gas storage and transmission include the following features [524]:

- Storage and park-and-loan service optimization, in which the pipeline company stores a producer's gas at its storage locations. A *park-and-loan service* is where the provider puts in gas for later use (parking) or takes out gas to meet a temporary imbalance, to be returned within a specified period (loaning). Using park-and-loan, shippers can generally avoid buying at high spot prices.

- Firm transportation revenue optimization, in which delivery is provided at the guaranteed delivery tariffs.

- Dynamic pricing of interruptible transportation tariffs (Table 10.9).

- Capacity contract optimization.

10.7 Electricity Generation and Transmission

The electricity industry has seen significant deregulation in the mid-1990s in the United States and many European Union countries. While in principle RM fits in nicely with these newly deregulated markets for electricity—and there has been considerable speculation on how RM could be applied in the industry [4]—to date we know of no electricity RM implementations per se, in the sense of market segmentation and capacity controls.

Still, the electricity industry has many characteristics that make it well suited to RM methods. Demand for energy is highly variable, varying by time of day, day of week, temperature, and season, yet generation and transmission capacity is relatively inflexible. Firms use a mix of generation technologies (hydro, nuclear, coal, and gas) in an attempt to respond to demand variations, but generating capacity has limits, and near-peak-capacity wholesale prices can rise to nearly 300 times the average price [426]—even in regions with 20% reserve capacity [125]. Finally, the industry has long used risk and demand management and sophisticated trading technologies, so it has the scientific and software culture to adapt RM.

10.7.1 Industry Structure

Much of the deregulation in the electricity industry has focused on separating the generation and transmission (distribution) functions, creating competitive wholesale markets for generators to sell to distributors and competitive retail markets for distributors to sell to end consumers. Figure 10.7 shows the four main models of electricity competition (Hunt and Shuttleworth [261]). In the complete monopoly model (10.7(i)), a single monopoly controls generation, transmission, and retail sales. In the purchasing-agency model, (10.7(ii)), there is competition in generation between the IPPs (independent power producers), but a single monopolist controls buying and distributing the electricity. In the wholesale competition model (10.7(iii)), there is competition in generation and among distribution companies (Distcos). In the model of retail competition, there is competition in the generation, wholesale, and retail markets for the final consumers (10.7(iv)). Models (i) and (ii) are generally followed by the traditional—usually state-owned or regulated—utility companies, while models (iii) and (iv) are a result of deregulation.

Industry Profiles

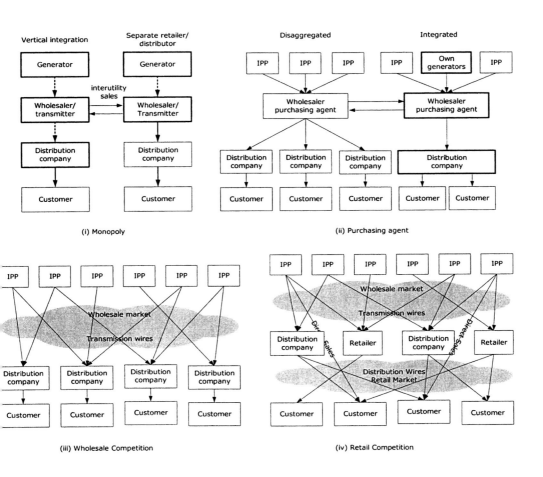

IPP stands for independent power producer

Figure 10.7. Four models of electricity industry structures and competition (*Source:* Hunt and Shuttleworth [261]).

10.7.2 Customers, Products, and Pricing

The pricing structures and the management of prices have also changed significantly in the wake of deregulation. Wholesale markets have moved toward dynamic, market-based pricing, while retail prices have, to a large extent, remained fixed (or have been capped by regulators). This has created often devastating problems for distributors that have to buy in wholesale markets and sell in retail markets.[8] Risk-management techniques are widely used by energy-sector companies in an attempt to manage such risks, but many of these problems are structural.

Wholesale markets in electricity are organized at various regional levels: state, country, pan-Europe, and global exchanges, such as the Amsterdam Power Exchange. The organization of these wholesale markets varies considerably from market to market. For example, trade may be through a central compulsory pool market or consist of bilateral trades between generators and consumers; prices may be set by bid/ask mechanisms, market makers, or auctions; the products traded may be electricity dispatch contracts, options or derivatives; settlement may be based on ex ante or ex post pricing; time of bid submission varies (real-time, 5 minutes ahead, 30-minute blocks) as do scheduling and dispatch rules; start-up costs and incumbent adjustments (price adjustments to account for the incumbents' sunk costs during the regulated period) may be factored in as well. Regardless of the exact form and rules of each market, the overall goal is to allow supply and demand to determine prices—often in real-time—and encourage efficient utilization and allocation of energy resources.

10.7.3 RM Practice

The electricity industry is making some preliminary attempts to implement dynamic pricing for end consumers—at least for larger industrial clients [545, 181]. Indeed, industrial customers have long paid differential prices depending on the quantity (nonlinear tariffs) and time (peak-load pricing) of their energy usage. However, such schemes require new technologies, such as time-of-usage meters, to enable real-time monitoring and accurate billing.

These same meters can, in principle, be used for individual households and small businesses. Radio control of the meters give utilities the capability to shut off large appliances (such as air-conditioners and swimming

---

[8]For example, only a few years after the deregulation of California's energy markets, one of the state's largest distributors, Pacific Gas and Electric, filed for Chapter 11 bankruptcy protection with accumulated debts of over $8.9 billion due to soaring wholesale power costs.

pool pumps) during periods of peak loads in exchange for lowered rates on electricity. Such practices have been reported in trials (see [545]) and exhorted by consultants (see [125] and [254]), but applications of such technologies and pricing schemes is not common yet.

The RM challenges in the electricity industry will undoubtedly be distinct from traditional RM industries. RM systems will need to determine the value of futures and long-term contracts for generators, evaluate complex contract conditions (such as preemptability) and handle new forecasting requirements (weather, economic condition, price sensitivity, market-price predictions).

Fortunately, the data available in the electricity industry is quite detailed and accurate. Historical hourly demand and price information by market is often publicly available. For example, the Electric Power Research Institute (EPRI) in California sells a database (StatsBank) of observed load responses to various types of risk-managed prices in the United States and United Kingdom. EPRI also has many ongoing projects on customer behavior and responses to electricity pricing. Load forecasting is also based on weather forecasts, which are commonly available. Publicly available macroeconomic factors (aggregate inventories, GDP, income data) can also be used for long-term energy forecasting.

Though traditional RM methods are not yet common in electricity markets, there is research on applying scientific methods to optimize the market pricing decisions of electric generators and distributors. For example, Anderson and Philpott [14] describe a body of work on optimizing "offer stacks" (price-quantity bids) in electricity pool markets. (See also Day et al. [149] and Neame et al. [274].) Researchers have also looked at how to optimally release power from hydroelectric dams (see Pritchard and Zakeri [430]) in response to dynamic, uncertain market prices. This work is very much in the spirit of RM, both in terms of the technical methods employed (such as large-scale stochastic optimization) and its philosophy of using scientific, model-based approaches to quantity and price-setting decisions.

10.8 Tour Operators

Tour operators sell packages of air and ground travel, cruises, and board. Some tour operators run their own charter air services, though most contract some amount of capacity from third-party suppliers. Tour operators share some of the same RM problems encountered by airlines and hotels. Yet their RM challenges are unique as they have to manage flexible capacity and multiple types of capacity with different costs and ownership.

10.8.1 Customers, Products, and Pricing

Tour operators almost exclusively target leisure and vacation customers. Sometimes a package is organized around a group—schools, businesses, associations, etc.—but many packages are unaccompanied, aimed at individuals and families, with the tour operator selling only a package of air travel, rental car, and hotel. Packages are published in catalogs or on the Internet and are offered for repeated dates for a season or a year. Distribution is through an operator's own retail offices, via the Internet, and through travel agencies.

To the leisure traveler, buying a package offers convenience (low search costs) and a low overall price for a trip, at the expense of some loss of flexibility. For the suppliers—airlines, car rental companies, and hotels - that sell their capacity to tour operators, tour operators offer them a chance to reach a very well targeted segment of demand.

A tour operator's product is therefore a complex mix of capacities of different types, and tour operators offer a large number of such products. The products are put together by negotiations with the airlines, ground transport operators, hotels, and rental-car companies. To price a product and plan its sales operations, tour operators either purchase blocks of capacity from their suppliers at fixed prices, or they negotiate just the rates and let their suppliers control the availability of capacity.

The planning and process for the tour operator consists of three stages: (1) capacity planning, where routes and capacities are fixed tentatively; (2) a pricing and purchasing stage, where capacity is purchased from various sources and prices for the packages are fixed; and (3) a RM stage, where discounts and promotions are used to stimulate demand during the booking period.

RM for tour operators—as in the rental car industry, which also has somewhat flexible capacity—is closely integrated with capacity management, which we describe briefly next.

10.8.2 Capacity Management and Base-Price Setting

As we mentioned, large tour operators use a mix of their own fleets and third-party carriers. The goal of capacity planning is to optimize the balance between own-fleet utilization and third-party purchases. Figure 10.8 shows an example of a capacity planning exercise of a tour operator.

Setting base prices by and large is guided by expected load factors, margins, competition, costs, and the previous year's prices. Figure 10.9 gives good insight into the price-setting process at a tour operator. It is

Industry Profiles

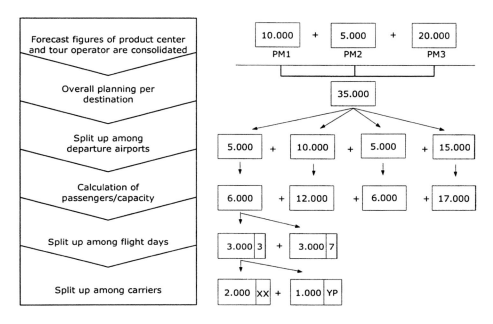

Figure 10.8. An example of capacity planning at a tour operator.

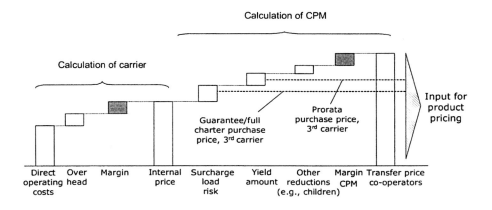

Figure 10.9. Purchase plan and price setting at a tour operator (*Source*: Modified from Remmers [439]).

essentially cost-based pricing allowing for margins and risks (load-factors as well as price-dilution risks). The planning process (purchases and price setting) is finalized around six weeks prior to the first departure. In the final five to six weeks of the booking period, a combination of

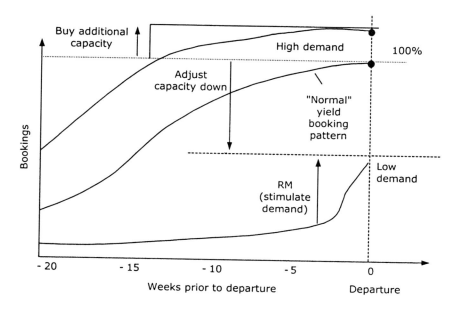

Figure 10.10. RM process for a tour operator [439].

capacity adjustments and RM is used to manage sales, which we describe next.

10.8.3 RM Practice

A tour operator can increase its revenue by (1) segmentation of its customer base (family holiday, weekend packages, exotic tours, back-packers tours, organized tours, school trips, cruise packages) marketing special packages to each segment; (2) allocating scarce capacity to more profitable packages by using capacity controls that make low-profit packages available only on unconstrained dates; and (3) using price-sensitivity estimation and dynamic pricing tools to adjust prices to stimulate demand.

Industry presentations indicate that to date only a few tour operators have implemented some form of RM [439, 557, 579, 264]. Current usage appears limited to long-term sales forecasting, booking-trends forecasting, and statistical reporting for package sales. Little optimization and control of availability appears to be used at present. The role of RM is to monitor demand during the booking period, and if demand exceeds the planned load factors, increase, or shift capacity from low-demand products, and if demand is below the planned load factors, advertise and run promotions to stimulate demand (Figure 10.10).

Part of the difficulty in implementing multiclass quantity-based RM in the tour industry is that there are no standard reservation systems for automating distribution and sales. Some industry standards are being developed specifically for the tour industry [518] by the EDI standards body Travel Technology Initiative (TTI) using XML, but these standards are not yet in widespread use.

10.9 Casinos

RM is applicable in casinos in two areas: renting out the casino's hotel rooms and managing capacity and pricing in the gaming area. We look at each in turn.

10.9.1 Customers, Products, and Pricing

The core business of a casino is gambling; renting rooms and meeting space is a secondary line of business in most casinos, whose main purpose is to attract gambling revenues. Two popular edicts in the industry are that gamers play more when they stay, and gamers play where they stay. The average daily gambling revenue from the different gamer types can range from $20 to $20,000 [422], so it is understandable that the revenue from rooms is not the highest priority for a casino. Indeed, many casinos give rooms away free to their top, "high-roller" customers.

10.9.2 RM Practice

The RM problem in casinos, therefore, is one of controlling availability based on a combination of room revenues and the amount a customer is expected to spend on the casino's gambling floor. To facilitate the latter, many casinos have instituted special loyalty-card programs for their repeat customers that are a cross between the debit cards of banks and the frequent flyer cards of airlines. These cards track how much the customer has spent on the floor. Hence, when a customer calls in to make a booking and gives his card number, the casino can assess how much revenue the customer will potentially bring in and the RM system quotes availability and price based on this information.

RM systems implement this customer value assessment through a *gaming value function*. The software recognizes and ranks repeat guests by their gambling history. Harrah's, a large U.S. casino, reports a modification in which its RM system was customized to work with their Total Rewards Player (TRP) program [92]. Guests in this program are ranked in tiers. High-rollers are identified and receive the lowest room rates, while first-time guests and nongamblers get the rack room rate. When a customer calls Harrah's for a reservation, the system automatically generates a customized room rate based on the customer segment of the caller. Harrah's has up to 64 segments and each segment has a code (such as AEP for avid experienced player [361]).

Another potential use of RM in casinos, mentioned in industry talks but not implemented yet, is to control capacity on the gaming floor itself—to determine the minimum bets at each table, and to adjust the number of tables of each game which are opened.

10.10 Cruise Ships and Ferry Lines

Even though both cruise and ferry lines have superficially similar operations (they transport people by ship), there are considerable differences between the two businesses when it comes to RM practices.

10.10.1 Customers, Products, and Prices

Cruise lines can vary from small harbor cruise boats to large international operators offering long multiport itineraries. Ferry lines operate regular scheduled transportation service to move both people and cargo. There are further subdivisions called *fast-ferry* and *cruise/ferry*.

Cruise ships are essentially floating hotels. There are a wide variety of cabin types and a large spread of fares, some of which are sold with advance-purchase restrictions. Sales are made either directly by the cruise company or through tour operators and travel agencies.

Unlike airlines and hotels, however, cruise customers are almost exclusively discretionary travelers. This has historically created some difficulties in maintaining price discipline in the industry. For example, there is a significant market for people willing to take a cruise on short notice at very low prices. Cruise lines facing an underutilized vessel are frequently tempted to lower prices and offer last minute bargain rates on some sailings. However, customers have, over time, learned of this practice and as a result are often unwilling to buy advance-purchase products.

Ferries, in contrast, have commercial, commuter, and leisure segments. Ferry traffic has more in common with the airlines or passenger railways than it does with hotels. They segment based on volume of purchases (monthly commuter passes), one-way and return purchase, peak and off-peak times, and customer type (student, child, senior citizen). Long-haul ferries also offer advance-purchase fares, and some offer private-cabin service.

10.10.2 RM Practice

Cruise lines have some special characteristics that distinguish their RM practices from hotels. First, all stays are of the same length; though some multiport cruises let customers join and leave at different ports. Also, overbooking for the ship as a whole is generally not practiced, as it is difficult to "walk" a cruise passenger; overbooking at a cabin-type level is more common.

Cruise operators also have to coordinate with airlines to bring customers to and from their various ports of departure. Cruise operators may purchase a block of seats from an airline and must therefore manage the air-travel capacity along with the ship capacity, similar to the problem faced by tour operators.

Finally, even if the cruise is all-inclusive, customers have many shopping opportunities on board, and the cruise lines consider shopping revenue in assessing the overall net revenue contribution of the different customer segments.

Ferry operators also face some unique RM challenges. For example, ferry lines have to manage passenger space and vehicle space jointly. Combination cruise/ferry lines have to manage inventories of cabins, passenger space, and vehicle space. Thus, they too are faced with multiple-resource RM problems.

Most large cruise lines, such as Norwegian Cruise Lines and Royal Caribbean, practice some form of RM for controlling sales on their ships. Smaller river and harbor cruise companies, like New York Cruise Lines, are also known to use GDSs and practice some simple RM. Ferry line RM implementation is more sporadic and less known, but there are a few firms, such as Transmediterranea, that are reported to have RM systems.

10.11 Passenger Railways

Passenger railroad RM is similar to airline or hotel RM, albeit with some differences.

10.11.1 Customers, Products, and Pricing

A large passenger railway company can have over 2,000 trains daily. Because many railways are nationalized, it is common to have only one passenger railway operating in a market. Consequently, they do not face direct price competition and have greater flexibility in pricing than do airlines or even hotels. Nevertheless, railways compete with the other modes of transportation such as airlines, automobiles, ferries, or buses,

Table 10.11. Amtrak accommodation and fare types.

Accommodation type	Fare Type
Superliner standard bedroom	Adult
Viewliner standard bedroom	Senior
Deluxe bedrooms	AAA adult
Family bedrooms	NARP
Viewliner accessible bedroom	Student Advantage
Superliner accessible bedroom	Veteran's Advantage

and hence pricing is influenced by the prices and availability of these transportation alternatives.

In addition to the prices of competing travel alternatives, pricing for trains also depends on the speed of the train, the time of operation, and the distance of travel. High-speed trains (such as the TGV in France) offer service that is comparable or better than airlines in terms of travel-time between city centers and on-board amenities and consequently are priced higher than ordinary trains.

The passenger mix varies depending on whether the route is a short-haul or long-haul and on the time of day. For example, a Washington-Philadelphia train at 6:00 PM will have a large number of business passengers, while a Washington-Chicago train carries primarily discretionary travelers. As a result, one sees multiple fares more often in short-haul markets. However, the number of fares offered by railways is typically small—two or three—with advance-purchase restrictions of five or fourteen days. For instance, Rail Canada sells seats in four classes Y, B, Q, and V with progressive advance-purchase discounts.

Long-haul trips rely more on product differentiation (type of accommodation) and identifiable customer types than on booking characteristics. Further segmentation and discounts are based on youth rail-passes, senior-citizen passes, and family packages with discounts for a family traveling together. Cancellation fees and other penalties normally apply for discount fares. Table 10.11 shows the product types and fare types for Amtrak's long-distance (overnight) train service. Prices for the accommodation part of the tariff range from $347 to $1,067 (standard versus deluxe bedroom going from Chicago to Emeryville). Dynamic pricing in the form of weekly promotions, which usually carry some sort of advance-purchase and cancellation restrictions, is becoming more prevalent.

10.11.2 RM Practice

AmTrak in the U.S, VIA Rail Canada, and Eurostar and Société Nationale des Chemins de Fer Francais (SNCF) in France are a few large passenger railways that are known to actively use RM techniques. The network structure for a railway is a cross between that of hotels and airlines. A single railway line resembles the length-of-stay network of a hotel; however, there are many connection points where passengers can switch trains, making it a more complicated network. Amtrak is reported to use five fare buckets, opening and closing them depending on demand to come (Johnston [271]), with the capacity decisions made jointly between the train or corridor manager and the central RM department.

10.12 Air Cargo

RM is a nearly universal mission-critical tool for passenger airlines, so cargo would seem a logical area for airlines to apply their RM skills. However, the use of RM in airline cargo at present is rather sporadic.

10.12.1 Customers, Products, and Pricing

Most passenger airlines accept cargo to fill the empty holds of their passenger aircraft. The bulk of the space is sold under long-term contracts to a small number of shippers, normally through a bidding process conducted only once or twice a year.

Still, the larger carriers have partially segmented their market and sell both premium (fast) and regular (slow) shipping. Ad-hoc requests and long-term contracts are some other segmentation criteria used. Restrictions or advance-purchase requirements are not prevalent at present. Pricing is primarily distance-based and the typical tariff structure does not segment the market significantly, though the potential exists to segment based on factors such as the quantity shipped and day of week.

10.12.2 RM Practice

Airline cargo RM is more complex than passenger airline RM. For one, the hold space of the carrier is limited both by the weight and volume it can carry, so shipments have to be controlled based both on their weight and volume. Also, the weight constraint of the aircraft depends on both the number of passengers carried and the cargo. Thus, the decisions for both passenger and cargo are interrelated and ideally should be coordinated by a single RM system.

The long-term nature of customer relationships also makes capacity control difficult. Air cargo carriers rarely—if ever—reject a shipment when they have available space. Long-term customer relations take priority because unlike passenger sales, which are anonymous and numerous, air cargo carriers work closely with a few important customers who ship large volumes. Still, there are occasional small-volume shippers in addition to large shippers, and some segmentation between the large, contract shippers and the smaller, "spot-market" shippers is possible.

To date, only the largest carriers practice some form of cargo RM, and even in these cases, the systems are not comparable in sophistication to the airline's passenger RM systems. Many cargo RM systems in operation do mostly simple reporting, overbooking, and aggregate forecasting. Unlike passenger systems, few make automated accept/reject or block allocation decisions.

The few airlines that have implemented or are currently implementing cargo RM are also cautious when evaluating the success of their systems. Edward R. O'Meara, senior director of cargo RM for Continental Airlines, cites the complexity of implementing RM in air cargo as one factor [483]:

> It's been a success story but it's not been a slam dunk either. It's a very difficult process because you're changing a lot of roles. I see it as an extra tool you don't use blindly. You always have to keep the customer in mind and that's the thing with RM: you don't want to do something stupid that hurts your customer.

According United Airline's Lung, the low status of air cargo in the industry is another factor [483]:

> Cargo is not as glamorous and is viewed more as a byproduct or an afterthought. Typically you have lots of investment opportunities within the airline and cargo is not viewed as your core business.

Another obstacle is data. Historical data is often collected only by weight or volume (usually weight), so many airlines don't have adequate data to implement a RM system. Legacy systems to manage cargo are also a significant barrier to implementing RM.

10.13 Freight

Shipping, trucking, railway, and intermodal companies transport billions of dollars worth of cargo every year. The emergence of global supply chains and just-in-time (JIT) manufacturing have increased the strategic importance of logistics, and companies are often willing to pay a premium to have reliable, fast deliveries. Freight transport is subject to significant capacity and time constraints. Together, these factors make freight transport a natural candidate for RM methods.

Table 10.12. Sample freight product differentiation.[a]

Exact Express™	Expedited, time-definite air and ground service with same-day, next-day and any-day service, featuring proactive notification.
Definite Delivery™	Guaranteed service with constant monitoring and proactive notification for standard transit time shipments.
Standard Ground™	Ground service.
Cross-Border™	Door-to-door transportation solutions between Canada, United States and Mexico
Global™	International via air, ocean, and land.

[a] *Source:* Yellow Freight, for year 2003.

10.13.1 Customers, Products, and Pricing

Most commercial freight today is moved in containers over one or more transport modes (ship, rail, and truck). (If the transportation involves more than one form of transport, it is called *intermodal shipping*.) Intermodal shipping offers a high level of service at a relatively low cost. It is often cheaper than moving a shipment entirely by truck, while offering comparable flexibility in routing and timing. Standard containerization also reduces transfer and handling costs considerably, as most rail yards and shipping docks today have specialized facilities for handling containerized freight.

Freight customers can have different service level requirements. For instance, package delivery companies and the U.S. Postal Service require strict service commitments and schedules tightly integrated with their own operations. Often a carrier has dedicated trucks and trains assigned to these customers. Other customers, such as *less-than-truckload (LTL) shippers* and freight forwarders, are less sensitive to delivery time and willing to accept longer delivery times and less reliable service for lower shipping charges. These time-price sensitivity differences provides a natural means for segmenting customers in the industry; a sample portfolio is shown in Table 10.12. Rates are also segmented based on the type of good being shipped, as well as by the weight and volume of each shipment. Many shippers purchase freight services through long-term contracts. The contracts require the freight company to guarantee capacity on demand. However, shippers have little obligation to use the services they contract for—or even to pay the prices agreed to in the contract. Indeed, if spot prices are cheaper, shippers commonly bypass their contract carriers and ship using the cheaper spot carriers [426]. These practices make it difficult for carriers to maintain pricing discipline.

Long-term contracts also make customer relations extremely important and—as shippers are reluctant to renegotiate long-term contracts—make RM type innovations hard to introduce. However, there is significant business from low-volume, smaller shippers, with whom the freight companies have some degree of pricing flexibility.

Tariffs for terminal-to-terminal shipping are usually different for wholesale agreement customers and retail customers and whether the shipping is domestic or international. Intermodal and rail pricing plans also differ considerably. For ocean carriers, the freight rate is normally a percentage of the value of the cargo. Ocean carriers can also be members of a conference that controls both the capacity members can introduce into the market, as well as monitors tariffs—but as the revenues depend on the cargo value, there is nothing to prevent the carrier from using RM-type controls to manage the cargo mix [360].

10.13.2 RM Practice

A few large freight companies, including Sea-Land [213] (Sea-Land is now integrated into CSX Intermodal), have implemented RM systems. However, RM is not yet widely practiced in freight, although the potential is significant. For example, low-value demand often fills up the weight limits of a ship or truck well before the volume limit is reached, forcing the carrier to reject low-volume/weight, high-value demand. Indeed, Maragos [360] reports that in the shipping industry, low-value customer demand tends to appear before the high value demand, creating conditions similar to those in the airlines.

Structurally, the RM problem in freight is similar to airline or hotel network problems, in that carriers provide a network of routes and scheduled service with capacity constraints on each link. In addition, orders are taken over a period of time, and there is a mix of customer types, ranging from high-value, low-weight items (PC components) to low-value, heavy-weight or large-volume items (raw materials). Internet websites (such as Rez1 and nacsfirst) are facilitating freight reservations.

Capacity conditions are somewhat different, however. For example, in the case of rail freight, there is considerable flexibility in adding or removing capacity to a train because the cars or car blocks are easy to add and remove. The capacity of the locomotive at some point prevents such additions, but the range is significant. Thus, at present it is rare for a railroad to reject a shipment because of capacity constraints. Trucking firms are also able to adjust capacity on routes by reallocating tractors and drivers. Ships, in contrast, have more of a hard capacity constraint, analogous to an airline flight.

10.14 Theaters and Sporting Events

Theaters and sporting events have many characteristics that seem well suited to RM methods. Indeed, the existence of "scalpers"—and the often exorbitant prices they charge for tickets for popular shows and events—is evidence of the demand management potential in this industry. Also, many firms are experimenting with auctions and dynamic pricing, driven mainly by the Internet.

10.14.1 Customers, Products, and Pricing

Tickets for events are purchased in advance or on location at the time of the event. There are many different customer segments—corporations, annual subscription customers, families, tourists—each with varying usage patterns and willingness to pay. There are few regulations on the prices that event sponsors can charge or on the sale conditions, with the exception of some local municipal laws against scalping.

Ticket prices for theaters and sporting events depend on factors such as the location relative to the stage, the expected demand for the event, group affiliation of customers, seasonalities, bulk-sales terms and advance-purchase restrictions—besides the draw of the performance itself. Table 10.13 gives an example of the different rate categories for a Broadway show [340].

Demand can be highly variable for events, and while historical data is sometimes available to make forecasts, there is often considerable uncertainty about the popularity of a new show or a particular sports event. For example, the success of a local sports team or the presence of a star player has a significant impact on attendance. For instance, A.T. Kearney [23] reports an Atlanta Hawks game with an overflow of 20,772 one day, and an attendance of 8,772 two days later—Michael Jordan played the first day; both days had the same prices with no capacity controls, a clear inefficiency in pricing.

The number of seats in the venue, of course, strictly limits capacity. However, there are different categories of seats based on location in the venue. In terms of the sales process, prices are sometimes published a year ahead (as in the case of opera or concert halls).

10.14.2 Ticket Scalping and Distribution

Unlike airline or hotel products, event tickets are almost always transferable (even many subscription tickets for opera/sports events can be passed on to others). A scalper ticket market shows both a missed opportunity to increase revenues for event sponsors as well as a potential

Table 10.13. An example of ticket categories for a broadway show *Seven Guitars*.[a]

Full-Price Sales	
Orchestra (543 seats)	The best seats in the house are located in the orchestra. By convention, for all Broadway shows, all full-price ; orchestra seats in a given theater have the same price, although different theaters may have different prices. This category is always the highest price category, and the price varies with the time of the week that the performance is held. Note however that many discount-price buyers will in fact be seated in the orchestra.
Front mezzanine (196 seats)	The best seats in the front mezzanine are preferable to some orchestra seats. For *Seven Guitars*, the front mezzanine is priced identically to the orchestra.
Rear mezzanine (126 seats)	For the first 16 weeks of performances, no distinction is made between the front and rear mezzanine. After week 16, the rear mezzanine price was lower than the front mezzanine price, which remained at the previous level.
Balcony (66 seats)	This is the least favorable seating in the house and also the cheapest, cheaper even than standing room.
Boxes (16 seats)	While the boxes are priced identically to the orchestra, the relative average seat quality of the boxes is unclear.
Standing room	The highest number of attendances, in a single performance, in this category is 13. It is possible that individuals purchase these tickets and then sit in vacant orchestra seats.

[a] *Source*: Leslie [340].

impediment to the practice of RM. We examine both aspects in this section, beginning with a brief look at the legality of ticket scalping.

10.14.2.1 Scalping Laws

Somewhat surprisingly, most legislation against scalping is at the state or municipal level, so there are many variations in the law. In the United States around half of the states have either no laws or give jurisdiction to municipalities; the other half have strict antiscalping laws. Most of the states that limit ticket resales (either a blanket law, or by type of event) specify that tickets cannot be resold above face value or face value plus a small amount.[9]

---

[9] *Source:* eBay Event Tickets Resale Policy.

Table 10.13 (continued)
An example of ticket categories for a broadway show *Seven Guitars*.[a]

Discount-Price Sales	
10% off	A catch-all category for 10% discounts given for an array of small-scale marketing initiatives.
Two-fer-one	Two tickets for the price of one. This is a marketing initiative during two disjoint time-periods. Coupons are available from locations such as dry cleaners and student cafeterias.
TKTS	50% discount for day of performance sales, with a $2.50 service charge to the buyer. Roughly 90% of TKTS sales are through the Times Square booth, where buyers must wait in an outdoor line to make their purchase.
MTC	The Manhattan Theater Club (MTC) is a subscriber organization that reached an agreement with the producer of *Seven Guitars* to give a discount to MTC members if they saw the show before the end of week 13.
AENY	Arts Entertainment of New York (AENY) is an organization that specializes in obtaining high-quality seats for various forms of live entertainment for its members. The producer of *Seven Guitars* gave AENY a 10% discount, but this is not passed on to the members who purchase the tickets. This category was only available during the final 7 weeks and never accounted for more than about 1% of the attendance in any given performance.
Direct mail	There were three direct-mail periods that are distinguished by pricing and the mailing list used.
Group	The group discount varies with the size of the group and is generally at least 10%.
Student	Approximately 50% discount for students.
TDF	The Theater Development Fund, a nonprofit company that also operates the TKTS booth, purchases tickets at a large discount operates (often around 70%) for use in various programs with an educational or audience-development emphasis.
Wheelchair	Approximately 50% discount for individuals in a wheelchair.
COMPL	Tickets given away for free.

[a] *Source*: Leslie [340].

The public reaction to scalping is generally not against individuals selling their tickets (on eBay or in newspaper ads) but against professional scalpers buying up large blocks of seats in anticipation of a popular event and also the nuisance caused by the presence of scalpers in front of event venues. To prevent scalping, many events in fact restrict the number of tickets they sell to each individual, though this is not always enforceable.

From an economics point of view, there is no reason to prevent a spot market just before the event time. There is also no reason to believe that scalpers will always charge a premium, as they do take a risk in buying up tickets ex ante. Indeed, Happel and Jennings [236] report the case of the Phoenix City Council, which allowed scalping by law but restricted scalpers to selling in a centralized lot in front of the Phoenix Suns stadium. The prices obtained for tickets went down as the event time approached, similar to markdown dynamic pricing. An econometric study by Williams [567] found that NFL teams in states with anti-scalping laws charged lower prices. His explanation is that secondary markets provide valuable information to event organizers if the event is underpriced (if it is overpriced, they know it themselves), allowing them to raise prices. So antiscalping laws may in fact hurt more than help event sponsors.

10.14.2.2 Primary Sales and Ticket Distribution

In contrast to ticket resales, there are few restrictions on the primary sale of tickets. (Primary sales are sales by the artist, team, promoter, or organizer of the event.) So in theory, there is little to prevent event organizers themselves from conducting auctions or using dynamic pricing if the public finds this acceptable.

At present most primary ticket sales in the United States are sold either by subscription, through an event website or box office, or through one of the electronic ticketing agents like TicketMaster (which bought up an earlier rival Ticketron). There are many Internet-only sellers such as Tickets.com and Ticketmall.com.

In addition, there are the ticket brokers. The U.S. National Association of Ticket Brokers, which boasts over 150 member firms, defines their role as to (1) provide tickets to events that are sold out through the primary market, (2) provide premium upfront seats which are the most desirable, and (3) to provide ticket holders a place to sell their unwanted or extra tickets.

There are a few auction marketplaces for event tickets currently, but as distribution moves increasingly to the Internet they are likely to gain in popularity. For instance, on a recent day eBay listed over 30,000 items in their Tickets category (including both primary and secondary sales). While this number is minuscule compared to what Ticketmaster sells on a given day (they recently *sold* more than a million tickets on a single day), it is growing. A recent *New York Times* article [157] reports that Ticketmaster is experimenting with online auctions and used them recently for a boxing match. Another online ticketing firm, StubHub, works with artists and entertainers to auction off front-row seats for charity.

10.14.2.3 Secondary Markets

Why don't event organizers create their own resale markets and extract the money that is now going to the scalpers? Some have. *USA Today* reports [251] that eight Major League Baseball teams have started online programs to facilitate the resale of seats, though the websites are only to facilitate sales, and they do not yet conduct auctions. In these systems, the holder of the ticket posts it for sale at a virtual exchange window. When he finds a buyer, the original ticket's bar code is invalidated and a ticket with a new bar code is created. This removes one of the main consumer risks in the scalper market—fraud.

The biggest impediment to dynamic pricing seems to be the fear of negative consumer reactions—the fear that along with the scalper's money, event organizers may also acquire the scalper's reputation. There is a concern that this could cause long-run damage that more than offsets any short-run boost in revenues. Indeed, in many cases, the ticket revenue for a single event is a small fraction of the total lifetime value of a customer. A performing artist, for instance, normally makes more money from album sales than from ticket sales, and a loss of fan goodwill could jeopardize future album sales. Sold out concerts also generate good publicity. Similarly, opera houses form long-term relationships with their clients and do not want to risk losing that patronage. For many sports teams, secondary spending in the arena is as important as the ticket revenue, so they would rather make sure *someone* gets in by facilitating the exchange of unused tickets than make money on the transaction per se. According to a survey on Fan Cost Index [440], an average major league baseball ticket costs $18.69, but a family of four spends an average of $148.66 for a single game, on souvenirs, snacks, drinks, and program. Yet none of this precludes some innovative applications of RM, as we discuss next.

10.14.3 RM Practice

Many microeconomic text books cite movie-ticket discounts for students and seniors as examples of third-degree price discrimination and discounts for midweek shows and matinees as examples of peak-load pricing. As can be seen from Table 10.13, for this Broadway event there are many rate categories with different prices even for the same inventory type (akin to airline fare classes in a cabin) and a price structure in the form of discounts off the top rate for the seating area (akin to percentage off the rack rate for hotels). Notice however that most of the rate categories in Table 10.13 are based on identifiable customer characteristics (third-degree price discrimination), so there is a fundamental difference between the segmentation as practiced in event ticketing and traditional airline RM.

Nevertheless, whenever the demand exceeds capacity for an event space, there is a need to manage the capacity intelligently. Not many venues manage their discounted demand in any systematic, scientific way. So although at first glance this seems like an industry tailor-made for RM, there are relatively few reported implementations in event ticketing. The few implementations include opera houses (San Francisco Opera [479], Washington Opera [137]), Internet event sellers (Tickets.com, [338]), sports teams (Mariners [22], Mets [454])—and RM has been proposed for movies [409]. However, even these reports point to somewhat tentative and limited implementations in this sector. Fear of negative customer reactions and consequent loss of customer goodwill are the main reasons firms seem to be avoiding bolder demand management strategies.

The Washington Opera is one of the pioneers of RM in the opera business. In 1994–1995 it initiated a rate plan consisting of nine categories based on the location of the seats (Table 10.14 shows the 2003–2004 season prices). Figure 10.11 shows the physical layout of the various categories. In addition, specially priced tickets and group tickets are also sold, the former at reduced prices, and the latter at normal prices. Subscription tickets are sold the earliest, with those subscribing to more shows given first priority. Next come group sales, around three months prior to the season beginning. After groups are booked, the box office opens for individual sales. Finally, excess demand is sold on the day of the performance at student or senior discounts or standing-only. Rather than lowering prices for low-demand shows (thereby upsetting the subscription customers), the preferred tactic is bundling where individual tickets to popular shows are sold only bundled with low-demand shows. So even though prices are fixed *and* there are no capacity controls, there is demand management going on—giving preference to customers who bring high value (lifetime value of the customer) and creating and pricing ticket bundles to smooth out demand and increase sales.

In baseball, the Mets pricing plan, shown in Table 10.15, varies prices by game rather than by segmenting customers and controlling capacity for each game. So it is more in the spirit of peak-load pricing than tactical RM. Other baseball teams, such as the Cubs, Yankees, and Giants have also implemented similar schemes, in which prices vary by game. Notice that significant efficiencies are lost, as prices are set ex ante, once, for the whole season, and there is little scope for demand management

Table 10.14. Washington Opera Kennedy Center pricing (2003–2004 season).

	Weeknights	Fri. and Sat. Evening, Sun. Matinee
Box	$260	$285
Premium orchestra	$160	$185
Prime orchestra	$135	$165
Orchestra	$117	$123
Rear orchestra	$90	$95
1st tier prime	$135	$165
1st tier	$93	$98
2nd tier 1	$65	$68
2nd tier 2	$41	$42

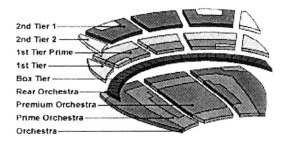

Figure 10.11. Washington Opera Kennedy Center layout.

on per-game basis. This can be considered a preliminary, experimental, step toward a full-fledged RM, and is a significant improvement over uniform pricing. However, the Mets' management ruled out, for now, the possibility of raising and lowering prices for individual games based on their popularity [454]. Theater and sporting events present some unique challenges for RM. For example, as mentioned, the popularity of an event or the success of a sports team must be factored into the forecasts of demand, which makes for a challenging forecasting. Also, seating in theaters suffers from bin-packing effects; having isolated empty seats in different locations is not appealing to couples or groups that want to sit together, so sales often drop off as a venue fills even though there are nominally many available seats. Assigning groups to seats and managing the configuration of available blocks of seats therefore becomes an important issue.

Table 10.15. The Mets four-tier pricing plan (year 2002).[a]

Gold Plan	Covers 17 dates, including opening day against the Chicago Cubs, weekend series against Atlanta, Seattle, the Yankees, and St. Louis, and a midweek series against San Francisco and Barry Bonds. The peak ticket price of $53, for inner-field boxes, is $10 more than it was in the previous season. The cheap seats rise to $16 from $12.
Silver Plan	For 21 games. Includes weekend series against Arizona, Philadelphia, Cincinnati, and Colorado and two midweek series against Atlanta. The top price rises $5 to $48, and the lowest price goes up $2 to $14.
Bronze Plan	For 27 dates, preserves the most expensive seat at $43 and the cheapest one at $12, for midweek games against Chicago, Los Angeles, Florida, Montreal, and Milwaukee and weekend games against Montreal and San Diego.
Value Plan	For 16 games, discounts the top seat to $38 and the least expensive one to $8, for midweek series against Houston, Philadelphia, Milwaukee, Florida, and Pittsburgh.

[a] Source: Sandomir [454].

10.15 Manufacturing

There is significant interest in applying RM in manufacturing. However, it is fair to say that to date there have been relatively few implementations. But manufacturing is clearly a vast sector of the economy and many SCM and ERP technology vendors are starting to offer pricing-optimization systems for manufacturers. It therefore warrants careful attention.

10.15.1 Customers, Products, and Pricing

Manufacturing spans a vast and diverse set of firms, so our discussion here is, of necessity, somewhat generic. A make-to-stock (MTS) manufacturer produce standardized products, typically in large volumes, based on forecasts of future demand. Most consumer goods (autos, electronics, food products, apparel) fall in this category. A key challenge in MTS manufacturers is to balance the need to meet demand, which is often variable and uncertain, against inventory and production costs. For most MTS manufacturers, pricing tends to be an aggregate decision and dynamic pricing is not routinely used to manage supply and demand. However, end-of-life-cycle discounting is a common practice in many sectors (such as automotive manufacturer rebates at the end of the model year). Also, trade promotions—discounts given by manufacturers to retailers and distributors—are a long-established practice among many MTS manufacturers. We mentioned one implementation of promotion and incentive optimization at Ford Motor Company earlier in Section 5.1.2 [135].

A make-to-order (MTO) manufacturer generally produces in smaller lots based on specific orders from end customers, who are often other manufacturers. MTO firms typically have to price a continuous stream of bids and request for quotes (RFQ). Most of these pricing decisions are made manually based on tactical factors, such as estimated costs, as well as strategic factors, such as the value of a long-term relationship with a buyer. Cost calculations are critical in MTO manufacturing. Such calculations are based on estimates of material costs, machine time, and labor rates, most often provided by a management accounting system. A variety of methods, such as activities-based costing (ABC), are used to arrive at these cost estimates. The volume of such request is high; a pricing department at a large manufacturing company may have to respond to over 250,000 RFQs every year.

Once a bid is accepted, the order is then scheduled into the firm's production planning and SCM system. SCM systems optimize the scheduling of current and new orders. Traditionally, they do not consider price explicitly as a mechanism to regulate orders, nor do they use price incentives to shift customer demand away from peak-load periods. Meeting the delivery deadlines at the lowest possible cost is the core objective of most SCM systems today. However, as mentioned above, many SCM vendors are currently working on incorporating price-optimization and demand-management functions into their systems, and some already offer pricing-optimization products. (See, for example, [262].)

10.15.2 RM Practice

Because variable costs and production planning are so important in manufacturing, manufacturing RM systems must coordinate their data and decisions with ERP, SCM, and management-accounting systems. Revenue has to be balanced with cost considerations in determining the profitability of accepting bids or adjusting prices. The interdependence of SCM and RM decisions is well recognized as the main reason why SCM vendors are leading the development of RM methods in the manufacturing sector.

Manufacturing RM differs from service RM in other important ways as well. For example, although idle capacity can be considered a perishable item, physical inventory of parts and raw materials can be stored for future use. This gives manufacturers more production flexibility. On the other hand, in high-tech manufacturing raw materials and parts lose valuable rapidly, so there is some degree of perishability. Another difference is that a manufacturing order need not be rejected outright; rather, it can be delayed, much as a bumped passenger on an overbooked flight can be flown on a later flight. As a result, RM can potentially be used for demand smoothing, delaying production of low-value demand to off-peak times, while ensuring prompt production and delivery of high-value demand during peak times.

The dynamic pricing tactic of choice for MTS manufacturers has been trade promotions. This is when the distributors and retailers are given a discount (which may or may not be passed on to consumers) if they buy a certain quantity or if they run a promotion special.

Applications of RM among MTO manufacturers are growing, with several vendors specializing in the area. Product configuration software tools (such as the ones used to order a PC online) incorporate pricing information and perform some segmentation (large business, small business, government, and so on), though most of the technology is rules-based. A form of RM where discounts are given based on advance purchase, lead-time, and delivery-time flexibility have been proposed in the literature, but we know of no major implementations at this stage. Applications of configuration-based pricing are reported in the trade literature, but few details have been published.

Finally, dynamic pricing, in the form of Internet auctions, has also had a significant impact in manufacturing recently. Excess inventory is now routinely auctioned off on the Internet, and procurement departments often source supplies (at least in-direct materials) from B2B exchanges. RM systems are likely to play a role in managing these surplus inventory auctions in the future.

10.16 Notes and Sources

The main source for the airline pricing description are a variety of ATPCO Rules and Footnotes documents. The discussion on airline operations draws on Barnhart and Talluri [28].

For a description of hotel operations, see Kimes [299, 300], Hanks, Cross and Noland [234], Orkin [411], Varini et al. [532], Burns [96], and Bitran and Mondschein [72]. Hadjinicola and Panayi [233] is one of the few papers that explicitly touches on hotel overbooking and tour-operator agreements. Pinchuk [422] gives a description of RM strategy for casinos, and Ladany and Arbel [319] for cruise lines.

Fellman [184] and Kuyumcu and Higbie [317] discuss media RM implementations. Our discussion of the television ad markets comes from Rust [451]. Bollapragada et al. [83] describe RM implementation at NBC.

Kasilingam [295] and Kasilingam and Hendricks [294] discuss cargo models and an implementation of cargo RM at American Airlines. The thesis of Maragos [360] advocates RM for ocean carriers, Campbell [97]

Industry Profiles 577

for intermodal, and a McKinsey article by Pompeo [426] argues the case for the freight industry in general.

In addition to the sources mentioned in the text on event pricing, Courty [133] provides a good survey of ticket pricing in the entertainment industry.

The two main sources for our description of rental car RM practice comes from Geraghty and Johnson [208], and Carroll and Grimes [100], describing RM implementations at National and Hertz respectively.

Passenger railway RM descriptions can be found in Kraft and Srikar [312], Di Pillo, Lucidi, and Palagi [158] and Ciancimino, Inzerillo, Lucidi, and Palagi [118].

Retail RM system information has been culled from vendor literature and also from Girard [214], Johnson [270] and Mantrala and Rao [359]. For background on the industry, see Leamon [333], Standard and Poor industry surveys [485], Subrahmanyan and Shoemaker [493] and Heching, Gallego, and van Ryzin [247].

For the natural-gas industry and RM descriptions, we have consulted Homes [255], Valkov and Secomandi [525] (the example comes from there), and Anthony and Harrington [20], and the tariff examples come from the Tennessee Gas Pipeline published tariffs for 2003. We consulted Hunt and Shuttleworth [261] and Wilson [569] for the electricity industry competition structure. See also Simth [481], Hirst [254], Wald [545], Colledge, Hicks, Robb, and Wagle [125] and Denton [153].

Harris and Pinder [238] describe an application of variable pricing in manufacturing at a repair facility, and Kay [297] describes dynamic pricing at Boise Cascade Office Products (retail) and Campbell Soups (manufacturing). The operations research literature has concentrated on models for joint inventory and pricing in manufacturing. (See the references at the end of Chapter 5.) Elimam and Dodin [176] and Kalyan [284] give examples of segmentation and RM applications from a ready-mix concrete plant and high-tech component procurement. See also Gray [219] for the potential of RM in manufacturing.

In addition to the industries mentioned in this chapter, RM applications have also been mentioned (although we are not aware of many implementations) in the following industries: bandwidth (Martin [363]), restaurants (Kimes, Barrash, and Alexander [298], and Bertsimas and Shioda [62]), golf courses (Kimes [303]), health care at the Duke University Diet and Fitness Center (DFC) (Chapman [106]), the nonprofit sector (Metters and Vargas [380]), and ISPs (Nair, Bapna and Brine [401]).

Chapter 11

IMPLEMENTATION

While models and science form the core of a revenue management system, in practice RM success hinges critically on the quality of implementation. Implementation involves more than software and hardware, the most visible aspects of the system. It requires designing products, aligning incentives and organizational structures, changing business processes to support the firm's RM objectives, and training employees properly. In this chapter, we examine these and other RM system-implementation issues.

11.1 Segmentation and Product Design

Segmentation and product design are the first steps in a RM implementation. *Segmentation* is the process of classifying customers into groups (segments) based on observed—or inferred—characteristics, behaviors, and preferences. The objective of segmentation is to understand who is buying the product, how they buy, what attributes they value (and don't value), and what price they are willing to pay—and then to classify them into groups based on these characteristics. Segmentation is then followed by *product design*, the objective of which is to construct bundles of product/service features to target each customer segment. Correlating customers' willingness to pay with their preferences and purchase behavior is the key to good product design.

Segmentation and product design are not operational processes, in the sense that they are seldom done on a day-to-day, routine basis (although there is scope for some fine-tuning of products after they are introduced). This affords more time for off-line analysis, but it also means that design should be robust and sustainable for a reasonably long period.

While both customer segmentation and product design are complex, data-intensive processes and therefore well suited to analytical methods, in practice they are still based primarily on managerial judgment and intuition. Yet RM managers are beginning to test analytical tools to support product design and to understand customer segments and preferences. Here we describe a few simple models and one such tool, conjoint analysis.

11.1.1 Segmentation

As mentioned, the goals of a segmentation analysis are to understand which customers are buying, how they buy, what they value, and how much they are willing to pay. Its aim is to uncover correlations between willingness to pay and segment characteristics and to exploit this segment behavior in some practical way that increases revenues. Toward this end, the following six criteria are widely used to evaluate a segmentation strategy [193, 558]:

- **Identifiability** Is it possible to identify customers as belonging to a segment, either prepurchase or postpurchase? If not, the segmentation, though perhaps conceptually valid, cannot be operationalized.

- **Substantiality** How large is the segment? If segments are too small, the costs of segmentation may not justify the benefits, though for some online channels the cost of customizing can be quite low.

- **Reachability** Can a segment be targeted by marketing techniques or product design? Or can the segment be induced to self-select their targeted product? If not, it may be impossible to reach the segment.

- **Stability** Do the segments change rapidly over time? If they do, it may be difficult to identify and estimate the characteristics of the segment. Stability is also necessary if a firm needs to design relatively static products.[1]

- **Responsiveness** Do customers in the same segment respond similarly to a product or marketing campaign? That is, are customers in each segment approximately homogenous in terms of their preferences and market response? If not, the resulting response from segmentation my be unpredictable or ineffective.

---

[1] Note that stability is not the same as saying a particular customer's behavior does not change: A customer who travels for both business and pleasure is one whose behavior changes depending on the occasion of purchase. However, his behavior fits either that of a business segment or leisure segment, and the behavior of the segment itself does not change.

Table 11.1. Customer segments and subsegments by industry.

Airline	Business, leisure, groups, negotiated
	Students, children, youth, seniors, military
Hotel	Leisure, conventions, meetings, contract accounts, single, double occupancy
	Location, facilities (chains)
Advertising	Upfront, scatter
Freight	Express, heavy weight, volume
Energy	Small, medium, Large power demand
	Office, retail, restaurant, grocery, school, lodging,
	Guaranteed, controlled lighting, space heating

- **Actionability** Is it feasible to price or market differently for each segment? That is, can we base product design or marketing decisions on the segmentation? Is the segmentation helpful in terms of suggesting practical ways to target customers?

A particular segmentation strategy should be evaluated along each of these criteria. For instance, consider a segmentation of families with more than two children. This segmentation might be difficult to reach in a direct-marketing campaign to households, but is reachable when selling a travel or hospitality product, for example, by giving a discount for three or more children. In the latter case, the segment is identifiable, substantial, reachable, and actionable. However, it may not necessarily prove to be responsive if families of size five or more do not have different price sensitivity than other customers.

11.1.1.1 Segmentation Bases for RM

A *segmentation basis* is the set of product attributes or customer characteristics that define a segment. For instance, the time of booking is a segmentation basis for many RM applications: a customer who books 21 days in advance is classified as a leisure customer (or more precisely, *likely* to be a leisure customer).

There could be more bases than there are segments. To give an example, a hotel could define only two broad customer segments (business and leisure) but use a number of different bases in combination (for instance, week day, time of booking, source of booking) to identify the customer with different degrees of certainty.

Note that some bases are observable (for instance, zip codes because of store location), and some are not (say, family size) and have to be inferred from data such as sales transactions, observed itinerary, and panel data. Table 11.2 gives a classification of segment bases as observable or unobservable and as customer or product-specific. For any basis that is

Table 11.2. Classification of segment bases.[a]

	Customer-Specific	Product-Specific
Observable	Geographic	Usage frequency
	Demographics	Store location
	Corporate affiliation	Purchase behavior
	Group booking	Channel
Unobservable	Values	Elasticities
	Life-style	Preferences
	Income	Reason for purchase

[a] *Source:* Frank, Massy and Wind [193].

observable, one can distinguish it further depending on whether it can be observed at the time of purchase or after purchase. For instance, a casino may classify some customers as high-rollers if they spend a lot on gambling, but the casino may not know this information at the time of purchase unless they used a separate mechanism to track customers (such as a loyalty card). Table 11.3 lists some common segmentation bases used in RM and their characteristics and purpose.

Quantity-based RM segmentation has some unique features: customers preferences are not necessarily based on the product itself[2] but rather on the conditions of purchase (such as advance-purchase restrictions, nonrefundability). In fact, one hesitates to use the term *preferences* in terms of these restrictions because, given the option, everyone would unequivocally prefer not to have restrictive conditions on purchases at all. Rather, some customers simply cannot meet certain purchase conditions while other customers can, or are willing to meet them in exchange for a lower price. The conditions of purchase can include delivery time (logistics, manufacturing), preemptability (advertising), and nonrefundability (airlines), among others.

11.1.1.2 Segmentation Mechanism

There are two basic approaches to constructing a segmentation mechanism. The first is to use an explicit *screening mechanism* based on observable characteristics—to restrict products and pricing to individuals based on their observed "types." Example of such mechanisms include age-based segments (child's and senior citizen's prices), geographical segments (zonal pricing), and group affiliation (corporate discounts); these are all mechanisms that use observable segment bases listed in

---

[2] For service industries, like hotels or airlines, we ignore minor differences in the product (such as view for hotel rooms and legroom for airline seats).

Implementation

Table 11.3. Some common segment bases used in RM.

Basis	Comments
Time of purchase	Used in the airline, media, retail, cruise line, and natural-gas industries. Separates low- and high-valuation customers depending on the value of buying early or late.
Time of reservation	Need not be the same as time of purchase. One can reserve early and purchase by a deadline. The firm takes the risk that you would not purchase by the deadline but cancel. Hotels, Airlines let you do this.
Day of week	Hotel, energy, and airline industries have peak demand during week-days. Weekends and weekdays also separate leisure from business customers.
Cancellation likelihood	In reservation-based industries, the service is used in the future. Customers differ in their uncertainty that they will use the service. This uncertainty is sometimes correlated with willingness to pay (as the case of business customers). As this is private information, customers are asked to self-select between a penalty for cancellation or a high price.
Senior and youth	Used in airline, movie theater, railways, and hotel industries. Directly observable and correlated with lower income and less willingness to pay.
Options and premptability	For instance, in media advertising, budgets may not be fixed when a firm buys ad space. So the firm buys options (similar to reservation without purchase). Premptability is also a common practice in the electricity, gas, and manufacturing industries. Customer gets a cheaper rate, but customer takes the risk that demand exceeds capacity and he will get preempted. Similar to airline standbys. Airlines are also often suspected that they use overbooking to preempt lower-paying customers, but this is not the case.
Channel	Easily identifiable. Different channels attract different types of customers. Far-away outlet stores separate low-search cost, low-valuation customers from the high-valuation customers. Sometimes, a channel offers the firm different informational content on the customers (Internet) and sale possibilities, allowing it to segment. For instance, last minute sales or auctions by airlines; customized pricing by e-tailers.
Trip length and length of Stay	Used by airlines, hotels, and rental-car companies. Longer stays (more than six days) or trip length signifies leisure customers.
Saturday-night stay	Businessmen prefer to return home by Friday night. So giving a discount only for trips that include a Saturday night stay will preclude businessmen purchasing at that rate.

Table 11.3 (continued)
Some common segment bases used in RM.

Basis	Comments
Group discounts	Groups are generally considered leisure customers. Beyond the segmentation aspect, it is volume discounting as the firm lessens the risk of unused inventory by selling a large block.
Package	Package holidays (e.g., airline, hotel room, and car) combined with some trip restrictions limit the product to leisure customers.
Business and individual	Used in retail, telecommunication, and energy industries. Identifiable at the time of the contract.
Size of business	Used in retail (e.g., PCs). Segmented as small, medium and large. The service requirements and sales effort vary by size of the client, so the firm can customize the product (or discounts).
Spend amount	Casinos and hotels track customer spend on food and beverages, gambling, and other services. Discounts may be tailored based on this quantity.
Loyalty	Repeat customers have a higher lifetime value for the firm. By using store discount cards or frequent-flyer cards, customers can be separated based on their loyalty.
Frequency	Frequent customers are not only loyal; they also provide more information about their preferences. Based on past purchasing habits and the frequency of purchases, the firm can separate frequent buyers from infrequent buyers.
Delivery time	Used in manufacturing, freight and package Delivery Industries. Customers with express orders are willing to pay more. The value of the service to the customer in a rush is much higher than normal. This is a easily implementable segmentation in most cases.

Table 11.2. Such segmentation is equivalent to third-degree price discrimination as discussed in Section 8.3.3.1.

The second approach is to use *self-selection segmentation*. This is necessary if a firm cannot observe or control which segment buys which product. It must attempt to induce customers to self-select the product targeted at them, which is the essence of second-degree price discrimination as discussed in Section 8.3.3.1. To give an everyday example, in the consumer packaged-goods industry, price-sensitive customers *choose* to use coupons more; and price-sensitive customers with large families *choose* to buy in bulk to get savings. When sales are anonymous or through third parties, then segmentation by self-selection is often the only viable alternative for a firm.

Using an explicit sorting mechanism to segment is relatively easy from an implementation standpoint, though one may face legal and customer acceptance issues in the process. With self-selection, legality is rarely an issue, but the firm risks a large amount of dilution—many customers in one segment ending up purchasing products not designed for them. Indeed, there is little a firm can do to prevent a rich consumer cutting out coupons or a business traveler staying over a Saturday night to get cheaper airfare. Still, the segmentation is considered successful if a *sufficient* number in each segment respond the way the firm had envisioned (substantiability). It is important, therefore, that the segment not be defined too narrowly as customer behavior is hard to predict except in an aggregate probabilistic sense. Most RM segmentation is of the self-selection type.

11.1.2 Product Design

Product design is the flip side of segmentation—"differentiating" the products to target the identified segments, with the idea of charging more for products targeted at customer segments with higher willingness to pay. As is the case for customer segmentation, there are very few models and analysis techniques currently used for RM product design. We present one methodology, conjoint analysis, that has found some success as a tool for physical product design and can conceivably used to design RM products also.

11.1.2.1 Product Design Using Conjoint Analysis

Conjoint analysis originated from statistical work by Luce and Tukey [348] and is widely used in marketing [224, 574]. Its role has expanded, from an initial positive goal of multiattribute utility measurement, to more normative uses such as new product design, segmentation, product positioning, and even pricing.

Many market-research firms now offer conjoint analysis as a service and commercial PC and Web-based software (Sawtooth Software, SAS) has become widely available to support it.

Table 11.4. Attributes and their levels for a hotel application.[a]

Levels	1	2	3	4
Breakfast	Premium	Standard	None	
Access to fitness	Fully equipped	Basic	None	
In-room tea- and coffee-making facilities	Yes	None		
Complimentary fresh fruit and bottled water in the room	Yes	None		
Bathroom amenities	A luxury selection of products	A selection of products	A single dispenser	
Complimentary pay TV	Yes	No		
Price: comparison to price paid on previous hotel stay (business travelers only)	−$15	Actual price	+$15	+$30
In-room computer equipment with free Internet and printer access (leisure travelers only)	Yes	No		
View from the room	Unique	Nice	None	

[a]Source: Varini et al. [532].

Conjoint studies are numerous in the CPG and automobile industries, and applications of conjoint analysis to design travel and hospitality industry facilities, products and amenities have been reported in [570, 102, 573, 574, 532, 216], but we know of no applications of conjoint analysis for quantity-based RM product "restriction" design as such. A second caveat is that current market environment often guides product design: what products the competition is providing and what prices they are charging. This presents an especially vexing problem for RM because (available) products and prices can change rapidly, making comparisons difficult. Barring these two caveats, the methodology has the potential to be useful in making rational product-design decisions.

The customer-behavior model of conjoint analysis is similar to the discrete-choice models discussed in Section 7.2.2, both having roots in microeconomic theories of preferences. There are M possible attributes of a product, and the firm can choose the *level* of attribute j in designing its product. Table 11.4 gives an example of the attributes and their levels for a hotel product. The choice of attributes and their levels is usually a result of management judgment or analysis of a survey among customers. It is advisable to be parsimonious and list only the most important attributes and the most reasonable range of their values.

A *profile* is a particular combination of the M attribute levels that can make up a potential product, $\mathbf{x} = (x_1, x_2, \ldots, x_M)$. Customers form a utility for a product that has the combination of attribute levels $\mathbf{x}$. The most common model[3] used for this utility formation is an additive composition of the *part-worths*, $u_j(\cdot)$ of attribute j:

$$U(x_1, x_2, \ldots, x_M) = \sum_{j=1}^{M} w_j u_j(x_j).$$

One can simplify this further and assume linear part-worth functions and a finite set of profiles. So if attribute k is assumed to have one of M_k levels, the utility of a customer i for a profile p is hypothesized to be

$$U_{ip} = \sum_{j=1}^{M} \sum_{k=1}^{M_k} u_{ijk} d_{jk}^{p}. \tag{11.1}$$

d_{jk}^{p} is an indicator variable equal to 1 if attribute j is at level k in the profile p and 0 otherwise, and u_{ijk} is the part-worth for customer i for the attribute j at level k. We would like to estimate these part-worth

---
[3] Other, more complicated, utility functions have been proposed, but we stick to this popular one for ease of exposition.

functions, so that the utility functions can be used in an optimization model for segmentation, product design or pricing.

The motivation behind conjoint analysis is the following. First, asking some sample customers to directly map their part-worths (say, for price) may be meaningless as they are convolved with all the other attributes. As an alternative, we can ask respondents to compare a discrete number of profiles, but the number of such profiles can quickly grow large even for a small number of attributes and levels. For instance, if there are five attributes, each with five different possible levels, then the number of possible profiles is 5^5, an immense number to test out in a survey. Even if many of these combinations can be eliminated as being unreasonable, we would still typically be left with too many profiles to test, since survey participants are unlikely to respond reliably to more than 5 to 10 rating questions at a time. To overcome this problem, conjoint surveys use experimental design techniques to construct an orthogonal subset of profiles for each respondent that is parsimonious yet ensures the results are statistically significant.

Next, even if presented a profile, most respondents are not able to give a utility value for each one, let alone break utilities down into part-worths. Rather, what a survey can meaningfully do is to ask respondents to compare one profile with another and rate which one they like better—or, given a set of profiles, pick the one they like best. The seminal work of Luce and Tukey [348] lays out the statistical methodology of extracting the part-worth measurements (or parameters of a hypothesized utility model) given only rankings data from a group of participants.

Current conjoint analysis software programs automate this analysis. A designer chooses a model and creates the relevant attributes and levels. The software then presents profiles to each survey participant and extracts part-worths based on statistical analysis of the resulting rankings data. The software is typically PC or Web-based and simulates a realistic choice environment by presenting graphics, images, or even videos.

Once the part-worth utilities have been extracted, they can be put to many uses. For example, the customer population can be segmented based on their part-worths (such as nonsmokers and exercise buffs for a hotel). One can also estimate the price sensitivity of each segment separately or design special marketing programs based on each segment's part-worths. A customer-behavior model can be fit using the utilities for various products under consideration, and their market shares and profitability can be estimated in a market-simulation model. Finally, product assortments can be designed for optimal positioning and overall profitability.

Our brief description here glosses over many details and decisions involved in a conjoint study: for example, whether to do pairwise comparison of attributes or full profiles; whether respondents should rate their preferences (say, on a scale of 1 to 10) or rank-order profiles (first choice, second choice, and so on). The interested reader should consult the references at the end of this chapter for such details.

11.1.2.2 RM Product Design Model

As mentioned, few firms currently utilize models to design their RM products. Nevertheless, it is conceptually useful to formulate the problem as an optimization problem to understand the many factors that impact product design.

Consider designing a set of K RM products for one particular resource with a capacity of C.[4] RM products are distinguished by the restrictions. Let there be M bases of restrictions (such as advance-purchase restrictions, min-stay, max-stay (see Table 11.3)). For each basis, there are multiple possibilities for creating a restriction. For example, for the advance-purchase basis, a RM product can use a restriction of three-day advance-purchase, seven-day advance-purchase and so on, or none at all. A RM product is composed of a set of restrictions, one along each basis.

Let $\mathcal{K}$ represent a collection of sets, with each set being a combination of M restrictions. Our design problem is then to pick K sets of restrictions from $\mathcal{K}$, fix prices for the K products, and in addition, decide on the portion $u_k, k = 1, \ldots, K$ of the capacity C to allocate to each of the K products (representing RM capacity controls).

$\mathcal{K}$, of course, could grow exponentially with the number of bases and the number of potential restriction values along each basis. In practice the number of bases would be small, and the number of values along each bases, four or five; hence, the size of $\mathcal{K}$ would be within reasonable limits.

The M bases and their values can be represented in an M-dimensional space. The potential restriction values create a grid in this M dimensional space with each block in the grid representing a potential product. Figure 11.1 gives an example of a 2-dimensional grid representing the product space with the advance-purchase and max-stay segmentation bases.

Let there be N customers. Each customer i has a set of valuations for the products in $\mathcal{K}$, This valuation could be represented by v_{ij}—

---

[4] K is considered an exogenous number fixed a priori by the firm. Alternatively, K could be an endogenous decision variable and we could model a fixed cost for introducing each additional product.

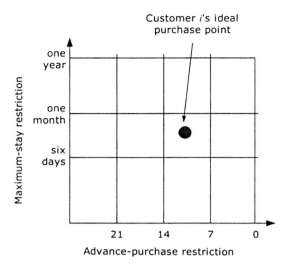

Figure 11.1. Advance-purchase and max-stay restrictions for an airline trip represented by a grid in a two-dimensional space. Customer's ideal point is the most restrictive point in this space where they have the highest valuation.

customer's i's valuation for product j. This however would create estimation difficulties, so below we use a simpler additive model in the spirit of the conjoint utility model of Section 11.1.

Let customer i have an *ideal point* defined as follows: it is the most restrictive product j for which he has his maximum valuation. For example, in Figure 11.1, customer i's uncertainty about his trip is resolved somewhere between 7 and 14 days before his trip date, and his trip takes more than 6 days and less than 1 month. Then his ideal point is represented by the potential product with restrictions of max-stay of 1 month and advance-purchase restriction of 7 days. Let his valuation for the product at this ideal point be v. If however, for reasons of price or availability, he is forced to purchase 14 days in advance, his willingness to pay would be lower. We model this as a reduction in valuation for purchasing less-than-ideal products.

Let w_{jk} represent the disutility for a customer whose ideal point is product j, if he has to purchase product k. (It is conceivable that this reduction is zero, especially when a customer purchases a less restrictive product.) If consumer i's ideal product is j_i, then his net utility for purchasing product k is given by $v - w_{j_i k} - p_k$, where p_k is the price charged by the firm for product k. So far we have assumed all the valuations v and w's are deterministic, but they could also be modeled as random variables to be more realistic. To keep the exposition simple, we assume

Implementation

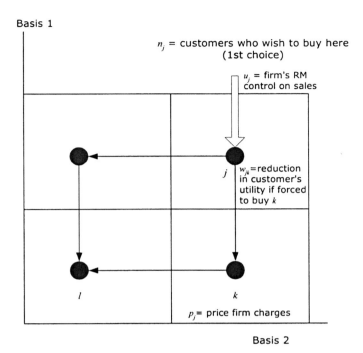

Figure 11.2. Products and customers utility reduction modeling.

deterministic valuations (see Talluri [502] for extension to stochastic private valuations). To keep things even simpler, the disutilities w_{jk}'s can be taken as the sum of the reductions along each basis between j and k. Figure 11.2 shows the graphic of the customer utility model.

If *all* the customers have the same deterministic valuation v (equivalently, there is a single segment), RM is not really necessary; the firm would just sell a single unrestricted product at a price v. This is because of the nature of RM products—customers will very likely have a disutility of zero for purchasing a less restrictive product, and the firm is better off not reducing the valuations by imposing restrictions.

So to justify designing multiple products the market should have multiple segments with different valuations and different disutilities (w's) for the restrictions. We can define segments as groups of customers who have identical valuations for their ideal products and identical disutility functions.[5] The firm can then use the multiple products to separate

---

[5]It would be more appropriate to model the valuations (for each segment) as random variables (from the firm's point of view), in which case the distinct segments will be groups with distinct

high valuation and low valuation customers. So, RM product design can be thought of as an optimal segmentation problem—taking a set of underlying segments (defined only by variations in their valuations and disutilities) and their purchase preferences (ideal points) and creating K segments (and, of course, the consequent operational problem of allocating capacity to the K resulting products). For the moment, assume that valuations are deterministic and there is a single segment. The description that follows generalizes to multiple segments in a transparent way.

Let n_j be the number of customers whose ideal point is product j. The firm's decision problem is to come up with prices p_j and allocations u_j, subject to $p_j \geq 0$ and $\sum u_j \leq C$.

$u_j = 0$ is taken to imply product j is not offered, so designing K RM products amounts to the restriction that at most K of the u_j's have non-negative values.

The firm's objective function then is given by

$$\max \sum_j p_j x_j \qquad (11.2)$$

$$\text{s.t.} \quad \sum_j u_j \leq C, \qquad (11.3)$$

$$0 \leq x_j \leq u_j, j \in \mathcal{K}. \qquad (11.4)$$

x_j represents the actual demand observed by the firm. We discuss how this is formed shortly. The restriction on number of products are captured by adding the following integer programming constraints to (11.2). Let y_j be a binary decision variable such that

$$y_j = 1 \quad \text{if} \quad u_j > 0.$$

Then, these restrictions are modeled by

$$y_j \geq \frac{u_j}{C}, \quad \sum_j y_j \leq K, \quad y_j \in \{0, 1\}.$$

Customers are utility maximizers in the sense that they will purchase the product (among the available products) that gives them the maximum positive net utility. The consumer's decision is of course influenced by the prices p_j and allocations u_j set by the firm.

Let x_{ijk} be a binary variable equal to 1 if a customer i with ideal point j purchases k and 0 otherwise ($\sum_k x_{ijk} \leq 1$). Let B be a sufficiently

---

distributions of valuations. The notion of "distinct" is necessarily vague. The definition is akin to cluster analysis, where we want to define distinct clusters that are similar within the cluster and as dissimilar as possible across the clusters.

large number (say $> vN$). The customer decision making is modeled as follows. Let $v_{jk} = v - w_{jk} - p_k$ be the net utility of a customer with ideal point j purchasing product k at price p_k. A customer with ideal point j ranks $v_{jk}, k \in \mathcal{K}$ and chooses the highest available one, provided his net utility is nonnegative. This is the equivalent to saying if $v_{jk} \geq v_{jl}$ then $x_{ijk} \geq x_{ijl}$.

Notice not all customers are assured of being able to buy the product (even if they have some product with positive net utility) because of rationing. We capture the utility maximization of customers by the following set of linear integer programming constraints added to (11.2).

$$\sum_k x_{ijk} \leq 1 \tag{11.5a}$$

$$1 + \frac{v_{jk}}{B} \geq x_{ijk} \tag{11.5b}$$

$$x_k = \sum_{i,j} x_{ijk} \tag{11.5c}$$

$$\sum_{i,k} x_{ijk} \leq n_j. \tag{11.5d}$$

$$1 + z_{il} + \frac{v_{jk} - v_{jl}}{B} \geq x_{ijk} - x_{ijl} \tag{11.5e}$$

$$1 - \frac{u_l - \sum_{m(m \neq i),j} x_{mjl}}{B} \geq z_{il} \tag{11.5f}$$

The set of constraints (11.5a–11.5f) model demand as follows: (11.5a) says that customer i buys at most one product; (11.5b) that if net utility for customer i (with ideal point j) to purchase k is less than zero, then he does not purchase k; (11.5c) gives the total demand for product k as sum of all the customers who choose k; (11.5d) sets the number of customers with ideal point j as n_j. Finally, (11.5e) and (11.5f) impose the utility-maximizing condition that for customer i if $v_{jk} < v_{jl}$ then $x_{ijk} < x_{ijl}$, unless there is no remaining capacity for l. z_{il} is a binary variable, equal to 1 if product l is not available for customer i (it had been sold out to other customers) and 0 if there is available capacity for product l. (11.5f) sets the values for z_{il}.

The above linear integer program captures many important elements of RM such as utility-maximizing customers, customer preferences, prices and restrictions (even if it is rather hopeless to solve—at least in its entirety—in practice). Note that customers are not strategic as in Section 5.5.2—they do not change their purchase behavior anticipating the firm's or other customers' actions. As with any posted price mechanism, the sequence of arrivals of the customers makes a difference (high valuation before low, etc.). The integer programming formulation here assumes the best possible ordering (from the firm's revenue-maximizing point of view).

11.2 System Architecture, Hardware, Software, and Interfaces

RM is a computationally intensive process. Huge amounts of data have to be collected and stored in databases. It must then be extracted and processed by the forecasting system to make thousands of forecasts on a daily basis. Finally, the optimization module uses the forecasts to come up with detailed quantity or price controls or both. Both forecasting and optimization can be computationally intensive tasks.

Figure 11.3 shows the flow of a nightly batch-process RM system. Figure 11.5 (repeated from Chapter 1) illustrates a prototypical process flow of RM. The steps involved in a forecasting module for an airline application (under the independent-demand model covered in Section 7.1) are shown in Figure 11.4, and the steps involved in the processing of a reservation request to a GDS are shown in Figure 11.8.

11.2.1 Hardware Requirements

Hardware requirements for RM systems can be immense.[6] That said, not every firm needs a million-dollar mainframe to run RM software. Some of the smaller, simpler applications (say, in a medium-size hotel) can be run on a PC. A multiprocessor database server and a powerful workstation for forecasting and optimization are usually sufficient for all but the largest RM systems. Current RM systems run on a large variety of platforms, from stand-alone PCs to Unix workstations and servers to mainframes. Reliability, redundancy, and good back-up procedures are important as RM is a mission-critical application; if a RM system is down, critical controls are not being set properly, which could lead to a significant loss of revenue.

11.2.2 User-Interface Design

The user interface (UI) is an important component of a RM system. They serve as the analysts' "window" on both market conditions and the RM system's response to these conditions. As mentioned in Chapter 1, RM is a man-machine process, with systems automating most of the routine decisions but under the oversight of analysts who intervene as necessary to respond to unusual market conditions or system errors. The

---

[6]United Airlines, to take a case in point, is reported to use several of IBM's "deep blue" supercomputers for portions of its RM system.

Implementation 595

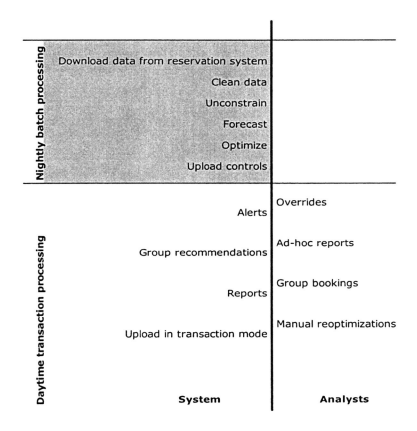

Figure 11.3. Nightly batch processing and daily activity.

ability of an analyst to interact effectively and productively with a RM system hinges on the quality of the UI design.

While the primary functionality of the UI is to allow analysts to analyze and monitor RM system controls and forecasts, the UI should also enforce security, so that only users with permissions have access to a specific functionality. For instance, analysts for a particular market should be prevented from overriding controls on markets outside their responsibility. Administrators should have the ability to maintain databases, set passwords, or otherwise monitor activity. In short, the UI supports all the basic functionality of the system (user administration, groups, forecasts, prices, availabilities, controls).

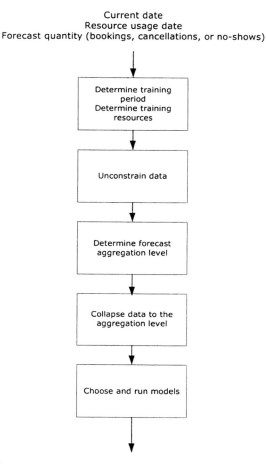

Figure 11.4. Forecasting process flow under the independent-class model.

Many hotel, rental-car, ferry, and cruise-line RM systems feature a color-coded calendar screen, with information on exceptions, booking curves, special events, and occupancy data. Alerts come with colors, sounds, and other notifications. As with any UI, ease-of-use and efficiency are important, as they reduce training costs and the chance that analysts will make mistakes.

At a technical level, the UI of most RM systems is written in one of any number of commercial programming languages including Visual Basic, Power Builder, C++, Motif (Unix) and Java/XML/HTML. Some older systems still use a DOS interface.

Implementation

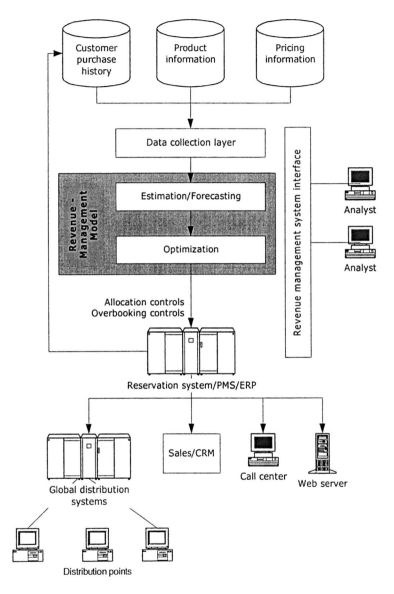

Figure 11.5. RM process flow.

The current trend is to have a Web interface for both analysts as well as system administrators, and therefore, Java/XML is the language of choice in UI design. With a Web interface, no special software needs to be installed on the client machines; any standard browser can be used

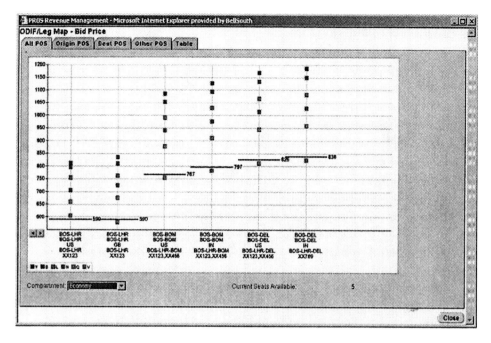

Figure 11.6. A quantity-based RM (airline RM) user interface (Courtesy: PROS Strat. Solns. Inc.).

to download and run the interface. Administration and control can be performed remotely as well, since only an Internet connection is needed. This is especially useful for a firm such as a hotel chain, which has to manage physically dispersed properties. Managers at the properties can access the system through a browser. In theory, the Web UI can be viewed from any browser, although in practice differences in browser versions, platforms, and capabilities create their own problems.

Figures 11.6 and 11.7 show two examples of RM software user interfaces—a screen shot of a quantity-based RM for the airline industry and a screen shot of a markdown pricing-optimization software system.

11.2.3 GDS, CRS, and PMS Interfaces

Hotels, airlines, and rental-car companies receive a substantial number of their bookings through global distribution systems (GDSs). The GDSs communicate with a firm's reservation system (host CRS for an airline or directly to the property management system (PMS) of a hotel), either periodically or for each booking request, to query for availability, retrieve passenger records, or process bookings (see Section 11.2.3.2 on seamless availability). Different GDSs are popular in different parts of

Implementation 599

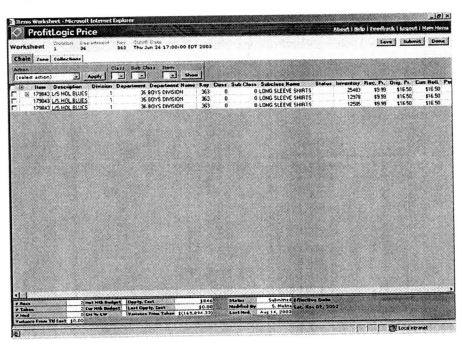

Figure 11.7. A screen-shot of a price-based RM (markdown) user interface (Courtesy: ProfitLogic Inc.).

the world. Table 11.5 gives a list of the major GDSs and their owners (as of 1998). GDSs charge per transaction (with different fees for different types of transactions) or per booking. GDS fees can quickly add range anywhere from 2% to 8% of distribution costs [191]. Because some GDSs are owned by airlines themselves and the order in which flights appear on travel agents' screens affects sales, there have been persistent concerns in the industry that GDSs are biased in favor of their owners [410]. In response to these concerns, today all GDSs are governed by regulations that attempt to ensure there is no such display bias. The governing rules also specify that any data generated by the system be available to all participants at equal and reasonable fees. The latter requirement is important because GDSs sell market share and sales data that can be used in RM systems.

11.2.3.1 GDS Interface Technology

The host CRS or PMS and GDSs interact thousands of times a day as travel agencies and other distributors (for instance, the switch companies in the hotel industry) query the CRS for availability and make

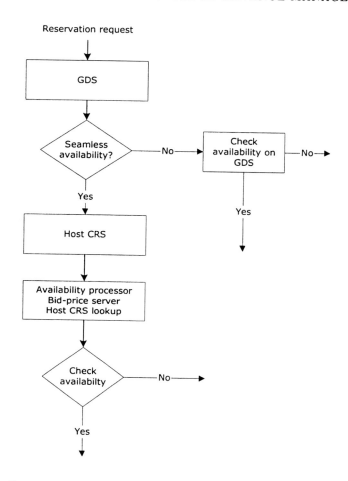

Figure 11.8. Processing of a reservation request to a GDS.

bookings. The communication to and from GDSs follow standard protocols (such as TCP/IP or X25) and can occur over dedicated or shared lines or even dial-up modems; essentially the client computer connects to the mainframe of the GDS and exchanges messages in each session. Table 11.6 gives a sample of an availability request for airline seats in its software form and in raw-data format. Hotel PMSs and rental-car reservation systems communicate in a similar fashion using industry-specific messages. Many of these message formats have EDI origins (Section 11.2.4.3), following data layouts similar to industry EDI standards, though each GDS can have significant variations or may even use its own proprietary message formats. The RM system may occasionally

Table 11.5. Major global distribution systems (GDSs) as of 1998.[a]

GDS	Shareholders	Outlets Number	Outlets %	Terminals Number	Terminals %
AMADEUS	Iberia	33,293	26.3	93,147	23.1
	Air France				
	Lufthansa				
	Continental				
	SAS				
GALILEO	United Airlines	30,161	23.82	115,454	28.62
	BA				
	Swiss				
	KLM				
	USAirways				
	Aer Lingus				
	Air Canada				
SABRE	American Airlines	29,277	23.13	119,546	29.64
WORLDSPAN	Delta Airlines	14,102	11.14	45,104	11.18
	Northwest				
	TWA				
AXESS	Japan Airlines	6,195	4.89	11,340	2.89
ABACUS	Singapore Airlines	4,200	3.32	10,500	2.6
	Thai Airways				
	Cathay Pacific				
INFINI	All Nippon Airways	6,195	4.89	7,700	1.9
GETS	SITA	3,150	2.49		

[a] *Source:* "Logistical Systems in Combined Transport", Working Document, Transport Series, TRAN 102, 1998.

need to communicate directly with the GDS, but more often, the RM system communicates only with the firm's own host reservation system.

Prior to the forecasting and optimization run, the RM system needs to download the total current reservations and remaining capacities. Table 11.7 shows some sample data tables from a hotel PMS used by a RM system. Similar, albeit more complex, tables exist in an airline CRS. Most PMS vendors license their data dictionaries to interface with RM systems. Table 11.8 shows the fields of a bid price implementation of a hotel PMS. The RM system periodically updates this table with new values. For interfaces, there has been an industry push recently to XML and other open messaging standards. This trend parallels the emergence of Internet sales channels. Internet travel sites may either query a GDS for availability or interface directly with the firm's CRS or PMS.

Table 11.6. An availability request message as software code and the same request as a message.

Availability{ Seats = '3',   //Number(2)

Class = 'Y',   //Text(1)

DepartureDate ='20020502',   //Text(8), YYYYMMDD

DepartureCity = 'BCN',   //Text(5)

ArrivalCity= 'JFK',   //Text(5)

⋮

}

Message Data String: "03Y20020502BCN JFK "

Table 11.7. Typical data tables provided by a hotel PMS.

Data Table	Description
Reservations table	List of all reservations for future inventory
Availability table	Remaining inventory in each category
User log table	User activity
Overbooking-limits table	Overbooking limits by day and room type
Room-categories table	Different room categories; mapping of rooms to categories
Reservation-types table	Different customer categories (can be the same as the RM classes or can be different)
Bid-price-control table	Table that the PMS uses to apply bid-price controls

11.2.3.2 Seamless Availability

Seamless availability is a technology for real-time communications between the host (internal) reservation system and (external) GDSs in the travel and transportation industry. It is a messaging standard developed under the auspices of IATA and is part of the EDI standards. The standards development body is a group called PADIS (Passenger and Airport Data Interchange Standards), and all new messaging standards have to be approved by its board.

The purpose of seamless availability is to replace periodic, batch uploads of static availability controls with real-time availability queries to

Table 11.8. Table BIDPRICE: A bid-price control implementation in a PMS. This particular PMS uses a delta approach to change the bid price after each booking. Note that this would constrain us to use a constant bid price with delta updates—a rather poor form of control.

Field Name	Type	Description
DATE	Date	Date of inventory
RMCATEG	Text	RM category classification (equivalent to airline coach and business compartments)
BP	Float	Bid price for this inventory on this date
DELTABP	Float	Bid price has to be added by this amount after each booking (or subtracted by this amount after each cancellation)
PMSCATEG	Text	The room category as stored in PMS (need not be the same as the RM classification
SOLDS	Integer	Number of rooms sold since last update (for applying delta calculation)
SELLLIMIT	Integer	Booking limit for this category (no rooms are sold in this category after this limit is reached)
DELTACEILING	Integer	If solds reach this number, delta is no longer valid (the table has to be updated by the optimization)

the host reservation system. Because the host reservation system has the most up-to-date information, by consulting it in real time the GDSs are able to provide travel agents and customers with accurate price and availability data.

Beyond the mechanics of encryption, data transfer, packaging, and hand-shakes, the main contents of a seamless-availability messages are (1) information on the travel request from the GDS to the airline and (2) information on price and availability from the airline to the GDS. Figure 11.9 shows two typical messages between an airline and GDS.

Seamless availability provides several advantages. For one, it allows the airline to base its accept/deny decisions on passenger characteristics (reducing the anonymity of the transaction somewhat) such as point of sale and frequent-flyer number. It also reduces errors in pricing and availability. Seamless availability also helps enforce *married logic*, a sys-

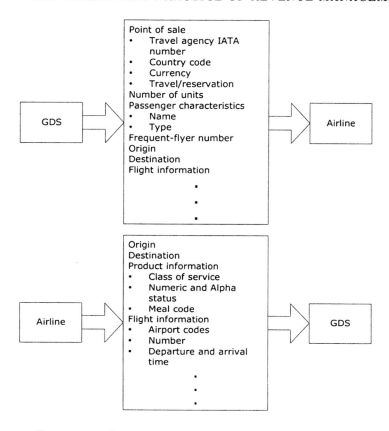

Figure 11.9. Two examples of seamless-availability messages

tem whereby flights can only be sold and cancelled as units.[7] Finally, it allows airlines to track rudimentary "click" behavior.

Seamless availability comes at a cost, though. To provide a response within the prescribed time limits, a firm usually needs to set up dedicated communication links and hardware. Also, GDSs charge more to implement this feature as it puts a heavy strain on their own systems.

11.2.3.3 GDS Abuse in the Travel Industry

While RM is not directly concerned with the inner workings of a GDS, limitations of GDSs often create challenges for a RM system implementation. One potential problem is abuse by travel agents. To give an example, if a GDS stores a multisegment itinerary as independent seg-

---

[7] *Married logic* is an airline industry practice to counter travel agents abusing the system by booking an itinerary to get cheaper fares and then cancelling part of the itinerary.

Implementation 605

ments, and if the fare for an O-D itinerary is lower than one of its constituent components (not uncommon in market-based pricing, as there may be competition on the O-D but not on the segments), a travel agent may be able to book the multisegment itinerary and cancel the unnecessary segment later. Or a travel agent may be able to book a block of seats on a fictional PNR to lock in cheap fares and then later fill in the passenger details or cancel them at will. A travel agent may be also able to make reservation on one GDS, then transfer it to another, and issue the tickets on the new GDS, doubling the GDS costs. And so on. The list of possible GDS abuses is endless and causes significant revenue leakage to airlines. Some of these are plain flaws in the product and restriction design, but if the GDS has limitations in implementation, a clever travel agent can always find ways around any restriction.

While the large airlines monitor travel-agency behavior and penalize violating firms (say, by cutting discounts), smaller airlines and hotels have little leverage to prevent these practices. In response, many RM systems have added functionality—and at times rather ad-hoc features—to prevent GDS flaws. For instance, as mentioned, airlines use married-segment logic to prevent agents from booking a through itinerary as two locals or from booking a through itinerary and then cancelling one segment to gain availability on a local flight. Managers in charge of the RM system must have a good working knowledge of the GDSs in order to intervene appropriately to prevent such abuse.

11.2.4   Retail Management Systems

Point-of-sale (POS) transaction databases are the central source of information for retail RM. These systems collect information from the point of sale, that—combined with product, inventory, price, and promotion information—gives a highly accurate picture of all shopping transactions in a store. A retail store management system (RMS) consists of a number of elements: POS terminals with attached bar-code readers, databases with product and inventory information, and EDI to connect to suppliers' ERP and supply-chain systems.

11.2.4.1   Bar Codes and POS Systems

Optical-scanner technology has revolutionized retail management. Almost all retail products today are encoded by a Universal Product Code (UPC),[8] a code consisting of 12 digits. The first digit represents the

---
[8]There are various flavors of UPC, but we describe the simplest and earliest one, what is called version A. Other standards include EAN (European Article Numbering), JAN (Japanese Article Numbering), ISBN, and code 39.

product type, the next five the manufacturer's code, the next five the product code and finally, a digit is reserved for check sum. The digits are encoded into bars and spaces that are read by bar-code readers. There have been many advances in bar-code technology, but essentially bar-code readers first scan and decode the bars and spaces to the correct UPC digits and then transmit the information to the retail system.

POS terminals are simply computer- and communication-enabled cash registers. They are connected to the bar-code and card (debit and credit and shop loyalty cards) readers on one end for fast, error-free check-out, and on the other end, to the store retail systems for real-time price lookups and updates to the inventory. As soon as the bar-code reader scans in an item, the POS system sends a PLU (price look up) query to the database to get unit price, tax, discount, and promotion information. Thus, the exact price (including shop discounts and coupons), time of sale, and shopping-basket composition are all recorded.

11.2.4.2 Pricing and Inventory Modules

The RMS database usually consists of the following tables: inventory, products, transactions, suppliers, purchasing, promotions, goods, customers, orders, contacts, and users, as well as others related to employees, and store layouts. On the pricing and promotions side, some tasks that a typical RMS can perform are given in Table 11.9.

11.2.4.3 Electronic Data Interchange (EDI)

In many business transactions, the output from one computer application is an input to another application. For instance, a purchase order by a retailer is an order entry for a manufacturer; an availability request by a reservation system is a query to a hotel bid price server. However, the formats and structure of the data required by each application can differ considerably. In order for applications to communicate, they must adopt a common format for data interchange. EDI provides such formatting standards.

Using EDI standards, a developer can create one single EDI interface that any application can understand. This promotes a low-cost, efficient way for multiple partners to implement automated transactions. EDI allows applications to "talk" to other applications, broadcast queries across multiple partners or otherwise communicate on a "many-to-many" basis.

EDI is a global standard, governed internationally by the United Nations Centre for Trade Facilitation and Electronic Business (UN/CEFACT) with participation by regional and national standards bodies (in the U.S., the Accredited Standards Committee (ASC)). EDI

Table 11.9. RMS pricing and inventory functions.

RMS Pricing Functions

"Buy X, get Y" discount capability
Lot pricing (single, six-pack, case, etc.)
Multiple price levels per customer (price-break tables)
Discount from retail
Mark-up from cost
Set profit margins
Discount by dollar amount or set percentage of discount
Put items on sale or promotion
Set up weekly sales schedules

RMS Inventory Functions

Track and manage all inventory
Automatically reorder inventory based on restock level, period, or rate of sale
Support links to suppliers ERP system
Generate purchase orders and receive shipments
Transfer inventory in and out
Store and track off-line inventory (items not for sale)

---

is technology independent and uses agreed-on message codes and structures to provide a secure and seamless exchange of data between trading partners. It is further extended by various industry-specific groups, each developing messaging standards suitable for their industry. There are standards for the retail, manufacturing, and travel and tourism industries. Most messaging between retailers and manufacturers, between reservation systems and travel providers occurs using EDI standards. EDI software is usually present as a module in enterprise applications and most retail systems and ERP applications have EDI modules built in.

The EDI standard for the travel, tourism, and leisure industry is called Unicorn. It is based on the traditional EDI "batch" message construction, although a more interactive version is also being developed by the standards body at the time of this writing.

Unicorn messages support the following business applications:

- Product information, enquiries, tariffs, schedules, and availability,

- Making of reservations,

- Enquiry on, amendment to, or cancellation of reservations,

- Pricing, ticketing, and production of similar documents, and
- Free text.

The Unicorn message set includes air, ferry, rail, most types of accommodation, car hire, package holidays, insurance, for sale of associated travel products and financial transactions such as statements and payment remittances. In addition, there are messages that allow the remote printing of travel documents, such as paper tickets and automated ticketing and boarding (ATB2). Any principal, agent, intermediary, or service provider is free to use a Unicorn message. Table 11.10 shows the functionality of Unicorn messaging.

11.3 Revenue-Opportunity Assessment and Revenue-Benefits Measurement

Because RM systems are expensive, time-consuming to implement, and require organizational changes that are disruptive to normal operations, it is natural for senior management to question whether the benefits justify the costs. It is important, therefore, to analyze a RM investment before implementation and then later, after the system is up and running, to validate the system benefits. The first type of analysis, performed during the preimplementation phase, is called the *revenue-opportunity assessment* and the latter, done post-implementation, is called *revenue-benefits measurement*. Most RM vendors perform a revenue-opportunity assessment as part of an engineering study phase and follow up with a benefits measurement study after system cutover.

11.3.1 Revenue-Opportunity Assessment

There are two basic approaches to revenue-opportunity assessment. The first is based on a *perfect-hindsight estimate* of revenue potential. This estimate is constructed as follows. First, historical data is analyzed, and corrections for censoring are made to estimate the control-unconstrained underlying demand. In a price-based RM setting, one fits demand functions directly to historical data based on observed price responses. Given this *a posteriori* estimate of demand, it is then possible to analyze the quantity or price controls that *would have* been optimal with perfect knowledge of demand. In quantity-based RM, this usually involves solving a deterministic linear integer program as discussed in Section 3.3.1 to optimally allocate capacity; in price-based RM, deterministic dynamic-pricing models of the type discussed in Section 5.2.1 can be used.

Table 11.10. Functionality of EDI for the travel and tourism industries.[a]

Functionality	Accommodation	Air	Car Hire	Ferry	General Sales	Insurance	Package	Rail
Information/tariffs	Yes	Yes	No	No	Yes	Yes	No	Yes
Inventory availability enquiry	Yes	Yes	No	Yes	Yes	Yes	No	Yes
Inventory alternatives enquiry	Yes	Yes	No	Yes	Yes	Yes	No	Yes
New reservation	Yes	Yes	No	Yes	Yes	Yes	No	Yes
Reservation recall	Yes	Yes	No	Yes	Yes	Yes	No	Yes
Reservation amendment	Yes	Yes	No	Yes	Yes	Yes	No	Yes
Reservation cancellation	Yes	Yes	No	Yes	Yes	Yes	No	Yes
Ticket (availability for issue) enquiry	Yes	Yes	Yes	Yes	Yes	Yes	Yes	Yes
Paper ticket issue	Yes	Yes	Yes	Yes	Yes	Yes	Yes	Yes
Tickets issued report	Yes	Yes	Yes	Yes	Yes	Yes	Yes	Yes
ATB2 pectab loading	Yes	Yes	Yes	Yes	Yes	Yes	Yes	Yes
ATB2 ticket issue	Yes	Yes	Yes	Yes	Yes	Yes	Yes	Yes
Ticket voiding	Yes	Yes	Yes	Yes	Yes	Yes	Yes	Yes
Ticket re-issue	Yes	Yes	Yes	Yes	Yes	Yes	Yes	Yes
General printing	Yes	Yes	Yes	Yes	Yes	Yes	Yes	Yes
Abort transmission	Yes	Yes	Yes	Yes	Yes	Yes	Yes	Yes
Connection time-up	Yes	Yes	Yes	Yes	Yes	Yes	Yes	Yes
RESCON message	Yes	Yes	Yes	Yes	Yes	Yes	Yes	Yes
Statement of accountant remittance advice	Yes	Yes	Yes	Yes	Yes	Yes	Yes	Yes

[a] Source: TTIR01 Version 2.4 February 1998.

These perfect-hindsight calculations then form an (estimated) upper bound on the revenue that might have been obtained in the past. Comparing this estimated maximum revenue to the historical revenue provides a measure of the potential gain. While this number clearly overestimates the revenue gains of a real system, it is common to assume some fraction of this potential gain is achievable. Often these estimated revenue gains are quite large, so even if the system achieves only a fraction of the gain, it provides more than enough justification for a RM investment.

The second methodology used for revenue-opportunity assessment is simulation. (See Section 11.4.) Simulation is more time-consuming but arguably more accurate in gauging potential revenue gains because in a simulation study it is possible to model consumer behavior, and replicate the exact forecasting and optimization methods being proposed. A simulation model can also model uncertainty and mimic salient features of the sales practices. Unlike historical perfect-hindsight studies, simulation can also be used to evaluate "what if" scenarios that have not been observed in the past. The disadvantage of simulation is that one must make a series of modeling assumptions, which may or may not reflect real-world conditions. Thus, it is important to get management approval of the model's validity prior to doing a detailed simulation study.

11.3.2 Revenue-Benefits Measurement

While in principle the benefits from a RM system should match the numbers given out by the revenue-opportunity assessment, this is rarely the case in practice. But this is to be expected. For one, business conditions change rapidly: recessions, economic shocks (wars) and currency changes all have a bigger impact than the effects of a RM system. Indeed, in RM one is often trying to measure benefits of the order of 1% to 2%, which can easily get washed out by even a mild demand shock or change in competitive conditions. Nevertheless, it is important to do such a study to attempt to validate the performance of a RM system. And it is best to aim for as unbiased a measurement as possible, ideally by a "neutral" internal team or an outside third-party.

Benefits measurement can and should be based on actual data. For this reason, it is important to collect and store all relevant data (prices and products, competitor prices and products, customer booking records, allocations) for a significant period—both prior to and after implementation of the RM system. The preimplementation data serves as the baseline for comparison. By collecting data over a long period, there is a better chance of being able to pick a period that is relatively stable or free from major outside shocks.

If there are major changes in the marketplace or economic conditions, benefits measurement becomes more difficult and potentially fraught with controversy. The data should ideally be "corrected" to account for these changes, but such corrections are difficult to make. So the judgment, fairness, and skills of the managers, statisticians, and consultants involved in the study play an important role in the credibility of the numbers.

One way to avoid such controversies is to perform a parallel test of the new system versus the old system. That is, selectively apply the new RM system to a sample of products and markets while continuing to manage the remaining products and markets with the old system and procedures. This allows for a controlled experiment of the revenue performance of the new and old systems under the same economic conditions. Many RM vendors use this approach, especially for first-time adopters of RM.

Of course, this approach has some drawbacks. One has to take care that the selected products and markets are representative of the total population and that analysts behave as "normally" as possible when using each system. And it may be technically or operationally difficult to try to run two systems in parallel. But the results of such side-by-side comparisons are often much more credible than those based on comparing performance of an entire system pre- and post-implementation.

Whichever approach is used for benefits measurement, it is important that such studies not be undertaken solely as a reward and validation process. Rather, they should be also viewed as an opportunity for process improvement—as a way to find and fix hidden bugs in the system or to determine areas where the system and models can be improved. As such, benefits measurement should ideally be part of a continuous improvement process.

In addition to revenue measurements, senior management is usually interested in tracking before and after measures along a number of dimensions. Table 11.11 lists some commonly used performance measures that are tracked pre- and post-implementation.

11.4 RM Simulation

As mentioned, simulation is a flexible and powerful methodology to evaluate RM system performance. It is also useful in basic R&D studies to evaluate the performance of new forecasting and optimization methods. Here, we look briefly at some RM-specific simulation issues.

The idea of simulation is to mimic both the customer demand process and the RM system responses to this simulated demand. A variety of questions can be addressed using a simulation study, such as: What is the potential revenue impact of changing to a new forecasting or opti-

Table 11.11. Commonly tracked RM system performance measures.

Performance Measure	Description
RevPAD	Revenue per available day (hospitality)
	Revenue per available ad (broadcasting)
RevPAR	Revenue per available room (hospitality industry, cruise lines)
RevPAF	Revenue per available square foot (retail, casinos)
RevPASM	Revenue per available seat mile (airlines)
Yield	Revenue per available unit of inventory (airline)
Occupancy	Percent of available inventory sold (hospitality)
ARR	Average room rate (hospitality)
Spoilage	Capacity that could have been sold but was not, due to a low overbooking pad
Load factor	Same as occupancy (airlines)
Spill	Estimate of demand that has been rejected
Group utilization ratio	Percentage of a group booking that shows up
ADY	Average daily yield
ADR	Average daily revenue
ATP	Average ticket price (events, theater)

mization system? What are the benefits of moving from a leg-level control to a network-level control? What is the revenue difference between different types of controls: Bid-price control or virtual nesting control? How robust is the system to errors in forecasting? How much revenue is lost by bad or biased forecasts? What are the revenue impacts of different types of customer-choice behavior (including sell-ups and revenue dilution)? And so on. The ability to provide detailed answers to such a wide range of questions is the main advantages of simulation.

For the results of a simulation study to be meaningful, the program has to model the current business and control processes and the planned processes as accurately as possible. Customer booking streams are normally generated using a pseudo-random number generator based on historical booking patterns. A simulation clock governs the progress of the simulation. At various points in time, one or more *events* occur. For example, events for a quantity-based RM simulation include booking requests, cancellations, no-shows, and optimization and forecasting runs. For a price-based RM simulation, events include price changes, customer arrivals, and purchase decisions (purchase, delay purchase, or no-purchase).

By carefully modeling the firm's sale practices and customer behavior, a reasonably accurate picture of revenue benefits can be obtained via simulation. But even at its best, it is important to remember that a simulation is only an abstraction of a real system—and it can only

Implementation

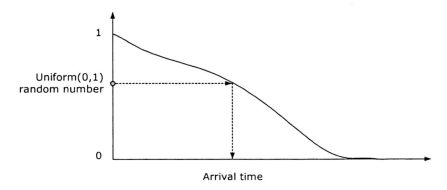

Figure 11.10. Generating arrivals over time.

represent phenomena that the analyst programs into it. Thus, the level of detail it provides can give a false sense of precision. Such caveats aside, simulation remains by far the most common method in practice for evaluating RM systems.

11.4.1 Generating Aggregate Number of Customers

Pseudo-random number generators are used for generating arrivals according to a specified distribution. While there are many subtleties involved in the algorithms used to generate pseudo-random numbers, a suite of well tested, fast, and stable algorithms have emerged over time that have good statistical properties. The basic pseudo-random number generators generate a random number uniformly distributed between 0 and 1. From this uniform random variable, a number of general techniques (such as transformation methods or rejection methods [429]) can be used to generate random numbers from a wide variety of distributions.

11.4.2 Generating the Customer-Arrival Pattern

In RM, the timing and order of customer arrivals has an impact on revenue gains. For instance, whether customers with a high willingness to pay arrive before those with low willingness to pay has an important effect on revenue as well as the effectiveness of the forecasting or optimization methods. For this reason, the RM simulation should be able to generate customer arrivals following the observed patterns for the various segments.

Mathematically, for each simulation run, we would like to generate a random number from a target probability distribution (such as a left-truncated normal distribution) representing the total number of customers of a particular segment, and then distribute these arrivals over the sale period, following a given arrival pattern. To avoid confusion with the distribution of the total number of customers, we call the distribution over time an *arrival pattern*.

The following procedure achieves this objective. First, generate a random number n from the (assumed) distribution for the total number of arrivals, representing the aggregate number of customers. Then for each one of these customers, generate their arrival time by considering the arrival-pattern curve as a cumulative distribution over time. That is, generate a uniform random variable between 0 and 1 and use the inverse of the given arrival pattern to generate an arrival time, as shown in Figure 11.10.

While this procedure gives a distribution of customers over time that conforms to both the aggregate demand distribution as well as the arrival pattern, it is important to note that the number of arrivals will not, in general, be independent across disjoint time intervals. Put another way, a RM forecasting system could gain an advantage by basing its forecasts on observed demand to date, since the expectation of future demand conditional on observed demand may be different from the unconditional expected demand. Whether this is as it should be is a matter of debate among RM professionals, but one should be aware that with arbitrary distributions, biases may be introduced into a simulation by using an unconditional distribution when forecasting demand to come.[9]

11.5 Customer Perceptions and Reactions

While firms, for understandable reasons, may be enthusiastic about RM, they may find that their customers are less so. Here we discuss strategies for making RM more palatable in the eyes of customers.

11.5.1 RM Perception Problems

One well-reported incident that highlights the potential customer perception problems with RM occurred in the fall of 2000 when Amazon.com conducted a price experiment, quoting different prices on DVDs to different customers. Several customers discovered the practice by logging onto Amazon.com's site at different times. What they found, which Amazon.com later confirmed, was that customers were given random discounts of between 20% and 40% on selected DVDs. For example, during one online price test conducted by the *E-Commerce Times*, the

---

[9] If aggregate demand is Poisson, however, then the unconditional distribution of demand to come is the same as the conditional distribution.

price for the DVD *Mission Impossible* was offered for sale at $17.99, a 40% discount off list price; several hours later, the price had risen to $20.99.[10]

The incident was picked up quickly by the press and created significant negative publicity for Amazon.com. The following are some customer reactions to Amazon.com's dynamic pricing experiment (*Source:* DVDtalk.com):

> This is a very strange business model, to charge customers more when they buy more or come back to the site more.

> I find this extremely sneaky and unethical.

> If you walk into a store, you aren't charged more based on how many times you pick up the DVD to look at the cover, are you?

> (Amazon.com is a) shyster.

Amazon.com denied setting prices based on customers' past purchases or demographic information. According to Amazon.com CEO Jeff Bezos:

> We've never tested and we never will test prices based on customer demographics. What we did was a random price test, and even that was a mistake because it created uncertainty for customers rather than simplifying their lives.[11]

And on September 27, 2000, Amazon issued a statement regarding the price test of early September:

> Price testing was not based on customer demographic information. These reports were incorrect and were not based on the facts. We have never tested and we never will test prices based on customer demographics. Contrary to these reports, Amazon varied the discount levels on a totally random basis. The purpose of the test was to determine how much sales are affected by lower prices. In retrospect, this random testing was a mistake, and we regret it because it created uncertainty and complexity for our customers, and our job is to simplify shopping for customers. That is why, more than two weeks ago in response to customer feedback, we changed our policy to protect customers should we ever do random price testing again (and currently we have no plans to do so). Now, if we ever do such a test again, we will automatically give customers who purchased a test item the lowest test price for that item at the conclusion of the test period—thereby ensuring that all customers

---

[10]Lori Enos, "Amazon Apologizes for Pricing Blunder," *E-Commerce Times*, September 28, 2000.
[11]Ibid.

pay the lowest available price. Under this new policy, by September 14, we had refunded to 6,896 customers an average of $3.10 as a result of the DVD random-price test.

This incident raises an interesting question: How can airlines get away with charging different prices to different customers while Amazon.com had such trouble doing, at least on the surface, the same thing? Certainly, the range of prices that airlines typically charge is much more than the range Amazon.com charged for its DVDs. (The highest ticket price can be ten times more than the lowest price on some flights.)

There are several explanations. For one, airline prices come with restrictions, so the products are in fact differentiated by sale conditions (even though the physical product is the same). Also, unlike in Amazon.com's case, customers self-select what airline fare they want to pay (second-degree price discrimination), whereas Amazon offered discounts to selected customers only (third-degree price discrimination). And while airline fares change over time as allocations close down, at any given point in time, everyone is given the same choices, and no one is treated differently. Airlines, moreover, do not control sales by changing fares directly; they change the allocation of capacity to each fare. So rather than prices rising, discounts "sell out," which seemingly creates a more acceptable perception among customers. Airline customers have also grown accustomed to frequent price changes based on many years of experience. In contrast, dynamic pricing in retail is a more recent practice, so it may take time before customers accept it. Finally, dynamic pricing is often perceived of as an irritating practice for low-priced items because the search cost for customers is high relative to the value of the item. In contrast, for expensive items like airline tickets or automobiles, customers are more willing to spend time and effort to shop around for a bargain. Jeff Bezos, the CEO of Amazon.com, acknowledged as such in a subsequent press statement.[12]

This list of differences should give a clue to what customers may or may not find acceptable about RM. If customers perceive that there is a tangible reason for price differences, they tend to be more accepting. Examples include early bird or advance-purchase discounts, volume discounts, a price that is related to dwindling or excess capacity (clearance sales), or prices related to the distress of the firm (going out of business sales). In short, if customers can "rationalize" the price differences they see, they are more accepting.

---

[12] Reuters news report, September 27, 2000.

Also, customers perceive RM as more fair if they can self-select options and all customer receive the same options at each point in time. We have already discussed this in the context of airline fares. As another example, consider auction mechanisms. They neither guarantee a uniform price to all customers nor offer stable prices over time. Yet most people regard auctions as "fair, " in the sense that the mechanism is transparent and customers win based purely on what they are willing to bid. Since each customer wins or loses based on his or her own self-determined bid, it is hard to argue that one is ever "cheated" out of winning as a result. If, however, the firm were to (hypothetically) perform the same allocation—that is, set prices and allocate units by customer valuation—then it is likely that customers would indeed question the fairness of the process.

Customers also appear more willing to tolerate differential prices if the product is expensive. In fact, for low-price products one can think of stable "fair" pricing as producing a beneficial equilibrium between customers and the firm, in which customers adopt a strategy of not searching for lower prices (thereby avoiding search costs) and in turn the firm adopts a strategy of not taking short-run advantage of loyal customers (thereby inducing loyal customers not to shop around). This perhaps explains the sense of "betrayal" among loyal Amazon.com customers to the price experiments. When the price of the product is high relative to search costs, however, this equilibrium is harder to sustain because customers have a much stronger incentive to comparison shop. Both parties seem to tolerate (or at least understand) "disloyal" behavior as a result.

When a firm raises prices because supply is low, it could easily be viewed as price gouging, especially for essential items. However, a firm can usually avoid this stigma if there are viable alternatives to the product and customers are made aware of these alternatives. So, for example, if demand is high for a particular resort hotel on a particular day, customers should be informed of other dates with lower prices. In this way, the *customer* makes the tradeoff between a higher price and convenience.

A sense of fairness can also be explained in terms of psychological and social factors. One theory from economics and marketing holds (see Appendix E) that customers form a reference price about an item based on past purchases, market prices and prices of comparable products. The principle of *dual entitlement*, another theory from economics and marketing [522, 275], says that customers believe that they are entitled to a reasonable price (say, the reference price) and firms in turn are entitled to a reasonable profit. Customers' views of fairness, so the theory goes, arise from these principles. Therefore, customers may perceive high prices during periods of high demand as "unfair" even though the economic law of supply and demand would justify such pricing [275, 276, 98, 296].

11.5.2 Managing Perceptions

The perception of unfairness is indeed a problem for RM. It is especially so if the customers come in contact (as in an airplane) and exchange information on the prices they paid and less of a problem when customers are more isolated (say a hotel). For example, it is reported that RM has often caused negative reactions in the cruise-line business because customers are confined together on a ship for many days and inevitably end up sharing information on the prices they paid.

This does not mean one should abandon thinking of implementing RM. Rather, it should suggest a need to identify strategies to manage customer perception on unfairness as part of an overall RM implementation.

How, then, can a firm appeal to the principle of dual entitlement to convince customers that it is not price gouging? Some possible strategies and tactics include the following.

- Clarify exactly why the firm charges different prices, whether limited supply, peak versus off-peak time, or advance-purchase discounts that reduce risk—and at the same time on what bases the firm does *not* charge different prices (such as knowledge of the customer). This enhances the sense of fairness of RM practices. In short, it is important to emphasize the rationale associated with discounts and not simply quote a price.

- Emphasize any differences in the product itself, no matter how small (room with a view, larger room). The additional value of these features may not justify the price differences, but if customers feel that the physical product itself has variations and that these variations are the reason the firm is charging different prices, they will tend to perceive the prices as fairer.

- Stress alternatives that the firm offers at a lower price. So, for example, if a discount is not available on a particular date, offer other dates on which the discount is available. This not only encourages efficient usage of capacity but also allows customers to decide, based on their individual valuations, which of the alternatives they want to buy.

- Emphasize self-selection wherever possible. Present a full menu of prices and products so that customers can choose for themselves.

- To the extent possible, discourage price disclosure among customers. For example, do not display price overtly on tickets and other paperwork, and keep sale interactions as private as possible.

11.5.3 Overbooking Perceptions

One area of RM where customer expectations have to be managed especially carefully is overbooking. Customers who make a reservation and pay for it expect to receive the product. At the same time, customers who reserve and cancel expect a full refund, since from their point of view they have not used the product. Overbooking, as discussed in Chapter 4, is the fine art of balancing these customer expectations while maximizing utilization of capacity.

Managing overbooking is certainly not limited to just the science and models. No matter how well the models perform, there will inevitably be a day when a customer with a guaranteed reservation cannot be accommodated. So policies and procedures need to be put in place to handle these inevitable denied-service situations.

Customer acceptance of overbooking depends to a large part on tradition and industry norms. In the airline industry in the United States for instance, it is pretty much accepted as part and parcel of airline travel. This can be attributed to a number of reasons. First, the voluntary denied-boarding system discussed in Section 4.1.1, whereby volunteers are requested to take an alternative flight in return for compensation, is widely perceived as a fair means of selecting who will be denied service. Second, airline travelers usually have a number of alternatives flights. Third, the process is transparent; all passengers are gathered together at one spot at the time of departure, so everyone realizes that the flight is oversold and that someone has to be left behind. (This cannot be said for a hotel, for instance, as an arriving customer sees no queue or clear evidence of a full hotel.) Finally, there is a long history and tradition of overbooking in the airline industry, and airline ticketing staff are well trained to handle denied boardings.

Planning, training of customer-relations staff, and well-established policies and procedures are the key factors in managing overbooking. Firms usually have advance notice that an overbooking situation is going to arise. By adequately making plans in advance, customers can often be accommodated with substitute arrangements.

If customers have to be denied service involuntarily, the procedures should be fair and minimize inconvenience to the customer as much as possible. Compensation should be flexible and geared toward the interests of the customer. If the customer expects cash (in lieu of, say, a 50% discount for the next stay at the hotel), the firm should specify how much cash can be offered. Customer-service staff also needs to be trained to manage customer expectations about the level of compensation.

It should be pointed out that in the airline industry, despite the highly refined practices and many years of experience, overbooking still ranks among the most prominent customer-service problems. Thus, the practice is something of an awkward compromise between service quality and efficiency.

11.6 Cultural, Organizational, and Training Issues

With all the sophisticated models and mathematical techniques employed in RM, it is easy lose sight of the human and organizational challenges involved in implementing and maintaining a RM system. The huge investments in technology also tend to overshadow important organizational issues. Yet poor organizational planning is often the reason cited for the failure of a RM implementation, and poor training is frequently blamed for subsequent inadequate performance. In this section, we look at the main RM organizational and training issues.

11.6.1 Changes in Responsibility by Function

Organizational and business process changes are usually required at the time of the introduction of RM, and once the RM system stabilizes, even broader organizational changes may be desirable. As a RM system implementation cuts across multiple departments and functional areas, it requires significant cross-functional coordination. Product design, capacity planning, pricing, inventory control, operations, IT, sales and marketing, and finance departments are all affected to some degree or the other by a RM system implementation. Let us first review how the main groups are affected.

11.6.1.1 Analysts

If a firm is managing pricing or inventory manually, the analysts who make the day-to-day decisions are the most affected by a RM system, in the sense that it is their work routines and roles that change the most. As the technology behind RM is alien to most analysts, they are often intimidated by the sheer complexity of a RM system and suspicious that an automated system could replace their own intuition and experience. They also often may feel their jobs are threatened by a move to automated decision making.

Therefore, top management ought to emphasize that RM is first and foremost a philosophy and a set of business practices and is secondarily a decision *support* technology. It is important to stress that the technology is a tool intended to help them make better decisions not a machine to replace them. It is worth emphasizing the vital role of analysts in both monitoring the system's recommendations and intervening when special conditions warrant intervention. Analysts should also be educated about the basics of the science behind the system's functionality, to demystify it as much as possible. They should understand its inputs, outputs, and the underlying assumptions behind the methodology. Emphasizing that they will be upgrading their skills and learning "leading-edge" technology also helps motivate the transition. Once the analysts understand and buy in into the RM concept, they will be more comfortable using the system.

The quality of analysts' jobs almost always improves once a RM system is implemented. Their role changes from making routine inventory or pricing decisions manually to monitoring the output relative to the current business situation. For example, if an unexpected event happens that affects demand, they can adjust forecasts manually to compensate; if business objectives require more or less aggressive overbooking, they can adjust the parameters to get the desired effect. Rarely do analysts need to override optimization outputs directly. However, if the business need arises, they can open up or close a particular booking class (for quantity-based RM) or set markdown rates manually across stores (in price-based RM). Analysts can also use the system to perform "what if" analysis, eliminating a large part of the guesswork involved in making decisions. Or they can let the system come up with the optimal decisions automatically based on revised inputs. In short, whereas the former role of analysts could be described as allocation or price setters, their new role becomes one of model calibrators, data analysts, problem solvers, system performance monitors, and business controllers.

11.6.1.2 Sales Teams

Sales teams are frequently effected by a RM implementation, especially if their compensation is based on sales volume. Volume of sales is not the primary objective for RM, of course; increased revenue is. And meeting this revenue objective may mean lower unit sales. More important, while the salesforce might have had the right to sell at their own discretion, once a RM system is in place they may be prevented from offering discounts. Sales representatives may view this as undermining their relationship with customers and limiting their ability to meet sales goals.

The expectations of the salesforce therefore also have to be managed. As with analysts, the salesforce has to be educated about the basic principles of RM. They should be trained to sell products using the forecasts and system allocations as guidelines. So instead of seeing closed allocation or discounts as a lost opportunity for selling, they should be encouraged to shift demand to products or period where discounts are open or persuade customers to upgrade. Their role should also include gathering market intelligence on private deals that competitors are offering, so products and discounts to match the competition can be introduced. Indeed, sales has always performed a market intelligence function, it is just made more challenging by RM because of the increase in the number of products and the complexity of the sale restrictions.

Finding the right incentive structure for sales teams in a RM framework is also an important challenge. Ideally, one would like to set incentives for the salesforce to generate profitable sales, while still maintaining the correct valuation for new incremental business. A few approaches have been tried using non-volume-based measures. For example, in airline RM group sales, the salesforce may be rewarded based on the revenue they generate *in excess* of the estimated bid prices (opportunity costs) for the capacity they sell. Such incentives have the potential to better align the saleforce's efforts with the objectives of the new RM system.

11.6.1.3 IT Department

Because of the massive technological development involved in RM, IT departments are frequently placed in charge of implementation. A RM system, however, is not like many other IT systems, in the sense that it is based on scientific models that are highly data-fragile. The system can be easily corrupted by data that is out of date, insufficient, or not cleaned properly, even while it keeps giving out reasonable-looking numbers.

RM systems also need data collected at the lowest possible level, and they need data stored for a relatively long time in operational databases. Storage costs have come down so much that it is no longer that expensive or difficult to store detailed customer data over many periods. Rather, the issue is more about developing the systems and procedures so that data is automatically stored, retrieved, and analyzed quickly. If the firm is not collecting the data suitable for RM, then the IT department must start working on such data collection far ahead of a RM implementation, so that the models have enough historical data to build on. Indeed, the IT department may need to start collecting data even prior to choosing a vendor or defining the need for a RM system, which may create budgetary and staffing conflicts that require high-level intervention to resolve. Also, there is the potential for something of a chicken-and-egg dilemma in this regard, as a RM system won't perform well without data, and the data is not worth gathering if the system is not performing well.

11.6.1.4 Other Functions

Other functional areas of the organization will be affected by a RM implementation as well. Pricing analysts and product-design groups have to input their knowledge on customer preferences and behavior so appropriate RM products can be created. Business rules and existing business contracts have to be respected and the RM system needs to conform to these rules.

Operations and customer-service divisions are affected by overbooking decisions made by a RM system. If the firm never practiced overbooking in the past, or did it in an ad-hoc way, then customer-service representatives and field managers have to be involved in the implementation and develop proper customer-service procedures to handle oversale situations.

11.6.2 Project and Organizational Structure

Given the interdepartment coordination and rapid communication that is required, any RM implementation should involve careful review and planning for the new organizational design, both at the project stage and at the operational stage.

11.6.2.1 Project Organization

The project leadership role could be taken by the operations research group—or lacking one, by the pricing or inventory-control group. However, IT and sales organizations have to be involved even from the project-conception stage. All this will be feasible, of course, only if senior management buys in and backs the project wholeheartedly. A senior officer has to be actively involved in the implementation to smooth over interdepartmental frictions and priorities. Indeed, most industry experiences of RM cite top-management leadership as a key success factor. It is also important to have all teams involved every step of the way—from initial proposal, to implementation, to training, to performance measurement. Table 11.12 shows a task list for a typical RM implementation.

11.6.2.2 Operational Organization

It is also important to think through the organizational design that will support RM on an ongoing basis. One important issue is how analysts are organized (by product, market, or resource). For example, in airline single-resource quantity-based RM, analysts are normally assigned collections of flights to manage. The flights may be related by geographical market (flights into or out of the same city) or by "type"

Table 11.12. Task list for a RM implementation.

	Task List
Pre-implementation	Form interdepartmental team (task force)
	Rough revenue opportunity model (industry and firm specific)
	Form business case
	Formulate strategic role
	Senior management buy-in
	RM audit
	Start polling vendors
	Start collecting data
	Buy or build decision
	Revenue opportunity analysis (simulations)
	Engineering study
	Formulate requirements
Implementation	Market segmentation/product design
	Database design
	System architecture design
	Model and system testing
	Data cleaning and validity testing
	Legacy system integration/phase-out
	Coordination with sales and product design
	Pricing feeds
	Hardware
	Software installation
	Analyst training and education
	Organizational changes
	Sales and customer-service agent training
	Testing phase
Operations	Cutover and ramp-up
	Monitoring and tracking
	Alerts and overrides monitoring
	Forecast error measurement
	Optimization performance measurement
	Overbooking and denied boardings
	Customer service complaints
	Groups performance
Benefits measurement	Before/after comparison
	Simulations

(leisure versus business markets). In retail RM, analysts (buyers) are typically organized by product category (women's casual sportswear), though they are occasionally organized by geographic region. Hotels are usually managed on an individual property basis, while car-rental ana-

Sample Airline RM Organizational Chart

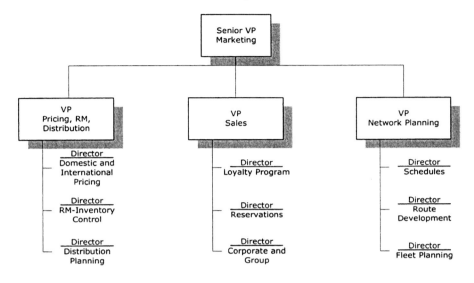

Sample Hotel RM Organizational Chart

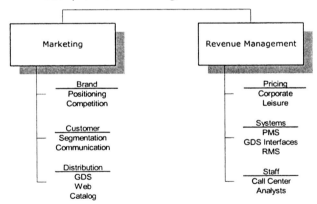

Figure 11.11. Sample organization charts for RM departments.

lysts are most often organized on a regional basis. Figure 11.11 shows some typical organizational charts in RM industries.

Each such organizational design involves a tradeoff; reducing the complexity of the decision-making task by decomposing the problem versus maintaining synergy and coordination among decisions by expanding the domain of control. There is no perfect solution to this problem, and different firms have had success with different organizational designs.

A second organizational issue is how to manage the various levels of decision making. For example, traditionally most airlines had separate marketing, pricing, and inventory-control groups. The marketing group was responsible for high-level decisions about branding, fare structure, and promotions; the pricing group was responsible for monitoring and reacting to fare changes in each O-D market; finally, the inventory-control group was responsible for day-to-day overbooking and capacity-control decisions on each flight. Today, however, many airlines are integrating all three functions into a single group, as the three sets of decisions are interrelated. Again, the main tradeoff here is between a desire for coordination versus a need to keep the analysts' job focused and manageable.

A final issue concerns which functional area has overall responsibility for RM. It is common to have both inventory control and pricing as marketing responsibility, although there are many firms where the finance department or even operations department is responsible for both, and some where pricing and inventory control are organized in a separate department reporting directly to the CEO.

11.6.2.3 Organizing for Network RM

Organizing for network RM presents special challenges. For one, if the current organization has analysts responsible for a limited number of markets, then as a network RM system makes decisions that cut across market boundaries, reorganizing analysts' responsibilities is almost always necessary to avoid conflicts. For example, if analysts make changes to a forecast, should they be allowed to reoptimize the network? What if multiple users are making changes at the same time? Such concurrency and locking issues frequently come up in a network RM system.

A network RM organizational structure normally has some analysts work at the market level and some at the network level, responsible for resolving any conflicts that arise across the different markets and for coordinating and batching forecast and parameter changes. Such network coordinators can also liaison with groups desks and network (product) designers.

If the RM system is implemented across multiple resources (a retail chain using price-based RM across multiple stores, a hotel chain implementing a RM system across all its properties, or a freight company managing inventory over many branch offices), then the question of centralized versus decentralized control arises. That is, should the system reside at head office and forecast and optimize across all the dispersed resources, or should each unit have its own decentralized RM system?

The argument for decentralization is that local field managers are in closer touch with their markets, so they are perhaps in a better position

Implementation 627

to manage the system's decisions. On the other hand, from a system's point of view, it is a lot easier to manage a centralized system. Moreover, a central system enables a firm to dedicate a set of skilled and experienced RM professionals focused solely on managing, developing, and maintaining the system. If the firm is using multiresource optimization, then RM by necessity has to be performed centrally as information has to be aggregated in real time. (For example, the hierarchical Bayes models of Section 9.3.6 requires information from multiple resources.)

With the increasing use of Web-enabled applications and interfaces, a compromise solution is to do the processing at a central office for all the units but let unit managers control the results, prices, or allocations, via a Web-enabled interface.

11.6.2.4 Operational Responsibility

Once the system is in operation, a different set of organizational issues comes up. To begin with, who should have operational responsibility of the system? Again, some interdepartmental coordination is unavoidable. While certainly the inventory control or pricing department "owns" the system as they are the daily users, they need the IT department to support the system, the operations research department to support the models and the science, and the pricing, sales, and marketing departments to coordinate prices, deals and promotions.

As mentioned, many firms with ongoing RM practices are finding that it is best to merge pricing and inventory-control departments. Pricing and RM are so strongly linked that unless one unit is responsible for both, no real coordination can be achieved. For instance, if prices are changed, forecasting based on historical data (as is done currently) can be unreliable. Analyst intervention is then required to manually adjust the forecasts. However, the processes for such close communication get complicated if the responsibility resides in different departments. Another reason for having pricing and inventory control under the same roof in the airline industry is the emergence of pricing decision support systems that complement RM systems.

11.6.3 Training

As regards training employees for RM, the abiding sentiment in industry is that there can never can be enough it. Most vendors offer on-site training programs. In addition, many consulting firms and professional schools have customized education and training programs by industry.

Training classes, both before and after implementation, should be conducted at various levels of management. For midlevel managers, the main emphasis should be on the principles of RM. For supervisors, more training is needed on the details of the technology (at least at a conceptual level) and how the system "thinks" in terms of coming up with its recommendations. Analysts and line managers need this same training, but they also need specific training in how to use the software effectively and when and when not to override the system recommendations.

Training should also be extended to groups that are not direct users of the system but are affected by it. This includes sales, customer service, pricing, and groups desks.

11.7 Notes and Sources

The standard technical book on market segmentation is Wedel and Kamakura [558]. A few more technical methods for segmentation can be found in Moorthy [389] and Kamakura and Russell [286] A nontechnical guide to market segmentation is McDonald and Dunbar [371].

Apart from the articles mentioned in the text, surveys on conjoint analysis and implementations can be found in Cattin and Wittink [102], Wittink, Vriens, and Burhenne [574], Wittink and Cattin [573], and Carroll and Green [99]. For applications of conjoint analysis for travel industry product design specifically, see Varini, Engelmann, Claessen and Schleusener [532] and Wind, Green, Schifflet, and Scarbrough [570]. Issues in estimation from conjoint analysis can be found in Green, Goldberg, and Montemayor [223] and Lenk, DeSarbo, Green and Young [339]. In addition, all the conjoint analysis software vendors' websites carry many articles on both technical and nontechnical aspects of conjoint analysis (e.g., www.sawtoothsoftware.com).

The product design model of Section 11.1.2 is due to Talluri [502]. A history of GDSs, PMSs and CRSs and their operations can be found in Gellman and Fitzgerald [204], Flint [191], Lee [334], and U.S. GAO report [410].

Further details on random number generation, with excellent computer codes, can be found in Press, Teukolosky, Vetterling, and Flannery [429]. Furthermore, there are many excellent books on simulations. We recommend Law and Kelton [331] for an introductory-level text.

Human-resources aspects of a RM implementation can be found in Donaghy and McMahon-Beattie [160, 159], Jones and Hamilton [272], Luciani [350], Peters and Riley [416], and Yeoman and Watson [585].

Discussions on fairness and consumer perceptions can be found in the following articles: Kimes [302], Campbell [98] and Kaufmann, Ortmeyer, and Smith [296]. Besides, the prospect theory of psychology and economics also has relevance on consumer perceptions of price discrimination (Kahneman and Tversky [278]).

The material on organizational issues has been gathered from many industry presentations and conversations with industry RM practitioners.

APPENDIX 11.A: Normal and Inverse Normal Approximations

Most spreadsheet programs and mathematical libraries have functions for the normal distribution and its inverse. The formulas in this appendix are useful for anyone programming a RM system (say, an EMSR algorithm). There are a number of such approximations circulating as folklore in the scientific programming community. The normal approximation we present is due from Abramowitz and Stegun [2] and Bagby [24] and is accurate up to four digits. The inverse normal distribution is based on Halley's method ([455]) and presented here as implemented by Acklam [3].

It is common in quantity-based RM practice to assume that the aggregate number of customers follows a normal distribution truncated to the left at zero. As is well-known, there are no closed-form expressions for the normal distribution and its inverse. In simulations (as well as in RM optimization algorithms such as the EMSRb) rational approximation functions to the normal distribution are used, which we describe next. Both are highly accurate and sufficient for most practical applications in RM.

The following approximation is accurate up to four digits.

Approximation formula for $\tilde{F}(x) \approx P(X \leq x)$, where $X \sim$ Normal$(0, 1)$:

if $x \geq 4.0$, $\tilde{F}(x) = 1.0$ stop
if $x \leq -4.0$, $\tilde{F}(x) = 0.0$ stop
let
$\quad a_1 = 0.319381530$
$\quad a_2 = -0.356563782$
$\quad a_3 = 1.781477937$
$\quad a_4 = -1.821255978$
$\quad a_5 = 1.330274429$
$\quad b = 0.2316419$
$\quad c = \frac{1}{\sqrt{2\pi}} \approx 0.398942208$

if $x > 0.0$, let
$\quad y = \frac{1}{1+bx}$
$\quad \tilde{F}(x) = 1 - ((e^{-0.5x^2} c)(y(a_1 + y(a_2 + y(a_3 + y(a_4 + a_5 y))))))$
else
$\quad y = \frac{1}{1-bx}$
$\quad \tilde{F}(x) = (e^{-0.5x^2} c)(y(a_1 + y(a_2 + y(a_3 + y(a_4 + a_5 y)))))$
end if

The next approximation method for inverse normal is considered accurate to nine digits.

Approximation formula for $\tilde{F}^{-1}(p) \approx F^{-1}(p) : (0, 1) \to \Re$, and $F^{-1}(p) = x$ if and only if $P(X \leq x) = p$, where $X \sim$ Normal$(0, 1)$:

let
$\quad a_1 = -3.969683028665376E + 01$

$a_2 = 2.209460984245205E + 02$
$a_3 = -2.759285104469687E + 02$
$a_4 = 1.383577518672690E + 02$
$a_5 = -3.066479806614716E + 01$
$a_6 = 2.506628277459239E + 00$
$b_1 = -5.447609879822406E + 01$
$b_2 = 1.615858368580409E + 02$
$b_3 = -1.556989798598866E + 02$
$b_4 = 6.680131188771972E + 01$
$b_5 = -1.328068155288572E + 01$
$c_1 = -7.784894002430293E - 03$
$c_2 = -3.223964580411365E - 01$
$c_3 = -2.400758277161838E + 00$
$c_4 = -2.549732539343734E + 00$
$c_5 = 4.374664141464968E + 00$
$c_6 = 2.938163982698783E + 00$
$d_1 = 7.784695709041462E - 03$
$d_2 = 3.224671290700398E - 01$
$d_3 = 2.445134137142996E + 00$
$d_4 = 3.754408661907416E + 00$
$P_l = 0.02425$
$P_h = 1 - P_l$

if $0 < p < P_l$,
    let $q = \sqrt{-2\ln(p)}$
    $\tilde{F}^{-1}(p) = \frac{(((((c_1 q + c_2)q + c_3)q + c_4)q + c_5)q + c_6)}{((((d_1 q + d_2)q + d_3)q + d_4)q + 1)}$
end if

if $P_l \leq p \leq P_h$,
    let $q = p - 0.5$, $r = q^2$
    $\tilde{F}^{-1}(p) = \frac{(((((a_1 r + a_2)r + a_3)r + a_4)r + a_5)r + a_6)q}{(((((b_1 r + b_2)r + b_3)r + b_4)r + b_5)r + 1)}$
end if

if $P_h < p < 1$,
    let $q = \sqrt{-2\ln(1-p)}$
    $\tilde{F}^{-1}(p) = \frac{-(((((c_1 q + c_2)q + c_3)q + c_4)q + c_5)q + c_6)}{((((d_1 q + d_2)q + d_3)q + d_4)q + 1)}$
end if

Appendix A
Notation

Scalars, Vectors, and Matrices

Scalars Scalars are denoted by plain (not boldface) characters, such as x, a, i, μ.

Vectors Vectors are denoted by boldface characters, so, for example, $\mathbf{x} = (x_1, x_2, \ldots, x_n)$.

Matrices Matrices are denoted by boldface uppercase characters, such as $\mathbf{A} = [a_{ij}]$ where a_{ij} denotes the element in the ith row and j^{th} column of $\mathbf{A}$. The ith row of a matrix $\mathbf{A}$ is denoted $\mathbf{A}^i$, and the jth column is denoted $\mathbf{A}_j$.

Inner products The inner product of two vectors $\mathbf{x}$ and $\mathbf{y}$ is denoted $\mathbf{x}^\top \mathbf{y}$ and is defined by

$$\mathbf{x}^\top \mathbf{y} = \sum_{i=1}^{n} x_i y_i.$$

The inner product of a matrix $\mathbf{A}$ and a vector $\mathbf{x}$ is denoted $\mathbf{A}^\top \mathbf{x}$ and is defined as the vector

$$\mathbf{A}^\top \mathbf{x} = (\mathbf{A}^{1\top}\mathbf{x}, \mathbf{A}^{2\top}\mathbf{x}, \ldots, \mathbf{A}^{n\top}\mathbf{x}).$$

The following is a list of variables along with a description of their typical meanings throughout the text.

Roman Variables

$\mathbf{A}$ Incidence matrix for a network model $\mathbf{A} = [a_{ij}]$, where $a_{ij} = 1$ if resource i is used by product j and $a_{ij} = 0$ otherwise; m rows, n columns.

$\mathbf{A}^i$ The ith row of the incidence matrix $\mathbf{A}$.

A^i The set of products that use resource i.

$\mathbf{A}_j$ The jth column of the incidence matrix $\mathbf{A}$. Also used to denote the set of resources used by product j.

A_j The set of resources used by product j.

$B_j(y, D)$ The jth "fill event."

b_j Booking limit or nested booking limit.

$c, c(x)$ Variable cost of production; cost function. Used in economics and overbooking models

$C_i, \mathbf{C}$ Initial capacity of resource i; vector of initial capacities. Also used to denote the jth complete set, $C_j = \{1, \ldots, j\}$.

$d_j, \mathbf{d}, d(p), \mathbf{d}(\mathbf{p})$ Demand (deterministic or mean) for product j; vector of demands. A demand function depending on price p; vector demand function.

$D_j, \mathbf{D}$ Demand (random variable) for product j; vector of demand random variables.

$h_{ij}, \mathbf{h}$ Cost parameters or vector of cost parameters in an overbooking models.

i Generally indexes resources but also used as a generic index.

j Generally indexes products but also used as a generic index.

$J(p), J(v)$ The marginal revenue as a function of price; the virtual value of a buyer with value v.

k Capacity cost in economics models; generic integer variable.

m The number of resources; generic integer variable.

n The number of products; generic integer variable.

N Population size or market potential in a pricing or an auction model.

$\mathcal{N}$ Denotes the set $\{1, 2, \ldots, n\}$ (e.g., set of n choice alternatives).

$p_j(t), \mathbf{p}(t), p_j, \mathbf{p}$ Price of product j at time t or vector of prices at time t; static price of product j; vectors of static prices.

$q_j, q_t, \mathbf{q}$ The probability that a customer shows up (e.g., the probability that class j does not cancel); vectors of probabilities.

$R(v)$ Expected revenue in an auction for buyer with value v.

S, S_k A subset of product classes or alternatives in a choice model; also used to represent a sum of random variables.

t Used to index time, either in discrete or continuous time.

T The number of periods in a discrete-time problem or the length of the horizon in a continuous-time problem. Also used to denote a generic set.

$u_j, \mathbf{u}, \mathbf{u}(t), u(x)$ Control variables in a dynamic program or other optimization problem, most often an accept or deny decision or a quantity decision. Also, u_j is used to denote the mean of a random-utility U_j in a random-utility model or to denote a utility function in economics models as in $u(x)$ is the utility of x.

$U_j, \mathbf{U}$ Random utility (random variable); vector of random utilities.

$v_j, \mathbf{v}$ Reservation price (private value) of customer j; vector of reservation price (private values).

$V_j(x), V_t(x)$ Optimal value function.

$V_t^M(\mathbf{x})$ A given approximation M to the optimal value function (e.g., $V_t^{DLP}(x)$ is the approximation of the value function produced by the deterministic linear program (DLP) model).

$x_i, \mathbf{x}$ Capacity variable; vector of capacities. For example, the remaining capacity of resource i in a dynamic program or the quantity of capacity chosen by firm i. Also used as the decision variable in overbooking models, where it represents the

APPENDIX A: Notation

overbooking limit (virtual capacity). Vector of such state variables or capacities. Finally, used as capacity- or quantity-choice variable in economic models.

$y, y_j, \mathbf{y}$ Allocation variable or protection level for product j; vector of allocations or protection levels. Used in models for finding partitioned or nested allocations. Also the state variable (number of reservations on hand) in overbooking models.

z_t Notation used in forecasting. Data value of a forecast observed at time t (realization of random variable Z_t).

$\hat{z}_t$ Notation used in forecasting. Forecast (point estimate) of time-series value at time t (estimate of unrealized value Z_t).

Z_t Notation used in forecasting. The tth random variable in a time series $Z_1, Z_2, \ldots$.

$Z(x), Z(y)$ Number of customers who show up (number of survivals) from a given number x, y of reservations on hand. Used in overbooking models.

$\bar{Z}(x)$ Number of customers who cancel from a given number x of reservations on hand; $\bar{Z}(x) = x - Z(x)$.

Greek Variables

λ, λ_j An arrival rate in a deterministic demand model and arrival intensity or arrival probability in a probabilistic-demand model.

Δ The first-difference operator; if $g(x)$ is a function, then $\Delta g(x) = g(x) - g(x-1)$.

$\epsilon(p), \epsilon_{ij}(\mathbf{p})$ The elasticity of demand; the cross-price elasticity of demand for product i with respect to the price of product j.

μ The mean of a random variable.

$\Omega, \Omega_p, \Omega_d$ A constraint set; the contraint set of prices p and demand rates d.

$\pi_i, \pi_i(x), \pi$ A bid price value or function—or a dual price from a math program.

σ The variance of a random variable.

θ A generic parameter of a distribution or a scaling parameter.

$\Phi(z)$ The standard normal distribution (i.e., $\Phi(z) = \int_{-\infty}^{x} \frac{1}{\sqrt{2\pi}} e^{-z/2} dz$).

$\phi(z)$ The standard normal density (i.e., $\phi(z) = \frac{1}{\sqrt{2\pi}} e^{-\frac{z}{2}}$).

$\psi_X(t)$ The moment-generating function of a random variable X.

ω An elementary outcome in a probability space (e.g., a random variable is $X(\omega)$).

Miscellaneous Symbols and Notation

$\Re, \Re_+, \Re^n, \Re^n_+$ The set of real numbers $(+\infty, +\infty)$; the set of nonnegative real numbers $[0, +\infty)$; the n-dimensional real plane and the n-dimensional positive orthant.

$\mathcal{Z}$ The set of integers, $\{\ldots, -2, -1, 0, 1, 2, \ldots\}$.

$\mathbf{x}^\top, \mathbf{A}^\top$ The transpose of a vector $\mathbf{x}$ or a matrix $\mathbf{A}$.

$x^+, (a-b)^+$ The positive part of x equal to $\max\{0, x\}$; the positive part of the quantity $(a-b)$.

$x^-, (a-b)^-$ The negative part of x equal to $\max\{0, -x\}$; the negative part of the quantity $(a-b)$.

e_j The jth unit vector; a vector with one in the jth component and zero in all other components.

$\mathbf{x}_{-j}$ The vector $\mathbf{x}$ without the jth component; that is, the vector $\mathbf{x}_{-j} = (x_1, \ldots, x_{j-1}, x_{j+1}, \ldots, x_n)$.

C^1, C^2 The class of continuously differentiable functions on $\Re^n$; the class of all twice-continuously differentiable functions on $\Re^n$.

Abbreviations

a.s. Almost surely.

c.d.f. Cumulative distribution function.

i.i.d. Independent and identically distributed.

p.d.f. Probability-density function.

p.m.f. Probability mass function.

Appendix B
Probability

Probability Spaces and Random Variables

A *probability space* is defined by a triple $(\Omega, \mathcal{F}, P)$, where Ω is a given set of elementary outcomes, $\mathcal{F}$ is a collection of subsets of Ω (each such subset B is called an *event*), and $P(\cdot)$ is a *probability measure* that assigns a nonnegative number $P(B)$ to each subset B in $\mathcal{F}$.

The collection of subsets $\mathcal{F}$ must satisfy

- If B is in $\mathcal{F}$, then so is its complement $\bar{B} = \{\omega \in \Omega : \omega \notin B\}$.
- If $B_1, B_2, \ldots$ are events in $\mathcal{F}$, then $\cup_k B_k$ and $\cap_k B_k$ are also in $\mathcal{F}$.

The probability measure must satisfy

- $P(B) \geq 0$ for all $B \in \mathcal{F}$.
- $P(\Omega) = 1$
- If $B_1, B_2, \ldots$ are disjoint events, the $P(\cup_k B_k) = \sum_k P(B_k)$.

A *random variable* is a function mapping elementary outcomes to real numbers, $X : \Omega \to \Re$ and is denoted $X(\omega)$—or simply X, where the dependence on ω is implicit. The *cumulative distribution function* (c.d.f.) of a random variable X—or just *distribution function* for short—is defined by

$$F(x) = P(X \leq x).$$

If X takes on only countable values, we define the *probability-mass function* (pmf) by the function

$$P(x) = P(X = x).$$

Such a random variable is said to be *discrete*. If F is differentiable, then the *probability-density function* is defined by

$$f(x) = \frac{\partial}{\partial x} F(x).$$

Such a random variable is said to be *continuous*.

Let $\mathbf{X} = (X_1, \ldots, X_n)$ denote a vector of random variables and $\mathbf{x} = (x_1, \ldots, x_n)$ a real vector. Then we define the *joint distribution* of X by

$$F(\mathbf{x}) = P(X_1 \leq x_1, \ldots, X_n \leq x_n).$$

The random variables $X_1, \ldots, X_n$ are said to be *independent random variables* if

$$F(\mathbf{x}) = \prod_{i=1}^{n} F_i(x_i),$$

where $F_i(x_i) = P(X_i \leq x_i)$ (which is referred to as the *marginal distribution* of X_i). If, in addition, $F_i(x_i) = F(x_i)$ for all i, then the random variables are said to be *independent and identically distributed*—or i.i.d. for short.

Expectations and Moment-Generating Functions

The *expected value*—or *mean*—of a random variable X is defined by the integral

$$E[X] = \int_{\Re} x \, dF(x),$$

where the right-hand side above is equal to $\int_{\Re} x f(x) dx$ if X is continuous and $\sum_x x P\{X = x\}$ when X is discrete. For a general function $g : \Re \to \Re$, the expected value of $g(X)$ is defined by

$$E[g(X)] = \int_{\Re} g(x) \, dF(x).$$

The *variance* of X is defined as

$$Var(X) = E[(X - E[X])^2].$$

If X and Y are two random variables, the *covariance* is defined by

$$Cov(X, Y) = E[(X - E[X])(Y - E[Y])].$$

The *moment-generating function* of X is defined as

$$\psi_X(t) = E[e^{tX}].$$

The nth *moment* of X is defined as $E[X^n]$. If the moment-generating function exists then one can determine the nth moment of X using the fact that

$$E[X^n] = \frac{\partial^n}{\partial t} \psi_X(t)|_{t=0},$$

where $\frac{\partial^n}{\partial t}$ denotes the nth derivative with respect to t.

The moment-generating function is also useful for analyzing sums of random variables. Indeed, if X and Y are two independent random variables with moment-generating functions $\psi_X(t)$ and $\psi_Y(t)$, respectively, and $Z = X + Y$, then

$$\psi_Z(t) = \psi_X(t)\, \psi_Y(t).$$

That is, the moment-generating function of a sum of independent random variables is simply the product of their individual moment-generating functions.

Inequalities

Jensen's inequality states that if g is a convex function, then

$$E[g(X)] \geq g(E[X]).$$

This is often useful in obtaining bounds on stochastic optimization problems.

Another useful bound in RM problems is due to Gallego [200] and involves a bound on the function $(X-x)^+ = \max\{X-x, 0\}$ (the positive part of $X-x$). It states that for any random variable X with mean μ and finite variance σ^2,

$$E[(X-x)^+] \leq \frac{\sqrt{\sigma^2 + (x-\mu)^2} - (x-\mu)}{2}.$$

For example, if X is demand and x is a capacity level, then $(X-x)^+$ is the rejected demand (spilled demand) and the above bound provides an upper bound on the expected spilled demand

Some Useful Distributions

We next provide the basic definitions of the most commonly used distributions in RM problems.

Discrete Distributions

Bernoulli

A random variable X has a Bernoulli distribution if it takes on only two values, 0 and 1. A Bernoulli distribution is characterized by a single parameter q (the probability that $X = 1$) with $0 \leq q \leq 1$. In RM, it is often used as the model of a single cancellation.

The basic definitions and properties are

$$P(x) = \begin{cases} q & x = 1 \\ 1-q & x = 0 \end{cases}$$
$$E[X] = q$$
$$Var(X) = q(1-q)$$
$$\psi(s) = qe^s + (1-q).$$

Binomial

A random variable X has a binomial distribution if it is the sum of n independent Bernoulli random variables. For example, the number of cancellations in a group of n reservations when each independently cancels with probability q. A binomial distribution is characterized by the two parameters q and n with $0 \leq q \leq 1$ and $n \geq 1$.

The basic definitions and properties are

$$P(x) = \binom{n}{x} q^x (1-q)^{n-x}, \quad x = 0, 1, \ldots, n$$
$$E[X] = nq$$

$$Var(X) = nq(1-q)$$
$$\psi(s) = (qe^s + (1-q))^n.$$

Poisson

In RM, the Poisson distribution is used as a model of demand or as a (continuous parameter) approximation to the Binomial distribution. It is characterized by a single nonnegative parameter λ (its mean).

The basic definitions and properties are

$$P(x) = e^{-\lambda}\frac{\lambda^x}{x!}, \; x = 0, 1, \ldots$$
$$E[X] = \lambda$$
$$Var(X) = \lambda$$
$$\psi(s) = e^{\lambda(e^s - 1)}.$$

Continuous Distributions
Uniform

A uniform distribution is defined by two constants $a < b$ and represents a case where the random variable is equally likely to assume any value in the interval $[a, b]$.

The basic definitions and properties are

$$f(x) = \frac{1}{b-a}, \; a \leq x \leq b$$
$$E[X] = \frac{a+b}{2}$$
$$Var(X) = \frac{(b-a)^2}{2}$$
$$\psi(s) = \frac{e^{sb} - e^{sa}}{s(b-a)}.$$

Exponential

The exponential distribution is defined by a single parameter λ.

The basic definitions and properties are

$$f(x) = \lambda e^{-\lambda x}, \; x \geq 0$$
$$E[X] = \frac{1}{\lambda}$$
$$Var(X) = \frac{1}{\lambda^2}$$
$$\psi(s) = \frac{\lambda}{\lambda - s}.$$

APPENDIX B: Probability

Normal

The normal (or Gaussian) distribution is frequently used as a model of demand. It is characterized by two parameters, its mean μ and its variance σ^2.
The basic definitions and properties are

$$f(x) = \frac{1}{\sqrt{2\pi}\sigma} e^{-\frac{(x-\mu)^2}{2\sigma^2}}, \quad -\infty < x < \infty$$
$$E[X] = \mu$$
$$Var(X) = \sigma^2$$
$$\psi(s) = e^{\mu s + \frac{\sigma^2 s}{2}}.$$

The normal has the property that if X and Y are two independent normal random variables, then the sum $X + Y$ also has a normal distribution (it is "closed under addition"). For example, if X and Y are independent with means μ_X, μ_Y and variances σ_X^2, σ_Y^2 (respectively), then their sum $X + Y$ has a normal distribution with mean $\mu_X + \mu_Y$ and variance $\sigma_X^2 + \sigma_Y^2$.

Gumbel

The Gumbel (or double-exponential) distribution is frequently used in discrete-choice models because it is "closed under maximization." That is, the maximum of two Gumbel random variables is also a Gumbel random variable. It is characterized by two parameters, a scale parameter μ and location parameter η.
The basic definitions and properties are

$$f(x) = \frac{1}{\mu} e^{-\frac{x-\eta}{\mu}} e^{-e^{-\frac{x-\eta}{\mu}}} \quad -\infty < x < \infty$$
$$E[X] = \eta + \frac{1}{\gamma\mu}$$
$$Var(X) = \frac{\mu^2 \pi^2}{6}$$
$$\psi(s) = e^{\eta s/\mu} \Gamma(1 + s\mu),$$

where $\gamma \approx 0.577$ is Euler's constant and $\Gamma(x)$ is the extension of the factorial function to real numbers

$$\Gamma(x) = \int_0^\infty t^{x-1} e^{-t} dt.$$

If X_1 and X_2 are two independent Gumbel random variables with parameters (η_1, μ) and (η_2, μ) respectively, then $\max\{X_1, X_2\}$ is a Gumbel random variable with parameters $(\mu(\ln(e^{\eta_1/\mu} + e^{\eta_2/\mu}), \mu)$.

Stochastic Monotonicity and Convexity

Consider a random variable X that depends on some parameter θ, so that $X = X(\theta)$. That is, $X(\theta)$ is a *random function* of θ. For example, X could be the number of customers who show up out of θ reservations, in which case

$$X(\theta) = \sum_{i=1}^{\theta} Y_i,$$

where Y_i are i.i.d. Bernoulli random variables with $P(Y_i = 1) = q$ and $P(Y_i = 0) = 1 - q$.

Given a function $g(x)$, suppose we are interested in determining properties of the expected value $E[g(X(\theta)]$ as a function of θ. For example, if g is increasing in x, is $E[g(X(\theta))]$ increasing in θ? If g is convex in x, is $E[g(X(\theta))]$ convex in θ? Stochastic monotonicity and convexity identify classes of random variables $X(\theta)$ for which such statements can be made. A good source for this material is the series of papers by Shaked and Shantikumar [460, 461] and their subsequent book [462].

DEFINITION B.1 *The random function $X(\theta)$ is* **stochastically increasing** *in θ if for all $\theta_1 \geq \theta_2$, $P(X(\theta_1) > x) \geq P(X(\theta_2) > x)$.*

A random function $X(\theta)$ is *stochastically decreasing* in θ if $-X(\theta)$ is stochastically increasing. An equivalent definition is provided by the following proposition:

PROPOSITION B.1 $X(\theta)$ *is stochastically increasing in θ if for any $\theta_1 \geq \theta_2$, there exists two random variables X_1 and X_2 defined on a common probability space $(\Omega, \mathcal{F}, P)$, such that X_1 and X_2 are equal in distribution to $X(\theta_1)$ and $X(\theta_2)$ (respectively), and they satisfy $X_1(\omega) \geq X_2(\omega)$ for all $\omega \in \Omega$.*

Continuing our example, we see that if $X(\theta) = \sum_{i=1}^{\theta} Y_i$, where Y_i are i.i.d. Bernoulli random variables, then $X(\theta)$ is stochastically increasing, since we can consider ω to define an infinite sequence $\{Y_1, Y_2, \ldots\}$ and consider $X(\theta)$ to be the sum of the first θ variables in this sequence. For every $\theta_1 \geq \theta_2$, the sums $X(\theta_1)$ and $X(\theta_2)$ will have the required distribution and $X(\theta_1) \geq X(\theta_2)$ for every such sequence ω.

The following proposition follows easily from this sample path definition of monotonicity:

PROPOSITION B.2 $X(\theta)$ *is stochastically increasing in θ if and only if for any real valued, increasing function $g(x)$, $E[g(X(\theta))]$ is increasing in θ.*

Similarly, one can define a notion of stochastic convexity for $X(\theta)$:

DEFINITION B.2 $X(\theta)$ *is* **stochastically convex** *(SCX) if for any real valued, convex function $g(x)$, $E[g(X(\theta))]$ is convex in θ.*

We say $X(\theta)$ is *stochastically concave* (SCV) if $-X(\theta)$ is stochastically convex, and we say $X(\theta)$ is *stochastically linear* if it is both stochastically convex and stochastically concave.

To verify whether the above holds is often difficult. However, two stronger notions of stochastic convexity are quite useful and both imply stochastic convexity. These are:

DEFINITION B.3 $X(\theta)$ *is said to be* **strongly stochastically convex** *(SSCX) if $X(\theta) = \psi(Z, \theta)$ where Z is a random variable independent of θ and ψ is convex in θ for every value of Z.*

APPENDIX B: Probability 641

For example, suppose $X(\theta) = \sigma Z + \theta$, where Z is a standard normal random variable. Then $X(\theta)$ is normal with mean θ and standard deviation σ, and $X(\theta)$ is strongly stochastically convex in θ.

A somewhat weaker version of stochastic convexity is the following:

DEFINITION B.4 $X(\theta)$ **is stochastically convex in the sample-path sense** *(SCX-sp) if for any four values $\theta_i, i = 1,2,3,4$ satisfying $\theta_2 - \theta_1 = \theta_4 - \theta_3$ and $\theta_4 \geq \max\{\theta_2, \theta_3\}$, there exist random variable $X_i, i = 1,2,3,4$ defined on a common probability space $(\Omega, \mathcal{F}, P)$, such that X_i is equal in distribution to $X(\theta_i)$, $i = 1,2,3,4$ and*

$$X_4(\omega) - X_3(\omega) \geq X_2(\omega) - X_1(\omega),$$

for all $\omega \in \Omega$.

To illustrate, we show that the sum of Bernoulli random variables is stochastically convex (and concave) in this sample path sense. To do so, let $\theta_i, i = 1,2,3,4$ be integers satisfying $\theta_2 - \theta_1 = \theta_4 - \theta_3$ and $\theta_4 \geq \max\{\theta_2, \theta_3\}$, and let ω define an infinite sequence $\{Y_1, Y_2, \ldots\}$ of i.i.d. Bernoulli random variables as before. Note that $\theta_1 \leq \min\{\theta_2, \theta_3\}$ (else $\theta_4 < \max\{\theta_2, \theta_3\}$), and define

$$X_1 = \sum_{i=1}^{\theta_1} Y_i$$

$$X_3 = \sum_{i=1}^{\theta_3} Y_i$$

$$X_4 = \sum_{i=1}^{\theta_4} Y_i$$

$$X_2 = X_1 + (X_4 - X_3).$$

Note X_i is equal in distribution to $X(\theta_i)$ since each is the sum of θ_i i.i.d. Bernoulli random variables, and by construction

$$X_4 - X_3 = X_2 - X_1,$$

so $X(\theta)$ is stochastically convex in the sample path sense.

The following proposition relates these versions of stochastic convexity:

PROPOSITION B.3 *SSCX $\Rightarrow$ SICX-sp $\Rightarrow$ SCX.*

So showing $X(\theta)$ is either strongly stochastically convex or stochastically convex in the sample path sense, implies that $X(\theta)$ is stochastically convex. Again, returning to our example, this implies that if $X(\theta)$ is the sum of θ i.i.d. Bernoulli random variables and $g(x)$ is a convex function, the $E[g(X(\theta))]$ is convex in θ.

Appendix C
Convexity and Optimization

Here we review the basic theory of optimization problems and associated definitions.

Convex Functions and Sets

Convexity is central to the theory of optimization. We begin with a definition of a convex function:

DEFINITION C.5 *A function $f : \Re^n \to R$ is* **convex** *on a set $X \subseteq \Re^n$ if, for all $\mathbf{x}, \mathbf{y} \in X$ and $\alpha \in [0,1]$*

$$f(\alpha \mathbf{x} + (1-\alpha)\mathbf{y}) \leq \alpha f(\mathbf{x}) + (1-\alpha)f(\mathbf{y})$$

If the inequality above is strict for all $\mathbf{x} \neq \mathbf{y}$, then f is said to be *strictly convex*. A function f is said to be *concave* if $-f$ is convex and *strictly concave* if $-f$ is strictly convex.

Convexity can also be defined for sets:

DEFINITION C.6 *A set $X \subseteq \Re^n$ is a* **convex set** *if, for all $\mathbf{x} \in X$, $\mathbf{y} \in X$ and $\alpha \in [0,1]$*

$$\alpha \mathbf{x} + (1-\alpha)\mathbf{y} \in X.$$

A point of the form $\alpha\mathbf{x}+(1-\alpha)\mathbf{y}$ is referred to as a *convex combination* of the points $\mathbf{x}$ and $\mathbf{y}$. A convex set, therefore, is one with the property that any convex combination of points in the set is also contained in the set. We also have

DEFINITION C.7 *A point $\mathbf{x}$ is said to be an* **extreme point** *of a convex set X if there are no two distinct points $\mathbf{y}, \mathbf{z} \in X$ with $\mathbf{y} \neq \mathbf{z}$ such that $\mathbf{x} = \alpha \mathbf{z} + (1-\alpha)\mathbf{y}$ for some $0 < \alpha < 1$.*

In other words, $\mathbf{x}$ cannot be expressed as the convex combination of two distinct points in X.

Let C^1 denotes the class of continuously differentiable functions on $\Re^n$ and C^2 denote the class of all twice-continuously differentiable functions on $\Re^n$. (See below.) Here are some properties of convex functions:

PROPOSITION C.4
(i) If f and g are convex, then $h(\mathbf{x}) = f(\mathbf{x}) + g(\mathbf{x})$ is convex.
(ii) If f is convex, then $h(\mathbf{x}) = af(\mathbf{x})$ is convex for all $a \geq 0$ and is concave for all $a \leq 0$.
(iii) If f is a convex function on a convex set X, then the level set $L(c) = \{\mathbf{x} : \mathbf{x} \in X, f(\mathbf{x}) \leq c\}$ is a convex set.
(iv) If $f \in C^1$, then f is convex over a convex set X if and only if $f(\mathbf{y}) \geq f(\mathbf{x}) + \nabla \mathbf{f}(\mathbf{x})^\top (\mathbf{y} - \mathbf{x})$ for all $\mathbf{x}, \mathbf{y} \in X$.
(v) If $f \in C^2$, then f is convex over a convex set X containing an interior point if and only if the Hessian, $\nabla^2 \mathbf{f}(\mathbf{x})$, is positive semidefinite throughout X.

Derivatives and Subderivatives

Let $\mathbf{e}_i$ denote the ith unit vector (the vector with all components zero except for the ith component, which is one). Then the ith partial derivative of a function $f : \Re^n \to \Re$ is defined by

$$\frac{\partial}{\partial x_i} f(\mathbf{x}) = \lim_{h \to 0} \frac{1}{h} \left[f(\mathbf{x} + h \mathbf{e}_i) - f(\mathbf{x}) \right],$$

provided the limit exists (where here $h \to 0$ denotes h tending to zero from above or below). If all partial derivatives exist, the *gradient* is defined as the (column) vector

$$\nabla \mathbf{f}(\mathbf{x}) = (\frac{\partial}{\partial x_1} f(\mathbf{x}), \ldots, \frac{\partial}{\partial x_n} f(\mathbf{x})).$$

If each of the partial derivatives of f at $\mathbf{x}$ is itself a differentiable function of x, then we define the *second partial derivatives* by

$$\frac{\partial}{\partial x_i \partial x_j} f(\mathbf{x}) = \lim_{h \to 0} \frac{1}{h} \left[\frac{\partial}{\partial x_i} f(\mathbf{x} + h \mathbf{e}_j) - \frac{\partial}{\partial x_i} f(\mathbf{x}) \right].$$

The $n \times n$ matrix of second partial derivatives is called the *Hessian* of f at x and is denoted

$$\nabla^2 \mathbf{f}(\mathbf{x}) = \left[\frac{\partial}{\partial x_i \partial x_j} f(\mathbf{x}) \right].$$

Consider a vector direction $\mathbf{d} \in \Re^n$. The *directional derivative* is defined by

$$D_f(\mathbf{x}; \mathbf{d}) = \lim_{h \downarrow 0} \frac{1}{h} \left[f(\mathbf{x} + h\mathbf{d}) - f(\mathbf{x}) \right], \qquad (C.1)$$

provide the limit exists. The function f is said to be *differentiable* at $\mathbf{x}$ if and only if $\nabla \mathbf{f}(\mathbf{x})$ exists and

$$D_f(\mathbf{x}; \mathbf{d}) = \nabla \mathbf{f}(\mathbf{x})^\top \mathbf{d}, \quad \forall \mathbf{d} \in \Re^n.$$

A function is said to be *continuously differentiable* on a set X if the gradient $\nabla \mathbf{f}(\mathbf{x})$ exists for all $\mathbf{x} \in X$ and is continuous on X. The class of all continuously differentiable functions on $\Re^n$ is denoted C^1; C^2 denotes the class of all functions with continuous second partial derivatives on $\Re^n$.

For a convex function f, the gradient exists almost everywhere (at all but a countable number of points in X). If f is convex but the gradient does not exist everywhere, it is useful to define a generalization of the gradient called a *subgradient* of f.

DEFINITION C.8 *Let $f : \Re^n \to \Re$ be a convex function defined on $\Re^n$. A* **subgradient** *of f at point $\mathbf{x}$ is a vector $\mathbf{d} \in \Re^n$ satisfying*

$$f(\mathbf{y}) \geq f(\mathbf{x}) + \mathbf{d}^\top(\mathbf{y} - \mathbf{x}), \quad \forall \mathbf{y} \in \Re^n.$$

The *subdifferential* of a f at $\mathbf{x}$, denoted $\partial f(\mathbf{x})$, is defined as the set of all subgradients of f at $\mathbf{x}$. The definition for concave functions simply has the above inequality reversed.

Optimization Problems

Let $\mathbf{x} \in \Re^n$ denote a vector of *decision variables*, $f : \Re^n \to R$ be a given *objective function* and $X \subseteq \Re^n$ be a *constraint set*. A point $\mathbf{x}$ in the set X is called *feasible*, and points not in X are called *infeasible*. In a *maximization problem*, we seek a feasible solution $\mathbf{x}^*$—called a *global maximum* (or *global optimum*)—such that

$$f(\mathbf{x}^*) \geq f(\mathbf{x}), \quad \forall \mathbf{x} \in X. \tag{C.2}$$

Equivalently, $\mathbf{x}^*$ solves

$$\max_{\mathbf{x} \in X} f(\mathbf{x}). \tag{C.3}$$

We say such an $\mathbf{x}^*$ is *globally optimal*. If X is the empty set, then the above optimization problem is said to be *infeasible*; otherwise, the problem is *feasible*. If $X = \Re^n$, the problem is said to be *unconstrained*. The problem is called *unbounded* if there exists a sequence of feasible points $\{\mathbf{x}^{(k)}; k = 1, 2, \ldots\}$ with $\mathbf{x}^{(k)} \in X$ for all k and $\lim_{k \to \infty} f(\mathbf{x}^{(k)}) = +\infty$.

A minimization problem reverses the inequality above and is equivalent to maximizing $-f(\mathbf{x})$. We focus here on only the maximization version. If f is concave and X is convex, then the problem (C.3) is called a *convex optimization problem*.

Let $N(\mathbf{x}, \epsilon) = \{\mathbf{y} : \|\mathbf{y} - \mathbf{x}\| \leq \epsilon\}$ denote the ball of radius ϵ about the point $\mathbf{x}$. A solution $\mathbf{x}^*$ is called a *local maximum* (or *local optimum*) if there exists an $\epsilon > 0$ such that

$$f(\mathbf{x}^*) \geq f(\mathbf{x}), \quad \forall \mathbf{x} \in X \cap N(\mathbf{x}^*, \epsilon).$$

We say such an $\mathbf{x}^*$ is *locally optimal*. Note all global optima are also locally optimal. In the convex case, local and global optima coincide:

PROPOSITION C.5 *If f is a concave function defined on a convex set X, then any local maximum is a global maximum. If f is strictly concave, then if a global maximum exists, it is unique.*

Optimality Conditions

Optimality conditions help identify and characterize optimal solutions. They are useful both theoretically and computationally.

Suppose $f \in C^1$. Then we have the following *first-order necessary conditions* for $\mathbf{x}^*$ to be an optimal solution:

PROPOSITION C.6 *If $f \in C^1$ and $\mathbf{x}^*$ is a local maximum, then there exists an $\epsilon > 0$ such that*

$$\nabla f(\mathbf{x}^*)^\top (\mathbf{x} - \mathbf{x}^*) \leq 0, \quad \forall \mathbf{x} \in X \cap N(\mathbf{x}^*, \epsilon).$$

In particular, if $X = \Re^n$ (the unconstrained case), then this condition reduces to

$$\nabla \mathbf{f}(\mathbf{x}^*) = 0. \tag{C.4}$$

If f is concave, then these conditions are also sufficient:

PROPOSITION C.7 *Suppose $f \in C^1$ is a concave function and X is a convex set. Then if a point $\mathbf{x}^*$ satisfies*

$$\nabla \mathbf{f}(\mathbf{x}^*)^\top (\mathbf{x} - \mathbf{x}^*) \leq 0, \ \forall \mathbf{x} \in X,$$

it is a global maximum.

In the nonconvex unconstrained case, local optimality is guaranteed by the following *second-order sufficiency conditions*:

PROPOSITION C.8 *If $f \in C^2$ and $X = \Re^n$, then if a point $\mathbf{x}^*$ satisfies*
(i) $\nabla \mathbf{f}(\mathbf{x}^*) = 0$
(ii) $\nabla^2 \mathbf{f}(\mathbf{x}^*)$ is positive definite,
it is a local maximum.

There are no general sufficient conditions for global optima in the nonconvex case.

Equality and Inequality Constraints

Suppose the set X is defined by a set of linear equalities. That is, $X = \{\mathbf{x} : h(\mathbf{x}) = \mathbf{b}\}$, where $h : \Re^n \to \Re^m$ (i.e., $\mathbf{h}(\mathbf{x}) = (h_1(\mathbf{x}), \ldots, h_m(\mathbf{x}))$) so the optimization problem to solve is

$$\begin{aligned} \max \quad & f(\mathbf{x}) \\ \text{s.t.} \quad & \mathbf{h}(\mathbf{x}) = \mathbf{b}. \end{aligned}$$

We require the following definition:

DEFINITION C.9 *A point $\mathbf{x}^*$ satisfying $\mathbf{h}(\mathbf{x}^*) = \mathbf{b}$ is said to be a **regular point** of the constraints $\mathbf{h}(\mathbf{x}) = \mathbf{b}$ if the vectors $\nabla \mathbf{h}_1(\mathbf{x}), \ldots, \nabla \mathbf{h}_m(\mathbf{x})$ are linearly independent.*

The assumption of regularity of x^* is an example of what is called a *constraint qualification*, a condition that ensures that the first-order conditions correctly identify a local optimum.

We then have the following first-order necessary conditions:

PROPOSITION C.9 *Suppose $f \in C^1$ and $\mathbf{x}^*$ is a local maximum of the function f over the constraint set $X = \{x : \mathbf{h}(\mathbf{x}) = \mathbf{b}\}$. Then if $\mathbf{x}^*$ is a regular point, there exist a vector $\boldsymbol{\pi} \in \Re^m$ such that*

$$\nabla \mathbf{f}(\mathbf{x}^*) - \boldsymbol{\pi}^\top \nabla \mathbf{h}(\mathbf{x}^*) = 0.$$

A vector $\boldsymbol{\pi}$ above is called a *Lagrange multiplier* of the constraints $\mathbf{h}(\mathbf{x}) = \mathbf{b}$.

If the constraint set is defined by inequalities, so the problem is

$$\begin{aligned} \max \quad & f(\mathbf{x}) \\ \text{s.t.} \quad & \mathbf{g}(\mathbf{x}) \leq \mathbf{d}, \end{aligned}$$

where $\mathbf{g} : \Re^n \to \Re^m$, then similar conditions apply. Indeed, the definition of a regular point in this case is:

APPENDIX C: Convexity and Optimization   647

DEFINITION C.10 *A point* $\mathbf{x}^*$ *satisfying* $\mathbf{g}(\mathbf{x}^*) \leq \mathbf{d}$ *is said to be a* **regular point** *of the constraints* $\mathbf{g}(\mathbf{x}) \leq \mathbf{d}$ *if the vectors in the set* $\{\nabla \mathbf{g}_j(\mathbf{x}^*) : g_j(\mathbf{x}^*) = d_j\}$ *are linearly independent.*

This leads to the following first-order necessary conditions (called the *Kuhn-Tucker conditions*):

PROPOSITION C.10 *Suppose* $f \in C^1$ *and* $\mathbf{x}^*$ *is a local maximum of the function* f *over the constraint set* $X = \{\mathbf{x} : \mathbf{g}(\mathbf{x}) \leq \mathbf{d}\}$. *Then if* $\mathbf{x}^*$ *is a regular point, there exist a vector* $\boldsymbol{\pi} \in \Re^m$ *with* $\boldsymbol{\pi} \geq 0$ *such that*

$$\nabla f(\mathbf{x}^*) - \boldsymbol{\pi}^T \nabla \mathbf{g}(\mathbf{x}^*) = 0$$
$$\boldsymbol{\pi}^T (\mathbf{d} - \mathbf{g}(\mathbf{x}^*)) = 0.$$

In the convex case, both Propositions C.9 and C.10 provide sufficient conditions for optimality. That is, if f is concave, the set X defined by the equality or inequality constraints is convex, and we find a feasible solution $\mathbf{x}^*$ and an associated multiplier $\boldsymbol{\pi}$ satisfy the conditions of Propositions C.9 (or Proposition C.10 in the inequality case), then $\mathbf{x}^*$ is a global maximum.

Sensitivity Analysis

The Lagrange multipliers have an interpretation as giving the rate of change of the objective function as a function of the right-hand side vectors. Indeed, let

$$v(\mathbf{b}) \equiv \max \quad f(\mathbf{x})$$
$$\text{s.t.} \quad \mathbf{h}(\mathbf{x}) = \mathbf{b}.$$

Then under some relatively mild regularity conditions (see Bertsekas [59]), one can show

$$\nabla_b v(\mathbf{b}) = \boldsymbol{\pi},$$

where $\boldsymbol{\pi}$ is the Lagrange multiplier associated with an equality-constrained optimal solution $\mathbf{x}^*$. Similarly, if

$$v(\mathbf{d}) \equiv \max \quad f(\mathbf{x})$$
$$\text{s.t.} \quad \mathbf{g}(\mathbf{x}) \leq \mathbf{d},$$

then

$$\nabla_d v(\mathbf{d}) = \boldsymbol{\pi},$$

where $\boldsymbol{\pi} \geq 0$ is the Lagrange multiplier associated with the corresponding optimal solution $\mathbf{x}^*$. The multipliers therefore measure the effect that small changes in the right-hand-sides have on the optimal objective function value.

Parametric Monotonicity

Paramteric monotonicity addresses the question of how optimal solutions vary as a function of the parameters of an optimization problem. These parametric monotonicity results are used, for example, in the analysis of the base-stock, list price policies of

Section 5.3.2, which show that the optimal list price in an inventory-pricing problem (the optimal solution) is decreasing in the inventory on hand (the parameter).

More abstractly, let $X \subset \Re^n$ be a constraint set, $\Theta \subset \Re^l$ be a set of parameter values and $f : X \times \Theta \to \Re$ be an objective function. We need the following definition:

DEFINITION C.11 *A function $f : X \times \Theta \to \Re$ is said to have increasing differences in $(\mathbf{x}, \boldsymbol{\theta})$, if for all $\mathbf{x}' \geq \mathbf{x}$ and $\boldsymbol{\theta}' \geq \boldsymbol{\theta}$,*

$$f(\mathbf{x}', \boldsymbol{\theta}') - f(\mathbf{x}, \boldsymbol{\theta}') \geq f(\mathbf{x}', \boldsymbol{\theta}) - f(\mathbf{x}, \boldsymbol{\theta}).$$

Define the component-wise minimum (the *meet*) of two vectors $\mathbf{x}$ and $\mathbf{y}$ in $\Re^n$ by

$$\mathbf{x} \wedge \mathbf{y} = (\min\{x_1, y_1\}, \ldots, \min\{x_n, y_n\})$$

and the component-wise maximum (the *join*) of the vectors by

$$\mathbf{x} \vee \mathbf{y} = (\max\{x_1, y_1\}, \ldots, \max\{x_n, y_n\}).$$

A set $X \subset \Re^n$ is called a *lattice* if for all $\mathbf{x}, \mathbf{y}$ in X, the meet and joint of $\mathbf{x}$ and $\mathbf{y}$ are also in X. If in addition X is compact (closed and bounded), then the set X is called a *compact sublattice*. A point $\mathbf{x}^*$ is said to be a *greatest element* (respectively, *least element*) of the sublattice X if $\mathbf{x}^* \geq x$ (respectively, $\mathbf{x}^* \leq \mathbf{x}$) for all $\mathbf{x} \in X$. We then have

PROPOSITION C.11 *If X is a nonempty, compact sublattice, then X has a greatest and least element.*

That is, if X is a compact sublattice, we can always identify a "largest" and "smallest" (component-wise) element of the set X.

Consider next the following definition:

DEFINITION C.12 *A function $f : \Re^n \to \Re$ is said to be supermodular in $\mathbf{z}$ if for all $\mathbf{z}$ and $\mathbf{z}'$ in $\Re^n$,*

$$f(\mathbf{z}) + f(\mathbf{z}') \leq f(\mathbf{z} \wedge \mathbf{z}') + f(\mathbf{z} \vee \mathbf{z}').$$

If f above is a C^2 function, then it is supermodular if and only if the cross-partial derivatives satisfy

$$\frac{\partial^2}{\partial z_i \partial z_j} f(\mathbf{z}) \geq 0, \quad \forall i, j, i \neq j.$$

So for C^2 functions, supermodularity corresponds to nonnegativity of the cross-partial derivatives.

Now consider the following optimization problem, for fixed $\boldsymbol{\theta} \in \Theta$,

$$\max_{\mathbf{x} \in X} \ f(\mathbf{x}, \boldsymbol{\theta}) \quad \text{(C.5)}$$

and define the *optimal action correspondence* (such as a set of optimal solutions)

$$D^*(\boldsymbol{\theta}) = \{\mathbf{x}^* : f(\mathbf{x}^*, \boldsymbol{\theta}) \geq f(\mathbf{x}, \boldsymbol{\theta}) \ \forall \mathbf{x} \in X\}.$$

We want to determine when these optimal solutions are in some sense "increasing" in $\boldsymbol{\theta}$.

APPENDIX C: Convexity and Optimization

The following theorem (see Sundaram [495] and Topkis [516] for proofs) shows that the property of decreasing differences and supermodularity can be used to establish such parametric monotonicity.

THEOREM C.1 *Suppose that (i) the optimization problem (C.5) has at least one optimal solution for each $\theta \in \Theta$, (ii) f satisfies decreasing differences in $(\mathbf{x}, \theta)$, and (iii) f is supermodular in $\mathbf{x}$ for each $\theta \in \Theta$. Then for each θ there exist a greatest optimal solution $\mathbf{x}^*(\theta) \in D^*(\theta)$. This greatest optimal solution is nondecreasing in the parameters θ; that is, $\mathbf{x}^*(\theta_1) \geq^* (\theta_2)$ for all $\theta_1 > \theta_2$.*

This result says that higher values of θ lead to higher optimal decisions $\mathbf{x}^*$. Corresponding definitions of decreasing difference and submodularity are used to show when optimal solutions are decreasing in a given parameter.

Linear Programs

An optimization problem is called a *linear program* if the objective function and all the equality and inequality constraints are defined by linear functions. That is, $f(\mathbf{x}) = \mathbf{c}^\top \mathbf{x}$ for some vector $\mathbf{c} \in \Re^n$, and the constraint set is of the form $X = \{\mathbf{Ax} = \mathbf{b}\}$ or $X = \{\mathbf{Ax} \leq \mathbf{b}\}$ (or combinations of inequality and equality constraints). Since we can always write an equality constraint $\mathbf{a}^\top \mathbf{x} = \mathbf{b}$ as two inequality constraints, $\mathbf{a}^\top \mathbf{x} \leq \mathbf{b}$ and $\mathbf{a}^\top \mathbf{x} \geq \mathbf{b}$, and we can always write a variable $\mathbf{x}$ as $\mathbf{x} = \mathbf{x}^+ - \mathbf{x}^-$, where $\mathbf{x}^+ \geq 0$ and $\mathbf{x}^- \geq 0$ - any linear program can be converted into the form

$$\begin{aligned} \max \quad & \mathbf{c}^\top \mathbf{x} \\ \text{s.t.} \quad & \mathbf{Ax} \leq \mathbf{b} \\ & \mathbf{x} \geq 0. \end{aligned} \quad (C.6)$$

Many problems can be expressed as linear programs, and there are specialized, highly efficient algorithms for solving them; hence, they warrant special attention. We have the following proposition, which shows that we can restrict our search for optimal solutions to extreme point solutions:

PROPOSITION C.12 *If the linear program (C.6) has an optimal solution, then it has an optimal solution that is an extreme point of the set $X = \{\mathbf{x} : \mathbf{Ax} \leq \mathbf{b}, \mathbf{x} \geq 0\}$.*

The popular *simplex algorithm* for solving linear programs is based on searching the extreme points of the set X.

The linear program (C.6) has an associated *dual* linear program (or dual problem) defined by

$$\begin{aligned} \min \quad & \pi^\top \mathbf{b} \\ \text{s.t.} \quad & \pi^\top \mathbf{A} \geq \mathbf{c}^\top \\ & \pi \geq 0. \end{aligned} \quad (C.7)$$

The original problem (C.6) is called the *primal* linear program (or primal problem). The primal and dual problems are related as follows:

PROPOSITION C.13 *If either the primal problem (C.6) or the dual problem (C.7) has a finite optimal solution, then so does the other, and the optimal objective function values are equal. Moreover, if the primal is unbounded, then the dual is infeasible, and if the dual is unbounded, then the primal is infeasible.*

A linear program is called a *network flow* problem if there is a directed graph $G = (N, A)$, where $N = \{1, \ldots, n\}$ is a set of *nodes* and A is a set of directed arcs connecting the nodes in N. That is, an arc is an ordered pair (i,j). Define x_{ij} to be the flow on arc (i,j). Then a minimum-cost network-flow problem is a linear program of the form

$$\max \sum_{i=1}^{n} \sum_{j=1}^{n} c_{ij} x_{ij}$$

$$\text{s.t.} \quad \sum_{j=1}^{n} x_{ij} - \sum_{k=1}^{n} x_{ki} = b_i$$

$$0 \leq x_{ij} \leq u_{ij}, \quad \forall i,j,$$

where c_{ij} are cost coefficients, u_{ij} are upper bounds on the flows (equal to zero if no arc (i,j) exists) and b_i are *source/sink* quantities satisfying $\sum_{i=1}^{n} b_i = 0$. Network-flow problems can be solved even more efficiently than general linear programs using specialized algorithms.

NonDifferentiable Optimization

In nondifferentiable-optimization problems, the gradient $\nabla \mathbf{f}(\mathbf{x})$ may not always exist everywhere on the constraint set X. For example, $f(x)$ may be a scalar continuous, piecewise linear function of x of the form

$$f(x) = \begin{cases} -x & x \geq 0 \\ 3x & x < 0. \end{cases}$$

This function has a corner point at $x = 0$ where the derivative does not exist. In such cases, we have the following necessary and sufficient condition for optimality in the unconstrained, convex case:

PROPOSITION C.14 *If f is concave, and $X = \Re^n$, then $\mathbf{x}^*$ is a global maximum if and only if $0 \in \partial f(\mathbf{x}^*)$.*

Note that for the example given above, f is concave and $\partial f(0) = [-1, 3]$, so zero is contained in the subdifferential at x^* and hence $x^* = 0$ is a maximum. Also observe that if f is differentiable at x^*, the above condition reduces to (C.4).

Appendix D
Dynamic Programming

Fundamentally, dynamic programming addresses how to make optimal decisions over time. While it can be applied to both deterministic and stochastic problems, our focus here is on stochastic problems because making decisions under uncertainty is central to revenue management. Our treatment largely follows that in Bertsekas [57] with some slight variations in notation. We summarize only the key results for the discrete-state, discrete-time, finite-horizon problem, again because it is the most frequently encountered one in RM. The reader is referred to Bertsekas [57] for an extensive treatment of other cases of dynamic programming and a discussion of further theoretical and computational issues.

Elements of a Dynamic Program

Dynamic programming involves the optimal control of a *system* over time. The system is dynamic and its *state* evolves over time as a function of both *control decisions* and *random disturbances* according to a *system equation*. The system generates *rewards* that are a function of both the state and the control decisions. The objective is to find a *control policy* that maximizes the total expected rewards from the system.

There are T time-periods. Time is indexed by t and the time indices run forward, so $t = 1$ is the first period and $t = T$ is the last period. The key elements of a dynamic program and related technical assumptions are

$\mathbf{x}(t)$ The *system state*. Assumed to be discrete and belonging to a finite-state space S_t.

$\mathbf{u}(t)$ The *control decision*. Assumed discrete and constrained to a finite set, $U_t(\mathbf{x}(t))$, that may depend on time t and the current state $\mathbf{x}(t)$.

$\mathbf{w}(t)$ The *random disturbance*. Assumed to be a discrete random variable (or vector) with known distribution, belonging to a countable state space W_t. The disturbances $\mathbf{w}(t), t = 1, \ldots, T$ are independent.

$\mathbf{f}_t(\mathbf{x}(t), \mathbf{u}(t), \mathbf{w}(t))$ A *system function*, which determines the next state as a function of the current state $\mathbf{x}(t)$, the decisions $\mathbf{u}(t)$ and disturbance $\mathbf{w}(t)$, according to the *system equation*:

$$\mathbf{x}(t+1) = \mathbf{f}_t(\mathbf{x}(t), \mathbf{u}(t), \mathbf{w}(t)).$$

$g_t(\mathbf{x}(t), \mathbf{u}(t), \mathbf{w}(t))$ A real-valued *reward function*, specifying the reward in period t as a function of the current state $\mathbf{x}(t)$, the decisions $\mathbf{u}(t)$, and disturbance $\mathbf{w}(t)$. The reward is assumed to be finite for all t. The total reward is additive,

$$g_{T+1}(\mathbf{x}(T+1)) + \sum_{t=1}^{T} g_t(\mathbf{x}(t), \mathbf{u}(t), \mathbf{w}(t)),$$

where $g_{T+1}(\mathbf{x}(T+1))$ is a *terminal reward*.

The objective is to maximize the total expected reward

$$E\left[g_{T+1}(\mathbf{x}(T+1)) + \sum_{t=1}^{T} g_t(\mathbf{x}(t), \mathbf{u}(t), \mathbf{w}(t))\right],$$

by choosing control actions $\mathbf{u}(1), \mathbf{u}(2), \ldots, \mathbf{u}(T)$. We will assume that the functions f_t, g_t and the disturbances $\mathbf{w}(t)$ are such that this expectation is always finite for any feasible sequence of control decisions.[1]

These control actions may be functions of the current state of the form $\mathbf{u}(t) = \boldsymbol{\mu}_t(\mathbf{x}(t))$.[2] A collection of such functions $\{\boldsymbol{\mu}_1, \boldsymbol{\mu}_2, \ldots, \boldsymbol{\mu}_T\}$ is called a *policy* and is denoted simply by μ. A policy is called *admissible* if $\mathbf{u}(t) \in U_t(\mathbf{x}(t))$ for all t and $\mathbf{x}(t) \in S_t$. The class of all admissible policies is denoted $\mathcal{M}$. For a given initial state $\mathbf{x}(1) = x$, the expected reward of a policy μ is

$$V_1^\mu(\mathbf{x}) = E\left[g_{T+1}(\mathbf{x}(T+1))) + \sum_{t=1}^{T} g_t(\mathbf{x}(t), \boldsymbol{\mu}_t(\mathbf{x}(t)), \mathbf{w}(t))\right]. \tag{D.1}$$

An optimal policy, denoted μ^*, is one for which

$$V_1^{\mu^*}(\mathbf{x}) = \sup_{\mu \in \mathcal{M}} V_1^\mu(\mathbf{x}).$$

The optimal expected reward is denoted simply $V_1(\mathbf{x})$, so

$$V_1(\mathbf{x}) = \sup_{\mu \in \mathcal{M}} V_1^\mu(\mathbf{x}).$$

The Principle of Optimality

The *principle of optimality*, due to Bellman [33], lies at the heart of dynamic programming. It is a strikingly simple idea; namely, that if a policy is optimal for the original problem stated above, then it must be optimal for any subproblem of this original problem as well. That is, define the reward-to-go for policy μ at time t by

$$V_t^\mu(\mathbf{x}) = E\left[g_{T+1}(\mathbf{x}(T+1))) + \sum_{s=t}^{T} g_s(\mathbf{x}(s), \boldsymbol{\mu}_s(\mathbf{x}(s)), \mathbf{w}(s)) \,\middle|\, \mathbf{x}(t) = \mathbf{x}\right].$$

---

[1] For example, the expectation is always finite if the reward function, state space, and disturbance space are all bounded.

[2] Note that we have explicitly assumed here that it is sufficient that the control depend only on the current state $\mathbf{x}(t)$ and the current time t, and it does not need to depend on any other information about the *history* of the process up to time t. Such controls are called *Markovian controls*. Since the disturbances are independent over time and the system function only depends on the current state, disturbance, and control, one can show that there always exists an optimal Markovian policy, so it is sufficient to consider only policies of this form.

APPENDIX D: Dynamic Programming

The function $V_t^\mu(\mathbf{x})$ gives the expected reward starting in state $\mathbf{x}$ at time t over the remainder of the truncated horizon $t, t+1, \ldots, T$. We call this truncated problem the *t-subproblem*. The principle of optimality then states the following:

THEOREM D.2 *If $\mu^* = \{\mu_1^*, \mu_2^*, \ldots, \mu_T^*\}$ is an optimal policy for the problem (D.1), then the truncated policy $\{\mu_t^*, \mu_{t+1}^*, \ldots, \mu_T^*\}$ is optimal for the t-subproblem. That is,*

$$V_t^{\mu^*}(\mathbf{x}) = \sup_{\mu \in \mathcal{M}} V_t^\mu(\mathbf{x}) \quad \forall \mathbf{x} \in S_t.$$

We omit a formal proof of this fact, but it is easy to see why it holds. Indeed, suppose $\{\mu_t^*, \mu_{t+1}^*, \ldots, \mu_T^*\}$ was not optimal for the t-subproblem. and another policy, $\hat{\mu}$, yields strictly greater expected reward. If this were true, then the policy

$$\{\mu_1^*, \ldots, \mu_{t+1}^*, \hat{\mu}_t, \hat{\mu}_{t+1}, \ldots, \hat{\mu}_T\}$$

would produce a strictly greater expected reward than does the policy μ^* for the original problem, which contradicts the optimality of μ^*. Hence, μ^* must be optimal for the t-subproblem.

The Dynamic Programming Recursion

The principle of optimality leads naturally to a recursive procedure for finding the optimal policy. First, for all $\mathbf{x} \in S_t$ and all $t = 1, \ldots, T$, define the optimal reward-to-go, called the *value function*, by

$$V_t(\mathbf{x}) = \sup_{\mu \in \mathcal{M}} V_t^\mu(\mathbf{x}).$$

The value function gives the optimal expected reward from time t onward given that we are in state $\mathbf{x}$ at time t. Note that $V_1(\mathbf{x})$ is the optimal expected reward for the original problem with initial state $\mathbf{x}$. The principle of optimality leads to the following recursive procedure for determining the value function:

PROPOSITION D.15 *The value function $V_t(\mathbf{x})$ is the unique solution to the recursion*

$$V_t(\mathbf{x}) = \max_{\mathbf{u} \in U_t(\mathbf{x})} E\left[g_t(\mathbf{x}, \mathbf{u}, \mathbf{w}(t)) + V_{t+1}(\mathbf{f}_t(\mathbf{x}, \mathbf{u}, \mathbf{w}(t)))\right], \tag{D.2}$$

for all $t = 1, \ldots, T$ and all $\mathbf{x} \in S_t$, with boundary conditions

$$V_{T+1}(\mathbf{x}) = g_{T+1}(\mathbf{x}), \quad \mathbf{x} \in S_{T+1}.$$

Moreover, if $\mathbf{u}^* = \mu_t^*(\mathbf{x})$ achieves the maximum in (D.2) for all t and $\mathbf{x} \in S_t$, then $\mu^* = \{\mu_1^*, \mu_2^*, \ldots, \mu_T^*\}$ is an optimal policy.

We omit a formal proof of this fact, but again the reasoning is quite intuitive—namely, since $V_{t+1}(\mathbf{x}(t+1))$ measures the optimal expected reward given state $\mathbf{x}(t+1)$ in the next time-period, $t+1$, the optimal value of the t-subproblem should be the result of maximizing the sum of the current expected reward, $E[g_t(\mathbf{x}, \mathbf{u}, \mathbf{w}(t))]$, and the expected reward from the $(t+1)$-subproblem, $E[V_{t+1}(\mathbf{x}(t+1))] = E[V_{t+1}(\mathbf{f}_t(\mathbf{x}, \mathbf{u}, \mathbf{w}(t)))]$. This is precisely what (D.2) does. The result yields the optimal t-subproblem value function $V_t(\mathbf{x})$ and the process is repeated.

The complexity of this recursion depends on the size of the state space S_t, control space U_t, and number of time-periods T. The worst-case complexity is $\sum_{t=1}^{T} |S_t||U_t|$, since for each time t and each state $\mathbf{x} \in S_t$, we have to search U_t for the control that maximizes the right-hand side of (D.2).

The usual difficulty with dynamic programming in practice is that the state space S_t can become quite large, making the recursion above computationally complex. For example, in a RM problem with n inventory classes, each with capacities in the range $0, 1, \ldots, C$, the size of the state space is C^n. For even moderate values of C and n, this becomes prohibitively large. This "curse of dimensionality" is the main drawback to dynamic programming. However, for problems with a moderate state space, dynamic programming provides a general procedure for computing and analyzing optimal decisions.

Systems with Observable Disturbances

We next consider a variation of this traditional dynamic programming formulation that helps simplify many RM models. Specifically, consider a case in which we can base our control action $\mathbf{u}$ on perfect knowledge of the disturbance $\mathbf{w}(t)$. In other words, we allow the control to be a function of both the state $\mathbf{x}$ and the disturbance $\mathbf{w}(t)$, so that $\mathbf{u} = \mathbf{u}(\mathbf{x}, \mathbf{w}(t))$. The idea here is that in such systems, we can observe the disturbance before making our control decision and therefore base our decision on the realized value of $\mathbf{w}(t)$.

In this case, the basic dynamic programming recursion becomes

$$V_t(\mathbf{x}) = \max_{\{\mathbf{u}(\mathbf{x},\mathbf{w}(t)) \in U_t(\mathbf{x})\}} E\left[g_t(\mathbf{x}, \mathbf{u}(\mathbf{x}, \mathbf{w}(t)), \mathbf{w}(t)) + V_{t+1}(\mathbf{f}_t(\mathbf{x}, \mathbf{u}(\mathbf{x}, \mathbf{w}(t)), \mathbf{w}(t)))\right],$$

where $\mathbf{u}(\mathbf{x}, \mathbf{w}(t))$ emphasizes that we can select a different control $\mathbf{u}$ for each value of $\mathbf{w}(t)$. However, since we can choose a control based on knowing $\mathbf{w}(t)$, the above recursion can be rewritten as

$$V_t(\mathbf{x}) = E\left[\max_{\{\mathbf{u} \in U_t(\mathbf{x})\}} \{g_t(\mathbf{x}, \mathbf{u}, \mathbf{w}(t)) + V_{t+1}(\mathbf{f}_t(\mathbf{x}, \mathbf{u}, \mathbf{w}(t)))\}\right]. \tag{D.3}$$

The recursion (D.3) can in fact be represented in the traditional form by expanding the state space. First, reindex the disturbances so that we have a new sequence of disturbance terms

$$\tilde{\mathbf{w}}(t) = \mathbf{w}(t+1), \quad t = 1, \ldots, T-1.$$

Consider adding the new system-state variable $\mathbf{y}(t)$, which along with $\mathbf{x}(t)$ is updated by the system equations

$$\begin{aligned} \mathbf{y}(t+1) &= \tilde{\mathbf{w}}(t) \\ \mathbf{x}(t+1) &= \mathbf{f}_t(\mathbf{x}(t), \mathbf{u}(t), \mathbf{y}(t)), \end{aligned}$$

where $\mathbf{f}_t$ is the same function as in (D.3). The initial state is $(\mathbf{x}(1), \mathbf{y}(1)) = (\mathbf{x}, \mathbf{w}(1))$ and the traditional dynamic programming recursion is

$$V_t(\mathbf{x}, \mathbf{y}) = \max_{\mathbf{u} \in U_t(\mathbf{x})} E\left[g_t(\mathbf{x}, \mathbf{u}, \mathbf{y}) + V_{t+1}(\mathbf{f}_t(\mathbf{x}, \mathbf{u}, \mathbf{y}), \tilde{\mathbf{w}}(t))\right],$$

for all $\mathbf{x} \in S_t$ and all $\mathbf{y} \in W_t$. To see this can be converted to the same form as (D.3), define

$$G_t(\mathbf{x}) = E[V_t(\mathbf{x}, \tilde{\mathbf{w}}(t))],$$

APPENDIX D: Dynamic Programming

where expectation is with respect to $\tilde{w}(t) = w(t+1)$, and note that

$$\begin{aligned} V_t(\mathbf{x}, \mathbf{y}) &= \max_{\mathbf{u} \in U_t(\mathbf{x})} E\left[g_t(\mathbf{x}, \mathbf{u}, \mathbf{y}) + V_{t+1}(\mathbf{f}_t(\mathbf{x}, \mathbf{u}, \mathbf{y}), \tilde{\mathbf{w}}(t))\right] \\ &= \max_{\mathbf{u} \in U_t(\mathbf{x})} \{g_t(\mathbf{x}, \mathbf{u}, \mathbf{y}) + E[V_{t+1}(\mathbf{f}_t(\mathbf{x}, \mathbf{u}, \mathbf{y}), \tilde{\mathbf{w}}(t))]\} \\ &= \max_{\mathbf{u} \in U_t(\mathbf{x})} \{g_t(\mathbf{x}, \mathbf{u}, \mathbf{y}) + G_{t+1}(\mathbf{f}_t(\mathbf{x}, \mathbf{u}, \mathbf{y}))\}. \end{aligned}$$

Now substituting $\mathbf{y} = \mathbf{w}(t)$ and taking expectations with respect to $\mathbf{w}(t)$ on both sides above we obtain

$$G_t(\mathbf{x}) = E\left[\max_{\mathbf{u} \in U_t(\mathbf{x})} \{g_t(\mathbf{x}, \mathbf{u}, \mathbf{w}(t)) + G_{t+1}(\mathbf{f}_t(\mathbf{x}, \mathbf{u}, \mathbf{w}(t)))\}\right],$$

which gives us a recursion exactly of the form (D.3).

Note, however, that by using this transformation we have reduced the original dynamic programming recursion from one with a state space $S_t \times W_t$ to one with only a state space of S_t. The function $G_t(\mathbf{x})$ has a similar interpretation as $V_t(\mathbf{x}, \mathbf{y})$ for this reduced state—namely, it is the optimal expected reward-to-go from time t onward given we are in the reduced state $\mathbf{x}(t)$ at time t, where $\mathbf{y} = \mathbf{w}(t)$ still uncertain (recall $G_t(\mathbf{x}) = E[V_t(\mathbf{x}, \mathbf{w}(t))]$). Indeed, one can think of this new recursion as propagating the system in two stages: first, the state $\mathbf{x}$ is realized but $\mathbf{y}$ remains uncertain. We measure the optimal expected reward at this point, yielding $G_t(\mathbf{x})$. Then the value $\mathbf{y} = \mathbf{w}(t)$ is realized, and we make our optimal decision. This takes us to a new state $\mathbf{x}(t+1)$, and the process repeats. Finally, note that this reduced-form recursion results in an optimization step of the form $E[\max\{\ \}]$ rather than the $\max E[\{\ \}]$ found in traditional dynamic programming formulations.

Here's a typical example of how this transformation arises in RM. Suppose $x(t)$ is a scalar capacity, $y(t)$ is the revenue of the request in period t, and $u(t) = 1$ if we decide to accept a request and zero otherwise. So the reward function is simply

$$y(t)u(t).$$

Capacity evolves according to the system equation

$$x(t+1) = x(t) - u(t),$$

and the revenue is driven by a random process

$$y(t+1) = w(t+1).$$

Formulated in traditional terms, we obtain

$$V_t(x, y) = \max_{u \in \{0,1\}} E\left[yu + V_{t+1}(x - u, w(t))\right].$$

However, with the transformation above, we can rewrite this in observable-disturbance form as

$$G_t(x) = E\left[\max_{u \in \{0,1\}} \{w(t)u + G_{t+1}(x - u)\}\right].$$

Since most dynamic programs in RM are of this observable-disturbance form, we typically use the simpler $E[\max\{\ \}]$ rather than the traditional $\max E[\{\ \}]$ form.

Appendix E
The Theory of Choice

In this appendix, we briefly review the theory of consumer choice. It is provided both as a background and reference on the core concepts of choice theory.

The most widely used theories of choice assume customers are *rational* decision makers who intelligently alter when, what, and how much to purchase to achieve the best possible outcome for themselves. This is a quite plausible assumption. Moreover, an important consequence of this rationality assumption is that customer behavior can be "predicted" by treating each customer as an agent that optimizes over possible choices and outcomes. Optimization theory can then be used to model their behavior. Indeed, for these reasons rational-customer models are the basis of most economic theory.

Yet despite the theoretical and intuitive appeal of the rationality assumption, instances of deviations from rational behavior are observed in experiments and in real life. Alternative theories of choice have emerged to explain such behavior. These models assume customers are not perfectly rational—that there are limits to how cleverly they behave or that they exhibit irrational biases in their choice decisions. These so-called *behavioral theories* are surveyed below as well.

Choice and Preference Relations

Given two alternatives, a *choice* corresponds to an expression of preference for one alternative over another. Here, "alternatives" may refer to different products, different quantities of the same product, bundles of different products or various uncertain outcomes (such as buying a house at the asking price versus waiting and bidding in an auction against other buyers). Similarly, given n alternatives, choice can be defined in terms of the preferences expressed for all pairwise comparisons between the n alternatives.

The mathematical construct that formalizes this notion of choice and preference is a *preference relation*. Customers are assumed to have a set of *binary preferences* over alternatives in a set X. That is, given any two alternatives x and y in X, customers can rank them and clearly say they prefer one over the other. This is represented by the notation $x \succeq y$. A customer strictly prefers x to y, denoted $x \succ y$, if he prefers x to y, but does not prefer y to x (that is, he is not indifferent between the two alternatives).

Consider a complete set of all such pairwise binary preferences between alternatives in X. The following two properties might be reasonably assumed about "rational" preferences:

- **Asymmetry** If x is strictly preferred to y, then y is not strictly preferred to x.

- **Negative transitivity** If x is not strictly preferred to y and y is not strictly preferred to z, then x is not strictly preferred to z.

Asymmetry and negative transitivity can be considered as "minimal consistency properties" for an expression of preference among a set of alternatives. A binary relation $\succ$ on a set X is called a *preference relation*, if it is asymmetric and negatively transitive. While asymmetry is quite plausible, negative transitivity is not a completely innocuous assumption, as illustrated by the following example:

Example E.1 Suppose you are choosing among jobs in three different cities. Suppose the two factors that matter most to you are income and the climate. The job in city x has a high salary of \$100,000, and the climate is average. The job in city y offers a salary of only \$50,000, but the climate is terrific. The job in city z offers a moderate salary of \$70,000 and the climate is poor. You might not strictly prefer x to y because although x offers a great salary, y offers a great climate. Likewise, you might not strictly prefer y to z because again, while y offers a great climate, z offers a higher salary. However, you may very well prefer x to z, since x has both a higher salary and a better climate than does z. These preferences would violate negative transitivity.

Despite such shortcomings, the properties of asymmetry and negative transitivity form the classical basis for modeling customer preferences. The following are some examples of preference relations:

Example E.2 (LEXICOGRAPHIC MODEL) This model of preferences, due to Tversky [521], assumes customers rank order various attributes of a product and then evaluate them using a lexicographic rule. For example, a tennis racquet comes in three models A, B, and C with the following features:

Product	Wide Body?	Graphite?	Black?
A	Yes	No	Yes
B	Yes	Yes	No
C	No	Yes	Yes

The customer's decision rule is to rank all attributes from most important to least important and then eliminate alternatives which do not possess the most important attributes. If more than one alternative remains, the next most important attribute is chosen as a criterion for elimination of alternatives, and so on.

For example, a customer may care most about whether a racquet has a wide body, then whether it is graphite, and lastly whether it is black. He would then prefer racquets with a wide body to all others without a wide body (regardless of the other attributes). Among all those with wide bodies, he would then select those that have graphite construction; among the remaining, he may select only the ones that are black, and so on. So for our three products above, this customer would prefer product in the order B, A, C. One can verify that the lexicographic model generates a preference relation among the alternatives.

Example E.3 (ADDRESS MODEL) Address models link attributes to preference without imposing the restriction that some attributes strictly dominate others as in the lexicographic model. Suppose we have n alternatives and each alternative has m attributes that take on real values. Alternatives can then be represented as n points, $z_1, \ldots, z_n$, in $\Re^m$, which is called *attribute space*. For example, in a travel context attributes may include departure time, arrival time, and price.

Each customer has an ideal point ("address") $y \in \Re^m$, reflecting his most preferred combination of attributes (such as an ideal departure time, arrival time, and price). A customer is then assumed to prefer the product closest to his ideal point in attribute space, where distance is defined by a metric ρ on $\Re^m \times \Re^m$ (such as Euclidean distance). These distances define a preference relation, in which $z_i \succ z_j$ if and only if $\rho(z_i, y) < \rho(z_j, y)$; that is, if z_i is "closer" to the ideal point y of the customer.

Utility Functions

Preference relations are intimately related to the existence of utility functions. Indeed, we have the following theorem (See Kreps [313] for a proof.):

THEOREM E.3 *If X is a finite set, a binary relation $\succ$ is a preference relation if and only if there exists a function $u : X \to \Re$ (called a **utility function**), such that*

$$x \succ y \quad \text{iff} \quad u(x) > u(y).$$

Intuitively, this theorem follows because if a consumer has a preference relation, then all products can be ranked (totally ordered) by his preferences; a utility function then simply assigns a numerical value corresponding to this ranking. Intuitively, one can think of utility as a measure of "value," though in a strict sense its numerical value need not correspond to any such tangible measure. Theorem E.3 applies to continuous sets X (such as travel times or continuous amounts of money) as well under mild regularity conditions, in which case the utility function $u(\cdot)$ is then continuous. The following examples illustrate the construct of utility:

Example E.4 A utility function corresponding to the lexicographic model of Example E.2 can be constructed as follows: Suppose there are n alternatives with m attributes each. Let the attributes be ordered so that 1 represents the highest-valued attribute and m the lowest. Let $a_k(x), k = 1, \ldots, m$, be binary digits representing whether alternative x, possesses attribute k. Then a utility satisfying Theorem E.3 is the binary number,

$$u(x) = a_{x1} a_{x2} \cdots a_{xm}.$$

Maximizing over these utilities leads to the same customer decisions as the lexicographic model.

Example E.5 Consider the address model of Example E.3. Again, Theorem E.3 guarantees that an equivalent utility maximization model exits that generates the same choices. In this case, it is easy to see that for customer y the continuous utilities

$$u(z) = c - \rho(z, y),$$

where c is an arbitrary constant, produce the same decision rule as the address model.

Utility for Money and Consumer Budgets

It is often convenient to narrow the choice of utilities further and express utility in monetary terms. To do so, one can pose the question: Given the customer's preference for n goods (purchase alternatives), a vector of market prices $\mathbf{p} = (p_1, \ldots, p_n)$ for these goods, and a level of monetary wealth w, how would a customer "spend" his wealth? To make matters simpler, we assume quantities x_i of each good i are continuous and our customer has a continuous utility function $u(\mathbf{x})$. Let $\mathbf{x} = (x_1, \ldots, x_n)$. The *consumer budget problem* can then be formulated as[1]

$$v(w) = \max \; u(\mathbf{x}) \quad \text{(E.1)}$$
$$\text{s.t.} \quad \mathbf{p}^\top \mathbf{x} \leq w$$
$$\mathbf{x} \geq 0.$$

In other words, customers purchase quantities x_i of each good i to maximize their total utility subject to the constraint that they can spend at most their total wealth w. The optimal solution gives the customer's utility for wealth (or money) $v(w)$; the optimal solution, $\mathbf{x}^*$, gives the customer's *demand* for each of the n goods.

Utility for money is increasing in w since one can always "not spend" the wealth w. Also, since the utility for money depends on the prices of goods, if prices change, both the demand x^* and the utility for money may change. The *marginal utility of money* $u'(w)$ also depends on the customer's wealth w. The utility for money $v(w)$ is concave if $u(\mathbf{x})$ is concave,[2] in which case the consumer has decreasing marginal utility for money. Intuitively, this is because at low levels of wealth only highly essential goods are purchased (food, water, clothing, shelter)—all of which have very high utility to most of us. As wealth rises, each marginal dollar is allocated to somewhat less important purchases.

If the function $u(\mathbf{x})$ is continuously differentiable and we let π denote the optimal Lagrange multiplier on the budget constraint in (E.1), then the marginal value of money is

$$v'(w) = \pi.$$

We can use this fact to redefine utilities in monetary terms. Indeed, since our customer's monetary utility for an additional dollar should be one dollar, we should have $v'(w) = \pi = 1$ if utilities are measured in dollars. This change of units can be accomplished by rescaling the customer's utility functions by $v'(w) = \pi$ to form the modified utilities

$$\tilde{u}(\mathbf{x}) = \frac{u(\mathbf{x})}{\pi}. \quad \text{(E.2)}$$

---

[1] Dynamic versions of this consumer budget problem can also be formulated by allowing customers to purchase over multiple periods and invest money at a given interest rate for future consumption. Other variations introduce wages and a utility for leisure time and allow customers to increase their monetary wealth by varying their time allocated to labor, and so on.

[2] This follows easily from the convexity of the budget constraint and the fact that (E.1) is a maximization problem. Concavity of the utility function corresponds to having decreasing marginal utility of consumption for goods, which is a natural assumption.

Reservation Prices

A *reservation price* is the monetary amount a consumer is willing to give up to acquire an extra marginal unit of some good. Reservation prices are also referred to as the customer's *willingness to pay*. Formally, if x^* denotes the optimal solution to (E.1), the reservation price, denoted v_i, for an additional unit of good i is given by

$$v_i \equiv \frac{\partial \tilde{u}(\mathbf{x}^*)}{\partial x_i}, \qquad (E.3)$$

where $\tilde{u}(\mathbf{x}^*)$ is the monetary utility (E.2). The first-order conditions of the budget problem imply $\frac{\partial \tilde{u}(\mathbf{x}^*)}{\partial x_i} = p_i$ since $\tilde{u}'(w) = \pi = 1$ when utilities are measured in dollars. Combining this with (E.3) implies that $v_i = p_i$. Thus, a customer's reservation price for goods that are *currently consumed* is simply the current market price. The reason for this equivalence, intuitively, is that if our customer valued another unit of good i at strictly more than its market price, then he would be able to increase his utility by reducing consumption of other goods and increasing his consumption of good i. Since our customer is assumed to be maximizing utility, this cannot occur.

On the other hand, for goods i that are not being consumed, so $x_i^* = 0$, the first-order conditions to (E.1) imply $\frac{\partial \tilde{u}(\mathbf{x}^*)}{\partial x_i} < p_i$, or equivalently $v_i < p_i$. In other words, by (E.3) the customer's reservation price for the first unit of good i is strictly less than its current market price. Moreover, the customer would change only his allocation and buy good i if its price p_i dropped below his reservation price v_i.

This formal analysis of reservation price is arguably less important in practice than the informal concept—namely, that the reservation price is the maximum amount a customer is willing to pay for an additional unit of good i. And to entice a customer to buy good i, the price must drop below his reservation price. Still, the analysis highlights the important fact that reservation prices are not "absolute" quantities. Like utility for money, they depend on customers' preferences, wealth, their current consumption levels, and the prices of other goods the customers may buy; change one of these factors, and customers' reservation price may change.

Lotteries and Stochastic Outcomes

Many choices in life involve uncertain outcomes, such as buying insurance, making investments or eating at a new restaurant. How do customers respond to these sorts of uncertainties? The theory of choice under uncertainty is a deep and extensive topic. Here, we outline the basic ideas and highlight the main concepts.

Consider again a discrete, finite set of n alternatives, $X = \{x_1, \ldots, x_n\}$. Let $\mathcal{P}$ be the class of all probability distributions $P(\cdot)$ defined on X. That is, $P \in \mathcal{P}$ is a function satisfying $\sum_i P(x_i) = 1$ and $P(x_i) \geq 0$ for $i = 1, \ldots, n$. One can think of each P as a "lottery," the outcome of which is that the customer is left with one of the alternatives x_i according to the distribution P.

What can we say about the customer's preference for these various lotteries? Specifically, when can we say that for any two lotteries P_1 and P_2, customers "prefer" one over the other (denoted by $P_1 \succ P_2$)?

To answer this question we again need to make some assumptions on customer preferences. First, we will assume there exists a preference relation $\succ$ on the n different outcomes x_i as before. Second, for any two lotteries P_1 and P_2, consider a compound lottery parameterized by α as follows:

STEP 1: A coin is flipped with probability of heads equal to α.

STEP 2: If the coin comes up heads, the customer enters lottery P_1; otherwise, the customer enters lottery P_2.

Denote this compound lottery by $\alpha P_1 + (1-\alpha) P_2$. Note this compound lottery is also contained in the set $\mathcal{P}$ (i.e., $\mathcal{P}$ is a convex set). We then require the following consistency properties on a customers preference for lotteries:

- **Substitution axiom** For all P_1, P_2, and P_3 in $\mathcal{P}$ and all $\alpha \in (0, 1]$, if $P_1 \succ P_2$, then $\alpha P_1 + (1-\alpha) P_3 \succ \alpha P_2 + (1-\alpha) P_3$.

- **Continuity axiom** For all P_1, P_2, and P_3 in $\mathcal{P}$ with $P_1 \succ P_2 \succ P_3$, there exist values $\alpha \in (0, 1)$ and $\beta \in (0, 1)$ such that $\alpha P_1 + (1-\alpha) P_3 \succ P_2 \succ \beta P_1 + (1-\beta) P_3$.

Roughly, the first axiom says that if one gamble produces strictly preferred outcomes for any realization of uncertainty, then the customer should strictly prefer it. The second axiom says that if a customer strictly prefers one gamble to another, then he should be willing to accept a sufficiently small risk of an even worse outcome to take the preferred gamble. Both are reasonable assumptions.

Under these two axioms, there exist utilities on outcomes such that the expected utility of each lottery defines a customer's preference relation among lotteries. Specifically,

THEOREM E.4 *A preference relation on the lotteries $\mathcal{P}$ exists that satisfies the substitution and continuity axioms if and only if there exists a utility function $u(\cdot)$ such that $P_1 \succ P_2$ if and only if*

$$\sum_{i=1}^{n} u(x_i) P_1(x_i) > \sum_{i=1}^{n} u(x_i) P_2(x_i).$$

That is, if and only if the expected utility from lottery P_1 exceeds the expected utility of lottery P_2. In addition, any two utility functions u and u' satisfying the above must be affine transformations of each other; that is,

$$u(x) = cu'(x) + d,$$

for some real $c > 0$ and d.

This result is due to von Neumann and Morgenstern [541] and is known as the *von Neumann-Morgenstern expected-utility theory*. Essentially, it allows us to extend utility as a model of customer preference to the case of uncertain outcomes, with expected utility replacing deterministic utility as the criterion for customer decision making. Since the original deterministic outcomes (e.g., outcome x_i occurs with probability $P(x_i) = 1$) are included in $\mathcal{P}$, the von Neumann-Morgenstern expected utilities also help us "narrow down" the list of possible utility functions for the customer.

Risk Preferences

An important special case of expected-utility theory is when outcomes represent different monetary amounts, so alternatives correspond to different levels of wealth and lotteries correspond to different gambles on a customer's ending wealth level. For

APPENDIX E: The Theory of Choice 663

this discussion, we assume the wealth levels are continuous and that the customer has preferences for wealth that satisfy the conditions of Theorem E.4. Also, assume the lotteries are now continuous distributions F on $\Re$.[3]

Consider now any given lottery F (a distribution on possible wealth outcomes) and μ_F denote the mean of the distribution. A customer is said to have *risk-averse preferences* if he prefers the certain wealth μ_F to the lottery F itself for all possible lotteries F. That is, the customer always prefers the certainty of receiving the expected wealth rather than a gamble with the same mean. The customer is said to have *risk-seeking preferences* if he prefers the gamble F to the certain outcome μ_F for all F. Finally, he has *risk-neutral preferences* if he is indifferent between the lottery F and the certain reward μ_F.[4] We then have the following result:

THEOREM E.5 *A customer's preference $\succ$ for lotteries exhibits risk-aversion (risk-seeking) behavior if and only if their von Neumann-Morgenstern utility function $u(w)$ is concave (convex). Their preference is risk-neutral if and only if $u(w)$ is affine.*

Thus, risk preferences are linked directly to concavity or convexity of the customer's utility function. The reason is quite intuitive; with a concave utility function for wealth, a customer gains less utility from a given increase in wealth than he loses in utility from the same decrease in wealth. Hence, the upside gains produced by the volatility in outcomes do not offset the downside losses, and customers therefore prefer the certain average to the uncertain outcomes of the lottery. Since most customers have a decreasing marginal utility for wealth, risk aversion is a good assumption in modeling customer behavior.

Still, the concept of risk aversion has to be addressed with care in operational modeling. While it is true that most customers are risk-averse when it comes to *large* swings in their wealth, often the gambles we face as consumers have a relatively small range of possible outcomes relative to our wealth. For example, a customer may face a price risk in buying a CD or book online. However, the differences in prices for such items are extremely small compared to his total wealth. In such cases, the utility function is "almost linear" in the range of outcomes affecting the decision and the customer tends to behave "as if" he were risk-neutral.[5] Similar statements apply to firms. Generally, they are risk-averse too, but for decisions and gambles that involve "small" outcomes relative to their total wealth and income, they tend to be approximately risk-neutral. Hence, risk-neutrality is a reasonable assumption in operational models and, indeed, is the standard assumption in RM practice.

---

[3]The extension of Theorem E.4 to the continuous case requires some additional technical conditions that are beyond the scope of this chapter. See Kreps [313].
[4]Note that a customer's preferences may not fall into any of these three categories. For example, many consumers take out fire insurance, preferring a certain loss in premium payments every year to the gamble between making no payments but potentially loosing their house, yet simultaneously play their local state lottery, which has an expected loss but provides a small probability of a large wealth pay-off. Such behavior violates a strict risk preference.
[5]Formally, one can see this by taking a Taylor series approximation of the utility function about the customer's current wealth w; the first-order approximation is affine, corresponding to risk-neutrality.

Information Asymmetry

Another important fact related to customer choice is that normally much of a customer's information is private, information that only the customer "knows" and information that cannot be directly observed by a firm. Normally, both a customer's preferences and wealth are private information. One can perhaps gain clues to a customer's preference by observing their purchase behavior over time (their so-called *revealed preferences*), and partial information on their wealth may be garnered from surveys and transactional data. But in general, much of the data affecting customers' choice behavior remains hidden.

This "information asymmetry" between customers and firms has implications for pricing and RM as discussed in detail in Chapters 6 and 8. To give a quick sense of the effect, consider how customers react to a posted price. Due to information asymmetry, the selling firm rarely knows a customer's true reservation price for their product. If they did, they could potentially offer the customer a price only marginally less than their reservation price and maximize the revenue obtained from each customer. Instead, most firms have to guess at each customer's reservation price. As a result, sometimes they price too high, and the customer does not purchase at all; other times they price too low, and although the customer may decide to purchase, they lose an opportunity for a revenue gain as the customer would have been willing to pay more. In this way, the private information of customers often allows them to retain some surplus, even from a monopoly seller.

Deviations from Rational Behavior

While rational behavior is the standard assumption underlying most of the theory and practice of RM, it is far from being completely accepted as a model of how an actual customer behaves. Indeed, much of the recent work in economics and customer behavior has centered on explaining observed, systematic deviations from rationality on the part of customers.

The seminal work in this area is that of Kahneman and Tversky [278, 277], who showed that customers often exhibit consistent biases when faced with simple choices in an experimental setting. Their key insight is that most individuals tend to evaluate choice in terms of losses and gains from their status quo wealth, rather than evaluating choices in terms of their terminal wealth as in classical utility theory. People also show a tendency toward "loss aversion" rather than risk aversion, and they have a strong preference for certainty of outcomes when evaluating choices. Finally, how gains and losses are expressed matter as well.

They showed that how questions of choice are "framed" have a large impact on customer choice. When choices are framed in terms of gains versus losses, customers typically care more about avoiding losses than about making gains. This is true even if the "gains" and "losses" amount to exactly the same choice. For example, if a public health policy choice is framed as a gain (200 of 800 diseased people will be saved) or as a loss (600 of 800 diseased people will die), most people respond differently, even though the outcomes are identical.

Other experiments revealed that people put a much higher value on a product they already own than one that they don't own because giving up a product they have feels like a loss. This behavior is part of the rationale behind the common marketing strategy of offering products on a "free 30-day trial"–that customers are much more willing to pay to "avoid losing" the trial product than they are willing to

APPENDIX E: The Theory of Choice

pay to acquire that same product initially. (Of course, other simpler explanations—such as reassuring the customer of the quality of the product—can also explain such guarantees.)

Another bias people exhibit is due to what is called *mental accounting*, in which customers tend to evaluate gains and losses for different categories of goods differently because they have "mental budgets" for each category of goods. For example, suppose you purchase a $1,000 watch and then immediately lose it. You might then be reluctant to replace it because in some sense your "budget" for purchasing watches has been exhausted. However, suppose you lost $1,000 in the stock market and you did not own a watch at the time. Then you might be willing to buy a new $1,000 watch because there is no direct association between the $1,000 dollar loss and the amount you might have "mentally allocated" to spend on a watch (for example, you might account for this as "an investment loss" not a "expensive-watch loss"). Such heuristic accounting again violates the rationality assumptions of classical consumer behavior.

Kahneman and Tversky [278] developed what they termed *prospect theory* to explain such effects. Prospect theory differs from expected-utility theory in several respects. For one, it handles the probabilities of outcomes differently, treating them as "decision weights" that may or may not correspond to actual probabilities. Indeed, prospect theory postulates that the subjective decision weights used by most customers tend to overweigh small probabilities and underweigh high probabilities. Prospect theory also uses the notion of "value" rather than "utility," where value is defined in terms of deviations from a reference point (the customer's status quo wealth). They postulate an S-shaped curve for the value function, which is convex for losses below the reference point and concave for gains above the reference point. Using this construct, Kahneman and Tversky [278] are able to model and explain many observed deviations from rational behavior.

Do such findings mean that expected utility theory is "dead"? Not really. In a gross sense, people do tend to behave in accordance with rationality assumptions. However, what this behavioral theory shows quite clearly is that the axioms of rational behavior, plausible as they are, do not apply uniformly and that there are situations in which deviations from rational behavior are systematic and substantial.

The main consequence of these findings for RM practice is that one should always understand the "environment" in which choices are made; the details of the buying situation matter in terms of customers' responses. How prices are presented, what "reference point" the customer perceives, the framing of the choice decision, their sense of "ownership" over the product—all can potentially influence their responses. While many of the tactics used to influence these factors lie in the domain of general marketing and are beyond the scope of this book, the general message that the choice environment matters is nevertheless an important one for RM practitioners and researchers to heed. Indeed, we expect these behavioral theories of demand to influence RM practice more directly in the years ahead.

Appendix F
Game Theory

This appendix provides some elementary facts about game theory and equilibrium concepts and should serve as a refresher for those readers with some background in game theory. However, for a proper and more complete explanation of the theory, the reader should refer to Fudenberg and Tirole [195], Myerson [399] or Mas-Colell, Whinston, and Green [365] (from which the material in this appendix is obtained). Also, Tirole [513] provides a User's Manual on Game Theory.

The Normal Form of Games

A game consists of a set of *players*, the *actions* that they can take (or in other words, the rules of the game), and the *information* that each player possesses at the time he takes his action. For each possible set of actions, the game defines a set of *outcomes* and *payoffs* for each player (such as how much profit or utility each player gets).

For instance, in the Bertrand pricing game (Section 8.4.1.4), there are n players (firms in the oligopoly). Their action space is the prices they set. Each possesses information that all the demand goes to the lowest-priced firm, and all have the same marginal costs. The outcome is that the demand goes to the lowest-priced firm. The payoffs are the revenues minus costs (profits).

Formally, let there be n players, let $\mathcal{H}_i$ be the collection of player i's information sets and $C(H) \subset \mathcal{A}$ be the set of actions possible for player i with information set H.

A *(pure) strategy* for player i is a function $s_i : \mathcal{H}_i \to \mathcal{A}$,—that is, the player has a mapping from each possible information set to a unique action. Moreover, the actions have to be feasible, so we assume that the strategy map is such that $s_i(H) \in C(H)$ for all $H \in \mathcal{H}_i$. Each player, given a set of pure strategies, can also randomize over these strategies (his strategy is to choose one of his pure strategies with a certain probability). This creates what are called *mixed strategies*.

A game's actions, outcomes, and payoffs can be defined by an extensive form or a normal form. Here we concentrate on the normal form. The normal form of the game is a specification of a set of possible strategies S_i for player i, and a payoff function $u_i(s_1, \ldots, s_N)$ if each player plays strategy $s_i \in S_i, i = 1, \ldots, N$. The game Γ is defined as the triple $\Gamma = [N, \{S_i\}, \{u_i(\cdot)\}]$.

For example, in the Bertrand pricing game with two players, the strategy space for player i is $(0, \infty)$, and the payoff if player 1 plays $p_1 \in (0, \infty)$ is $p_1 d_1(p_1, p_2)$, where

$$d_1(p_1, p_2) = \begin{cases} d(p_1) & \text{if } p_1 < p_2 \\ d(p_1)/2 & \text{if } p_1 = p_2 \\ 0 & \text{if } p_1 > p_2 \end{cases} \tag{F.1}$$

and $d(\cdot)$ is the market-demand function.

Simultaneous Move Static Games

In the simultaneous move static game, all players move exactly once and make their moves simultaneously. Hence, no player knows what the other players' moves are going to be, nor do they have any information on past moves of their opponents (as it is a one-move game).

These are rather restrictive assumptions. Nevertheless, such games are applicable in some situations (for instance, a sealed-bid auction), and they serve as the basis for the study of more complicated repeated games.

Game theory is concerned with predicting the outcomes of a game assuming the players are rational (utility-maximizing players). To this end, we define the concept of *equilibrium*, essentially a prediction of the possible outcomes of the game. There are many equilibrium concepts, depending on the nature of the information, and the assumptions on players' behavior.

We assume that players have *complete information* about the game. Each player knows the strategy sets, utility functions, and any other relevant parameters for all other players, and they also know that all the other players are rational and, like themselves, have complete information.

Dominant Strategies

A strategy $s_i \in S_i$ is a *dominant strategy* for player i if his payoff from playing s_i is no less than that from playing any other of his strategies, for all possible strategies of the other players.

Formally, let the the vector $\mathbf{s}_{-i} = (s_1, \ldots, s_{i-1}, s_{i+1}, \ldots, s_N)$ represent strategies of all players other than i, and S_{-i}, the set of all vectors of all possible strategies of all the players other than i.

Then, for a game $\Gamma = [N, \{S_i\}, \{u_i(\cdot)\}]$, $s_i \in S_i$ is a dominant strategy if for all $s_i' \neq s_i$,

$$u_i(s_i, s_{-i}) \geq u_i(s_i', s_{-i}), \quad \forall s_{-i} \in S_{-i}.$$

If rational player i has a strictly dominant strategy, it is reasonable to predict he would always play that strategy. There are very few games, however, where such dominant strategies exist.

Nash Equilibrium

Perhaps the most important and widely accepted notion on the outcome of games with rational players is the Nash equilibrium.

Nash equilibrium, definition A strategy vector $\mathbf{s}$ is a *(pure strategy) Nash equilibrium* for the game $\Gamma = [N, \{S_i\}, \{u_i(\cdot)\}]$ if for every player $i = 1, \ldots, N$, given the strategies of the other players $\mathbf{s}_{-i}$, his strategy s_i is optimal, that is,

$$u_i(s_i, \mathbf{s}_{-i}) \geq u_i(s_i', \mathbf{s}_{-i}), \quad \forall s_i' \in S_i, \ i = 1, \ldots, N.$$

APPENDIX F: Game Theory

If we allow players to randomize over their strategies, then a vector of mixed strategies, $\sigma = (\sigma_1, \ldots, \sigma_N)$ is a mixed-strategy Nash equilibrium if for every player i, given a profile of mixed strategies of the other players σ_{-i}, player i's mixed strategy σ_i is optimal. A game can have no Nash equilibrium, a unique Nash equilibrium, or many equilibria (pure-strategy or mixed).

Here are two basic results on the existence of pure and mixed-strategy Nash equilibria.

PROPOSITION F.16 *Every game with finite strategy sets for all the players has a mixed-strategy Nash equilibrium.*

PROPOSITION F.17 *If the strategy sets for the players S_i are nonempty, convex and compact subsets of $\Re^m$, $u_i(\mathbf{s})$ is continuous in $\mathbf{s}$ and quasiconcave in s_i for all i, then the game has a pure-strategy Nash equilibrium.*

However, in either case, there is no guarantee that the equilibrium is unique.

Bayesian Nash Equilibrium

Games with *incomplete information* model situations where the players do not know with certainty what the other players' strategy sets, parameters, and utility functions are. Each forms a probabilistic view of the other players' private information (akin to a Bayesian prior; this probabilistic view may be updated in a repeated game as the game reveals more information to the players).

The model is as follows: Player i's payoff function is now given by $u_i(\mathbf{s}, \theta_i)$, where $\theta_i \in \Theta_i$ is a random variable whose realization is observed only by player i. Let $\theta = (\theta_1, \ldots, \theta_N)$ and $\Theta = \Theta_1 \times \Theta_2 \times \cdots \times \Theta_N$. However, the joint probability distribution of $\theta \in \Theta$, $F(\theta)$ is common knowledge among the players. The Bayesian game is then $\Gamma = [N, \{S_i\}, \{u_i(\cdot)\}, \Theta, F(\cdot)]$.

A pure-strategy for player i is in this case a decision rule $s_i(\theta_i)$. His strategy is a function of the realization of his θ_i. Given a vector of pure strategies for all the players $(s_1(), \ldots, s_N(\cdot))$, player i's payoff is given by the expectation over the θ's:

$$\tilde{u}_i(s_i(\cdot), \mathbf{s}_{-i}(\cdot)) = E_\theta[u_i(s_i(\cdot), \mathbf{s}_{-i}(\cdot))].$$

The extension to the Nash equilibrium concept is then as follows. A *(pure-strategy) Bayesian Nash equilibrium* is a vector of decision rules $\mathbf{s}(\cdot) = (s_1(\cdot), \ldots, s_N(\cdot))$ such that

$$\tilde{u}_i(s_i(\cdot), \mathbf{s}_{-i}(\cdot)) \geq \tilde{u}_i(s_i'(\cdot), \mathbf{s}_{-i}(\cdot)).$$

Repeated Games

Finite repeated games are one-shot games that are repeated over a number of periods. At the beginning of each period, firms are aware of the others' past moves and make their decisions simultaneously and noncooperatively for that period.

For example, a repeated Bertrand game would have firms setting prices simultaneously at the beginning of each period, and a repeated Cournot game would have firms deciding how much to produce at the beginning of each period. For instance, when all the firms in the market post their prices on a centralized industrywide reservation system every day, the repeated game's period is one day, and at the beginning of each day firms set prices simultaneously without knowing how the other firms will choose their prices that day.

Models where moves don't occur simultaneously but have a leader-follower structure are called *Stackelberg* games. It can be shown that the first mover in a Stackelberg game has an advantage under certain scenarios [195].

To analyze repeated games, we need a refinement of the Nash equilibrium concept known as a *subgame-perfect equilibrium*. Roughly, a subgame-perfect equilibrium is one in which the initial equilibrium is simultaneously a Nash equilibrium for any subgame (the game from any subsequent stage assuming all the information on actions from the previous stages) of the initial game.

The idea is best illustrated by an example. Consider a T period, two-player, Bertrand pricing game where prices are the strategic variables. Then $[(p^1(1), p^2(1)), \ldots, (p^1(T), p^2(T))]$ is subgame-perfect equilibrium if (i) it is a Nash-equilibrium and (ii) for all $1 < t \leq T$, the decisions $[(p^1(t), p^2(t)), \ldots, (p^1(T), p^2(T))]$ is a Nash equilibrium for the subgame starting from period t to period T.

The subgame-perfect equilibrium refinement allows one to restrict attention to strategies that only contain credible threats or promises. For instance, in a two-player, two-period Bertrand pricing game, suppose firm one adopts a strategy of pricing high in period one and promises to continue to price high in period two provided the other firm does not undercut its price in period one. While this may result in a Nash equilibrium with both firms pricing high in each period, it does not constitute a subgame-perfect equilibrium because once the firms reach period two, it is in firm 1's interest to deviate from it's announced strategy and undercut its rival's price. Thus, the promise to continue to price high is not credible.

Infinitely repeated games (called *supergames*) provide a richer set of results than do finite repeated games. The assumption of infinite interaction may seem excessive, but in situations where there are many opportunities for frequent interactions or when the end of a game is uncertain, it is a reasonable modeling assumption.

References

[1] B. Abraham and J. Ledolter. *Statistical Methods for Forecasting*. Wiley, New York, NY, 1983.

[2] M. Abramowitz and I. A. Stegun. *Functions with Formulas, Graphs, and Mathematical Tables*. Dover, New York, NY, 1972.

[3] P. J. Acklam. An algorithm for computing the inverse normal cumulative distribution function. website, 2003. http://home.online.no/ pjacklam/.

[4] Thinking Networks AG. Revenue management in liberalized energy markets. Technical report, March 2001.

[5] K. Ailawadi and S. A. Neslin. The effect of promotion on consumption: Buying more and consuming it faster. *Journal of Marketing Research*, 35:390–398, 1998.

[6] J. Albert and S. Chib. Bayesian analysis of binary and polychotomous data. *Journal of the American Statistical Association*, 88:669–679, 1993.

[7] S. Algers and M. Besser. Modeling choice of flight and booking class: A study using stated preference and revealed preference data. *International Journal of Services Technology and Management*, 2:28–45, 2001.

[8] G. M. Allenby, N. Arora, and J. L. Ginter. On the heterogeneity of demand. *Journal of Marketing Research*, 35:384–89, 1998.

[9] G. M. Allenby and J. L. Ginter. Using extremes to design products and segment markets. *Journal of Marketing Research*, 32:392–403, November 1995.

[10] G. Allenby and P. Rossi. A Bayesian approach to estimating household parameters. *Journal of Marketing Research*, 30:171–182, 1993.

[11] G. Allenby and P. Rossi. Marketing models of consumer heterogeneity. *Journal of Econometrics*, 89:57–78, 1999.

[12] D. M. Allen. The relationship between variable selection and data augmentation and a method for prediction. *Technometrics*, 16:125–127, 1974.

[13] R. Amir. Cournot oligopoly and the theory of supermodular games. *Games and Economic Behavior*, 15:132–148, 1996.

[14] E. J. Anderson and A. B. Philpott. On supply-function bidding in electricity markets. In C. Greengard and A. Ruszczynski, editors, *Decision Making Under Uncertainty: Energy and Power*, pages 115–134. Springer-Verlag, 2002.

[15] S. P. Anderson, A. de Palma, and Y. Nesterov. Oligopolistic competition and the optimal provision of products. *Econometrica*, 63:1281–1301, 1995.

[16] S. P. Anderson, A. de Palma, and J-F. Thisse. *Discrete-Choice Theory of Product Differentiation*. MIT Press, Cambridge, MA, 1992.

[17] S. P. Anderson and A. de Palma. Multiproduct firms: A nested logit approach. *Journal of Industrial Economics*, 40:261–276, 1992.

[18] S. E. Andersson. Operational planning in airline business—Can science improve efficiency? Experiences from SAS. *European Journal of Operations Research*, 43:3–12, 1989.

[19] S. E. Andersson. Passenger choice analysis for seat capacity control: A pilot project in Scandinavian Airlines. *International Transaction in Operational Research*, 5:471–486, 1998.

[20] D. Anthony and J. R. Harrington. An overview of emerging IT and the gas pipeline industry. *Pipeline & Gas Journal*, March 2001.

[21] L. M. Ausubel and R. J. Deneckere. Reputation in bargaining and durable goods monopoly. *Econometrica*, 57:511–531, 1989.

[22] M. Axelrod. Mariners testing out web-bidding system for ticket sales. ABCNews.com, July 21, 2000.

[23] A. T. Kearney, Inc. Let the games begin. A. T. Kearney report, May 2002.

[24] R. J. Bagby. Calculating normal probabilities. *American Mathematical Monthly*, 102:46–49, 1995.

[25] M. Bagnoli and T. Bergstrom. Log-concave probability and its applications. Working Paper, University of Michigan, 1989.

[26] M. Bagnoli, S. W. Salant, and J. E. Swierzbinski. Durable-goods monopoly with discrete demand. *Journal of Political Economy*, 97:1459–1478, 1989.

[27] R. Bamon and J. Fraysee. Existence of Cournot equilibrium in large markets. *Econometrica*, 53:587–597, 1985.

[28] C. Barnhart and K. T. Talluri. Airline Operations Research. In C. ReVelle and A. McGarrity, editors, *Systems for Civil and Environmental Engineering: An Advanced Text Book*. Wiley, New York, NY, 1997.

[29] M. Basgall. Hotel help: Fuzzy thinking helps clear up tradeoffs on hotel reservations, June 1998. Duke University News, News release.

REFERENCES

[30] J. M. Bates and C. W. J. Granger. The combination of forecasts. *Operational Research Quarterly*, 20:451–468, 1969.

[31] M. J. Beckmann. Decision and team problems in airline reservations. *Econometrica*, 26:134–145, 1958.

[32] M. J. Beckman and F. Bobkowski. Airline demand: An analysis of some frequency distributions. *Naval Research Quarterly*, 43:43–51, 1958.

[33] R. Bellman. *Applied Dynamic Programming*. Princeton University Press, Princeton, NJ, 1957.

[34] D. R. Bell, J. Chiang, and V. Padmanabhan. The decomposition of promotional response: An empirical generalization. *Marketing Science*, 18:504–526, 1999.

[35] D. R. Bell, G. Iyer, and V. Padmanabhan. Price competition under stockpiling and flexible consumption. *Journal of Marketing Research*, 39:292–303, 2002.

[36] P. P. Belobaba and S. Lee. PODS update: Large network O-D control results. In *2000 AGIFORS Reservations and Yield Management Study Group Symposium Proceedings*, New York, NY, 2000.

[37] P. P. Belobaba and L. R. Weatherford. Comparing decision rules that incorporate customer diversion in perishable asset revenue management situations. *Decision Sciences*, 27:343–363, 1996.

[38] P. P. Belobaba. Airline yield management: An overview of seat inventory control. *Transportation Science*, 21:63–73, 1987.

[39] P. P. Belobaba. *Air Travel Demand and Airline Seat Inventory Management*. PhD thesis, Flight Transportation Laboratory, MIT, Cambridge, MA, 1987.

[40] P. P. Belobaba. Application of a probabilistic decision model to airline seat inventory control. *Operations Research*, 37:183–197, 1989.

[41] P. P. Belobaba. Optimal vs. heuristic methods for nested seat allocation. Presentation at ORSA/TIMS Joint National Meeting, November 1992.

[42] P. P. Belobaba. PODS results update: Impacts of forecasting on O-D control methods. In *1998 AGIFORS Reservations and Yield Management Study Group Symposium Proceedings*, Melbourne, Australia, 1998.

[43] P. P. Belobaba. Revenue and competitive impacts of O-D control: Summary of PODS results. In *First Annual INFORMS Revenue Management Section Meeting*, New York, NY, 2001.

[44] D. A. Belsley, E. Kuh, and R. E. Welsch. *Regression Diagnostics: Identifying Influential Data and Sources of Collinearity*. Wiley, New York, NY, 1980.

[45] A. C. Bemmaor and D. Mouchoux. Measuring the short-term effect of in-store promotion and retail advertising on brand sales: A factorial experiment. *Journal of Marketing Research*, 28:202–214, 1991.

[46] J. P. Benoit and V. Krishna. Dynamic duopoly: Prices and quantities. *Review of Economic Studies*, 54:23–35, 1987.

[47] D. Bensanko and W. L. Winston. Optimal price skimming by a monopolist facing rational consumers. *Management Science*, 36:555–567, 1990.

[48] M. Ben-Akiva and S. Lerman. *Discrete-Choice Analysis: Theory and Application to Travel Demand*. MIT Press, Cambridge, MA, 1985.

[49] M. Ben-Akiva. *Structure of Passenger Travel Demand Models*. PhD thesis, Department of Civil Engineering, MIT, Cambridge, MA, 1973.

[50] M. E. Berge and C. A. Hopperstad. Demand driven dispatch: A method for dynamic aircraft assignment, models and algorithms. *Operations Research*, 41:153–168, 1993.

[51] T. Bergstrom and J. K. MacKie-Mason. Some simple analytics of peak-load pricing. *RAND Journal of Economics*, 22:241–249, 1991.

[52] S. Berry, J. Levinsohn, and A. Pakes. Automobile prices in market equilibrium. *Econometrica*, 63:841–890, July 1995.

[53] S. Berry. Estimating discrete-choice models of product differentiation. *RAND Journal of Economics*, 25:242–262, 1994.

[54] J. Bertrand. Theorie math ématique de la richesse sociale. *Journal des Savants*, pages 499–508, 1983.

[55] D. P. Bertsekas and J. N. Tsitsiklis. *Neuro-Dynamic Programming*. Athena Scientific, Belmont, MA, 1996.

[56] D. P. Bertsekas and J. N. Tsitsiklis. *Parallel and Distributed Computing*. Athena Scientific, Belmont, MA, 2nd edition, 1997.

[57] D. P. Bertsekas. *Dynamic Programming and Optimal Control*, volume 1. Athena Scientific, Belmont, MA, 1995.

[58] D. P. Bertsekas. *Constrained Optimization and Lagrange Multiplier Methods*. Athena Scientific, Belmont, MA, 1996.

[59] D. P. Bertsekas. *Nonlinear Programming*. Athena Scientific, Belmont, MA, 2nd edition, 1999.

[60] D. J. Bertsimas and S. de Boer. A stochastic booking limit policy for airline network revenue management. Technical report, Operations Research Center, MIT, Cambridge, MA, 2001.

[61] D. J. Bertsimas and I. Popescu. Revenue management in a dynamic network environment. *Transportation Science*, 37:257–277, 2003.

[62] D. Bertsimas and R. Shioda. Restaurant revenue management. *Operations Research*, 51:472–486, 2003.

REFERENCES

[63] D. Besanko, S. Gupta, and D. Jain. Logit demand estimation under competitive pricing behavior: An equilibrium framework. *Management Science*, 44:1533–1547, 1998.

[64] A. V. Bhatia and S. C. Parekh. Optimal allocation of seats by fare. Presentation to AGIFORS Reservations Study Group, Trans World Airlines, 1973.

[65] R. R. Bhat. *Managing the Demand for Fashion Items*. UMI Research Press, Ann Arbor, MI, 1985.

[66] H. Bierman Jr. and J. Thomas. Airline overbooking strategies and bumping procedures. *Public Policy*, 21:601–606, 1975.

[67] S. Biller, L. M. A. Chan, D. Simchi-Levi, and J. Swann. Dynamic pricing and the direct-to-customer model in the automotive industry. Technical report, Northwestern University, Evanston, IL, 2000. Working Paper.

[68] J. R. Birge and F. Louveaux. *Introduction to Stochastic Programming*. Springer-Verlag, Berlin, Germany, 1997.

[69] C. Bishop. *Neural Networks for Pattern Recognition*. Oxford University Press, Oxford, UK, 1995.

[70] G. R. Bitran, R. Caldentey, and S. V. Mondschein. Coordinating clearance markdown sales of seasonal products in retail chains. *Operations Research*, 46:609–624, 1998.

[71] G. R. Bitran and S. Gilbert. Managing hotel reservations with uncertain arrivals. *Operations Research*, 44:35–49, 1996.

[72] G. R. Bitran and S. V. Mondschein. An application of yield management to the hotel industry considering multiple day stays. *Operations Research*, 43:427–444, 1995.

[73] G. R. Bitran and S. V. Mondschein. Periodic pricing of seasonal products in retailing. *Management Science*, 43:61–79, 1997.

[74] R. C. Blattberg, R. Briesch, and E. J. Fox. How promotions work. *Marketing Science*, 14:G122–G132, 2001.

[75] R. C. Blattberg and E. I. George. Shrinkage estimation of price and promotional elasticities: Seemingly unrelated equations. *Journal of the American Statistical Association*, 86:304–315, 1991.

[76] R. C. Blattberg and S. A. Neslin. *Sales Promotion: Concepts, Methods, and Strategies*. Prentice-Hall, Englewood Cliffs, NJ, 1990.

[77] R. C. Blattberg and K. J. Wisniewski. How retail price promotions work: Empirical results. Technical Report 43, University of Chicago, 1987.

[78] R. C. Blattberg and K. J. Wisniewski. Price-induced patterns of competition. *Marketing Science*, 8:291–309, 1989.

[79] H. D. Block and J. Marschak. Random orderings and stochastic theories of responses. In I. Olkin, editor, *Contributions to Probability and Statistics*. Stanford University Press, California, CA, 1960.

[80] S. E. Bodily and P. E. Pfeifer. Overbooking decision rules. *Omega*, 20:129–133, 1992.

[81] C. H. Bohutinsky. The sell-up potential of airline demand. Master's thesis, Flight Transportation Lab, MIT, Cambridge, MA, 1990.

[82] M. Boiteux. Peak-load pricing. *Journal of Business*, 33:157–179, 1960.

[83] S. Bollapragada, H. Cheng, M. Phillips, M. Garbiras, M. Scholes, T. Gibbs, and M. Humphreville. NBC's optimization systems increase revenue and productivity. *Interfaces*, 32:47–60, 2002.

[84] R. N. Bolton. The relationship between market characteristics and promotional price elasticities. *Marketing Science*, 8:153–169, 1989.

[85] G. E. P. Box and G. M. Jenkins. *Time Series Analysis: Forecasting and Control*. Holden-Day, Oakland, CA, 1976.

[86] A. Boyd. Airline alliances. *OR/MS Today*, 25:28–31, 1998.

[87] S. Bratu. Network value concept in airline revenue management. Master's thesis, Department of Aeronautics and Astronautics, MIT, Cambridge, MA, 1999.

[88] W. Brock and J. A. Scheinkman. Price setting supergames with capacity constraints. *Review of Economic Studies*, 52:371–382, 1985.

[89] J. K. Brueckner and P. T. Spiller. Competition and mergers in airline networks. *International Journal of Industrial Organization*, 9:323–343, 1991.

[90] S. L. Brumelle, J. I. McGill, T. H. Oum, K. Sawaki, and M. W. Tretheway. Allocation of airline seat between stochastically dependent demands. *Transportation Science*, 24:183–192, 1990.

[91] S. L. Brumelle and J. I. McGill. Airline seat allocation with multiple nested fare classes. *Operations Research*, 41:127–137, 1993.

[92] R. Bruns. Mastering the yield. *Hospitality Technology*, October 2001.

[93] J. I. Bulow. Durable-goods monopolists. *Journal of Political Economy*, 90:314–332, 1982.

[94] J. Bulow and P. Klemperer. Rational frenzies and crashes. *Journal of Political Economy*, 102:1–23, 1994.

[95] J. Bulow and J. Roberts. The simple economics of optimal auctions. *Journal of Political Economy*, 97:1060–1090, 1989.

[96] J. Burns. Understanding and maximizing a hotel's electronic distribution options. Hospitality Technology Consulting, 2000.

REFERENCES

[97] K. C. Campbell. The applicability of revenue management to intermodal transport. Technical report, Systems Engineering Department, University of Pennsylvania, 1996.

[98] M. C. Campbell. Perceptions of price unfairness: Antecedents and consequences. *Journal of Marketing Research*, 36:187–199, 1999.

[99] J. D. Carroll and P. E. Green. Psychometric methods in marketing research: Part 1, Conjoint analysis. *Journal of Marketing Research*, 32:385–391, 1995.

[100] W. J. Carroll and R. C. Grimes. Evolutionary change in product management: Experiences in the car rental industry. *Interfaces*, 25:84–104, 1995.

[101] G. Casella and E. George. Explaining the Gibbs sampler. *American Statistician*, 46:167–174, 1992.

[102] P. Cattin and D. R. Wittink. Commercial use of conjoint analysis: A survey. *Journal of Marketing*, 46:44–53, 1982.

[103] E. Chamberlin. *The Theory of Monopolistic Competition*. Harvard University Press, Cambridge, MA, 1933.

[104] P. Chandon and B. Wansink. When are stockpiled products consumed faster? A convenience-salience framework of postpurchase consumption incidence and quantity. *Journal of Marketing Research*, 39:321–335, 2002.

[105] L. M. A. Chan, D. Simchi-Levi, and J. Swann. Flexible pricing strategies to improve supply chain performance. Technical report, University of Toronto, 2000. Working Paper.

[106] S. N. Chapman and J. Carmel. Demand capacity management in health care: An application of yield management. *Health Care Management Review*, 17:45–54, 1992.

[107] R. E. Chatwin. *Optimal Airline Overbooking*. PhD thesis, Stanford University, Palo Alto, CA, 1993.

[108] R. E. Chatwin. Continuous-time airline overbooking with time-dependent fares and refunds. *Transportation Science*, 33:805–819, 1999.

[109] R. E. Chatwin. Multiperiod airline overbooking with a single fare class. *Operations Research*, 46:805–819, 1999.

[110] B. Cheng and D. M. Titterington. Neural networks: A review from a statistical perspective. *Statistical Science*, 9:2–54, 1994.

[111] V. C. P. Chen, D. Gunther, and E. L. Johnson. A Markov decision problem-based approach to the airline YM problem. Technical report, Georgia Institute of Technology, Atlanta, GA, 1998. Working Paper.

[112] V. C. P. Chen, D. Gunther, and E. L. Johnson. A bid price heuristic. Technical report, Georgia Institute of Technology, Atlanta, GA, 1999. Working Paper.

[113] X. Chen and D. Simchi-Levi. Coordinating inventory control and pricing strategies with random demand and fixed ordering costs: The finite horizon case. Technical report, MIT, Cambridge, MA, 2002. Working Paper.

[114] M. Chevalier and R. C. Curhan. Retail promotions as a function of trade promotions: A descriptive analysis. *Sloan Management Review*, pages 19–32, Winter 1976.

[115] S. Chib and E. Greenberg. Markov chain Monte Carlo simulation methods in econometrics. *Econometric Theory*, 12:409–431, 1996.

[116] P. K. Chintagunta, V. Kadiyali, and N. J. Vilcassim. Endogeneity and simultaneity in competitive pricing and advertising: A logit demand analysis. Technical report, Graduate School of Business, University of Chicago, Chicago, IL, January 1999.

[117] P. K. Chintagunta. Heterogenous logit model implications for brand positioning. *Journal of Marketing Research*, 31:304–311, 1994.

[118] A. Ciancimino, G. Inzerillo, S. Lucidi, and L. Palagi. A methematical programming approach for the solution of the railway yield management problem. *Transportation Science*, 33:168–181, 1999.

[119] *Civil Aeronautics Board Economic Regulations Docket 16563*. Washington, DC, January 10, 1967.

[120] E. Clarke. Multipart pricing of public goods. *Public Choice*, 2:19–33, 1971.

[121] E. W. Clemens. Price discrimination and the multiproduct firm. *Review of Economic Studies*, 19:1–11, 1951.

[122] R. T. Clemen and R. L. Winkler. Limits on the precision and value of information from dependent sources. *Operations Research*, 33:427–442, 1985.

[123] R. T. Clemen. Combining overlapping information. *Management Science*, 33:373–380, 1987.

[124] R. H. Coase. Durability and monopoly. *Journal of Law and Economics*, 15:143–149, 1972.

[125] J. A. Colledge, J. Hicks, J. B. Robb, and D. Wagle. Power by the minute. *McKinsey Quarterly*, (1), 2002.

[126] McKinsey & Company. The benefits of demand-side management and dynamic pricing programs. Technical report, March 2001.

[127] L. G. Cooper and M. Nakanishi. *Market-share analysis*. Kluwer, Norwell, MA, 1988.

[128] L. G. Cooper. Competitive maps: The structure underlying asymmetric cross elasticities. *Management Science*, 34:707–723, 1988.

[129] W. L. Cooper and R. P. Menich. Airline ticket auctions: Revenue management and the pivotal mechanism. Technical report, University of Minnesota, Minneapolis, MI, 1998. Working Paper.

REFERENCES

[130] W. L. Cooper. Pathwise properties and performance bounds for a perishable inventory system. *Operations Research*, 49:455–466, 2001.

[131] W. L. Cooper. Asymptotic behavior of an allocation policy for revenue management. *Operations Research*, 50:720–727, 2002.

[132] A. Cournot. *Recherches sur les Principes Mathématiques de la Théorie des Richesses*. (English Version) Macmillan, Paris, France, 1838.

[133] P. Courty. An economic guide to ticket pricing in the entertainment industry. *Louvain Economic Review*, 66:167–189, 1994.

[134] D. R. Cox and D. Oakes. *Analysis of Survival Data*. Chapman and Hall, London, UK, 1984.

[135] P. Coy. The power of smart pricing: Companies are fine-tuning their price strategies—and it's paying off. *Business Week*, pages 160–164, April 10, 2000.

[136] M. Crew, C. Fernando, and P. Kleindorfer. The theory of peak-load pricing. *Journal of Regulatory Economics*, 8:215–248, 1995.

[137] R. G. Cross. *Revenue Management: Hardcore Tactics for Market Domination*. Broadway Books, New York, NY, 1997.

[138] R. C. Curhan and R. J. Kopp. Factors influencing grocery retailer's support of trade promotions. Technical Report 86-114, Marketing Science Institute, Cambridge, MA, 1986.

[139] R. E. Curry. Optimal airline seat allocation with fare classes nested by origins and destinations. *Transportation Science*, 24:193–204, 1990.

[140] R. E. Curry. Real-time revenue management: Bid price strategies for origins/destinations and legs. *Scorecard, Aeronomics Inc.*, Atlanta, 2Q, 1992.

[141] J. D. Dana. Advance-purchase discounts and price discrimination in competitive markets. *Journal of Political Economy*, 106:395–422, 1998.

[142] J. D. Dana. Equilibrium price dispersion under demand uncertainty: The roles of costly capacity and market structure. *RAND Journal of Economics*, 30:632–660, 1999.

[143] J. D. Dana. Using yield management to shift demand when the peak time is unknown. *RAND Journal of Economics*, 30:456–474, 1999.

[144] C. G. Dasgupta, G. S. Dispensa, and S. Ghose. Comparing the predictive performance of a neural network model with some traditional market response models. *International Journal of Forecasting*, 10:235–244, 1994.

[145] G. Das Varma and N. Vettas. Optimal dynamic pricing with inventories. *Economics Letters*, 72:335–340, 2001.

[146] S. Daudel and G. Vialle. *Yield Mangement: Applications to Air Transport and Other Service Industries*. Les Presses de L'Institut du Transport Aerien, Paris, France, 1994.

[147] C. Davidson and R. Deneckere. Long-run competition in capacity, short-run competition in price and the Cournot model. *RAND Journal of Economics*, 17:404–415, 1986.

[148] A. C. Davison and D. V. Hinkley. *Bootstrap Methods and Their Application*. Cambridge University Press, Cambridge, UK, 1997.

[149] C. J. Day, B. F. Hobbs, and J. S. Pang. Oligopolistic competition in power networkds: A conjectured supply function approach. Technical Report PWP-090, Program for Workable Energy Regulation (POWER), University of California, Berkeley, February 2002.

[150] G. Debreu. Review of R. D. Luce, individual choice behavior: A theoretical analysis'. *American Economic Review*, 50:186–188, 1960.

[151] M. H. DeGroot. *Probability and Statistics*. Addison Wesley, Reading, MA, 2nd edition, 1985.

[152] A. P. Dempster, N. M. Laird, and D. B. Rubin. Maximum likelihood from incomplete data via the EM algorithm. *Journal of the Royal Statistics Society, B*, 39:1–38, 1977.

[153] M. J. Denton. Trading & risk management focus in north america. *World Power*, 2001.

[154] S. V. de Boer, R. Freling, and N. Piersma. Stochastic programming for multiple-leg network revenue management. *European Journal of Operational Research*, 137:72–92, 2002.

[155] S. V. de Boer. *Advances in Airline Revenue Management and Pricing*. PhD thesis, Sloan School of Management, MIT, Cambridge, MA, June 2003.

[156] S. de Vries and R. Vohra. Combinatorial auctions: A survey. *INFORMS Journal on Computing*, 15:284–309, 2003.

[157] T. Di Nome. Hot tickets, hawked legitimately online. *New York Times*, July 3, 2003.

[158] G. Di Pillo, S. Lucidi, and L. Palagi. An algorithm for the nonlinear programming problem of the railway yield management. Technical report, Dipartimento di Informatica e Sistemistica, Universita di Roma "La Sapienza", 2002.

[159] K. Donaghy and U. McMahon-Beattie. Implementing yield management: Lessons from the hotel sector. *International Journal of Contemporary Hospitality Management*, 9:50–54, 1997.

[160] K. Donaghy and U. McMahon-Beattie. The impact of yield management on the role of the hotel general manager. *Progress in Tourism and Hospitality Research*, 4:217–228, 1998.

[161] N. R. Draper and H. Smith. *Applied Regression Analysis*. Wiley, New York, NY, 1980.

[162] M. Dror, P. Trudeau, and S. P. Ladany. Network models for seat allocation on flights. *Transportation Research*, 22B:239–250, 1988.

[163] E. D'Sylva. OD seat assignment to maximize expected revenue. Technical report, Boeing Commercial Airplane Company, Seattle, WA, 1982.

[164] M. Dudey. Dynamic Edgeworth-Bertrand competition. *Quarterly Journal of Economics*, 107:1461–1477, November 1992.

[165] H. N. Dunleavy. Airline passenger overbooking. In D. Jenkins, editor, *The Handbook of Airline Economics*, pages 469–482. McGraw-Hill, New York, NY, 1995.

[166] J. Durbin and G. S. Watson. Testing for serial correlation in least squares regression, I. *Biometrika*, 37:407–428, 1950.

[167] J. Durbin and G. S. Watson. Testing for serial correlation in least squares regression, II. *Biometrika*, 38:159–178, 1951.

[168] J. Durbin and G. S. Watson. Testing for serial correlation in least squares regression, III. *Biometrika*, 58:1–42, 1971.

[169] J. Durbin. Errors in variables. *Review of International Statistics*, 22:23–32, 1954.

[170] J. Eaton and M. Engers. Intertemporal price competition. *Econometrica*, 58:637–659, 1990.

[171] B. Efron. Censored data and the bootstrap. *Journal of the American Statistical Association*, 76:312–319, 1981.

[172] B. Efron. Empirical Bayes methods for combining likelihoods. *Journal of the American Statistical Association*, 83:414–425, 1988.

[173] B. Efron. Logistic regression, survival analysis, and the Kaplan-Meier curve. *Journal of the American Statistical Association*, 83:414–425, 1988.

[174] J. Eliashberg and G. L. Lilien, editors. *Management Science in Marketing*. North Holland, Amsterdam, The Netherlands, 1991.

[175] J. Eliashberg and R. Steinberg. Marketing-production joint decision making. In J. Eliashberg and J. D. Lilien, editors, *Management Science in Marketing, Handbooks in Operations Research and Management Science*. North Holland, Amsterdam, The Netherlands, 1991.

[176] A. A. Elimam and B. M. Dodin. Incentives and yield management in improving productivity of manufacturing facilities. *IIE Transactions*, 33:449–462, 2001.

[177] W. J. Elmaghraby and P. Keskinocak. Dynamic pricing: Research overview, current practices, and future directions. *Management Science*, 2003. (To appear).

[178] Y. Ermoliev. Stochastic quasigradient methods. In Y. Ermoliev and R. J. B Wets, editors, *Numerical Techniques for Stochastic Optimization*. Springer-Verlag, New York, NY, 1988.

[179] P. S. Fader and J. M. Lattin. Accounting for heterogeneity and nonstationarity in a cross-sectional model of consumer purchase behavior. *Marketing Science*, 12:304–317, 1991.

[180] L. M. Falkson. Airline overbooking: Some comments. *Journal Transport Economics and Policy*, 3:352–354, 1969.

[181] A. Farqui, S. Shaffer, J. P. Seiden, S. Black, and J. H. Chamberlin. Customer response to rate options. Technical report, EPRI Palo Alto: EPRI Demand-Side Planning Division, 1991.

[182] A. Federgruen and H. Groenevelt. The greedy procedure for resource allocation problems: Necessary and sufficient conditions. *Operations Research*, 34:909–918, 1986.

[183] A. Federgruen and A. Heching. Combined pricing and inventory control under uncertainty. *Operations Research*, 47:454–475, 1999.

[184] C. L. Fellman. Television advertising—a spot is a spot is a spot—or is it? In *Proceedings of the Fifth IATA Revenue Management Conference and Technical Briefing*, 1993.

[185] Y. Feng and G. Gallego. Optimal starting times for end-of-season sales and optimal stopping times for promotional fares. *Management Science*, 41:1371–1391, 1995.

[186] Y. Feng and G. Gallego. Perishable asset revenue management with Markovian time dependent demand intensities. *Management Science*, 46:941–956, 2000.

[187] Y. Feng and B. Xiao. Maximizing revenue of perishable assets with a risk factor. *Operations Research*, 47:337–341, 1999.

[188] Y. Feng and B. Xiao. A continuous-time yield management model with multiple prices and reversible price changes. *Management Science*, 46:644–657, 2000.

[189] Y. Feng and B. Xiao. Optimal policies of yield management with multiple predetermined prices. *Operations Research*, 48:332–343, 2000.

[190] H. Feyen and C. Hüglin. Data mining to improve forecast accuracy in the airline business. In *Reservations and Yield Management Study Group Annual Meeting Proceedings*, Bangkok, Thailand, 2001. AGIFORS.

[191] P. Flint. End the CRS oligopoly. *Air Transport World*, 35(4), 1998.

[192] D. P. Foster and R. V. Vohra. A randomization rule for selecting forecasts. *Operations Research*, 41:704–801, 1993.

[193] R. Frank, W. Massy, and Y. Wind. *Market Segmentation*. Prentice-Hall, Englewood Cliffs, NJ, 1972.

[194] S. C. Friend and P. H. Walker. Welcome to the new world of merchandising. *Harvard Business Review*, 79, November 2001.

[195] D. Fudenberg and J. Tirole. *Game Theory*. MIT Press, Cambridge, MA, 1991.

REFERENCES

[196] I. L. Gale and T. J. Holmes. Advance-purchase discounts and monopoly allocation of capacity. *American Economic Review*, 83:135–146, 1993.

[197] G. Gallego, G. Iyengar, and R. Phillips. Network revenue optimization with flexible products. Conference Presentation, 2003. The Third Annual INFORMS Pricing and Revenue Management Conference, Columbia University.

[198] G. Gallego and G. J. van Ryzin. Optimal dynamic pricing of inventories with stochastic demand over finite horizons. *Management Science*, 40:999–1020, 1994.

[199] G. Gallego and G. J. van Ryzin. A multi-product dynamic pricing problem and its applications to network yield management. *Operations Research*, 45:24–41, 1997.

[200] G. Gallego. A minmax distribution free procedure for the (Q, R) inventory model. *Operations Research Letters*, 11:55–60, 1992.

[201] A. Garcia-Diaz and A. Kuyumcu. A cutting-plane method for maximizing revenues in yield management. *Computers and Industrial Engineering*, 33:51–54, 1997.

[202] R. Geary. Determination of linear relations between systematic parts of variables with errors of observation the variance of which is unknown. *Econometrica*, 17:30–58, 1949.

[203] A. E. Gelfand, A. F. M. Smith, and T. M. Lee. Bayesian analysis of constrained parameter and truncated data problems using Gibbs sampling. *Journal of the American Statistical Association*, 87:523–532, 1992.

[204] A. Gellman and C. Fitzgerald. The history and outlook of travel distribution in the PC-based internet environment. Technical report, Global Aviation Associates, Chicago, IL, 2001.

[205] S. Geman and D. Geman. Stochastic relaxation, Gibbs distributions and the Bayesian restoration of images. *IEEE Transactions on Pattern Analysis and Machine Intelligence*, 12:609–628, 1984.

[206] D. Gensch and P. Shaman. Models of competitive television ratings. *Journal of Marketing Research*, 17:307–315, 1980.

[207] E. I. George and R. E. McCulloch. Variable selection via Gibbs sampling. *Journal of the American Statistical Association*, 88:881–889, 1993.

[208] M. K. Geraghty and E. Johnson. Revenue management saves National Car Rental. *Interfaces*, 27:107–127, 1997.

[209] Y. Gerchak, M. Parlar, and T. K. M. Ye. Optimal rationing policies and production quantities for products with several demand classes. *Canadian Journal of Administration Science*, 2:161–176, 1985.

[210] Y. Gerchak and M. Parlar. A single period inventory problem with partially controlled demand. *Computers and Operations Research*, 14:1–9, 1987.

[211] E. Gerstner and J. D. Hess. A theory of channel price promotions. *American Economic Review*, 81:872–886, 1991.

[212] W. R. Gilks, S. Richardson, and D. J. Spiegelhalter, editors. *Markov Chain Monte Carlo in Practice*. Chapman and Hall, London, UK, 1996.

[213] P. J. Gill. Sea-Land sets new course with IT. *UniForum's ITSolutions*, March 1996.

[214] G. Girard. Revenue management: The price can't be right if the tools aren't. Technical report, AMR Research Inc., Boston, MA, September 2000.

[215] F. Glover, R. Glover, J. Lorenzo, and C. McMillan. The passenger mix problem in the scheduled airlines. *Interfaces*, 12:73–79, 1982.

[216] S. M. Goldberg, P. E. Green, and Y. Wind. Conjoint analysis of price premiums for hotel amenities. *Journal of Business*, 57(1):111–132, 1984.

[217] T. Gorin. Revenue benefits of sell-up models. In *Reservations and Yield Management Study Group Annual Meeting Proceedings*, New York, NY, 2000. AGIFORS.

[218] C. W. J. Granger and P. Newbold. *Forecasting Economic Time Series*. Academic Press, New York, NY, 1977.

[219] D. Gray. Revenue based capacity management, a survival technique developed by the airline industry, enters manufacturing sector. *OR/MS Today*, October 1994.

[220] W. Greene. *Econometric Analysis*. Prentice-Hall, Englewood Cliffs, NJ, 2nd edition, 1997.

[221] E. A. Greenleaf. The impact of reference price effects on the profitability of price promotions. *Marketing Science*, 14:82–104, 1995.

[222] E. J. Green and R. H. Porter. Noncooperative collusion under imperfect price information. *Econometrica*, 52:87–100, 1984.

[223] P. E. Green, S. M. Goldberg, and M. Montemayor. A hybrid utility estimation model for conjoint analysis. *Journal of Marketing*, 45:33–41, 1981.

[224] P. E. Green. On the design of choice experiments involving multifactor alternatives. *Journal of Consumer Research*, 1:61–68, 1974.

[225] R. Grover and V. Srinivasan. Evaluating the multiple effects of retail promotions on brand loyal and brand switching segments. *Journal of Marketing Research*, 29:76–89, 1992.

[226] T. Groves. Incentives in teams. *Econometrica*, 41:617–631, 1973.

[227] P. M. Guadagni and J. D. C. Little. A logit model of brand choice calibrated on scanner data. *Marketing Science*, 3:203–238, 1983.

[228] F. Gul, H. Sonnenschein, and R. B. Wilson. Foundations of dynamic monopoly and the Coase conjecture. *Journal of Economic Theory*, 39:155–190, 1986.

[229] E. J. Gumbel. *Statistics of Extremes*. Columbia University Press, New York, NY, 1958.

[230] S. Gupta and P. C. Wilton. Combination of forecasts: An extension. *Management Science*, 33:356–372, 1987.

[231] I. Guttman. *Linear Models: An Introduction*. Wiley, New York, NY, 1982.

[232] J. D. Gwartney and R. L. Stroup. *Economics: Private and Public Choice*. Harcourt, Stamford, CT, 8th edition, 1997.

[233] G. C. Hadjinicola and C. Panayi. The overbooking problem in hotels with multiple tour operators. *International Journal of Operations and Production Management*, 17:874–889, 1997.

[234] R. D. Hanks, R. G. Cross, and R. P. Noland. Discounting in the hotel industry: A new approach. *Cornell Hotel and Restaurant Administration Quarterly*, 33:40–45, 1992.

[235] D. M. Hanssens, L. J. Parsons, and R. L. Schultz. *Market Response Models: Econometric And Time Series Analysis*. Kluwer, Norwell, MA, 1990.

[236] S. K. Happel and M. M. Jennings. Creating a futures market for major ticket events: Problems and prospects. *Cato Journal*, 21(3), Winter 2002.

[237] B. G. S. Hardie, E. J. Johnson, and P. S. Fader. Modeling loss aversion and reference dependent effects on brand choice. *Marketing Science*, 12:378–394, 1993.

[238] F. H. Harris and P. Peacock. Hold my place, please. *Marketing Management*, 4:34–46, 1995.

[239] M. Harris and A. Raviv. A theory of monopoly pricing schemes with demand uncertainity. *American Economic Review*, 71:347–365, 1981.

[240] M. Harris and R. M. Townsend. Resource allocation under asymmetric information. *Econometrica*, 49:33–64, 1981.

[241] J. Harsanyi. Games with incomplete information played by Bayesian players, I, II, and III. *Management Science*, 14:159–182, 320–334, 486–503, 1967-8.

[242] A. C. Harvey. *Forecasting, Structural Time Series Models and the Kalman Filter*. Cambridge University Press, Cambridge, UK, 1989.

[243] A. C. Harvey. *Time Series Models*. MIT Press, Cambridge, MA, 1993.

[244] J. Hausman and D. McFadden. A specification test for the multinomial logit model. *Econometrica*, 52:1219–1240, 1984.

[245] J. Hausman. Specification tests in econometrics. *Econometrica*, 46:1251–1272, 1978.

[246] A. Y. Ha. Stock-rationing policy for a make-to-stock production system with two priority classes and backordering. *Naval Research Logistics*, 44:457–472, 1997.

[247] A. Heching, G. Gallego, and G. J. van Ryzin. A theoretical and empirical investigation of markdown pricing in apparel retailing. *Journal of Revenue Managment and Pricing*, 2002.

[248] J. Heckman and B. Singer. A method for minimizing the impact of distributional assumptions in econometric models for duration data. *Econometrica*, 52:271–320, 1984.

[249] A. L. Hempenius. *Monopoly with Random Demand.* Rotterdam University Press, Rotterdam, The Netherlands, 1970.

[250] C. Heun. Dynamic pricing boosts bottom line. *Information Week. Com*, (29), October 2001.

[251] M. Hiestand. Online sales allow leagues to gain money, ticketholder information. USA Today, May 1, 2003.

[252] T. Hill, L. Marquez, M. O'Connor, and W. Remus. Artifical neural network models for forecasting and decision making. *International Journal of Forecasting*, 10:5–15, 1994.

[253] T. Hill, M. O'Connor, and W. Remus. Neural network models for time series forecasting. *Management Science*, 42:1082–1091, 1994.

[254] E. Hirst. Real-time pricing could tame the wholesale market. *Electric Perspectives*, March 2001.

[255] S. Holmes, L. Hyde, and M. Rothkopf. Auctions and their use in the natural gas markets. Technical report, Office of Economic Policy, FERC, October 1998.

[256] C. Hopperstad. The application of path preference and stochastic demand modeling to market based forecasting. In *AGIFORS Reservations and Yield Management Study Group Proceedings*, Hong Kong, 1994.

[257] J. H. Horen. Scheduling of network television programs. *Management Science*, 26:354–370, 1980.

[258] R. A. Horn and C. R. Johnson. *Matrix Analysis.* Cambridge University Press, Cambridge, UK, 1994.

[259] H. Hruska. Determining market response functions by neural network modeling: A comparison to econometric techniques. *European Journal of Operational Research*, 66:27–35, 1993.

[260] J. Huber and K. Train. On the similarity of classical and Bayesian estimates of individual mean partworths. *Marketing Letters*, 12:257–267, 2001.

[261] S. Hunt and G. Shuttleworth. *Competition and choice in electricity.* Wiley, Chicester, UK, 1996.

[262] Manugistics Inc. Pricing and revenue optimization. White Paper, Manugistics Inc., available at www.manugistics.com, 2001.

REFERENCES

[263] A. Ingold, U. McMahon-Beattie, and I. Yeoman, editors. *Yield Management: Strategies for the Service Sector.* Continuum, London, UK, 2nd edition, 2000.

[264] IS Report. Revenue management with the world's largest travel agency. Technical Report 5, Thinking Networks AG, 2001.

[265] S. S. Iyengar and M. Lepper. When choice is demotivating: Can one desire too much of a good thing? *Journal of Personality and Social Psychology,* 76:995–1006, 2000.

[266] D. Jain, F. M. Bass, and Y. M. Chen. Estimation of latent class models with heterogenous choice probabilities: An application to market structuring. *Journal of Marketing Research,* 27:94–101, 1990.

[267] D. Jain, N. J. Vilcassim, and P. K. Chintagunta. A random-coefficients logit brand-choice model applied to panel data. *Journal of Business and Economic Studies,* 12:317–328, 1994.

[268] D. Jenkins, editor. *Handbook of Airline Economics.* McGraw-Hill, New York, NY, 1995.

[269] C. A. Johnson. Pricing gets personal. Technical report, Forrester Research Inc., Cambridge, MA, April 2000.

[270] C. A. Johnson. Retail revenue optimization: Timely and rewarding. Technical report, Forrester Research Inc., Cambridge, MA, July 23, 2001.

[271] B. Johnston. The scoop on sleeper pricing. *Trains Magazine,* 2003.

[272] P. Jones and D. Hamilton. Yield management: Putting people in the big picture. *Cornell Hotel and Restaurant Administration Quarterly,* 33:89–95, 1992.

[273] G. G. Judge, W. E. Griffiths, R. C. Hill, H. Lütkepohl, and T. C. Lee. *The Theory and Practice of Econometrics.* Wiley, New York, NY, 2nd edition, 1985.

[274] Neame P. J., Philpott A. B., and G. Pritchard. Offer stack optimization in electricity pool markets. *Operations Research,* 2002.

[275] D. Kahneman, J. L. Knetsch, and R. H. Thaler. Fairness and the assumption of economics. *Journal of Business,* 59:285–300, 1986.

[276] D. Kahneman, J. L. Knetsch, and R. H. Thaler. Fairness as a constraint on profit seeking: Entitlements in the market. *American Economic Review,* 76:728–741, 1986.

[277] D. Kahneman, P. Slovic, and A. Tversky, editors. *Judgement Under Uncertainty: Heuristics and Biases.* Cambridge University Press, Cambridge, UK, 1982.

[278] D. Kahneman and A. Tversky. Prospect theory: An analysis of decision under risk. *Econometrica,* 47:263–292, 1979.

[279] C. Kalish. Monopolist pricing with dynamic demand and production costs. *Marketing Science*, 2:135–159, 1983.

[280] K. U. Kalka and K. Weber. PNR-based no-show forecast. In *Reservations and Yield Management Study Group Annual Meeting Proceedings*, New York, NY, 2000. AGIFORS.

[281] M. U. Kalwani, H. J. Rinne, Y. Sugita, and C. K. Yim. A price expectations model of customer brand choice. *Journal of Marketing Research*, 27:251–262, 1990.

[282] M. U. Kalwani and C. K. Yim. Consumer price and promotion expectations: An experimental study. *Journal of Marketing Research*, 29:90–100, 1992.

[283] K. Kalyanam. Pricing decisions under demand uncertainty: A Bayesian mixture model approach. *Marketing Science*, 15:207–221, 1996.

[284] V. Kalyan. Dynamic customer value management. *Information Systems Frontiers*, 4:101–119, 2002.

[285] W. A. Kamakura, B. D. Kim, and J. Lee. Modeling preference and structural heterogeneity in consumer choice. *Marketing Science*, 15:152–172, 1996.

[286] W. A. Kamakura and G. J. Russell. A probabilistic choice model for market segmentation and elasticity structure. *Journal of Marketing Research*, 26:379–390, 1989.

[287] P. K. Kannan and G. P. Wright. Modeling and testing structured markets: A nested logit approach. *Marketing Science*, 10:58–82, 1991.

[288] A. Kaplan. Stock rationing. *Management Science*, 15:260–267, 1969.

[289] E. L. Kaplan and P. Meier. Nonparametric estimates from incomplete observations. *American Statistical Association Journal*, 53:457–481, 1958.

[290] I. Karaesmen and G. J. van Ryzin. Overbooking with substitutable inventory classes. *Operations Research*, 1998. (To appear).

[291] I. Karaesmen and G. J. van Ryzin. Combining overbooking and fare class allocation on a network. Technical report, Graduate School of Business, Columbia University, New York, NY, 2001. Working paper.

[292] I. Karaesmen. *Three Essays in Revenue Management*. PhD thesis, Graduate School of Business, Columbia University, New York, NY, 2001.

[293] S. Karlin and C. R. Carr. Prices and optimal inventory policies. In K. J. Arrow, S. Karlin, and H. Scarf, editors, *Studies in Applied Probability and Management Science*. Stanford University Press, California, CA, 1962.

[294] R. G. Kasilingam and G. L. Hendricks. Cargo revenue management at American Airlines. In *Proceedings of the Cargo Study Group Meeting*, Rome, Italy, 1993. AGIFORS.

[295] R. G. Kasilingam. Air cargo revenue management: Characteristics and complexities. *European Journal of Operations Research*, 96:36–44, 1996.

REFERENCES

[296] P. J. Kaufmann, G. Ortmeyer, and N. C. Smith. Fairness in consumer pricing. *Journal of Consumer Policy*, 14:117–140, 1991.

[297] E. Kay. Flexed pricing. *Datamation*, 44:58–62, 1998.

[298] S. E. Kimes, D. I. Barrash, and J. E. Alexander. Developing a restaurant revenue management strategy. *Cornell Hotel and Restaurant Administration Quarterly*, 40:18–29, 1999.

[299] S. E. Kimes. The basics of yield management. *Cornell Hotel and Restaurant Administration Quarterly*, 30:14–19, 1989.

[300] S. E. Kimes. A tool for capacity-constrained service firms. *Journal Operations Management*, 8:348–363, 1989.

[301] S. E. Kimes. Yield management: A tool for capacity-constrained service firms. *Journal of Operations Management*, 8:348–363, 1989.

[302] S. E. Kimes. Perceived fairness of yield management. *Cornell Hotel and Restaurant Administration Quarterly*, 35:22–29, 1994.

[303] S. E. Kimes. Applying yield management to the golf course industry. *Cornell Hotel and Restaurant Administration Quarterly*, 41:120–127, 2001.

[304] W. M. Kincaid and D. Darling. An inventory pricing problem. *Journal of Mathematical Analysis and Applications*, 7:183–208, 1963.

[305] P. Klemperer. Auction theory: A guide to the literature. *Journal of Economic Surveys*, 13:227–286, 1999.

[306] P. Klemperer, editor. *The Economic Theory of Auctions*, volume 1–2. Elgar, Cheltenham, UK, 2000.

[307] A. J. Kleywegt and J. D. Papastavrou. The dynamic and stochastic knapsack problem. *Operations Research*, 46:17–35, 1998.

[308] J. Köhler. Revealed preferences for logit modeling of SAS booking data. Master's thesis, Royal Institute of Technology, Stockholm, Sweden, 1993.

[309] P. K. Kopalle, C. F. Mela, and L. Marsh. The dynamic effect of discounting on sales: Empirical analysis and normative pricing implications. *Marketing Science*, 18:317–332, 1999.

[310] P. K. Kopalle, A. Rao, and J. L. Assunção. Asymmetric reference price effects and dynamic pricing policies. *Marketing Science*, 15:60–85, 1996.

[311] S. Kotz and N. L. Johnson, editors. *Encyclopedia of Statistical Sciences*, volume 6. Wiley, New York, NY, 1985.

[312] E. R. Kraft and B. N. Srikar. Revenue management in railroad applications. *Transportation Quarterly*, 54:157–177, 2000.

[313] D. M. Kreps. *Notes on the Theory of Choice*. Westview Press, London, UK, 1988.

[314] D. Kreps and J. Scheinkman. Quantity precommitment and Bertrand competition yield Cournot outcomes. *Bell Journal of Economics*, 14:326–337, 1983.

[315] L. Krishnamurthi and S. P. Raj. A model of brand choice and purchase quantity price sensitivities. *Marketing Science*, 7:1–20, 1988.

[316] V. Kumar and S. P. Leone. Measuring the effect of retail store promotions on brand and store substitution. *Journal of Marketing Research*, 25:178–185, 1988.

[317] A. Kuyumcu and J. Higbie. A framework for media pricing and revenue optimization problems. In *First Annual INFORMS Revenue Management Section Conference*, New York, NY, June 2001. Columbia University.

[318] A. H. Kvanli, C. S. Guynes, and R. J. Pavur. *Introduction to Business Statistics*. West, St. Paul, MN, 2nd edition, 1989.

[319] S. P. Ladany and A. Arbel. Optimal cruise-liner passenger cabin pricing policy. *European Journal of Operations Research*, 55:136–147, 1991.

[320] S. P. Ladany. Dynamic operating rules for motel reservations. *Decision Sciences*, 7:829–841, 1976.

[321] S. P. Ladany. Bayesian dynamic operating rules for optimal hotel reservations. *Zeitschrift Operations Research*, 21:B165–B176, 1977.

[322] R. Lal and J. M. Villas-Boas. Exclusive dealing and price promotions. *Journal of Business*, 69:159–172, 1996.

[323] R. Lal and J. M. Villas-Boas. Price promotions and trade deals with multi-product retailers. *Management Science*, 44:247–262, 1998.

[324] R. Lal. Manufacturer trade deals and retail price competition. *Journal of Marketing Research*, 39:428–444, 1990.

[325] R. Lal. Price promotions: Limiting competitive encroachment. *Marketing Science*, 9:247–262, 1990.

[326] C.U. Lambert, J. M. Lambert, and T. P. Cullen. The overbooking question: A simulation. *Cornell Hotel and Restaurant Administration Quarterly*, 29:7–8, 1989.

[327] P. Lamb and D. Logue. Implementation of a gas load forecaster at Williams Gas pipeline. In *2001 Pipeline Simulation Interest Group Proceedings*. PSIG, 2001.

[328] K. J. Lancaster. The economics of product variety: A survey. *Marketing Science*, 9:189–206, 1990.

[329] J. M. Lattin and R. E. Bucklin. Reference effects of price and promotion on brand-choice behavior. *Journal of Marketing Research*, 26:299–310, 1989.

[330] C. J. Lautenbacher and S. J. Stidham. The underlying Markov decision process in the single-leg airline yield management problem. *Transportation Science*, 34:136–146, 1999.

[331] A. M. Law and W. D. Kelton. *Simulation Modeling and Analysis*. McGraw-Hill, New York, NY, 1982.

[332] E. P Lazear. Retail pricing and clearance sales. *American Economic Review*, 76:14–32, 1986.

[333] A. Leamon. A note on the retailing industry. *Harvard Business School Case, Number 9-598-148*, 2, 1998.

[334] A. Lee. Computer reservations systems: An industry of its own. Case HKU055, CABC, School of Business, The University of Hong Kong, 2000.

[335] K. Lee, T. Choi, C. Ku, and J. Park. Neural network architectures for short term load forecasting. In *Proceedings of the IEEE International Conference on Neural Networks*, Orlando, FL, 1994. IEEE ICNN.

[336] T. C. Lee and M. Hersh. A model for dynamic airline seat inventory control with multiple seat bookings. *Transportation Science*, 27:252–265, 1993.

[337] M. M. Lefever. The gentle art of overbooking. *Cornell Hotel and Restaurant Administration Quarterly*, 30:15–20, 1988.

[338] S. Leibs. Ford heeds the profits. *CFO Magazine*, August 2001.

[339] P. J. Lenk, W. S. DeSarbo, P. E. Green, and M. R. Young. Hierarchical Bayes conjoint analysis: Recovery of partworth heterogeneity from reduced experimental designs. *Marketing Science*, 15:173–191, 1996.

[340] P. Leslie. A structural econometric analysis of price discrimination in Broadway Theatre. Yale University, 1997.

[341] R. Levitan and M. Shubik. Price duopoly and capacity constraints. *International Economic Review*, 13:111–122, 1972.

[342] Y. Liang. Solution to the continuous time dynamic yield management model. *Transportation Science*, 33:117–123, 1999.

[343] V. Liberman and U. Yechiali. Hotel overbooking problem—Inventory system with stochastic cancellations. *Management Science*, 24:1117–1126, 1977–78.

[344] V. Liberman and U. Yechiali. On the hotel overbooking problem. *Management Science*, 24:1117–1126, 1978.

[345] D. Lindley and A. F. M. Smith. Bayes estimates for the linear model. *Journal of the Royal Statistical Association, Ser. B*, 34:1–41, 1972.

[346] S. A. Lippman and K. F. McCardle. The competitive newsboy. *Operations Research*, 45:54–65, 1997.

[347] K. Littlewood. Forecasting and control of passenger bookings. In *Proceedings of the Twelfth Annual AGIFORS Symposium*, Nathanya, Israel, 1972.

[348] R. D. Luce and J. W. Tukey. Simultaneous conjoint measurement: A new type of fundamental measurement. *Journal of Mathematical Psychology*, 1:1–27, 1964.

[349] R. Luce. *Individual Choice Behavior: A Theoretical Analysis.* Wiley, New York, NY, 1959.

[350] S. Luciani. Implementing yield management in small and medium sized hotels: An investigation of obstacles and success factors in Florence hotels. *International Journal of Hospitality Management,* 18:129–142, 1999.

[351] D. Lucking-Reily. Vickrey auctions in practice: From nineteenth century philately to twenty-first-century E-commerce. *Journal of Economic Perspectives,* 13:183–192, 2000.

[352] J. G. MacKinnon. Model specification tests and artificial regressions. *Journal of Economic Literature,* 30:102–146, 1992.

[353] G. Maddala. *Introduction to Econometrics.* Macmillan, New York, NY, 2nd edition, 1997.

[354] C. Maglaras and J. Meissner. Dynamic pricing strategies for multi-product revenue management problems. Working Paper, 2003.

[355] S. Mahajan and G. J. van Ryzin. Inventory competition under dynamic consumer substitution. *Operations Research,* 49:646–657, 2001.

[356] S. Makridakis and R. L. Winkler. Averages of forecast: Some empirical results. *Management Science,* 29:987–996, 1983.

[357] C. F. Manski and D. McFadden, editors. *Structural Analysis of Discrete Data with Econometric Applications.* MIT Press, Cambridge, MA, 1981.

[358] C. Manski. The structure of random utility models. *Theory and Decisions,* 8:229–254, 1977.

[359] M.K. Mantrala and S. Rao. A decision-support system that helps retailers decide order quantities and markdowns for fashion goods. *Interfaces,* 31:146–165, 2001.

[360] S. A. Maragos. Revenue management for ocean carriers: Optimal capacity allocation with multiple nested freight rate classes. Master's thesis, Department of Ocean Engineering, MIT, Cambridge, MA, 1994.

[361] D. L. Margulius. Priced to sell ... to you. *Infoworld,* February 2002.

[362] R. Martinez and M. Sanchez. Automatic booking level control. In *Proceedings of the Tenth AGIFORS Symposium,* 1970.

[363] H. Martin. Broadband meets yield management. *Lightwave,* June 2002.

[364] E. Maskin and J. Riley. Optimal multi-unit auctions. In P. Klemperer, editor, *The Economic Theory of Auctions,* volume 2, pages 312–336. Elgar, 2000.

[365] A. Mas-Collel, M. Whinston, and J. Green. *Microeconomic Theory.* Oxford University Press, 1995.

REFERENCES

[366] S. Matthews. A technical primer on auction theory I: Independent private values. Discussion Paper No. 1096, CMS-EMS Northwestern University, May 1995.

[367] G. E. Mayhew and R. Winer. An empirical analysis of internal and external reference prices using scanner data. *Journal of Consumer Research*, 19:62–70, 1992.

[368] S. G. Maynes and D. J. Wood. Estimating sell-up factors for American Airline's DINAMO seat inventory control model using censored regression techniques. In *Reservations and Yield Management Study Group Annual Meeting Proceedings*. AGIFORS, 1992.

[369] R. P. McAfee and J. McMillan. Auctions and bidding. *Journal of Economic Literature*, 25:699–738, 1987.

[370] R. P. McAfee and T. Wiseman. Capacity choice counters the Coase conjecture. Technical report, Department of Economics, University of Texas at Austin, April 2003.

[371] M. McDonald and I. Dunbar. *Market Segmentation*. Macmillan, London, UK, 2nd edition, 1998.

[372] D. McFadden. Conditional logit analysis of qualitative choice behavior. In P. Zarembka, editor, *Frontiers in Econometrics*, pages 105–142. Academic Press, New York, NY, 1974.

[373] D. McFadden. Econometric models of probabilistic choice among products. *Journal of Business*, 53:513–529, 1980.

[374] J. I. McGill and G. J. van Ryzin. Revenue management: Research overview and prospects. *Transportation Science*, 33:233–256, 1999.

[375] J. I. McGill. *Optimization and Estimation Problems in Airline Yield Management*. PhD thesis, Faculty of Commerce and Business Administration, University of British Columbia, Vancouver, Canada, 1989.

[376] J. I. McGill. Censored regression analysis of multiclass demand data subject to joint capacity constraints. *Annals of Operations Research*, 60:209–240, 1995.

[377] G. McLachlan and T. Krishnan. *The EM Algorithm and Extensions*. Wiley, New York, NY, 1996.

[378] M. McManus. Equilibrium, numbers and size in Cournot oligopoly. *Yorkshire Bulletin of Social and Economic Research*, 16:68–75, 1964.

[379] A. Merrick. Priced to move: Retailers attempt to get a leg up on markdowns with new software. *Wall Street Journal*, 2001.

[380] R. Metters and V. Vargas. Yield management for the nonprofit sector. *Journal of Service Research*, 1:215–226, 1999.

[381] P. Milgrom. Auctions and bidding: A primer. *Journal of Economic Perspectives*, 3:3–22, 1989.

[382] A. J. Miller. *Subset selection in regression.* Chapman and Hall, New York, NY, 1990.

[383] R. G. Miller, Jr. *Survival Analysis.* Wiley, New York, NY, 1981.

[384] R. G. Miller, Jr. What price Kaplan-Meier? *Biometrics*, 39:1077–1081, 1983.

[385] R. G. Miller. The jackknife: A review. *Biometrika*, 61:1–15, 1974.

[386] E. S. Mills. Uncertainty and price theory. *Quarterly Journal of Economics*, 73:117–130, 1959.

[387] T. J. Mitchell and J. J. Beauchamp. Bayesian variable selection in linear regression. *Journal of the American Statistical Association*, 83:1023–1036, 1993.

[388] D. C. Montgomery, L. A. Johnson, and J. S. Gardiner. *Forecasting and Time Series Analysis.* McGraw-Hill, New York, NY, 2nd edition, 1990.

[389] K. S. Moorthy. Market segmentation, self-selection and product line design. *Marketing Science*, 3:288–307, 1984.

[390] M. M. Moriarty. Retail promotional effects on intra and interbrand sales performance. *Journal of Retailing*, 61:27–47, 1985.

[391] D. G. Morrison and D. C. Schmittlein. How many forecasters do you really have? Mahalanobis provides the intuition for the surprising Clemen and Winkler result. *Operations Research*, 39:519–523, 1991.

[392] J. Morris and P. Maes. Negotiating beyond the bid price. Technical report, Media Lab, MIT, Cambridge, MA, 2001.

[393] J. Morris, P. Ree, and P. Maes. Sardine: Dynamic seller strategies in an auction marketplace. In *ACM Conference on Electronic Commerce (EC '00)*. ACM, October 2000.

[394] F. J. Mulhern and R. J. Caprara. A nearest neighbor model for forecasting market response. *International Journal of Forecasting*, 10:191–207, 1994.

[395] F. J. Mulhern and R. P. Leone. Implicit price bundling of retail products: A multiproduct approach to maximizing store profitability. *Journal of Marketing*, 55:63–76, 1991.

[396] F. J. Mulhern. *An econometric analysis of consumer response to retail price promotions.* PhD thesis, University of Texas, Austin, TX, 1989.

[397] B. Müller, J. Reinhardt, and M. T. Strickland. *Neural Networks: An Introduction.* Springer-Verlag, Berlin, Germany, 1995.

[398] R. Myerson. Optimal auction design. *Mathematics of Operations Research*, 6:58–73, 1981.

[399] R. Myerson. *Game Theory: Analysis of Conflict.* Harvard University Press, Cambridge, MA, 1991.

REFERENCES

[400] N. T. Nagle and R. K. Holden (contributor). *The Strategy and Tactics of Pricing: A Guide to Profitable Decision Making*. Prentice-Hall, Englewood Cliffs, NJ, 2nd edition, 1994.

[401] S. K. Nair, R. Bapna, and L. Brine. An application of yield management for internet service providers. *Naval Research Logistics*, 48:348–362, 2001.

[402] J. Nash. Non-cooperative games. *Annals of Mathematics*, 54:286–295, 1951.

[403] J. Neter and W. Wasserman. *Applied Linear Statistical Models*. Irwin, Homewood, IL, 1974.

[404] S. Netessine and R. A. Shumsky. Revenue management games. Technical report, W. E. Simon Graduate School of Business Administration, University of Rochester, Rochester, NY, 2000.

[405] S. Netessine and R. A. Shumsky. Private communication, 2003.

[406] A. Nevo. A practitioner's guide to estimation of random-coefficients logit models of demand. *Journal of Economics and Management Strategy*, 9:513–548, 2000.

[407] P. Newbold and C. W. J. Granger. Experience with forecasting univariate time series and the combination of forecasts. *Journal of the Royal Statistical Society*, 137:131–146, 1974.

[408] W. Novshek. On the existence of Cournot equilibrium. *Review of Economic Studies*, 52:86–98, 1985.

[409] R. Oberwetter. Building blockbuster business: Can revenue management land a starring role in the movie theater industry? *OR/MS Today*, June 2001.

[410] United States General Accounting Office. Airline competition: Impact of computerized reservation systems. Report to Congressional Requesters, May 1986.

[411] E. B. Orkin. Boosting your bottom line with yield management. *Cornell Hotel and Restaurant Administration Quarterly*, 28:52–56, 1988.

[412] T. H. Oum. A warning on the use of linear logit models in transport mode choice studies. *Bell Journal of Economics*, 10:374–388, 1979.

[413] I. C. Paschalidis and J. N. Tsitiklis. Congestion-dependent pricing of network services. *IEEE/ACM Transactions on Networking*, 8:171–184, 2000.

[414] P. P. B. Pashigan and B. Bowen. Why are products sold on sale? Explanations of pricing regularities. *Quarterly Journal of Economics*, pages 1015–1038, November 1991.

[415] P. P. B. Pashigan. Demand uncertainty and sales. *American Economic Review*, 78:936–953, December 1988.

[416] S. Peters and J. Riley. Yield management transition: A case example. *International Journal of Contemporary Hospitality Management*, 9:89–91, 1997.

[417] N. C. Petruzzi and M. Dada. Pricing and the newsvendor problem: A review with extensions. *Operations Research*, 47:183–194, 1999.

[418] R. L. Phillips, D. W. Boyd, and T. A. Grossman. An algorithm for calculating consistent itinerary flows. *Transportation Science*, 25:225–239, 1991.

[419] R. L. Phillips. A marginal value approach to airline origin and destination revenue management. In J. Henry and P. Yvon, editors, *Proceedings of the Sixteenth Conference on System Modeling and Optimization*, New York, NY, 1994. Springer-Verlag.

[420] L. Phlips. *The Economics of Price Discrimination*. Cambridge University Press, Cambridge, UK, 1980.

[421] A. C. Pigou. *The Economics of Welfare*. Macmillan, London, UK, 1920.

[422] S. Pinchuk. Casino hotel RM and distribution. Presentation at the EyeForTravel RM and Distribution Conference, Amsterdam, March, 2003.

[423] I. Poli and R. D. Jones. A neural net model for prediction. *Journal of the American Statistical Association*, 89:117–121, 1994.

[424] S. Pölt. Forecasting is difficult—especially if it refers to the future. In *Reservations and Yield Management Study Group Annual Meeting Proceedings*, Melbourne, Australia, 1998. AGIFORS.

[425] S. Polt. Back to the roots: New results on leg optimization. In *1999 AGIFORS Reservations and Yield Management Study Group Symposium*, London, UK, 1999.

[426] L. Pompeo and T. Sapountzis. Freight expectations. *McKinsey Quarterly*, (2), 2002.

[427] R. H. Porter. On the incidence and duration of price wars. *Journal of Industrial Economics*, 33:415–426, 1985.

[428] E. C. Prescott. Efficiency of the natural rate. *Journal of Political Economy*, 83:1229–1236, 1976.

[429] W. H. Press, S. A. Teukolosky, W. T. Vetterling, and B. P. Flannery. *Numerical Recipes in C: The Art of Scientific Computing*. Cambridge University Press, Cambridge, UK, 1992.

[430] G. Pritchard and G. Zakeri. Market offering strategies for hydro-electric generators. *Operations Research*, 2002. (Forthcoming).

[431] M. H. Quenouille. Approximate tests of correlation of time-series. *Journal of the Royal Statistical Society B*, 11:68–84, 1949.

[432] J. R. Quinlan. *Programs for Machine Learning*. Morgan Kaufmann, San Mateo, CA, 1993.

[433] A. Rajan, Rakesh, and R. Steinberg. Dynamic pricing and ordering decisions by a monopolist. *Management Science*, 38:240–262, 1992.

REFERENCES

[434] J. S. Raju. The effect of price promotions on variability in product category sales. *Marketing Science*, 11:207–220, 1992.

[435] C. R. Rao and Y. Wu. A strongly consistent procedure for model selection in a regression problem. *Biometrika*, 76:369–374, 1989.

[436] R. C. Rao, R. V. Arjuni, and B. P. S. Murthi. Game theory and empirical generalization concerning competitive promotions. *Marketing Science*, 14:G89–G100, 1995.

[437] S. K. Reddy, J. E. Aronson, and A. Stam. SPOT: Scheduling programs optimally for television. *Management Science*, 44:83–102, 1998.

[438] O. Reiersøl. Confluence analysis by means of instrumental sets of variables. *Arkiv for Mathematik, Astronomi och Fysik*, 32, 1945.

[439] J. Remmers. Revenue management at an integrated tour operator. In *EyeForTravel Conference*, Amsterdam, The Netherlands, 2003.

[440] Team Marketing Report. 2002 MLB fan costindex (FCI) survey. Team Marketing Report, April 2002.

[441] J. Riley and W. Samuelson. Optimal auctions. *American Economic Review*, 71:381–392, 1981.

[442] J. Rissanen. A predictive least squares principle. *Journal of Mathematical Control Information*, 3:211–222, 1986.

[443] H. Robbins and S. Monro. A stochastic approximation method. *Annals of Mathematical Statistics*, 22:400–407, 1951.

[444] J. Roberts and H. Sonnenshein. On the existence of Cournot equilibrium without concave profit functions. *Journal of Economic Theory*, 13:112–117, 1976.

[445] L. W. Robinson. Optimal and approximate control policies for airline booking with sequential nonmonotonic fare classes. *Operations Research*, 43:252–263, 1995.

[446] M. Rothstein and A. W. Stone. Passenger booking levels. In *Proceedings of the Seventh AGIFORS Symposium*, 1967.

[447] M. Rothstein. Airline overbooking: The state of the art. *Journal Transport Economics and Policy*, 5:96–99, 1971.

[448] M. Rothstein. Airline overbooking: Fresh approaches are needed. *Transportation Science*, 2:169–173, 1975.

[449] M. Rothstein. O.R. and the airline overbooking problem. *Operations Research*, 33:237–248, 1985.

[450] K. M. Ruppenthal and R. Toh. Airline deregulation and the no-show/overbooking problem. *Logistics and Transportation Review*, 19:111–121, 1983.

[451] R. T. Rust. *Advertising Media Models: A Practical Guide.* Lexington, Lexington, MA, 1986.

[452] S. Salop and J. E. Stiglitz. A theory of sales: A simple model of equilibrium price dispersion with identical agents. *American Economic Review,* 72:1121–1130, 1982.

[453] L. Samulson. *Evolutionary Games and Equilibrium Selection (Economic Learning and Social Evolution).* MIT Press, Cambridge, MA, 1998.

[454] R. Sandomir. Ticket prices for mets tailored to the occasion. New York Times (nytimes.com), November 27, 2002.

[455] T. R. Scavo and J. B. Thoo. On the geometry of Halley's method. *American Mathematical Monthly,* 102:417–426, 1995.

[456] J. L. Schafer. *Analysis of Incomplete Multivariate Data.* Chapman and Hall, London, UK, 1997.

[457] D. C. Schmittlein, J. Kimm, and D. G. Morrison. Combining forecasts: Operational adjustments to theoretically optimal rules. *Management Science,* 36:1044–1056, 1990.

[458] R. Sethuraman and V. Srinivasan. The asymmetric share effect: An empirical generalization of cross-price effects. *Journal of Marketing Research,* 39:379–386, 2002.

[459] A. Shaked and J. Sutton. Relaxing price competition through product differentiation. *Review of Economic Studies,* 58:3–13, 1982.

[460] M. Shaked and J. G. Shantikumar. Stochastic convexity and its applications. *Advances in Applied Probability,* 20:427–446, 1988.

[461] M. Shaked and J. G. Shantikumar. Convexity of a set of stochastically ordered random variables. *Advances in Applied Probability,* 22:160–177, 1990.

[462] M. Shaked and J. G. Shantikumar. *Stochastic Orders and Their Applications.* Academic Press, San Diego, CA, 1994.

[463] C. Shapiro. Theories of oligopoly behavior. In R. Schmalensee and R. D. Willig, editors, *Handbook of Industrial Organization,* pages 329–414. Elsevier Science, New York, NY, 1989.

[464] R. Shibata. Approximate efficiency of a selection procedure for the number of regressor variables. *Biometrika,* 71:43–49, 1984.

[465] E. Shlifer and Y. Vardi. An airline overbooking policy. *Transportation Science,* 9:101–114, 1975.

[466] S. M. Shugan and R. Desiraju. Strategic service pricing and yield management. *Journal of Marketing,* 63:44–56, 1999.

[467] S. M. Shugan and J. Xie. Advance pricing of services and other implications of separating purchase and consumption. *Journal of Service Research,* 2:227–239, 2000.

REFERENCES

[468] S. M. Shugan. Editorial: Marketing science, models, monopoly models, and why we need them. *Marketing Science*, 21:223–228, 2002.

[469] O. Shy. *Industrial Organization: Theory and Applications*. MIT Press, Cambridge, MA, 1995.

[470] J. M. Silva-Risso, R. E. Bucklin, and D. G. Morrison. A decision support system for planning manufacturer's sales promotion calendars. *Marketing Science*, 18:274–300, 1999.

[471] H. Simon. *Price Management*. North-Holland, Amsterdam, The Netherlands, 1989.

[472] J. L. Simon. An almost practical solution to airline overbooking. *Journal of Transport Economics and Policy*, 2:201–202, 1968.

[473] J. L. Simon. Airline overbooking: The state of the art—a reply. *Journal Transport Economics and Policy*, 6:255–256, 1972.

[474] J. L. Simon. The airline oversales auction plan: How it was adopted and how it has fared. In *Fifth IATA Revenue Management Conference*, Montreal, Canada, 1993.

[475] J. L. Simon. The airline oversales auction plan: The results. *Journal Transport Economics & Policy*, 28:319–323, 1994.

[476] R. W. Simpson. Using network flow techniques to find shadow prices for market and seat inventory control. Technical Report Memorandum, M89-1, Flight Transportation Laboratory, MIT, Cambridge, MA, 1989.

[477] B. C. Smith, J. F. Leimkuhler, and R. M. Darrow. Yield management at American Airlines. *Interfaces*, 22:8–31, 1992.

[478] B. C. Smith and C. W. Penn. Analysis of alternative origin-destination control strategies. In *Proceedings of the Twenty Eighth Annual AGIFORS Symposium*, New Seabury, MA, 1988.

[479] R. Smith. Revenue management: Hotels, airlines, opera houses hope this tool will help them maximize sales and profits. *San Francisco Chronicle*, May 25, 1999.

[480] S. A. Smith and D. D. Achabal. Clearance pricing and inventory policies for retail chains. *Management Science*, 44:285–300, 1998.

[481] S. A. Smith. A linear programming model for real-time pricing of electric power service. *Operations Research*, 41:470–483, 1993.

[482] J. Sobel. The timing of sales. *Review of Economic Studies*, LI:353–368, 1984.

[483] B. Sobie. Freight's yield signs. *Air Cargo World*, July 2000.

[484] W. Stadje. A full information pricing problem for the sale of several identical commodities. *Zeitschriftfur Operations Reserach*, 34:161–181, 1990.

[485] Standard and Poor. *Industry Surveys: Retailing*. Standard and Poor, New York, NY, 1998.

[486] P. Steiner. Peak-loads and efficient pricing. *Quarterly Journal of Economics*, 71:585–610, 1957.

[487] S. Stidham. Socially and individually optimal control of arrivals to a GI/GI/1 queue. *Management Science*, pages 1598–1610, 1978.

[488] G. Stigler. A theory of oligopoly. *Journal of Political Economy*, 72:44–61, 1964.

[489] G. Stigler. *Theory of Price*. Macmillan, London, UK, 1987.

[490] N. Stokey. Intertemporal price descrimination. *Quarterly Journal of Economics*, 94:355–371, 1979.

[491] N. Stokey. Rational expectations and durable goods pricing. *Bell Journal of Economics*, 12:112–128, 1981.

[492] M. Stone. Cross-validation choice and assessment for statistical predictions. *Journal of the Royal Statistical Society, B*, 36:111–147, 1974.

[493] S. Subrahmanyan and R. Shoemaker. Developing optimal pricing and inventory policies for retailers who face uncertain demand. *Journal of Retailing*, 72:7–30, 1996.

[494] J. Subramanian, S. Stidham Jr., and C. Lautenbacher. Airline yield management with overbooking, cancellations and no-shows. *Transportation Science*, 33:147–167, 1999.

[495] R. K. Sundaram. *A First Course in Optimization Theory*. Cambridge University Press, Cambridge, UK, 1996.

[496] S. Sun. The fine-tuned learning enhancement to the standard backpropagation algorithm. In *Reservations and Yield Management Study Group Annual Meeting Proceedings*, Zurich, Switzerland, 1996. AGIFORS.

[497] J. Swann. *Dynamic Pricing Models to Improve Supply Chain Performance*. PhD thesis, Department of Industrial Engineering and Management Science, Northwestern University, Evanston, IL, 2001.

[498] K. T. Talluri and G. J. van Ryzin. An analysis of bid-price controls for network revenue management. *Management Science*, 44:1577–1593, 1999.

[499] K. T. Talluri and G. J. van Ryzin. A randomized linear programming method for computing network bid prices. *Transportation Science*, 33:207–216, 1999.

[500] K. T. Talluri and G. J. van Ryzin. Revenue management under a general discrete choice model of consumer behavior. *Management Science*, January 2004.

[501] K. T. Talluri. Airline revenue management with passenger routing control: A new model with solution approaches. *International Journal of Services Technology and Management*, 2:102–115, 2001.

[502] K. T. Talluri. Design and positioning of revenue management products. Technical report, UPF, 2003.

[503] K. T. Talluri. On equilibria in duopolies with finite strategy spaces. Technical Report WP-701, UPF, 2003.

[504] C. J. Taylor. The determination of passenger booking levels. In *Proceedings of the Second AGIFORS Symposium*, 1962.

[505] B. Tedeschi. Using discounts to build a client base. *New York Times*, May 31, 1999.

[506] B. Tedeschi. Scientifically priced retail goods. *New York Times (nytimes.com)*, September 3, 2002.

[507] G. J. Tellis and S. Zufryden. Tackling the retailer decision maze: Which brands to discount, how much, when and why? *Marketing Science*, 14:271–289, 1995.

[508] H. R. Thompson. Statistical problems in airline reservation control. *Operations Research Quarterly*, 12:167–185, 1961.

[509] G. T. Thowsen. A dynamic, nonstationary inventory problem for a price/quantity setting firm. *Naval Research Logistics*, 22:461–476, 1975.

[510] L. L. Thurstone. A law of comparative judgement. *Psychology Review*, 34:273–286, 1927.

[511] L. L. Thurstone. Psychological analysis. *American Journal of Psychology*, 38:368–389, 1927.

[512] S-C Ting and G-H. Tzeng. Fuzzy multi-objective programming approach to allocating containership slots for liner shipping revenue management. In *MCDM Winter Conference*, Semmering, Austria, 2002.

[513] J. Tirole. *The Theory of Industrial Organization*. MIT Press, Cambridge, MA, 1993.

[514] B. Titze and R. Griesshaber. Realistic passenger booking behaviors and the simple low-fare/high-fare seat allotment model. In *Proceedings of the Twenty Third Annual AGIFORS Symposium*, 1983.

[515] D. M. Topkis. Optimal ordering and rationing policies in a nonstationary dynamic inventory model with n demand classes. *Management Science*, 15:160–176, 1968.

[516] D. M. Topkis. Minimizing a submodular function on a lattice. *Operations Research*, 26:305–321, 1978.

[517] J. Tschirhart and F. Jen. Behavior of a monopoly offering interruptible service. *Bell Journal of Economics*, 10:244–258, 1979.

[518] TTI. Working together. TTI Newsletter, 2001.

[519] J. W. Tukey. Bias and confidence in not-quite large samples. *Annals of Mathematical Statistics*, 29:614, 1958.

[520] J. Tukey. *Exploratory Data Analysis.* Addison Wesley, Reading, MA, 1979.

[521] A. Tversky. Choice by elimination. *Journal of Mathematical Psychology*, 9:341–367, 1972.

[522] J. E. Urban, T. J. Madden, and P. R. Dickson. All's not fair in pricing: An initial look at the dual entitlement principle. *Marketing Letters*, 1:17–25, 1989.

[523] United States Department of Transportation, Office of the Secretary, *14 C.F.R. Chapter II*, 1-1-00, part 250 edition, 2000.

[524] T. V. Valkov and N. Secomandi. Revenue management for the natural gas industry. *EN Energy Industry Management*, 1(1), July 2000.

[525] T. V. Valkov and N. Secomandi. Revenue management for the natural gas industry. *Power & Gas Marketing*, July 2000.

[526] G. J. van Ryzin and J. I. McGill. Revenue management without forecasting or optimization: An adaptive algorithm for determining seat protection levels. *Management Science*, 46:760–775, 2000.

[527] G. J. van Ryzin and G. Vulcano. Optimal auctioning and ordering in an infinite horizon inventory-pricing system. Technical report, Graduate School of Business, Columbia University, New York, NY, 2002. Working Paper.

[528] G. J. van Ryzin and G. Vulcano. A stochastic subgradient algorithm for network capacity control. Technical report, Graduate School of Business, Columbia University, New York, NY, 2002. Working Paper.

[529] G. J. van Ryzin. The brave new world of pricing. In Survey: Mastering management, Financial Times, October 16, 2000.

[530] R. Van Slyke and Y. Young. Finite horizon stochastic knapsacks with applications to yield management. *Operations Research*, 48:155–172, 2000.

[531] H. R. Varian. Price discrimination. In R. Schmalensee and R. D. Willig, editors, *Handbook of Industrial Organization*, pages 597–654. Elsevier Science, New York, NY, 1989.

[532] K. Varini, R. Engelmann, B. Claessen, and M. Schleusener. Evaluation of the price-value perception of customers in Swiss hotels. Technical report, Ecole Hôtelière Lausanne, Lausanne, Switzerland, 2002.

[533] W. Vickrey. Counterspeculation, auctions and competitive sealed tenders. *Journal of Finance*, 16:8–37, 1961.

[534] W. Vickrey. Airline overbooking: Some further solutions. *Journal Transport Economics and Policy*, 6:257–270, 1972.

[535] N. J. Vilcassim and P. K. Chintagunta. Modeling purchase timing and brand-switching behavior incorporating explanatory variables and unobserved heterogeneity. *Journal of Marketing Research*, 28:29–41, 1991.

[536] J. M. Villas-Boas and R. S. Winer. Endogeneity in brand-choice models. *Management Science*, 45:1324–1338, 1999.

REFERENCES

[537] B. Vinod. A set partitioning algorithm for virtual nesting indexing using dynamic programming. Technical report, Internal Technical Report, SABRE Decision Technologies, 1989.

[538] B. Vinod. Origin-and-destination yield management. In D. Jenkins, editor, *The Handbook of Airline Economics*, pages 459–468. McGraw-Hill, New York, NY, 1995.

[539] X. Vives. *Oligopoly Pricing: Old Ideas and New Tools*. MIT Press, Cambridge, MA, 1998.

[540] N-H. M. von der Fehr and K-U. Kühn. Coase versus Pacman: Who eats whom in the durable-goods monopoly? *Journal of Political Economy*, 103:785–812, 1995.

[541] J. von Neumann and O. Morgenstern. *Theory of Games and Economic Behavior*. Princeton University Press, Princeton, NJ, 1944.

[542] G. Vulcano, G. J. van Ryzin, and C. Maglaras. Optimal dynamic auctions for revenue management. *Management Science*, 48(11).

[543] D. Walczak and S. Brumelle. Semi-Markov partial information model for revenue management and dynamic pricing. Working Paper, Faculty of Commerce and Business Administration, University of British Columbia, 2001.

[544] D. Walczak and S. Brumelle. Dynamic airline revenue management with multiple semi-Markov demand. *Operations Research*, 51:137–148, 2003.

[545] M. Wald. Utilities trying new approaches to pricing energy. *New York Times*, (7), July 2000.

[546] R. G. Walters and S. B. MacKenzie. A structural equations analysis of the impact of price promotions on store performance. *Journal of Marketing Research*, 25:51–63, 1988.

[547] R. G. Walters and H. J. Rinne. An empirical investigation into the impact of price promotions on retail store performance. *Journal of Retailing*, 62:237–266, 1986.

[548] R. G. Walters. Retail promotions and retail store performance: A test of some key hypothesis. *Journal of Retailing*, 64:153–180, 1988.

[549] R. G. Walters. Assessing the impact of retail price promotions on product substitution, complementary purchase, and interstore sales displacement. *Journal of Marketing*, 55:17–28, 1991.

[550] K. Wang. Optimum seat allocation for multi-leg flights. In *Proceedings of the Twenty-Third AGIFORS Symposium*, Memphis, TN, 1983.

[551] R. Wang. Auctions versus posted-price selling. *American Economic Review*, 83:838–851, 1993.

[552] R. Wang. Auctions versus posted-price selling: The case of correlated private valuations. *Canadian Journal of Economics*, 31(2):395–410, 1998.

[553] B. Wansink and R. Deshpande. Out of sight, out of mind: Pantry stockpiling and brand-usage frequency. *Marketing Letters*, 5:91–100, 1994.

[554] E. J. Warner and R. B. Barsky. The timing and magnitude of retail store markdowns: Evidence from weekends and holidays. *Quarterly Journal of Economics*, pages 321–352, May 1995.

[555] L. R. Weatherford, S. E. Bodily, and P. E. Pfeifer. Modeling the customer arrival process and comparing decision rules in perishable asset revenue management situations. *Transportation Science*, 27:239–251, 1993.

[556] L. R. Weatherford and S. E. Bodily. A taxonomy and research overview of perishable-asset revenue management: Yield management, overbooking, and pricing. *Operations Research*, 40:831–844, 1992.

[557] K. Weber. From O&D bid-price control to package bid-price control. In *Proceedings*, Bangkok, Thailand, 2001. AGIFORS Reservation and Yield Management Study Group.

[558] M. Wedel and W. Kamakura. *Market segmentation: Conceptual and Methodological Foundations*. Kluwer, Norwell, MA, 2nd edition, 2000.

[559] C. Z. Wei. On predictive least squares principles. *Annals of Statistics*, 20:1–42, 1992.

[560] W. S. Wei. *Time Series Analysis*. Addison-Wesley, Redwood City, CA, 1990.

[561] A. Westerhof. CO_2 in the air. In *Reservations and Yield Management Study Group Annual Meeting Proceedings*, Melbourne, Australia, 1998. AGIFORS.

[562] P. M. West, P. L. Brockett, and L. L. Golden. A comparative analysis of neural networks and statistical methods for predicting consumer choice. *Marketing Science*, 16:370–391, 1997.

[563] H. White. A heteroskedasticity-consistent covariance matrix estimator and a direct test for heteroskedasticity. *Econometrica*, 48:817–838, 1980.

[564] T. M. Whitin. Inventory control and price theory. *Management Science*, 2:61–68, 1955.

[565] E. L. Williamson. Comparison of optimization techniques for origin-destination seat inventory control. Master's thesis, Flight Transportation Laboratory, MIT, Cambridge, MA, 1988.

[566] E. L. Williamson. *Airline Network Seat Inventory Control: Methodologies and Revenue Impacts*. PhD thesis, Flight Transportation Laboratory, MIT, Cambridge, MA, 1992.

[567] A. T. Williams. Do anti-scalping laws make a difference? *Managerial and Decision Economics*, 15:503–509, 1994.

[568] C. Wilson. On the optimal pricing of a monopolist. *Journal of Political Economy*, 96:164–176, 1988.

[569] R. B. Wilson. *Nonlinear Pricing*. Oxford University Press, Oxford, UK, 1993.

REFERENCES

[570] J. Wind, P. E. Green, D. Schifflet, and M. Scarbrough. Courtyard by Marriott: Designing a hotel facility with consumer-based marketing models. *Interfaces*, 19:25–27, 1989.

[571] W. L. Winston. *Operations Research: Applications and Algorithms*. Duxbury, Belmont, CA, 3rd edition, 1994.

[572] D. R. Wittink, M. Addona, W. Hawkes, and J. C. Porter. SCAN*PRO: The estimation, validation and use of promotional effects based on scanner data. Technical report, Johnson Graduate School of Management, Cornell University, Ithaca, NY, 1988.

[573] D. R. Wittink and P. Cattin. Commercial use of conjoint analysis in Europe: An update. *Journal of Marketing*, 53:91–96, 1989.

[574] D. R. Wittink, M. Vriens, and W. Burhenne. Commercial use of conjoint analysis in Europe: Results and critical reflections. *International Journal of Research in Marketing*, 11:41–52, 1994.

[575] R. D. Wollmer. A hub-and-spoke seat management model. Company report, Douglas Aircraft Company, McDonnell Douglas Corporation, 1986.

[576] R. D. Wollmer. An airline seat management model for a single leg route when lower fare classes book first. *Operations Research*, 40:26–37, 1992.

[577] J. T. Wong, F. S. Koppelman, and M. S. Daskin. Flexible assignment approach to itinerary seat allocation. *Transportation Research*, 27B:33–48, 1993.

[578] J. T. Wong. *Airline Network Seat Allocation*. PhD thesis, Northwestern University, Evanston, IL, 1990.

[579] T. Woodall. Travel suppliers seeking to catch up with airlines in areas of revenue management. In *New Frontiers, Third-Fourth Annual Convention*, Salt Lake City, UT, 2002.

[580] A. G. Woodside and G. L. Waddle. Sales effects of in-store advertising. *Journal of Advertising Research*, 15:29–33, 1975.

[581] J. Woolridge. *Introductory Econometrics: A Modern Approach*. South-Western, Columbus, OH, 2nd edition, 2003.

[582] C. F. J. Wu. On the convergence properties of the EM algorithm. *Annals of Statistics*, 11:95–103, 1983.

[583] D-M. Wu. Alternative tests of independence between stochastic regressors and disturbances. *Econometrica*, 41:733–750, 1973.

[584] R. Wysong. A simplified method for including network effects in capacity control. In *Proceedings of the Twenty-Eighth AGIFORS Symposium*, New Seabury, MA, 1988.

[585] I. Yeoman and S. Watson. Yield management: A human activity system. *International Journal of Contemporary Hospitality Management*, 9:80–83, 1997.

[586] P. S. You. Dynamic pricing in airline seat management for flights with multiple legs. *Transportation Science*, 34:192–206, 1999.

[587] R. H. Zeni. *Improved Forecast Accuracy in Airline Revenue Management by Unconstraining Demand Estimates from Censored Data*. PhD thesis, Graduate School, Rutgers, State University of New Jersey, 2001. Also published by dissertation.com.

[588] G. Zhang, B. E. Patuwo, and M. Y. Hu. Forecasting with artificial neural networks: The state of the art. *International Journal of Forecasting*, 14:35–62, 1998.

[589] W. Zhao and Y.-S. Zheng. Optimal dynamic pricing for perishable assets with nonhomogeneous demand. *Management Science*, 46:375–388, 2000.

[590] A. Ziegler and E. P. Lazear. The dominance of retail stores. Technical Report 9795, NBER, June 2003.

[591] S. Ziya, A. Hayriye, and R. D. Foley. On assumptions ensuring a strictly unimodal revenue function. Technical report, School of Industrial and Systems Engineering, Georgia Institute of Technology, Atlanta, GA, 2003. Working Paper.

Index

3G, 243

Activities-based costing (ABC), 575
Ad-hoc forecasting methods, 434
Adjusted sales rate, 196–197
Advance-purchase restrictions, 516
Advance-purchase restrictions, 516
Advance-purchases, 345
Advance purchases, 372
Air cargo RM, 563
Airline itinerary, 517
Airline RM, 515
Algorithm
  adaptive, 50
  Robbins-Monro, 50
  stochastic gradient, 169
Alternating-direction method, 170
American Airlines, 516
Amsterdam Power Exchange, 554
APEX fares, 516
Asymptotically unbiased estimator, 421
ATPCO, 520
Auction mechanism, 254
Auction
  ask offers, 243
  bid offers, 243
  common-value model, 266
  discriminatory, 270
  Dutch, 245
  English, 245
  independent private-value model, 247
  mechanism, 241
  multiunit, 246
  network, 288
  online, 244
  open ascending, 245
  open descending, 245
  optimal dynamic single-resource, 272
  optimal dynamic with replenishment, 280
  oversale, 136
  procurement, 243
  reverse, 136
  risk aversion in, 268
  sealed-bid first-price, 245
  sealed-bid second-price, 245
  symmetric equilibrium, 255
  Vickrey, 245
Augmented Lagrangian, 170
Authorization levels, 139
Autocorrelation, 441
  ACF, 441
  PACF, 441
Autocovariance, 441
Autoregressive integrated moving-average process, 448

Base-stock
  list price policy, 285
  posted-price policy, 214
Bayesian forecasting, 450
Bayesian Nash equilibrium, 669
Bias detection and correction, 487, 497
Biased estimator, 421
Bid-price control, 31, 86, 89, 292
  heuristic method, 101
  in network auction, 291
  nonoptimality example, 90
  relation to opportunity cost, 91
Bidding ring, 270
Bidding strategy, 254
Bin-packing, 57
Binary logit, 305
Binary preferences, 657
Binary probit, 305
Binomial model, 163–164, 167
  normal approximation to, 147
  of cancellations, 139
  validation, 141

Blocked-seat allotment, 122
Booking-curve, 410
Booking-profile forecasting, 410
Booking limit, 28
  nested, 28
  partitioned, 28
  with no-shows, 157
Bottom-up forecasting strategy, 418
Bounded optimization problem, 645
Broken-assortment effect, 196
Buckets, 104
Buy-up, 62
Buy-up factors, 62

Cancellations, 138
Capacity-controlled fares, 8
Capacity, virtual, 139
Casino RM, 559
Category management, 538
Censored-data forecasting, 473
Christie's, 242
Civil Aeronautics Board, 131
Closed-to-arrival control, 530
Coase conjecture, 364–365
Code-sharing, 121
Coefficient of determination, 488
Coefficient of skewness, 148
Collusion, 270
Combinability rules, 518
Commodity, 336
Complements, 321
Component-wise concavity, 164
Concave function, 643
Consistent, 422
Consolidating system, 529
Constant-elasticity-demand function, 325
Constraint set, 645
Constructed fares, 518
Consumer budget problem, 660
Continuously differentiable, 644
Contraction mapping, 110
Convex combination, 643
Convex function, 643
Convex set, 643
Covariance, 636
Covisint, 244
CPM, 544
CRM, 17
Cross-price elasticity, 321
CRS, 598
Cruise ship RM, 560
Cumulative distribution function, 635
Customer-relationship management (CRM) system, 412

Data collection points (DCPs), 18

DAVN, 108
Delphi method, 472
Demand-driven dispatch (D^3), 4
Demand function
  constant elasticity, 325
  linear, 323
  log-linear (exponential), 324
  logit, 326
Demand, elastic and inelastic, 313
Denied service
  compensation, 135
  involuntary, 132
  oversale auction, 136
  selection criteria, 135
  voluntary, 132, 137
Deterministic linear programming model, 166
Direct-revelation mechanism, 256, 369
Directional derivative, 644
Discrete-choice model, in single-resource model, 64
Displacement-adjusted revenue, 104, 107
Displacement-adjusted virtual nesting, 103
Displacement cost, 33
Diversion, 62
Dominant strategy, 250, 668
  equilibrium, 250
Double-log, 507
Double auctions, 243
Duke University Diet and Fitness Center (DFC), 577
Durable good, 184, 364
Dynamic programming decomposition, 107

E-commerce, 178
EBay, 244
Efficient estimator, 422
Efficient set, 67
Elastic demand, 313
Electric Power Research Institute (EPRI), 555
Electricity RM, 551
Electronic Data Interchange (EDI), 606
Empirical Bayes approximation, 455
EMSR
  EMSR-a, 45
  EMSR-b, 47, 52
    with buy-up, 63, 74
  prorated, 102
EnronOnline (EOL), 550
Equality-constrained optimization problem, 646
Equilibrium strategy, 255
ERP, 17
Error back-propagation method, 469
Expectation-maximization (EM) method, 474

INDEX

Expected value, 636
Exponential-demand function, 324
Exponential model, 507
Exponential smoothing, 436
Extreme point, 643

Fare-basis code, 522
Fare class, 521–522
Feasible solution, 645
Federal Energy Regulatory Commission, 547
Ferry RM, 560
Finite-population model, 184–185, 223
First-order necessary conditions, 645–647
Forecasting, bottom-up versus top-down, 418
Free individual traveler (FIT), 524
Free sale, 122
FreeMarkets, 244
Freight RM, 564
Friedman, Milton, 137

Geometric distributed-lag model, 507
Gibbs sampling, 481
Global distributed system (GDS), 86, 523, 598
Global optimum, 645
Gradient, 644
 estimator, 170
Gram-Charlier series, 148
Group size
 empirical distribution, 150
Group size, in overbooking models, 150
GRP, 544
Gutenberg model, 507

Hausman-type test, 491
Hazard rate, 316
Hessian, 644
Holt-Winter's method, 438
Homoscedasticity, 423
Hotel RM, 524

i.i.d., 636
IATA, 516, 518
Imputed cost, 146
Incentive compatibility constraint, 256
Incidence matrix, 166
Incidence vector, 87
Increasing-differences property, 648
Independent-demand model, 301
Independent random variables, 636
Indexing, 85, 104
Inelastic demand, 313
Inequality constrained optimization problem, 646

Infinite-population model, 184–185, 225
Infinite overbooking, 158
Information asymmetry, 664
Information rent, 258
Instrumental-variables (IV) techniques, 429
Intercontinental Exchange (ICE), 550
Interlining, 518
Intermodal shipping, 565
Interpool, 532
Intrapool, 532
Inventory-depletion effect, 195
Inverse-demand function, 313
Iterative DAVN, 108
Iterative decomposition methods, 108, 110
Iterative prorated EMSR, 108–109

Jackknife estimator, 487
Jensen's inequality, 637
Joint distribution, 635

Kahn, Alfred, 137
Kalman filter, 458
Kaplan-Meier estimator, 52, 483
Kuhn-Tucker conditions, 647

Lagrange multiplier, 646
Law of one price, 337
Length-of-stay control
 maximum stay, 530
 minimum stay, 530
Less-than-truckload (LTL) shippers, 565
Likelihood ratio test, 489
Linear-demand function, 323
Linear-regression estimator, 423
Littlewood's rule, 36
Local optimum, 645
Log-linear-demand function, 324
Log-linear, 507
Log-reciprocal, 507
Logistic, 507

M-period moving-average forecast, 435
Machine learning, 464
Make-to-order (MTO) manufacturer, 575
Make-to-stock (MTS) manufacturer, 574
Manufacturing RM, 574
Manufacturing, dynamic pricing in, 178
Marginal distribution, 636
Marginal revenue, 315
 monotone, 316
Market-basket effects, 538
Market information data tapes (MIDT), 413
Markov-chain Monte Carlo (MCMC), 481
Maximum-likelihood (ML) estimator, 425
Maximum concave envelope, 195

Mean, 636
Media and broadcasting RM, 542
Mental accounting, 665
Menu costs, 539
Method of moments, 427
Minimum Square Error (MSE) estimator, 422
Moment-generating functions, 151
Monopoly, 349
Monopoly model, of demand in dynamic pricing, 186
Monopoly
  advance purchases, 372
  capacity-constrained, 351
  optimal mechanism design, 369
  peak-load pricing, 372
  price discrimination, 352
  single-price, 350
  strategic customers, 363
Monte Carlo integration, 42
Multinomial logit (MNL), 306
  finite-mixture, 309
  independence from irrelevant alternatives (IIA) property, 307
  linear-in-attributes model, 305
Multiplicative or power model, 506
Myopic customers, 182, 228, 366
  price skimming, 223

Nader, Ralph, 132
Nash equilibrium
  definition, 668
  subgame-perfect, 670
Natural gas RM, 546
Negative bias, 421
Nested booking limits, 85
Nesting
  by revenue order, 72
  standard, 30
  theft, 30
Net bookings, 140, 154
Network-capacity control
  model definition, 87
  optimal policy, 89
Network-flow problem, 162
Neural-network methods, 464
New York Cruise Lines, 561
Newsvendor problem, 378
No-shows, 138
Nonlinear tariffs, 554
Nonparametric estimation, 416
Normal distribution, 147

Objective function, 645
OD factors, 101

ODIF (origin-destination itinerary fare-class), 81
Oligopoly, 375
  model of demand in dynamic pricing, 186
  Bertrand-Edgeworth model, 385
  Bertrand model, 382
  Cournot model, 376
  dynamic models, 388
  static models, 376
Opportunity cost, 33
Ordinary least-squares (OLS) estimator, 423
Overbooking level
  Gram-Charlier series approximation, 148
  normal approximation, 147
Overbooking
  airline history, 131
  average service level, 143
  class-dependent cancellation refunds, 160
  class-dependent no-show refunds, 158
  combined with capacity controls, 155
  cost and revenue parameters, 145
  customer class mix, 149
  deterministic approximation, 147
  dynamic model, 152
  dynamic models, 152
  economic criteria, 144
    use in network overbooking, 166
  Gram-Charlier series approximation, 152
  group cancellations, 150
  groups, 156
  imputed cost, 146
  infinite, 158
  limit, 138, 141–142
    optimal dynamic, 153–154
  moment-generating functions, 151
  network, 166
  normal approximation, 152
  notification, 134
  pad, 142
  Poisson model, 163
  static model, 138
  substitutable capacity, 161
  type 1 service level, 141, 145
  type 2 service level, 142

Pacific Gas and Electric, 554
Panel data, 542
Parametric estimation, 416
Partial-booking data, 415
Partial autocorrelation function (PACF), 441
Partial derivative, 644
Participation constraint, 257
Partitioned booking-limit, 84
Passenger name record (PNR), 413, 523
Passenger rail RM, 561
Peak-load pricing, 341, 372, 554

INDEX

Perfect competition, 336
  advance-purchases, 345
  demand uncertainty, 338
  peak-load pricing, 341
Perfectly competitive models, 186
Pick-up forecasting methods, 470
Plotting procedures for censored data, 484
Point-of-sale (POS)
  data, 542
  system, 412, 606
Point estimate, 407
Poisson distribution, 163
Poisson model, 167, 171
Positive bias, 421
Preference relation, 657–658
Prescott equilibrium, 340
Price discrimination, 226, 352
  classification of types, 352
  first-degree, 352, 354
  perfect, 355
  second-degree, 223, 353, 357
  third-degree, 353–354
Price elasticity, 313
Price skimming, 223
  myopic customers, 223
  strategic customers, 226
Price taker, 187, 337
Priceline.com, 244, 520
Priority-pricing, 371
Private fares, 520
Probability-density function, 635
Probability-mass function, 635
Probability measure, 635
Probability space, 635
Projection-detruncation method, 485
Property management system (PMS), 527, 598
Prospect theory, 665
Protection level, 30
Protection level
  nested
    optimality in discrete-choice model, 71
Public fares, 520
Purchase restrictions, 8

Quantile estimators, 427

Reservation price, 661
Rack rate, 525
Random-coefficients logit model, 309
Random variable, 635
Rate bands, 525
Rationing rules, 330
Recapture, 411
Reduced revenue, 161
Regression estimator, 422

Regular point, 646
Rental car RM, 531
Request for quotes (RFQ), 575
Reservation-period, 163
Reservation period, 166
Retail management system (RMS), 605
Retail RM, 533
Retailing, dynamic pricing in, 177
Revealed preferences, 664
Revenue-benefits measurement, 608
Revenue-maximizing price, 264, 318
Revenue-opportunity assessment, 608
Revenue function, 315
Risk aversion, in auctions, 268
Risk preference, 662
Robbins-Monro algorithm, 50
Run-out price, 264

S-shaped model, 507
Saturday-night stay, 516
SCAN*PRO model, 508
Sea-Land, 566
Seamless availability, 86
Second-order sufficiency conditions, 646
Second partial derivative, 644
Sell-up, 62
Semilogarithmic model, 506
Sensitivity analysis, 647
Service level
  Type 1, 147
  Type 2, 146, 148
Service period, 163, 166
Show demand, 138, 140, 144, 147, 156
Simon, Julian, 136
Single-resource
  approximations and heuristics, 44
  dynamic model, 57
    optimal policy, 59
    with cancellations and no-shows, 160
    with discrete-choice behavior, 64
    with no-shows, 157
  group arrivals, 56
Single-resource
  static model
    adaptive methods, 50
    computing optimal policies, 41
    multiple classes, 36
    optimal policy, 38
    optimality conditions for continuous demand, 40
    two-classes, 35
    with cancellations and no-shows, 159
    with no-shows, 157
SITA, 520
Société Nationale des Chemins de Fer Francais (SNCF), 563
Sorting mechanism, 353

Sotheby's, 242
Specification errors, 488
Specification tests, 488
Spill, 411
Sporting event RM, 567
Stackelberg game, 670
State-space forecasting methods, 458
Stationary time series, 440
Steepest decent algorithm, 111
Stochastic approximation, 50
Stochastic convexity, 156, 160
Stochastic gradient, 111, 169
  algorithm, 170
  method, 111
Stock-clearing price, 193
Stock-keeping unit (SKU), 535
Strategic customer, 182, 228
  durable goods monopolist, 363
  price skimming, 226
Subdifferential, 645
Subgame-perfect equilibrium, 227
Subgradient, 645
Submodular, 164
Substitutes, 321
Super saver fares, 516
Supergames, 670
Supermodular function, 213, 648
Switch companies, 529

Texas Air Corporation, 516
Theater RM, 567
THISCO, 529
Time series, 439
Top-down forecasting strategy, 418
Tour operator RM, 555

Transmediterranea, 561
Transportation problem, 163
Travel-agent fees, 529
Travel Technology Initiative (TTI), 559

U.S. Postal Service, 565
Unadjusted sales rate, 196
Unbiased estimator, 421
Unbounded optimization problem, 645
Unconstraining data, 473
Unicorn message standard, 607
Unimodal function, 317
User influences, 408, 412
Utilization ratio, 151

Variance, 636
VIA Rail Canada, 563
Vickrey-Clarke-Groves mechanism, 288
Virtual capacity, 167
Virtual class, 85, 104
Virtual nesting, 85
  displacement-adjusted (DAVN), 103
Virtual utility, 316
Virtual value, 258, 370
Von Neumann-Morgenstern expected-utility, 662

Willingness to pay, 13, 182, 184, 188, 661
Winner's curse, 267
WizCom, 529

XML, 559

Early Titles in the
INTERNATIONAL SERIES IN
OPERATIONS RESEARCH & MANAGEMENT SCIENCE
Frederick S. Hillier, Series Editor, *Stanford University*

Saigal/ *A MODERN APPROACH TO LINEAR PROGRAMMING*
Nagurney/ *PROJECTED DYNAMICAL SYSTEMS & VARIATIONAL INEQUALITIES WITH APPLICATIONS*
Padberg & Rijal/ *LOCATION, SCHEDULING, DESIGN AND INTEGER PROGRAMMING*
Vanderbei/ *LINEAR PROGRAMMING*
Jaiswal/ *MILITARY OPERATIONS RESEARCH*
Gal & Greenberg/ *ADVANCES IN SENSITIVITY ANALYSIS & PARAMETRIC PROGRAMMING*
Prabhu/ *FOUNDATIONS OF QUEUEING THEORY*
Fang, Rajasekera & Tsao/ *ENTROPY OPTIMIZATION & MATHEMATICAL PROGRAMMING*
Yu/ *OR IN THE AIRLINE INDUSTRY*
Ho & Tang/ *PRODUCT VARIETY MANAGEMENT*
El-Taha & Stidham/ *SAMPLE-PATH ANALYSIS OF QUEUEING SYSTEMS*
Miettinen/ *NONLINEAR MULTIOBJECTIVE OPTIMIZATION*
Chao & Huntington/ *DESIGNING COMPETITIVE ELECTRICITY MARKETS*
Weglarz/ *PROJECT SCHEDULING: RECENT TRENDS & RESULTS*
Sahin & Polatoglu/ *QUALITY, WARRANTY AND PREVENTIVE MAINTENANCE*
Tavares/ *ADVANCES MODELS FOR PROJECT MANAGEMENT*
Tayur, Ganeshan & Magazine/ *QUANTITATIVE MODELS FOR SUPPLY CHAIN MANAGEMENT*
Weyant, J./ *ENERGY AND ENVIRONMENTAL POLICY MODELING*
Shanthikumar, J.G. & Sumita, U./ *APPLIED PROBABILITY AND STOCHASTIC PROCESSES*
Liu, B. & Esogbue, A.O./ *DECISION CRITERIA AND OPTIMAL INVENTORY PROCESSES*
Gal, T., Stewart, T.J., Hanne, T. / *MULTICRITERIA DECISION MAKING: Advances in MCDM Models, Algorithms, Theory, and Applications*
Fox, B.L. / *STRATEGIES FOR QUASI-MONTE CARLO*
Hall, R.W. / *HANDBOOK OF TRANSPORTATION SCIENCE*
Grassman, W.K. / *COMPUTATIONAL PROBABILITY*
Pomerol, J-C. & Barba-Romero, S. / *MULTICRITERION DECISION IN MANAGEMENT*
Axsäter, S. / *INVENTORY CONTROL*
Wolkowicz, H., Saigal, R., & Vandenberghe, L. / *HANDBOOK OF SEMI-DEFINITE PROGRAMMING: Theory, Algorithms, and Applications*
Hobbs, B.F. & Meier, P. / *ENERGY DECISIONS AND THE ENVIRONMENT: A Guide to the Use of Multicriteria Methods*
Dar-El, E. / *HUMAN LEARNING: From Learning Curves to Learning Organizations*
Armstrong, J.S. / *PRINCIPLES OF FORECASTING: A Handbook for Researchers and Practitioners*
Balsamo, S., Personé, V., & Onvural, R./ *ANALYSIS OF QUEUEING NETWORKS WITH BLOCKING*
Bouyssou, D. et al. / *EVALUATION AND DECISION MODELS: A Critical Perspective*
Hanne, T. / *INTELLIGENT STRATEGIES FOR META MULTIPLE CRITERIA DECISION MAKING*
Saaty, T. & Vargas, L. / *MODELS, METHODS, CONCEPTS and APPLICATIONS OF THE ANALYTIC HIERARCHY PROCESS*
Chatterjee, K. & Samuelson, W. / *GAME THEORY AND BUSINESS APPLICATIONS*
Hobbs, B. et al. / *THE NEXT GENERATION OF ELECTRIC POWER UNIT COMMITMENT MODELS*
Vanderbei, R.J. / *LINEAR PROGRAMMING: Foundations and Extensions, 2nd Ed.*
Kimms, A. / *MATHEMATICAL PROGRAMMING AND FINANCIAL OBJECTIVES FOR SCHEDULING PROJECTS*
Baptiste, P., Le Pape, C. & Nuijten, W. / *CONSTRAINT-BASED SCHEDULING*
Feinberg, E. & Shwartz, A. / *HANDBOOK OF MARKOV DECISION PROCESSES: Methods and Applications*

* *A list of the more recent publications in the series is at the front of the book* *

Printed in the United States
81129LV00003B/1-2